Applied Probability and Statistics (Continued)

BARTHOLOMEW · Stochastic Models for Social Processes, *Second Edition*
BENNETT and FRANKLIN · Statistical Analysis in Chemistry and the Chemical Industry
BHAT · Elements of Applied Stochastic Processes
BOX and DRAPER · Evolutionary Operation: A Statistical Method for Process Improvement
BROWNLEE · Statistical Theory and Methodology in Science and Engineering, *Second Edition*
BURY · Statistical Models in Applied Science
CHERNOFF and MOSES · Elementary Decision Theory
CHIANG · Introduction to Stochastic Processes in Biostatistics
CHOW · Analysis and Control of Dynamic Economic Systems
CLELLAND, deCANI, BROWN, BURSK, and MURRAY · Basic Statistics with Business Applications, *Second Edition*
COCHRAN · Sampling Techniques, *Second Edition*
COCHRAN and COX · Experimental Designs, *Second Edition*
COX · Planning of Experiments
COX and MILLER · The Theory of Stochastic Processes, *Second Edition*
DANIEL and WOOD · Fitting Equations to Data
DAVID · Order Statistics
DEMING · Sample Design in Business Research
DODGE and ROMIG · Sampling Inspection Tables, *Second Edition*
DRAPER and SMITH · Applied Regression Analysis
DUNN and CLARK · Applied Statistics: Analysis of Variance and Regression
ELANDT-JOHNSON · Probability Models and Statistical Methods in Genetics
FLEISS · Statistical Methods for Rates and Proportions
GOLDBERGER · Econometric Theory
GROSS and HARRIS · Fundamentals of Queueing Theory
GUTTMAN, WILKS and HUNTER · Introductory Engineering Statistics, *Second Edition*
HAHN and SHAPIRO · Statistical Models in Engineering
HALD · Statistical Tables and Formulas
HALD · Statistical Theory with Engineering Applications
HARTIGAN · Clustering Algorithms
HOEL · Elementary Statistics, *Third Edition*
HOLLANDER and WOLFE · Nonparametric Statistical Methods
HUANG · Regression and Econometric Methods
JOHNSON and KOTZ · Distributions in Statistics
Discrete Distributions
Continuous Univariate Distributions-1
Continuous Univariate Distributions-2
Continuous Multivariate Distributions
JOHNSON and LEONE · Statistics and Experimental Design: In Engineering and the Physical Sciences, Volumes I and II
LANCASTER · The Chi Squared Distribution
LANCASTER · An Introduction to Medical Statistics

continued on back

The Statistical Analysis of Time Series

Also by T. W. Anderson

An Introduction to Multivariate Statistical Analysis
John Wiley & Sons, Inc., 1958

The Statistical Analysis of Times Series

John Wiley & Sons, 1971

With S. D. Gupta and G. P. H. Styan

Bibliography of Multivariate Statistical Analysis

A Halsted Press Publication, John Wiley & Sons, 1972

The Statistical Analysis of Time Series

T. W. ANDERSON
Professor of Statistics and Economics
Stanford University

JOHN WILEY & SONS, INC.
New York London Sydney Toronto

 This work was supported by the Logistics and Mathematical Statistics Branch of the Office of Naval Research under Contracts Nonr-3279(00), NR 042-219, Nonr-266(33), NR 042-034; Nonr-4195(00), NR 042-233; Nonr-4259(08), NR 042-034; Nonr-225(53), NR 042-002; N 00014-67-A-0112-0030, NR 042-034.

Library of Congress Catalog Card Number: 70-126222
ISBN 0 471 02900 9

Printed in the United States of America

10 9 8 7 6 5 4

To

DOROTHY

Preface

In writing a book on the statistical analysis of time series an author has a choice of points of view. My selection is the mathematical theory of statistical inference concerning probabilistic models that are assumed to generate observed time series. The probability model may involve a deterministic trend and a random part constituting a stationary stochastic process; the statistical problems treated have to do with aspects of such trends and processes. Where possible, the problem is posed as one of finding an optimum procedure and such procedures are derived. The statistical properties of the various methods are studied; in many cases they can be developed only in terms of large samples, that is, on information from series observed a long time. In general these properties are derived on a rigorous mathematical basis.

While the theory is developed under appropriate mathematical assumptions, the methods may be used where these assumptions are not strictly satisfied. It can be expected that in many cases the properties of the procedures will hold approximately. In any event the precisely stated results of the theorems give some guidelines for the use of the procedures. Some examples of the application of the methods are given, and the uses, computational approaches, and interpretations are discussed, but there is no attempt to put the methods in the form of programs for computers.

This book grew out of a graduate course that I gave for many years at Columbia University, usually for one semester and occasionally for two semesters. By now the material included in the book cannot be covered completely in a two-semester course; an instructor using this book as a text will select the material that he feels most interesting and important. Many exercises are given. Some of these are applications of the methods described; some of the problems are to work out special cases of the general theory; some of the exercises fill in details in complicated proofs; and some extend the theory.

Besides serving as a text book I hope this book will furnish a means by which statisticians and other persons can learn about time series analysis without resort to a formal course. Reading this book and doing selected exercises will lead to a considerable knowledge of statistical methodology useful for the analysis of time series. This book may also serve for reference. Much material which has not been assembled together before is presented here in a fairly coherent fashion. Some new theorems and methods are presented. In other cases the assumptions of previously stated propositions have been weakened and conclusions strengthened.

Since the area of time series analysis is so wide, an author must select the topics he will include. I have described in the introduction (Chapter 1) the material included as well as the limitations, and the Table of Contents also gives an indication. It is hoped that the more basic and important topics are treated here, though to some extent the coverage is a matter of taste. New methods are constantly being introduced and points of view are changing; the results here can hardly be definitive. In fact, some material included may at the present time be rather of historical interest.

In view of the length of this book a few words of advice to readers and instructors may be useful in selecting material to study and teach. Chapter 2 is a self-contained summary of the methods of least squares; it may be largely redundant for many statisticians. Chapters 3 and 4 deal with models with independent random terms (known sometimes as "errors in variables"); some ideas and analysis are introduced which are used later, but the reader interested mainly in the later chapters can pass over a good deal (including much of Sections 3.4, 4.3, and 4.4). Autoregressive processes, which are useful in applications and which illustrate stationary stochastic processes, are treated in Chapter 5; Sections 5.5 and 5.6 on large-sample theory contain relevant theorems, but the proofs involve considerable detail and can be omitted. Statistical inference for these models is basic to analysis of stationary processes "in the time domain." Chapter 6 is an extensive study of serial correlation and tests of independence; Sections 6.3 and 6.4 are primarily of theoretical statistical interest; Section 6.5 develops the algebra of quadratic forms and ratios of them; distributions, moments, and approximate distributions are obtained in Sections 6.7 and 6.8, and tables of significance points are given for tests. The first five sections of Chapter 7 constitute an introduction to stationary stochastic processes and their spectral distribution functions and densities. Chapter 8 develops the theory of statistics pertaining to stationary stochastic processes. Estimation of the spectral density is treated in Chapter 9; it forms the basis of analyzing stationary processes "in the frequency domain." Section 10.2 extends regression analysis (Chapter 2) to stationary random terms; Section

10.3 extends Chapters 8 and 9 to this case; and Section 10.4 extends Chapter 6 to the case of residuals from fitted trends. Parts of the book that constitute units which may be read somewhat independently from other parts are (i) Chapter 2, (ii) Chapters 3 and 4, (iii) Chapter 5, (iv) Chapter 6, (v) Chapter 7, and (vi) Chapters 8 and 9.

The statistical analysis of time series in practical applications will also invoke less formal techniques (which are now sometimes called "data analysis"). A graphical presentation of an observed time series contributes to understanding the phenomenon. Transformations of the measurement and relations to other data may be useful. The rather precisely stated procedures studied in this book will not usually be used in isolation and may be adapted for various situations. However, in order to investigate statistical methods rigorously within a mathematical framework some aspects of the analysis are formalized. For instance, the determination of whether an effect is large enough to be important is sometimes formalized as testing the hypothesis that a parameter is 0.

The level of this book is roughly that of my earlier book, *An Introduction to Multivariate Statistical Analysis*. Some knowledge of matrix algebra is needed. (The necessary material is given in the appendix of my earlier book; additional results are developed in the text and exercises of this present book.) A general knowledge of statistical methodology is useful; in particular, the reader is expected to know the standard material of univariate analysis such as t-tests and F-statistics, the multivariate normal distribution, and the elementary ideas of estimation and testing hypotheses. Some more sophisticated theory of testing hypotheses, estimation, and decision theory that is referred to is developed in the exercises. [The reader is referred to Lehmann (1959) for a detailed and rigorous treatment of testing hypotheses.] A moderate knowledge of advanced calculus is assumed. Although real-valued time series are treated, it is sometimes convenient to write expressions in terms of complex variables; actually the theory of complex variables is not used beyond the simple fact that $e^{i\theta} = \cos\theta + i\sin\theta$ (except for one problem). Probability theory is used to the extent of characteristic functions and some basic limit theorems. The theory of stochastic processes is developed to the extent that it is needed.

As noted above, there are many problems posed at the end of each chapter except the first which is the introduction. Solutions to these problems have been prepared by Paul Shaman. Solutions which are referred to in the text or which demonstrate some particularly important point are printed in Appendix B of this book. Solutions to most other problems (except solutions that are straightforward and easy) are contained in a Solutions Manual which is issued as a separate booklet. This booklet is available free of charge by writing to the publisher.

I owe a great debt of gratitude to Paul Shaman for many contributions to this book in matters of exposition, selection of material, suggestions of references and problems, improvements of proofs and exposition, and corrections of errors of every magnitude. He has read my manuscript in many versions and drafts. The conventional statement that an acknowledged reader of a manuscript is not responsible for any errors in the publication I feel is usually superfluous because it is obvious that anyone kind enough to look at a manuscript assumes no such responsibility. Here such a disclaimer may be called for simply because Paul Shaman corrected so many errors that it is hard to believe any remain. However, I admit that in this material it is easy to generate errors and the reader should throw the blame on the author for any he finds (as well as inform him of them).

My appreciation also goes to David Hinkley, Takamitsu Sawa, and George Styan, who read all or substantial parts of the manuscript and proofs and assisted with the preparation of the bibliography and index. There are many other colleagues and students to thank for assistance of various kinds. They include Selwyn Gallot, Joseph Gastwirth, Vernon Johns, Ted Matthes, Emily Stong Myers, Emanuel Parzen, Lloyd Rosenberg, Ester Samuel, and Morris Walker as well as Anupam Basu, Nancy David, Ronald Glaser, Elizabeth Hinkley, Raul Mentz, Fred Nold, Arthur V. Peterson, Jr., Cheryl Schiffman, Kenneth Thompson, Roger Ward, Larry Weldon, and Owen Whitby. No doubt I have forgotten others. I also wish to thank J. M. Craddock, C. W. J. Granger, M. G. Kendall, A. Stuart, and Herman Wold for use of some material.

In preparing the typescript my greatest debt is to Pamela Oline Gerstman, my secretary for four years, who patiently went through innumerable drafts and revisions. (Among other tribulations her office was used as headquarters for the "liberators" of Fayerweather Hall in the spring of 1968.) The manuscript also bore the imprints of Helen Bellows, Shauneen Nelson, Carol Andermann Novak, Katherine Cane, Carol Hallett Robbins, Alexandra Mills, Susan Parry, Noreen Browne Ettl, Sandi Hochler Frost, Judi Campbell, and Polly Bixler.

An important factor in my writing this book was the sustained financial support of the Office of Naval Research over a period of some ten years. The Logistics and Mathematical Statistics Branch has been helpful, encouraging, accommodating and patient in its sponsorship.

Stanford University
Stanford, California
February, 1970

T. W. ANDERSON

Contents

CHAPTER 4

Cyclical Trends 92

CHAPTER 5

Linear Stochastic Models with Finite Numbers of Parameters 164

CHAPTER 6

Serial Correlation 254

CHAPTER 1

Introduction

A time series is a sequence of observations, usually ordered in time, although in some cases the ordering may be according to another dimension. The feature of time series analysis which distinguishes it from other statistical analyses is the explicit recognition of the importance of the order in which the observations are made. While in many problems the observations are statistically independent, in time series successive observations may be dependent, and the dependence may depend on the positions in the sequence. The nature of a series and the structure of its generating process may also involve in other ways the sequence in which the observations are taken.

In almost every area there are phenomena whose development and variation with the passing of time are of interest and importance. In daily life one is interested in aspects of the weather, in prices that one pays, and in features of one's health; these change in time. There are characteristics of a nation, affecting many individuals, such as economic conditions and population, that evolve and fluctuate over time. The activity of a business, the condition of an industrial process, the level of a person's sleep, and the reception of a television program vary chronologically. The measurement of some particular characteristic over a period of time constitutes a time series. It may be an hourly record of temperature at a given place or the annual rainfall at a meteorological station. It may be a quarterly record of gross national product; an electrocardiogram may compose several time series.

There are various purposes for using time series. The objective may be the prediction of the future based on knowledge of the past; the goal may be the control of the process producing the series; it may be to obtain an understanding of the mechanism generating the series; or simply a succinct description of the salient features of the series may be desired. As statisticians we shall be interested in statistical inference; on the basis of a limited amount of information, a time series of finite length, we wish to make inferences about the probabilistic mechanism that produced the series; we want to analyze the underlying structure.

In principle the measurement of many quantities, such as temperature and voltage, can be made continuously and sometimes is recorded continuously in the form of a graph. In practice, however, the measurements are often made discretely in time; in other cases, such as the annual yield of grain per acre, the measurement must be made at definite intervals of time. Even if the data are recorded continuously in time only the values at discrete intervals can be used for digital computations. In this book we shall confine ourselves to time series that are recorded discretely in time, that is, at regular intervals, such as barometric pressure recorded each hour on the hour. Although the effect of one quantity on another and the interaction of several characteristics over time are often of consequence, in many studies a great deal of knowledge may be gained by the investigation of a single time series; this book (except with respect to autoregressive systems) is concerned with statistical methods for analyzing a univariate time series, that is, one type of measurement made repeatedly on the same object or individual. We shall, furthermore, suppose that the measurement is a real number, such as temperature, which is not limited to a finite (or denumerable) number of values; such a measurement is often called a continuous variable. Some measurements we treat mathematically as if they were continuous in time; for example, annual national income can at best be measured to the nearest penny, but the number of values that this quantity can take on is so large that there is no serious slight to reality in considering the variable as continuous. Moreover, we shall consider series which are rather stable, that is, ones which tend to stay within certain bounds or at least are changing slowly, not explosively or abruptly; we would include many meteorological variables, for instance, but would exclude shock waves.

Let an observed time series be $y_1, y_2, \ldots, y_T$. The notation means that we have T numbers, which are observations on some variable made at T equally distant time points, which for convenience we label $1, 2, \ldots, T$. A fairly general mathematical, statistical, or probabilistic model for the time series can be written as follows:

$$y_t = f(t) + u_t, \qquad t = 1, 2, \ldots, T. \tag{1}$$

The observed series is considered as made up of a completely determined sequence $\{f(t)\}$, which may be called the systematic part, and of a random or stochastic sequence $\{u_t\}$, which obeys a probability law. (Signal and noise are sometimes used for these two components.) These two components of the observed series are not observable; they are theoretical quantities. For example, if the measurements are daily rainfall, the $f(t)$ might be the climatic norm, which is the long-run average over many years, and the u_t would represent those vagaries and irregularities in the weather that describe fluctuations from the climatic norm. Exactly what the decomposition means depends not only on the data, but, in part, depends on what is thought of as repetitions of the experiment giving rise to the data. The interpretation of the random part made here is the so-called "frequency" interpretation. In principle one can repeat the entire situation, obtaining a new set of observations; the $f(t)$ would be the same as before, but the random terms would be different, as a new realization of the stochastic process. The random terms may include errors of observation. [In effect $f(t) = \mathscr{E}y_t$.]

We have some intuitive ideas of what time should mean in such a model or process. One notion is that time proceeds progressively in one direction. Another is that events which are close together in time should be relatively highly related and happenings farther apart in time should not be as strongly related. The effect of time in the mathematical model (1) can be inserted into specifications of the function or sequence $f(t)$; it can be put into the formulation of the probability process that defines the random term u_t; or it can be put into both components. The first part of this book will be devoted to time series represented by "error" models, in which the observations are considered to be independent random deviations from some function representing trend. In the second part we shall be concerned with sequences of dependent random variables, in general stationary stochastic processes with particular emphasis on autoregressive processes. Finally, we shall treat models in which there is a trend and the random terms constitute a stationary stochastic process. Stationary stochastic processes are explained in Chapter 7.

In many cases the model may be completely specified except for a finite number of parameters; in such a case the problems of statistical inference concern these parameters. In other cases the model may be more loosely defined and the corresponding methods are nonparametric. The model is to represent the mechanism generating the relevant series reasonably well, but as a mathematical abstraction the model is only an approximation to reality. How precisely the model can be determined depends on the state of knowledge about the process being studied, and, correspondingly, the information that can be supplied by statistical analysis depends on this state of knowledge. In this book many methods and their properties will be described, but, of course, these are only a selection from the many methods which are useful and available. Here the emphasis is on statistical inference and its mathematical basis.

The early development of time series analysis was based on models in which the effect of time was made in the systematic part, but not in the random part.

For convenience this case might be termed the classical model, because in a way it goes back to the time when Gauss and others developed least squares theory and methods for use in astronomy and physics. In this case we assume that the random part does not show any effect of time. More specifically, we assume that the mathematical expectation (that is, the mean value) is zero, that the variance is constant, and that the u_t are uncorrelated at different points in time. These specifications essentially force any effects of time to be made in the systematic part $f(t)$. The sequence $f(t)$ may depend on unknown coefficients and known quantities which vary over time. Then $f(t)$ is a "regression function." Methods of inference for the coefficients in a regression function are useful in many branches of statistics. The cases which are peculiar to time series analysis are those cases in which the quantities varying over time are known functions of t.

Within the limitations set out we may distinguish two types of sequences in time, $f(t)$. One is a slowly moving function of time, which is often called a trend, particularly by economists, and is exemplified by a polynomial of fairly low degree. Another type of sequence is cyclical; this is exemplified by a finite Fourier series, which is a finite sum of pairs of sine and cosine terms. A pair may be $\alpha \cos \lambda t + \beta \sin \lambda t$ $(0 < \lambda < \pi)$, which can also be written as a cosine function, say $\rho \cos (\lambda t - \theta)$. The period of such a function of time is $2\pi/\lambda$; that is, the function repeats itself after t has gone this amount; the frequency is the reciprocal of the period, namely $\lambda/(2\pi)$. The coefficient $\rho = \sqrt{\alpha^2 + \beta^2}$ is the amplitude and θ is the phase. The observed series is considered to be the sum of such a series $f(t)$ and a random term. Figure 1.1 presents $y_t = 5 + 2 \cos 2\pi t/6 + \sin 2\pi t/6 + u_t$, where u_t is normally distributed with mean 0 and variance 1. [The function $f(t)$ is drawn as a function of a continuous variable t.] The successive values of y_t are scattered randomly above and below $f(t)$. If we know this curve and the error distribution, information about $y_1, \ldots, y_{t-1}$ gives us no help in predicting y_t; the plot of $f(s)$ for $s > t - 1$ does not depend on $y_1, \ldots, y_{t-1}$.

Such a model may be appropriate in certain physical or economic problems. In astronomy, for example, $f(t)$ might be one coordinate in space of a certain planet at time t. Because telescopes are not perfect, and because of atmospheric variations, the observation of this coordinate at time t will have a small error. This error of observation does not affect later positions of the planet nor our observations of them. In the case of a freely swinging pendulum the displacement of the pendulum is a trigonometric function $\rho \cos (\lambda t - \theta)$ when measured from its lowest point.

One general model with the effect of time represented in the random part is a

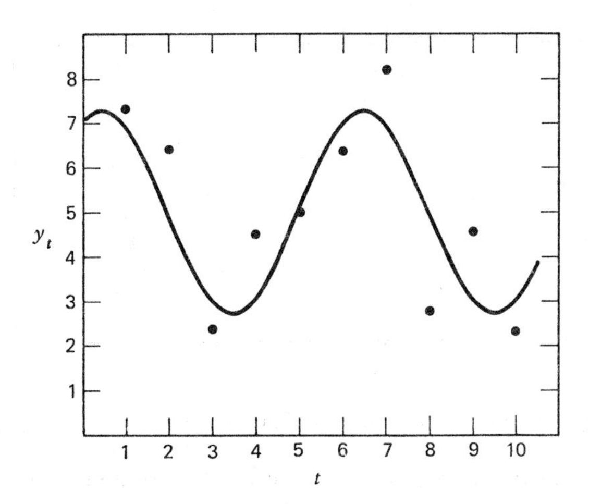

Figure 1.1. Series with a trigonometric trend.

stationary stochastic process; we can illustrate this with an autoregressive process. Suppose y_1 has some distribution with mean 0; let y_1 and y_2 have the joint distribution of y_1 and $\rho y_1 + u_2$, where u_2 is distributed with mean 0, independently of y_1. We write $y_2 = \rho y_1 + u_2$. Define in turn the joint distribution of $y_1, y_2, \ldots, y_{t-1}, y_t$ as the joint distribution of $y_1, y_2, \ldots, y_{t-1}$, $\rho y_{t-1} + u_t$, where u_t is distributed with mean 0, independently of $y_1, \ldots, y_{t-1}$, $t = 3, 4, \ldots$. When the (marginal) distributions of $u_2, u_3, \ldots$ are identical and the distribution of y_1 is chosen suitably, $\{y_t\}$ is a stationary stochastic process, in fact, an autoregressive process, and

$$y_t = \rho y_{t-1} + u_t \tag{2}$$

is a stochastic difference equation of first order. This construction is illustrated in Figure 1.2 for $\rho = \frac{1}{2}$. In this model the "disturbance" u_t has an effect on y_t and all later y_r's. It follows from the construction that the conditional expectation of y_t, given $y_1, \ldots, y_{t-1}$, is

$$\mathscr{E}(y_t \mid y_1, \ldots, y_{t-1}) = \rho y_{t-1}. \tag{3}$$

(In fact, for a first-order process y_t and $y_{t-2}, \ldots, y_1$ are conditionally independent given y_{t-1}.) If we want to predict y_t from $y_1, \ldots, y_{t-1}$ and know the parameter ρ, our best prediction (in the sense of minimizing the mean square error) is ρy_{t-1}; in this model knowledge of earlier observations assists in predicting y_t.

An autoregressive process of second order is obtained by taking the joint

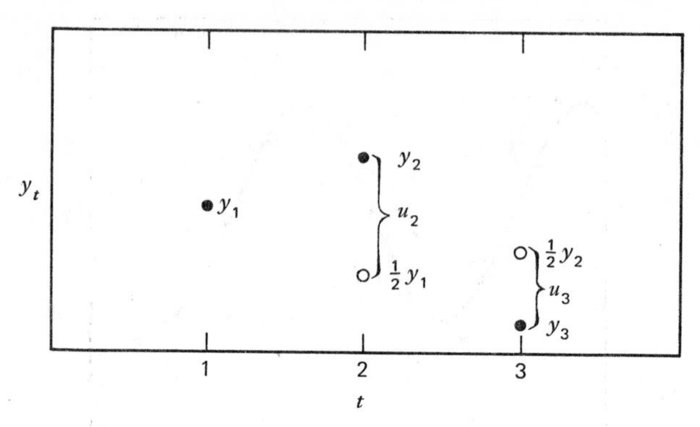

Figure 1.2. Construction of an autoregressive series.

distribution of $y_1, y_2, \ldots, y_{t-1}, y_t$ as the joint distribution of $y_1, y_2, \ldots, y_{t-1}$, $\rho_1 y_{t-1} + \rho_2 y_{t-2} + u_t$, where u_t is independent of $y_1, y_2, \ldots, y_{t-1}$, $t = 3, 4, \ldots$, and the distribution of y_1 and y_2 is suitably chosen; graphs of such series are given in Appendix A.2. Graphs of other randomly generated series are given by Kendall and Stuart (1966), Chapter 45, and Wold (1965), Chapter 1. The variable y_t may represent the displacement of a swinging pendulum when it is subjected to random "shocks" or pushes, u_t. The values of y_t tend to be a trigonometric function $\rho \cos(\lambda t - \theta)$ but with varying amplitude, varying frequency, and varying phase. An autoregressive process of order 4, generated by $y_t = \sum_{s=1}^{4} \rho_s y_{t-s} + u_t$, tends to be like the sum of two trigonometric functions with varying amplitudes, frequencies, and phases.

A general stationary stochastic process can be approximated by an autoregressive process of sufficiently high order, or it can be approximated by a process

$$\sum_{j=1}^{q} (A_j \cos \lambda_j t + B_j \sin \lambda_j t), \tag{4}$$

where $A_1, B_1, \ldots, A_q, B_q$ are independently distributed with $\mathscr{E}A_j = \mathscr{E}B_j = 0$ and $\mathscr{E}A_j^2 = \mathscr{E}B_j^2 = \phi(\lambda_j)$. The process is the sum of q trigonometric functions, whose amplitudes and phases are random variables. On the average the importance of the trigonometric function with frequency $\lambda_j/(2\pi)$ is proportional to the expectation of its squared amplitude, which is $2\phi(\lambda_j)$. In these terms a stationary stochastic process (in a certain class) may be characterized by a *spectral density* $f(\lambda)$ such that $\int_a^b f(\lambda)\, d\lambda$ is approximated by the sum of $\phi(\lambda_j)$ over λ_j such that $a \leq \lambda_j < b$. A feature of stationary stochastic processes is that the covariance $\mathscr{E}(y_t - \mathscr{E}y_t)(y_s - \mathscr{E}y_s)$ only depends on the difference in time $|t - s|$ and hence can be denoted by $\sigma(t - s)$. The covariance sequence and the spectral density (when it exists) are alternative ways of describing the second-order moment structure of a stationary stochastic process. The covariance sequence may be more appropriate and instructive when the time sequence is most relevant, which is the case of many economic series. The spectral density may be more suitable for other analyses. This is particularly true in the physical sciences because the harmonic or trigonometric function of time frequently gives the basic description of a phenomenon. Since the fluctuation of air pressure of a pure tone is given by a cosine function, the natural analysis of sound is in Fourier terms. In particular, the ear operates in this manner in the way it senses pitch.

The effect of time may be present both in the systematic part $f(t)$ as a trend in time and in the random part u_t as a stationary stochastic process. For example, an economic time series may consist of a long-run movement and seasonal variation, which together constitute $f(t)$, and an oscillatory component and other irregularities which constitute u_t and which may be described by an autoregressive process.

When the trend $f(t)$ has a specified structure and depends on a finite number of parameters, we consider problems of inference concerning these parameters. For example, we may estimate the coefficients of powers of t in polynomial trends and the coefficients of sines and cosines in trigonometric trends. In the former case we may want to decide the highest power of t to include, and in the latter case we may want to decide which of several terms to include. When the trend is not specified so exactly, we may use nonparametric methods, such as smoothing, to estimate the trend.

When the stochastic process is specified in terms of a finite number of parameters, such as an autoregressive process, we may wish to estimate the coefficients, test hypotheses about them, or decide what order process to use. A null hypothesis of particular interest is the independence of the random terms; this hypothesis may be tested by means of a serial correlation coefficient. When the process is stationary, but specified more loosely, we may estimate the covariances $\{\sigma(h)\}$ or the spectral density; these procedures are rather nonparametric.

The methods presented here are mainly for inference concerning the structure of the mechanism producing the process. Methods of prediction of the future values of the process when the structure is known are indicated; when the structure is not known, it may be inferred from the data and that inferred structure used for forecasting.

REFERENCES

Kendall and Stuart (1966), Wold (1965).

CHAPTER 2

The Use of Regression Analysis

2.1. INTRODUCTION

Many of the statistical techniques used in time series analysis are those of regression analysis (classical least squares theory) or are adaptations or analogues of them. The independent variables may be given functions of time, such as powers of time t or trigonometric functions of t. We shall first summarize statistical procedures when the random terms are uncorrelated (Sections 2.2 and 2.3); these results are used in trend analysis (Chapters 3 and 4). Then these procedures are modified for the case of the random terms having an arbitrary covariance matrix, which is known to within a proportionality factor (Section 2.4). Statistical procedures for analyzing trends when the random terms constitute a stationary stochastic process are studied in Chapter 10. Some large-sample theory for regression analysis which holds when the random terms are not necessarily normally distributed is developed (Section 2.6); generalizations of these results are useful for estimates of coefficients of stochastic difference equations (Chapter 5) because exact distribution theory is unmanageable; other generalizations are needed when the random terms constitute a more general stationary process (Section 10.2).

2.2. THE GENERAL THEORY OF LEAST SQUARES

Consider random variables $y_1, y_2, \ldots, y_T$ which are uncorrelated and have means and variances

$$\mathscr{E}y_t = \sum_{i=1}^{p} \beta_i z_{it}, \qquad t = 1, \ldots, T, \tag{1}$$

$$\mathscr{E}(y_t - \mathscr{E}y_t)^2 = \sigma^2, \qquad t = 1, \ldots, T, \tag{2}$$

where the z_{it}'s are given numbers. The z_{it}'s are called *independent* variables and the y_t's are called *dependent* variables. If we introduce the following vector notation

$$\boldsymbol{\beta} = \begin{pmatrix} \beta_1 \\ \beta_2 \\ \cdot \\ \cdot \\ \cdot \\ \beta_p \end{pmatrix}, \qquad \boldsymbol{z}_t = \begin{pmatrix} z_{1t} \\ z_{2t} \\ \cdot \\ \cdot \\ \cdot \\ z_{pt} \end{pmatrix}, \qquad t = 1, \dots, T, \tag{3}$$

the means (1) can be written

$$\mathscr{E} y_t = \boldsymbol{\beta}' \boldsymbol{z}_t, \qquad t = 1, \dots, T. \tag{4}$$

(The transpose of any vector or matrix $\boldsymbol{a}$ is indicated by $\boldsymbol{a}'$.)

The estimate of $\boldsymbol{\beta}$, denoted by $\boldsymbol{b}$, is the solution to the normal equation

$$\boldsymbol{A}\boldsymbol{b} = \boldsymbol{c}, \tag{5}$$

where

$$\boldsymbol{A} = \sum_{t=1}^{T} \boldsymbol{z}_t \boldsymbol{z}_t', \qquad \boldsymbol{c} = \sum_{t=1}^{T} y_t \boldsymbol{z}_t, \tag{6}$$

and $\boldsymbol{A}$ is assumed nonsingular (and thus $T \geq p$). The vector $\boldsymbol{b} = \boldsymbol{A}^{-1}\boldsymbol{c}$ minimizes $\sum_{t=1}^{T} (y_t - \tilde{\boldsymbol{b}}' \boldsymbol{z}_t)^2$ with respect to vectors $\tilde{\boldsymbol{b}}$, and is called the *least squares estimate* of $\boldsymbol{\beta}$. An unbiased estimate s^2 of σ^2 may be obtained (when $T > p$) from

$$\begin{aligned} (T - p)s^2 &= \sum_{t=1}^{T} (y_t - \boldsymbol{b}' \boldsymbol{z}_t)^2 \\ &= \sum_{t=1}^{T} y_t^2 - \boldsymbol{b}' \boldsymbol{A} \boldsymbol{b}. \end{aligned} \tag{7}$$

The least squares estimate $\boldsymbol{b}$ is an unbiased estimate of $\boldsymbol{\beta}$,

$$\mathscr{E} \boldsymbol{b} = \boldsymbol{\beta}, \tag{8}$$

with covariance matrix

$$\mathscr{E}(\boldsymbol{b} - \boldsymbol{\beta})(\boldsymbol{b} - \boldsymbol{\beta})' = \sigma^2 \boldsymbol{A}^{-1}. \tag{9}$$

The Gauss-Markov Theorem states that the elements of $\boldsymbol{b}$ are the best linear unbiased estimates (BLUE) of the corresponding components of $\boldsymbol{\beta}$ in the sense that each element of $\boldsymbol{b}$ has the minimum variance of all unbiased estimates of the corresponding element of $\boldsymbol{\beta}$ that are linear in $y_1, \dots, y_T$.

If the y_t's are independently normally distributed, $\boldsymbol{b}$ is the maximum likelihood estimate of $\boldsymbol{\beta}$ and $(T - p)s^2/T$ is the maximum likelihood estimate of σ^2. Then $\boldsymbol{b}$ is distributed according to $N(\boldsymbol{\beta}, \sigma^2 \boldsymbol{A}^{-1})$, where $N(\boldsymbol{\beta}, \sigma^2 \boldsymbol{A}^{-1})$ denotes the multivariate normal distribution with mean $\boldsymbol{\beta}$ and covariance matrix $\sigma^2 \boldsymbol{A}^{-1}$, and $(T - p)s^2/\sigma^2$ is distributed according to a χ^2-distribution with $T - p$ degrees of freedom independently of $\boldsymbol{b}$. The vector $\boldsymbol{b}$ and s^2 form a sufficient set of statistics for $\boldsymbol{\beta}$ and σ^2.

On the assumption of normality we can develop tests of hypotheses about the β_i's and confidence regions for the β_i's. Let

$$\boldsymbol{\beta} = \begin{pmatrix} \boldsymbol{\beta}^{(1)} \\ \boldsymbol{\beta}^{(2)} \end{pmatrix}, \tag{10}$$

where

$$\boldsymbol{\beta}^{(1)} = \begin{bmatrix} \beta_1 \\ \cdot \\ \cdot \\ \cdot \\ \beta_r \end{bmatrix}, \qquad \boldsymbol{\beta}^{(2)} = \begin{bmatrix} \beta_{r+1} \\ \cdot \\ \cdot \\ \cdot \\ \beta_p \end{bmatrix}, \tag{11}$$

and suppose that z_t and $\boldsymbol{b}$ are partitioned similarly. [Partitioned vectors and matrices are treated in T. W. Anderson (1958), Appendix 1, Section 3.] Then to test the hypothesis H: $\boldsymbol{\beta}^{(2)} = \overline{\boldsymbol{\beta}}^{(2)}$, where $\overline{\boldsymbol{\beta}}^{(2)}$ is a specified vector of numbers, we may use the F-statistic

$$\frac{(\boldsymbol{b}^{(2)} - \overline{\boldsymbol{\beta}}^{(2)})'(A^{22})^{-1}(\boldsymbol{b}^{(2)} - \overline{\boldsymbol{\beta}}^{(2)})}{(p - r)s^2} = \frac{(\boldsymbol{b}^{(2)} - \overline{\boldsymbol{\beta}}^{(2)})'(A_{22} - A_{21}A_{11}^{-1}A_{12})(\boldsymbol{b}^{(2)} - \overline{\boldsymbol{\beta}}^{(2)})}{(p - r)s^2}, \tag{12}$$

where A and A^{-1} have been partitioned into r and $p - r$ rows and columns

$$A = \begin{pmatrix} A_{11} & A_{12} \\ A_{21} & A_{22} \end{pmatrix}, \qquad A^{-1} = \begin{pmatrix} A^{11} & A^{12} \\ A^{21} & A^{22} \end{pmatrix}. \tag{13}$$

[The equality $A^{22} = (A_{22} - A_{21}A_{11}^{-1}A_{12})^{-1}$ is indicated in Problem 8.] This statistic has the F-distribution with $p - r$ and $T - p$ degrees of freedom if the normality assumption holds and the null hypothesis is true. In general if the normality assumption holds, this statistic has the noncentral F-distribution with noncentrality parameter

$$\frac{(\boldsymbol{\beta}^{(2)} - \overline{\boldsymbol{\beta}}^{(2)})'(A_{22} - A_{21}A_{11}^{-1}A_{12})(\boldsymbol{\beta}^{(2)} - \overline{\boldsymbol{\beta}}^{(2)})}{\sigma^2}. \tag{14}$$

These results follow from the fact that $\boldsymbol{b}^{(2)}$ is distributed according to $N(\boldsymbol{\beta}^{(2)}, \sigma^2 A^{22})$. [See Section 2.4 of T. W. Anderson (1958), for example.]

If $\overline{\boldsymbol{\beta}}^{(2)} = \mathbf{0}$, the null hypothesis means that $z_t^{(2)}$ does not enter into the regression function and one says that the y_t's are independent of the $z_t^{(2)}$'s. In this important case the numerator of (12) is simply $\boldsymbol{b}^{(2)\prime}(A_{22} - A_{21}A_{11}^{-1}A_{12})\boldsymbol{b}^{(2)}$.

A confidence region for $\boldsymbol{\beta}^{(2)}$ with confidence coefficient $1 - \varepsilon$ is given by the set

$$(15) \qquad \left\{\boldsymbol{\beta}^{(2)} \,\middle|\, \frac{(\boldsymbol{b}^{(2)} - \boldsymbol{\beta}^{(2)})'(A_{22} - A_{21}A_{11}^{-1}A_{12})(\boldsymbol{b}^{(2)} - \boldsymbol{\beta}^{(2)})}{(p - r)s^2} \leq F_{p-r,T-p}(\varepsilon)\right\},$$

where $F_{p-r,T-p}(\varepsilon)$ is the upper ε significance point of the F-distribution with $p - r$ and $T - p$ degrees of freedom.

If one is interested in only one component of $\boldsymbol{\beta}$, these inferences can be based on a t-statistic (instead of an F-statistic). Suppose the component is β_p; then $A^{22} = a^{pp}$, say, is a scalar. The essential fact is that $(b_p - \beta_p)/(s\sqrt{a^{pp}})$ has a t-distribution with $T - p$ degrees of freedom.

We note that the residuals $\hat{y}_t = y_t - \boldsymbol{b}'z_t$ are uncorrelated with the independent variables z_t in the sample

$$(16) \qquad \sum_{t=1}^{T} \hat{y}_t z_t = \sum_{t=1}^{T} y_t z_t - \sum_{t=1}^{T} z_t z_t' \boldsymbol{b} = \mathbf{0},$$

and the set of residuals is uncorrelated with the sample regression vector in the population. (See Problem 7.)

The geometric interpretation in Figure 2.1 is enlightening. Let $\boldsymbol{y} = (y_1, \ldots, y_T)'$ be a vector in a T-dimensional Euclidean space, let the r columns of $\boldsymbol{Z}_1 = (z_1^{(1)}, \ldots, z_T^{(1)})'$ be r vectors in the space, let the $p - r$ columns of $\boldsymbol{Z}_2 = (z_1^{(2)}, \ldots, z_T^{(2)})'$ be $p - r$ vectors, and let $\boldsymbol{Z} = (\boldsymbol{Z}_1\ \boldsymbol{Z}_2)$. Then the expectation of $\boldsymbol{y}$ is $\boldsymbol{Z\beta}$, which is a vector in the p-dimensional space spanned by the columns of $\boldsymbol{Z}$. The sample regression $\boldsymbol{Zb}$ is the projection of $\boldsymbol{y}$ on this p-dimensional space; the vector of residuals $\boldsymbol{y} - \boldsymbol{Zb}$ is orthogonal to every vector in the p-dimensional space; it is (parallel to) the projection of $\boldsymbol{y}$ on the $(T - p)$-dimensional space orthogonal to the columns of $\boldsymbol{Z}$.

Figure 2.2 represents the p-dimensional space spanned by the columns of $\boldsymbol{Z}$. The projection of the vector $\boldsymbol{Zb} - \boldsymbol{Z\beta} = \boldsymbol{Z}(\boldsymbol{b} - \boldsymbol{\beta})$ on the r-dimensional subspace spanned by the columns of $\boldsymbol{Z}_1$ is $\boldsymbol{Z}_1\boldsymbol{b}^{*(1)} - \boldsymbol{Z}_1\boldsymbol{\beta}^{*(1)} = \boldsymbol{Z}_1(\boldsymbol{b}^{*(1)} - \boldsymbol{\beta}^{*(1)})$, where $\boldsymbol{b}^{*(1)} = (\boldsymbol{Z}_1'\boldsymbol{Z}_1)^{-1}\boldsymbol{Z}_1'\boldsymbol{y}$. (See Section 2.3.) The projection on the $(p - r)$-dimensional space of $\boldsymbol{Z}$ orthogonal to $\boldsymbol{Z}_1$ is $\boldsymbol{Z}(\boldsymbol{b} - \boldsymbol{\beta}) - \boldsymbol{Z}_1(\boldsymbol{b}^{*(1)} - \boldsymbol{\beta}^{*(1)}) = (\boldsymbol{Z}_2 - \boldsymbol{Z}_1A_{11}^{-1}A_{12})(\boldsymbol{b}^{(2)} - \overline{\boldsymbol{\beta}}^{(2)})$. The numerator of the F-statistic (12) is the squared length of this last vector, and the denominator is proportional to the squared length of $\boldsymbol{y} - \boldsymbol{Zb}$.

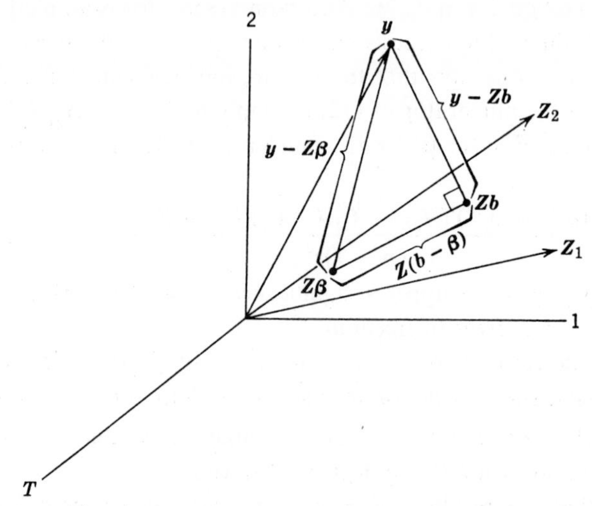

Figure 2.1. Geometry of least squares estimation.

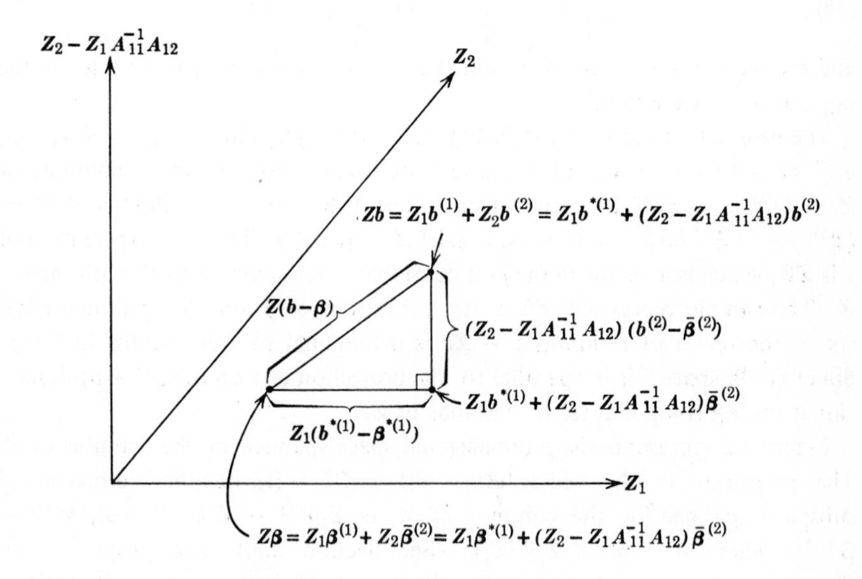

Figure 2.2. Geometry of testing $\beta^{(2)} = \bar{\beta}^{(2)}$.

2.3. LINEAR TRANSFORMATIONS OF INDEPENDENT VARIABLES; ORTHOGONAL INDEPENDENT VARIABLES

The regression function can be expressed in terms of different p-coordinate systems for the independent variables. Some coordinate systems may be more convenient than others.

Let $z_t^* = Gz_t$, $t = 1, \ldots, T$, for an arbitrary nonsingular matrix G, and let $\boldsymbol{\beta} = G'\boldsymbol{\beta}^*$. Then the expected value $\mathscr{E}y_t$ can be written

$$\boldsymbol{\beta}'z_t = \boldsymbol{\beta}^{*\prime}GG^{-1}z_t^* = \boldsymbol{\beta}^{*\prime}z_t^*, \qquad t = 1, \ldots, T. \tag{1}$$

The components of the z_t^*'s are coordinates of the vectors z_t referred to another coordinate system. The estimates of $\boldsymbol{\beta}^*$ and σ^2, based on observations $y_1, \ldots, y_T, z_1, \ldots, z_T$, are given by

$$\begin{aligned} b^* &= A^{*-1}\sum_{t=1}^{T} z_t^* y_t = (GAG')^{-1}G\sum_{t=1}^{T} z_t y_t \\ &= (G')^{-1}b, \end{aligned} \tag{2}$$

$$\begin{aligned} (T - p)s^{*2} &= \sum_{t=1}^{T}(y_t - b^{*\prime}z_t^*)^2 = \sum_{t=1}^{T}(y_t - b'z_t)^2 \\ &= (T - p)s^2, \end{aligned} \tag{3}$$

since

$$\begin{aligned} b^{*\prime}z_t^* &= [(G')^{-1}b]'Gz_t \\ &= b'z_t. \end{aligned} \tag{4}$$

The estimated regression functions $b'z_t = b^{*\prime}z_t^*$ in the two coordinate systems give the same values.

The independent variables may be divided into two sets, $z_t' = (z_t^{(1)\prime}\ z_t^{(2)\prime})$, with the latter set $z_t^{(2)}$ receiving special attention. For instance, one may wish to test the null hypothesis $\boldsymbol{\beta}^{(2)} = \mathbf{0}$. In such a case it may be convenient to transform the independent variables by

$$G = \begin{pmatrix} G_{11} & 0 \\ G_{21} & G_{22} \end{pmatrix}, \tag{5}$$

where G_{11} is square. Then

$$\begin{aligned} G^{-1} &= \begin{pmatrix} G^{11} & 0 \\ G^{21} & G^{22} \end{pmatrix} \\ &= \begin{pmatrix} G_{11}^{-1} & 0 \\ -G_{22}^{-1}G_{21}G_{11}^{-1} & G_{22}^{-1} \end{pmatrix} \end{aligned} \tag{6}$$

and

$$\boldsymbol{\beta}^* = \begin{pmatrix} \boldsymbol{\beta}^{*(1)} \\ \boldsymbol{\beta}^{*(2)} \end{pmatrix} = \begin{pmatrix} (G_{11}^{-1})' & -(G_{22}^{-1}G_{21}G_{11}^{-1})' \\ \mathbf{0} & (G_{22}^{-1})' \end{pmatrix} \begin{pmatrix} \boldsymbol{\beta}^{(1)} \\ \boldsymbol{\beta}^{(2)} \end{pmatrix} \tag{7}$$

$$= \begin{pmatrix} (G_{11}^{-1})'\boldsymbol{\beta}^{(1)} - (G_{22}^{-1}G_{21}G_{11}^{-1})'\boldsymbol{\beta}^{(2)} \\ (G_{22}^{-1})'\boldsymbol{\beta}^{(2)} \end{pmatrix}.$$

The hypothesis H: $\boldsymbol{\beta}^{(2)} = \mathbf{0}$ is therefore equivalent to the hypothesis H: $\boldsymbol{\beta}^{*(2)} = \mathbf{0}$. The transformation of z_t is

$$\begin{pmatrix} z_t^{*(1)} \\ z_t^{*(2)} \end{pmatrix} = \begin{pmatrix} G_{11} & \mathbf{0} \\ G_{21} & G_{22} \end{pmatrix} \begin{pmatrix} z_t^{(1)} \\ z_t^{(2)} \end{pmatrix} \tag{8}$$

$$= \begin{pmatrix} G_{11}z_t^{(1)} \\ G_{21}z_t^{(1)} + G_{22}z_t^{(2)} \end{pmatrix}.$$

Since $b^{*(2)} = (G_{22}^{-1})'b^{(2)}$ and

$$A_{22}^* - A_{21}^*(A_{11}^*)^{-1}A_{12}^* = G_{22}(A_{22} - A_{21}A_{11}^{-1}A_{12})G_{22}', \tag{9}$$

the F-criteria for testing the null hypothesis are equal.

If G is

$$\begin{pmatrix} G_{11} & \mathbf{0} \\ G_{21} & G_{22} \end{pmatrix} = \begin{pmatrix} I & \mathbf{0} \\ -A_{21}A_{11}^{-1} & I \end{pmatrix}, \tag{10}$$

then $z_t^{*(1)} = z_t^{(1)}$ and $z_t^{*(2)} = z_t^{(2)} - A_{21}A_{11}^{-1}z_t^{(1)}$ are orthogonal; that is,

$$A_{21}^* = A_{12}^{*\prime} = \sum_{t=1}^{T} z_t^{*(2)} z_t^{*(1)\prime} \tag{11}$$

$$= A_{21} - A_{21}A_{11}^{-1}A_{11} = \mathbf{0}.$$

Since

$$A_{22}^* = A_{22} - A_{21}A_{11}^{-1}A_{12}, \tag{12}$$

the F-criterion is $b^{*(2)\prime}A_{22}^*b^{*(2)}/[(p-r)s^2]$.

The linear transformation G can be chosen so that every pair of components of z_t^* is orthogonal. Let

$$z_t^* = \boldsymbol{\Gamma} z_t, \tag{13}$$

where

$$\Gamma = \begin{bmatrix} 1 & 0 & 0 & \cdots & 0 \\ \gamma_{21} & 1 & 0 & \cdots & 0 \\ \gamma_{31} & \gamma_{32} & 1 & \cdots & 0 \\ \cdot & \cdot & \cdot & & \cdot \\ \cdot & \cdot & \cdot & & \cdot \\ \cdot & \cdot & \cdot & & \cdot \\ \gamma_{p1} & \gamma_{p2} & \gamma_{p3} & \cdots & 1 \end{bmatrix}. \tag{14}$$

The form of $\boldsymbol{\Gamma}$ implies that $z_{1t}^* = z_{1t}$ and

$$z_{kt}^* = z_{kt} + \sum_{j=1}^{k-1} \gamma_{kj} z_{jt}, \qquad k = 2, \ldots, p; \tag{15}$$

that is, z_{kt}^* depends on z_{jt} with index j no greater than k. The conditions that $(z_{k1}^*, \ldots, z_{kT}^*)$ is orthogonal to $(z_{11}^*, \ldots, z_{1T}^*), \ldots, (z_{k-1,1}^*, \ldots, z_{k-1,T}^*)$ are the same as that $(z_{k1}^*, \ldots, z_{kT}^*)$ is orthogonal to $(z_{11}, \ldots, z_{1T}), \ldots, (z_{k-1,1}, \ldots, z_{k-1,T})$, namely

$$-(\gamma_{k1}, \ldots, \gamma_{k,k-1}) \begin{bmatrix} a_{11} & \cdots & a_{1,k-1} \\ \cdot & & \cdot \\ \cdot & & \cdot \\ \cdot & & \cdot \\ a_{k-1,1} & \cdots & a_{k-1,k-1} \end{bmatrix} = (a_{k1}, \ldots, a_{k,k-1}). \tag{16}$$

Then z_{kt}^* is the residual of z_{kt} from its formal regression on $z_{1t}, \ldots, z_{k-1,t}$, $t = 1, \ldots, T$. Thus

$$\sum_{t=1}^{T} z_t^* z_t^{*\prime} = A^* = \boldsymbol{\Gamma} A \boldsymbol{\Gamma}' = \begin{bmatrix} a_{11}^* & 0 & \cdots & 0 \\ 0 & a_{22}^* & \cdots & 0 \\ \cdot & \cdot & & \cdot \\ \cdot & \cdot & & \cdot \\ \cdot & \cdot & & \cdot \\ 0 & 0 & \cdots & a_{pp}^* \end{bmatrix} \tag{17}$$

is diagonal. In components we have

$$\sum_{t=1}^{T} z_{it}^* z_{jt}^* = 0, \qquad i \neq j. \tag{18}$$

(If the orthogonal variables are normalized to have sum of squares of unity, that is, transformed to $z^*_{jt}/\sqrt{a^*_{jj}}$, this orthogonalization procedure is often known as the Gram-Schmidt orthogonalization.)

The formulas and computations are considerably simplified when the independent variables are orthogonal. Since

$$
(19)\qquad A^{*-1} = \begin{bmatrix} a_{11}^{*-1} & 0 & \cdots & 0 \\ 0 & a_{22}^{*-1} & \cdots & 0 \\ \cdot & \cdot & & \cdot \\ \cdot & \cdot & & \cdot \\ \cdot & \cdot & & \cdot \\ 0 & 0 & \cdots & a_{pp}^{*-1} \end{bmatrix},
$$

the elements of $\boldsymbol{b}^*$ are uncorrelated and the variance of b_i^* is σ^2/a^*_{ii}, $i = 1, \ldots, p$. In this case the normal equations have the simple form

$$
(20)\qquad b_i^* = \frac{c_i^*}{a_{ii}^*} = \frac{1}{a_{ii}^*}\sum_{t=1}^{T} y_t z_{it}^*, \qquad i = 1, \ldots, p,
$$

and the formula for the estimate of the variance is

$$
(21)\qquad (T - p)s^2 = \sum_{t=1}^{T} y_t^2 - \sum_{i=1}^{p} a_{ii}^* b_i^{*2}.
$$

The F-statistic (12) of Section 2.2 for testing $\boldsymbol{\beta}^{*(2)} = \overline{\boldsymbol{\beta}}^{*(2)}$ is

$$
(22)\qquad \frac{\sum_{i=r+1}^{p} a_{ii}^*(b_i^* - \bar{\beta}_i^*)^2}{(p - r)s^2} = F_{p-r,\,T-p}.
$$

In some cases independent variables are chosen at the outset so as to be orthogonal. As we shall see in Chapter 4 the trigonometric sequences $\{\cos 2\pi jt/T\}$, $j = 0, 1, \ldots, [\frac{1}{2}T]$, and $\{\sin 2\pi kt/T\}$, $k = 1, \ldots, [\frac{1}{2}(T - 1)]$ are orthogonal for $t = 1, \ldots, T$ and can be used as components of z_t. Although any set of independent variables can be orthogonalized, it is usually not worthwhile to do so if the given set is to be used only once. However, if the same set of independent variables is to be used many times, then the transformation to orthogonal z_t^*'s can save considerable computational effort.

The case of polynomial regression provides an example of the use of orthogonalized independent variables. Suppose $z_{it} = t^{i-1}$; let

$$
(23)\qquad \mathscr{E} y_t = \beta_1 + \beta_2 t + \cdots + \beta_p t^{p-1}, \qquad t = 1, \ldots, T.
$$

The powers of t can be replaced by orthogonal polynomials $\phi_{0T}(t) = 1$, $\phi_{1T}(t), \ldots, \phi_{p-1,T}(t)$. These have the form

$$\phi_{kT}(t) = t^k + C_{k-1}(k, T)t^{k-1} + \cdots + C_1(k, T)t + C_0(k, T), \tag{24}$$

where the C's depend on the length of the series T and the degree of the polynomial k and are determined by

$$\sum_{t=1}^{T} \phi_{iT}(t)\phi_{jT}(t) = 0, \qquad i \neq j. \tag{25}$$

Then

$$\mathscr{E}y_t = \gamma_0\phi_{0T}(t) + \gamma_1\phi_{1T}(t) + \cdots + \gamma_{p-1}\phi_{p-1,T}(t). \tag{26}$$

Orthogonal polynomials will be considered in detail in Section 3.2.

Most of the computational methods for solving the normal equations $\boldsymbol{Ab} = \boldsymbol{c}$ consist of a so-called forward solution and a backward solution. The forward solution consists of operations involving the rows of $(\boldsymbol{A}\ \boldsymbol{c})$ which end up with $\boldsymbol{A}$ in triangular form; that is, it amounts to $\boldsymbol{D}(\boldsymbol{A}\ \boldsymbol{c}) = (\bar{\boldsymbol{A}}\ \bar{\boldsymbol{c}})$, namely

$$\begin{bmatrix} 1 & 0 & \cdots & 0 \\ d_{21} & 1 & \cdots & 0 \\ \cdot & \cdot & & \cdot \\ \cdot & \cdot & & \cdot \\ \cdot & \cdot & & \cdot \\ d_{p1} & d_{p2} & \cdots & 1 \end{bmatrix} (\boldsymbol{A}\ \boldsymbol{c}) = \begin{bmatrix} \bar{a}_{11} & \bar{a}_{12} & \cdots & \bar{a}_{1p} & \bar{c}_1 \\ 0 & \bar{a}_{22} & \cdots & \bar{a}_{2p} & \bar{c}_2 \\ \cdot & \cdot & & \cdot & \cdot \\ \cdot & \cdot & & \cdot & \cdot \\ \cdot & \cdot & & \cdot & \cdot \\ 0 & 0 & \cdots & \bar{a}_{pp} & \bar{c}_p \end{bmatrix}. \tag{27}$$

The matrix $\boldsymbol{D}$ is of the form of $\boldsymbol{\Gamma}$. Consideration of (27) shows that each row of $\boldsymbol{D}$ is identical to the corresponding row of $\boldsymbol{\Gamma}$ and hence that $\boldsymbol{D} = \boldsymbol{\Gamma}$. [Different methods triangularize $\boldsymbol{A}$ by different sequences of operations but they are algebraically equivalent, given the ordering of the variables; note that the forward solution of (27) for one k is part of the forward solution for each subsequent k.] Thus

$$\bar{a}_{kk} = \sum_{t=1}^{T} z_{kt}^* z_{kt} = \sum_{t=1}^{T} (z_{kt}^*)^2 = a_{kk}^*, \tag{28}$$

$$\bar{c}_k = \sum_{t=1}^{T} z_{kt}^* y_t = c_k^*. \tag{29}$$

The sample regression coefficients for the orthogonalized variables $b_k^* = c_k^*/a_{kk}^*$ can be obtained from the forward solution of the normal equations as

$b_k^* = \bar{c}_k/\bar{a}_{kk}$. In fact, in many procedures such as the Doolittle method each row of $(\bar{A}\ \bar{c})$ is divided by the leading nonzero element and recorded; the last element in each row is then the regression coefficient of the corresponding orthogonal variable. Thus we see that the forward solution of the normal equations involves the same algebra as defining the orthogonal variables; the important difference, of course, is that the computation of the orthogonal variables involves the calculation of the pT numbers z_{kt}^*. We note in passing that $\boldsymbol{b}^{(2)\prime}(\boldsymbol{A}_{22} - \boldsymbol{A}_{21}\boldsymbol{A}_{11}^{-1}\boldsymbol{A}_{12})\boldsymbol{b}^{(2)}$ can be obtained from the forward solution as $\sum_{k=r+1}^{p} b_k^* c_k^* = \sum_{k=r+1}^{p} \bar{c}_k^2/\bar{a}_{kk}$. (See Problem 12.)

2.4. CASE OF CORRELATION

If the dependent variables are correlated and the covariance matrix is known except (possibly) for a multiplying factor, the foregoing theory can be modified. Suppose

$$\mathscr{E} y_t = \boldsymbol{\beta}' \boldsymbol{z}_t, \qquad t = 1, \ldots, T, \tag{1}$$

$$\mathscr{E}(y_t - \boldsymbol{\beta}'\boldsymbol{z}_t)(y_s - \boldsymbol{\beta}'\boldsymbol{z}_s) = \sigma_{ts}, \qquad t, s = 1, \ldots, T, \tag{2}$$

where $\boldsymbol{z}_t$ is a known (column) vector of p numbers, $t = 1, \ldots, T$, and $\sigma_{ts} = \sigma^2\psi_{ts}$, $t, s = 1, \ldots, T$, where the ψ_{ts}'s are known. It is convenient to put this model in a more complete matrix notation. Let $\boldsymbol{y} = (y_1, \ldots, y_T)'$, $\boldsymbol{Z} = (\boldsymbol{z}_1, \ldots, \boldsymbol{z}_T)'$, $\boldsymbol{\Sigma} = (\sigma_{ts})$, and $\boldsymbol{\Psi} = (\psi_{ts})$. Then (1) and (2) are

$$\mathscr{E}\boldsymbol{y} = \boldsymbol{Z}\boldsymbol{\beta}, \tag{3}$$

$$\mathscr{E}(\boldsymbol{y} - \boldsymbol{Z}\boldsymbol{\beta})(\boldsymbol{y} - \boldsymbol{Z}\boldsymbol{\beta})' = \boldsymbol{\Sigma}, \tag{4}$$

and $\boldsymbol{\Sigma} = \sigma^2\boldsymbol{\Psi}$, where $\boldsymbol{\Psi}$ is known. Let $\boldsymbol{D}$ be a matrix such that

$$\boldsymbol{D}\boldsymbol{\Psi}\boldsymbol{D}' = \boldsymbol{I}. \tag{5}$$

Let $\boldsymbol{D}\boldsymbol{y} = \boldsymbol{x} = (x_1, \ldots, x_T)'$ and $\boldsymbol{D}\boldsymbol{Z} = \boldsymbol{W} = (\boldsymbol{w}_1, \ldots, \boldsymbol{w}_T)'$. Multiplication of (3) on the left by $\boldsymbol{D}$ and multiplication of (4) on the left by $\boldsymbol{D}$ and on the right by $\boldsymbol{D}'$ gives

$$\mathscr{E}\boldsymbol{x} = \boldsymbol{W}\boldsymbol{\beta}, \tag{6}$$

$$\mathscr{E}(\boldsymbol{x} - \boldsymbol{W}\boldsymbol{\beta})(\boldsymbol{x} - \boldsymbol{W}\boldsymbol{\beta})' = \sigma^2\boldsymbol{I}, \tag{7}$$

which is the model treated in Section 2.2. The normal equation $\boldsymbol{A}\boldsymbol{b} = \boldsymbol{c}$, where $\boldsymbol{A} = \boldsymbol{W}'\boldsymbol{W}$ and $\boldsymbol{c} = \boldsymbol{W}'\boldsymbol{x}$, is equivalent to

$$\boldsymbol{Z}'\boldsymbol{D}'\boldsymbol{D}\boldsymbol{Z}\boldsymbol{b} = \boldsymbol{Z}'\boldsymbol{D}'\boldsymbol{D}\boldsymbol{y}. \tag{8}$$

From (5) we see that $\boldsymbol{\Psi} = \boldsymbol{D}^{-1}(\boldsymbol{D}')^{-1} = (\boldsymbol{D}'\boldsymbol{D})^{-1}$ and $\boldsymbol{\Psi}^{-1} = \boldsymbol{D}'\boldsymbol{D}$; the solution of (8) is

$$b = (Z'\Psi^{-1}Z)^{-1}Z'\Psi^{-1}y. \tag{9}$$

Then

$$\mathscr{E}b = (Z'\Psi^{-1}Z)^{-1}Z'\Psi^{-1}\mathscr{E}y = \boldsymbol{\beta}, \tag{10}$$

(11)

$$\begin{aligned}\mathscr{E}(b - \boldsymbol{\beta})(b - \boldsymbol{\beta})' &= (Z'\Psi^{-1}Z)^{-1}Z'\Psi^{-1}\mathscr{E}(y - Z\boldsymbol{\beta})(y - Z\boldsymbol{\beta})'\Psi^{-1}Z(Z'\Psi^{-1}Z)^{-1}\\ &= \sigma^2(Z'\Psi^{-1}Z)^{-1}.\end{aligned}$$

Each element of $\boldsymbol{b}$ is the best linear unbiased estimate of the corresponding component of $\boldsymbol{\beta}$ (Gauss-Markov Theorem); $\boldsymbol{b}$ is called the Markov estimate of $\boldsymbol{\beta}$. The vector $\boldsymbol{b}$ minimizes $(y - Z\tilde{b})'\Psi^{-1}(y - Z\tilde{b})$. If $y_1, \ldots, y_T$ have a joint normal distribution, $\boldsymbol{b}$ and $\hat{\sigma}^2 = (T - p)s^2/T$ are the maximum likelihood estimates of $\boldsymbol{\beta}$ and σ^2 and constitute a sufficient set of statistics for $\boldsymbol{\beta}$ and σ^2.

The least squares estimate

$$b_L = (Z'Z)^{-1}Z'y \tag{12}$$

is unbiased,

$$\mathscr{E}b_L = (Z'Z)^{-1}Z'\mathscr{E}y = \boldsymbol{\beta}, \tag{13}$$

with covariance matrix

$$\begin{aligned}\mathscr{E}(b_L - \boldsymbol{\beta})(b_L - \boldsymbol{\beta})' &= (Z'Z)^{-1}Z'\mathscr{E}(y - Z\boldsymbol{\beta})(y - Z\boldsymbol{\beta})'Z(Z'Z)^{-1}\\ &= \sigma^2(Z'Z)^{-1}Z'\Psi Z(Z'Z)^{-1}.\end{aligned} \tag{14}$$

Unless the columns of $\boldsymbol{Z}$ have a specific relation to $\boldsymbol{\Psi}$ the variance of a linear combination of $\boldsymbol{b}_L$, say $\boldsymbol{\gamma}'\boldsymbol{b}_L$, will be greater than the variance of the corresponding linear combination of $\boldsymbol{b}$, namely $\boldsymbol{\gamma}'\boldsymbol{b}$; then the difference between the matrix (14) and the matrix (11) is positive semidefinite.

THEOREM 2.4.1. *If $Z = V^*C$, where the p columns of V^* are p linearly independent characteristic vectors of Ψ and C is a nonsingular matrix, then the least squares estimate* (12) *is the same as the Markov estimate* (9).

PROOF. The condition on V^* may be summarized as

$$\Psi V^* = V^*\Lambda^*, \tag{15}$$

where $\mathbf{\Lambda}^*$ is diagonal, consisting of the (positive) characteristic roots of $\mathbf{\Psi}$ corresponding to the columns of V^*. Then

$$V^{*\prime}\mathbf{\Psi}^{-1} = \mathbf{\Lambda}^{*-1}V^{*\prime}, \tag{16}$$

and $V^{*\prime}\mathbf{\Psi}^{-1}V^* = \mathbf{\Lambda}^{*-1}V^{*\prime}V^*$. The least squares estimate is

$$\boldsymbol{b}_L = C^{-1}(V^{*\prime}V^*)^{-1}V^{*\prime}\boldsymbol{y}, \tag{17}$$

and the Markov estimate is

$$\begin{aligned} \boldsymbol{b} &= (C'V^{*\prime}\mathbf{\Psi}^{-1}V^*C)^{-1}C'V^{*\prime}\mathbf{\Psi}^{-1}\boldsymbol{y} \\ &= (C'\mathbf{\Lambda}^{*-1}V^{*\prime}V^*C)^{-1}C'\mathbf{\Lambda}^{*-1}V^{*\prime}\boldsymbol{y}, \end{aligned} \tag{18}$$

which is identical to (17). Q.E.D.

In Section 10.2.1 it will be shown that the condition in Theorem 2.4.1 is necessary as well as sufficient. The condition may be stated equivalently that there are p linearly independent linear combinations of the columns of $\boldsymbol{Z}$ that are characteristic vectors of $\mathbf{\Psi}$. The point of the theorem is that when the condition is satisfied, the least squares estimates (which are available when $\mathbf{\Psi}$ is not known) have minimum variances among unbiased linear estimates. In Section 10.2.1 we shall also take up the case that there are an arbitrary number of linearly independent linear combinations of the columns of $\boldsymbol{Z}$ that are characteristic vector S of $\mathbf{\Psi}$; the least squares estimates of the coefficients of these linear combinations are the same as the Markov estimates when the other independent variables are orthogonal to these. That the condition of Theorem 2.4.1 implies that the least squares estimates are maximum likelihood under normality was shown by T. W. Anderson (1948).

2.5. PREDICTION

Let us consider prediction of y_τ, the observation at $t = \tau$. If we know the regression function, then we know $\mathscr{E}y_\tau$ and this is the best prediction of y_τ in the sense of minimizing the mean square error of prediction.

Suppose now that we have observations $y_1, \ldots, y_T$ and wish to use these to predict y_τ $(\tau > T)$, where $\mathscr{E}y_\tau = \sum_{k=1}^{p}\beta_k z_{k\tau} = \boldsymbol{\beta}'\boldsymbol{z}_\tau$ and where $\boldsymbol{\beta}$ is unknown and $\boldsymbol{z}_\tau$ is known. It seems reasonable that we would estimate $\boldsymbol{\beta}$ by the least squares estimate $\boldsymbol{b}$ and predict y_τ by $\boldsymbol{b}'\boldsymbol{z}_\tau$. Let us justify this procedure. We shall consider linear predictors $\sum_{t=1}^{T} d_t y_t$, where the d_t's may depend on $\boldsymbol{z}_1, \ldots, \boldsymbol{z}_T$ and $\boldsymbol{z}_\tau$. First we ask that the predictor be unbiased, that is, that

$$\mathscr{E}\left(\sum_{t=1}^{T} d_t y_t - y_\tau\right) = 0. \tag{1}$$

Thus the predictor is to be an unbiased estimate of $\mathscr{E}y_\tau = \boldsymbol{\beta}'z_\tau$. We ask to minimize the variance (or equivalently the mean square error of prediction)

$$(2) \qquad \mathscr{E}\left(\sum_{t=1}^{T} d_t y_t - y_\tau\right)^2 = \mathscr{E}\left[\left(\sum_{t=1}^{T} d_t y_t - \boldsymbol{\beta}'z_\tau\right) - (y_\tau - \mathscr{E}y_\tau)\right]^2$$
$$= \mathscr{E}\left(\sum_{t=1}^{T} d_t y_t - \boldsymbol{\beta}'z_\tau\right)^2 + \sigma^2.$$

Thus the problem is to find the minimum variance linear unbiased estimate of $\boldsymbol{\beta}'z_\tau$, a linear combination of the regression coefficients. The Gauss-Markov Theorem implies that this is $\boldsymbol{b}'z_\tau$. This follows from the earlier discussion since the model can be transformed so that $\boldsymbol{\beta}'z_\tau$ is a component of $\boldsymbol{\beta}^*$ when one column of $\boldsymbol{G}^{-1}$ is z_τ. The variance of $\boldsymbol{b}'z_\tau$ is

$$(3) \qquad \mathscr{E}(\boldsymbol{b}'z_\tau - \boldsymbol{\beta}'z_\tau)^2 = \mathscr{E}z_\tau'(\boldsymbol{b} - \boldsymbol{\beta})(\boldsymbol{b} - \boldsymbol{\beta})'z_\tau$$
$$= \sigma^2 z_\tau' \boldsymbol{A}^{-1} z_\tau.$$

The mean square error of prediction is

$$(4) \qquad \mathscr{E}(\boldsymbol{b}'z_\tau - y_\tau)^2 = \sigma^2(1 + z_\tau'\boldsymbol{A}^{-1}z_\tau).$$

The property of this predictor can alternatively be stated as the property of minimizing the mean square error among linear predictors with bounded mean square error. (See Problem 15.)

If we assume normality, we can give a prediction interval for y_τ. We have that $\boldsymbol{b}'z_\tau - y_\tau$ is normally distributed with mean 0 and variance (4) and independently of s^2. Thus

$$(5) \qquad \frac{\boldsymbol{b}'z_\tau - y_\tau}{s\sqrt{1 + z_\tau'\boldsymbol{A}^{-1}z_\tau}}$$

has a t-distribution with $T - p$ degrees of freedom. Then a prediction interval for y_τ with confidence $1 - \varepsilon$ is

$$(6) \qquad \boldsymbol{b}'z_\tau - t_{T-p}(\varepsilon)s\sqrt{1 + z_\tau'\boldsymbol{A}^{-1}z_\tau} \leq y_\tau \leq \boldsymbol{b}'z_\tau + t_{T-p}(\varepsilon)s\sqrt{1 + z_\tau'\boldsymbol{A}^{-1}z_\tau},$$

where $t_{T-p}(\varepsilon)$ is the number such that probability of the t-distribution with $T - p$ degrees of freedom between $\pm t_{T-p}(\varepsilon)$ is $1 - \varepsilon$.

Now let us study the error of prediction when some independent variables are neglected. Suppose the components of z_t are numbered in a suitable way, that z_t' is partitioned as before into $(z_t^{(1)\prime}\ z_t^{(2)\prime})$, and that the effect of $z_t^{(2)}$ is neglected, that is, that the regression is erroneously assumed to be a linear function of $z_t^{(1)}$ instead of a function of z_t. It is convenient to write $\boldsymbol{\beta}'z_t$ as

$\boldsymbol{\beta}^{*\prime}z_t^*$, where $\boldsymbol{\beta}^*$ is defined by (7) of Section 2.3 with $G_{11} = I$, $G_{21} = -A_{21}A_{11}^{-1}$, and $G_{22} = I$ and

$$z_t^* = Gz_t = \begin{pmatrix} I & 0 \\ -A_{21}A_{11}^{-1} & I \end{pmatrix}\begin{pmatrix} z_t^{(1)} \\ z_t^{(2)} \end{pmatrix} = \begin{pmatrix} z_t^{(1)} \\ z_t^{(2)} - A_{21}A_{11}^{-1}z_t^{(1)} \end{pmatrix} = \begin{pmatrix} z_t^{*(1)} \\ z_t^{*(2)} \end{pmatrix}. \tag{7}$$

Then $z_t^{*(1)}$ and $z_t^{*(2)}$ are orthogonal, $t = 1, \ldots, T$, and

$$A^* = \begin{pmatrix} A_{11}^* & 0 \\ 0 & A_{22}^* \end{pmatrix} = \begin{pmatrix} A_{11} & 0 \\ 0 & A_{22} - A_{21}A_{11}^{-1}A_{12} \end{pmatrix}. \tag{8}$$

The estimate of $\boldsymbol{\beta}^{*(1)}$ is

$$b^{*(1)} = A_{11}^{-1}c^{(1)} = A_{11}^{-1}\sum_{t=1}^{T} z_t^{(1)}y_t. \tag{9}$$

The statistical properties of the estimate are

$$\begin{aligned} \mathscr{E}b^{*(1)} &= A_{11}^{-1}\sum_{t=1}^{T} z_t^{(1)}(z_t^{(1)\prime}\boldsymbol{\beta}^{*(1)} + z_t^{*(2)\prime}\boldsymbol{\beta}^{*(2)}) \\ &= A_{11}^{-1}(A_{11}\boldsymbol{\beta}^{*(1)} + 0\boldsymbol{\beta}^{*(2)}) \\ &= \boldsymbol{\beta}^{*(1)}, \end{aligned} \tag{10}$$

$$\mathscr{E}(b^{*(1)} - \boldsymbol{\beta}^{*(1)})(b^{*(1)} - \boldsymbol{\beta}^{*(1)})' = \sigma^2 A_{11}^{-1}. \tag{11}$$

The prediction at time τ ($> T$) is $b^{*(1)\prime}z_\tau^{*(1)} = b^{*(1)\prime}z_\tau^{(1)}$. The bias in prediction at τ is

$$\begin{aligned} \mathscr{E}(b^{*(1)\prime}z_\tau^{(1)} - y_\tau) &= \boldsymbol{\beta}^{*(1)\prime}z_\tau^{(1)} - \boldsymbol{\beta}'z_\tau \\ &= -\boldsymbol{\beta}^{*(2)\prime}z_\tau^{*(2)} \\ &= -\boldsymbol{\beta}^{(2)\prime}(z_\tau^{(2)} - A_{21}A_{11}^{-1}z_\tau^{(1)}). \end{aligned} \tag{12}$$

The variance of $b^{*(1)\prime}z_\tau^{(1)}$ is

$$\mathscr{E}(b^{*(1)\prime}z_\tau^{(1)} - \boldsymbol{\beta}^{*(1)\prime}z_\tau^{(1)})^2 = \sigma^2 z_\tau^{(1)\prime}A_{11}^{-1}z_\tau^{(1)} \tag{13}$$

and the mean square error of prediction is

$$\begin{aligned} &\mathscr{E}(b^{*(1)\prime}z_\tau^{(1)} - y_\tau)^2 \\ &\quad = \boldsymbol{\beta}^{(2)\prime}(z_\tau^{(2)} - A_{21}A_{11}^{-1}z_\tau^{(1)})(z_\tau^{(2)\prime} - z_\tau^{(1)\prime}A_{11}^{-1}A_{12})\boldsymbol{\beta}^{(2)} + \sigma^2 z_\tau^{(1)\prime}A_{11}^{-1}z_\tau^{(1)} + \sigma^2. \end{aligned} \tag{14}$$

In general, neglecting of $z_t^{(2)}$ causes a bias in the prediction but decreases the variance. (See Problem 18.)

In some instances one might expect to predict in situations where the z_τ's of the future are like the z_t's of the past. The squared bias averaged over the z_t's already observed is

$$(15) \quad \sum_{t=1}^{T} [\boldsymbol{\beta}^{(2)\prime}(z_t^{(2)} - A_{21}A_{11}^{-1}z_t^{(1)})]^2$$
$$= \sum_{t=1}^{T} \boldsymbol{\beta}^{(2)\prime}(z_t^{(2)} - A_{21}A_{11}^{-1}z_t^{(1)})(z_t^{(2)\prime} - z_t^{(1)\prime}A_{11}^{-1}A_{12})\boldsymbol{\beta}^{(2)}$$
$$= \boldsymbol{\beta}^{(2)\prime}(A_{22} - A_{21}A_{11}^{-1}A_{12})\boldsymbol{\beta}^{(2)},$$

which is proportional to the noncentrality parameter in the distribution of the F-statistic for testing the hypothesis $\boldsymbol{\beta}^{(2)} = \mathbf{0}$. This fact may give another reason for preferring the F-test to other tests of the hypothesis.

2.6. ASYMPTOTIC THEORY

The procedures of testing hypotheses and setting up confidence regions have been based on the assumption that the observations are normally distributed. When the assumption of normality is not satisfied, these procedures can be justified for large samples on the basis of asymptotic theory.

THEOREM 2.6.1. *Let* $y_t = \boldsymbol{\beta}' z_t + u_t$, $t = 1, 2, \ldots,$ *where the* u_t*'s are independent and* u_t *has mean* 0, *variance* σ^2, *and distribution function* $F_t(u)$, $t = 1, 2, \ldots$. *Let* $A_T = \sum_{t=1}^{T} z_t z_t'$, $c_T = \sum_{t=1}^{T} y_t z_t$, $b_T = A_T^{-1} c_T$,

$$(1) \quad D_T = \begin{bmatrix} \sqrt{a_{11}^T} & 0 & \cdots & 0 \\ 0 & \sqrt{a_{22}^T} & \cdots & 0 \\ \vdots & \vdots & & \vdots \\ 0 & 0 & \cdots & \sqrt{a_{pp}^T} \end{bmatrix},$$

and

$$(2) \quad \mathbf{R}_T = D_T^{-1} A_T D_T^{-1}.$$

Suppose (i) $a_{ii}^T \to \infty$ *as* $T \to \infty$, $i = 1, \ldots, p$, (ii) $z_{i,T+1}^2 / a_{ii}^T \to 0$ *as* $T \to \infty$, $i = 1, \ldots, p$, (iii) $\mathbf{R}_T \to \mathbf{R}_\infty$ *as* $T \to \infty$, (iv) $\mathbf{R}_\infty$ *is nonsingular, and* (v)

$$(3) \quad \sup_{t=1,2,\ldots} \int_{|u|>c} u^2 \, dF_t(u) \to 0$$

as $c \to \infty$. *Then* $D_T(b_T - \beta)$ *has a limiting normal distribution with mean* $\mathbf{0}$. *and covariance matrix* $\sigma^2 \mathbf{R}_\infty^{-1}$.

PROOF. We have

$$D_T(b_T - \beta) = D_T(A_T^{-1} c_T - \beta) \tag{4}$$
$$= D_T\left[A_T^{-1} \sum_{t=1}^{T} z_t(u_t + z_t'\beta) - \beta\right]$$
$$= (D_T^{-1} A_T D_T^{-1})^{-1} D_T^{-1} \sum_{t=1}^{T} z_t u_t.$$

We shall prove that $D_T^{-1} \sum_{t=1}^{T} z_t u_t$ has a limiting normal distribution with mean $\mathbf{0}$ and covariance matrix $\sigma^2 \mathbf{R}_\infty$ by showing that $\alpha' D_T^{-1} \sum_{t=1}^{T} z_t u_t$ has a limiting normal distribution with mean 0 and variance $\sigma^2 \alpha' \mathbf{R}_\infty \alpha$ for every arbitrary vector α ($\neq \mathbf{0}$). (See Theorem 7.7.7.) Let $\gamma_t^T = \alpha' D_T^{-1} z_t = \sum_{j=1}^{p} (\alpha_j z_{jt}/\sqrt{a_{jj}^T})$, $t = 1, \ldots, T$, $T = 1, 2, \ldots$. Then

$$\operatorname{Var}\left(\sum_{t=1}^{T} \gamma_t^T u_t\right) = \sigma^2 \sum_{t=1}^{T} (\gamma_t^T)^2 = \sigma^2 \alpha' \mathbf{R}_T \alpha. \tag{5}$$

Let $w_t^T = \gamma_t^T u_t / (\sigma \sqrt{\alpha' \mathbf{R}_T \alpha})$. Then $\mathscr{E} w_t^T = 0$, $\sum_{t=1}^{T} \operatorname{Var}(w_t^T) = 1$, and

$$\sum_{t=1}^{T} \int_{|w| > \delta} w^2 \, dF_t^T(w) = \sum_{t=1}^{T} \frac{(\gamma_t^T)^2}{\sigma^2 \alpha' \mathbf{R}_T \alpha} \int_{u^2 > \delta^2 [\alpha' \mathbf{R}_T \alpha / (\gamma_t^T)^2] \sigma^2} u^2 \, dF_t(u) \tag{6}$$
$$\leq \frac{1}{\sigma^2} \sup_{t=1,2,\ldots} \int_{u^2 > \delta^2 [\alpha' \mathbf{R}_T \alpha / \max_{t=1,\ldots,T} (\gamma_t^T)^2] \sigma^2} u^2 \, dF_t(u) \to 0,$$

where $F_t^T(w)$ is the distribution of w_t^T, $t = 1, \ldots, T$. The limit in (6) follows by (3) since

$$\max_{t=1,\ldots,T} |\gamma_t^T| = \max_{t=1,\ldots,T} \left| \sum_{j=1}^{p} \alpha_j \frac{z_{jt}}{\sqrt{a_{jj}^T}} \right| \leq \sum_{j=1}^{p} |\alpha_j| \max_{t=1,\ldots,T} \frac{|z_{jt}|}{\sqrt{\sum_{s=1}^{T} z_{js}^2}}, \tag{7}$$

which converges to 0 by (i), (ii), and Lemma 2.6.1 following. The Lindeberg-Feller condition (6) implies that $\alpha' D_T^{-1} \sum_{t=1}^{T} z_t u_t$ has a limiting normal distribution. [See Loève (1963), Section 21.2, or Theorem 7.7.2.] Therefore $D_T^{-1} \sum_{t=1}^{T} z_t u_t$ has a limiting normal distribution with mean $\mathbf{0}$ and covariance matrix $\sigma^2\mathbf{R}_\infty$ and $\boldsymbol{D}_T(\boldsymbol{b}_T - \boldsymbol{\beta})$ has a limiting normal distribution with mean $\mathbf{0}$ and covariance matrix $\sigma^2\mathbf{R}_\infty^{-1}$. Q.E.D.

LEMMA 2.6.1. *(i) $a_{ii}^T \to \infty$ and (ii) $z_{i,T+1}^2/a_{ii}^T \to 0$ imply*

$$\frac{\max\limits_{t=1,\ldots,T} z_{it}^2}{a_{ii}^T} \to 0 \tag{8}$$

as $T \to \infty$. Conversely (8) implies (i) and (ii).

PROOF. Let $t(T)$ be the largest index t ($t = 1, \ldots, T$) such that $z_{it(T)}^2 = \max_{t=1,\ldots,T} z_{it}^2$. Then $\{t(T)\}$ is a nondecreasing sequence of integers. If $\{t(T)\}$ is bounded, let τ be the maximum; then

$$\frac{\max\limits_{t=1,\ldots,T} z_{it}^2}{a_{ii}^T} \leq \frac{z_{i\tau}^2}{a_{ii}^T} \to 0 \tag{9}$$

by (i). If $t(T) \to \infty$ as $T \to \infty$, then

$$\frac{\max\limits_{t=1,\ldots,T} z_{it}^2}{a_{ii}^T} \leq \frac{z_{it(T)}^2}{a_{ii}^{t(T)-1}} \to 0 \tag{10}$$

by (ii). Proof of the converse is left to the reader.

COROLLARY 2.6.1. *Let $y_t = \boldsymbol{\beta}' z_t + u_t$, $t = 1, 2, \ldots$, where the u_t's are independently and identically distributed with mean 0 and variance σ^2. Suppose (i), (ii), (iii), and (iv) of Theorem 2.6.1 are satisfied. Then $\boldsymbol{D}_T(\boldsymbol{b}_T - \boldsymbol{\beta})$ has a limiting normal distribution with mean $\mathbf{0}$ and covariance matrix $\sigma^2\mathbf{R}_\infty^{-1}$.*

COROLLARY 2.6.2. *Let $y_t = \boldsymbol{\beta}' z_t + u_t$, $t = 1, 2, \ldots$, where the u_t's are independent and u_t has mean 0 and variance σ^2. Suppose (i), (ii), (iii), and (iv) of Theorem 2.6.1 are satisfied. If there exist $\delta > 0$ and $M > 0$ such that $\mathscr{E}\,|u_t|^{2+\delta} \leq M$, $t = 1, 2, \ldots$, then $\boldsymbol{D}_T(\boldsymbol{b}_T - \boldsymbol{\beta})$ has a limiting normal distribution with mean $\mathbf{0}$ and covariance matrix $\sigma^2\boldsymbol{R}_\infty^{-1}$.*

Theorem 2.6.1 in a slightly different form was given by Eicker (1963). The two corollaries use stronger sufficient conditions. Corollary 2.6.2 with the condition (ii) replaced by $z_t' z_t$ uniformly bounded can be proved directly by use of the Liapounov central limit theorem. (See Problem 20.) However, the boundedness of $z_t' z_t$ is not sufficiently general for our purposes since polynomials in t do not satisfy the condition.

THEOREM 2.6.2. *Under the conditions of Theorem 2.6.1, Corollary 2.6.1, or Corollary 2.6.2 s^2 converges stochastically to σ^2.*

Proof. Since

$$\sum_{t=1}^{T}(y_t - \boldsymbol{\beta}'z_t)^2 = \sum_{t=1}^{T}[(y_t - b_T'z_t) + (b_T - \boldsymbol{\beta})'z_t]^2 \tag{11}$$

$$= \sum_{t=1}^{T}(y_t - b_T'z_t)^2 + (b_T - \boldsymbol{\beta})'A_T(b_T - \boldsymbol{\beta}),$$

we have

$$s^2 = \frac{T}{T-p}\,\frac{\sum_{t=1}^{T}(y_t - \boldsymbol{\beta}'z_t)^2}{T} - (b_T - \boldsymbol{\beta})'\frac{1}{T-p}A_T(b_T - \boldsymbol{\beta}). \tag{12}$$

The second term in (12) is nonnegative and has expected value

$$\frac{1}{T-p}\mathscr{E}(b_T - \boldsymbol{\beta})'A_T(b_T - \boldsymbol{\beta}) = \frac{1}{T-p}\operatorname{tr}\mathscr{E}A_T(b_T - \boldsymbol{\beta})(b_T - \boldsymbol{\beta})' \tag{13}$$

$$= \frac{p}{T-p}\sigma^2,$$

which goes to 0 as $T \to \infty$. Hence, by Tchebycheff's inequality the second term converges stochastically to 0. The theorem will follow from the appropriate law of large numbers as

$$\operatorname*{plim}_{T\to\infty}\frac{\sum_{t=1}^{T}(y_t - \boldsymbol{\beta}'z_t)^2}{T} = \operatorname*{plim}_{T\to\infty}\frac{\sum_{t=1}^{T}u_t^2}{T} = \operatorname*{plim}_{T\to\infty}\frac{\sum_{t=1}^{T}x_t}{T} = \sigma^2, \tag{14}$$

where $x_t = u_t^2$ and $\mathscr{E}x_t = \mathscr{E}u_t^2 = \sigma^2$. For the conditions of Corollary 2.6.1 the x_t's are identically distributed; for the conditions of Corollary 2.6.2, $\mathscr{E}\,|x_t|^{1+\frac{1}{2}\delta} < M$ [Loève (1963), Section 20.1]; and for Theorem 2.6.1

$$\sup_{t=1,2,\ldots}\int_{x>d} x\,dG_t(x) \to 0 \tag{15}$$

as $d \to \infty$, where $G_t(x)$ is the distribution function of $x_t = u_t^2$. [See Loève (1963), Section 20.2, and Problem 21.] Q.E.D.

The implication of the theorems is that if the usual normal theory is used in the case of large samples it will be approximately correct even if the observations are not normally distributed. We shall see in Section 5.5 that in autoregressive processes, where $\boldsymbol{\beta}'z_t$ is replaced by a linear combination of lagged values of y_t, there is further asymptotic theory justifying the procedures for large samples when the assumptions are not satisfied. The asymptotic theory will be generalized and presented in more detail in Section 5.5.

In Section 10.2.4 similar theorems will be proved for $\{u_t\}$ a stationary stochastic process of the moving average type.

REFERENCES

The theory of regression is treated more fully in several textbooks; for example, *Applied Regression Analysis* by N. R. Draper and H. Smith; *An Introduction to Linear Statistical Models,* Vol. I, by Franklin A. Graybill; *The Design and Analysis of Experiments* by Oscar Kempthorne; *The Advanced Theory of Statistics*, Vol. II, by Maurice G. Kendall and Alan Stuart; *Principles of Regression Analysis* by R. L. Plackett; *The Analysis of Variance* by Henry Scheffé; *Mathematical Statistics* by S. S. Wilks; and *Regression Analysis* by E. J. Williams.

Section 2.2. T. W. Anderson (1958).
Section 2.4. T. W. Anderson (1948).
Section 2.6. Eicker (1963), Loève (1963).

PROBLEMS

1. (Sec. 2.2.) Prove that $\boldsymbol{b}$ defined by (5) minimizes $\sum_{t=1}^T(y_t - \tilde{\boldsymbol{b}}'\boldsymbol{z}_t)^2$ with respect to $\tilde{\boldsymbol{b}}$. [*Hint:* Show that

$$\sum_{t=1}^T (y_t - \tilde{\boldsymbol{b}}'\boldsymbol{z}_t)^2 = \sum_{t=1}^T (y_t - \boldsymbol{b}'\boldsymbol{z}_t)^2 + (\boldsymbol{b} - \tilde{\boldsymbol{b}})'\boldsymbol{A}(\boldsymbol{b} - \tilde{\boldsymbol{b}}).]$$

2. (Sec. 2.2.) Verify (8) and (9).

3. (Sec. 2.2.) Prove the Gauss-Markov Theorem. [*Hint:* Show that if the components of $\sum_{t=1}^T \boldsymbol{a}_t y_t$ are unbiased estimates of the components of $\boldsymbol{\beta}$, where $\boldsymbol{a}_t = \boldsymbol{A}^{-1}\boldsymbol{z}_t + \boldsymbol{d}_t$, $t = 1, \ldots, T$, and $\boldsymbol{a}_1, \ldots, \boldsymbol{a}_T$ and $\boldsymbol{d}_1, \ldots, \boldsymbol{d}_T$ are vectors of p components, then $\sum_{t=1}^T \boldsymbol{d}_t\boldsymbol{z}_t' = \boldsymbol{0}$. Show that the variances of such estimates are the diagonal elements of $\sigma^2\boldsymbol{A}^{-1} + \sigma^2\sum_{t=1}^T \boldsymbol{d}_t\boldsymbol{d}_t'$.]

4. (Sec. 2.2.) Show that $\boldsymbol{b}$ and $\hat{\sigma}^2 = (T-p)s^2/T$ are the maximum likelihood estimates of $\boldsymbol{\beta}$ and σ^2 if $y_1, \ldots, y_T$ are independently normally distributed.

5. (Sec. 2.2.) Show that $\boldsymbol{b}$ and s^2 are a sufficient set of statistics for $\boldsymbol{\beta}$ and σ^2 if $y_1, \ldots, y_T$ are independently normally distributed. [*Hint:* Show that the density is

$$(2\pi\sigma^2)^{-\frac{1}{2}T} \exp\{-\tfrac{1}{2}[(\boldsymbol{b} - \boldsymbol{\beta})'\boldsymbol{A}(\boldsymbol{b} - \boldsymbol{\beta}) + (T-p)s^2]/\sigma^2\}.]$$

6. (Sec. 2.2.) If $\mathscr{E}\boldsymbol{y} = \boldsymbol{Z\beta}$ and $\mathscr{E}(\boldsymbol{y} - \boldsymbol{Z\beta})(\boldsymbol{y} - \boldsymbol{Z\beta})' = \sigma^2\boldsymbol{I}$, prove that the covariance matrix of residuals $\boldsymbol{y} - \boldsymbol{Zb}$ is $\sigma^2(\boldsymbol{I} - \boldsymbol{ZA}^{-1}\boldsymbol{Z}')$.

7. (Sec. 2.2.) In the formulation of Problem 6, prove $\mathscr{E}(\boldsymbol{y} - \boldsymbol{Zb})(\boldsymbol{b} - \boldsymbol{\beta})' = \boldsymbol{0}$.

8. (Sec. 2.2.) Let $\boldsymbol{A}$ (nonsingular) and $\boldsymbol{B} = \boldsymbol{A}^{-1}$ be partitioned similarly as

$$\boldsymbol{A} = \begin{pmatrix} \boldsymbol{A}_{11} & \boldsymbol{A}_{12} \\ \boldsymbol{A}_{21} & \boldsymbol{A}_{22} \end{pmatrix}, \qquad \boldsymbol{B} = \begin{pmatrix} \boldsymbol{B}_{11} & \boldsymbol{B}_{12} \\ \boldsymbol{B}_{21} & \boldsymbol{B}_{22} \end{pmatrix},$$

where $\boldsymbol{A}_{11}$ is nonsingular. By solving $\boldsymbol{AB} = \boldsymbol{I}$ according to the partitioning, prove

$$\text{(i)} \quad \boldsymbol{B}_{12}\boldsymbol{B}_{22}^{-1} = -\boldsymbol{A}_{11}^{-1}\boldsymbol{A}_{12},$$

$$\text{(ii)} \quad \boldsymbol{B}_{22} = (\boldsymbol{A}_{22} - \boldsymbol{A}_{21}\boldsymbol{A}_{11}^{-1}\boldsymbol{A}_{12})^{-1}.$$

9. (Sec. 2.3.) Show that if $\boldsymbol{G} = (g_{ij})$ is nonsingular and lower triangular (that is, $g_{ij} = 0,\ i < j$), then $\boldsymbol{G}^{-1}$ is lower triangular.

10. (Sec. 2.3.) Let $\tilde{y}_t$ be the residual of y_t from the sample regression on $\boldsymbol{z}_t^{(1)}$, and let $\tilde{\boldsymbol{z}}_t^{(2)}$ be the residuals of $\boldsymbol{z}_t^{(2)}$ from the formal sample regression on $\boldsymbol{z}_t^{(1)}$.

(a) Show $\mathscr{E}\tilde{y}_t = \boldsymbol{\beta}^{(2)\prime}\tilde{\boldsymbol{z}}_t^{(2)}$.

(b) Show that the least squares estimate of $\boldsymbol{\beta}^{(2)}$ based on $\tilde{y}_t$ and $\tilde{\boldsymbol{z}}_t^{(2)}$ according to (a) is identical to that based on y_t and $\boldsymbol{z}_t$.

11. (Sec. 2.3.) Prove explicitly and completely that $\boldsymbol{D} = \boldsymbol{\Gamma}$.

12. (Sec. 2.3.) Let z_{kt}^* be the orthogonalization of z_{kt}, $t = 1, \ldots, T$, $\boldsymbol{\beta}^{*\prime} = (\boldsymbol{\beta}^{*(1)\prime}\ \boldsymbol{\beta}^{*(2)\prime})$ be the corresponding transform of $\boldsymbol{\beta}' = (\boldsymbol{\beta}^{(1)\prime}\ \boldsymbol{\beta}^{(2)\prime})$. Prove that criterion (12) of Section 2.2 for testing the hypothesis H: $\boldsymbol{\beta}^{(2)} = \boldsymbol{0}$ is

$$\frac{\sum_{i=r+1}^{p} a_{ii}^*(b_i^*)^2}{(p-r)s^2} = \frac{\sum_{i=r+1}^{p} b_i^* c_i^*}{(p-r)s^2} = \frac{\sum_{i=r+1}^{p} (c_i^*)^2/a_{ii}^*}{(p-r)s^2}.$$

13. (Sec. 2.4.) Show algebraically that the difference between (14) and (11) is positive semidefinite. [*Hint:* The difference is σ^2 times

$$(\boldsymbol{Z}'\boldsymbol{Z})^{-1}[\boldsymbol{Z}'\boldsymbol{\Psi Z} - \boldsymbol{Z}'\boldsymbol{Z}(\boldsymbol{Z}'\boldsymbol{\Psi}^{-1}\boldsymbol{Z})^{-1}\boldsymbol{Z}'\boldsymbol{Z}](\boldsymbol{Z}'\boldsymbol{Z})^{-1},$$

which is positive semidefinite if

$$\begin{pmatrix} \boldsymbol{Z}'\boldsymbol{\Psi}^{-1}\boldsymbol{Z} & \boldsymbol{Z}'\boldsymbol{Z} \\ \boldsymbol{Z}'\boldsymbol{Z} & \boldsymbol{Z}'\boldsymbol{\Psi Z} \end{pmatrix} = \begin{pmatrix} \boldsymbol{Z}'\boldsymbol{\Psi}^{-1} \\ \boldsymbol{Z}' \end{pmatrix} \boldsymbol{\Psi}(\boldsymbol{\Psi}^{-1}\boldsymbol{Z} \quad \boldsymbol{Z})$$

is positive semidefinite.]

14. (Sec. 2.5.) Verify that the Gauss-Markov Theorem implies that $\boldsymbol{b}'\boldsymbol{z}_\tau$ is the best linear unbiased estimate of $\boldsymbol{\beta}'\boldsymbol{z}_\tau$.

15. (Sec. 2.5.) Show that if a linear estimate $\sum_{t=1}^T k_t y_t$ of an expected value is biased, then the mean square error is unbounded. [*Hint:* If $\mathscr{E} \sum_{t=1}^T k_t y_t \neq \boldsymbol{\beta}' \boldsymbol{z}_\tau$ for $\boldsymbol{\beta} = \boldsymbol{\gamma}$, then consider $\boldsymbol{\beta} = k\boldsymbol{\gamma}$ as $k \to \infty$.]

16. (Sec. 2.5.) Show that one-sided prediction intervals for y_τ with confidence $1 - \varepsilon$ are given alternatively by

$$y_\tau \leq \boldsymbol{b}'\boldsymbol{z}_\tau + t_{T-p}(2\varepsilon)s\sqrt{1 + \boldsymbol{z}_\tau' \boldsymbol{A}^{-1}\boldsymbol{z}_\tau}$$

and

$$\boldsymbol{b}'\boldsymbol{z}_\tau - t_{T-p}(2\varepsilon)s\sqrt{1 + \boldsymbol{z}_\tau' \boldsymbol{A}^{-1}\boldsymbol{z}_\tau} \leq y_\tau.$$

17. (Sec. 2.5.) Show that a prediction region for y_τ and y_ρ ($\tau > T$, $\rho > T$, $\tau \neq \rho$) with confidence $1 - \varepsilon$ is given by

$$(\boldsymbol{b}'\boldsymbol{z}_\tau - y_\tau \quad \boldsymbol{b}'\boldsymbol{z}_\rho - y_\rho)\begin{pmatrix} 1 + \boldsymbol{z}_\tau'\boldsymbol{A}^{-1}\boldsymbol{z}_\tau & \boldsymbol{z}_\tau'\boldsymbol{A}^{-1}\boldsymbol{z}_\rho \\ \boldsymbol{z}_\rho'\boldsymbol{A}^{-1}\boldsymbol{z}_\tau & 1 + \boldsymbol{z}_\rho'\boldsymbol{A}^{-1}\boldsymbol{z}_\rho \end{pmatrix}^{-1}\begin{pmatrix} \boldsymbol{b}'\boldsymbol{z}_\tau - y_\tau \\ \boldsymbol{b}'\boldsymbol{z}_\rho - y_\rho \end{pmatrix} \leq 2s^2 F_{2,T-p}(\varepsilon).$$

18. (Sec. 2.5.) Prove

$$\boldsymbol{z}_\tau^{(1)\prime}\boldsymbol{A}_{11}^{-1}\boldsymbol{z}_\tau^{(1)} \leq \boldsymbol{z}_\tau'\boldsymbol{A}^{-1}\boldsymbol{z}_\tau.$$

[*Hint:* Use (7).]

19. (Sec. 2.6.) Prove Corollary 2.6.2 using Theorem 2.6.1.

20. (Sec. 2.6.) Prove Corollary 2.6.2 (without use of Theorem 2.6.1) with condition (ii) replaced by (ii′) there exists a constant L such that $\boldsymbol{z}_t'\boldsymbol{z}_t \leq L$, $t = 1, 2, \ldots$.

21. (Sec. 2.6.) Prove that (15) implies

(i)
$$\sum_{t=1}^T \int_{x \geq T} dG_t(x) \to 0,$$

(ii)
$$\frac{1}{T}\sum_{t=1}^T \int_{x<T} x\, dG_t(x) \to \sigma^2,$$

(iii)
$$\frac{1}{T^2}\sum_{t=1}^T \left\{ \int_{x<T} x^2\, dG_t(x) - \left[\int_{x<T} x\, dG_t(x) \right]^2 \right\} \to 0$$

as $T \to \infty$ [which is the condition of Loève (1963), Section 20.2]. [*Hint:* For an arbitrary constant d' and $T > d'$ show that the first term above is

$$\frac{1}{T^2}\sum_{t=1}^T \int_{0 \leq x \leq d'} x^2\, dG_t(x) + \frac{1}{T^2}\sum_{t=1}^T \int_{d'<x<T} x^2\, dG_t(x) \leq \frac{\sigma^2 d'}{T} + \frac{1}{T}\sum_{t=1}^T \int_{x>d'} x\, dG_t(x).\Big]$$

CHAPTER 3

Trends and Smoothing

3.1. INTRODUCTION

We shall study so-called error models in which the observed time series is treated as the sum of systematic parts or trend and random parts or error. In this chapter we consider error models in which the underlying trend has a smooth movement increasing or decreasing with time but not regularly repeating; in the next chapter we consider trends which are periodic and which behave roughly the same in various ranges of time. The random terms are assumed to have equal variances at all time points and to be uncorrelated; they may be errors of observation or other kinds of irregularities. The assumptions of equal variances and lack of correlation are approximations to reality. In Chapter 10 we shall consider the difficulties due to these assumptions not being satisfied. Sometimes the assumptions of equal variances and of the additive nature of the trend and error are approximated well by measuring the quantity studied suitably; for example, in some econometric studies analysis is made of the logarithms of prices. Linear combinations of the variable at successive time points may be more nearly uncorrelated; if the coefficients of the linear combination are known (for example, of differences), the linear combination may be treated.

In some cases the underlying trend is a known function of time or other observable quantities and of (possibly unknown) parameters. If the function is linear in the parameters, we have the familiar regression case reviewed in Chapter 2. However, some functions of time such as growth curves are not linear in the parameters; in such cases the estimation of parameters and testing hypotheses about them is not easy. We consider these problems briefly in Section 3.5.

In another category of cases the trend or general movement is a function of time or of unspecified quantities which is unknown but which can be adequately represented by a linear combination of known functions of time. If the trend is periodic, it can be represented by a linear combination of trigonometric terms (that is, a finite Fourier series); the statistical problems of this approach are considered in Chapter 4. If the trend is fluctuating with broad movements up or down, it can often be adequately approximated by a polynomial; inference in this case is treated in Section 3.2.

Sometimes the trend is an unknown function of time which moves gradually with time but the movement over long ranges is such that it is not represented well by a polynomial of low degree or by a short Fourier series. Methods which involve "smoothness" properties are treated in Sections 3.3 and 3.4. These are nonparametric methods.

3.2. POLYNOMIAL TRENDS

3.2.1. Orthogonal Polynomials

In many studies the time series has a gradual movement in general, though it also has fluctuations and irregularities. For example, many economic time series have been roughly going up. This broad movement may be called a *trend.* In many instances it may be desired to infer this trend from the single time series studied, though it may be known that the trend is due to certain other factors such as the growth of an economic time series being due to growth of population. When there is no theory to specify the trend as a certain function of time, it may be possible to approximate it by a polynomial in time of low degree. In the simplest nontrivial case of a uniform movement up or down, a first-degree polynomial, that is, a straight line, may give an adequate representation.

The polynomial trend is primarily a descriptive device; it summarizes succinctly the over-all characteristics of the series. To do this usefully the polynomial must be of fairly low degree. In many cases there is no substantive meaning which can be given to the coefficients of the polynomial; the polynomial is a substitute for a more sophisticated and complicated (but unknown) function of time. The fitted polynomial can usually be used for interpolation, but must be used cautiously for extrapolation since there is a question of how good an approximation the polynomial is to the underlying trend outside the range of the given data.

In some studies the primary interest is in short-term cycles, fluctuations, and irregularities. Then one wants to fit a trend in order to obtain a base-line from which to measure the aspects of the series that change in small units of time.

Our basic "error model" specifies that the observable y_t is the sum of a trend in time $f(t)$ and an (unobservable) error u_t; that is, $y_t = f(t) + u_t$, where the

u_t's are uncorrelated and have means $\mathscr{E}u_t = 0$ and variances $\mathscr{E}u_t^2 = \sigma^2$. In this section we assume that the trend is a polynomial of degree q

$$f(t) = \alpha_0 + \alpha_1 t + \cdots + \alpha_q t^q. \tag{1}$$

A first-degree function represents a uniform movement upward or downward, a second-degree function can display a decrease followed by an increase or vice versa, etc. Usually q will be small relative to T. Of course, the customary techniques of regression analysis can be used.

As noted in Chapter 2, we can transform the so-called independent variates, $1, t, \ldots, t^q$ in this case, to orthogonal independent variates $\phi_{0T}(t), \ldots, \phi_{qT}(t)$ here. Since the polynomials are used again and again, it is worthwhile to orthogonalize them once and for all; then all subsequent computations are simpler. The powers of t have a natural ordering. In almost all cases if a given power is to be included in the regression function all lower powers will also be included.

The orthogonal polynomial of degree k will be

$$\phi_{kT}(t) = t^k + C_{k-1}(k, T)t^{k-1} + \cdots + C_1(k, T)t + C_0(k, T), \qquad k = 1, \ldots, T-1, \tag{2}$$

and $\phi_{0T}(t) = 1$. For $\phi_{kT}(t)$ to be orthogonal to $\phi_{0T}(t), \ldots, \phi_{k-1,T}(t)$ or equivalently to $1, t, \ldots, t^{k-1}$, we must have

$$\sum_{t=1}^{T} \phi_{kT}(t)t^i = 0, \qquad i = 0, 1, \ldots, k-1; \tag{3}$$

that is,

$$C_0(k, T)\sum_{t=1}^{T} t^i + C_1(k, T)\sum_{t=1}^{T} t^{i+1} + \cdots + C_{k-1}(k, T)\sum_{t=1}^{T} t^{i+k-1} = -\sum_{t=1}^{T} t^{i+k}, \qquad i = 0, 1, \ldots, k-1. \tag{4}$$

These equations constitute a special case of (16) of Section 2.3 (with obvious changes in notation). For $k \leq T-1$ the set of equations (4) has a unique solution for $C_0(k, T), \ldots, C_{k-1}(k, T)$. For example, we obtain

$$\begin{aligned} \phi_{0T}(t) &= 1, \\ \phi_{1T}(t) &= t - \tfrac{1}{2}(T+1), \\ \phi_{2T}(t) &= t^2 - (T+1)t + (T+1)(T+2)/6. \end{aligned} \tag{5}$$

We have defined the orthogonal polynomials so the leading term is simply the highest power of t. We could multiply each $\phi_{kT}(t)$ by a constant without affecting the orthogonality property

$$\sum_{t=1}^{T} \phi_{iT}(t)\phi_{kT}(t) = 0, \qquad i \neq k, \quad i, k = 0, 1, \ldots, T-1. \tag{6}$$

The full set, $\phi_{0T}(t), \ldots, \phi_{T-1,T}(t)$, constitutes a nonsingular linear transformation of $1, t, \ldots, t^{T-1}$ (which is the maximum number of linearly independent polynomials over the range $t = 1, \ldots, T$).

In terms of the orthogonal polynomials the trend (1) is written

$$f(t) = \gamma_0 + \gamma_1\phi_{1T}(t) + \cdots + \gamma_q\phi_{qT}(t). \tag{7}$$

Then

$$\alpha_k = \gamma_k + C_k(k+1, T)\gamma_{k+1} + \cdots + C_k(q, T)\gamma_q, \qquad k = 0, 1, \ldots, q. \tag{8}$$

The estimates of the coefficients in (7) based on observations $y_1, \ldots, y_T$ are

$$c_k = \frac{\sum_{t=1}^{T} y_t\phi_{kT}(t)}{\sum_{t=1}^{T} \phi_{kT}^2(t)}, \qquad k = 0, 1, \ldots, q. \tag{9}$$

The unbiased estimate of the variance is

$$\begin{aligned} s^2 &= \frac{\sum_{t=1}^{T} [y_t - c_0 - c_1\phi_{1T}(t) - \cdots - c_q\phi_{qT}(t)]^2}{T-q-1} \\ &= \frac{\sum_{t=1}^{T} y_t^2 - c_0^2 T - c_1^2 \sum_{t=1}^{T} \phi_{1T}^2(t) - \cdots - c_q^2 \sum_{t=1}^{T} \phi_{qT}^2(t)}{T-q-1}. \end{aligned} \tag{10}$$

The numerical computation of (9) and (10) is usually done in desk calculation by means of tables of the orthogonal polynomials. Fisher and Yates (1963) give them up to degree 5 from $T = 3$ to $T = 52$ in Table 23, and R. L. Anderson and Houseman (1942) give them to $T = 104$. For a given value of T (denoted by n in Fisher and Yates) there is tabulated

$$\xi'_{kt} = \lambda_k\phi_{kT}(t), \tag{11}$$

where λ_k is the smallest rational number so that the values of $\lambda_k \phi_{kT}(t)$ are integers. [It will be noted that the coefficients of $C_0(k, T), \ldots, C_{k-1}(k, T)$ in (4) are integers and hence $C_0(k, T), \ldots, C_{k-1}(k, T)$ are rational numbers.] For example, $\phi_{1T}(t) = t - \frac{1}{2}(T + 1)$. When T is odd, $\phi_{1T}(t)$ is an integer and $\lambda_1 = 1$; when T is even, $\phi_{1T}(t)$ is an integer plus $\frac{1}{2}$ and $\lambda_1 = 2$. At the foot of each column of the tables is given

$$\sum_{t=1}^{T} (\xi'_{kt})^2 = \lambda_k^2 \sum_{t=1}^{T} \phi_{kT}^2(t) \tag{12}$$

and λ_k. The purpose of displaying the polynomials in integers is that then there is no rounding error in the computation until the division in (9) by the sum of squares. The computations with the tabulated values give c_k/λ_k, which are estimates of γ_k/λ_k (which are the coefficients of ξ'_{kt}). A numerical example is given at the end of this section.

When high-speed computers are used, tables of orthogonal polynomials are not necessary or convenient. As was observed in Chapter 2, the fitted regression is the same for powers of t and for orthogonal polynomials to the same degree, and the coefficients of the highest degree are the same. Any method of pivotal condensation yields the coefficients of the orthogonal polynomials as well as the coefficients of powers of t. The coefficients of the orthogonal polynomials (normalized by division by the square roots of the sums of squares of orthogonal polynomials) are easily interpreted as indicating the effect of that degree in the regression beyond the effect of lower-degree terms.

3.2.2. Determining the Degree of a Polynomial Trend

From the theory of least squares we know that c_k is an unbiased estimate of γ_k and s^2 is an unbiased estimate of σ^2. The variance of c_k is

$$\mathscr{E}(c_k - \gamma_k)^2 = \frac{\sigma^2}{\sum_{t=1}^{T} \phi_{kT}^2(t)}, \tag{13}$$

and the c_k's are uncorrelated. We shall assume the y_t's are normally distributed. Then $c_0, c_1, \ldots, c_q$ are independently normally distributed and $(T - q - 1)s^2/\sigma^2$ is distributed as χ^2 with $T - q - 1$ degrees of freedom independently of $c_0, c_1, \ldots, c_q$. The usual tests and confidence-interval procedures can be used. For example, if one wants to test the hypothesis H: $\gamma_q = 0$ against alternatives

$\gamma_q \neq 0$ at significance level ε one rejects the hypothesis if

$$\frac{|c_q| \sqrt{a_{qq}}}{s} > t_{T-q-1}(\varepsilon), \tag{14}$$

where

$$a_{qq} = \sum_{t=1}^{T} \phi_{qT}^2(t) \tag{15}$$

and $t_{T-q-1}(\varepsilon)$ is the (two-sided) ε-significance point of the t-distribution with $T - q - 1$ degrees of freedom. The null hypothesis is that the polynomial is of degree less than q given that its degree is at most q. This test of hypothesis H: $\gamma_q = 0$ under normality is best at significance level ε in the following senses: It is the (i) uniformly most powerful unbiased test, (ii) uniformly most powerful invariant test (invariant under scale transformations $c_i \to kc_i$, $i = 0, 1, \ldots, q$, $s^2 \to k^2 s^2$ and location transformations† $c_i \to c_i + a_i$, $i = 0, 1, \ldots, q - 1$), (iii) uniformly most powerful test with power depending on only γ_q^2/σ^2, (iv) uniformly most powerful symmetric similar-region test (that is, symmetric in c_q), and (v) uniformly most powerful similar-region test with power depending on $|\gamma_q|$ (that is, not depending on the sign of γ_q). [See Lehmann (1959), Chapter 5, Section 2, and Chapter 6, Problem 11.]

Frequently the investigator does not know in advance the degree of polynomial to fit and he does not merely ask whether the degree should be a specified q or should be less. He wants to decide the appropriate degree within some range of possibilities. It is an advantage to represent the trend by a polynomial of low degree because the curve is smoother, the presumed "explanation" is simpler, and the function is more economical. However, if the underlying mean value of the observed variable is not approximately a polynomial of low degree the investigator may want to use a polynomial of higher degree. (In statistical terms a disadvantage of too low a degree is a bias in estimating the trend, and a disadvantage of too high a degree is an unnecessarily large variability in the estimation of the trend.) We shall assume in general that there is some number m (perhaps $m = 0$) which is the lowest possible degree and some maximum degree q; presumably the investigator has some a priori information of this kind.

The investigator is confronted with a multiple-decision problem, namely, deciding whether the degree is $m, m + 1, \ldots, q - 1$, or q. We can formalize this problem by saying that he wants to decide to which of the following mutually

† The scale transformation is equivalent to $y_t \to ky_t$, $t = 1, \ldots, T$, and the location transformations are equivalent to adding to $\{y_t\}$ an arbitrary polynomial of degree $q - 1$.

exclusive sets the parameter point $(\gamma_{m+1}, \ldots, \gamma_q)$ belongs:

$$
\begin{aligned}
H_q:&\quad \gamma_q \neq 0,\\
H_{q-1}:&\quad \gamma_q = 0, \qquad \gamma_{q-1} \neq 0,\\
&\quad \vdots\\
H_{m+1}:&\quad \gamma_q = \cdots = \gamma_{m+2} = 0, \qquad \gamma_{m+1} \neq 0,\\
H_m:&\quad \gamma_q = \cdots = \gamma_{m+1} = 0.
\end{aligned}
\tag{16}
$$

The set H_i implies that the polynomial is of degree i. An alternative formulation is that he wishes to decide which (if any) of the following null hypotheses are true:

$$
\begin{aligned}
H_q^*:&\quad \gamma_q = 0,\\
H_{q,q-1}^*:&\quad \gamma_q = \gamma_{q-1} = 0,\\
&\quad \vdots\\
H_{q,q-1,\ldots,m+1}^*:&\quad \gamma_q = \cdots = \gamma_{m+1} = 0.
\end{aligned}
\tag{17}
$$

If any hypothesis of (17) is true, the preceding hypotheses must be true and if any hypothesis is false succeeding ones must be false; that is, $H_{q,q-1,\ldots,m+1}^* \subset \cdots \subset H_q^*$. These two families of sets are related by

$$
\begin{aligned}
H_q^* &= H_{q-1} \cup \cdots \cup H_m,\\
H_{q,q-1}^* &= H_{q-2} \cup \cdots \cup H_m,\\
&\ \vdots\\
H_{q,q-1,\ldots,m+1}^* &= H_m.
\end{aligned}
\tag{18}
$$

See Figure 3.1.

We shall suppose that the investigator wants to control directly the probabilities of errors of saying that coefficients are not zero when they are zero or correspondingly of choosing a higher degree than suitable, and that given these probabilities he wants to minimize the probabilities of saying coefficients are zero when they are not, or correspondingly of choosing a lower degree than suitable. We suppose that the investigator assigns a significance level to

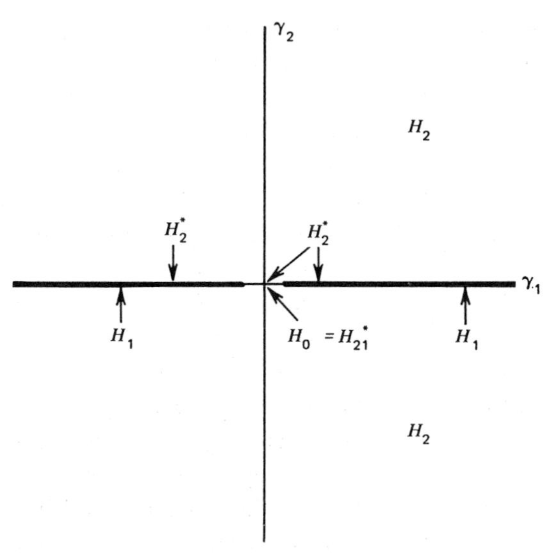

Figure 3.1. Families of multiple-decision sets.

each null hypothesis:

$$
\begin{aligned}
p_q &= \Pr\{\text{reject } H_q^* \mid H_q^*\}, \\
p_q + p_{q-1} &= \Pr\{\text{reject } H_{q,q-1}^* \mid H_{q,q-1}^*\}, \\
&\;\;\vdots \\
p_q + \cdots + p_{m+1} &= \Pr\{\text{reject } H_{q,q-1,\ldots,m+1}^* \mid H_{q,q-1,\ldots,m+1}^*\} \\
&= \Pr\{\text{reject } H_m \mid H_m\},
\end{aligned} \tag{19}
$$

where $p_i \geq 0$ and $p_q + \cdots + p_{m+1} \leq 1$. Since one null hypothesis includes the next (that is, each subsequent null hypothesis is more stringent), the probabilities of rejection are taken monotonically nondecreasing (that is, the probability of rejecting a more restricted null hypothesis when it is true is not less than that of rejecting a less restricted one when it is true). In terms of the mutually exclusive categories the specification is

$$
\begin{aligned}
p_q &= \Pr\{\text{accept } H_q \mid H_{q-1} \cup \cdots \cup H_m\}, \\
p_{q-1} &= \Pr\{\text{accept } H_{q-1} \mid H_{q-2} \cup \cdots \cup H_m\}, \\
&\;\;\vdots \\
p_{m+1} &= \Pr\{\text{accept } H_{m+1} \mid H_m\}.
\end{aligned} \tag{20}
$$

Let $p_m = 1 - p_q - \cdots - p_{m+1} = \Pr\{\text{accept } H_m \mid H_m\}$. In terms of (20) the investigator specifies the probability of deciding to use a polynomial of degree i when the degree of the polynomial needed is less than i (for each i).

A nonrandomized statistical procedure for this multiple-decision problem consists of a set of $q - m + 1$ (mutually exclusive and exhaustive) regions in the space of $c_0, c_1, \ldots, c_q$ and $S = (T - q - 1)s^2$ (or in the original space of $y_1, \ldots, y_T$), which we denote $R_m, R_{m+1}, \ldots, R_q$; if the sample point falls in R_i, then H_i is accepted. The assignment of significance levels implies that these regions are "similar regions" in the sense that when $\gamma_i = \cdots = \gamma_q = 0$ the probabilities of the sample point falling in $R_i, \ldots, R_q$ are $p_i, \ldots, p_q$, respectively (independent of $\gamma_0, \gamma_1, \ldots, \gamma_{i-1}$, and σ^2); that is, when the degree of polynomial is less than i the probability of making an error by saying the degree is i does not depend on what that (lower-degree) polynomial is.

In many cases the investigator wishes only to determine whether a coefficient is zero or nonzero regardless of it being positive or negative; we ask that the probabilities associated with the procedure be unaffected by changing the sign of any relevant coefficient; that is, the probabilities depend on the parameters through $|\gamma_{m+1}|, \ldots, |\gamma_q|$. (It will be seen below that this requirement could be replaced by an assignment of signs to nonzero coefficients.)

Subject to the above requirements, we ask for the best regions in the sense that we want to maximize the probability of R_i when H_i is true, $i = m + 1, \ldots, q$. It should be noticed that we are trying to maximize simultaneously the probabilities of $q - m$ different regions (each for all nonzero values of the corresponding parameter). It will be shown that under the above conditions the maximized probabilities of one region are not affected by the choice of another. This fact permits us to optimize $R_{m+1}, \ldots, R_q$ simultaneously. It might be emphasized that these developments are based on the y_t's being normally distributed.

As noted above, the test with rejection region (14) is optimum for testing the hypothesis H: $\gamma_q = 0$ which is H_q^*, the complement of H_q. Thus the best choice of R_q is (14) with $\varepsilon = p_q$. Indeed the best procedure for deciding H: $\gamma_i = 0$ at a specified significance level ε_i when we assume $\gamma_{i+1} = \cdots = \gamma_q = 0$ is to reject this hypothesis if

$$\frac{c_i^2 a_{ii}}{c_{i+1}^2 a_{i+1,i+1} + \cdots + c_q^2 a_{qq} + S} > \frac{t_{T-i-1}^2(\varepsilon_i)}{T - i - 1}, \tag{21}$$

where $S = (T - q - 1)s^2$ and

$$a_{ii} = \sum_{t=1}^{T} \phi_{iT}^2(t). \tag{22}$$

Secondly, we note that if T_i is a similar region of size ε_i for testing the hypothesis $\gamma_i = 0$, that is, if

(23) $$\Pr\{T_i \mid \gamma_i = \gamma_{i+1} = \cdots = \gamma_q = 0\} = \varepsilon_i$$

for all $\gamma_0, \gamma_1, \ldots, \gamma_{i-1}$, and σ^2 (> 0), then T_i cuts out relative probability ε_i on almost every combination of specified values of $c_0, c_1, \ldots, c_{i-1}$, and $c_i^2 a_{ii} + \cdots + c_q^2 a_{qq} + S$ (the sufficient statistics for $\gamma_0, \gamma_1, \ldots, \gamma_{i-1}$, and σ^2 when $\gamma_i = \cdots = \gamma_q = 0$); that is,

(24)
$$\Pr\{T_i \mid c_0, c_1, \ldots, c_{i-1}, c_i^2 a_{ii} + \cdots + c_q^2 a_{qq} + S; \gamma_i = \cdots = \gamma_q = 0\} = \varepsilon_i$$

almost everywhere (or with probability 1). In fact some of the optimum properties of the t-test noted above are based on this "Neyman structure" of similar regions. [See Lehmann (1959), Chapter 4, or Problem 10.] The requirements (20) [or equivalently (19)] imply that R_i has such a structure; in fact, $R_i \cup R_{i+1} \cup \cdots \cup R_q$, the region of rejection of $H^*_{q,q-1,\ldots,i}$, given $H^*_{q,q-1,\ldots,i}$, has such a structure. We make use of this fact to show that the choice of $R_q, \ldots, R_{i+1}$ [subject to (20)] does not affect the choice of R_i in the sense that the probability of R_i (a function of $|\gamma_i|$) when $\gamma_{i+1} = \cdots = \gamma_q = 0$ does not depend on how $R_q, \ldots, R_{i+1}$ are chosen. Note that we are interested in γ_i only when $\gamma_{i+1} = \cdots = \gamma_q = 0$, and if one of $\gamma_{i+1}, \ldots, \gamma_q$ is not 0 (that is, the degree is greater than i), we are not interested in whether γ_i is 0.

LEMMA 3.2.1. *Let S_{i+1} be a set in the space of $c_0, \ldots, c_q$, and $S = (T - q - 1)s^2$ such that*

(25) $$\Pr\{S_{i+1} \mid \gamma_{i+1} = \cdots = \gamma_q = 0\} = p_q + \cdots + p_{i+1},$$

and let T_i be a set defined by $c_0, \ldots, c_i$, and $c_{i+1}^2 a_{i+1,i+1} + \cdots + c_q^2 a_{qq} + S$. Then

(26) $$\Pr\{\bar{S}_{i+1} \cap T_i \mid \gamma_{i+1} = \cdots = \gamma_q = 0\}$$
$$= (1 - p_q - \cdots - p_{i+1}) \Pr\{T_i \mid \gamma_{i+1} = \cdots = \gamma_q = 0\},$$

where $\bar{S}_{i+1}$ is the complement of S_{i+1}.

PROOF. The requirement (25) is that S_{i+1} be a similar region (with respect to $\gamma_0, \ldots, \gamma_i$, and σ^2) and therefore

(27) $$\Pr\{S_{i+1} \mid c_0, \ldots, c_i, c_{i+1}^2 a_{i+1,i+1} + \cdots + c_q^2 a_{qq} + S; \gamma_{i+1} = \cdots = \gamma_q = 0\}$$
$$= p_q + \cdots + p_{i+1}$$

for almost all $c_0, \ldots, c_i$, and $c_{i+1}^2 a_{i+1,i+1} + \cdots + c_q^2 a_{qq} + S$. Let $f_i(c_0, \ldots, c_i, c_{i+1}^2 a_{i+1,i+1} + \cdots + c_q^2 a_{qq} + S)$ be the characteristic set function of T_i. ($f_i = 1$ if the point is in the set T_i and is 0 otherwise.) Then

$$(28)\quad \Pr\{\bar{S}_{i+1} \cap T_i\}$$
$$= \mathscr{E}[\Pr\{\bar{S}_{i+1} \mid c_0, \ldots, c_i, c_{i+1}^2 a_{i+1,i+1} + \cdots + c_q^2 a_{qq} + S\}$$
$$\times f_i(c_0, \ldots, c_i, c_{i+1}^2 a_{i+1,i+1} + \cdots + c_q^2 a_{qq} + S)]$$
$$= \mathscr{E}[(1 - p_q - \cdots - p_{i+1}) f_i(c_0, \ldots, c_i, c_{i+1}^2 a_{i+1,i+1} + \cdots + c_q^2 a_{qq} + S)],$$

which is (26), proving the lemma. [Note that this proves the t-tests in (21) are independent.]

The point of the lemma is that however $R_q, \ldots, R_{i+1}$ are chosen subject to (20), which implies (25) for $S_{i+1} = R_q \cup \cdots \cup R_{i+1}$, the probability of a region R_i defined as an intersection $\bar{S}_{i+1} \cap T_i$ depends only on T_i and not on S_{i+1} (when $H^*_{q,q-1,\ldots,i+1}$ is true).

Now let T_i^* be the region defined by (21) for $\varepsilon_i = p_i/(1 - p_q - \cdots - p_{i+1})$.

LEMMA 3.2.2. *Given S_{i+1} satisfying* (25) *and any R_i disjoint with S_{i+1} such that*

$$(29)\qquad \Pr\{R_i \mid \gamma_i = \gamma_{i+1} = \cdots = \gamma_q = 0\} = p_i,$$

$$(30)\quad \Pr\{R_i \mid \gamma_i, \gamma_{i+1} = \cdots = \gamma_q = 0\} = \Pr\{R_i \mid -\gamma_i, \gamma_{i+1} = \cdots = \gamma_q = 0\},$$

then

$$(31)\quad \Pr\{\bar{S}_{i+1} \cap T_i^* \mid \gamma_{i+1} = \cdots = \gamma_q = 0\} \geq \Pr\{R_i \mid \gamma_{i+1} = \cdots = \gamma_q = 0\}.$$

PROOF. Suppose there were some value of γ_i so that the inequality (31) were violated. We shall show that this contradicts the previous assertion that $T_i^* = (S_{i+1} \cap T_i^*) \cup (\bar{S}_{i+1} \cap T_i^*)$ is the uniformly most powerful similar-region test of the hypothesis $\gamma_i = 0$ with power independent of the sign of γ_i. For T_i^* and $R_i \cup (S_{i+1} \cap T_i^*)$ are critical regions of the same size, but the power of the second at this special value of γ_i would be greater than that of T_i^*. Since this is false, the lemma must be true.

The implication of the two lemmas is that whatever $R_q, \ldots, R_{i+1}$ are, the best choice of R_i is the part of T_i^* disjoint from $R_q, \ldots, R_{i+1}$ and for such a choice

$$\text{(32)}\quad \Pr\{R_i \mid \gamma_{i+1} = \cdots = \gamma_q = 0\} \\ = (1 - p_q - \cdots - p_{i+1}) \Pr\{T_i^* \mid \gamma_{i+1} = \cdots = \gamma_q = 0\},$$

which does not depend on the choice of $R_q, \ldots, R_{i+1}$.

THEOREM 3.2.1. *Let $R_m, R_{m+1}, \ldots, R_q$ be $q - m + 1$ exhaustive disjoint regions in the sample space such that*

$$\text{(33)}\qquad \Pr\{R_i \mid \gamma_i = \gamma_{i+1} = \cdots = \gamma_q = 0\} = p_i, \qquad i = m+1, \ldots, q,$$

where $p_{m+1} + \cdots + p_q \leq 1$, and

$$\text{(34)}\quad \Pr\{R_i \mid \gamma_i, \gamma_{i+1} = \cdots = \gamma_q = 0\} = \Pr\{R_i \mid -\gamma_i, \gamma_{i+1} = \cdots = \gamma_q = 0\}, \\ i = m+1, \ldots, q.$$

Then for every value of γ_i (34) *is maximized by R_i defined by the intersection of* (21) *for $\varepsilon_i = p_i/(1 - p_q - \cdots - p_{i+1})$ and the complement of $R_q \cup \cdots \cup R_{i+1}$, $i = m+1, \ldots, q$.*

The optimum procedure is, therefore,

$$\begin{aligned} R_q &= T_q^*, \\ R_{q-1} &= \bar{T}_q^* \cap T_{q-1}^*, \\ &\cdot \\ &\cdot \\ &\cdot \\ R_i &= \bar{T}_q^* \cap \bar{T}_{q-1}^* \cap \cdots \cap \bar{T}_{i+1}^* \cap T_i^*, \qquad \text{(35)} \\ &\cdot \\ &\cdot \\ &\cdot \\ R_{m+1} &= \bar{T}_q^* \cap \bar{T}_{q-1}^* \cap \cdots \cap \bar{T}_{m+2}^* \cap T_{m+1}^*, \\ R_m &= \bar{T}_q^* \cap \bar{T}_{q-1}^* \cap \cdots \cap \bar{T}_{m+2}^* \cap \bar{T}_{m+1}^*. \end{aligned}$$

What the procedure amounts to is that one tests $\gamma_q = 0$, $\gamma_{q-1} = 0, \ldots$ in turn until either one rejects such a hypothesis, say rejects $\gamma_i = 0$ and decides H_i, or one accepts all such hypotheses and eventually H_m. Thus the procedure is a sequential one; this results because of the requirement that the probability of deciding that the degree of the polynomial is less than a given integer when that is the case should not depend on what the polynomial is.

The t-test of the hypothesis $\gamma_i = 0$ has several optimum properties; five of them were stated earlier in this section. Several of these lead to corresponding statements of the theorem. The requirements of similarity (33) can be replaced by requirements of unbiasedness, namely that

$$\Pr\{\text{reject } H^*_{q,q-1,\ldots,i} \mid H^*_{q,q-1,\ldots,i}\} \le p_q + \cdots + p_i,\ i = m+1, \ldots, q, \tag{36}$$

for unbiasedness implies that the regions are similar and that the probabilities of the regions are independent of the signs of $\gamma_q, \ldots, \gamma_{m+1}$. [See Lehmann (1959), Chapter 4, Section 1.]

We have not stated here how $p_{m+1}, \ldots, p_q$ are to be chosen. If ε_i is fixed, say ε, then $p_q = \varepsilon$, $p_{q-1} = \varepsilon(1-\varepsilon), \ldots, p_i = \varepsilon(1-\varepsilon)^{q-i}, \ldots, p_{m+1} = \varepsilon(1-\varepsilon)^{q-m-1}$, $p_m = (1-\varepsilon)^{q-m}$. In general one has to balance the desirability of not overestimating the degree with sensitivity of the procedure to nonzero coefficients. A reasonable approach is to set q fairly large, but make ε_i very small for i near q. If the effect of the ith degree is relatively large, there is a chance of learning that fact; if high degrees are not needed, the probability is small that a high degree is decided on. An alternative is to pick q large and let ε_j be small for large j. For example, let $\varepsilon_j = \varepsilon^*(q+1-j)/(q-m)$, $j = m+1, \ldots, q$. Then p_m is approximately $e^{-\frac{1}{2}\varepsilon^*(q-m+1)} \sim 1 - \frac{1}{2}(q-m+1)\varepsilon^*$.

The procedure derived here is not a new one, but the approach is different. Another procedure that has been suggested for deciding the degree of a polynomial regression is also a sequence of significance tests, but in the reverse order. First, test $\gamma_{m+1} = 0$ by the t-test; then if this hypothesis is rejected test $\gamma_{m+2} = 0$; and continue until some hypothesis is accepted or until $\gamma_q = 0$ has been rejected. A disadvantage of this procedure is that if some γ_i is very large, the probability of deciding on a polynomial of too low a degree is large. For example, if γ_2 is large and γ_1 is small, the probability of

$$\frac{c_1^2 a_{11}}{c_2^2 a_{22} + \cdots + c_q^2 a_{qq} + S} < k \tag{37}$$

is relatively large, that is, of accepting $\gamma_1 = 0$ and deciding that the degree of the polynomial is 0. [It may be noted that the procedure does not satisfy (19) or (20) for $\gamma_3 = \cdots = \gamma_q = 0$.]

A practical disadvantage of the procedure which is theoretically best in our formulation is that it requires computation of S and hence of $c_0, c_1, \ldots, c_q$ for a value of q chosen in advance, whereas in the other sequential procedure one computes $c_1, c_2, \ldots$ in sequence (with $c_2^2 a_{22} + \cdots + c_q^2 a_{qq} + S = \sum_{t=1}^T y_t^2 - c_0^2 T - c_1^2 a_{11}$, etc.), and one needs to compute only as long as the hypotheses are rejected. However, this disadvantage is limited because the regression coefficients of orthogonal polynomials are relatively easy to compute since tables of the polynomials are available and one would usually choose q small since orthogonal polynomials are not very useful if the situation calls for a high degree. (In practice if one accepted H_q one might be tempted to compute c_{q+1} as a check on the choice of q.) As indicated earlier, with high-speed computers one uses regression on powers of t; the coefficients of the orthogonal polynomials which are used in the procedure are available in any method of pivotal condensation.

As was noted in Section 2.3, the test of the hypothesis that a particular regression coefficient is zero is unchanged if a linear transformation of independent variables is made that leaves that coefficient unchanged when the regression function is expressed in terms of the new independent variables; a particular case of such a transformation is one which orthogonalizes the variables in a sequence so the variable of interest is the last to be orthogonalized. Thus the test that $\gamma_i = 0$ when $\gamma_{i+1} = \cdots = \gamma_q = 0$ is assumed is exactly the same as the test that $\alpha_i = 0$ when $\alpha_{i+1} = \cdots = \alpha_q = 0$ is assumed [where the α_j's are the coefficients of the powers of t in the polynomial trend (1)]. The multiple-decision procedure outlined above can, therefore, be carried out on the basis of the regression analysis of polynomial trend. (In fact, the procedure applies to any set of independent variables that are ordered.) As indicated in Section 2.3, the test of a coefficient being zero requires only the calculation of the forward solution of the normal equations; in carrying out the multiple-decision procedure one only needs the *forward* solution for the full set of variables since the forward solution of each subset obtained by deleting the last variable is a part of that forward solution.

Prediction of y_t for $t > T$ can be made as indicated in Section 2.5; that is, one predicts y_t by $\sum_k c_k\phi_{kT}(t)$. However, usually prediction by use of a fitted polynomial trend is risky because the polynomial is merely an approximation to the real trend and one cannot expect that the approximation will remain good outside of the range of t over which the polynomial was fitted.

Example 3.1. In Table 3.1 the quantity of meat consumed per year per person in the United States from 1919 to 1941 is tabulated† along with the values of the orthogonal polynomials up to the fifth degree [from R. L. Anderson and Houseman (1942)]. Here $T = 23$. We also give the multiplier λ_k. The calculations of regression coefficients and t^2-ratios are displayed in Table 3.2.

† Tintner (1952), pp. 195–198. The preparation of the series is described on p. 177; "meat" includes meat, poultry, and fish.

TABLE 3.1

ANNUAL CONSUMPTION OF MEAT IN THE UNITED STATES, 1919–1941

t	y_t	ξ'_{1t}	ξ'_{2t}	ξ'_{3t}	ξ'_{4t}	ξ'_{5t}	Y_t
1919	171.5	−11	77	−77	1463	−209	165.805
1920	167.0	−10	56	−35	133	76	169.456
1921	164.5	−9	37	−3	−627	171	171.927
1922	169.3	−8	20	20	−950	152	173.350
1923	179.4	−7	5	35	−955	77	173.859
1924	179.2	−6	−8	43	−747	−12	173.585
1925	172.6	−5	−19	45	−417	−87	172.662
1926	170.5	−4	−28	42	−42	−132	171.223
1927	168.6	−3	−35	35	315	−141	169.399
1928	164.7	−2	−40	25	605	−116	167.325
1929	163.0	−1	−43	13	793	−65	165.132
1930	162.1	0	−44	0	858	0	162.954
1931	160.2	1	−43	−13	793	65	160.923
1932	161.2	2	−40	−25	605	116	159.172
1933	165.8	3	−35	−35	315	141	157.833
1934	163.5	4	−28	−42	−42	132	157.040
1935	146.7	5	−19	−45	−417	87	156.925
1936	160.2	6	−8	−43	−747	12	157.620
1937	156.8	7	5	−35	−955	−77	159.260
1938	156.8	8	20	−20	−950	−152	161.975
1939	165.4	9	37	3	−627	−171	165.900
1940	174.7	10	56	35	133	−76	171.167
1941	178.7	11	77	77	1463	209	177.908
λ_k		1	1	1/6	7/12	1/60	

t denotes the year, y_t denotes per capita consumption of meat (in pounds) in year t, and Y_t denotes the value of the fitted third-degree polynomial.

TABLE 3.2

COMPUTATIONS WITH ORTHOGONAL POLYNOMIALS

Degree of Polynomial k	Sum of Squares $\sum_{t=1}^{T} \xi_{kt}'^2$	Sum of Products $\sum_{t=1}^{T} y_t \xi_{kt}'$	Coefficient c_k $\overline{\lambda_k}$	Square† $c_k^2 a_{kk}$
0	23		166.1913	
1	1,012	−383.6	−0.379,051	145.404
2	35,420	2,606.0	0.073,574	191.735
3	32,890	4,366.0	0.132,745,5	579.567
4	13,123,110	17,252.4	0.001,314,66	22.681
5	340,860	17.8	0.000,052,22	0.000,929

k	Residual Sum of Squares $c_{k+1}^2 a_{k+1,k+1} + \cdots + S$	Mean Residual Sum of Squares $\frac{c_{k+1}^2 a_{k+1,k+1} + \cdots + S}{T - k - 1}$	Ratio t^2
0	1369.538	62.2517	
1	1224.413	58.3054	2.494
2	1032.400	51.6200	3.714
3	452.833	23.8333	24.318
4	430.152	23.8973	0.949
5	430.151	25.3030	0.000,037

$\sum_{t=1}^{T} y_t = 3822.4$, $\sum_{t=1}^{T} y_t^2 = 636{,}619.18$, $S = 430.150{,}796$.

† Calculated as $(\sum_{t=1}^{T} y_t \xi_{kt}')^2 / \sum_{t=1}^{T} \xi_{kt}'^2$.

In this case it is not hard to determine the appropriate degree of polynomial. We take $q = 5$. Unless p_5 is taken infinitesimally close to 1 we decide $\gamma_5 = 0$; unless $p_4/(1 - p_5)$ is greater than about 1/3 we decide $\gamma_4 = 0$. We take $\gamma_3 \neq 0$ [unless $p_3/(1 - p_5 - p_4)$ is infinitesimal]. Thus the degree is 3. The last column of Table 3.1 gives the fitted cubic.

The observed series and the fitted trend are graphed in Figure 3.2. It will be observed that the curve gives a good fit, most of the points being close to the curve. We interpret the curve as the expected or normal consumption of meat if it were not affected by year-to-year irregularities. It should be clear that the fitted third-degree polynomial cannot be good for prediction—at least not very far in the future. Far to the right of these data this polynomial increases and

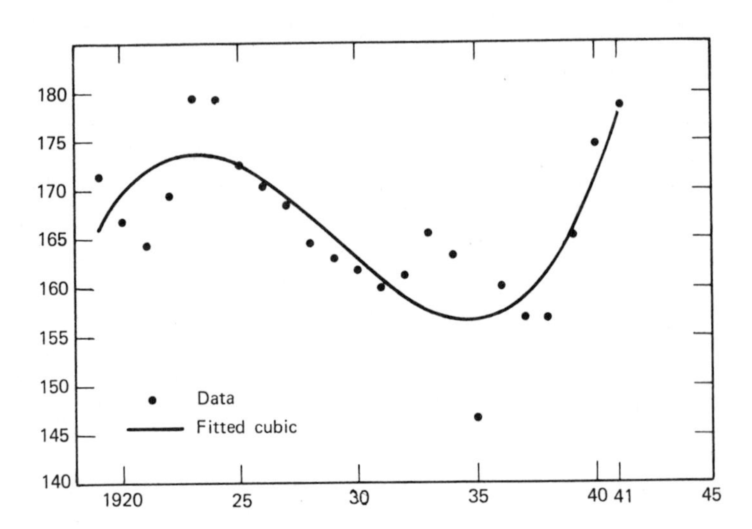

Figure 3.2. Annual per capita consumption of meat in the United States, 1919–1941, and fitted cubic.

with an increasing slope; even without the effect of the war it does not seem reasonable that per capita meat consumption will increase indefinitely and at an increasing rate of increase.

3.3. SMOOTHING

3.3.1. Smoothing Procedures

Sometimes the trend is a smooth function of time and does not fluctuate greatly in any small interval of time, but still is not closely approximated by a simple function of time over the entire range of time under consideration. However, just the simple idea that the trend is smooth but the random error terms are usually irregular leads to the statistical technique of smoothing. Smoothing a time series means representing the trend at a given point in time by a weighted average of the observed values near that point. The observed value is considered to be the sum of the trend and a random error. The weighted average of the trend is roughly the same as the trend at that point, and the weighted average of the random terms will tend to be numerically small (the random terms being independent and having expected values 0). Hence, the weighted average of the observed values will estimate the trend. At each point in time (except the first few and the last few) one applies this same weighted average of neighboring values. Thus the rather irregular graph of observed points is replaced by a smooth graph of the moving average.

We can give a more definite idea of what we mean by smooth by saying that a smooth function can be adequately represented by a polynomial of fairly low degree over some span of time not too small. This idea is somewhat justified by Taylor's theorem. (We are actually interested here only in the approximation at a few integer values of the argument.) The polynomial that approximates a function in one interval may not be the same as the polynomial approximation in other intervals; if a polynomial is fitted to one interval it may be quite different from the smooth function at another part of the range. As an example, think of a trend that is first a parabola concave downward and then a parabola concave upward, for example,

$$\begin{aligned} f(t) &= a[1 - (t-k)^2/k^2], & t &= 0, 1, \ldots, 2k, \\ &= -a[1 - (t-3k)^2/k^2], & t &= 2k, \ldots, 4k, \end{aligned} \tag{1}$$

for an integer k. The two parabolas have the same value and tangent at $t = 2k$. This function is smooth. It is exactly a polynomial of degree 1 or 2 at every three successive points, and except at the center is a parabola over a longer span. However, the parabola constituting the function over the first half of the entire interval is far from the function at the end of the interval. In fact, the function is fairly close to a third-degree polynomial from 0 to $4k$. (See Problem 20.) The point of this discussion is that the assumption of smoothness is a local property whereas the assumption of a polynomial trend concerns the entire time interval $t = 1, \ldots, T$. Correspondingly, the assumption of smoothness permits using only observations near a given time point to estimate the trend at that point, while the assumption of a polynomial trend implies that all observations are used to estimate the polynomial which represents the trend over the entire interval.

Given the observations $y_1, \ldots, y_T$ we shall estimate the trend at t by

$$y_t^* = \sum_{s=-m}^{m} c_s y_{t+s}, \qquad t = m+1, \ldots, T-m, \tag{2}$$

which is a weighted average of the observed y_t's in a time interval within m time units of t. The c_s's are normalized so

$$\sum_{s=-m}^{m} c_s = 1. \tag{3}$$

(There is no loss of generality in taking the limits of the sum symmetric about 0 since some c_s's can be 0.) The derived sequence, $y_{m+1}^*, \ldots, y_{T-m}^*$, is called a *moving average* of the original $\{y_t\}$ sequence. When $y_t = f(t) + u_t$, then

$$y_t^* = \sum_{s=-m}^{m} c_s f(t+s) + u_t^*, \tag{4}$$

where

$$u_t^* = \sum_{s=-m}^{m} c_s u_{t+s}. \tag{5}$$

As before, we assume $\mathscr{E}u_t = 0$, $\mathscr{E}u_t^2 = \sigma^2$ and $\mathscr{E}u_t u_s = 0$, $t \neq s$; thus u_t^* has variance $\sigma^2 \sum_{s=-m}^{m} c_s^2$. The c_s's are chosen so that the variance of u_t^* is considerably smaller than that of u_t. If the values of $f(t+s)$, $s = \pm 1, \ldots, \pm m$, are close to $f(t)$, then $\mathscr{E}y_t^*$ is again approximately $f(t)$. Thus the moving average or smoothed series $\{y_t^*\}$ has about the same sequence of expected values as $\{y_t\}$ but a smaller (common) variance. However, successive terms in the smoothed series are in general correlated. We have

$$\begin{aligned} \mathscr{E}u_t^* u_{t+h}^* &= \sum_{s=-m}^{m} \sum_{r=-m}^{m} c_s c_r \mathscr{E} u_{t+s} u_{t+h+r} \\ &= \sigma^2 \sum_{s=-m+h}^{m} c_s c_{s-h}, \qquad h = 0, 1, \ldots, 2m, \\ &= 0, \qquad h = 2m+1, \ldots . \end{aligned} \tag{6}$$

A particular case is the arithmetic average when $c_s = 1/(2m+1)$. Then the smoothed value is

$$\begin{aligned} y_t^* &= \frac{1}{2m+1} \sum_{s=-m}^{m} y_{t+s} \\ &= \frac{1}{2m+1} \sum_{s=-m}^{m} f(t+s) + \frac{1}{2m+1} \sum_{s=-m}^{m} u_{t+s}. \end{aligned} \tag{7}$$

The averaged error term has the variance of the mean of $2m+1$ uncorrelated variates, namely $\sigma^2/(2m+1)$, and the covariances of successive u_t^*'s are

$$\begin{aligned} \mathscr{E}u_t^* u_{t+h}^* &= \frac{2m+1-h}{(2m+1)^2}\sigma^2, \qquad h = 0, 1, \ldots, 2m, \\ &= 0, \qquad h = 2m+1, \ldots . \end{aligned} \tag{8}$$

The general basis for most smoothing formulas is in effect to fit a polynomial to $2m+1$ successive observed values and use this fitted polynomial to estimate the trend at the middle value. Since the estimates of the coefficients of the polynomial are linear in the observed values, so is the estimate of the trend and hence has the form (2). We now suppose that the trend $f(t+s)$ at the time points $t+s = t-m, \ldots, t+m$ can be approximated by

$$f_t(s) = \alpha_0 + \alpha_1 s + \cdots + \alpha_q s^q, \qquad s = -m, \ldots, m. \tag{9}$$

(The α's depend on t, but we shall not indicate that fact in the notation.) In particular, then $f(t)$ is approximately $f_t(0) = \alpha_0$. We can estimate the coefficients of the polynomial on the basis of $y_{t-m}, \ldots, y_{t+m}$ by using the method of least squares. The normal equations for the estimates $a_0, a_1, \ldots, a_q$ are

$$a_0 \sum_{s=-m}^{m} s^j + a_1 \sum_{s=-m}^{m} s^{j+1} + \cdots + a_q \sum_{s=-m}^{m} s^{j+q} = \sum_{s=-m}^{m} s^j y_{t+s}, \qquad j = 0, 1, \ldots, q. \tag{10}$$

By symmetry the sum of any odd power of s is 0. Hence, for even j the coefficients of $a_1, a_3, \ldots$ in (10) are 0, and for odd j the coefficients of $a_0, a_2, \ldots$ in (10) are 0. Since our estimate of $f(t)$ shall be our estimate of $f_t(0) = \alpha_0$, we want to solve (10) for a_0; these are the equations for j even, namely,

$$a_0 \sum_{s=-m}^{m} s^{2i} + a_2 \sum_{s=-m}^{m} s^{2i+2} + \cdots + a_{2[q/2]} \sum_{s=-m}^{m} s^{2i+2[q/2]} = \sum_{s=-m}^{m} s^{2i} y_{t+s}, \qquad i = 0, 1, \ldots, [\tfrac{1}{2}q], \tag{11}$$

where

$$\begin{aligned} [\tfrac{1}{2}q] &= \tfrac{1}{2}q, && q \text{ even}, \\ &= \tfrac{1}{2}(q-1), && q \text{ odd}. \end{aligned} \tag{12}$$

We notice that the equations to be solved for a_0 for odd q are the same as for the next lower (even) value of q. (We need study only the degrees 0, 2, 4, etc.) Let $[\frac{1}{2}q] = k$. Then (11) can be written

$$\begin{aligned} (2m+1)a_0 + 2\sum_{s=1}^{m} s^2 a_2 + \cdots + 2\sum_{s=1}^{m} s^{2k} a_{2k} &= \sum_{s=-m}^{m} y_{t+s}, \\ 2\sum_{s=1}^{m} s^{2i} a_0 + 2\sum_{s=1}^{m} s^{2i+2} a_2 + \cdots + 2\sum_{s=1}^{m} s^{2i+2k} a_{2k} &= \sum_{s=1}^{m} s^{2i}(y_{t-s} + y_{t+s}), \qquad i = 1, \ldots, k. \end{aligned} \tag{13}$$

We see that each coefficient on the left of these normal equations depends only on m, and on the right the coefficient of y_{t-s} is the same as that of y_{t+s}, $s = 1, \ldots, m$. The solution of (13) for a_0 is

$$a_0 = \sum_{s=-m}^{m} c_s y_{t+s} \tag{14}$$

with $c_{-s} = c_s$. We note that the coefficients are functions of m and k and are polynomials in s. Then $y_t^* = a_0$ as determined by (13). That $\sum_{s=-m}^{m} c_s = 1$ can be verified from (14) and observing that if $y_{t+s} = \alpha$, $s = -m, \ldots, m$, then $f_t(s) \equiv \alpha$ is the best fitting polynomial.

As an example, let $m = 2$ and $q = 2$; that is, we use $y_{t-2}, y_{t-1}, y_t, y_{t+1}, y_{t+2}$ to estimate $\mathscr{E}y_t$ on the assumption that a parabola is a good approximation to $f(t-2), f(t-1), f(t), f(t+1), f(t+2)$. The normal equations for a_0 and a_2 are

$$(15)\qquad \begin{aligned} 5a_0 + 10a_2 &= \sum_{s=-2}^{2} y_{t+s}, \\ 10a_0 + 34a_2 &= \sum_{s=1}^{2} s^2(y_{t-s} + y_{t+s}). \end{aligned}$$

Then

$$(16)\qquad \begin{aligned} a_0 &= \frac{34}{70}\sum_{s=-2}^{2} y_{t+s} - \frac{10}{70}\sum_{s=1}^{2} s^2(y_{t-s} + y_{t+s}) \\ &= \tfrac{1}{35}(-3y_{t-2} + 12y_{t-1} + 17y_t + 12y_{t+1} - 3y_{t+2}). \end{aligned}$$

When $k = 0$ ($q = 0$ or 1) there is only one equation in (13) in the unknown a_0, and the solution is $a_0 = \sum_{s=-m}^{m} y_{t+s}/(2m+1)$; that is, $c_s = 1/(2m+1)$, $s = -m, \ldots, m$. In other words the moving average with equal weights is the special case of polynomial smoothing when the polynomial is of degree 0 or degree 1.

When $k = 1$ ($q = 2$ or 3) the normal equations are (see Problem 7)

$$(17)\qquad \begin{aligned} (2m+1)a_0 + \frac{(2m+1)m(m+1)}{3}a_2 &= \sum_{s=-m}^{m} y_{t+s}, \\ \frac{(2m+1)m(m+1)}{3}a_0 + \frac{(2m+1)m(m+1)(3m^2+3m-1)}{15}a_2 &= \sum_{s=1}^{m} s^2(y_{t-s} + y_{t+s}). \end{aligned}$$

The solution is

$$(18)\qquad a_0 = \frac{\sum_{s=-m}^{m} [3(3m^2+3m-1) - 15s^2]y_{t+s}}{(2m-1)(2m+1)(2m+3)}.$$

Table 3.3 lists the values of c_s's for $k = 1$ and 2 and some values of m. The case $k = 1$ ($q = 2$ or 3) was treated above, and the case $k = 2$ ($q = 4$ or 5) is assigned as Problem 22. Kendall and Stuart (1966) in Section 46.5 give coefficients for other cases, as well as discuss related smoothing formulas.

TABLE 3.3

COEFFICIENTS IN SMOOTHING FORMULAS, $k = 1, 2$

m	$c_{-5} = c_5$	$c_{-4} = c_4$	$c_{-3} = c_3$	$c_{-2} = c_2$	$c_{-1} = c_1$	c_0
	$k = 1$					
2				$-\frac{3}{35}$	$\frac{12}{35}$	$\frac{17}{35}$
3			$-\frac{2}{21}$	$\frac{3}{21}$	$\frac{6}{21}$	$\frac{7}{21}$
4		$-\frac{21}{231}$	$\frac{14}{231}$	$\frac{39}{231}$	$\frac{54}{231}$	$\frac{59}{231}$
5	$-\frac{36}{429}$	$\frac{9}{429}$	$\frac{44}{429}$	$\frac{69}{429}$	$\frac{84}{429}$	$\frac{89}{429}$
	$k = 2$					
3			$\frac{5}{231}$	$-\frac{30}{231}$	$\frac{75}{231}$	$\frac{131}{231}$
4		$\frac{15}{429}$	$-\frac{55}{429}$	$\frac{30}{429}$	$\frac{135}{429}$	$\frac{179}{429}$
5	$\frac{18}{429}$	$-\frac{45}{429}$	$-\frac{10}{429}$	$\frac{60}{429}$	$\frac{120}{429}$	$\frac{143}{429}$

It might be noted that if $m < k$ the coefficient a_0 is undetermined. If $m = k$, one is fitting a $(2k + 1)$-degree polynomial to $2m + 1 = 2k + 1$ points; the fit is perfect and hence $a_0 = y_t$. If $m > k$, the moving average is nontrivial, that is, it involves several values of y_{t+s}. The simplest case is $m = k + 1$; then†

$$1 - c_0 = C\binom{2k+2}{k+1}, \tag{19}$$

$$c_{-s} = c_s = (-1)^{s+1}C\binom{2k+2}{k+1+s}, \qquad s = 1, \ldots, m, \tag{20}$$

where

$$C = \frac{\binom{2k+2}{k+1}}{\binom{4k+4}{2k+2}} = \frac{\dfrac{(2k+2)!}{(k+1)!^2}}{\dfrac{(4k+4)!}{(2k+2)!^2}} = \frac{(2k+2)!^3}{(k+1)!^2\,(4k+4)!}. \tag{21}$$

We shall verify this in Section 3.4 after we develop some results concerning sequences of differences.

3.3.2. Properties of Smoothing Procedures

Now let us consider some of the properties of these smoothing methods. One of the primary purposes of smoothing is to reduce the random error, that

† $\binom{h}{g} = \dfrac{h!}{g!\,(h-g)!}$, $g = 0, 1, \ldots, h$, with $0! = 1$.

is, make the variance of the smoother sequence small relative to the variance of the original sequence.

THEOREM 3.3.1. *The variance of* $y_t^* = a_0$ *is*

$$\sigma^2 b^{00} = \sigma^2 c_0, \tag{22}$$

where b^{00} *is the upper left-hand element in the inverse of the matrix* $\boldsymbol{B}$ *of coefficients in* (13), *where*

$$\begin{aligned} b_{00} &= 2m + 1, \\ b_{ij} &= 2\sum_{s=1}^{m} s^{2(i+j)}, \qquad i + j > 0, \quad i, j = 0, 1, \ldots, k. \end{aligned} \tag{23}$$

PROOF. If we denote the right-hand side of (13) by Y_i, $i = 0, 1, \ldots, k$, the equations can be written

$$\sum_{j=0}^{k} b_{ij} a_{2j} = Y_i, \qquad i = 0, 1, \ldots, k, \tag{24}$$

and the solution for a_0 is

$$\begin{aligned} a_0 &= \sum_{j=0}^{k} b^{0j} Y_j \\ &= b^{00} \sum_{s=-m}^{m} y_{t+s} + \sum_{j=1}^{k} b^{0j} Y_j, \end{aligned} \tag{25}$$

where $(b^{ij}) = \boldsymbol{B}^{-1}$. Since Y_j does not involve y_t for $j > 0$, we see that b^{00} is the coefficient of y_t. However, the general theory of the least squares method states that the variance of a_0 is $\sigma^2 b^{00}$. Q.E.D.

If $k = 0$, the variance is $\sigma^2/(2m + 1)$. If $k = 1$, the variance is

$$\frac{3(3m^2 + 3m - 1)}{(2m - 1)(2m + 1)(2m + 3)} \sigma^2. \tag{26}$$

If $m = k + 1$, the variance is

$$\sigma^2\left[1 - C\binom{2k+2}{k+1}\right] = \sigma^2\left[1 - \frac{(2k+2)!^4}{(4k+4)!\,(k+1)!^4}\right]. \tag{27}$$

In Table 3.4 we give the variances of some smoothed values.

For given k the variance goes down as the number of points used increases. For a given number of points (that is, given m) the variance increases with k.

TABLE 3.4

VARIANCES OF SMOOTHED VALUES

($\sigma^2 = 1$)

$2m+1$	$k=0$, $q=0,1$	$k=1$, $q=2,3$	$k=2$, $q=4,5$	$k=3$, $q=6,7$
3	$\frac{1}{3} = 0.333$			
5	$\frac{1}{5} = 0.200$	$\frac{17}{35} = 0.486$		
7	$\frac{1}{7} = 0.143$	$\frac{1}{3} = 0.333$	$\frac{131}{231} = 0.567$	
9	$\frac{1}{9} = 0.111$	$\frac{59}{231} = 0.255$	$\frac{179}{429} = 0.417$	$\frac{797}{1287} = 0.619$
11	$\frac{1}{11} = 0.091$	$\frac{89}{429} = 0.207$	$\frac{1}{3} = 0.333$	

In fact for given $m - k$ (which is half the excess of points over implicitly fitted constants) the variance increases with k.

We notice also that $y_t - y_t^*$, the residual of the observed value from the fitted value, is uncorrelated with y_t^* because in general the estimated regression coefficients are uncorrelated with the residuals. (See Problem 7 of Chapter 2.) Thus

$$\begin{aligned} \text{Var}\,(y_t - y_t^*) &= \sigma^2 - \text{Var}\, y_t^* \\ &= \sigma^2(1 - c_0). \end{aligned} \tag{28}$$

As indicated earlier, successive smoothed values are correlated. As an example, the correlations of y_t^* with y_{t-1}^*, y_{t-2}^*, y_{t-3}^*, and y_{t-4}^* in the case of $k = 1$ and $m = 2$ are

$$\tfrac{336}{595} = 0.565, \qquad \tfrac{42}{595} = 0.071, \qquad -\tfrac{72}{595} = -0.121, \qquad \tfrac{9}{595} = 0.015, \tag{29}$$

respectively. We shall study this effect more in Chapter 7 after we have developed more powerful mathematical tools.

When $y_t = f(t) + u_t$ and the smoothing formula with coefficients c_s is used, the systematic error in the smoothed value is

$$\mathscr{E}(y_t - y_t^*) = f(t) - \sum_{s=-m}^{m} c_s f(t+s). \tag{30}$$

If the smoothing formula is based on a polynomial of degree q and the trend is a polynomial of this degree (or less), the systematic error is 0; otherwise there will be some systematic error. Suppose $k = 0$ ($q = 0$ or 1) and the coefficients are the same; then the systematic error is

$$\mathscr{E}(y_t - y_t^*) = f(t) - \frac{1}{2m+1} \sum_{s=-m}^{m} f(t+s), \tag{31}$$

the difference between $f(t)$ and the arithmetic mean of the neighboring values. Suppose $f(t+s)$, $s = -m, \ldots, m$, is written in terms of the orthogonal polynomials to degree $2m+1$ (orthogonal over the range $-m, \ldots, m$) as

$$f(t+s) = \gamma_0 + \gamma_1 \phi^*_{1,2m+1}(s) + \cdots + \gamma_{2m+1}\phi^*_{2m+1,2m+1}(s), \qquad s = -m, \ldots, m. \tag{32}$$

Then use of the smoothing formula based on a polynomial of degree $2k$ or $2k+1$ has systematic error

$$\mathscr{E}(y_t - y_t^*) = \gamma_{2k+2}\phi^*_{2k+2,2m+1}(0) + \gamma_{2k+4}\phi^*_{2k+4,2m+1}(0) + \cdots + \gamma_{2m}\phi^*_{2m,2m+1}(0), \tag{33}$$

since the fitted polynomial consists of the terms of (32) to degree $2k$ or $2k+1$ and $\phi^*_{i,2m+1}(0) = 0$ for i odd. (See Problem 30.)

A mean value function we shall study in Chapter 4 is $f(t) = \cos(\lambda t - \theta)$. This is a cosine function with period $2\pi/\lambda$. If the coefficients are $c_s = 1/(2m+1)$, the expected value of the smoothed variate is

$$\begin{aligned} \mathscr{E} y_t^* &= \frac{1}{2m+1}\sum_{s=-m}^{m} \cos[\lambda(t+s) - \theta] \\ &= \frac{\sin \tfrac{1}{2}\lambda(2m+1)}{(2m+1)\sin\tfrac{1}{2}\lambda}\cos(\lambda t - \theta) \\ &= \frac{\sin \tfrac{1}{2}\lambda(2m+1)}{(2m+1)\sin\tfrac{1}{2}\lambda} f(t), \qquad 0 < \lambda < 2\pi. \end{aligned} \tag{34}$$

(See Problem 31.) That is, the smoothing operation simply reduces the magnitude of $f(t)$. If λ is small (that is, the period is large), the reduction is small (Problem 32). For given λ, the larger m [satisfying $(2m+1)\lambda < 2\pi$] the smaller is the factor (Problem 33). If $2m+1 = 2\pi/\lambda$ (the length of the moving average is the period), the smoothed value is 0.

The primary purpose of smoothing is to estimate the trend or expected value of y_t with minimum error. The error consists of the bias (30) and the random part $u_t^* = \sum_{s=-m}^{m} c_s u_{t+s}$. The first part can be measured by its square and the second by its variance $\sigma^2 b^{00} = \sigma^2 c_0$. For given k, the bias goes up (in most cases) and the variance goes down with increasing m, while for given m the bias goes down and the variance goes up with increasing k. The statistician who wants to use a smoothing formula must choose values of k and m. He might use as his measure of error the mean square error, which is the sum of the variance and the average of the squared bias. If σ^2, the variance of u_t, were known and if the average squared bias were known for each combination of k

and m, the statistician could select the combination of k and m minimizing the criterion. It is hard to give any generalizations about this process because the variance and the average squared bias behave oppositely with respect to k and m. If σ^2 is small, one can be satisfied with m relatively small. The smoother $f(t)$ is the smaller one can take k for given m (or the larger m for given k). In point of fact, of course, these characteristics are not known, but must be estimated from the data; hence the selection of k and m is a statistical multiple-decision problem. It is difficult to formulate this and give a mathematical statistical solution. The practitioner, thus, must proceed on the basis of general experience and intuition.

Another approach is to ask what is the smallest k such that the average squared bias is 0 or close to 0 when m is given or when m is a given function of k, for example, $m = k + 1$. We shall consider this in the next section.

The advantage of smoothing to estimate the trend is its flexibility in the sense that the assumptions for its use (that is, smoothness) are not stringent. However, because the method is not based on an explicit probability model the properties are not well-defined and statistical inference is limited. For example, the trend is not determined by a small number of parameters for which a confidence region can be given. One cannot test a hypothesis about the trend. One cannot directly relate the estimated trend to a theory or model for the generation of the observed series. Smoothing leads to an estimated trend that is descriptive rather than analytic or explanatory. Because it is not based on an explicit probabilistic model, the method cannot be treated fully and rigorously in terms of mathematical statistics (at least not succinctly).

There is another serious practical difficulty in smoothing and that is that to obtain y_t^*, the estimated trend at t, one uses $y_{t-m}, \ldots, y_{t+m}$. Since this procedure is based on $y_1, \ldots, y_T$, the first smoothed value is y_{m+1}^* and the last is y_{T-m}^*. Thus one does not have an estimate of the trend at the beginning of the period or at the end. Some other considerations must be involved to estimate the trend at these extremes.

Smoothing in itself, of course, does not provide a means of prediction. Extrapolation of the estimated trend is precarious, particularly because the trend is not estimated for the last m time points.

We have based our smoothing on an odd number of terms with symmetric weights. If one uses an even number of terms with symmetric weights one interprets the smoothed value as an estimate of the trend midway between the two middle time points. This may be awkward.

A moving average with equal weights ($k = 0$) can be carried out easily on a desk calculator, for the sum $\sum_{s=-m}^{m} y_{t+s}$ changes at each t by subtraction of one term and addition of another. The sums are recorded and then each divided by $2m + 1$ [or multiplied by $1/(2m + 1)$]. There has been considerable interest in approximating a smoothing procedure corresponding to unequal weights by a succession of smoothing procedures using equal weights. Of course, with high-speed computers there is no need to use simple coefficients.

Graduation by use of moving averages has a long history and has also been approached from a nonstatistical point of view. [See Whittaker and Robinson (1926).] Sometimes one wants to use a table of data for time points not observed, that is, one wants to interpolate between observed time points. Interpolation formulas use successive differences; smoothing formulas are used to give a graduation for which successive differences are smooth. From this point of view two smoothing procedures are equivalent to a certain order if differences of that order agree for each pair of smoothed series resulting from the two procedures. (See Section 3.4.) A procedure is said to be correct to differences of a given order if differences of this order of polynomials are not affected by the procedure. One popular procedure, that is correct to third-order differences, is Spencer's 15-point formula, which consists of

$$y_t^* = \tfrac{1}{4}(-3y_{t-2} + 3y_{t-1} + 4y_t + 3y_{t+1} - 3y_{t+2}), \tag{35}$$

then averaging (with equal weights) 5 successive y_t^*'s, then averaging 4 successive terms of the resulting series, and finally averaging 4 successive terms of this last series; another procedure which maintains third-order differences is Spencer's 21-point formula which consists of

$$y_t^* = \tfrac{1}{2}(-y_{t-3} + y_{t-1} + 2y_t + y_{t+1} - y_{t+3}) \tag{36}$$

and averaging 7 and 5 and 5 terms in turn. These two procedures are relatively easy to carry out.

3.3.3. Seasonal Variation

Another consideration comes in when there is a regular periodic variation imposed on the time series. For example, many monthly economic time series will have a seasonal factor. The underlying function of time may be written

$$\mathscr{E} y_t = f(t) = g(t) + h(t), \tag{37}$$

where $g(t)$ is a periodic function of period n (12 for monthly data, 4 for quarterly data, etc.); that is,

$$g(t + n) = g(t), \qquad t = 1, \ldots, T - n. \tag{38}$$

We can normalize $g(t)$ so

$$\sum_{t=1}^{n} g(t) = 0; \tag{39}$$

the periodicity implies

$$\sum_{t=1}^{n} g(t+s) = 0, \qquad s = 0, \ldots, T-n. \tag{40}$$

Usually T is a multiple of n, say $T = hn$. (For example, $T = 12h$ for h years of monthly data.) Given an arbitrary $f(t)$ the above specifications do not permit a unique determination of $g(t)$ until conditions have been put on $h(t)$, which is generally considered to be a slowly moving trend or cycle.

A moving average of n terms with equal coefficients will eliminate the seasonal variation of $g(t)$ in the sense that

$$\mathscr{E}\frac{1}{n}\sum_{s=1}^{n} y_{t+s} = \frac{1}{n}\sum_{s=1}^{n} h(t+s). \tag{41}$$

If n is even, say $n = 2m$, which is usually the case in economic data, we use

$$y_t^* = \frac{1}{2m}\left[\sum_{s=-(m-1)}^{m-1} y_{t+s} + \tfrac{1}{2}y_{t-m} + \tfrac{1}{2}y_{t+m}\right], \qquad t = m+1, \ldots, T-m. \tag{42}$$

Then

$$\begin{aligned}\mathscr{E}y_t^* &= \frac{1}{2m}\left[\sum_{s=-(m-1)}^{m-1} f(t+s) + \tfrac{1}{2}f(t-m) + \tfrac{1}{2}f(t+m)\right] \\ &= \frac{1}{2m}\left[\sum_{s=-(m-1)}^{m-1} h(t+s) + \tfrac{1}{2}h(t-m) + \tfrac{1}{2}h(t+m)\right].\end{aligned} \tag{43}$$

If the long-term trend $h(t)$ changes slowly, (43) will be close to $h(t)$. In particular, if $h(t)$ is linear, (43) will equal $h(t)$.

When $T = hn$, we can define $g(t)$ uniquely by

$$g(t) = \frac{1}{h}\sum_{j=0}^{h-1} f(t+nj) - \frac{1}{T}\sum_{s=1}^{T} f(s), \qquad t = 1, \ldots, n. \tag{44}$$

For example, the seasonal effect of December is the difference between the average of all Decembers and the over-all average. Then an estimate of $g(t)$ is

$$\frac{1}{h}\sum_{j=0}^{h-1} y_{t+nj} - \frac{1}{T}\sum_{s=1}^{T} y_s, \qquad t = 1, \ldots, n. \tag{45}$$

This is an unbiased estimate. The variance is

$$\left[h\left(\frac{1}{h}-\frac{1}{T}\right)^2+(T-h)\left(-\frac{1}{T}\right)^2\right]\sigma^2=\frac{n-1}{T}\sigma^2. \tag{46}$$

An alternative estimation procedure is to take deviations from the smoothed values. When $n = 2m$, the estimate of $g(t)$ is

$$\begin{aligned} &\frac{1}{h-1}\sum_{j=1}^{h-1}(y_{t+nj}-y^*_{t+nj}), \qquad t=1,\ldots,m,\\ &\frac{1}{h-1}\sum_{j=0}^{h-2}(y_{t+nj}-y^*_{t+nj}), \qquad t=m+1,\ldots,2m. \end{aligned} \tag{47}$$

Because the smoothed series is lacking m terms at the beginning and m terms at the end, the averages (47) must be based on $h - 1$ deviations. They are

$$\begin{aligned} &\frac{1}{h-1}\left[\sum_{j=1}^{h-1}y_{t+nj}-\frac{1}{2m}\left(\sum_{s=t+m+1}^{T-m-1+t}y_s+\tfrac{1}{2}y_{t+m}+\tfrac{1}{2}y_{T-m+t}\right)\right],\\ &\qquad\qquad t=1,\ldots,m,\\ &\frac{1}{h-1}\left[\sum_{j=0}^{h-2}y_{t+nj}-\frac{1}{2m}\left(\sum_{s=t-m+1}^{T-3m-1+t}y_s+\tfrac{1}{2}y_{t-m}+\tfrac{1}{2}y_{T-3m+t}\right)\right],\\ &\qquad\qquad t=m+1,\ldots,2m. \end{aligned} \tag{48}$$

The variance of (48) is

$$\left[\frac{n-1}{T-n}-\frac{1}{2(T-n)^2}\right]\sigma^2. \tag{49}$$

The estimate (48) may seem to have some appeal because a moving average is more flexible than a specified trend, but as is seen above the difference between (48) and (45) is only in the treatment of the end terms, namely omitting $2m - 1$ end terms and using $\frac{1}{2}$ of the other two end terms. It is easily verified that (49) is greater than (46). [See Durbin (1963).]

Economic statisticians frequently consider effects in many economic series to be multiplicative; that is,

$$Y(t) = G(t)H(t)U(t), \tag{50}$$

where $H(t)$ is the trend, $G(t)$ the seasonal factor, and $U(t)$ the random factor; all of these are positive. The so-called "ratio-to-moving average" involves forming the moving average series (42) and then the ratios $Y(t + nj)/Y^*(t + nj)$

instead of the differences which are summed in (47). The ratios are averaged for each t (or the median of the ratios is found) to give an estimate of a seasonal factor, corresponding to the sums (47). If the product of the $2m$ average (or median) ratios is not 1, as is usually the case [whereas the sum of the estimates (48) must be 0 except for end-term effects], each ratio is multiplied by the same factor so that the resulting product is 1. An alternative to this procedure is to use logarithms, that is, let

$$(51) \qquad \begin{aligned} y_t &= \log Y(t), & g(t) &= \log G(t), \\ h(t) &= \log H(t), & u_t &= \log U(t). \end{aligned}$$

If the resulting series and its components satisfy the assumptions above, then (45) or (48) may be used to estimate $g(t) = \log G(t)$. This seems to be a preferable method. The ratio-to-moving average method is not justified on a mathematical statistical basis and intuitively is unappealing (mathematically) because of the use of additive methods on multiplicative factors (showing up in the need to modify the estimates to force them to multiply to 1).

We defined the seasonal variation $g(t)$ in (44) in a purely formal manner. The economist, however, usually thinks of the seasonal variation as the effect of certain behavior due to the seasons and the trend† as a long-term movement due to more stable effects. Thus $f(t)$ is the sum of the seasonal effect $g(t)$ and the long-term trend $h(t)$, where these two functions are (at least conceptually) independently defined. In this sense $g(t)$ may not be strictly periodic; it might change slowly over time. For example, one might assume

$$(52) \qquad g(t + nj) = g^*(t)g^{**}(j),$$

where $g^{**}(j)$ changes slowly. We shall not go further into this matter here. Wald (1936) has proposed a method of seasonal analysis based on a model somewhat like this. Hannan (1964) and Box and Jenkins (1970) have further considered changing seasonal variation.

Economic statisticans sometimes consider an economic time series as made up of a long-run trend, a cyclical movement, a seasonal effect, and an irregular movement. The trend is a long-term movement due to population growth, technological changes, and other rather permanent effects, while the cyclical movement is the fluctuation known as the business cycle. From this point of view the cyclical movement is not periodic as was the seasonal effect in (38). Nevertheless, a moving average is sometimes proposed to eliminate the effect

† We use "trend" here to mean any systematic movement of the expected value not due to the seasons. Economists often distinguish between "cyclical movements" which are short-run variations and "trends" which are long-run.

of this cyclical movement in estimating the trend. The efficacy of this procedure is questionable; we shall return to this question later.

Smoothing formulas can be based on fitting functions other than polynomials. Also the function may be fitted by using weights other than all equal; that is, the values at $s = -m, \ldots, m$ may be treated as if the values at $s = 0, \pm 1, \ldots, \pm m$ have increasing variance.

The problem of providing smoothed values at the ends of the series may be solved in several ways; the values at the beginning of the series may be unimportant, but values at the immediate past are usually most relevant. If the smoothing formula is based on fitting a polynomial of degree q to $2m + 1$ points, the polynomial fitted to $y_{T-2m}, y_{T-2m+1}, \ldots, y_T$ can be used to provide values y_t^* of the smoothed series for $t = T - m + 1, T - m + 2, \ldots, T$ as well as for $t = T - m$; these smoothed values of y_t^* are also linear combinations of $y_{T-2m}, \ldots, y_T$. [Cowden (1962) supplies the coefficients for $q = 1, \ldots, 5$ and $m = 1, \ldots, 12$.] Another approach is to apply smoothing procedures based on smaller values of m and k; such an approach implies that $y_T^* = y_T$ since m and k must be 0, which is rather unsatisfactory. Either method yields smoothed values which have more variability at the end of the series than in the middle.

3.4. THE VARIATE DIFFERENCE METHOD

3.4.1. Introduction

The variate difference method has been proposed for estimating the variance of the error term when the trend is smooth. An extension of its use is to answer questions about the degree of smoothness of the trend. These statistical problems are treated in terms of the model $y_t = f(t) + u_t$, where the u_t's have means 0, variances σ^2 and are uncorrelated, and $f(t)$ is smooth in the sense of being approximated well by a polynomial of low degree in consecutive intervals of time.

Another use of the variate differences is to test lack of correlation. Such problems will be treated in Chapter 6 with respect to a different probability model.

In this section we develop some general theory, consider the estimation of σ^2 and tests of hypotheses of smoothness, and relate the method to smoothing. In studying serial correlation later we shall use some of these results.

The use of differences seems to have first been made by Cave-Browne-Cave (1904) and Hooker (1905) for correlation between two series; the method was extended by Student (1914). O. Anderson (1929) and Tintner (1940) studied intensively the method for estimating variances.

3.4.2. Differencing Sequences

The variate difference method is based on successive differences of elements in a time series. We want to establish a notation for such computations and record some properties. Let $\mathscr{P}$ be an operator such that

$$\mathscr{P}u_t = u_{t+1}, \qquad t = \ldots, -1, 0, 1, \ldots. \tag{1}$$

That is, given any sequence $\ldots, u_{-1}, u_0, u_1, \ldots$, the operator† $\mathscr{P}$ gives a new sequence in which the subscripts are shifted by 1. We can also write $\mathscr{P}\{u_t\} = \{u_{t+1}\}$.

An operator of this kind is a function whose argument is a sequence and whose value is a sequence. An operator $\mathscr{O}$ is a *linear operator* if for each sequence $\{u_t\}$ and each real number c

$$\mathscr{O}\{cu_t\} = c\mathscr{O}\{u_t\} \tag{2}$$

and for each pair of sequences $\{u_t\}$ and $\{v_t\}$

$$\mathscr{O}\{u_t + v_t\} = \mathscr{O}\{u_t\} + \mathscr{O}\{v_t\}. \tag{3}$$

It is obvious that $\mathscr{P}$ is a linear operator. We shall use (1) to mean alternatively the operator acting on the entire sequence and the effect of the operation on the particular element u_t.

We define $\mathscr{P}^0 = 1$ as the identity operator ($\mathscr{P}^0 u_t = u_t$) and we define $\mathscr{P}^n$ recursively as $\mathscr{P}(\mathscr{P}^{n-1})$; that is,

$$\mathscr{P}^n u_t = \mathscr{P}(\mathscr{P}^{n-1}u_t), \qquad n = 2, 3, \ldots. \tag{4}$$

One can verify by induction that

$$\mathscr{P}^n u_t = u_{t+n}. \tag{5}$$

The definition of $c\mathscr{P}$ is

$$(c\mathscr{P})u_t = c(\mathscr{P}u_t). \tag{6}$$

Furthermore, we define addition of the operators as

$$(c_1\mathscr{P}^{n_1} + \cdots + c_k\mathscr{P}^{n_k})u_t = c_1\mathscr{P}^{n_1}u_t + \cdots + c_k\mathscr{P}^{n_k}u_t. \tag{7}$$

It follows that (7) is

$$(c_1\mathscr{P}^{n_1} + \cdots + c_k\mathscr{P}^{n_k})u_t = c_1u_{t+n_1} + \cdots + c_ku_{t+n_k}. \tag{8}$$

† We do not use the more customary symbol E in order to avoid confusion with expectation.

Now a polynomial in $\mathscr{P}$ (with real coefficients) is well-defined; the operations (of multiplication, etc.) with these polynomials agree with the operations with polynomials in an indeterminate. [A sum of an infinite number of operators is a limit of the left-hand side of (8) if the right-hand side converges in some sense.]

A particular polynomial of interest to us is the first-order difference $\Delta = \mathscr{P} - 1$; that is,

$$\Delta u_t = (\mathscr{P} - 1)u_t = \mathscr{P}u_t - u_t = u_{t+1} - u_t. \tag{9}$$

The second-order difference is

$$\begin{aligned} \Delta^2 u_t &= \Delta(\Delta u_t) = \Delta u_{t+1} - \Delta u_t \\ &= u_{t+2} - 2u_{t+1} + u_t. \end{aligned} \tag{10}$$

Equivalently,

$$\Delta^2 u_t = (\mathscr{P} - 1)^2 u_t = (\mathscr{P}^2 - 2\mathscr{P} + 1)u_t. \tag{11}$$

Algebraically, we can consider the higher-order differences

$$\begin{aligned} \Delta^r u_t = (\mathscr{P} - 1)^r u_t &= \sum_{j=0}^{r} (-1)^{r-j} \binom{r}{j} \mathscr{P}^j u_t \\ &= \sum_{j=0}^{r} (-1)^{r-j} \binom{r}{j} u_{t+j}. \end{aligned} \tag{12}$$

Computationally, we may prefer to use $\Delta^r u_t = \Delta(\Delta^{r-1}u_t)$. Given a sequence $\{u_t\}$, we would compute Δu_t, $\Delta^2 u_t = \Delta(\Delta u_t)$, ... in turn.

An important property of differencing is its effect on a sequence which is made up of a polynomial in the index t. We have

$$\begin{aligned} &\Delta(\alpha_0 + \alpha_1 t + \cdots + \alpha_k t^k) \\ &= \alpha_0 + \alpha_1(t+1) + \cdots + \alpha_k(t+1)^k - [\alpha_0 + \alpha_1 t + \cdots + \alpha_k t^k] \\ &= \alpha_1 + \alpha_2[(t+1)^2 - t^2] + \alpha_3[(t+1)^3 - t^3] + \cdots + \alpha_k[(t+1)^k - t^k] \\ &= k\alpha_k t^{k-1} + \left[\binom{k}{2}\alpha_k + \binom{k-1}{1}\alpha_{k-1}\right] t^{k-2} \\ &\quad + \cdots + \left[\binom{k}{k-1}\alpha_k + \binom{k-1}{k-2}\alpha_{k-1} + \cdots + \binom{2}{1}\alpha_2\right] t \\ &\qquad + [\alpha_k + \alpha_{k-1} + \cdots + \alpha_2 + \alpha_1]. \end{aligned} \tag{13}$$

The important point is that differencing a polynomial reduces its degree by 1. It follows that if $f(t)$ is a polynomial of degree k

$$\Delta^r f(t) = 0, \qquad r = k+1, k+2, \ldots . \tag{14}$$

We have considered previously trends which were polynomials or approximated by polynomials in intervals; differencing such trends will reduce them to 0 or approximately so. [For a development of the difference calculus see Jordan (1939) or Miller (1960).]

3.4.3. Differencing Observed Series

Consider an observed time series $\{y_t\}$ which is assumed to be composed of a trend $f(t)$ and random error u_t. Since Δ is a linear operator

$$\Delta y_t = \Delta[f(t) + u_t] = \Delta f(t) + \Delta u_t. \tag{15}$$

More generally,

$$\Delta^r y_t = \Delta^r f(t) + \Delta^r u_t. \tag{16}$$

If $f(t)$ is a polynomial in t of degree less than r, then $\Delta^r f(t) = 0$ and $\mathscr{E}\Delta^r y_t = 0$. In any case we can write

$$\mathscr{E}\Delta^r y_t = \Delta^r f(t). \tag{17}$$

Now we want the variance of $\Delta^r y_t$, which is the variance of

$$\begin{aligned} \Delta^r u_t &= (\mathscr{P} - 1)^r u_t \\ &= \left[\mathscr{P}^r - \binom{r}{1}\mathscr{P}^{r-1} + \cdots + (-1)^r\right] u_t \\ &= u_{t+r} - \binom{r}{1} u_{t+r-1} + \cdots + (-1)^r u_t. \end{aligned} \tag{18}$$

The following lemma is useful.

LEMMA 3.4.1. *Let $P(x)$ and $Q(x)$ be polynomials of degree p and q, respectively. Then $\mathscr{E}P(\mathscr{P})u_t Q(\mathscr{P})u_t$ is σ^2 times the coefficient of x^q in $x^q P(x)Q(x^{-1})$ or equivalently the coefficient of x^p in $x^p P(x^{-1})Q(x)$.*

PROOF. Let

$$\begin{aligned} P(x) &= \sum_{i=0}^{p} a_i x^i, \\ Q(x) &= \sum_{j=0}^{q} b_j x^j. \end{aligned} \tag{19}$$

Then

$$(20)\qquad \mathscr{E}P(\mathscr{P})u_tQ(\mathscr{P})u_t = \sum_{i=0}^{p}\sum_{j=0}^{q} a_ib_j\mathscr{E}u_{t+i}u_{t+j}$$

$$= \sigma^2 \sum_{i=0}^{\min(p,q)} a_ib_i,$$

since $\mathscr{E}u_tu_s = 0$, $t \neq s$. However, this is σ^2 times the coefficient of x^q in

$$(21)\qquad P(x)x^qQ(x^{-1}) = \sum_{i=0}^{p} a_ix^i \sum_{j=0}^{q} b_jx^{q-j}$$

$$= \sum_{i=0}^{p}\sum_{j=0}^{q} a_ib_jx^{i-j+q}.$$

The proof is completed by writing out the other polynomial.

The variance of $\Delta^r u_t$ is then

$$(22)\qquad \mathscr{E}(\Delta^r u_t)^2 = \mathscr{E}(\mathscr{P}-1)^r u_t(\mathscr{P}-1)^r u_t$$

$$= \sigma^2\binom{2r}{r} = \sigma^2\frac{(2r)!}{(r!)^2},$$

which is σ^2 times the coefficient of x^r in

$$(23)\qquad (x-1)^r(1-x)^r = (-1)^r(x-1)^{2r}$$

$$= \sum_{j=0}^{2r}(-1)^{j+r}\binom{2r}{j}x^j.$$

The covariance between $\Delta^r u_t$ and $\Delta^r u_{t+s}$ is

$$(24)\qquad \mathscr{E}\Delta^r u_t\Delta^r u_{t+s} = \mathscr{E}(\mathscr{P}-1)^r u_t(\mathscr{P}-1)^r\mathscr{P}^s u_t$$

$$= \sigma^2(-1)^s\binom{2r}{r+s}, \qquad s = 0, 1, \ldots, r,$$

$$= 0, \qquad s = r+1, \ldots .$$

A moving average is a linear operator and can be expressed as a polynomial in $\mathscr{P}$,

$$(25)\qquad \sum_{s=-m}^{m} c_sy_{t+s} = \left(\sum_{s=-m}^{m} c_s\mathscr{P}^{m+s}\right)y_{t-m}.$$

In particular the smoothing formulas of Section 3.3.1 can be written in this fashion as well as the residual from a moving average

$$y_t - \sum_{s=-m}^{m} c_s y_{t+s} = \left(\mathscr{P}^m - \sum_{s=-m}^{m} c_s \mathscr{P}^{m+s}\right) y_{t-m}. \tag{26}$$

When the smoothing formula (25) is based on a polynomial of degree $2k + 1$, the operator (26) annihilates (that is, reduces to 0) every polynomial of degree $2k + 1$ or less. Let us show that this implies that the effect of the operator (except for a shift in time) is a linear combination of the differences of the sequence, $\Delta^{2k+2}, \ldots, \Delta^{2m}$.

Lemma 3.4.2. *If $Q(x)$ is a polynomial in x of degree n and $Q(\mathscr{P})$ annihilates every polynomial sequence of degree p $(< n)$, then $Q(\mathscr{P})$ can be expressed as a linear combination of $\Delta^{p+1}, \ldots, \Delta^n$.*

Proof. Let $Q(y + 1) = d_n y^n + d_{n-1} y^{n-1} + \cdots + d_1 y + d_0$. Since $\mathscr{P} = \Delta + 1$,

$$Q(\mathscr{P}) f(t) = d_n \Delta^n f(t) + d_{n-1} \Delta^{n-1} f(t) + \cdots + d_1 \Delta f(t) + d_0 f(t). \tag{27}$$

If $f(t) = t^q$, $0 \leq q \leq p$, then

$$Q(\mathscr{P}) f(t) = d_q \Delta^q t^q + d_{q-1} \Delta^{q-1} t^q + \cdots + d_1 \Delta t^q + d_0 t^q = 0 \tag{28}$$

since $\Delta^r t^q = 0$ for $r > q$. Consideration of (28) for $q = 0, 1, \ldots, p$ in turn shows $d_0 = 0$, $d_1 = 0, \ldots, d_p = 0$, respectively. This proves the lemma.

Corollary 3.4.1. *The residual (26) of a smoothing formula of $2m + 1$ terms based on a polynomial of degree $2k$ or $2k + 1$ is a linear combination of $\Delta^{2k+2}, \ldots, \Delta^{2m}$ operating on y_{t-m}.*

In the case of $m = k + 1$, the operator is of degree $2m = 2k + 2$ and hence must be a constant multiple of $(\mathscr{P} - 1)^{2k+2} = \Delta^{2k+2}$, say $C' \Delta^{2k+2}$. We now evaluate the constant by comparing two formulas for the variance of

$$u_t - \sum_{s=-(k+1)}^{k+1} c_s u_{t+s} = C' \Delta^{2k+2} u_{t-k-1}. \tag{29}$$

We have

$$\mathscr{E}(C' \Delta^{2k+2} u_{t-k-1})^2 = C'^2 \binom{4k+4}{2k+2} \sigma^2 \tag{30}$$

from (22). Furthermore

$$\mathscr{E}\left(u_t - \sum_{s=-(k+1)}^{k+1} c_s u_{t+s}\right)^2 = \mathscr{E}u_t^2 - \mathscr{E}\left(\sum_{s=-(k+1)}^{k+1} c_s u_{t+s}\right)^2 \tag{31}$$

$$= \sigma^2 - \sigma^2 c_0$$

follows from Theorem 3.3.1 and the fact that the estimated regression and the residual from the estimated regression are uncorrelated. Since $1 - c_0$ is the coefficient of x^{k+1} in $C'(x-1)^{2k+2}$ we have

$$(-1)^{k+1}C'\binom{2k+2}{k+1} = C'^2\binom{4k+4}{2k+2}. \tag{32}$$

Thus

$$C' = (-1)^{k+1}\frac{\binom{2k+2}{k+1}}{\binom{4k+4}{2k+2}}, \tag{33}$$

$$1 - c_0 = \frac{\binom{2k+2}{k+1}^2}{\binom{4k+4}{2k+2}}, \tag{34}$$

$$c_{-s} = c_s = (-1)^{s+1}\frac{\binom{2k+2}{k+1}\binom{2k+2}{k+1+s}}{\binom{4k+4}{2k+2}}, \tag{35}$$

which were given in (19), (20), and (21) of Section 3.3.

3.4.4. Estimating the Error Variance

We have shown that if the trend $f(t)$ is approximately a polynomial of low degree in short intervals a difference sequence has mean approximately 0. If these means are exactly 0,

$$V_r = \frac{\sum_{t=1}^{T-r}(\Delta^r y_t)^2}{(T-r)\binom{2r}{r}} \tag{36}$$

is an unbiased estimate of σ^2. In fact, whatever the trend

$$(37)\qquad \mathscr{E}V_r = \sigma^2 + \frac{\sum_{t=1}^{T-r}[\Delta^r f(t)]^2}{(T-r)\binom{2r}{r}}.$$

Let $K_r = 1\Big/\left[(T-r)\binom{2r}{r}\right]$. Then the variance of V_r when $\Delta^r f(t) = 0$, $t = 1, \ldots, T-r$, is K_r^2 times

(38)

$$\mathscr{E}\left[\sum_{t=1}^{T-r}(\Delta^r u_t)^2 - (T-r)\binom{2r}{r}\sigma^2\right]^2 = \sum_{t,s=1}^{T-r}\mathscr{E}(\Delta^r u_t)^2(\Delta^r u_s)^2 - (T-r)^2\binom{2r}{r}^2\sigma^4.$$

LEMMA 3.4.3. *Let* $\sum_{i,j=1}^n a_{ij}u_iu_j = \boldsymbol{u}'A\boldsymbol{u}$, *where* A *is symmetric and* $\mathscr{E}u_i = 0$, $\mathscr{E}u_i^2 = \sigma^2$, $\mathscr{E}u_iu_j = 0$, $i \neq j$, $\mathscr{E}u_i^4 = \kappa_4 + 3\sigma^4$, $\mathscr{E}u_i^2u_j^2 = \sigma^4$, $i \neq j$, *and* $\mathscr{E}u_iu_ju_ku_l = 0$ *unless the subscripts are equal in pairs. Then*

$$(39)\qquad \mathscr{E}\sum_{i,j=1}^n a_{ij}u_iu_j = \sigma^2\sum_{i=1}^n a_{ii} = \sigma^2 \operatorname{tr} A,$$

$$(40)\qquad \mathscr{E}\left(\sum_{i,j=1}^n a_{ij}u_iu_j\right)^2 = \sum_{i,j,k,l=1}^n a_{ij}a_{kl}\mathscr{E}u_iu_ju_ku_l$$

$$= \kappa_4\sum_{i=1}^n a_{ii}^2 + \sigma^4\left(\sum_{i=1}^n a_{ii}\right)^2 + 2\sigma^4\sum_{i,j=1}^n a_{ij}^2.$$

LEMMA 3.4.4. *The variance of* $\sum_{i,j=1}^n a_{ij}u_iu_j$ *under the conditions of Lemma* 3.4.3 *is*

$$(41)\qquad \operatorname{Var}\left(\sum_{i,j=1}^n a_{ij}u_iu_j\right) = \kappa_4\sum_{i=1}^n a_{ii}^2 + 2\sigma^4\sum_{i,j=1}^n a_{ij}^2$$

$$= \kappa_4\sum_{i=1}^n a_{ii}^2 + 2\sigma^4 \operatorname{tr} A^2.$$

If the u_i's are normally distributed, $\kappa_4 = 0$ and the first term in the variance is 0. Let

$$(42)\qquad S_r = \sum_{t=1}^{T-r}(\Delta^r y_t)^2$$

$$= \sum_{t,s=1}^{T} a_{ts}^{(r)}y_ty_s,$$

and let $A_r = (a_{ts}^{(r)})$. Then $A_0 = I$,

$$A_1 = \begin{bmatrix} 1 & -1 & 0 & \cdots & 0 \\ -1 & 2 & -1 & \cdots & 0 \\ 0 & -1 & 2 & \cdots & 0 \\ \cdot & \cdot & \cdot & & \cdot \\ \cdot & \cdot & \cdot & & \cdot \\ \cdot & \cdot & \cdot & & \cdot \\ 0 & 0 & 0 & \cdots & 1 \end{bmatrix}, \tag{43}$$

$$A_2 = \begin{bmatrix} 1 & -2 & 1 & 0 & \cdots & 0 \\ -2 & 5 & -4 & 1 & \cdots & 0 \\ 1 & -4 & 6 & -4 & \cdots & 0 \\ 0 & 1 & -4 & 6 & \cdots & 0 \\ \cdot & \cdot & \cdot & \cdot & & \cdot \\ \cdot & \cdot & \cdot & \cdot & & \cdot \\ \cdot & \cdot & \cdot & \cdot & & \cdot \\ 0 & 0 & 0 & 0 & \cdots & 1 \end{bmatrix}, \tag{44}$$

$$A_3 = \begin{bmatrix} 1 & -3 & 3 & -1 & 0 & \cdots & 0 \\ -3 & 10 & -12 & 6 & -1 & \cdots & 0 \\ 3 & -12 & 19 & -15 & 6 & \cdots & 0 \\ -1 & 6 & -15 & 20 & -15 & \cdots & 0 \\ 0 & -1 & 6 & -15 & 20 & \cdots & 0 \\ \cdot & \cdot & \cdot & \cdot & \cdot & & \cdot \\ \cdot & \cdot & \cdot & \cdot & \cdot & & \cdot \\ \cdot & \cdot & \cdot & \cdot & \cdot & & \cdot \\ 0 & 0 & 0 & 0 & 0 & \cdots & 1 \end{bmatrix}, \tag{45}$$

$$
(46) \quad A_4 = \begin{bmatrix}
1 & -4 & 6 & -4 & 1 & 0 & \cdots & 0 \\
-4 & 17 & -28 & 22 & -8 & 1 & \cdots & 0 \\
6 & -28 & 53 & -52 & 28 & -8 & \cdots & 0 \\
-4 & 22 & -52 & 69 & -56 & 28 & \cdots & 0 \\
1 & -8 & 28 & -56 & 70 & -56 & \cdots & 0 \\
0 & 1 & -8 & 28 & -56 & 70 & \cdots & 0 \\
\cdot & \cdot & \cdot & \cdot & \cdot & \cdot & & \cdot \\
\cdot & \cdot & \cdot & \cdot & \cdot & \cdot & & \cdot \\
\cdot & \cdot & \cdot & \cdot & \cdot & \cdot & & \cdot \\
0 & 0 & 0 & 0 & 0 & 0 & \cdots & 1
\end{bmatrix}.
$$

Also $V_r = K_r S_r$. When $f(t) = 0$, we compute

$$(47) \quad \operatorname{Var} V_0 = \frac{1}{T}\{\kappa_4 + 2\sigma^4\},$$

$$(48) \quad \operatorname{Var} V_1 = \frac{1}{T-1}\left\{\left[1 - \frac{1}{2(T-1)}\right]\kappa_4 + 2\left[\frac{3}{2} - \frac{1}{2(T-1)}\right]\sigma^4\right\},$$

$$(49) \quad \operatorname{Var} V_2 = \frac{1}{T-2}\left\{\left[1 - \frac{5}{9(T-2)}\right]\kappa_4 + 2\left[\frac{35}{18} - \frac{2}{2(T-2)}\right]\sigma^4\right\},$$

$$(50) \quad \operatorname{Var} V_3 = \frac{1}{T-3}\left\{\left[1 - \frac{69}{100(T-3)}\right]\kappa_4 + 2\left[\frac{231}{100} - \frac{3}{2(T-3)}\right]\sigma^4\right\},$$

$$(51) \quad \operatorname{Var} V_4 = \frac{1}{T-4}\left\{\left[1 - \frac{194}{245(T-4)}\right]\kappa_4 + 2\left[\frac{1287}{490} - \frac{4}{2(T-4)}\right]\sigma^4\right\},$$

where κ_4 is the fourth cumulant.

To find a general formula for the variance of V_r is difficult because the elements on each diagonal of A_r at the two ends are different from the elements nearer the middle. We have

$$
\begin{aligned}
(52) \quad S_r = \sum_{t=1}^{T-r} (\Delta^r y_t)^2 &= \sum_{t=1}^{T-r} \sum_{\alpha,\beta=0}^{r} (-1)^{\alpha+\beta} \binom{r}{\alpha}\binom{r}{\beta} y_{t+\alpha} y_{t+\beta} \\
&= \sum_{s,u=1}^{T} (-1)^{s+u} \sum_{t=\max(s-r,u-r,1)}^{\min(s,u,T-r)} \binom{r}{s-t}\binom{r}{u-t} y_s y_u .
\end{aligned}
$$

Then $a_{ss}^{(r)}$ is the coefficient of y_s^2 in (52), which is

$$a_{ss}^{(r)} = \sum_{\alpha=0}^{r} \binom{r}{\alpha}^2 = \binom{2r}{r}, \qquad s = r+1, \ldots, T-r, \tag{53}$$

$$= \sum_{\alpha=0}^{s-1} \binom{r}{\alpha}^2 = a_{T-s+1,T-s+1}^{(r)}, \qquad s = 1, \ldots, r.$$

The first sum is evaluated from the identity

$$\sum_{j=0}^{2r} (-1)^{j+r} \binom{2r}{j} x^j = (-1)^r (x-1)^{2r} \tag{54}$$

$$= (x-1)^r (1-x)^r$$

$$= \sum_{\alpha,\beta=0}^{r} (-1)^{\alpha+\beta} \binom{r}{\alpha}\binom{r}{\beta} x^{r-\alpha+\beta},$$

being the coefficient of x^r. Other nonzero elements are

$$a_{s,s+k}^{(r)} = (-1)^k \sum_{\alpha=0}^{r-k} \binom{r}{\alpha}\binom{r}{\alpha+k} = (-1)^k \binom{2r}{r+k}, \tag{55}$$

$$s = r-k+1, \ldots, T-r,$$

$$= (-1)^k \sum_{\alpha=0}^{s-1} \binom{r}{\alpha}\binom{r}{\alpha+k} = a_{T-s-k+1,T-s+1}^{(r)},$$

$$s = 1, \ldots, r-k,$$

for $k = 1, \ldots, r$.

The variances can now be evaluated, but the calculation is tedious. We have for even r, say $r = 2n$,

$$\sum_{s=1}^{T} [a_{ss}^{(r)}]^2 = \sum_{s=1}^{n} \{[a_{ss}^{(r)}]^2 + [a_{T-s+1,T-s+1}^{(r)}]^2\} \tag{56}$$

$$+ \sum_{s=n+1}^{2n} \{[a_{ss}^{(r)}]^2 + [a_{T-s+1,T-s+1}^{(r)}]^2\} + (T-2r)\binom{2r}{r}^2$$

$$= (T-r)\binom{2r}{r}^2 + 4\sum_{s=1}^{n} a_{ss}^{(r)}\left[a_{ss}^{(r)} - \binom{2r}{r}\right],$$

because $a^{(r)}_{T-s+1,T-s+1} = a^{(r)}_{ss}$ and

$$a^{(r)}_{r+1-s,r+1-s} = \binom{2r}{r} - a^{(r)}_{ss}, \qquad s = 1, \ldots, r. \tag{57}$$

Similarly, if $r = 2n + 1$, we have

$$\begin{aligned}\sum_{s=1}^{T} [a^{(r)}_{ss}]^2 &= (T - 2r)\binom{2r}{r}^2 + 2\sum_{s=1}^{n} [a^{(r)}_{ss}]^2 + 2[a^{(r)}_{n+1,n+1}]^2 + 2\sum_{s=n+2}^{2n+1} [a^{(r)}_{ss}]^2 \\ &= (T - r)\binom{2r}{r}^2 + 4\sum_{s=1}^{n} a^{(r)}_{ss}\left[a^{(r)}_{ss} - \binom{2r}{r}\right] - \frac{1}{2}\binom{2r}{r}^2\end{aligned} \tag{58}$$

because

$$a^{(r)}_{n+1,n+1} = \frac{1}{2}\binom{2r}{r}, \qquad r = 2n + 1. \tag{59}$$

In each case the leading term is $(T - r)\binom{2r}{r}^2$.

We also find that for $k = 1, \ldots, r$

$$\begin{aligned}&\sum_{s=1}^{T-k} [a^{(r)}_{s,s+k}]^2 \\ &= (T - 2r + k)\binom{2r}{r+k}^2 + 2\sum_{s=1}^{r-k} [a^{(r)}_{s,s+k}]^2 \\ &= (T - r)\binom{2r}{r+k}^2 + 4\sum_{s=1}^{\frac{1}{2}(r-k)} a^{(r)}_{s,s+k}\left[a^{(r)}_{s,s+k} - \binom{2r}{r+k}\right], \qquad r - k \text{ even}, \\ &= (T - r)\binom{2r}{r+k}^2 + 4\sum_{s=1}^{\frac{1}{2}(r-k-1)} a^{(r)}_{s,s+k}\left[a^{(r)}_{s,s+k} - \binom{2r}{r+k}\right] - \frac{1}{2}\binom{2r}{r+k}^2, \\ &\qquad r - k \text{ odd}.\end{aligned} \tag{60}$$

(See Problem 55.) Then the leading term in $\sum_{s,t=1}^{T} [a^{(r)}_{st}]^2$ is

$$\begin{aligned}(T - r)\left[\binom{2r}{r}^2 + 2\sum_{k=1}^{r}\binom{2r}{r+k}^2\right] &= (T - r)\sum_{k=-r}^{r}\binom{2r}{r+k}^2 \\ &= (T - r)\binom{4r}{2r}.\end{aligned} \tag{61}$$

For $r = 2n$ the other term is

$$(62)\quad 4\sum_{s=1}^{n} a_{ss}^{(r)}\left[a_{ss}^{(r)} - \binom{2r}{r}\right] + 8\sum_{s=1}^{n-1} a_{s,s+1}^{(r)}\left[a_{s,s+1}^{(r)} - \binom{2r}{r+1}\right] - \binom{2r}{r+1}^2$$
$$+ 8\sum_{s=1}^{n-1} a_{s,s+2}^{(r)}\left[a_{s,s+2}^{(r)} - \binom{2r}{r+2}\right]$$
$$+ \cdots + 8a_{1,r-1}^{(r)}\left[a_{1,r-1}^{(r)} - \binom{2r}{2r-2}\right] - \binom{2r}{2r-1}^2.$$

Kendall and Stuart [(1966), p. 389] give this other term as $-\beta_r = -\dfrac{r}{2}\dbinom{2r}{r}^2$. Thus we have [for $\Delta^r f(t) = 0$]

$$(63)\quad \operatorname{Var} S_r = \left[(T-r)\binom{2r}{r}^2 - \alpha_r\right]\kappa_4 + 2\left[(T-r)\binom{4r}{2r} - \beta_r\right]\sigma^4,$$

$$(64)\quad \operatorname{Var} V_r = \frac{1}{T-r}\left\{\left[1 - \frac{\alpha_r}{(T-r)\binom{2r}{r}^2}\right]\kappa_4 + 2\left[\frac{\binom{4r}{2r}}{\binom{2r}{r}^2} - \frac{\beta_r}{(T-r)\binom{2r}{r}^2}\right]\sigma^4\right\},$$

where κ_4 is the fourth cumulant and α_r is deduced from (56) or (58). [Kendall and Stuart (1966), p. 389, also give an expression for α_r.]

If T is large, the variance of V_r is approximately

$$(65)\quad \frac{1}{T-r}\left\{\kappa_4 + 2\frac{\binom{4r}{2r}}{\binom{2r}{r}^2}\sigma^4\right\}.$$

If the u_t's are normally distributed, $\kappa_4 = 0$.

Since the variance goes to 0 as T increases, V_r is a consistent estimate of σ^2. It can be shown that when κ_4 exists and the u_t's are independently and identically distributed

$$(66)\quad \sqrt{T-r}\,\frac{V_r - \sigma^2}{\sqrt{\kappa_4 + 2\sigma^4\binom{4r}{2r}\Big/\binom{2r}{r}^2}}$$

has a limiting normal distribution with mean 0 and variance 1. (See Section 3.4.5.) If $\kappa_4 = 0$, the variance of $\sqrt{T-r}\,V_r$ is asymptotically $2\sigma^4\binom{4r}{2r}\Big/\binom{2r}{r}^2$, and this is a measure of the efficiency of V_r as an estimate of σ^2. As is to be expected, the variance of V_r increases with r.

LEMMA 3.4.5. *The covariance of the two (symmetric) quadratic forms* $\sum_{i,j=1}^n a_{ij}u_iu_j = u'Au$ *and* $\sum_{k,l=1}^n b_{kl}u_ku_l = u'Bu$ *under the conditions of Lemma* 3.4.3 *is*

$$\text{(67)} \qquad \operatorname{Cov}\left(\sum_{i,j=1}^n a_{ij}u_iu_j, \sum_{k,l=1}^n b_{kl}u_ku_l\right) = \kappa_4\sum_{i=1}^n a_{ii}b_{ii} + 2\sigma^4\sum_{i,j=1}^n a_{ij}b_{ij}$$

$$= \kappa_4\sum_{i=1}^n a_{ii}b_{ii} + 2\sigma^4 \operatorname{tr} \boldsymbol{AB}.$$

It can be shown that the covariance between V_r and V_q is approximately

$$\text{(68)} \qquad \frac{1}{T-r}\left\{\kappa_4 + 2\frac{\binom{2r+2q}{r+q}}{\binom{2r}{r}\binom{2q}{q}}\sigma^4\right\}.$$

The exact formula differs from this one by terms which result from the fact that the coefficients of terms near the ends of the series are different from those in the middle. Kendall and Stuart [(1966), p. 389] give the covariance exactly for the case of $q = r + 1$.

Quenouille (1953) has pointed out that a linear combination $c_pV_p + \cdots + c_qV_q$ with $c_p + \cdots + c_q = 1$ is also an unbiased estimate of σ^2 [if $\Delta^p f(t) = 0$]. If $p = 0$, then $V_0 = \sum_{t=1}^T y_t^2/T$ is the best unbiased estimate of σ^2 because then $f(t) \equiv 0$ and V_0 is a sufficient statistic for σ^2 under normality. [If $p = 1$, and hence $f(t)$ is constant, then $\bar{y}$ and $\sum_{t=1}^T (y_t - \bar{y})^2$ are a sufficient set of statistics under normality.] For $p \geq 1$ Quenouille raised the question of what linear combination has minimum variance for various p and q when $\kappa_4 = 0$. If $p = 1$ and $q = 4$, the "best" coefficients are $c_1 = 7.5$, $c_2 = -20.25$, $c_3 = 22.50$, and $c_4 = -8.75$; the variance of this is 3/4 the variance of V_1. It turns out that these linear combinations have coefficients that alternate in sign. Thus there is no assurance that the estimate will be positive; in fact, since the coefficients are numerically large relative to 1, one can expect that the probability of a negative estimate of the variance is not at all negligible.

In formulating this problem one is led to the question of why the limit on the number of V_r's and—more important—why restrict the statistician only to sums of squares of the variate differences? What is desired is the best estimate of σ^2 when the trend is "smooth." The difficulty is in giving a precise enough definition of "smooth" so that the problem of best estimate is mathematically defined. If smoothness is defined so that it implies the trend is a polynomial of degree q, then the best estimate of σ^2 is the one given in Section 3.2. Any other definition of smoothness is either too vague or too complicated.

Kendall [in Problem 30.8 (1946a)] has proposed modifying the variate difference procedure by inserting dummy variables $y_{-r+1} = y_{-r+2} = \cdots = y_0 = 0$ and $y_{T+1} = \cdots = y_{T+r} = 0$ and computing the sum of squares of $\Delta^r y_t$ from $t = -r + 1$ to T. Then there are simple relations between these sums of squares and sums of cross-products $\sum_{t=1}^{T-j} y_t y_{t+j}$. Quenouille (1953) has proposed another modification that involves changing the end-terms. His mth modified sum, say, is an average of two $(m - 1)$st modified sums as

$$\sum_{t=1\,(m)}^{T} w_t = \frac{1}{2}\left[\sum_{t=1\,(m-1)}^{T-1} w_t + \sum_{t=2\,(m-1)}^{T} w_t\right], \tag{69}$$

where

$$\sum_{t=1\,(0)}^{T} w_t = \sum_{t=1}^{T} w_t. \tag{70}$$

Modified sums of $(\Delta^r y_t)^2$ and $y_t y_{t+j}$ can then be used and related. Another modification that leads to some simplification is using A_1^r instead of A_r. (See Chapter 6.)

3.4.5. Determination of the Degree of Smoothness of a Trend

One may be interested in questions of whether the trend has a certain degree of smoothness. For instance, the choice of a smoothing formula depends on this feature. That is, one may want to know whether a trend is smooth in the sense that polynomials of specified degree q are adequate to represent the trend in each interval of time. This corresponds to a problem of testing the hypothesis that a given degree is suitable against the alternative hypothesis that a greater degree is needed. We can also consider the multiple-decision problem of determining the appropriate degree (between a specified m and a specified q) of a polynomial as an approximation of the trend.

If one considers an exact fit the only adequate representation, then these questions are equivalent to those treated in Section 3.2 on polynomial trends. If one permits a more generous, but unspecified, interpretation of adequate representation these problems are not mathematically defined. It is conceivable that one specify the maximum allowable discrepancy between the actual trend and the polynomial approximation, but the resulting mathematical statistical problems would be intractable. We might note that the general theory of multiple-decision problems, which is the generalization of that given in Section 3.2, is not applicable here.

We shall now consider these problems restricted to the use of the sums of squares of variate differences. From (37) we see that the expected value of V_r depends on $\sum_{t=1}^{T-r}[\Delta^r f(t)]^2$. This is 0 if $f(t)$ is a polynomial of degree less than r and is close to 0 if $f(t)$ is approximately a polynomial of degree less than r for every succession of $r+1$ values of t. These facts suggest that the V_r's are suitable for the problems of statistical inference mentioned. To decide whether $\Delta^r f(t)$ is approximately 0 for all available t [that is, that $(r-1)$ degree polynomials are adequate representation] against the alternative that $\Delta^{r+1}f(t)$ is approximately 0 one could see whether V_r was not much larger than V_{r+1}.

The smoothing formulas considered in Sections 3.3 and 3.4 were based on the assumption that a polynomial of degree $2k$ or $2k+1$ was an adequate representation of the trend over $2m+1$ successive time points. In particular we noted that when $m=k+1$ the bias in estimating the trend was $C'\Delta^{2k+2}f(t)$. Thus the questions we are studying here correspond to problems about the suitability of smoothing formulas.

Consider testing the hypothesis $\Delta^r f(t)=0$ under the assumption that $\Delta^q f(t)=0$ $(q>r)$. We shall reject this hypothesis if V_r is much larger than V_q. The procedure can be based on

$$\sqrt{T-q}\,\frac{V_r-V_q}{V_q}=\sqrt{T-q}\left(\frac{V_r}{V_q}-1\right). \tag{71}$$

From (68) we see that the variance of the numerator $\sqrt{T-q}(V_r-V_q)$ is approximately

$$\left\{\kappa_4+2\frac{\binom{4r}{2r}}{\binom{2r}{r}^2}\sigma^4\right\}+\left\{\kappa_4+2\frac{\binom{4q}{2q}}{\binom{2q}{q}^2}\sigma^4\right\}-2\left\{\kappa_4+2\frac{\binom{2q+2r}{q+r}}{\binom{2q}{q}\binom{2r}{r}}\sigma^4\right\} \tag{72}$$

$$=2\sigma^4\left[\frac{\binom{4r}{2r}}{\binom{2r}{r}^2}+\frac{\binom{4q}{2q}}{\binom{2q}{q}^2}-2\frac{\binom{2q+2r}{q+r}}{\binom{2q}{q}\binom{2r}{r}}\right].$$

When the hypothesis is true, the expected value of the numerator of (71) is 0, and otherwise it is positive. We note that we can write

$$(73)\quad V_r - V_q = \frac{1}{T-q}\sum_{t=1}^{T-q}\left[\frac{(\Delta^r y_t)^2}{\binom{2r}{r}} - \frac{(\Delta^q y_t)^2}{\binom{2q}{q}}\right] + \left(\frac{1}{T-r} - \frac{1}{T-q}\right)\sum_{t=1}^{T-q}\frac{(\Delta^r y_t)^2}{\binom{2r}{r}} + \frac{1}{T-r}\sum_{t=T-q+1}^{T-r}\frac{(\Delta^r y_t)^2}{\binom{2r}{r}}.$$

Each of the last two terms multiplied by $\sqrt{T-q}$ converges stochastically to 0. (Note the factor in front of the second term is of order $1/T^2$.) The first term is the average of $T - q$ terms. If the u_t's are independently and identically distributed these terms are identically distributed but not independently distributed. However, the sequence of terms forms a stationary stochastic process (see Chapter 7) and terms more than q apart are independent; this is called a finitely dependent stationary stochastic process. Theorem 7.7.5 asserts that $\sqrt{T-q}\,(V_r - V_q)$ has a limiting normal distribution. Since V_q is a consistent estimate of σ^2 (whether the null hypothesis is true or not), (71) has a limiting normal distribution with variance given by (72) with σ^4 omitted.

THEOREM 3.4.1. *If $\Delta^r f(t) = 0$, $t = 1, \ldots, T - r$, and $u_1, u_2, \ldots, u_T$ are independently and identically distributed with $\mathscr{E}u_t = 0$, and $\mathscr{E}u_t^4 < \infty$, then*

$$(74)\quad \sqrt{\frac{T-q}{2\left[\dfrac{\binom{4r}{2r}}{\binom{2r}{r}^2} + \dfrac{\binom{4q}{2q}}{\binom{2q}{q}^2} - 2\dfrac{\binom{2q+2r}{q+r}}{\binom{2q}{q}\binom{2r}{r}}\right]}}\cdot\frac{V_r - V_q}{V_q}$$

has a limiting normal distribution with mean 0 *and variance* 1.

For large samples we reject the null hypothesis at significance level ε if (74) computed for the sample values exceeds $t(2\varepsilon)$, where

$$(75)\quad \int_{t(2\varepsilon)}^{\infty}\frac{1}{\sqrt{2\pi}}e^{-\frac{1}{2}u^2}\,du = \varepsilon.$$

In this large-sample theory we have not used the exact variance because the difference between that and (72) goes to 0 as $T \to \infty$ and the central limit theorem that justifies the procedure neglects such differences. It is possible that use of such terms may make the limiting distribution a better approximation to the exact distribution for a given T; that is not known. Tintner (1940) has used more exact moments (involving the consistent estimation of κ_4) and has given tables that facilitate the computation of V_r as well as the use of the limiting distribution.

If the u_t's are normally distributed, the criterion (74) has a distribution free of nuisance parameters when the null hypothesis is true. The distributions of quadratic forms in normally distributed variables and ratios of such quadratic forms will be discussed in detail in Chapter 6 on serial correlation. As will be seen there, the distribution of $(V_r - V_q)/V_q$, or equivalently V_r/V_q, is not simple (it cannot be reduced to a simple canonical form). Several modifications have been proposed to simplify the distribution problem. Tintner [(1940), Chapter 8] has suggested replacing V_r and V_q by sums of $(\Delta^r y_t)^2$ and $(\Delta^q y_t)^2$, respectively, where the terms are selected so that no y_t appears twice; then the numerator and denominator are sums of squares of independent normal variables, and a multiple of the ratio has an F-distribution. This proposed method, however, is extremely inefficient since the number of terms in each sum is about $1/(q + r)$ of the total possible. Another modification suggested by Tintner (1955) is to use a circular definition (see Chapter 6); this simplifies the distribution problem but may cause considerable bias since the trend at the beginning of the series may be quite different from that at the end. (In fact, the bias may increase with T.) Kamat (1955) and Geisser (1956) have suggested omitting one or two terms from the middle of V_r (or V_q) to simplify its distribution; another possibility is to replace A_r by A_1^r. We shall discuss these problems further in Chapter 6 (where A_1 is used to denote other matrices).

Now consider the multiple-decision problem of selecting one of the following:

$$
\begin{aligned}
&H_q: && \Delta^q f(t) \neq 0, && \\
&H_{q-1}: && \Delta^q f(t) = 0, && \Delta^{q-1} f(t) \neq 0, \\
& && \vdots && \\
&H_{m+1}: && \Delta^{m+2} f(t) = 0, && \Delta^{m+1} f(t) \neq 0, \\
&H_m: && \Delta^{m+1} f(t) = 0,
\end{aligned}
\tag{76}
$$

where $\Delta^r f(t) = 0$ means equality for all t and $\Delta^r f(t) \neq 0$ means inequality for at least one t. This set of possibilities is like (16) of Section 3.2. Our decision procedures here are to be based on $V_{m+1}, \ldots, V_{q+1}$. We shall assume $\Delta^{q+1} f(t) = 0$.

As pointed out, the general theory of Section 3.2 does not now apply because the V_r's are not sufficient statistics. However, the procedure of Section 3.2 suggests that the investigator proceed sequentially, testing first $\Delta^q f(t) = 0$ by use of $(V_q - V_{q+1})/V_{q+1}$ as indicated above. If this hypothesis is accepted, test $\Delta^{q-1} f(t) = 0$ by use of $(V_{q-1} - V_q)/V_q$. At any particular step use $(V_r - V_{r+1})/V_{r+1}$. If $\Delta^{m+1} f(t) = 0$ is accepted, stop. For a long-run series (that is, large T) these tests can be based on the asymptotic theory given above. However, the different numerators are not asymptotically independent. Hence the limiting probability of deciding $\Delta^q f(t) = 0$ and then $\Delta^{q-1} f(t) = 0$ is given by a bivariate normal distribution. The limiting probability of r decisions involves an r-variate normal distribution. In practice, then, one would have to ignore this difficulty and simply perform each test by itself.

It has been suggested that one proceed in the reverse order. First consider $(V_{m+1} - V_{m+2})/V_{m+2}$; if this is large, decide $\Delta^{m+1} f(t) \neq 0$. Then consider $(V_{m+2} - V_{m+3})/V_{m+3}$, and so on. In practice the procedure may be carried out informally. One computes the sequence $\{\Delta y_t\}$ by differencing the original series, and calculates V_1. Then one computes the sequence $\{\Delta^2 y_t\}$ by differencing the first differences, and calculates V_2. As the V_r's are successively calculated, each is compared with the preceding ones. Actual significance tests may be applied only when the sequence levels off. Besides the difficulty of controlling the probabilities of errors because the component significance tests are not independent, there is the possibility that, say, $\sum_{t=1}^{T-r} [\Delta^r f(t)]^2$ may be small and $\sum_{t=1}^{T-r} [\Delta^{r+1} f(t)]^2$ may be large, thus tending to lead to the erroneous conclusion that $\Delta^r f(t) = 0$.

Tintner [(1952), p. 320] has given the differences up to order 5 of the series in Table 3.1. (The last number in each column of Tintner's table is incorrect.)

TABLE 3.5

VARIANCES OF DIFFERENCES OF TABLE 3.1

Order	S_r	$(T-r)\binom{2r}{r}$	V_r
0			62.2517
1	905.64	44	20.5827
2	1,860.23	126	14.7637
3	5,411.77	400	13.5294
4	17,321.25	1330	13.0235
5	58,446.06	4536	12.8849

The statistics V_r calculated from these are given in Table 3.5. V_r for $r = 0$ is $\sum_{t=1}^{23} (y_t - \bar{y})^2/22$ and is the same as the mean residual sum of squares for $k = 0$ of the lower half of Table 3.2. Note that the estimate V_r of σ^2 based on an approximate polynomial trend of degree $r - 1$ is considerably smaller than the estimate which is the mean sum of squares of residuals about a fitted polynomial of degree $r - 1$.

Tintner uses V_r instead of V_{r+1} as a consistent estimate of σ^2 with the difference $V_r - V_{r+1}$. The latter seems preferable because it is consistent when $\Delta^r f(t) \neq 0$ and $\Delta^{r+1} f(t) = 0$.

3.5. NONLINEAR TRENDS

Time series sometimes exhibit trends which are best described by functions that are not linear in the parameters which are to be estimated. For example, population studies often reveal a characteristic trend in the growth of populations which may be reasonably well described by the so-called "logistic" growth curve. [See Davis (1941), pp. 247–271.] The formula for the logistic curve as a function of time is

$$f(t; \alpha_1, \alpha_2, \alpha_3) = \frac{\alpha_1}{1 + \alpha_2 e^{-\alpha_3 t}}. \tag{1}$$

In the error model formulation the observed values of population size $y_1, y_2, \ldots, y_T$ are assumed to differ from the trend by uncorrelated random amounts, and we may establish a least squares criterion for estimating the parameters α_1, α_2, and α_3 as was done in the linear case. Thus we require that the expression

$$\sum_{t=1}^{T} [y_t - f(t; a_1, a_2, a_3)]^2 = S(a_1, a_2, a_3) \tag{2}$$

be minimized with respect to a_1, a_2, and a_3. The minimizing values of a_1, a_2, and a_3 will then be the least squares estimates of the parameters α_1, α_2, and α_3, respectively. This computational problem may often be solved iteratively by considering the first terms of the Taylor series expansion of $f(t; a_1, a_2, a_3)$ about suitable values a_1^0, a_2^0, and a_3^0,

$$f(t; a_1, a_2, a_3) = f(t; a_1^0, a_2^0, a_3^0) + \sum_{j=1}^{3} (a_j - a_j^0) \left. \frac{\partial f(t; \boldsymbol{a})}{\partial a_j} \right|_{a_1^0, a_2^0, a_3^0} + R, \tag{3}$$

where R is a remainder term. Given initial estimates a_1^0, a_2^0, and a_3^0, a linear least squares technique can be applied to (3) to obtain estimates a_1^1, a_2^1, and a_3^1.

The resulting linear equations for $a_i - a_i^0$ upon neglect of the remainder term are

$$(4) \qquad \sum_{t=1}^{T} \frac{\partial f(t; \boldsymbol{a}^0)}{\partial \boldsymbol{a}^0} \left[\frac{\partial f(t; \boldsymbol{a}^0)}{\partial \boldsymbol{a}^0}\right]' (\boldsymbol{a} - \boldsymbol{a}^0) = \sum_{t=1}^{T} [y_t - f(t; \boldsymbol{a}^0)] \frac{\partial f(t; \boldsymbol{a}^0)}{\partial \boldsymbol{a}^0},$$

where $f(t; \boldsymbol{a}^0) = f(t; a_1^0, a_2^0, a_3^0)$, $\boldsymbol{a} - \boldsymbol{a}^0$ is the (column) vector with elements $a_i - a_i^0$, and $\partial f(t; \boldsymbol{a}^0)/\partial \boldsymbol{a}^0$ is the vector whose elements are the partial derivatives with respect to a_1, a_2, and a_3 evaluated at the initial points. The procedure can then be repeated using a_1^1, a_2^1, and a_3^1 in place of a_1^0, a_2^0, and a_3^0 to obtain more exact solutions, and so on. In many cases the sequence of solutions so obtained will converge to the values minimizing (2).

Another way of developing iterative procedures is to treat (2) directly. The partial derivatives of (2) with respect to a_1, a_2, and a_3 set equal to 0 yield

$$(5) \qquad \sum_{t=1}^{T} f(t; a_1, a_2, a_3) \frac{\partial f(t; \boldsymbol{a})}{\partial a_i} = \sum_{t=1}^{T} y_t \frac{\partial f(t; \boldsymbol{a})}{\partial a_i}, \qquad i = 1, 2, 3.$$

If we now expand the partial derivatives as well as the original function in a Taylor series about a point (a_1^0, a_2^0, a_3^0) and omit all second-degree terms (in $a_i - a_i^0$), the resulting equations are

$$(6) \qquad \left\{\sum_{t=1}^{T} \frac{\partial f(t; \boldsymbol{a}^0)}{\partial \boldsymbol{a}^0} \left[\frac{\partial f(t; \boldsymbol{a}^0)}{\partial \boldsymbol{a}^0}\right]' - \sum_{t=1}^{T} [y_t - f(t; \boldsymbol{a}^0)] \frac{\partial^2 f(t; \boldsymbol{a}^0)}{\partial \boldsymbol{a}^0 \, \partial \boldsymbol{a}^{0\prime}}\right\} (\boldsymbol{a} - \boldsymbol{a}^0) = \sum_{t=1}^{T} [y_t - f(t; \boldsymbol{a}^0)] \frac{\partial f(t; \boldsymbol{a}^0)}{\partial \boldsymbol{a}^0},$$

where

$$(7) \qquad \frac{\partial^2 f(t; \boldsymbol{a}^0)}{\partial \boldsymbol{a}^0 \, \partial \boldsymbol{a}^{0\prime}} = \left(\frac{\partial^2 f(t; \boldsymbol{a})}{\partial a_i \, \partial a_j}\bigg|_{a_1^0, a_2^0, a_3^0}\right).$$

If the random terms are considered normally and independently distributed with constant variance, the method of least squares is the method of maximum likelihood, and this iterative method is equivalent to the Newton-Raphson method to obtain the maximum. [If (7) is replaced by its expected value the procedure is the method of scoring. See C. R. Rao (1952), Section 4c.2.]

Other means can be used to minimize (2). At an initial point $\boldsymbol{a}^0$ the function $S(a_1, a_2, a_3)$ decreases most rapidly in the direction $-\partial S/\partial a_1$, $-\partial S/\partial a_2$, $-\partial S/\partial a_3$; a change in the vector $\boldsymbol{a}$ in this direction will produce a decrease in $S(a_1, a_2, a_3)$. Methods of steepest descent are designed to estimate the size of the move that makes the greatest decrease. In practice the partial derivatives may be approximated by finite differences. From an initial point (a_1^0, a_2^0, a_3^0), S

is calculated for the values of $a_i = -\lambda\, \partial S/\partial a_i$, $i = 1, 2, 3$, for increasing λ until S no longer decreases; this procedure establishes the second point $\boldsymbol{a}^1$, and the procedure is repeated. Draper and Smith (1966) in Section 10.3 discuss a method ("Marquardt's Compromise") that is based on both the linearization of the function $f(t;\boldsymbol{a})$ and a method of steepest descent. [Williams (1959) also discusses nonlinear regression.]

We have indicated some general approaches to finding iterative solutions to least squares problems; writing the details for specifically three parameters is simply a convenience. A great deal of work has been done in this area; it does not seem appropriate to attempt to summarize that work here.

There has also been a considerable amount of work on the fitting of special functions such as the logistic [Davis (1941), pp. 250-254, for example]. We shall not attempt to outline that work either. Some of the methods make use of the ordering of the observations in time; for instance,

$$\sum_{t=1}^{T-1} [\Delta y_t - \Delta f(t;\, a_1, a_2, a_3)]^2 \tag{8}$$

may be minimized (with perhaps one function of the estimate determined by other means). Such methods may not be based on the model with uncorrelated error components, but may be based on other models, which are more suitable in many cases.

3.6. DISCUSSION

The models treated in this chapter have assumed that the observations y_t are made up of a function of time $f(t)$ plus uncorrelated errors u_t. The function of time is considered to be unaffected by the irregularities. This may be reasonable when the major irregularities or randomness are in the measuring process. One of the classical areas of time series study is astronomy; a sequence of observations is made on the position of a planet. The irregularities observed are due mainly to variations in the earth's atmosphere and settings in the telescope. These random effects do not influence the course of the planet. Moreover, from day to day or week to week these effects may be taken to be uncorrelated.

In other areas these assumptions may not be fulfilled. The quantity of meat consumed in the United States each year is subject to error of measurement. Some meat will not be reported; some consumption will be assigned to the wrong year; and so on. To some extent these errors will be correlated. In

Chapter 10 we shall consider the effect of the errors being correlated; a large-sample theory is developed for the general case of the errors constituting a stationary stochastic process.

However, there are other irregularities in the time series of meat consumption which may make us dissatisfied with representing actual meat consumption as a simple function of time. There are effects of general economic conditions which influence the income of the consumers and the prices of meat; these effects are neither functions of time nor uncorrelated random terms. If such effects are not taken into account explicitly (as regression variables, say) then they are taken account of implicitly [in $f(t)$ or u_t]. These are some irregularities that become incorporated into the quantity studied. If meat consumption unexpectedly goes up one year due to some random conditions, the consumers may shift their eating habits, tending to increase permanently consumption of meat. This behavior is not reflected in the models studied in this chapter but is in models studied later. In Chapter 5 we combine the regression model for the systematic part and the autoregressive process for the random part.

REFERENCES

Section 3.2. R. L. Anderson and Houseman (1942), Fisher and Yates (1963), Lehmann (1959), Tintner (1952).

Section 3.3. Box and Jenkins (1970), Cowden (1962), Durbin (1963), Hannan (1964), Kendall and Stuart (1966), Wald (1936), Whittaker and Robinson (1926).

Section 3.4. O. Anderson (1929), Cave-Browne-Cave (1904), Geisser (1956), Hooker (1905), Jordan (1939), Kamat (1955), Kendall (1946a), Kendall and Stuart (1966), Miller (1960), Quenouille (1953), Student (1914), Tintner (1940), (1952), (1955).

Section 3.5. Davis (1941), Draper and Smith (1966), C. R. Rao (1952), Williams (1959).

PROBLEMS

1. (Sec. 3.2.1.) In the polynomial $f(t)$ given by (1) substitution of t by $t^* + \tau$ amounts to a change in the origin of time. Give the coefficients of the resulting polynomial in t^*. (Note that if $\alpha_q \neq 0$ no power of t^* less than q can be absent from every polynomial resulting from such substitutions.)

2. (Sec. 3.2.1.) Find the orthogonal polynomial of third degree for $t = 1, \ldots, T$.

3. (Sec. 3.2.1.) Find the orthogonal polynomial of third degree for $t = 1, \ldots, 17$.

4. (Sec. 3.2.1.) Find the orthogonal polynomials to third degree for $t = 1, \ldots, 23$.

5. (Sec. 3.2.1.) Let $\psi_k(T) = \sum_{t=1}^{T} t^k$. Show that

$$\sum_{k=0}^{p} \binom{p+1}{k} \psi_k(T) = (T+1)^{p+1} - 1.$$

[*Hint:* Show that both sides of the above equation are $\sum_{t=1}^{T} (t+1)^{p+1} - \sum_{t=1}^{T} t^{p+1}$.]

6. (Sec. 3.2.1.) Show that

$$2\sum_{i=0}^{q} \binom{2q+1}{2i} \psi_{2i}(T) = (T+1)^{2q+1} + T^{2q+1} - 1,$$

where $\psi_k(T)$ is defined in Problem 5. [*Hint:* Show that both sides of the above equation are $\sum_{t=1}^{T} (t+1)^{2q+1} - \sum_{t=1}^{T} (t-1)^{2q+1}$.]

7. (Sec. 3.2.1.) Verify the following table for $\psi_k(T) = \sum_{t=1}^{T} t^k$, $k = 0, 1, \ldots, 8$, by using Problem 5:

$$\sum_{t=1}^{T} 1 = T,$$

$$\sum_{t=1}^{T} t = \frac{T(T+1)}{2},$$

$$\sum_{t=1}^{T} t^2 = \frac{(2T+1)T(T+1)}{6},$$

$$\sum_{t=1}^{T} t^3 = \frac{T^2(T+1)^2}{4},$$

$$\sum_{t=1}^{T} t^4 = \frac{(2T+1)T(T+1)(3T^2+3T-1)}{30},$$

$$\sum_{t=1}^{T} t^5 = \frac{T^2(T+1)^2(2T^2+2T-1)}{12},$$

$$\sum_{t=1}^{T} t^6 = \frac{(2T+1)T(T+1)(3T^4+6T^3-3T+1)}{42},$$

$$\sum_{t=1}^{T} t^7 = \frac{T^2(T+1)^2(3T^4+6T^3-T^2-4T+2)}{24},$$

$$\sum_{t=1}^{T} t^8 = \frac{T(T+1)(2T+1)(5T^6+15T^5+5T^4-15T^3-T^2+9T-3)}{90}.$$

8. (Sec. 3.2.1.) Find $\psi_{2h}(T) = \sum_{t=1}^{T} t^{2h}$ for $h = 1, 2, 3, 4$ by using Problem 6.

9. (Sec. 3.2.2.) Let $x_1, \ldots, x_N$ be independently normally distributed with variance σ^2 and expected values $\mathscr{E}x_i = \mu_i$, $i = 1, \ldots, n$, and $\mathscr{E}x_i = 0$, $i = n+1, \ldots, N$. Prove that for R a (measurable) set in the N-dimensional space of $x_1, \ldots, x_N$

$$\text{(i)} \qquad \Pr\left\{R \mid x_1, \ldots, x_n, \sum_{i=n+1}^{N} x_i^2\right\} = \varepsilon$$

implies

$$\text{(ii)} \qquad \Pr\{R\} = \varepsilon.$$

[*Hint:* Take the expectation of (i) relative to the distribution of $x_1, \ldots, x_n$ and $\sum_{i=n+1}^{N} x_i^2$.]

10. (Sec. 3.2.2.) In the formulation of Problem 9 prove that (ii) (identically in $\mu_1, \ldots, \mu_n$ and σ^2) implies (i) almost everywhere (that is, with probability 1). [*Hint:* (ii) is

$$\text{(iii)} \qquad 1 = \int_0^\infty \int_{-\infty}^\infty \cdots \int_{-\infty}^\infty \frac{\Pr\{R \mid x_1, \ldots, x_n, v\}}{\varepsilon} \prod_{i=1}^{n} n(x_i \mid \mu_i, \sigma^2) h\left(\frac{v}{\sigma^2}\right) \prod_{i=1}^{n} dx_i \frac{dv}{\sigma^2}$$

identically in $\mu_1, \ldots, \mu_n$, and σ^2, where h is the density of χ^2_{N-n} and $n(x_i \mid \mu_i, \sigma^2)$ is the normal density with mean μ_i and variance σ^2. If $1/\sigma^2$ is replaced by $(1/\sigma^2) - 2\theta_0$ and μ_i by $[(\mu_i/\sigma^2) + \theta_i]/[(1/\sigma^2) - 2\theta_0]$ in (iii), the moment generating function of the integrand can be deduced; this is the moment generating function of the density of $x_1, \ldots, x_n$, and v.]

11. (Sec. 3.2.2.) Let $x_1, \ldots, x_N$ be independently normally distributed with variance σ^2 and expected values $\mathscr{E}x_i = \mu_i$, $i = 1, \ldots, n+1$, and $\mathscr{E}x_i = 0$, $i = n+2, \ldots, N$. Prove $x_1, \ldots, x_{n+1}$ and $\sum_{i=n+2}^{N} x_i^2$ form a sufficient set of statistics for $\mu_1, \ldots, \mu_{n+1}$, and σ^2.

12. (Sec. 3.2.2.) Prove that in the formulation of Problem 11 the best test of the hypothesis $\mu_{n+1} = 0$ at significance level ε (that is, probability ε of rejection uniformly in $\mu_1, \ldots, \mu_n$, and σ^2 when $\mu_{n+1} = 0$) against a particular alternative $\mu_{n+1} > 0$ has a rejection region of the form

$$\text{(iv)} \qquad \frac{x_{n+1}}{\sqrt{x_{n+1}^2 + \cdots + x_N^2}} > k.$$

[*Hint:* Since Problem 10 shows that the rejection region satisfies (i), consider the density at given $x_1, \ldots, x_n$, and $\sum_{i=n+1}^{N} x_i^2$. Compare the intersection of $x_1 = c_1, \ldots, x_n = c_n, x_{n+1}^2 + \cdots + x_N^2 = c$ with the region R above and the intersection with any other region R^* satisfying (i).]

13. (Sec. 3.2.2.) Prove that in the formulation of Problem 11 the uniformly most powerful test of the hypothesis $\mu_{n+1} = 0$ at significance level ε against alternatives $\mu_{n+1} > 0$ has a rejection region (iv) of Problem 12.

14. (Sec. 3.2.2.) Let $x_1, \ldots, x_n$ be normally independently distributed with means 0 and variances 1, and let S be independently distributed as χ^2 with m degrees of freedom. Let

$$t_i = \frac{x_i}{\sqrt{\dfrac{x_{i+1}^2 + \cdots + x_n^2 + S}{m + n - i}}}, \qquad i = 1, \ldots, n.$$

Prove that $t_1, \ldots, t_n$ are independently distributed according to t-distributions, the distribution of t_i having $m + n - i$ degrees of freedom. [*Hint:* Transform the density of $x_1, \ldots, x_n$, and S to that of $t_1, \ldots, t_n$, and $u = x_1^2 + \cdots + x_n^2 + S$.]

15. (Sec. 3.2.2.) Show that (19) and (20) are equivalent.

16. (Sec. 3.2.2.) Show that a_{ii} defined by (22) is

$$\frac{(i!)^4 T(T^2 - 1)(T^2 - 4)(T^2 - 9) \cdots (T^2 - i^2)}{(2i)!(2i + 1)!}.$$

17. (Sec. 3.2.2.) Give the fitted trend of Example 3.1 as a polynomial in t (where t is the year less 1918).

18. (Sec. 3.2.2.)

Dow Jones Industrial Stock Price Averages

Year	Price	Year	Price	Year	Price
1897	45.5	1903	55.5	1909	92.8
1898	52.8	1904	55.1	1910	84.3
1899	71.6	1905	80.3	1911	82.4
1900	61.4	1906	93.9	1912	88.7
1901	69.9	1907	74.9	1913	79.2
1902	65.4	1908	75.6		

Using these data do the following:

(a) Graph the data.

(b) Assuming the least squares model with $\mathscr{E}y_t = \gamma_0 + \gamma_1\phi_{1T}(t) + \gamma_2\phi_{2T}(t)$, where $\phi_{iT}(t)$ are the orthogonal polynomials, find the estimates of γ_0, γ_1, and γ_2.

(c) Find a confidence region for γ_2 with a confidence coefficient $1 - \varepsilon = .95$.

(d) Test the hypothesis H: $\gamma_2 = 0$ at the .05 level of significance.

(e) If $\mathscr{E}y_t = \beta_0 + \beta_1 t + \beta_2 t^2$, find the least squares estimates of β_0, β_1, and β_2. [*Hint:* These estimates can be found from (b).]

(f) Compute and graph the residuals from the estimated linear trend $c_0 + c_1\phi_{1T}(t)$.

19. (Sec. 3.2.2.) Suppose the polynomial regression is

$$\mathscr{E}y_t = \beta_0 + \beta_1 t + \cdots + \beta_q t^q$$

and the investigator wants to determine the degree of polynomial according to the method of Section 3.2.2. Show how the computation can be done efficiently by using the forward solution of the normal equations.

20. (Sec. 3.3.1.) In (1) let $a = 1$ and $k = 3$.

(a) Tabulate and graph $f(t)$.

(b) Verify that the two constituent functions define df/dt consistently at $t = 2k = 6$.

(c) Find the parabola that approximates $f(t)$ best over the range $t = 0, 1, \ldots, 12$ in the sense of minimizing the sum of squared deviations.

(d) Tabulate and graph this parabola.

(e) Tabulate the deviations.

(f) What parabola fits best at $t = 5, 6, 7$ and what are its values at these points?

(g) What is the value of the first constituent parabola at $t = 12$?

(h) What is the best fitting parabola at $t = 4, 5, 6, 7, 8$ and what are its values at these points?

(i) What is the best fitting cubic over the range $t = 0, 1, \ldots, 12$?

(j) Tabulate and graph the cubic.

21. (Sec. 3.3.1.) If $\sum_{s=-m}^{m} c_s = 1$, show that

$$\sum_{s=-m}^{m} c_s^2 \geq \frac{1}{2m+1}.$$

If, furthermore, $c_s \geq 0$, show that

$$\sum_{s=-m}^{m} c_s^2 \leq 1$$

and that equality holds only if all the c_s's but one are 0.

22. (Sec. 3.3.1.) Find a_0 when $k = 2$. [*Hint:* Find the orthogonal polynomials $\phi^*_{i,2m+1}(s)$, $i = 0, 2, 4$, defined over $s = -m, \ldots, m$ and represent a_0 in terms of the fitted linear combination of these.]

23. (Sec. 3.3.1.) Smooth the series in Example 3.1 with the procedure based on $k = 1$ and $m = 2$. Compare with the fitted cubic polynomial.

24. (Sec. 3.3.1.)

(a) Smooth the series in Problem 18 with the procedure based on $k = 0$ and $m = 3$.

(b) Smooth with the procedure based on $k = 1$ and $m = 2$.

(c) Compare the smoothed series with the fitted trend in Problem 18.

25. (Sec. 3.3.1.) Let

$$C = \begin{pmatrix} 1 & 1 & 1 & \cdots & 1 \\ 1 & 2^2 & 3^2 & \cdots & m^2 \\ 1 & 2^4 & 3^4 & \cdots & m^4 \\ \cdot & \cdot & \cdot & & \cdot \\ \cdot & \cdot & \cdot & & \cdot \\ \cdot & \cdot & \cdot & & \cdot \\ 1 & 2^{2k} & 3^{2k} & \cdots & m^{2k} \end{pmatrix}.$$

Prove that the matrix of coefficients in (13) is

$$\begin{bmatrix} 1 & 0 & \cdots & 0 \\ 0 & 0 & \cdots & 0 \\ \cdot & \cdot & & \cdot \\ \cdot & \cdot & & \cdot \\ \cdot & \cdot & & \cdot \\ 0 & 0 & \cdots & 0 \end{bmatrix} + 2CC'.$$

26. (Sec. 3.3.1.) In Problem 25 let $D = 2CC'$ and let D_{00} be the cofactor of d_{00}. Prove

$$b^{00} = \frac{D_{00}}{D_{00} + |D|} = \frac{d^{00}}{1 + d^{00}}.$$

27. (Sec. 3.3.1.) Verify the entries in Table 3.3.

28. (Sec. 3.3.2.) Find a confidence interval for α_0 at t given an independent estimate s^2 of σ^2, where s^2/σ^2 has a χ^2-distribution with n degrees of freedom.

29. (Sec. 3.3.2.) Let the orthogonal polynomials $\phi^*_{i,2m+1}(s)$ be defined over $-m, \ldots, m$. Find $\phi^*_{1,5}(s)$ and $\phi^*_{2,5}(s)$.

30. (Sec. 3.3.2.) Let the orthogonal polynomials $\phi^*_{i,2m+1}(s)$ be defined over $-m, \ldots, m$. Prove $\phi^*_{i,2m+1}(0) = 0$ for i odd. [*Hint:* Let $\phi^*(s) = c_0 + c_1 s + \cdots + c_{2r}s^{2r} + s^{2r+1}$. Prove that $\sum_{s=-m}^{m} \phi^*(s)s^{2j} = 0$, $j = 0, 1, \ldots, r$, yield a set of $r + 1$ homogeneous equations in the $r + 1$ unknowns $c_0, c_2, \ldots, c_{2r}$ with nonsingular coefficient matrix.]

31. (Sec. 3.3.2.) Verify (34). [*Hint:* Let $\cos[\lambda(t + s) - \theta] = \mathscr{R} \exp[i(\lambda s + \lambda t - \theta)]$, where $\mathscr{R} \exp[iy]$ is the real part of e^{iy}, and use the formula for a geometric sum $\sum_{s=0}^{m-1} x^s = (1 - x^m)/(1 - x)$, $x \neq 1$, for $x = e^{iy}$.]

32. (Sec. 3.3.2.) Prove

$$\lim_{x \to 0} \frac{\sin ax}{\sin x} = a.$$

33. (Sec. 3.3.2.) Prove that $(\sin y)/y$ is a decreasing function of y $(0 \leq y \leq \pi)$.

34. (Sec. 3.3.2.) Let

$$y_t^* = \frac{1}{2m + 1} \sum_{s=-m}^{m} y_{t+s},$$

$$y_t^{**} = \frac{1}{2n + 1} \sum_{r=-n}^{n} y_{t+r}^*.$$

Find the coefficients c_s in

$$y_t^{**} = \sum_{s=-(m+n)}^{m+n} c_s y_{t+s}.$$

35. (Sec. 3.3.3.) When $h(t) = \cos(\lambda t - \theta)$ show (43) is

$$\frac{\sin \frac{1}{2}\lambda(2m+1) + \sin \frac{1}{2}\lambda(2m-1)}{4m \sin \frac{1}{2}\lambda} \cos(\lambda t - \theta) = \frac{\sin \lambda m \cos \frac{1}{2}\lambda}{2m \sin \frac{1}{2}\lambda} \cos(\lambda t - \theta).$$

[*Hint:* See Problem 31.]

36. (Sec. 3.3.3.) Verify (49).

37. (Sec. 3.4.2.) Verify (5).

38. (Sec. 3.4.2.) Show that if $\mathcal{O}_1$ and $\mathcal{O}_2$ are linear operators $c_1\mathcal{O}_1 + c_2\mathcal{O}_2$ is a linear operator.

39. (Sec. 3.4.2.) Show that if $\Delta\phi_1(t) = \Delta\phi_2(t)$ for $t = 1, 2, \ldots$, where $\Delta\phi_1(t) = \phi_1(t+1) - \phi_1(t)$, then $\phi_1(t) - \phi_2(t)$ is a constant.

40. (Sec. 3.4.2.) Let $x^{(0)} = 1$, $x^{(r)} = x(x-1)\cdots(x-r+1)$, $r = 1, 2, \ldots$. Show that

$$\Delta x^{(r)} = rx^{(r-1)},$$

where $\Delta x^{(r)} = (x+1)^{(r)} - x^{(r)}$.

41. (Sec. 3.4.2.) Show

$$\sum_{t=1}^{T} t^{(r)} = \frac{(T+1)^{(r+1)}}{r+1}, \qquad r = 1, 2, \ldots.$$

42. (Sec. 3.4.2.) For $x^{(r)}$ defined in Problem 40,

$$x^n = \sum_{r=1}^{n} S_r^n x^{(r)}, \qquad n = 1, 2, \ldots,$$

where S_r^n are Stirling numbers of the second kind. Derive the following table:

Table of Stirling Numbers of the Second Kind S_r^n

n \ r	1	2	3	4	5	6	7	8
1	1							
2	1	1						
3	1	3	1					
4	1	7	6	1				
5	1	15	25	10	1			
6	1	31	90	65	15	1		
7	1	63	301	350	140	21	1	
8	1	127	966	1701	1050	266	28	1

43. (Sec. 3.4.2.) Evaluate $\psi_k(T) = \sum_{t=1}^{T} t^k$ by use of Problems 41 and 42 for $k = 1, \ldots, 8$.

44. (Sec. 3.4.2.) Show that if $\Delta\phi(t)$ is a polynomial of degree $k - 1$, then $\phi(t)$ is a polynomial of degree k. [*Hint:* Use (13) to show a solution exists and Problem 39 to show no other type of solution exists.]

45. (Sec. 3.4.2.) Show $\psi_k(T) = \sum_{t=1}^{T} t^k$ is a polynomial of degree $k + 1$.

46. (Sec. 3.4.2.) Use Problem 45 to find $\psi_k(T) = \sum_{t=1}^{T} t^k$, $k = 1, \ldots, 8$, as given in Problem 7.

47. (Sec. 3.4.3.) Prove

$$f(t + 2) - \frac{1}{5}\sum_{s=-2}^{2} f(t + 2 + s) = -\frac{1}{5}[\Delta^4 + 5\Delta^3 + 5\Delta^2]f(t).$$

48. (Sec. 3.4.3.) Show that the residual $y_t - y_t^*$ of an observed series from a moving average based on fitting a polynomial of degree $q = 2k + 1$ to $2m + 1$ points can be written

$$y_{t+k+1} - y_{t+k+1}^* = \Delta^{2(k+1)}\left[\sum_{s=-(m-k-1)}^{m-k-1} \beta_s y_{t+s}\right]$$

and show $\beta_s = \beta_{-s}$, $s = 1, \ldots, m - k - 1$. Show how the β_r's can be calculated from the c_r's. [*Hint:* Write

$$\mathscr{P}^{k+1}\left(1 - \sum_{s=-m}^{m} c_s \mathscr{P}^s\right)y_t = (\mathscr{P} - 1)^{2(k+1)}\left(\sum_{r=-(m-k-1)}^{m-k-1} \beta_r \mathscr{P}^r y_t\right).\Bigg]$$

49. (Sec. 3.4.3.) In the notation of Problem 48 show that

$$\begin{aligned}\sum_{s=-1}^{1} \beta_s y_{t+s} &= \tfrac{1}{21}(2y_{t-1} + 5y_t + 2y_{t+1}) \\ &= \tfrac{1}{21}(2\Delta^2 + 9\mathscr{P})y_{t-1}\end{aligned}$$

for $q = 2k + 1 = 3$ and $m = 3$.

50. (Sec. 3.4.3.) In the notation of Problem 48 show that

$$\begin{aligned}\sum_{s=-2}^{2} \beta_s y_{t+s} &= \tfrac{1}{231}(21y_{t-2} + 70y_{t-1} + 115y_t + 70y_{t+1} + 21y_{t+2}) \\ &= (\tfrac{1}{11}\Delta^4 + \tfrac{2}{3}\Delta^2\mathscr{P} + \tfrac{9}{7}\mathscr{P}^2)y_{t-2}\end{aligned}$$

for $q = 2k + 1 = 3$ and $m = 4$.

51. (Sec. 3.4.3.) Verify that if the rth order difference operator annihilates the trend then the trend is a polynomial of degree at most $r - 1$.

52. (Sec. 3.4.3.) Show that if the trend $f(t)$ is *exactly* a polynomial of degree q in every set of $q + 2$ consecutive points, then $f(t)$ is a polynomial of degree q over the whole range.

53. (Sec. 3.4.4.) When the trend is a polynomial of degree q and the residuals are normally and independently distributed with variance σ^2, show that s^2 given by (10) of Section 3.2 is the minimum variance unbiased estimate of σ^2.

54. (Sec. 3.4.4.) Prove that if the trend is exactly annihilated by differencing of order $q + 1$ and the residuals are normally and independently distributed with variance σ^2, the minimum variance unbiased estimate of σ^2 is s^2 given by (10) of Section 3.2. [*Hint:* Reduce to Problem 53.]

55. (Sec. 3.4.4.) Prove

$$a^{(r)}_{r+1-s-k,r+1-s} = (-1)^k \binom{2r}{r+k} - a^{(r)}_{s,s+k}, \quad s = 1, \ldots, r-k, \quad k = 1, \ldots, r-1,$$

$$a^{(r)}_{n+1,n+1+k} = \tfrac{1}{2}(-1)^k \binom{2r}{r+k}, \qquad r - k = 2n + 1.$$

56. (Sec. 3.4.4.) Prove (66) has a limiting normal distribution with mean 0 and variance 1. [*Hint:* $\{(\Delta^r u_t)^2\}$ forms a finitely dependent stationary stochastic process; apply Theorem 7.7.5.]

57. (Sec. 3.4.4.) Prove Lemma 3.4.5.

58. (Sec. 3.4.4.) Show that

$$\lim_{T\to\infty} \frac{1}{T-q} \sum_{s,t=1}^{T} a^{(r)}_{st} a^{(q)}_{st} = \binom{2r+2q}{r+q}.$$

[*Hint:* Show

$$\sum_{k=-p}^{p} \binom{2r}{r+k}\binom{2q}{q+k} = \binom{2r+2q}{r+q},$$

where $p = \min(r, q)$, by comparing coefficients of x^{r+q} in alternative expansions of $(x-1)^{2r}(1-x)^{2q}$.]

59. (Sec. 3.4.4.) Prove (68).

60. (Sec. 3.4.4.) Using (68) with $\kappa_4 = 0$, find c_1 and c_2 subject to $c_1 + c_2 = 1$ to minimize the (approximate) variance of $c_1 V_1 + c_2 V_2$. Give the resulting variance.

61. (Sec. 3.4.4.) Show $\sum_{t=1}^{T-r} (\Delta^r y_t)^2$ is

$$\sum_{t=r+1}^{T-r} \sum_{j=-r}^{r} (-1)^j \binom{2r}{r+j} y_t y_{t+j}$$

except for end terms.

62. (Sec. 3.4.5.) Find the limiting joint distribution of $\sqrt{T-r-2}\,(V_{r+1} - V_{r+2})/V_{r+2}$ and $\sqrt{T-r-1}\,(V_r - V_{r+1})/V_{r+1}$ when $\Delta^r f(t) = 0$, $t = 1, 2, \ldots$.

63. (Sec. 3.4.5.) Let $u_1, \ldots, u_T$ be independently distributed with $\mathscr{E}u_t = 0$, $\mathscr{E}u_t^2 = \sigma^2$, and $\mathscr{E}u_t^4 = \kappa_4 + 3\sigma^4$; let $Q_1 = \sum_{s,t=1}^{T} a_{st}u_s u_t$ and $Q_2 = \sum_{s,t=1}^{T} b_{st}u_s u_t$.

Show that

$$\operatorname{Var}\left(Q_1 - \frac{\mathscr{E}Q_1}{\mathscr{E}Q_2} Q_2\right) = \kappa_4 \sum_{t=1}^{T} \left(a_{tt} - \frac{\operatorname{tr} \boldsymbol{A}}{\operatorname{tr} \boldsymbol{B}} b_{tt}\right)^2 + 2\sigma^4 \operatorname{tr} \left(\boldsymbol{A} - \frac{\operatorname{tr} \boldsymbol{A}}{\operatorname{tr} \boldsymbol{B}} \boldsymbol{B}\right)^2.$$

64. (Sec. 3.4.5.) Let $\mathscr{E}\boldsymbol{u}'\boldsymbol{A}_T\boldsymbol{u} = \alpha\sigma^2$ and $\mathscr{E}\boldsymbol{u}'\boldsymbol{B}_T\boldsymbol{u} = \beta\sigma^2$, $\beta > 0$. Suppose that $\sqrt{T}(\boldsymbol{u}'\boldsymbol{A}_T\boldsymbol{u} - \alpha\sigma^2)$ and $\sqrt{T}(\boldsymbol{u}'\boldsymbol{B}_T\boldsymbol{u} - \beta\sigma^2)$ have a limiting bivariate normal distribution. Show that $\sqrt{T}[\boldsymbol{u}'\boldsymbol{A}_T\boldsymbol{u}/\boldsymbol{u}'\boldsymbol{B}_T\boldsymbol{u} - \alpha/\beta]$ has a limiting normal distribution with variance

$$\lim_{T\to\infty} T\left[\kappa_4 \sum_{t=1}^{T} \left(a_{tt}^{(T)} - \frac{\alpha}{\beta} b_{tt}^{(T)}\right)^2 + 2\sigma^4 \operatorname{tr}\left(\boldsymbol{A}_T - \frac{\alpha}{\beta}\boldsymbol{B}_T\right)^2\right] \bigg/ (\beta^2\sigma^4).$$

Show that the variance of the limiting distribution does not depend on κ_4 if $a_{11}^{(T)}/b_{11}^{(T)} = \cdots = a_{TT}^{(T)}/b_{TT}^{(T)}$. [*Hint:* Show

$$\operatorname{Var}\left(\boldsymbol{u}'\boldsymbol{A}_T\boldsymbol{u} - \frac{\alpha}{\beta}\boldsymbol{u}'\boldsymbol{B}_T\boldsymbol{u}\right) = \kappa_4 \sum_{t=1}^{T} \left(a_{tt}^{(T)} - \frac{\alpha}{\beta} b_{tt}^{(T)}\right)^2 + 2\sigma^4 \operatorname{tr}\left(\boldsymbol{A}_T - \frac{\alpha}{\beta}\boldsymbol{B}_T\right)^2.\bigg]$$

65. (Sec. 3.4.5.) Verify that the sequence $\Delta^r y_1, \Delta^q y_{r+2}, \Delta^r y_{r+q+3}, \ldots$ is a sequence of statistically independent variables. Use this fact to construct an F-test of the hypothesis $\Delta^r f(t) = 0$ under the assumptions of normality and that $\Delta^q f(t) = 0$, $r < q$.

66. (Sec. 3.4.5.) For Table 3.5 calculate $(V_r - V_{r+1})/V_{r+1}$, $r = 0, 1, \ldots, 4$, convert to normalized variables, and consider whether $\Delta^r f(t) = 0$.

67. (Sec. 3.4.5.) Find Δy_t, $\Delta^2 y_t$, and $\Delta^3 y_t$ for the data of Problem 18. Calculate V_1, V_2, and V_3.

68. (Sec. 3.5.) Write out the equations (4) explicitly for the logistic function (1).

69. (Sec. 3.5.) The population of the United States (in millions) at 30-year intervals was

1820	9.6
1850	23.2
1880	50.2
1910	92.0
1940	131.4

Fit the logistic to these data. [*Hint:* Try out several guessed triplets (a_1, a_2, a_3) by graphing; then use such a triplet as the basis for one iteration of (4).]

CHAPTER 4

Cyclical Trends

4.1. INTRODUCTION

Time series often show more or less regular fluctuations as well as broad long-run movements. These up-and-down movements may be strictly periodic or nearly so, such as hours of sunshine per day. On the other hand, they may fluctuate irregularly; typical of the latter case are economic time series, the fluctuations of which reflect or constitute the business cycle. In this chapter we treat the model in which the observed time series y_t is considered to be composed of a periodic trend $f(t)$ plus a random error u_t. That the trend is periodic means that it repeats itself after a certain period; if the period is ϕ, then

$$f(t+\phi) = f(t). \tag{1}$$

This repetition is perfectly regular and periodic; given the function on any interval of length ϕ it is determined over its entire range. Since we assume the trend to be a given function of time, the irregular effects u_t do not affect it. This model is to be contrasted with other models we shall study later in which the fluctuations are not regular and in which random effects are incorporated into the sequence.

As we shall see, the analysis in this chapter is naturally in terms of linear combinations of sine and cosine functions of time, the coefficients being parameters. Later on in models of stationary stochastic processes we again treat linear combinations of sines and cosines, but the coefficients will be random variables. Some of the mathematical theory and many of the computational procedures are common to the two approaches.

4.2. TRANSFORMATIONS AND REPRESENTATIONS

4.2.1. Orthogonal Periodic Functions

The trigonometric functions $\cos t$ and $\sin t$ are periodic with period 2π; that is,

$$\cos(t + 2\pi) = \cos t, \qquad \sin(t + 2\pi) = \sin t. \tag{1}$$

It follows that

$$\cos(t + 2\pi k) = \cos t, \qquad \sin(t + 2\pi k) = \sin t, \qquad k = \pm 1, \pm 2, \ldots . \tag{2}$$

We can make a linear transformation of the argument and retain the property of periodicity. Cos $(\lambda t - \theta)$ and $\sin(\lambda t - \theta)$ are periodic with *period* $2\pi/\lambda$; that is,

$$\begin{aligned} \cos[\lambda(t + 2\pi/\lambda) - \theta] &= \cos[\lambda t + 2\pi - \theta] = \cos[\lambda t - \theta], \\ \sin[\lambda(t + 2\pi/\lambda) - \theta] &= \sin[\lambda t + 2\pi - \theta] = \sin[\lambda t - \theta]. \end{aligned} \tag{3}$$

The reciprocal of the period, namely, $\lambda/(2\pi)$, is the *frequency* because this is the number of periods (not necessary integral) in the unit interval; that is, the function goes through its pattern this number of times. The effect of the multiplication by λ is to expand or contract the time scale. The effect of subtracting θ is to translate the cosine or sine curve. The maximum of $\cos(\lambda t - \theta)$ occurs at $\lambda t = \theta + 2\pi k$, $k = 0, \pm 1, \ldots$; that is, $t = (\theta + 2\pi k)/\lambda$. The angle θ is called the *phase*. We usually take θ so the first maximum occurs at $t = \theta/\lambda$; then $0 \leq \theta < 2\pi$. At $t = 0$ the trigonometric function is $\cos\theta$ or $-\sin\theta$.

A translated cosine or sine curve is a linear combination of a cosine and sine curve and vice versa. From the trigonometric formula $\cos(a - b) = \cos a \cos b + \sin a \sin b$ we have

$$\begin{aligned} \rho\cos(\lambda t - \theta) &= \rho(\cos\lambda t\cos\theta + \sin\lambda t\sin\theta) \\ &= \alpha\cos\lambda t + \beta\sin\lambda t, \end{aligned} \tag{4}$$

where

$$\alpha = \rho\cos\theta, \qquad \beta = \rho\sin\theta, \tag{5}$$

and equivalently

$$\begin{aligned} \rho &= \sqrt{\alpha^2 + \beta^2}, \\ \tan\theta &= \frac{\beta}{\alpha}, \qquad \theta = \arctan\frac{\beta}{\alpha}. \end{aligned} \tag{6}$$

The coefficient ρ, which is the maximum of $\rho\cos(\lambda t - \theta)$, is called the *amplitude* of the function. The expression (4) can also be written as $\rho\sin(\lambda t + \phi)$, where $\tan\phi = \alpha/\beta$, but usually the cosine function is preferred.

The trigonometric functions are convenient to work with because of certain orthogonality properties. We consider here orthogonality properties for sums over the range $1, \ldots, T$; these correspond to the orthogonality of the polynomials treated in Section 3.2.1. Consider frequencies $\lambda/(2\pi) = j/T$, $j = 0, 1, \ldots, [\frac{1}{2}T]$, where $[\frac{1}{2}T] = \frac{1}{2}T$ for T even and $[\frac{1}{2}T] = \frac{1}{2}(T-1)$ for T odd. A period is $2\pi/\lambda = T/j$; there are j of these in the length of observation T. The T cosine and sine functions of these frequencies are orthogonal. To show this it is convenient to use

$$\begin{aligned} e^{i\lambda} &= \cos\lambda + i\sin\lambda, \\ \cos\lambda &= \tfrac{1}{2}(e^{i\lambda} + e^{-i\lambda}), \\ \sin\lambda &= \frac{1}{2i}(e^{i\lambda} - e^{-i\lambda}). \end{aligned} \tag{7}$$

Then

$$\begin{aligned} &\sum_{t=1}^{T} \cos\frac{2\pi j}{T}t\left[\cos\frac{2\pi k}{T}t + i\sin\frac{2\pi k}{T}t\right] \\ &= \sum_{t=1}^{T} \tfrac{1}{2}[e^{i2\pi jt/T} + e^{-i2\pi jt/T}]e^{i2\pi kt/T} \\ &= \tfrac{1}{2}e^{i2\pi(k+j)/T}\left[\sum_{t=0}^{T-1} e^{i2\pi(k+j)t/T}\right] + \tfrac{1}{2}e^{i2\pi(k-j)/T}\left[\sum_{t=0}^{T-1} e^{i2\pi(k-j)t/T}\right] \\ &= \begin{cases} \tfrac{1}{2}e^{i2\pi(k+j)/T}\left[\dfrac{1-e^{i2\pi(k+j)}}{1-e^{i2\pi(k+j)/T}}\right] + \tfrac{1}{2}e^{i2\pi(k-j)/T}\left[\dfrac{1-e^{i2\pi(k-j)}}{1-e^{i2\pi(k-j)/T}}\right], & 0 \le k \ne j \le [\tfrac{1}{2}T], \\ \tfrac{1}{2}e^{i2\pi(k+j)/T}\left[\dfrac{1-e^{i2\pi(k+j)}}{1-e^{i2\pi(k+j)/T}}\right] + \tfrac{1}{2}T, & 0 < k = j < \tfrac{1}{2}T, \\ T, & k = j = 0, \tfrac{1}{2}T, \end{cases} \\ &= \begin{cases} 0, & 0 \le k \ne j \le [\tfrac{1}{2}T], \\ \tfrac{1}{2}T, & 0 < k = j < \tfrac{1}{2}T, \\ T, & k = j = 0, \tfrac{1}{2}T. \end{cases} \end{aligned} \tag{8}$$

Hence equating the real and imaginary parts of the left and right sides of (8) we obtain

$$\sum_{t=1}^{T} \cos \frac{2\pi j}{T} t \cos \frac{2\pi k}{T} t = \begin{cases} 0, & 0 \leq k \neq j \leq [\tfrac{1}{2}T], \\ \tfrac{1}{2}T, & 0 < k = j < \tfrac{1}{2}T, \\ T, & k = j = 0, \tfrac{1}{2}T, \end{cases} \tag{9}$$

$$\sum_{t=1}^{T} \cos \frac{2\pi j}{T} t \sin \frac{2\pi k}{T} t = 0, \qquad k, j = 0, 1, \ldots, [\tfrac{1}{2}T]. \tag{10}$$

A similar argument shows that

$$\sum_{t=1}^{T} \sin \frac{2\pi j}{T} t \sin \frac{2\pi k}{T} t = \begin{cases} 0, & 0 \leq k \neq j \leq [\tfrac{1}{2}T], \\ \tfrac{1}{2}T, & 0 < k = j < \tfrac{1}{2}T, \\ 0, & k = j = 0, \tfrac{1}{2}T. \end{cases} \tag{11}$$

Also, letting $j = 0$ in (9) and (10), we obtain

$$\sum_{t=1}^{T} \cos \frac{2\pi k}{T} t = 0, \qquad k = 1, \ldots, [\tfrac{1}{2}T], \tag{12}$$

$$\sum_{t=1}^{T} \sin \frac{2\pi k}{T} t = 0, \qquad k = 0, 1, \ldots, [\tfrac{1}{2}T]. \tag{13}$$

If T is odd, then $\cos 2\pi 0t/T = 1$ and $\cos 2\pi kt/T$ and $\sin 2\pi kt/T$, $k = 1, \ldots, \frac{1}{2}(T-1)$, form a set of T sequences of T numbers, each pair of which is orthogonal. If T is even, the set is 1, $\cos 2\pi kt/T$, $\sin 2\pi kt/T$, $k = 1, \ldots, \frac{1}{2}T - 1$, and $\cos 2\pi \frac{1}{2}Tt/T = \cos \pi t = (-1)^t$. The sum of squares of each sequence is $\frac{1}{2}T$ except for the sequences 1 and $(-1)^t$, for which each sum of squares is T.

4.2.2. Representation of a Finite Sequence in Terms of its Frequencies

Consider a sequence of T numbers, say $y_1, \ldots, y_T$ (which is not necessarily an observed time series). This sequence is a set of coordinates of a point in a space of T dimensions. The point can be referred to another coordinate system; sometimes the coordinates in another system are more meaningful. We shall use the orthogonal trigonometric functions properly normalized to define an orthogonal matrix which will transform the coordinates $y_1, \ldots, y_T$

to another set. If T is even, let the $T \times T$ matrix $\boldsymbol{M}$ be defined by

$$\boldsymbol{M} = \sqrt{\frac{2}{T}} \begin{bmatrix} \frac{1}{\sqrt{2}} & \cos\frac{2\pi}{T} & \sin\frac{2\pi}{T} & \cos\frac{4\pi}{T} & \cdots & \sin\frac{2\pi(\frac{1}{2}T-1)}{T} & -\frac{1}{\sqrt{2}} \\ \frac{1}{\sqrt{2}} & \cos\frac{4\pi}{T} & \sin\frac{4\pi}{T} & \cos\frac{8\pi}{T} & \cdots & \sin\frac{4\pi(\frac{1}{2}T-1)}{T} & \frac{1}{\sqrt{2}} \\ \vdots & \vdots & \vdots & \vdots & & \vdots & \vdots \\ \frac{1}{\sqrt{2}} & 1 & 0 & 1 & \cdots & 0 & \frac{1}{\sqrt{2}} \end{bmatrix}. \tag{14}$$

From the orthogonality relations (9) to (13) we deduce

$$\boldsymbol{M}'\boldsymbol{M} = \boldsymbol{I}; \tag{15}$$

that is, $\boldsymbol{M}$ is an orthogonal matrix. Now let

$$\boldsymbol{y} = \begin{bmatrix} y_1 \\ \cdot \\ \cdot \\ \cdot \\ y_T \end{bmatrix} \tag{16}$$

and

$$\boldsymbol{x} = \begin{bmatrix} x_1 \\ \cdot \\ \cdot \\ \cdot \\ x_T \end{bmatrix} = \boldsymbol{M}'\boldsymbol{y}; \tag{17}$$

that is,

$$\begin{aligned} x_1 &= \frac{1}{\sqrt{T}} \sum_{t=1}^{T} y_t, \\ x_{2k} &= \sqrt{\frac{2}{T}} \sum_{t=1}^{T} y_t \cos\frac{2\pi k}{T}\, t, \qquad k = 1, \ldots, \tfrac{1}{2}T - 1, \\ x_{2k+1} &= \sqrt{\frac{2}{T}} \sum_{t=1}^{T} y_t \sin\frac{2\pi k}{T}\, t, \qquad k = 1, \ldots, \tfrac{1}{2}T - 1, \\ x_T &= \frac{1}{\sqrt{T}} \sum_{t=1}^{T} y_t(-1)^t. \end{aligned} \tag{18}$$

Then, since $\boldsymbol{MM}' = \boldsymbol{I}$,

$$\boldsymbol{y} = \boldsymbol{Mx}, \tag{19}$$

$$y_t = \sqrt{\frac{2}{T}}\left(\frac{1}{\sqrt{2}}\,x_1 + x_2 \cos\frac{2\pi}{T}\,t + \cdots + x_T\frac{(-1)^t}{\sqrt{2}}\right), \qquad t = 1, \ldots, T. \tag{20}$$

Sometimes (20) is called the Fourier representation of $y_1, \ldots, y_T$ with coefficients $x_1, \ldots, x_T$.

If T is odd, the matrix $\boldsymbol{M}$ is defined by

$$M = \sqrt{\frac{2}{T}}\begin{bmatrix} \frac{1}{\sqrt{2}} & \cos\frac{2\pi}{T} & \sin\frac{2\pi}{T} & \cos\frac{4\pi}{T} & \cdots & \sin\frac{2\pi\frac{1}{2}(T-1)}{T} \\ \frac{1}{\sqrt{2}} & \cos\frac{4\pi}{T} & \sin\frac{4\pi}{T} & \cos\frac{8\pi}{T} & \cdots & \sin\frac{4\pi\frac{1}{2}(T-1)}{T} \\ \cdot & \cdot & \cdot & \cdot & & \cdot \\ \cdot & \cdot & \cdot & \cdot & & \cdot \\ \cdot & \cdot & \cdot & \cdot & & \cdot \\ \frac{1}{\sqrt{2}} & 1 & 0 & 1 & \cdots & 0 \end{bmatrix}. \tag{21}$$

Then the Fourier representation is

$$y_t = \sqrt{\frac{2}{T}}\left(\frac{1}{\sqrt{2}}\,x_1 + x_2 \cos\frac{2\pi}{T}\,t + \cdots + x_T \sin\frac{2\pi\frac{1}{2}(T-1)}{T}\,t\right), \qquad t = 1, \ldots, T. \tag{22}$$

[In some ways the unitary matrix $\boldsymbol{N} = (n_{st}) = (e^{i2\pi st/T}/\sqrt{T})$, satisfying $\boldsymbol{N\bar{N}}' = \boldsymbol{I}$, where $\boldsymbol{\bar{N}} = (\bar{n}_{st})$ and $\bar{n}_{st}$ is the conjugate of n_{st}, is easier to work with. Note $\boldsymbol{N} = \boldsymbol{N}'$.]

4.2.3. Representation of a Periodic Sequence when the Period is an Integer

Suppose the sequence of numbers $y_1, \ldots, y_T$ is periodic with period n, where n is an integer; that is,

$$y_{t+n} = y_t, \qquad t = 1, \ldots, T-n. \tag{23}$$

We shall show how this sequence can be represented in terms of n trigonometric functions. Let

$$\boldsymbol{y}^* = \begin{bmatrix} y_1 \\ \cdot \\ \cdot \\ \cdot \\ y_n \end{bmatrix}. \tag{24}$$

Then

$$\begin{bmatrix} y_1 \\ y_2 \\ \cdot \\ \cdot \\ \cdot \\ y_{hn} \end{bmatrix} = \begin{bmatrix} \boldsymbol{y}^* \\ \boldsymbol{y}^* \\ \cdot \\ \cdot \\ \cdot \\ \boldsymbol{y}^* \end{bmatrix}, \tag{25}$$

where $\boldsymbol{y}^*$ is repeated h times and $hn \leq T < (h + 1)n$. (There may be $T - hn$ terms which form only a partial cycle.) Now we define the $n \times n$ matrix $\boldsymbol{M}^*$ as in Section 4.2.2 so that

$$\boldsymbol{y}^* = \boldsymbol{M}^*\boldsymbol{x}^*, \qquad \boldsymbol{x}^* = \boldsymbol{M}^{*\prime}\boldsymbol{y}^*; \tag{26}$$

that is,

$$\begin{aligned}
x_1^* &= \frac{1}{\sqrt{n}} \sum_{t=1}^{n} y_t, \\
x_{2k}^* &= \sqrt{\frac{2}{n}} \sum_{t=1}^{n} y_t \cos \frac{2\pi k}{n}\, t, \quad k = 1, \ldots, [\tfrac{1}{2}(n-1)], \\
x_{2k+1}^* &= \sqrt{\frac{2}{n}} \sum_{t=1}^{n} y_t \sin \frac{2\pi k}{n}\, t, \quad k = 1, \ldots, [\tfrac{1}{2}(n-1)], \\
x_n^* &= \frac{1}{\sqrt{n}} \sum_{t=1}^{n} y_t (-1)^t, \qquad n \text{ even.}
\end{aligned} \tag{27}$$

Then for $t = 1, \ldots, n$ (even), we have

$$y_t = \sqrt{\frac{2}{n}} \left(\frac{1}{\sqrt{2}}\, x_1^* + x_2^* \cos \frac{2\pi}{n}\, t + \cdots + x_n^* \frac{(-1)^t}{\sqrt{2}} \right). \tag{28}$$

We note that for $t \leq n$

$$\begin{aligned}
y_{t+n} = y_t &= \sqrt{\frac{2}{n}} \left(\frac{1}{\sqrt{2}}\, x_1^* + x_2^* \cos \frac{2\pi}{n}\, t + \cdots + x_n^* \frac{(-1)^t}{\sqrt{2}} \right), \\
&= \sqrt{\frac{2}{n}} \left(\frac{1}{\sqrt{2}}\, x_1^* + x_2^* \cos \frac{2\pi}{n}\, (t+n) + \cdots + x_n^* \frac{(-1)^{t+n}}{\sqrt{2}} \right),
\end{aligned} \tag{29}$$

because $\cos 2\pi k\,(t+n)/n = \cos 2\pi k\,t/n$ and $\sin 2\pi k\,(t+n)/n = \sin 2\pi k\,t/n$. In a similar way we show that (28) holds for every $t = 1, \ldots, T$. If n is odd, the term involving $(-1)^t$ is omitted.

We have demonstrated how to represent an arbitrary periodic sequence of integral period n as a linear combination of n trigonometric sequences. It will be noted that the result applies to any periodic sequence. However, if the given periodic sequence is a cosine-sine function with period n (that is, a multiple of the minimum period is n) then only one or two trigonometric sequences (the cosine and the sine) are needed.

The trigonometric functions with period n will not necessarily be orthogonal over $1, \ldots, T$ unless T is a multiple of n, say $T = hn$. In that case

$$\cos \frac{2\pi j}{n} t = \cos \frac{2\pi(jh)}{T} t, \qquad \sin \frac{2\pi j}{n} t = \sin \frac{2\pi(jh)}{T} t. \tag{30}$$

Thus the trigonometric functions in (28) are a subset of those in (20) or (22); then some (or many) of the coefficients in the Fourier representation are 0. The full representation is convenient because certain periodic sequences are simple special cases.

In Section 3.3 we considered a model in which the trend contained a periodic component $g(t)$; this was interpreted as the seasonal variation in a series. The period n is 2 for semiannual data, 4 for quarterly data, and 12 for monthly data. The seasonal variation $g(t)$ can be represented as a linear combination of n trigonometric terms.

A periodic sequence $y_1, \ldots, y_T$, where n is the period and $T = hn$, can be arranged in a matrix

$$\begin{bmatrix} y_1 & y_2 & \cdots & y_n \\ y_{n+1} & y_{n+2} & \cdots & y_{2n} \\ \cdot & \cdot & & \cdot \\ \cdot & \cdot & & \cdot \\ \cdot & \cdot & & \cdot \\ y_{(h-1)n+1} & y_{(h-1)n+2} & \cdots & y_T \end{bmatrix}, \tag{31}$$

where the rows are identical.

4.2.4. Representation of a Periodic Function

We now consider a periodic function $f(t)$ defined for all real t. If the period is ϕ, then

$$f(t) = f(t + \phi) = f(t + 2\phi) = \cdots . \tag{32}$$

We might expect to represent this function in terms of trigonometric functions which are periodic with the same period ϕ, namely, 1, $\cos 2\pi t/\phi$, $\sin 2\pi t/\phi$, $\cos 4\pi t/\phi$, $\sin 4\pi t/\phi, \ldots$. Consider an infinite series in these pairs of functions

$$\alpha_0 + \left(\alpha_1 \cos \frac{2\pi}{\phi} t + \beta_1 \sin \frac{2\pi}{\phi} t\right) + \left(\alpha_2 \cos \frac{4\pi}{\phi} t + \beta_2 \sin \frac{4\pi}{\phi} t\right) + \cdots . \tag{33}$$

If the infinite series converges to $f(t)$ for a value t, then it does for $t + \phi$ since

$$\begin{aligned} \cos \frac{2\pi k}{\phi} (t + \phi) &= \cos \frac{2\pi k}{\phi} t, \\ \sin \frac{2\pi k}{\phi} (t + \phi) &= \sin \frac{2\pi k}{\phi} t, \end{aligned} \tag{34}$$

and the sum of the series is periodic with period ϕ. The trigonometric functions are normalized and are orthogonal as follows:

$$\int_0^\phi \cos^2 \frac{2\pi j}{\phi} t \, dt = \tfrac{1}{2}\phi, \qquad j \neq 0, \tag{35}$$

$$\int_0^\phi \sin^2 \frac{2\pi j}{\phi} t \, dt = \tfrac{1}{2}\phi, \qquad j \neq 0, \tag{36}$$

$$\int_0^\phi \cos \frac{2\pi j}{\phi} t \cos \frac{2\pi k}{\phi} t \, dt = 0, \qquad j \neq k, \tag{37}$$

$$\int_0^\phi \sin \frac{2\pi j}{\phi} t \sin \frac{2\pi k}{\phi} t \, dt = 0, \qquad j \neq k, \tag{38}$$

$$\int_0^\phi \cos \frac{2\pi j}{\phi} t \sin \frac{2\pi k}{\phi} t \, dt = 0, \qquad \text{all } j, k. \tag{39}$$

These can be verified by using trigonometric identities. (See Problems 1, 2, 8, and 9.) For example,

$$\begin{aligned} \int_0^\phi \cos^2 \frac{2\pi j}{\phi} t \, dt &= \int_0^\phi \left(\tfrac{1}{2} + \tfrac{1}{2} \cos \frac{4\pi j}{\phi} t\right) dt \\ &= \tfrac{1}{2}\phi + \frac{\phi}{8\pi j} \int_0^{4\pi j} \cos s \, ds \\ &= \tfrac{1}{2}\phi, \qquad j \neq 0. \end{aligned} \tag{40}$$

If the infinite series (33) converges to $f(t)$ and if term-by-term integration is permissible, then

$$(41)\quad \int_0^\phi f(t)\cos\frac{2\pi k}{\phi}t\,dt$$

$$= \alpha_0\int_0^\phi \cos\frac{2\pi k}{\phi}t\,dt + \int_0^\phi \sum_{j=1}^{\infty}\left(\alpha_j\cos\frac{2\pi j}{\phi}t + \beta_j\sin\frac{2\pi j}{\phi}t\right)\cos\frac{2\pi k}{\phi}t\,dt$$

$$= \tfrac{1}{2}\phi\alpha_k, \qquad k \neq 0.$$

This determines α_k as

$$(42)\qquad \alpha_k = \frac{2}{\phi}\int_0^\phi f(t)\cos\frac{2\pi k}{\phi}t\,dt, \qquad k \neq 0.$$

Similarly

$$(43)\qquad \beta_k = \frac{2}{\phi}\int_0^\phi f(t)\sin\frac{2\pi k}{\phi}t\,dt, \qquad k \neq 0,$$

$$(44)\qquad \alpha_0 = \frac{1}{\phi}\int_0^\phi f(t)\,dt.$$

When the coefficients are chosen according to (42), (43), and (44), the series is said to *represent* $f(t)$. There are a number of theorems about the convergence of the series under different conditions. One of these states that if $f(t)$ is of bounded variation in the closed interval $[0, \phi]$, then the series (33) converges to $f(t)$ at every continuity point of $f(t)$. [See Whittaker and Watson (1943), pp. 174-179, for example.] It is not our purpose at this time to develop the theory of Fourier analysis, but simply to show how a (periodic) mathematical function can be developed in these trigonometric terms.

When a function $f(t)$ is used as a trend function in an error model, only its values at $t = 1, \ldots, T$ are relevant. As was observed previously, if $f(t)$ is periodic with period ϕ and if ϕ is an integer n, then only n values of the function appear, namely, $f(1), f(2), \ldots, f(n)$. (If T/n is not an integer, then these n values do not appear exactly the same number of times.) In such a case, the function can be represented at $t = 1, \ldots$ by a linear combination of n trigonometric functions. However, if ϕ is not an integer, then as many as T terms might be included to represent $f(t)$ at T points. The import of the above discussion is that when ϕ is not an integer the periodic $f(t)$ may still be approximated by a relatively small number of trigonometric terms.

4.2.5. Representation of a Function

We now turn to representing a real-valued function over the entire real line. A case familiar to statisticians is the relationship between a probability density

$p(x)$ and its characteristic function $\phi(t)$ defined as

$$\begin{aligned}(45)\qquad \phi(t) &= \int_{-\infty}^{\infty} e^{itx} p(x)\, dx \\ &= \int_{-\infty}^{\infty} [\cos tx + i \sin tx] p(x)\, dx \\ &= \phi_1(t) + i\phi_2(t),\end{aligned}$$

where $\phi_1(t)$ and $\phi_2(t)$ are real functions. From the definition we see that $\phi_1(t) = \phi_1(-t)$ is an even function and $\phi_2(t) = -\phi_2(-t)$ is an odd function. The inverse transform is

$$\begin{aligned}(46)\qquad p(x) &= \frac{1}{2\pi}\int_{-\infty}^{\infty} e^{-itx}\phi(t)\, dt \\ &= \frac{1}{2\pi}\int_{-\infty}^{\infty} [\cos(-tx) + i\sin(-tx)][\phi_1(t) + i\phi_2(t)]\, dt \\ &= \frac{1}{2\pi}\int_{-\infty}^{\infty} [\cos tx\, \phi_1(t) + \sin tx\, \phi_2(t)]\, dt.\end{aligned}$$

To summarize the representations or transforms: A finite sequence can be transformed into a finite set of Fourier coefficients by use of the orthogonal matrix $\boldsymbol{M}$. A periodic function can be represented by an infinite sum of Fourier terms whose coefficients (under certain conditions) are trigonometric integrals; conversely (under certain conditions) the Fourier series defines a periodic function. Finally, a (nonperiodic, integrable) function can be expressed as a Fourier integral, the integrand being the reciprocal Fourier integral.

4.3. STATISTICAL INFERENCE WHEN THE PERIODS OF THE TREND ARE INTEGRAL DIVISORS OF THE SERIES LENGTH

4.3.1. Least Squares Estimates of the Coefficients and the Variance

We consider now the model $y_t = f(t) + u_t$, $t = 1, \ldots, T$, where $\mathscr{E}u_t = 0$, $\mathscr{E}u_t^2 = \sigma^2$, $\mathscr{E}u_t u_s = 0$, $t \neq s$, and where $f(t)$ is periodic with known periods which divide T; that is, the periods of $f(t)$ are given by the numbers T/k_j, $j = 1, \ldots, q$, where $(k_1, \ldots, k_q)$ is a subset of the integers $1, \ldots, \frac{1}{2}(T-1)$ if T is odd, or of $1, \ldots, \frac{1}{2}T - 1$ if T is even. In the latter case period 2 may also be included. Hence $f(t)$ may be represented by†

$$(1)\qquad f(t) = \alpha_0 + \sum_{j=1}^{q}\left[\alpha(k_j)\cos\frac{2\pi k_j}{T}t + \beta(k_j)\sin\frac{2\pi k_j}{T}t\right],$$

† The notation $\alpha(k_j)$, etc., is used to avoid a subscript to a subscript.

when the term with period 2 is not to be included; if T is even and period 2 is desired, $f(t)$ may be represented by

$$(2)\qquad f(t) = \alpha_0 + \sum_{j=1}^{q} \left[\alpha(k_j) \cos \frac{2\pi k_j}{T} t + \beta(k_j) \sin \frac{2\pi k_j}{T} t \right] + \alpha_{\frac{1}{2}T}(-1)^t.$$

The trigonometric functions are a subset of those tabled in the matrix $\boldsymbol{M}$. It is possible that all T terms be included, but often fewer would be. For instance, if one had h years of monthly data, $T = 12h$, and one wanted to account for seasonal variation, one would take the constant and the 11 terms with minimum periods

$$(3)\qquad \begin{array}{lll} \dfrac{T}{\frac{1}{2}T} = \dfrac{12}{6} = 2, & \dfrac{T}{\frac{5}{12}T} = \dfrac{12}{5}, & \dfrac{T}{\frac{1}{3}T} = \dfrac{12}{4} = 3, \\ \dfrac{T}{\frac{1}{4}T} = \dfrac{12}{3} = 4, & \dfrac{T}{\frac{1}{6}T} = \dfrac{12}{2} = 6, & \dfrac{T}{\frac{1}{12}T} = 12; \end{array}$$

the frequencies are $1/2 = 6/12$ (one term), $5/12$, $1/3 = 4/12$, $1/4 = 3/12$, $1/6 = 2/12$, and $1/12$. In the case of period 2 there is only one term; this term and the constant completely account for a cycle of period 2 (that is, alternating between two values). For each other minimum period there are two terms, cosine and sine; the pair of terms accounts for a cosine function possibly translated. We can write (1) and (2) as

$$(4)\qquad f(t) = \alpha_0 + \sum_{j=1}^{q} \rho(k_j) \cos \left[\frac{2\pi k_j}{T} t - \theta(k_j) \right],$$

$$(5)\qquad f(t) = \alpha_0 + \sum_{j=1}^{q} \rho(k_j) \cos \left[\frac{2\pi k_j}{T} t - \theta(k_j) \right] + \alpha_{\frac{1}{2}T}(-1)^t,$$

respectively, where

$$(6)\qquad \rho(k_j) = \sqrt{\alpha^2(k_j) + \beta^2(k_j)}, \qquad \theta(k_j) = \arctan \left[\frac{\beta(k_j)}{\alpha(k_j)} \right],$$

$$(7)\qquad \alpha(k_j) = \rho(k_j) \cos \theta(k_j), \qquad \beta(k_j) = \rho(k_j) \sin \theta(k_j).$$

We now consider the problem of estimating α_0, $\alpha_{\frac{1}{2}T}$, $\alpha(k_j)$ and $\beta(k_j)$, $j = 1, \ldots, q$, by the method of least squares. Let a_0, $a_{\frac{1}{2}T}$, $a(k_j)$, and $b(k_j)$ be the least squares estimates of α_0, $\alpha_{\frac{1}{2}T}$, $\alpha(k_j)$, and $\beta(k_j)$, respectively. Then the normal equations for the a's and b's are

$$
(8)\qquad
\begin{bmatrix}
T & 0 & 0 & \cdots & 0 \\
0 & \tfrac{1}{2}T & 0 & \cdots & 0 \\
0 & 0 & \tfrac{1}{2}T & \cdots & 0 \\
\vdots & \vdots & \vdots & & \vdots \\
0 & 0 & 0 & \cdots & \tfrac{1}{2}T
\end{bmatrix}
\begin{bmatrix}
a_0 \\ a(k_1) \\ b(k_1) \\ \vdots \\ b(k_q)
\end{bmatrix}
=
\begin{bmatrix}
\sum_{t=1}^{T} y_t \\
\sum_{t=1}^{T} y_t \cos \frac{2\pi k_1}{T} t \\
\sum_{t=1}^{T} y_t \sin \frac{2\pi k_1}{T} t \\
\vdots \\
\sum_{t=1}^{T} y_t \sin \frac{2\pi k_q}{T} t
\end{bmatrix}
$$

when $f(t)$ is represented by (1), or

$$
(9)\qquad
\begin{bmatrix}
T & 0 & 0 & \cdots & 0 & 0 \\
0 & \tfrac{1}{2}T & 0 & \cdots & 0 & 0 \\
0 & 0 & \tfrac{1}{2}T & \cdots & 0 & 0 \\
\vdots & \vdots & \vdots & & \vdots & \vdots \\
0 & 0 & 0 & \cdots & \tfrac{1}{2}T & 0 \\
0 & 0 & 0 & \cdots & 0 & T
\end{bmatrix}
\begin{bmatrix}
a_0 \\ a(k_1) \\ b(k_1) \\ \vdots \\ b(k_q) \\ a_{\frac{1}{2}T}
\end{bmatrix}
=
\begin{bmatrix}
\sum_{t=1}^{T} y_t \\
\sum_{t=1}^{T} y_t \cos \frac{2\pi k_1}{T} t \\
\sum_{t=1}^{T} y_t \sin \frac{2\pi k_1}{T} t \\
\vdots \\
\sum_{t=1}^{T} y_t \sin \frac{2\pi k_q}{T} t \\
\sum_{t=1}^{T} y_t(-1)^t
\end{bmatrix}
$$

when $f(t)$ is represented by (2). Because the trigonometric functions are orthogonal, the matrix of coefficients of the estimates is diagonal. Thus the solutions to these normal equations are

$$
(10)\qquad a_0 = \frac{1}{T} \sum_{t=1}^{T} y_t = \bar{y},
$$

$$a(k_j) = \frac{2}{T}\sum_{t=1}^{T} y_t \cos\frac{2\pi k_j}{T} t, \qquad j = 1, \ldots, q, \tag{11}$$

$$b(k_j) = \frac{2}{T}\sum_{t=1}^{T} y_t \sin\frac{2\pi k_j}{T} t, \qquad j = 1, \ldots, q, \tag{12}$$

and in the case of (9)

$$a_{\frac{1}{2}T} = \frac{1}{T}\sum_{t=1}^{T} y_t(-1)^t. \tag{13}$$

It is interesting to note that expressions (10) through (13) may also be obtained by substituting the a's and b's for the α's and β's and the y_t's for $f(t)$ in equations (1) or (2) and applying the Fourier inversion formulas giving the coefficients in terms of the y_t's.

The least squares estimate of σ^2 is given by

$$s^2 = \frac{\sum_{t=1}^{T} y_t^2 - T\bar{y}^2 - \frac{1}{2}T\sum_{j=1}^{q}[a^2(k_j) + b^2(k_j)]}{T - (2q+1)} \tag{14}$$

when $f(t)$ is represented by (1), or

$$s^2 = \frac{\sum_{t=1}^{T} y_t^2 - T(\bar{y}^2 + a_{\frac{1}{2}T}^2) - \frac{1}{2}T\sum_{j=1}^{q}[a^2(k_j) + b^2(k_j)]}{T - (2q+2)}, \tag{15}$$

when $f(t)$ is represented by (2).

The estimates of $\rho(k_j)$ and $\theta(k_j)$ are

$$R(k_j) = \sqrt{a^2(k_j) + b^2(k_j)}, \qquad \hat{\theta}(k_j) = \arctan\frac{b(k_j)}{a(k_j)}, \quad j = 1, \ldots, q, \tag{16}$$

respectively. These estimates could be obtained directly by minimizing $\sum_{t=1}^{T}[y_t - f(t)]^2$ with respect to $\rho(k_j)$ and $\theta(k_j)$, $j = 1, \ldots, q$, and α_0 when $f(t)$ is given by (4) or with respect to these and $\alpha_{\frac{1}{2}T}$ when $f(t)$ is given by (5).

Example 4.1. Consider the data [taken from United States Department of Agriculture (1939), p. 390] in Table 4.1 on the receipts of butter (in units of 1,000,000 pounds) at five markets (Boston, Chicago, San Francisco, Milwaukee, and St. Louis).

TABLE 4.1

RECEIPTS OF BUTTER AT FIVE MARKETS

	Year				
Month	1935	1936	1937	Total m_t	Average
Jan.	48.9	48.3	42.4	139.6	46.5333
Feb.	43.4	47.1	41.4	131.9	43.9667
March	43.8	52.4	49.0	145.2	48.4000
April	50.8	55.3	50.8	156.9	52.3000
May	67.6	64.7	65.8	198.1	66.0333
June	83.7	79.5	85.9	249.1	83.0333
July	82.7	62.6	70.6	215.9	71.9667
Aug.	60.8	51.3	55.8	167.9	55.9667
Sept.	55.4	51.0	49.1	155.5	51.8333
Oct.	48.4	54.0	45.7	148.1	49.3667
Nov.	37.7	45.2	43.8	126.7	42.2333
Dec.	41.0	44.9	46.7	132.6	44.2000
Total	664.2	656.3	647.0	1967.5	655.8333
Average	55.350000	54.691667	53.916667	163.958333	54.652778

We want to estimate the constant α_0, the coefficient $\alpha_{\frac{1}{2}T} = \alpha_{18}$ corresponding to period 2 (frequency $\frac{1}{2} = 6/12$) and the pairs of coefficients corresponding to (minimum) periods of 12/5, 3, 4, 6, and 12 (frequencies 5/12, 1/3 = 4/12, 1/4 = 3/12, 1/6 = 2/12, and 1/12, respectively). Let

$$m_t = \sum_{j=0}^{2} y_{t+12j} \tag{17}$$

be the sum of the observations for the tth month over the 3 years. Because of the periodicity of the trigonometric functions

$$
\begin{aligned}
a_0 &= \bar{y} = \frac{1}{36}\sum_{t=1}^{36} y_t = \frac{1}{36}\sum_{t=1}^{12} m_t, \\
a_{3k} &= \frac{1}{18}\sum_{t=1}^{36} y_t \cos\frac{2\pi 3k}{36}t = \frac{1}{18}\sum_{t=1}^{12} m_t \cos\frac{2\pi k}{12}t, \qquad k = 1, 2, 3, 4, 5, \\
b_{3k} &= \frac{1}{18}\sum_{t=1}^{36} y_t \sin\frac{2\pi 3k}{36}t = \frac{1}{18}\sum_{t=1}^{12} m_t \sin\frac{2\pi k}{12}t, \qquad k = 1, 2, 3, 4, 5, \\
a_{18} &= \frac{1}{36}\sum_{t=1}^{36} y_t(-1)^t = \frac{1}{36}\sum_{t=1}^{12} m_t(-1)^t.
\end{aligned} \tag{18}
$$

Tables of the trigonometric functions are given by Kendall (1946b). Table 4.1, arranged according to the period and including the totals, is called the Buijs-Ballot table [Buijs-Ballot (1847)].

The coefficients computed from the data in Table 4.1 are given in Table 4.2. The sum of squared deviations is 474.51 and $s^2 = 19.771$. The largest sample amplitude in Table 4.2 is $R_3 = 14.96$ corresponding to the pair of terms with minimum period 12; this reflects the tendency in the series m_t to have one fluctuation roughly like a cosine-sine function.

TABLE 4.2

COEFFICIENTS OF CYCLICAL TRENDS

k	0	3	6	9	12	15	18
a_k	54.65	−14.82	6.60	−3.98	2.21	−0.61	0.15
b_k		−2.02	1.23	0.30	1.73	0.60	
Minimum period		12	6	4	3	12/5	2
Frequency		1/12	1/6	1/4	1/3	5/12	1/2
R_k		14.96	6.71	3.99	2.81	0.86	0.15
R_k^2		223.7	45.1	15.9	7.88	0.73	0.02

4.3.2. The Periodogram and Spectrogram

The graph of $R_k^2 = a_k^2 + b_k^2$ against the period T/k is called the *periodogram* (Figure 4.1). The use of orthogonal functions above defines the periodogram at $T/k = 2$ (if T is even), . . . , $T/4$, $T/3$, $\frac{1}{2}T$, T.

The graph of R_k^2 against the frequency k/T is called the *spectrogram* (Figure 4.2). The graph against frequency is more convenient because the squared amplitudes computed for the orthogonal terms are equally spaced with the range not depending on T.

We can define the same quantities for the population squared amplitude ρ_k^2. Sometimes the squared amplitude is called the *intensity*. The terms periodogram and spectrogram are used freely and loosely in the literature and are sometimes interchanged. Various factors are used to multiply R_k^2 in the definitions.

It will be observed that a_k and b_k are the values of a and b that minimize

$$\sum_{t=1}^{T}\left[y_t - \left(a \cos \frac{2\pi k}{T} t + b \sin \frac{2\pi k}{T} t\right)\right]^2, \tag{19}$$

and, hence, $R_k = \sqrt{a_k^2 + b_k^2}$ and $\hat{\theta}_k = \arctan\,(b_k/a_k)$ are the values of R and θ that minimize

$$\sum_{t=1}^{T}\left[y_t - R \cos\left(\frac{2\pi k}{T} t - \theta\right)\right]^2. \tag{20}$$

The minimum value of the sum of squares is

$$\sum_{t=1}^{T} y_t^2 - \tfrac{1}{2}T(a_k^2 + b_k^2) = \sum_{t=1}^{T} y_t^2 - \tfrac{1}{2}TR_k^2. \tag{21}$$

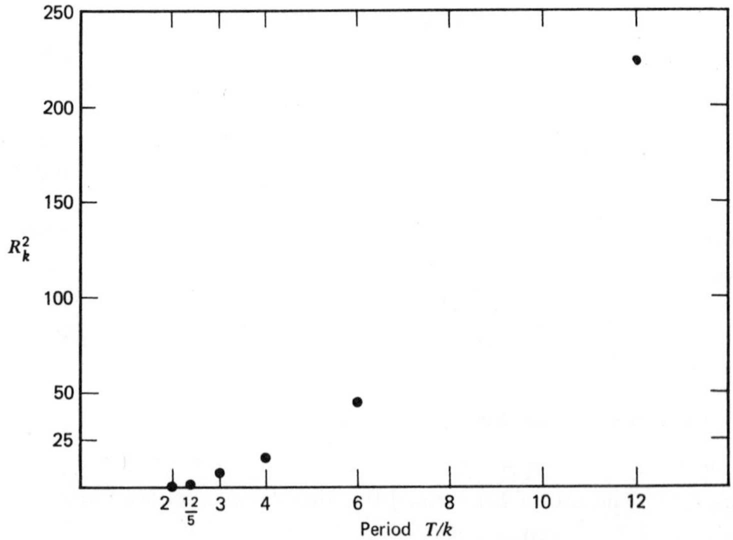

Figure 4.1. Periodogram for Example 4.1.

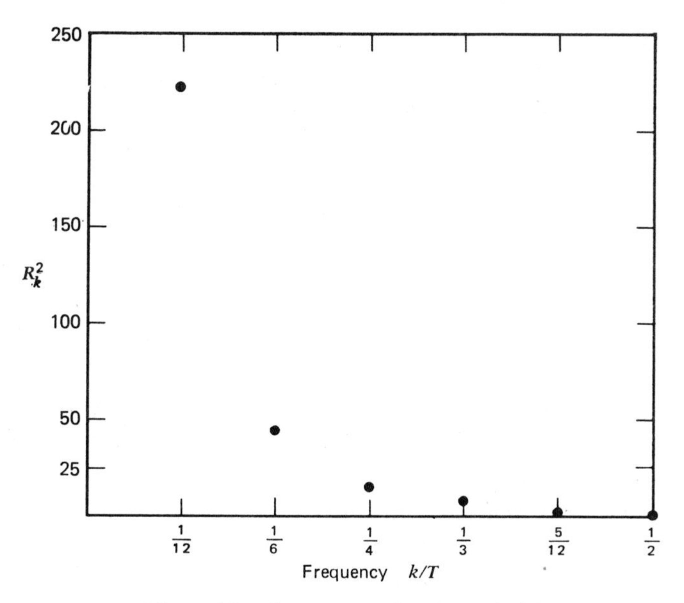

Figure 4.2. Spectrogram for Example 4.1.

In this sense R_k^2 is a measure of how closely the suitably chosen trigonometric function with frequency k/T fits the observed data.

The correlation coefficient between two sequences, $y_1, \ldots, y_T$ and $z_1, \ldots, z_T$ is

$$\text{(22)} \qquad \frac{\sum_{t=1}^{T}(y_t - \bar{y})(z_t - \bar{z})}{\sqrt{\sum_{t=1}^{T}(y_t - \bar{y})^2}\sqrt{\sum_{t=1}^{T}(z_t - \bar{z})^2}} = \frac{\sum_{t=1}^{T} y_t z_t - T\bar{y}\bar{z}}{\sqrt{\sum_{t=1}^{T} y_t^2 - T\bar{y}^2}\sqrt{\sum_{t=1}^{T} z_t^2 - T\bar{z}^2}},$$

where $\bar{y} = \sum_{t=1}^{T} y_t/T$ and $\bar{z} = \sum_{t=1}^{T} z_t/T$, and is a measure of the similarity between the two sequences disregarding location and scale factors. If $z_t = a \cos 2\pi kt/T + b \sin 2\pi kt/T = R \cos (2\pi kt/T - \theta)$, the correlation is

$$\text{(23)} \qquad \frac{\sum_{t=1}^{T} y_t\left(a \cos \frac{2\pi k}{T} t + b \sin \frac{2\pi k}{T} t\right)}{\sqrt{\sum_{t=1}^{T}(y_t - \bar{y})^2}\sqrt{\tfrac{1}{2}T(a^2 + b^2)}} = \frac{\sum_{t=1}^{T} y_t \cos\left(\frac{2\pi k}{T} t - \theta\right)}{\sqrt{\sum_{t=1}^{T}(y_t - \bar{y})^2}\sqrt{\tfrac{1}{2}T}}$$

$$= \sqrt{\frac{\tfrac{1}{2}T}{\sum_{t=1}^{T}(y_t - \bar{y})^2}}\,[a_k \cos \theta + b_k \sin \theta].$$

This correlation is maximized with respect to θ by $\hat{\theta}_k = \arctan b_k/a_k$, and (23) is $[\frac{1}{2}T/\sum_{t=1}^{T}(y_t - \bar{y})^2]^{\frac{1}{2}}R_k$. Thus the correlation between $\{y_t\}$ and $\{\cos(2\pi kt/T - \theta)\}$ maximized with respect to θ is proportional to R_k.

4.3.3. Tests of Hypotheses and Confidence Regions for the Coefficients

The estimated coefficients defined by (10) through (13) are unbiased estimates; that is,

$$\begin{aligned} &\mathscr{E}a_0 = \alpha_0, && \mathscr{E}a_{\frac{1}{2}T} = \alpha_{\frac{1}{2}T}, \\ &\mathscr{E}a(k_j) = \alpha(k_j), && \mathscr{E}b(k_j) = \beta(k_j), \qquad j = 1, \ldots, q. \end{aligned} \tag{24}$$

The variances of a_0 and $a_{\frac{1}{2}T}$ are σ^2/T, and the variances of $a(k_j)$ and $b(k_j)$ are $2\sigma^2/T$, $j = 1, \ldots, q$. These estimates are mutually uncorrelated. If the y_t's are normally distributed, these estimates are normally distributed and are independent.

The estimate of variance s^2 is an unbiased estimate of σ^2. If the y_t's are normally distributed, then s^2/σ^2 is proportional to a χ^2-variable, and it is distributed independently of the estimated coefficients. If $f(t)$ is given by (1) and s^2 by (14), then $(T - 2q - 1)s^2/\sigma^2$ has a χ^2-distribution with $T - 2q - 1$ degrees of freedom; if $f(t)$ is given by (2) and s^2 by (15) (that is, T is even and $a_{\frac{1}{2}T}$ is included), then $(T - 2q - 2)s^2/\sigma^2$ has a χ^2-distribution with $T - 2q - 2$ degrees of freedom.

Under the assumption of normality the general tests of hypotheses and confidence regions summarized in Chapter 2 are available. Because of the orthogonality and normalization properties of the regression variables here, the procedures take particularly simple forms. One null hypothesis we may be interested in is whether there is a cyclic term with given minimum period, say T/k_j. The null hypothesis is

$$H:\ \alpha(k_j) = \beta(k_j) = 0. \tag{25}$$

The question includes both the sine and cosine terms because together they can give a translated cosine function; we assume here that the question is not about the timing, that is, the phase $\theta(k_j)$. Thus the hypothesis is equivalently $H:\ \rho(k_j) = 0$.

When the null hypothesis is true, $a(k_j)$ and $b(k_j)$ are independently and normally distributed with means 0 and variances $2\sigma^2/T$; then

$$\frac{T}{2}\cdot\frac{a^2(k_j) + b^2(k_j)}{\sigma^2} = \frac{T}{2}\cdot\frac{R^2(k_j)}{\sigma^2} \tag{26}$$

has a χ^2-distribution with 2 degrees of freedom. In general, this quantity has the noncentral χ^2-distribution with noncentrality parameter $T\rho^2(k_j)/(2\sigma^2)$.

[See (49) below.] Let p be the number of estimated coefficients ($2q + 1$ or $2q + 2$). Then under the null hypothesis

$$\frac{\frac{T}{2} \cdot \frac{R^2(k_j)}{\sigma^2}}{2 \frac{s^2}{\sigma^2}} = \frac{TR^2(k_j)}{4s^2} \tag{27}$$

has the F-distribution with 2 and $T - p$ degrees of freedom; in general (27) has the noncentral F-distribution with noncentrality parameter $T\rho^2(k_j)/(2\sigma^2)$. The null hypothesis is rejected at significance level ε if (27) is greater than the ε-significance point of the F-distribution. The cumulative distribution function is

$$1 - \left(1 + \frac{2}{T - p} F\right)^{-\frac{1}{2}(T-p)}. \tag{28}$$

The hypothesis-testing problem is left invariant by the following transformations of the estimates: $a_0^* = ka_0 + c_0$, $a^*(k_i) = ka(k_i) + c_i$, $b^*(k_i) = kb(k_i) + d_i$, $i \neq j$, $a_{\frac{1}{2}T}^* = ka_{\frac{1}{2}T} + c_{\frac{1}{2}T}$ (if T is even), $a^*(k_j) = k[a(k_j)\cos\theta + b(k_j)\sin\theta]$, $b^*(k_j) = k[-a(k_j)\sin\theta + b(k_j)\cos\theta]$, and $s^{*2} = k^2 s^2$ for arbitrary $k \neq 0$, c_0, c_i, d_i, $i \neq j$, $c_{\frac{1}{2}T}$, and θ. The corresponding transformation on the observations, $y_1, \ldots, y_T$, is

$$\begin{aligned} y_t^* = ky_t + c_0 + \sum_{\substack{i=1 \\ i \neq j}}^{q} \left(c_i \cos\frac{2\pi k_i}{T} t + d_i \sin\frac{2\pi k_i}{T} t\right) \\ + \frac{2}{T} \sum_{s=1}^{T} \left\{\cos\left[\frac{2\pi k_j}{T}(s - t) - \theta\right] - \cos\frac{2\pi k_j}{T}(s - t)\right\} ky_s + (-1)^t c_{\frac{1}{2}T}, \\ t = 1, \ldots, T; \end{aligned} \tag{29}$$

the term $(-1)^t c_{\frac{1}{2}T}$ does not appear if T is odd. (See Problems 11 and 12.) The distribution of $y_1^*, \ldots, y_T^*$ is the distribution of $y_1, \ldots, y_T$ with transformed parameters. Any function of the parameters invariant under the induced transformations is a function of $\rho^2(k_j)/\sigma^2$, and any function of the sufficient statistics invariant under the transformations is a function of $R^2(k_j)/s^2$. Since the family of distributions is complete, the only invariant tests based on sufficient statistics with assigned significance level are tests based on $F = TR^2(k_j)/(4s^2)$. Hence, the uniformly most powerful invariant test at level ε is the test which rejects if the observed F exceeds $F_{2,T-p}(\varepsilon)$. The test is also uniformly most powerful among tests whose power depends only on $\rho^2(k_j)/\sigma^2$.

Hypotheses concerning α_0 and $\alpha_{\frac{1}{2}T}$ may be tested by t-tests. (See Problems 14 and 15.) If σ^2 is known, the F-tests can be replaced by χ^2-tests and t-tests by normal tests.

Suppose now that $T = hn$, n is even, and it is assumed a priori that $f(t) = f(t+n)$. Then to test the hypothesis that the trend of the series has no periodic variation, we must test the hypothesis that the coefficients of all of the orthogonal trigonometric functions with this period (except the constant 1) are zero. This may be done by using the F-statistic

$$\frac{\frac{1}{2}T \sum_{k=1}^{\frac{1}{2}n-1} [a^2(kh) + b^2(kh)] + Ta_{\frac{1}{2}T}^2}{(n-1)s^2}, \tag{30}$$

which has the F-distribution with $n - 1$ and $T - n$ degrees of freedom under the null hypothesis that $f(t)$ is a constant. In general [when $f(t)$ is periodic with period n], (30) has the noncentral F-distribution with noncentrality parameter

$$\frac{\frac{1}{2}T \sum_{k=1}^{\frac{1}{2}n-1} [\alpha^2(kh) + \beta^2(kh)] + T\alpha_{\frac{1}{2}T}^2}{\sigma^2}, \tag{31}$$

where $\alpha(kh)$ and $\beta(kh)$ are the coefficients of $\cos 2\pi kht/T = \cos 2\pi kt/n$ and $\sin 2\pi kht/T = \sin 2\pi kt/n$, $k = 1, \ldots, \frac{1}{2}n - 1$, in $f(t)$. (Here $k_j = jh$.)

We can arrange the observations in the Buijs-Ballot table

$$\begin{array}{cccc|c}
y_1 & y_{n+1} & \cdots & y_{(h-1)n+1} & \bar{y}_1 \\
y_2 & y_{n+2} & \cdots & y_{(h-1)n+2} & \bar{y}_2 \\
\cdot & \cdot & & \cdot & \cdot \\
\cdot & \cdot & & \cdot & \cdot \\
\cdot & \cdot & & \cdot & \cdot \\
y_n & y_{2n} & \cdots & y_{hn} & \bar{y}_n
\end{array} \tag{32}$$

where

$$\bar{y}_t = \frac{1}{h} \sum_{j=0}^{h-1} y_{t+nj}, \qquad t = 1, \ldots, n. \tag{33}$$

Let

$$\bar{y} = \frac{1}{n} \sum_{t=1}^{n} \bar{y}_t = \frac{1}{T} \sum_{t=1}^{T} y_t. \tag{34}$$

We have

$$\mathscr{E} y_{t+nj} = \mathscr{E} \bar{y}_t = f(t), \qquad t = 1, \ldots, n, \ j = 0, \ldots, h-1. \tag{35}$$

The null hypothesis is $f(1) = f(2) = \cdots = f(n)$. This is the problem of equality of means in the analysis of variance with equal numbers of observations in each class of a one-way classification. The usual F-statistic for testing the null hypothesis is

$$\frac{h \sum_{t=1}^{n} (\bar{y}_t - \bar{y})^2}{\sum_{t=1}^{n} \sum_{j=0}^{h-1} (y_{t+nj} - \bar{y}_t)^2} \cdot \frac{n(h-1)}{n-1}, \tag{36}$$

which has the F-distribution with $n - 1$ and $n(h - 1) = T - n$ degrees of freedom under the null hypothesis and the corresponding noncentral F-distribution with noncentrality parameter

$$\frac{h \sum_{t=1}^{n} [f(t) - \bar{f}]^2}{\sigma^2}, \tag{37}$$

where $\bar{f} = \sum_{t=1}^{n} f(t)/n$, in general. Using the properties of the trigonometric functions given in Section 4.2.1, one can show that the two F-statistics (30) and (36) are the same and the two noncentrality parameters (31) and (37) are the same.

Confidence regions for the parameters can be given in the usual way. For example, a confidence region for $\alpha(k_j)$ and $\beta(k_j)$ with confidence coefficient $1 - \varepsilon$ consists of the pairs of numbers (α^*, β^*) satisfying

$$[\alpha^* - a(k_j)]^2 + [\beta^* - b(k_j)]^2 \leq \frac{4s^2}{T} F_{2,T-p}(\varepsilon), \tag{38}$$

where $F_{2,T-p}(\varepsilon)$ is the ε-significance point of the F-distribution with 2 and $T - p$ degrees of freedom. This confidence region consists of the circumference and interior of the circle with center $[a(k_j), b(k_j)]$ and radius $\sqrt{4s^2 F_{2,T-p}(\varepsilon)/T}$. The points of the circumference and interior of this circle in polar coordinates, $\rho^* = \sqrt{\alpha^{*2} + \beta^{*2}}$ and $\theta^* = \arctan(\beta^*/\alpha^*)$, constitute a confidence region for the amplitude $\rho(k_j)$ and phase $\theta(k_j)$. The minimum and maximum ρ^* in the circle are the end points of the interval

$$\left(R(k_j) - \sqrt{\frac{4s^2}{T} F_{2,T-p}(\varepsilon)}, \quad R(k_j) + \sqrt{\frac{4s^2}{T} F_{2,T-p}(\varepsilon)}\right); \tag{39}$$

the interval (39) constitutes a confidence interval for $\rho(k_j)$ with confidence greater than $1 - \varepsilon$. [If the lower limit of (39) is negative, it can be replaced by

0.] Using the noncentral F-distribution of (27), we could obtain a confidence interval on the noncentrality parameter $T\rho^2(k_j)/(2\sigma^2)$, but this is not very useful since σ^2 is unknown.

If $R(k_j) < \sqrt{4s^2F_{2,T-p}(\varepsilon)/T}$, then the origin is in the confidence circle and $\rho^* = 0$ is admitted; only the upper limit in (39) is effective. In this case the null hypothesis $\rho(k_j) = 0$ is not rejected at significance level ε. For every angle θ^* there are points in the confidence circle. However, if the origin is not in the confidence circle, one can define a confidence interval for $\theta(k_j)$ by including all angles θ^* in the circle; that is, one draws the two tangents to the circle that go through the origin and the angles of these are the endpoints of the confidence interval for $\theta(k_j)$. In this case the length of the interval for θ is less than π. The procedure yields confidence greater than $1 - \varepsilon$.

Another approach is to use the fact that $a(k_j)\tan\theta - b(k_j)$ is normally distributed with mean $\alpha(k_j)\tan\theta - \beta(k_j)$ and variance $2\sigma^2(1 + \tan^2\theta)/T = 2\sigma^2\sec^2\theta/T$. If $\theta = \theta(k_j)$, the mean is 0; then

$$\Pr\left\{\frac{[a(k_j)\tan\theta - b(k_j)]^2}{\frac{2}{T}s^2(1+\tan^2\theta)} \le F_{1,T-p}(\varepsilon)\right\} = 1 - \varepsilon. \tag{40}$$

The event in the braces is

$$\left[a^2(k_j) - \frac{2}{T}s^2F_{1,T-p}(\varepsilon)\right]\left[\tan\theta - \frac{a(k_j)b(k_j)}{a^2(k_j) - \frac{2}{T}s^2F_{1,T-p}(\varepsilon)}\right]^2 \le \frac{\frac{2}{T}s^2F_{1,T-p}(\varepsilon)\left[R^2(k_j) - \frac{2}{T}s^2F_{1,T-p}(\varepsilon)\right]}{a^2(k_j) - \frac{2}{T}s^2F_{1,T-p}(\varepsilon)}. \tag{41}$$

Inversion of (41) to obtain limits on $\tan\theta$ gives

$$\frac{a(k_j)b(k_j) - \sqrt{\frac{2}{T}s^2F_{1,T-p}(\varepsilon)\left[R^2(k_j) - \frac{2}{T}s^2F_{1,T-p}(\varepsilon)\right]}}{a^2(k_j) - \frac{2}{T}s^2F_{1,T-p}(\varepsilon)} \le \tan\theta \le \frac{a(k_j)b(k_j) + \sqrt{\frac{2}{T}s^2F_{1,T-p}(\varepsilon)\left[R^2(k_j) - \frac{2}{T}s^2F_{1,T-p}(\varepsilon)\right]}}{a^2(k_j) - \frac{2}{T}s^2F_{1,T-p}(\varepsilon)} \tag{42}$$

if the denominator is positive (for then the quantity under the radical is nonnegative), and

$$
(43) \qquad \tan\theta \geq -\frac{a(k_j)b(k_j) + \sqrt{\frac{2}{T}s^2F_{1,T-p}(\varepsilon)\left[R^2(k_j) - \frac{2}{T}s^2F_{1,T-p}(\varepsilon)\right]}}{a^2(k_j) - \frac{2}{T}s^2F_{1,T-p}(\varepsilon)} \quad \text{or}
$$

$$
\tan\theta \leq -\frac{a(k_j)b(k_j) - \sqrt{\frac{2}{T}s^2F_{1,T-p}(\varepsilon)\left[R^2(k_j) - \frac{2}{T}s^2F_{1,T-p}(\varepsilon)\right]}}{a^2(k_j) - \frac{2}{T}s^2F_{1,T-p}(\varepsilon)},
$$

if the denominator is negative and the quantity under the radical is nonnegative. If the quantity under the radical is negative, the two ratios are imaginary. In that case (41) is satisfied by all θ since $a^2(k_j) < 2s^2F_{1,T-p}(\varepsilon)/T$. The confidence region for $\tan\theta$ then is (42) when the denominator is positive and (43) when the denominator is negative in case the quantity under the radical is nonnegative; it is the entire line when the denominator is negative in case the quantity under the radical is negative. The probability that the interval be the whole line is the probability that

$$
(44) \qquad \frac{TR^2(k_j)}{4s^2} < \tfrac{1}{2}F_{1,T-p}(\varepsilon);
$$

the left-hand side has a noncentral F-distribution with 2 and $T - p$ degrees of freedom and noncentrality parameter $T\rho^2(k_j)/(2\sigma^2)$. The probability of (44) is small if $T\rho^2(k_j)/(2\sigma^2)$ is large. Note that the confidence interval based on (38) is trivial when the left-hand side of (44) is less than $F_{2,T-p}(\varepsilon)$, which is about 3/2 of the right-hand side of (44). [Scheffé (1970) has proposed a procedure that yields the confidence set (38) for $\alpha(k_j)$ and $\beta(k_j)$ when that set includes the origin and yields intervals similar to (42) and (43) with $F_{1,T-p}(\varepsilon)$ replaced by a suitably chosen, decreasing function of $F_{2,T-p}(\varepsilon)$].

The three cases are (42) when $2s^2F_{1,T-p}(\varepsilon)/T \leq a^2(k_j)$, (43) when $a^2(k_j) \leq 2s^2F_{1,T-p}(\varepsilon)/T \leq a^2(k_j) + b^2(k_j)$, and the entire line when $a^2(k_j) + b^2(k_j) \leq 2s^2F_{1,T-p}(\varepsilon)/T$. These three cases occur when $a^2(k_j)$ is large, when $a^2(k_j)$ is small and $b^2(k_j)$ is large, and when both are small, respectively.

The test procedures and confidence regions given in this chapter are based on the observations being independently normally distributed. However, the results hold asymptotically for fixed periods T/k_j if the y_t's are independently distributed with uniformly bounded (absolute) moments of order $2 + \delta$ for some $\delta > 0$ or if a Lindeberg-Feller type condition is satisfied. (See Section 2.6.)

4.3.4. Deciding About Inclusion of Trigonometric Terms

A difficult but pertinent problem is what trigonometric terms to include in a cyclical trend. This is a multiple-decision problem that also arises in other areas where regression methods are used, but there are some special features that occur here. We assume there are p terms in the trend function, including α_0 and possibly $\alpha_{\frac{1}{2}T}(-1)^t$, if T is even, and $\frac{1}{2}(p-2)$ or $\frac{1}{2}(p-1)$ pairs of trigonometric terms with periods $T/k_j, j = 1, \ldots, \frac{1}{2}(p-2)$ or $\frac{1}{2}(p-1)$; then there is available an estimate of σ^2 with $T-p$ degrees of freedom. Suppose there are q pairs of trigonometric terms whose inclusions are in question; that is, $p-2q$ terms are certain to be included in the trend. For each pair of trigonometric terms we wish to decide whether the amplitude is 0; that is, $\rho(k_j) = 0$ for each j. The possible decisions are

$$
\begin{aligned}
&H_0: \quad \rho(k_1) = \cdots = \rho(k_q) = 0, \\
&H_i: \quad \rho(k_i) > 0, \quad \rho(k_j) = 0, \qquad j \neq i, j = 1, \ldots, q, \\
&H_{hi}: \quad \rho(k_h) > 0, \rho(k_i) > 0, \rho(k_j) = 0, \quad j \neq i,\ j \neq h,\ j = 1, \ldots, q, \qquad (45)\\
&\quad \cdot \\
&\quad \cdot \\
&\quad \cdot \\
&H_{12\cdots q}: \quad \rho(k_h) > 0, \qquad h = 1, \ldots, q.
\end{aligned}
$$

The multiple-decision problem here is basically different from the one treated in Section 3.2.2 because there is no a priori meaningful ordering of the periods as there is an ordering of degrees of polynomials.

We assume that there is no interest in the phases and no prior knowledge of them. Hence, our procedures will be based on $R^2(k_j)$. These procedures are invariant under transformations $a_0^* = a_0 + c_0$, $a_{\frac{1}{2}T}^* = a_{\frac{1}{2}T} + c_{\frac{1}{2}T}$ (if T is even), $a^*(k_i) = a(k_i) + c_i$, $b^*(k_i) = b(k_i) + d_i$, $i = q+1, \ldots, \frac{1}{2}(p-2)$ or $\frac{1}{2}(p-1)$, and $a^*(k_j) = a(k_j)\cos\theta_j + b(k_j)\sin\theta_j$, $\quad b^*(k_j) = -a(k_j)\sin\theta_j + b(k_j)\cos\theta_j$, $j = 1, \ldots, q$; that is,

$$
(46)\quad y_t^* = y_t + c_0 + \sum_{i=q+1}^{[\frac{1}{2}p-1]} \left(c_i \cos\frac{2\pi k_i}{T} t + d_i \sin\frac{2\pi k_i}{T} t \right)
$$

$$
+ \frac{2}{T} \sum_{s=1}^{T} \sum_{j=1}^{q} \left\{ \cos\left[\frac{2\pi k_j}{T}(s-t) - \theta_j\right] - \cos\frac{2\pi k_j}{T}(s-t) \right\} y_s + c_{\frac{1}{2}T}(-1)^t.
$$

Also we assume that there is no a priori preference for any one periodic function among the q. Hence, it seems reasonable to impose symmetry on any procedure to be considered.

Let us consider first the case when σ^2 is known. This case is of some historical interest and is the limiting case of some other cases. Under the assumption of normality a_0, $a(k_j)$, $b(k_j)$, $j = 1, \ldots, \frac{1}{2}(p-2)$ or $\frac{1}{2}(p-1)$, and possibly $a_{\frac{1}{2}T}$ form a sufficient set of statistics for α_0, $\alpha(k_j)$, $\beta(k_j)$, $j = 1, \ldots, \frac{1}{2}(p-2)$ or $\frac{1}{2}(p-1)$, and possibly $\alpha_{\frac{1}{2}T}$. We ask that procedures be invariant with respect to transformations (46). [Changes of phases are rotations of pairs $a(k_j)$, $b(k_j)$.] Thus the procedures should be based on $R^2(k_1), \ldots, R^2(k_q)$. If $\rho(k_j) = 0$,

$$z_j = \frac{T}{2\sigma^2} R^2(k_j) \tag{47}$$

has a χ^2-distribution with 2 degrees of freedom; the density and distribution functions are

$$\begin{aligned} k(z) &= \tfrac{1}{2} e^{-\frac{1}{2}z}, \\ K(z) &= \int_0^z k(v)\, dv = 1 - e^{-\frac{1}{2}z}, \end{aligned} \tag{48}$$

respectively. In general z_j has a noncentral χ^2-distribution with 2 degrees of freedom and noncentrality parameter $T\rho^2(k_j)/(2\sigma^2) = \tau^2$, say; the density† is

$$\begin{aligned} k(z \mid \tau^2) &= \frac{1}{2\sqrt{\pi}} e^{-\frac{1}{2}(\tau^2+z)} \sum_{\gamma=0}^{\infty} \frac{(\tau^2)^\gamma z^\gamma \Gamma(\gamma + \frac{1}{2})}{(2\gamma)!\, \Gamma(\gamma+1)} \\ &= \tfrac{1}{2} e^{-\frac{1}{2}(\tau^2+z)} \sum_{\gamma=0}^{\infty} \frac{(\tau^2)^\gamma z^\gamma}{2^{2\gamma}(\gamma!)^2} \\ &= \tfrac{1}{2} e^{-\frac{1}{2}(\tau^2+z)} I_0(\tau\sqrt{z}), \end{aligned} \tag{49}$$

where

$$I_0(z) = \sum_{\gamma=0}^{\infty} \frac{z^{2\gamma}}{2^{2\gamma}(\gamma!)^2} \tag{50}$$

is the Bessel function of the first kind with purely imaginary argument of order 0. Because $I_0(z)$ is a power series with positive coefficients, it is a monotonically increasing function of z $(0 \le z < \infty)$, and therefore $k(z \mid \tau^2)/k(z) = e^{-\frac{1}{2}\tau^2} I_0(\tau\sqrt{z})$ is a monotonically increasing function of z $(0 \le z < \infty)$ for every $\tau^2 > 0$.

Consider testing the null hypothesis that a specified population amplitude is zero using only the corresponding sample amplitude.

† See T. W. Anderson (1958), p. 113, for example. The noncentrality parameter is defined to be the quadratic form having the χ^2-distribution with the normal variables replaced by their expected values. Some authors divide this expression by 2 to obtain a different definition of the noncentrality parameter.

THEOREM 4.3.1. *The uniformly most powerful invariant test of H_j^*: $\rho(k_j) = 0$ against alternatives $\rho(k_j) > 0$ at significance level ε_j has the rejection region*

$$\frac{TR^2(k_j)}{2\sigma^2} > -2 \log \varepsilon_j. \tag{51}$$

PROOF. The probability of (51) under the null hypothesis is ε_j since the cumulative distribution function of $z_j = TR^2(k_j)/(2\sigma^2)$ is $1 - e^{-\frac{1}{2}z_j}$. That the test of H_j^* which is most powerful against any specified alternative, say $\rho^2(k_j) = 2\sigma^2\tau^2/T$, has rejection region (51) is a consequence of the Neyman-Pearson Fundamental Lemma. To prove the result in this case we let the set (51) be A^* and A be any other rejection region with $\Pr\{A \mid H_j^*\} = \varepsilon_j$. The inequality (51) is equivalent to

$$\frac{k(z_j \mid \tau^2)}{k(z_j)} = e^{-\frac{1}{2}\tau^2} I_0(\tau\sqrt{z_j}) > e^{-\frac{1}{2}\tau^2} I_0(\tau\sqrt{-2 \log \varepsilon_j}) = d_j, \tag{52}$$

say. Then

$$\begin{aligned} \Pr\{A^* \mid \tau^2\} &= \Pr\{A^* \cap \bar{A} \mid \tau^2\} + \Pr\{A^* \cap A \mid \tau^2\} \\ &= \int_{A^* \cap \bar{A}} k(z_j \mid \tau^2)\, dz_j + \Pr\{A^* \cap A \mid \tau^2\} \\ &> d_j \int_{A^* \cap \bar{A}} k(z_j)\, dz_j + \Pr\{A^* \cap A \mid \tau^2\} \\ &= d_j \int_{\bar{A}^* \cap A} k(z_j)\, dz_j + \Pr\{A^* \cap A \mid \tau^2\} \\ &\geq \int_{\bar{A}^* \cap A} k(z_j \mid \tau^2)\, dz_j + \Pr\{A^* \cap A \mid \tau^2\} \\ &= \Pr\{A \mid \tau^2\}. \end{aligned} \tag{53}$$

The theorem follows because (53) holds for every $\tau^2 > 0$.

This procedure is known as the *Schuster test* [Schuster (1898)].

The multiple-decision problem with which we are concerned can be considered as a combination of problems involving hypotheses about the individual amplitudes; that is,

$$
\begin{aligned}
H_0 &= H_1^* \cap H_2^* \cap \cdots \cap H_q^* = \bigcap_{i=1}^{q} H_i^*, \\
H_i &= \bar{H}_i^* \cap \bigcap_{\substack{j=1 \\ j\neq i}}^{q} H_j^*, \qquad i = 1, \ldots, q, \\
H_{hi} &= \bar{H}_h^* \cap \bar{H}_i^* \cap \bigcap_{\substack{j=1 \\ j\neq i,\, j\neq h}}^{q} H_j^*, \qquad h < i,\ h, i = 1, \ldots, q, \\
&\ \ \vdots \\
H_{12\cdots q} &= \bigcap_{i=1}^{q} \bar{H}_i^*.
\end{aligned} \tag{54}
$$

Let R_0, R_i, etc., denote the regions in the sample space of $R^2(k_1), \ldots, R^2(k_q)$ for making the decisions that H_0, H_i, etc., are true, respectively, and let R_i^*, $i = 1, \ldots, q$, be the regions for accepting H_i^*, respectively. Then these two sets of regions in the sample space are related in exactly the same way as the two sets of regions in the parameter space are related, as indicated in (54). Suppose we ask that

$$
\Pr\{R_j^* \mid \rho(k_1), \ldots, \rho(k_q); \rho(k_j) = 0\} = 1 - \varepsilon_j; \tag{55}
$$

that is, that the probability of deciding $\rho(k_j) = 0$ does not depend on what the other amplitudes are. We shall call this property independence of irrelevant parameters. Then an argument similar to that of Section 3.2.2 (which we shall outline) demonstrates that the intersection of R_j^* and $R^2(k_i) = c_i$, $i \neq j$, has probability $1 - \varepsilon_j$ for almost every set of $q - 1$ nonnegative $c_1, \ldots, c_{j-1}, c_{j+1}, \ldots, c_q$. Let $h[R^2(k_1), \ldots, R^2(k_q)]$ be 1 in R_j^* and 0 outside. (This is the critical function and could be generalized to a randomized critical function.) The expectation of h relative to $R^2(k_j)$ is a bounded function of the other $q - 1$ arguments. Condition (55) is that the expectation of this quantity minus $1 - \varepsilon_j$ is 0 identically in the noncentrality parameters. The result follows from the bounded completeness of the noncentral χ^2-distributions. (See Problems 25 and 26.)

Then in effect the decision about $\rho(k_j)$ is based on $R^2(k_j)$. When we invoke the symmetry condition, we see that $\varepsilon_1 = \cdots = \varepsilon_q = \varepsilon$, say. Given ε, the best procedure for H_j^* is given by Theorem 4.3.1 (for $\varepsilon_j = \varepsilon$). We note that

$$\text{(56)} \qquad \begin{aligned} \Pr\{R_0 \mid H_0\} &= \Pr\left\{\bigcap_{j=1}^{q} R_j^* \mid \rho(k_1) = \cdots = \rho(k_q) = 0\right\} \\ &= \prod_{j=1}^{q} \Pr\{R_j^* \mid \rho(k_j) = 0\} \\ &= (1-\varepsilon)^q. \end{aligned}$$

If this probability is specified to be, say $1 - \delta$, then $1 - \varepsilon = (1-\delta)^{1/q}$ or $\log(1-\varepsilon) = [\log(1-\delta)]/q$.

THEOREM 4.3.2. *The uniformly most powerful symmetric invariant decision procedure for selection of positive amplitudes given the probability* $(1-\varepsilon)^q$ *of deciding all are zero when that is true and given independence of irrelevant parameters is to decide* $\rho(k_j)$ *is positive when* $TR^2(k_j)/(2\sigma^2) > -2\log\varepsilon$.

If every $\rho(k_j) = 0$, an error is deciding that at least one $\rho(k_j) > 0$. Since the procedure is to decide $\rho(k_i) > 0$ if $z_i > C$, the probability of an error is

$$\text{(57)} \qquad \Pr\{\bar{R}_0 \mid H_0\} = 1 - (1 - e^{-\frac{1}{2}C})^q.$$

This function has been tabulated by Davis [(1941), Table 1], writing our $\frac{1}{2}C$ as his K and our q as his $\frac{1}{2}N$. The procedure as a test of significance of H_0 is known as the *Walker test* [G. T. Walker (1914)]. This general type of decision problem has been treated in a different way by Lehmann (1957). We note that if $q = [\frac{1}{2}(T-1)]$

$$\text{(58)} \qquad \begin{aligned} \lim_{q\to\infty} \Pr\{z_j < C + 2\log q, j = 1, \ldots, q\} &= \lim_{q\to\infty} [1 - \exp(-\tfrac{1}{2}C - \log q)]^q \\ &= \exp(-e^{-\frac{1}{2}C}). \end{aligned}$$

Other formulations of the decision problem can be given. Suppose it is assumed that at most one amplitude is positive. Then the hypotheses to be considered are

$$\text{(59)} \qquad \begin{aligned} &H_0\colon \ \rho(k_1) = \cdots = \rho(k_q) = 0, \\ &H_j\colon \ \rho(k_j) > 0,\ \rho(k_i) = 0, \qquad i \neq j,\ i = 1, \ldots, q, \\ &\qquad\qquad\qquad\qquad\qquad\qquad\quad j = 1, \ldots, q. \end{aligned}$$

The joint density of the normalized sample amplitudes is

$$\text{(60)} \qquad \prod_{j=1}^{q} k(z_j \mid \tau_j^2),$$

where at most one $\tau_j^2 = T\rho^2(k_j)/(2\sigma^2)$ is positive. If, for instance, H_j is true, that is, $\rho(k_j) > 0$ and $\rho(k_i) = 0$, $i \neq j$, then the likelihood ratio for H_j and H_0 is

$$(61)\qquad \frac{k(z_j \mid \tau_j^2) \prod_{i \neq j} k(z_i)}{\prod_{i=1}^{q} k(z_i)} = \frac{k(z_j \mid \tau_j^2)}{k(z_j)} = e^{-\frac{1}{2}\tau_j^2} I_0(\tau_j \sqrt{z_j}),$$

which is a monotonic increasing function of z_j. This suggests that H_j should be preferred to H_0 if z_j is large. The likelihood ratio for H_j and H_i, $i \neq j$, $i > 0$, is

$$(62)\qquad \frac{k(z_j \mid \tau_j^2) \prod_{h \neq j} k(z_h)}{k(z_i \mid \tau_i^2) \prod_{g \neq i} k(z_g)} = \frac{k(z_j \mid \tau_j^2) k(z_i)}{k(z_j) k(z_i \mid \tau_i^2)} = \frac{e^{-\frac{1}{2}\tau_j^2} I_0(\tau_j \sqrt{z_j})}{e^{-\frac{1}{2}\tau_i^2} I_0(\tau_i \sqrt{z_i})}.$$

If $\tau_j^2 = \tau_i^2 = \tau^2$, say, the likelihood ratio is $I_0(\tau\sqrt{z_j})/I_0(\tau\sqrt{z_i})$. This is greater than 1 or less than 1 according to the ratio z_j/z_i being greater than 1 or less than 1. Consider for a moment deciding between H_j and H_i.

LEMMA 4.3.1. *The uniformly most powerful symmetric invariant procedure for deciding between* H_j *and* H_i *is to choose* H_j *if* $z_j > z_i$ *and* H_i *if* $z_j < z_i$.

PROOF. Here symmetric means symmetric between z_j and z_i; that is, B_j and B_i are symmetric if interchanging z_j and z_i in the definition of B_j gives a definition of B_i. Let $A_j^* = \{(z_1, \ldots, z_q) \mid z_j > z_i\}$ and $A_i^* = \{(z_1, \ldots, z_q) \mid z_j < z_i\}$ and let A_j and A_i be two other mutually exclusive, exhaustive, and symmetric sets. [We ignore sets of probability 0, such as $\{(z_1, \ldots, z_q) \mid z_j = z_i\}$.] Then

$$\begin{aligned}
(63)\quad & \Pr\{A_j^* \mid \tau_j^2 = \tau^2, \tau_i^2 = 0\} \\
&= \Pr\{A_j^* \cap A_i \mid \tau_j^2 = \tau^2, \tau_i^2 = 0\} + \Pr\{A_j^* \cap A_j \mid \tau_j^2 = \tau^2, \tau_i^2 = 0\} \\
&= \int \cdots \int_{A_j^* \cap A_i} \prod_{h=1}^{q} k(z_h) e^{-\frac{1}{2}\tau^2} I_0(\tau\sqrt{z_j}) \prod_{h=1}^{q} dz_h + \Pr\{A_j^* \cap A_j \mid \tau_j^2 = \tau^2, \tau_i^2 = 0\} \\
&\geq \int \cdots \int_{A_j^* \cap A_i} \prod_{h=1}^{q} k(z_h) e^{-\frac{1}{2}\tau^2} I_0(\tau\sqrt{z_i}) \prod_{h=1}^{q} dz_h + \Pr\{A_j^* \cap A_j \mid \tau_j^2 = \tau^2, \tau_i^2 = 0\} \\
&= \int \cdots \int_{A_j \cap A_i^*} \prod_{h=1}^{q} k(z_h) e^{-\frac{1}{2}\tau^2} I_0(\tau\sqrt{z_j}) \prod_{h=1}^{q} dz_h + \Pr\{A_j^* \cap A_j \mid \tau_j^2 = \tau^2, \tau_i^2 = 0\} \\
&= \Pr\{A_j \mid \tau_j^2 = \tau^2, \tau_i^2 = 0\}
\end{aligned}$$

because $A_j^* \cap A_i$ and $A_j \cap A_i^*$ are symmetric. The above inequality holds for every value of $\tau^2 > 0$. Thus this symmetric procedure is most powerful uniformly in τ^2.

Theorem 4.3.2 and Lemma 4.3.1 suggest that in the multiple-decision problem one should decide H_0 if all of the sample amplitudes are small and decide H_j if $R^2(k_j)$ is the largest sample amplitude and is sufficiently large. This procedure is symmetric in the sense that permuting the indices of $H_1, \ldots, H_q$ and $R^2(k_1), \ldots, R^2(k_q)$ does not change the procedure.

THEOREM 4.3.3. *The uniformly most powerful symmetric invariant decision procedure for deciding among* $H_0, H_1, \ldots, H_q$ *given* $\Pr\{R_0 \mid H_0\} = (1-\varepsilon)^q$ *and given independence of irrelevant parameters is to decide* H_0 *if* $z_j = TR^2(k_j)/(2\sigma^2) \leq -2\log\varepsilon$, $j = 1, \ldots, q$, *and otherwise to decide* H_j *when* $z_j > z_i$, $i \neq j$, $i = 1, \ldots, q$.

PROOF. Let the mutually exclusive and exhaustive regions in the space of $z_1, \ldots, z_q$ for deciding $H_0, H_1, \ldots, H_q$ be $R_0, R_1, \ldots, R_q$, respectively, for an arbitrary symmetric procedure with $\Pr\{R_0 \mid H_0\} = (1-\varepsilon)^q$. Let $R_0' = R_0$ and

$$R_j' = \bar{R}_0 \cap \{(z_1, \ldots, z_q) \mid z_j > z_i, i \neq j\}, \qquad j = 1, \ldots, q. \tag{64}$$

Then $\Pr\{R_0' \mid H_0\} = (1-\varepsilon)^q$ and

$$\begin{aligned}
(65)\quad & \Pr\{R_j' \mid \tau_j^2 = \tau^2, \tau_h^2 = 0, h \neq j\} \\
&= \Pr\{R_j' \cap \bar{R}_j \mid \tau_j^2 = \tau^2, \tau_h^2 = 0, h \neq j\} + \Pr\{R_j' \cap R_j \mid \tau_j^2 = \tau^2, \tau_h^2 = 0, h \neq j\} \\
&= \sum_{i \neq j} \Pr\{R_j' \cap R_i \mid \tau_j^2 = \tau^2, \tau_h^2 = 0, h \neq j\} + \Pr\{R_j' \cap R_j \mid \tau_j^2 = \tau^2, \tau_h^2 = 0, h \neq j\} \\
&\geq \sum_{i \neq j} \Pr\{R_j \cap R_i' \mid \tau_j^2 = \tau^2, \tau_h^2 = 0, h \neq j\} + \Pr\{R_j' \cap R_j \mid \tau_j^2 = \tau^2, \tau_h^2 = 0, h \neq j\} \\
&= \Pr\{R_j \mid \tau_j^2 = \tau^2, \tau_h^2 = 0, h \neq j\}, \qquad j = 1, \ldots, q,
\end{aligned}$$

because

$$\begin{aligned}
(66)\quad \Pr\{R_j' \cap R_i \mid \tau_j^2 = \tau^2, \tau_h^2 = 0, h \neq j\} \\
\geq \Pr\{R_j \cap R_i' \mid \tau_j^2 = \tau^2, \tau_h^2 = 0, h \neq j\}
\end{aligned}$$

by symmetry and the argument of Lemma 4.3.1.

Now let

$$R_0^* = \{(z_1, \ldots, z_q) \mid z_i < -2\log\varepsilon, i = 1, \ldots, q\}, \tag{67}$$

$$R_j^* = \bar{R}_0^* \cap \{(z_1, \ldots, z_q) \mid z_j > z_i, i \neq j\}, \qquad j = 1, \ldots, q. \tag{68}$$

Then $\Pr\{R_0^* \mid H_0\} = (1-\varepsilon)^q$ and

$$(69)\quad \begin{aligned} &\Pr\{R_j^* \mid \tau_j^2 = \tau^2, \tau_h^2 = 0, h \neq j\} \\ &= \Pr\{R_j^* \cap R_0' \mid \tau_j^2 = \tau^2, \tau_h^2 = 0, h \neq j\} + \Pr\{R_j^* \cap R_j' \mid \tau_j^2 = \tau^2, \tau_h^2 = 0, h \neq j\} \\ &\geq \Pr\{R_j' \cap R_0^* \mid \tau_j^2 = \tau^2, \tau_h^2 = 0, h \neq j\} + \Pr\{R_j^* \cap R_j' \mid \tau_j^2 = \tau^2, \tau_h^2 = 0, h \neq j\} \\ &= \Pr\{R_j' \mid \tau_j^2 = \tau^2, \tau_h^2 = 0, h \neq j\} \end{aligned}$$

by the argument of Theorem 4.3.1. Since (65) and (69) hold for every τ^2, the theorem follows.

THEOREM 4.3.4. *Any symmetric Bayes solution is a procedure to accept H_0 if $TR^2(k_j)/(2\sigma^2) \leq C$, $j = 1, \ldots, q$, and otherwise to accept H_j if $z_j > z_i$, $i \neq j$, $i = 1, \ldots, q$.*

PROOF. Any Bayes solution is defined by [see Section 6.6, T. W. Anderson (1958), for example]

$$(70)\qquad R_0:\quad p_0 \prod_{j=1}^{q} k(z_j) > p_i k(z_i \mid \tau^2) \prod_{j \neq i} k(z_j), \qquad i = 1, \ldots, q,$$

$$(71)\qquad R_h:\quad \begin{cases} p_h k(z_h \mid \tau^2) \prod_{j \neq h} k(z_j) > p_i k(z_i \mid \tau^2) \prod_{j \neq i} k(z_j), \\ \qquad\qquad\qquad\qquad i = 1, \ldots, q,\ i \neq h, \\ p_h k(z_h \mid \tau^2) \prod_{j \neq h} k(z_j) > p_0 \prod_{j=1}^{q} k(z_j), \end{cases}$$

where $p_i \geq 0$, $i = 0, 1, \ldots, q$, and $\sum_{i=0}^{q} p_i = 1$. [These regions are defined to be disjoint; a point on the boundary of two regions (of probability 0) can be assigned to either region.] The first set of inequalities ($p_i > 0$) is

$$(72)\qquad R_0:\quad \frac{k(z_i \mid \tau^2)}{k(z_i)} < \frac{p_0}{p_i}, \qquad i = 1, \ldots, q,$$

the left-hand side being a monotonic increasing function of z_i. This region is symmetric if and only if $p_1 = \cdots = p_q$. Then R_0 is of the form given in the theorem and is identical when p_0 $(= 1 - \sum_{i=1}^{q} p_i)$ is chosen suitably. Then the R_i's agree with the theorem. Q.E.D.

We note that this procedure amounts to the test of significance given before with the further proviso that if the null hypothesis H_0 is rejected, one decides that the largest sample amplitude indicates the one population amplitude which is positive. This type of problem has been treated in a general way by Karlin and Truax (1960) and by Kudo (1960).

Since the class of Bayes solutions is the same as the class of admissible solutions in this problem [see Theorem 6.6.4 of T. W. Anderson (1958), for example], Theorem 4.3.3 and Theorem 4.3.4 are equivalent.

Now let us turn to the more important case of σ^2 being unknown. We must distinguish between the situations when there is available an estimate of σ^2 based on $n = T - p$ degrees of freedom and attention is focused on some q given frequencies, not involved in the estimate of σ^2, and the situation when there is no estimate of σ^2 that does not involve the frequencies of interest. In the first situation there is an estimate s^2, such that $ns^2/\sigma^2 = S/\sigma^2$ has a χ^2-distribution with n degrees of freedom independent of the q pairs $a(k_1)$, $b(k_1)$, . . . , $a(k_q)$, $b(k_q)$.

In this first situation we may ask for symmetric procedures satisfying (55); that is, that the probability of deciding whether $\rho(k_j) = 0$ does not depend on the other amplitudes and the probability of deciding all the amplitudes are 0 when that is the case does not depend on σ^2. Then the best procedure for testing H_j^* is to reject H_j^* if $F_j = TR^2(k_j)/(4s^2) > c_j$. [This follows by consideration of the sufficient statistics invariant under the transformations (29) and the completeness of the family of distributions; see the remarks below (29).] For a symmetric procedure we require $c_1 = \cdots = c_q = c$, say. Then

$$\Pr\{R_0 \mid H_0\} = \Pr\left\{z_j < 2c\frac{s^2}{\sigma^2}, j = 1, \ldots, q \mid H_0\right\} \tag{73}$$

$$= \int_0^\infty (1 - e^{-cu/n})^q \frac{u^{\frac{1}{2}n-1}e^{-\frac{1}{2}u}}{2^{\frac{1}{2}n}\Gamma(\frac{1}{2}n)}\,du$$

$$= \int_0^\infty \sum_{j=0}^{q} (-1)^j \binom{q}{j} \frac{u^{\frac{1}{2}n-1}e^{-\frac{1}{2}[1+2cj/n]u}}{2^{\frac{1}{2}n}\Gamma(\frac{1}{2}n)}\,du$$

$$= \sum_{j=0}^{q} (-1)^j \binom{q}{j}\left(1 + \frac{2cj}{n}\right)^{-\frac{1}{2}n}.$$

[This approaches 1 minus (57) for $c = \frac{1}{2}C$ as $n \to \infty$.] One can adjust c so (73) is a preassigned number.

THEOREM 4.3.5. *When σ^2 is unknown, the uniformly most powerful symmetric invariant decision procedure for selection of positive amplitudes given the probability of deciding all are zero when that is the case and given independence of irrelevant parameters is to decide $\rho(k_j)$ is positive when $TR^2(k_j)/(4s^2) > c$, where c is determined so (73) is the given probability.*

When σ^2 is unknown, the procedure above is possible only if there is an estimate s^2 of σ^2 whose distribution does not depend on $\alpha(k_1), \ldots, \beta(k_q)$ which are involved in the hypotheses. Otherwise one cannot make the requirement of independence of irrelevant parameters. This latter situation occurs when one is unwilling to assume that any coefficients $\alpha(k_j)$, $\beta(k_j)$ are 0; that is, one may want to permit trigonometric terms of any frequency (of the form j/T, $j = 0, 1, \ldots, [\frac{1}{2}T]$). However, one may assume that there are at most only a small number of these nonzero but without delineating them further. We treat the simplest case in some detail.

Suppose at most one of q population amplitudes is positive. The problem is to decide which of the hypotheses $H_0, H_1, \ldots, H_q$ is true. We ask that the procedure be symmetric and that $\Pr\{R_0 \mid H_0\}$ be specified independent of the unknown σ^2. We shall further assume that there are $r - q$ sample amplitudes, $R^2(k_{q+1}), \ldots, R^2(k_r)$, corresponding to zero population amplitudes, though $r - q$ may be 0. That leaves $T - 2r$ coefficients in the trend which may be nonzero and which we do not question. (This formulation leads to a more general problem than those considered by other authors.) When H_0 is true, $\sum_{j=1}^{r} R^2(k_j)$, a_0, and any sample coefficients corresponding to other population coefficients possibly nonzero form a sufficient set of statistics for the parameters. Hence, for almost every set of values of $\sum_{j=1}^{r} R^2(k_j)$, a_0, and the other coefficients, the conditional probability of R_0 must be the assigned $\Pr\{R_0 \mid H_0\}$. (See Problem 10 of Chapter 3.) The joint density of $z_j = TR^2(k_j)/(2\sigma^2)$, $j = 1, \ldots, r - 1$, given $\sum_{j=1}^{r} z_j = c$, that is, $z_r = c - \sum_{j=1}^{r-1} z_j$, is

$$\prod_{j=1}^{r} k(z_j) = (\tfrac{1}{2})^r e^{-\frac{1}{2}c}, \qquad z_j \geq 0, \; j = 1, \ldots, r, \tag{74}$$

$$= 0, \qquad \text{otherwise},$$

when H_0 is true and is

$$k(z_j \mid \tau_j^2) \prod_{i \neq j} k(z_i) = (\tfrac{1}{2})^r e^{-\frac{1}{2}c} e^{-\frac{1}{2}\tau_j^2} I_0(\tau_j \sqrt{z_j}), \; z_i \geq 0, \; i = 1, \ldots, r, \tag{75}$$

$$= 0, \qquad \text{otherwise},$$

when H_j is true. We use the same argument as in proving Theorem 4.3.3. The intersection of the region R_0 and $\sum_{j=1}^{r} z_j = c$ is defined by $z_i < g(c)$, $i = 1, \ldots, q$, where $g(c)$ is a number so the conditional probability is the assigned $\Pr\{R_0 \mid H_0\}$. The intersection of R_h, $h = 1, \ldots, q$, and $\sum_{j=1}^{r} z_j = c$ is defined by $z_h > g(c)$ and $z_h > z_i$, $i \neq h$, $i = 1, \ldots, q$.

The conditional joint distribution of $z_1, \ldots, z_r$ given $\sum_{j=1}^r z_j = c$ is uniform over the region in r-dimensional space where $z_j \geq 0$, $j = 1, \ldots, r$, and $\sum_{j=1}^r z_j = c$. Thus the conditional distribution of $z_1/c, \ldots, z_r/c$ given $\sum_{j=1}^r z_j/c = 1$ is uniform over the region $z_j/c \geq 0$, $j = 1, \ldots, r$, and $\sum_{j=1}^r z_j/c = 1$. If $g(c) = gc$ for each positive g,

$$(76)\quad \Pr\left\{z_j \leq gc, j = 1, \ldots, q \,\middle|\, \sum_{j=1}^r z_j = c, H_0\right\} = \Pr\left\{\frac{z_j}{c} \leq g, j = 1, \ldots, q \,\middle|\, \sum_{j=1}^r \frac{z_j}{c} = 1, H_0\right\}$$

does not depend on c, and hence is the same as the unconditional probability

$$(77)\quad \Pr\left\{\frac{z_j}{\sum_{i=1}^r z_i} \leq g,\ j = 1, \ldots, q \mid H_0\right\}.$$

THEOREM 4.3.6. *When σ^2 is unknown, the uniformly most powerful symmetric invariant decision procedure for deciding among $H_0, H_1, \ldots, H_q$ given* $\Pr\{R_0 \mid H_0\}$ *is to decide H_0 if*

$$(78)\quad x_j = \frac{R^2(k_j)}{\sum_{i=1}^r R^2(k_i)} \leq g, \qquad j = 1, \ldots, q,$$

and otherwise decide H_j when $R^2(k_j) > R^2(k_i)$, $i \neq j$, $i = 1, \ldots, q$, with g chosen so that the probability of (78) *under H_0 is the specified* $\Pr\{R_0 \mid H_0\}$.

Now let us evaluate 1 minus (77) which is

$$(79)\quad \Pr\left\{\max_{j=1,\ldots,q} x_j > g \mid H_0\right\} = \Pr\{\text{at least one } x_j > g, j = 1, \ldots, q \mid H_0\}.$$

Let A_j be the event $\{x_j > g\}$. Then under H_0

$$(80)\quad \Pr\left\{\max_{j=1,\ldots,q} x_j > g\right\} = \Pr\left\{\bigcup_{j=1}^q A_j\right\} = \sum_{j=1}^q \Pr\{A_j\} - \sum_{j<i} \Pr\{A_i \cap A_j\} + \cdots + (-1)^{q-1} \Pr\left\{\bigcap_{j=1}^q A_j\right\}.$$

[See Feller (1968), Section 1 of Chapter IV, for example.] Now by the equidistribution of the x_j's we have

$$\Pr\left\{\bigcup_{j=1}^{q} A_j\right\} = q \Pr\{A_1\} - \binom{q}{2} \Pr\{A_1 \cap A_2\} + \cdots + (-1)^{q-1} \Pr\left\{\bigcap_{j=1}^{q} A_j\right\}. \tag{81}$$

Since $\sum_{j=1}^{r} x_j = 1$, and $x_j \geq 0, j = 1, \ldots, r$, we have

$$\Pr\left\{\bigcap_{j=1}^{k} A_j\right\} = \Pr\{x_1 > g, \ldots, x_k > g\} = 0 \tag{82}$$

whenever $kg \geq 1$. Hence

(83)

$$\Pr\left\{\bigcup_{j=1}^{q} A_j\right\} = q \Pr\{A_1\} - \binom{q}{2} \Pr\{A_1 \cap A_2\} + \cdots + (-1)^{n-1}\binom{q}{n} \Pr\left\{\bigcap_{j=1}^{n} A_j\right\},$$

where n is the minimum of q and $[1/g]$, which is the largest integer not greater than $1/g$.

The conditional distribution of $x_1, \ldots, x_r$, given $\sum_{j=1}^{r} x_j = 1$, is uniform over the region where $x_j \geq 0$, $j = 1, \ldots, r$, and $\sum_{j=1}^{r} x_j = 1$ (and zero elsewhere). The unconditional distribution of the x_j's is the same as the conditional distribution since the condition is automatically satisfied by the way x_j is defined. The r coordinates in the $(r-1)$-dimensional space $\sum_{j=1}^{r} x_j = 1$ are *barycentric* coordinates. The probability of a set is proportional to its $(r-1)$-dimensional volume. We see

$$\begin{aligned}
(84)\quad \Pr\{A_1\} &= \Pr\{x_1 > g\} \\
&= \Pr\left\{g < x_1 \leq 1,\quad 0 \leq x_j < 1 - g,\quad j \neq 1 \,\middle|\, \sum_{j=1}^{r} x_j = 1\right\} \\
&= \frac{\text{Volume}\left\{g < x_1 \leq 1,\quad 0 \leq x_j < 1 - g,\quad j \neq 1, \sum_{j=1}^{r} x_j = 1\right\}}{\text{Volume}\left\{0 \leq x_j \leq 1, \text{all } j, \quad \sum_{j=1}^{r} x_j = 1\right\}}
\end{aligned}$$

$$= \frac{\text{Volume}\left\{0 < x_1 - g \leq 1 - g, 0 \leq x_j < 1 - g, j \neq 1, x_1 - g + \sum_{j=2}^{r} x_j = 1 - g\right\}}{\text{Volume}\left\{0 \leq x_j \leq 1, \text{all } j, \quad \sum_{j=1}^{r} x_j = 1\right\}}.$$

Letting $z_1 = x_1 - g, z_2 = x_2, \ldots, z_r = x_r$ in the numerator, we have

$$\Pr\{A_1\} = \frac{\text{Volume}\left\{0 \le z_j \le 1 - g, \text{all } j, \sum_{j=1}^{r} z_j = 1 - g\right\}}{\text{Volume}\left\{0 \le x_j \le 1, \text{all } j, \sum_{j=1}^{r} x_j = 1\right\}}. \tag{85}$$

The right-hand side of (85) is the ratio of the volumes of two similar figures in $(r - 1)$-dimensional subspaces and the linear dimensions of the figures are in ratio $1 - g$: 1. Therefore

$$\Pr\{A_1\} = (1 - g)^{r-1}. \tag{86}$$

To find

$$\Pr\{A_1 \cap A_2\} = \Pr\{x_1 > g, x_2 > g\}, \tag{87}$$

we consider

(88)

$$\text{Volume}\left\{g < x_j < 1 - g, j = 1, 2, 0 \le x_j < 1 - 2g, j = 3, \ldots, r, \sum_{j=1}^{r} x_j = 1\right\}$$

$$= \text{Volume}\left\{0 < x_j - g < 1 - 2g, j = 1, 2, 0 \le x_j < 1 - 2g, j = 3, \ldots, r, \right.$$

$$\left. (x_1 - g) + (x_2 - g) + \sum_{j=3}^{r} x_j = 1 - 2g\right\}.$$

Letting $z_1 = x_1 - g, z_2 = x_2 - g, z_3 = x_3, \ldots, z_r = x_r$, we have for (88)

$$\text{Volume}\left\{0 \le z_j \le 1 - 2g, \text{all } j, \sum_{j=1}^{r} z_j = 1 - 2g\right\}. \tag{89}$$

Hence by the same argument as before we have

$$\Pr\{A_1 \cap A_2\} = (1 - 2g)^{r-1}. \tag{90}$$

Similarly,

$$\Pr\left\{\bigcap_{j=1}^{k} A_j\right\} = (1 - kg)^{r-1}, \qquad k \le \frac{1}{g}. \tag{91}$$

Finally we have from (83) and the above

$$(92)\quad \Pr\left\{\max_{j=1,\ldots,q} x_j > g\right\} = \Pr\left\{\bigcup_{j=1}^{q} A_j\right\}$$
$$= q(1-g)^{r-1} - \binom{q}{2}(1-2g)^{r-1} + \cdots + (-1)^{n-1}\binom{q}{n}(1-ng)^{r-1},$$

where n is the minimum of q and $[1/g]$.

This distribution was given by Fisher (1929) with tables, and the procedure is known as *Fisher's test*. A geometric representation for $q = r = 3$ is studied in Problem 27. Whittle (1951) used the method of characteristic functions; Irwin (1955) discussed various methods.

We can extend this result on probabilities. Consider

$$(93)\quad \Pr\{\text{second largest } x_j > g,\ j = 1, \ldots, q \mid H_0\}$$
$$= \Pr\{\text{at least two } x_j > g,\ j = 1, \ldots, q \mid H_0\}$$
$$= \Pr\left\{\bigcup_{i<j}(A_i \cap A_j)\right\}$$
$$= \sum_{i<j} \Pr\{A_i \cap A_j\} - 2\sum_{i<j<k} \Pr\{A_i \cap A_j \cap A_k\}$$
$$+ 3\sum_{h<i<j<k} \Pr\{A_h \cap A_i \cap A_j \cap A_k\} - \cdots + (-1)^q(q-1)\Pr\left\{\bigcap_{j=1}^{q} A_j\right\}$$
$$= \binom{q}{2}\Pr\{A_1 \cap A_2\} - 2\binom{q}{3}\Pr\{A_1 \cap A_2 \cap A_3\}$$
$$+ 3\binom{q}{4}\Pr\{A_1 \cap A_2 \cap A_3 \cap A_4\} - \cdots + (-1)^q(q-1)\Pr\left\{\bigcap_{j=1}^{q} A_j\right\}$$
$$= \binom{q}{2}(1-2g)^{r-1} - 2\binom{q}{3}(1-3g)^{r-1}$$
$$+ 3\binom{q}{4}(1-4g)^{r-1} - \cdots + (-1)^n(n-1)\binom{q}{n}(1-ng)^{r-1}$$

for n the minimum of q and $[1/g]$. The expansion of $\Pr\{\bigcup_{i<j}(A_i \cap A_j)\}$ is given in Section 5 of Chapter IV of Feller (1968).

THEOREM 4.3.7. *If* $x_1, \ldots, x_r$ *are uniformly distributed in the* $(r-1)$-*dimensional region* $x_j \geq 0, j = 1, \ldots, r$, *and* $\sum_{j=1}^r x_j = 1$,

$$\begin{aligned}
(94)\quad &\Pr\{\text{at least } p \ \ x_j \geq g,\ j = 1, \ldots, q\} \\
&= \binom{q}{p}(1 - pg)^{r-1} - \binom{q}{p+1}\binom{p}{p-1}[1 - (p+1)g]^{r-1} \\
&\quad + \binom{q}{p+2}\binom{p+1}{p-1}[1 - (p+2)g]^{r-1} - \cdots \\
&\quad + (-1)^{n-p}\binom{q}{n}\binom{n-1}{p-1}(1 - ng)^{r-1} \\
&= q\binom{q-1}{p-1}\left\{\frac{1}{p}(1 - pg)^{r-1} - \frac{1}{p+1}\binom{q-p}{1}[1 - (p+1)g]^{r-1}\right. \\
&\quad + \frac{1}{p+2}\binom{q-p}{2}[1 - (p+2)g]^{r-1} - \cdots \\
&\quad \left. + \frac{(-1)^{n-p}}{n}\binom{q-p}{n-p}(1 - ng)^{r-1}\right\},
\end{aligned}$$

where $p \leq q$ *and* n *is the minimum of* q *and* $[1/g]$.

Stevens (1939) derived this probability as the solution to another problem and gave tables for $p = 2$. Fisher (1940) related Stevens' problem to the one we consider and also gave some tables.

The application of this result, however, is limited. This distribution, for example, for $p = 2$ gives the probability of two ratios x_j being larger than a given number g under H_0 and this is relevant to a procedure which specifies acceptance of H_0 unless at least *two* ratios are larger. Fisher (1940) suggests that this procedure might be suitable when it is anticipated that a real effect would show up in two amplitudes, rather than one. We may rephrase this proposal in our terminology as follows: accept H_0 unless at least two $x_j \geq g$ and in that case accept H_{ij} where $R^2(k_i)$ and $R^2(k_j)$ are the largest sample amplitudes, $i, j = 1, \ldots, q$.

Let us now formulate the Bayes problem of deciding among $H_0, H_1, \ldots, H_q$, $H_{12}, H_{13}, \ldots, H_{q-1,q}$ when σ^2 is known. In view of the symmetry of the problem we shall take the a priori probabilities symmetric and consider the hypotheses at symmetric values of $\tau_1^2, \ldots, \tau_q^2$. Let p_0 be the a priori probability of H_0, p_{I} be the a priori probability that $\tau_j^2 = \tau_{\mathrm{I}}^2$, $\tau_i^2 = 0$, $i \neq j$, say H_j', $j = 1, \ldots, q$, and p_{II} be the a priori probability that $\tau_j^2 = \tau_h^2 = \tau_{\text{II}}^2$, $\tau_i^2 = 0$, $i \neq j$, $i \neq h$, say H'_{jh}, $j < h$, $j, h = 1, \ldots, q$. We have $p_0 + qp_{\text{I}} + \frac{1}{2}q(q-1)p_{\text{II}} = 1$. Then the a posteriori probabilities of H_0, H'_j, and H'_{jh} are, respectively, proportional to

$$p_0 \prod_{g=1}^{q} k(z_g), \tag{95}$$

$$p_{\text{I}} e^{-\frac{1}{2}\tau_{\text{I}}^2} I_0(\tau_{\text{I}}\sqrt{z_j}) \prod_{g=1}^{q} k(z_g), \qquad j = 1, \ldots, q, \tag{96}$$

$$p_{\text{II}} e^{-\tau_{\text{II}}^2} I_0(\tau_{\text{II}}\sqrt{z_j}) I_0(\tau_{\text{II}}\sqrt{z_h}) \prod_{g=1}^{q} k(z_g), \quad j < h, j, h = 1, \ldots, q. \tag{97}$$

Then R_0 is defined by the inequalities

$$I_0(\tau_{\text{I}}\sqrt{z_j}) < \frac{p_0}{p_{\text{I}}} e^{\frac{1}{2}\tau_{\text{I}}^2}, \qquad j = 1, \ldots, q, \tag{98}$$

$$I_0(\tau_{\text{II}}\sqrt{z_j}) I_0(\tau_{\text{II}}\sqrt{z_h}) < \frac{p_0}{p_{\text{II}}} e^{\tau_{\text{II}}^2}, \quad j < h, j, h = 1, \ldots, q. \tag{99}$$

The inequalities (98) are equivalent to $z_j < c$, $j = 1, \ldots, q$, for

$$I_0(\tau_{\text{I}}\sqrt{c}) = \frac{p_0}{p_{\text{I}}} e^{\frac{1}{2}\tau_{\text{I}}^2}. \tag{100}$$

Since $I_0(\tau_{\text{I}}\sqrt{c}) \geq 1$, (100) implies $p_0 e^{\frac{1}{2}\tau_{\text{I}}^2} \geq p_{\text{I}}$. If

$$I_0^2(\tau_{\text{II}}\sqrt{c}) \leq \frac{p_0}{p_{\text{II}}} e^{\tau_{\text{II}}^2}, \tag{101}$$

(98) implies (99). [When $\tau_{\text{I}}^2 = \tau_{\text{II}}^2$, (101) is $p_0 p_{\text{II}} \leq p_{\text{I}}^2$.] Since $I_0(\tau_{\text{II}}\sqrt{z_h}) \geq 1$, (99) implies

$$I_0(\tau_{\text{II}}\sqrt{z_j}) < \frac{p_0}{p_{\text{II}}} e^{\tau_{\text{II}}^2}; \tag{102}$$

thus if

$$I_0(\tau_{\text{II}}\sqrt{c}) \geq \frac{p_0}{p_{\text{II}}} e^{\tau_{\text{II}}^2}, \tag{103}$$

(99) implies (98). [When $\tau_{\text{I}}^2 = \tau_{\text{II}}^2$, (103) is $p_{\text{II}} \geq p_{\text{I}} e^{\frac{1}{2}\tau_{\text{I}}^2}$.] The set R_j is defined by $z_j > c$,

$$I_0(\tau_{\text{I}}\sqrt{z_j}) > I_0(\tau_{\text{I}}\sqrt{z_h}), \qquad h \neq j, h = 1, \ldots, q, \tag{104}$$

$$p_{\text{I}} e^{-\frac{1}{2}\tau_{\text{I}}^2} I_0(\tau_{\text{I}}\sqrt{z_j}) > p_{\text{II}} e^{-\tau_{\text{II}}^2} I_0(\tau_{\text{II}}\sqrt{z_i}) I_0(\tau_{\text{II}}\sqrt{z_h}), \tag{105}$$

$$i \neq h, i, h = 1, \ldots, q.$$

[These regions are defined to be disjoint; a point on the boundary of any region (of probability 0) can be assigned to either region.] The inequalities (104) are equivalent to $z_j > z_h$, $h \neq j$, $h = 1, \ldots, q$. An inequality (105) for $i = j$ is equivalent to

$$\frac{I_0(\tau_{\rm I}\sqrt{z_j})}{I_0(\tau_{\rm II}\sqrt{z_j})} > \frac{p_{\rm II}}{p_{\rm I}}\, e^{\frac{1}{2}\tau_{\rm I}^2-\tau_{\rm II}^2} I_0(\tau_{\rm II}\sqrt{z_h}). \tag{106}$$

If $\tau_{\rm I}^2 = \tau_{\rm II}^2$, (106) is equivalent to

$$I_0(\tau_{\rm I}\sqrt{z_h}) < \frac{p_{\rm I}}{p_{\rm II}}\, e^{\frac{1}{2}\tau_{\rm I}^2}, \tag{107}$$

which is equivalent to $z_h < d$, where

$$I_0(\tau_{\rm I}\sqrt{d}) = \frac{p_{\rm I}}{p_{\rm II}}\, e^{\frac{1}{2}\tau_{\rm I}^2}. \tag{108}$$

If $d \leq c$ ($p_{\rm I}^2 \leq p_0 p_{\rm II}$), then $z_j > c$ and $z_h < d$ imply $z_h < z_j$, which is (104). If $\tau_{\rm I}^2 > \tau_{\rm II}^2$, the ratio $I_0(\tau_{\rm I}\sqrt{z_j})/I_0(\tau_{\rm II}\sqrt{z_j})$ is monotonically increasing and the boundary (106) has a derivative dz_h/dz_j that is positive. If $\tau_{\rm I}^2 < \tau_{\rm II}^2$, the ratio $I_0(\tau_{\rm I}\sqrt{z_j})/I_0(\tau_{\rm II}\sqrt{z_j})$ is monotonically decreasing and the boundary (106) has a derivative dz_h/dz_j that is negative.

The region R_{jh}, $j < h$, is defined by

$$I_0(\tau_{\rm II}\sqrt{z_j})I_0(\tau_{\rm II}\sqrt{z_h}) > \frac{p_0}{p_{\rm II}}\, e^{\tau_{\rm II}^2}, \tag{109}$$

$$p_{\rm II}e^{-\tau_{\rm II}^2}I_0(\tau_{\rm II}\sqrt{z_j})I_0(\tau_{\rm II}\sqrt{z_h}) > p_{\rm I}e^{-\frac{1}{2}\tau_{\rm I}^2}I_0(\tau_{\rm I}\sqrt{z_i}), \quad i = 1, \ldots, q, \tag{110}$$

and $z_j > z_i$, $z_h > z_i$, $i \neq j$, $i \neq h$, $i = 1, \ldots, q$. If $\tau_{\rm I}^2 = \tau_{\rm II}^2$, (110) for $i = h$ is equivalent to $z_j > d$ and (110) for $i = j$ is equivalent to $z_h > d$.

Figure 4.3 indicates R_0, R_1, R_2, and R_{12} when $\tau_{\rm I}^2 = \tau_{\rm II}^2$. The shapes of the regions depend on p_0, $p_{\rm I}$, $p_{\rm II}$, and $\tau_{\rm I}^2$. [Inequality in (103) is impossible when $\tau_{\rm I}^2 = \tau_{\rm II}^2$.] If $\tau_{\rm I}^2 > \tau_{\rm II}^2$, the lines $z_1 = d$ and $z_2 = d$ are replaced by curves sloping upwards. If $\tau_{\rm I}^2 < \tau_{\rm II}^2$, the lines $z_1 = d$ and $z_2 = d$ are replaced by curves with negative slopes. The shapes of the regions depend on $p_0, p_{\rm I}, p_{\rm II}, \tau_{\rm I}^2$, and $\tau_{\rm II}^2$. There is no procedure that is optimal uniformly in $\tau_{\rm I}^2$ and $\tau_{\rm II}^2$.

These symmetric procedures may be characterized by the probabilities of correct decision $\Pr\{R_0 \mid H_0\}$, $\Pr\{R_1 \mid \tau_1^2 = \tau_{\rm I}^2, \tau_i^2 = 0, i \geq 2\}$, and $\Pr\{R_{12} \mid \tau_1^2 = \tau_2^2 = \tau_{\rm II}^2, \tau_i^2 = 0, i \geq 3\}$. An alternative formulation is to set $\Pr\{R_0 \mid H_0\}$ and $\Pr\{R_1 \mid \tau_1^2 = \tau_{\rm I}^2, \tau_i^2 = 0, i \geq 2\}$ for a specified value of $\tau_{\rm I}^2$ and maximize the third probability for a specified value of $\tau_{\rm II}^2$.

If σ^2 is unknown, we can require that $\Pr\{R_0 \mid H_0\}$ not depend on σ^2. That requirement leads to the above considerations conditional on $\sum_{j=1}^{r} R^2(k_j)$. The regions for each value of $\sum_{j=1}^{r} z_j$ will depend on p_0, p_{I}, p_{II}, τ_{I}^2, and τ_{II}^2. Birnbaum (1959), (1961) has considered this approach in a somewhat similar problem.

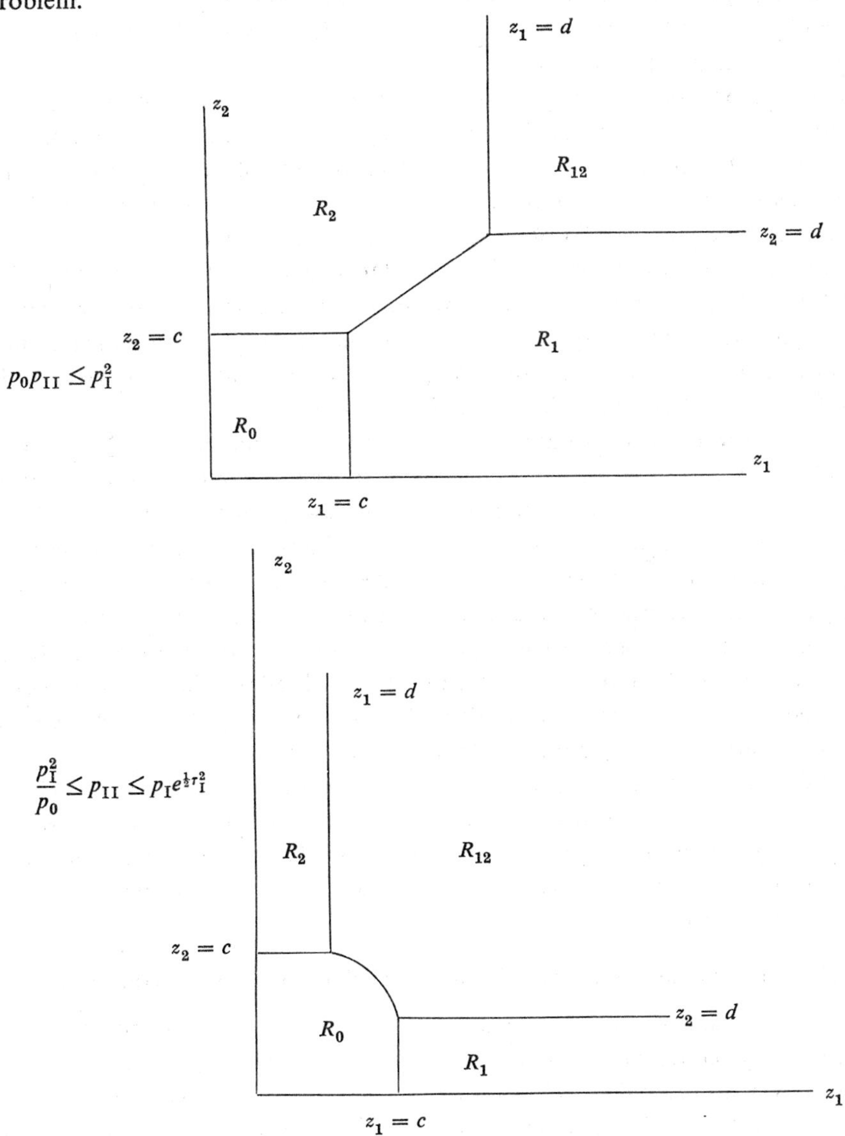

Figure 4.3. Decision regions for H_0, H_1', H_2', H_{12}' when $\tau_{\mathrm{I}}^2 = \tau_{\mathrm{II}}^2$.

Another suggestion [for example, by Whittle (1952)] is to decide H_0 if $R^2(k_j) < g \sum_{i=1}^r R^2(k_i), j = 1, \ldots, q$, where g is such that (92) is ε, and decide $\rho(k_j) > 0$ if $R^2(k_j) > g \sum_{i=1}^r R^2(k_i)$ and $R^2(k_j) > R^2(k_i)$, $i \neq j$. Then treat the remaining $r - 1$ $R^2(k_i)$'s as constituting a new problem of the same sort with q replaced by $q - 1$ and r replaced by $r - 1$. The procedure may be continued step by step until the remaining $\rho(k_i)$'s are decided to be 0. (The procedure for $q = r = 3$ is graphed in Problem 28.)

The procedure is not very satisfactory from a theoretical point of view. The specification of the probability $\Pr\{R_0 \mid H_0\} = 1 - \varepsilon$ controls the probability of one error. If one $\rho(k_j) > 0$, then the probability of correctly deciding this increases with the value of $\rho(k_j)$ (which is satisfactory). The procedure, however, also permits deciding that more than one amplitude is positive. If two amplitudes are positive and one is much larger than the other, then the probability of deciding in turn that the two amplitudes are positive may be high. If two amplitudes are positive and about equal in magnitude, the probability may not be high that the larger sample amplitude is sufficiently large relative to the sum to warrant deciding H_0 is not true.

If H_0 is true, the distribution of the x_j's is uniform on the plane $\sum_{j=1}^r x_j = 1$, $x_j \geq 0$. It can be seen from the geometry that the conditional distribution of any $r - 1$ of the x_j's, given their sum is 1 minus the other and that the other is larger than g, is again a uniform distribution (if g is large enough). Hence, the probability that the ratio of the largest of the $r - 1$ to the sum is greater than a specified number can be found in a similar fashion. However, when H_0 is not true, then the distributions will depend on the $\rho^2(k_j)/\sigma^2$ which are not 0. If one is different from 0 and it is so large that the probability is high that the corresponding x_j is the largest, then again the conditional distribution of the other $r - 1$ is approximately uniform. Other cases are possible.

We note that to use the procedure of Theorem 4.3.6 we do not need to find g so (92) is the desired probability and then see whether the maximum observed x_j is larger than this number. Instead we make the comparison by computing (92) for g equal to the maximum x_j and see whether (92) is smaller than the desired probability.

4.3.5. Calculation of Fourier Coefficients—the Fast Fourier Transform

In most situations use of relations between trigonometric functions can drastically reduce the computation of the coefficients a_j and b_j. To see the idea let us consider $T = T_1 T_2$. Then write $t = t_1 + (t_2 - 1)T_1$, $t_1 = 1, \ldots, T_1$ and $t_2 = 1, \ldots, T_2$, and write $j = j_1 + j_2 T_2$, $j_1 = 0, \ldots, T_2 - 1$ and $j_2 = 0, \ldots, T_1 - 1$. Note $a_{T-j} = a_j$ and $b_{T-j} = -b_j$. We want the real and imaginary parts of

(111)
$$\sum_{t=1}^{T} y_t e^{i2\pi jt/T} = \sum_{t_1=1}^{T_1}\sum_{t_2=1}^{T_2} y_{t_1+(t_2-1)T_1} \exp\left[i2\pi\left(\frac{j_1t_1 + j_2t_1T_2}{T} + \frac{j_1(t_2-1)}{T_2}\right)\right]$$
$$= \sum_{t_1=1}^{T_1} \exp\left[i2\pi\left(\frac{j_1 + j_2T_2}{T}\right)t_1\right] \sum_{t_2=1}^{T_2} \exp\left[i2\pi\frac{j_1(t_2-1)}{T_2}\right] y_{t_1+(t_2-1)T_1}.$$

Thus one can first calculate

$$c_{j_1,t_1} = \sum_{t_2=1}^{T_2} y_{t_1+(t_2-1)T_1} \cos\frac{2\pi j_1}{T_2}(t_2-1), \tag{112}$$

$$s_{j_1,t_1} = \sum_{t_2=1}^{T_2} y_{t_1+(t_2-1)T_1} \sin\frac{2\pi j_1}{T_2}(t_2-1), \tag{113}$$

for each pair $(j_1, t_1), j_1 = 0, \ldots, T_2 - 1$ and $t_1 = 1, \ldots, T_1$. Since $c_{T_2-j_1,t_1} = c_{j_1,t_1}$ and $s_{T_2-j_1,t_1} = -s_{j_1,t_1}$, there are approximately $\frac{1}{2}T_1T_2 = \frac{1}{2}T$ such pairs to compute. Next one calculates as the real and imaginary parts of (111)

$$\sum_{t_1=1}^{T_1} c_{j_1,t_1} \cos 2\pi\frac{j_1 + j_2T_2}{T}t_1 - \sum_{t_1=1}^{T_1} s_{j_1,t_1} \sin 2\pi\frac{j_1 + j_2T_2}{T}t_1, \tag{114}$$

$$\sum_{t_1=1}^{T_1} c_{j_1,t_1} \sin 2\pi\frac{j_1 + j_2T_2}{T}t_1 + \sum_{t_1=1}^{T_1} s_{j_1,t_1} \cos 2\pi\frac{j_1 + j_2T_2}{T}t_1 \tag{115}$$

for $j = 0, 1, \ldots, [\frac{1}{2}T]$. The number of products in the first sums is about T_2T, and the number of products in the second sums is about $2T_1T$. The number of products in the entire computation is about $T(T_2 + 2T_1)$ as compared to about T^2 involved in the definitions of a_j and b_j, $j = 0, 1, \ldots, [\frac{1}{2}T]$.

If 2 is a factor of T, there are special forms of the computation. More generally, if $T = T_1T_2T_3$, we can write

$$t = t_1 + (t_2 - 1)T_1 + (t_3 - 1)T_1T_2, \qquad t_i = 1, \ldots, T_i,\ i = 1, 2, 3, \tag{116}$$

$$j = j_1 + j_2T_3 + j_3T_3T_2, \quad j_1 = 0, 1, \ldots, T_3 - 1,\ j_2 = 0, \ldots, T_2 - 1,\ j_3 = 0, \ldots, T_1 - 1. \tag{117}$$

Then

$$\sum_{t=1}^{T} y_t e^{i2\pi jt/T} = \sum_{t_1=1}^{T_1} e^{i2\pi(j_1+j_2T_3+j_3T_3T_2)t_1/T} \times \sum_{t_2=1}^{T_2} e^{i2\pi(j_1+j_2T_3)(t_2-1)/(T_2T_3)} \sum_{t_3=1}^{T_3} e^{i2\pi j_1(t_3-1)/T_3} y_{t_1+(t_2-1)T_1+(t_3-1)T_1T_2}. \tag{118}$$

We shall not go further into these methods. A detailed description has been given by Cooley, Lewis, and Welch (1967). The method has been developed by Tukey and collaborators. [See Cooley and Tukey (1965).] The idea seems to go back to Runge (1903).

4.4. STATISTICAL INFERENCE WHEN THE PERIODS OF THE TREND ARE NOT INTEGRAL DIVISORS OF THE SERIES LENGTH

4.4.1. Definitions of Trigonometric Coefficients and Amplitudes for Arbitrary Frequencies and their Moments

The assumption that an underlying cyclical trend has periods that are integral divisors of the series length, the case considered in the previous section, is reasonable in some cases where prior information is available about the possible periods and the length of the series is suited to these periods (for instance, a certain number of years of monthly data when seasonal variation is considered). However, in many cases the length of the series is determined on some basis not connected with the possible periods, and the investigator will not want to limit his consideration only to periods which are integral divisors of the series length. For example, this model of a periodic trend has been suggested for economic series displaying the fluctuations of the so-called business cycle. The data may be annual, and the length of the series is determined by the availability of material (such as when a certain agency started collecting certain data) or external events (such as world wars). The possible periods of the business cycle may be of any length; the econometrician may want to consider any possible period greater than or equal to 2. (We shall consider later the criticism regarding the applicability of this model to economic time series.)

A general formulation of the model is that the trend is

$$f(t) = f_1(t) + \cdots + f_m(t), \tag{1}$$

where $f_i(t)$ has (minimum) period n_i, but this formulation is too general to be considered on the basis of simply T observations. We shall discuss some more restricted formulations. We shall consider the simplest case where $f(t)$ consists of a single periodic function and this function is trigonometric. Specifically, we assume $y_t = f(t) + u_t$, where $\mathscr{E}u_t = 0$, $\mathscr{E}u_t^2 = \sigma^2$, $\mathscr{E}u_t u_s = 0$, $t \neq s$, and

$$\begin{aligned} f(t) &= \alpha \cos \lambda t + \beta \sin \lambda t \\ &= \rho \cos (\lambda t - \theta), \qquad 0 \leq \lambda \leq \pi, \end{aligned} \tag{2}$$

where

$$\rho^2 = \alpha^2 + \beta^2, \qquad \tan\theta = \frac{\beta}{\alpha}. \tag{3}$$

The observations $y_1, \ldots, y_T$ may be replaced by the coefficients $a_0 = \bar{y}$,

$$a_j = \frac{2}{T}\sum_{t=1}^{T} y_t \cos\frac{2\pi j}{T}t, \qquad j = 1, \ldots, [\tfrac{1}{2}(T-1)], \tag{4}$$

$$b_j = \frac{2}{T}\sum_{t=1}^{T} y_t \sin\frac{2\pi j}{T}t, \qquad j = 1, \ldots, [\tfrac{1}{2}(T-1)], \tag{5}$$

$$a_{\frac{1}{2}T} = \frac{1}{T}\sum_{t=1}^{T} y_t(-1)^t, \tag{6}$$

the last being included if T is even. As has been pointed out before, this set of T is simply a transformation of the data. The variances and covariances of these coefficients are, as before,

$$\begin{aligned} \operatorname{Var}(a_0) &= \frac{1}{T}\sigma^2 = \operatorname{Var}(a_{\frac{1}{2}T}), \\ \operatorname{Var}(a_j) &= \frac{2}{T}\sigma^2 = \operatorname{Var}(b_j), && j \neq 0, \tfrac{1}{2}T, \\ \operatorname{Cov}(a_j, a_k) &= 0 = \operatorname{Cov}(b_j, b_k), && j \neq k, \\ \operatorname{Cov}(a_j, b_k) &= 0. \end{aligned} \tag{7}$$

We may wish to consider more generally

$$A(\nu) = \frac{2}{T}\sum_{t=1}^{T} y_t \cos\nu t, \qquad 0 \leq \nu \leq \pi, \tag{8}$$

$$B(\nu) = \frac{2}{T}\sum_{t=1}^{T} y_t \sin\nu t, \qquad 0 \leq \nu \leq \pi, \tag{9}$$

for all ν in the interval $[0, \pi]$. Then $a_j = A(2\pi j/T)$ and $b_j = B(2\pi j/T)$. The means, variances, and covariances of these coefficients involve sums of products of trigonometric functions.

LEMMA 4.4.1. *For $\nu + \lambda \neq 2\pi k$ and $\nu - \lambda \neq 2\pi k$, $k = 0, \pm 1, \ldots,$*

$$(10)\quad \sum_{t=1}^{T} \cos \lambda t \cos \nu t$$

$$= \frac{1}{2} \frac{\sin \frac{1}{2}(\nu + \lambda)T}{\sin \frac{1}{2}(\nu + \lambda)} \cos \tfrac{1}{2}(\nu + \lambda)(T + 1) + \frac{1}{2} \frac{\sin \frac{1}{2}(\nu - \lambda)T}{\sin \frac{1}{2}(\nu - \lambda)} \cos \tfrac{1}{2}(\nu - \lambda)(T + 1),$$

$$(11)\quad \sum_{t=1}^{T} \cos \lambda t \sin \nu t$$

$$= \frac{1}{2} \frac{\sin \frac{1}{2}(\nu + \lambda)T}{\sin \frac{1}{2}(\nu + \lambda)} \sin \tfrac{1}{2}(\nu + \lambda)(T + 1) + \frac{1}{2} \frac{\sin \frac{1}{2}(\nu - \lambda)T}{\sin \frac{1}{2}(\nu - \lambda)} \sin \tfrac{1}{2}(\nu - \lambda)(T + 1),$$

$$(12)\quad \sum_{t=1}^{T} \sin \lambda t \sin \nu t$$

$$= -\frac{1}{2} \frac{\sin \frac{1}{2}(\nu + \lambda)T}{\sin \frac{1}{2}(\nu + \lambda)} \cos \tfrac{1}{2}(\nu + \lambda)(T + 1) + \frac{1}{2} \frac{\sin \frac{1}{2}(\nu - \lambda)T}{\sin \frac{1}{2}(\nu - \lambda)} \cos \tfrac{1}{2}(\nu - \lambda)(T + 1).$$

For $\nu \neq \pi k$, $k = 0, \pm 1, \ldots,$

$$(13)\quad \sum_{t=1}^{T} \cos^2 \nu t = \tfrac{1}{2}T + \frac{1}{2} \frac{\sin \nu T}{\sin \nu} \cos \nu(T + 1),$$

$$(14)\quad \sum_{t=1}^{T} \cos \nu t \sin \nu t = \frac{1}{2} \frac{\sin \nu T}{\sin \nu} \sin \nu(T + 1),$$

$$(15)\quad \sum_{t=1}^{T} \sin^2 \nu t = \tfrac{1}{2}T - \frac{1}{2} \frac{\sin \nu T}{\sin \nu} \cos \nu(T + 1).$$

The equalities are verified by considering the sums $\sum_{t=1}^{T} \cos \lambda t e^{i\nu t}$ for (10), (11), (13), and (14) and $\sum_{t=1}^{T} \sin \nu t e^{i\lambda t}$ for (11), (12), (14), and (15) and by using the method of (8) of Section 4.2. (See Problem 29.)

The means, $\mathscr{E}A(\nu) = \alpha(\nu)$ and $\mathscr{E}B(\nu) = \beta(\nu)$, say, and the variances and covariances of $A(\nu)$ and $B(\nu)$ are given in Theorem 4.4.1.

THEOREM 4.4.1. *When* $\mathscr{E}y_t = \alpha \cos \lambda t + \beta \sin \lambda t = \rho \cos (\lambda t - \theta)$,

$$
\begin{aligned}
(16)\quad \mathscr{E}A(\nu) = \alpha(\nu) &= \frac{\alpha}{T}\left[\frac{\sin \frac{1}{2}(\lambda + \nu)T}{\sin \frac{1}{2}(\lambda + \nu)} \cos \tfrac{1}{2}(\lambda + \nu)(T + 1)\right. \\
&\qquad \left. + \frac{\sin \frac{1}{2}(\lambda - \nu)T}{\sin \frac{1}{2}(\lambda - \nu)} \cos \tfrac{1}{2}(\lambda - \nu)(T + 1)\right] \\
&\qquad + \frac{\beta}{T}\left[\frac{\sin \frac{1}{2}(\lambda + \nu)T}{\sin \frac{1}{2}(\lambda + \nu)} \sin \tfrac{1}{2}(\lambda + \nu)(T + 1)\right. \\
&\qquad \left. + \frac{\sin \frac{1}{2}(\lambda - \nu)T}{\sin \frac{1}{2}(\lambda - \nu)} \sin \tfrac{1}{2}(\lambda - \nu)(T + 1)\right] \\
&= \frac{\rho}{T}\left\{\frac{\sin \frac{1}{2}(\lambda + \nu)T}{\sin \frac{1}{2}(\lambda + \nu)} \cos [\tfrac{1}{2}(\lambda + \nu)(T + 1) - \theta]\right. \\
&\qquad \left. + \frac{\sin \frac{1}{2}(\lambda - \nu)T}{\sin \frac{1}{2}(\lambda - \nu)} \cos [\tfrac{1}{2}(\lambda - \nu)(T + 1) - \theta]\right\}, \\
&\qquad\qquad \nu \neq \lambda, 0 < \nu < \pi, \\
&= \alpha + \frac{\rho}{T}\frac{\sin \lambda T}{\sin \lambda} \cos [\lambda(T + 1) - \theta], \qquad 0 < \lambda = \nu < \pi,
\end{aligned}
$$

$$
\begin{aligned}
(17)\quad \mathscr{E}B(\nu) = \beta(\nu) &= \frac{\alpha}{T}\left[\frac{\sin \frac{1}{2}(\lambda + \nu)T}{\sin \frac{1}{2}(\lambda + \nu)} \sin \tfrac{1}{2}(\lambda + \nu)(T + 1)\right. \\
&\qquad \left. - \frac{\sin \frac{1}{2}(\lambda - \nu)T}{\sin \frac{1}{2}(\lambda - \nu)} \sin \tfrac{1}{2}(\lambda - \nu)(T + 1)\right] \\
&\qquad + \frac{\beta}{T}\left[-\frac{\sin \frac{1}{2}(\lambda + \nu)T}{\sin \frac{1}{2}(\lambda + \nu)} \cos \tfrac{1}{2}(\lambda + \nu)(T + 1)\right. \\
&\qquad \left. + \frac{\sin \frac{1}{2}(\lambda - \nu)T}{\sin \frac{1}{2}(\lambda - \nu)} \cos \tfrac{1}{2}(\lambda - \nu)(T + 1)\right] \\
&= \frac{\rho}{T}\left\{\frac{\sin \frac{1}{2}(\lambda + \nu)T}{\sin \frac{1}{2}(\lambda + \nu)} \sin [\tfrac{1}{2}(\lambda + \nu)(T + 1) - \theta]\right. \\
&\qquad \left. - \frac{\sin \frac{1}{2}(\lambda - \nu)T}{\sin \frac{1}{2}(\lambda - \nu)} \sin [\tfrac{1}{2}(\lambda - \nu)(T + 1) - \theta]\right\}, \\
&\qquad\qquad \nu \neq \lambda, 0 < \nu < \pi, \\
&= \beta + \frac{\rho}{T}\frac{\sin \lambda T}{\sin \lambda} \sin [\lambda(T + 1) - \theta], \qquad 0 < \lambda = \nu < \pi,
\end{aligned}
$$

$$
(18)\qquad \operatorname{Var}[A(\nu)] = \frac{2\sigma^2}{T}\left[1 + \frac{\sin \nu T}{T \sin \nu} \cos \nu(T + 1)\right], \quad 0 < \nu < \pi,
$$

$$
(19)\qquad \operatorname{Var}[B(\nu)] = \frac{2\sigma^2}{T}\left[1 - \frac{\sin \nu T}{T \sin \nu} \cos \nu(T + 1)\right], \quad 0 < \nu < \pi,
$$

$$
(20)\qquad \operatorname{Cov}[A(\nu), B(\nu)] = \frac{2\sigma^2}{T}\frac{\sin \nu T}{T \sin \nu} \sin \nu(T + 1), \qquad 0 < \nu < \pi.
$$

If $\nu = 2\pi j/T$, the variances and covariances reduce to (7). The expected values also simplify because $\sin(x \pm \pi j) = (-1)^j \sin x$ and $\cos(x \pm \pi j) = (-1)^j \cos x$.

COROLLARY 4.4.1.

$$(21)\quad \mathscr{E}A\left(\frac{2\pi j}{T}\right) = \mathscr{E}a_j = \alpha\left(\frac{2\pi j}{T}\right)$$

$$= \frac{\rho}{T}\sin\tfrac{1}{2}\lambda T\left\{\frac{\cos\left[\tfrac{1}{2}\lambda(T+1) + \dfrac{\pi j}{T} - \theta\right]}{\sin\dfrac{1}{2}\left(\lambda + \dfrac{2\pi j}{T}\right)} + \frac{\cos\left[\tfrac{1}{2}\lambda(T+1) - \dfrac{\pi j}{T} - \theta\right]}{\sin\dfrac{1}{2}\left(\lambda - \dfrac{2\pi j}{T}\right)}\right\},$$

$$\lambda \neq \frac{2\pi j}{T}, j = 1, \ldots, [\tfrac{1}{2}(T-1)],$$

$$= \alpha, \qquad \lambda = \frac{2\pi j}{T}, j = 1, \ldots, [\tfrac{1}{2}(T-1)].$$

$$(22)\quad \mathscr{E}B\left(\frac{2\pi j}{T}\right) = \mathscr{E}b_j = \beta\left(\frac{2\pi j}{T}\right)$$

$$= \frac{\rho}{T}\sin\tfrac{1}{2}\lambda T\left\{\frac{\sin\left[\tfrac{1}{2}\lambda(T+1) + \dfrac{\pi j}{T} - \theta\right]}{\sin\dfrac{1}{2}\left(\lambda + \dfrac{2\pi j}{T}\right)} - \frac{\sin\left[\tfrac{1}{2}\lambda(T+1) - \dfrac{\pi j}{T} - \theta\right]}{\sin\dfrac{1}{2}\left(\lambda - \dfrac{2\pi j}{T}\right)}\right\},$$

$$\lambda \neq \frac{2\pi j}{T}, j = 1, \ldots, [\tfrac{1}{2}(T-1)],$$

$$= \beta, \qquad \lambda = \frac{2\pi j}{T}, j = 1, \ldots, [\tfrac{1}{2}(T-1)].$$

We further note that

$$(23)\qquad \tfrac{1}{2}\mathscr{E}A(0) = \mathscr{E}a_0 = \alpha_0 = \frac{\rho}{T}\frac{\sin\tfrac{1}{2}\lambda T}{\sin\tfrac{1}{2}\lambda}\cos\left[\tfrac{1}{2}\lambda(T+1) - \theta\right], \qquad \lambda \neq 0,$$

and $\tfrac{1}{2}\mathscr{E}A(0) = \mathscr{E}a_0 = \alpha$ for $\lambda = 0$. Also

$$(24)\quad \tfrac{1}{2}\mathscr{E}A(\pi) = \mathscr{E}a_{\frac{1}{2}T} = \alpha_{\frac{1}{2}T} = \frac{\rho}{T}\frac{\sin\tfrac{1}{2}\lambda T}{\sin\tfrac{1}{2}(\lambda+\pi)}\cos\left[\tfrac{1}{2}\lambda(T+1) + \tfrac{1}{2}\pi - \theta\right], \qquad \lambda \neq \pi,$$

for T even, and $\tfrac{1}{2}\mathscr{E}A(\pi) = \mathscr{E}a_{\frac{1}{2}T} = \alpha$ for $\lambda = \pi$. We shall assume $0 \leq \lambda \leq \pi$ because if $\pi < \lambda < 2\pi$, we could replace it by $\lambda^* = 2\pi - \lambda$ since

$$\begin{aligned}(25)\quad \alpha\cos\lambda t + \beta\sin\lambda t &= \alpha\cos\left[2\pi t - (2\pi-\lambda)t\right] + \beta\sin\left[2\pi t - (2\pi-\lambda)t\right]\\ &= \alpha\cos\left[-\lambda^* t\right] + \beta\sin\left[-\lambda^* t\right]\\ &= \alpha\cos\lambda^* t + (-\beta)\sin\lambda^* t.\end{aligned}$$

The sample intensity function is

$$
\begin{aligned}
(26)\qquad R^2(\nu) &= A^2(\nu) + B^2(\nu) \\
&= \frac{4}{T^2}\sum_{s,t=1}^{T} y_s y_t(\cos \nu s \cos \nu t + \sin \nu s \sin \nu t) \\
&= \frac{4}{T^2}\sum_{s,t=1}^{T} y_s y_t \cos \nu(s-t) \\
&= \left|\frac{2}{T}\sum_{t=1}^{T} y_t e^{i\nu t}\right|^2, \qquad 0 \le \nu \le \pi.
\end{aligned}
$$

If $\nu = 2\pi j/T$, $R^2(2\pi j/T) = R_j^2 = a_j^2 + b_j^2, j = 1, \ldots, [\frac{1}{2}(T-1)]$. The population quantity corresponding to $R^2(\nu)$ is

$$
\begin{aligned}
(27)\qquad \rho^2(\nu) &= \alpha^2(\nu) + \beta^2(\nu) \\
&= \frac{\rho^2}{T^2}\left\{\frac{\sin^2 \frac{1}{2}(\lambda+\nu)T}{\sin^2 \frac{1}{2}(\lambda+\nu)} + \frac{\sin^2 \frac{1}{2}(\lambda-\nu)T}{\sin^2 \frac{1}{2}(\lambda-\nu)}\right. \\
&\qquad \left. + 2\frac{\sin \frac{1}{2}(\lambda+\nu)T}{\sin \frac{1}{2}(\lambda+\nu)}\frac{\sin \frac{1}{2}(\lambda-\nu)T}{\sin \frac{1}{2}(\lambda-\nu)} \cos[\lambda(T+1) - 2\theta]\right\}, \\
&\qquad\qquad \nu \ne \lambda,\ 0 < \nu < \pi, \\
&= \rho^2\left\{1 + \frac{\sin^2 \lambda T}{T^2 \sin^2 \lambda} + 2\frac{\sin \lambda T}{T \sin \lambda}\cos[\lambda(T+1) - 2\theta]\right\}, \\
&\qquad\qquad 0 < \lambda = \nu < \pi.
\end{aligned}
$$

If $\nu = 2\pi j/T$,

$$
\begin{aligned}
(28)\qquad \rho^2\left(\frac{2\pi j}{T}\right) = \rho_j^2 &= \frac{\rho^2 \sin^2 \frac{1}{2}\lambda T}{T^2}\left\{\frac{1}{\sin^2 \frac{1}{2}\left(\lambda + \frac{2\pi j}{T}\right)} + \frac{1}{\sin^2 \frac{1}{2}\left(\lambda - \frac{2\pi j}{T}\right)}\right. \\
&\qquad \left. + 2\frac{\cos[\lambda(T+1) - 2\theta]}{\sin \frac{1}{2}\left(\lambda + \frac{2\pi j}{T}\right)\sin \frac{1}{2}\left(\lambda - \frac{2\pi j}{T}\right)}\right\}, \\
&\qquad\qquad \lambda \ne \frac{2\pi j}{T},\ j = 1, \ldots, [\tfrac{1}{2}(T-1)], \\
&= \rho^2, \qquad\qquad \lambda = \frac{2\pi j}{T},\ j = 1, \ldots, [\tfrac{1}{2}(T-1)].
\end{aligned}
$$

It will be observed that if the trigonometric trend has a period that divides the series length, that is, if $\lambda = 2\pi j/T, j = 1, \ldots, [\frac{1}{2}(T-1)]$, then the expected values of the sample trigonometric coefficients with periods that divide the series length, that is, $A(2\pi k/T)$, $B(2\pi k/T)$, $k = 1, \ldots, [\frac{1}{2}(T-1)]$, are 0 if the period of the coefficients is different from that of the trend, $k \neq j$, and the sum of squares of the expected values, namely, the square of the population amplitude, ρ_k^2, is 0 in this case. The square of the population amplitude is the square of the amplitude of the trend ρ, if $k = j$; also $2\alpha_0^2 = 0$ and, if T is even, $2\alpha_{\frac{1}{2}T}^2 = 0$. (If $\lambda = 0$, $\rho_k^2 = 0$, $k = 1, \ldots, [\frac{1}{2}(T-1)]$, $2\alpha_0^2 = 2\alpha^2$ and, if T is even, $2\alpha_{\frac{1}{2}T}^2 = 0$; if $\lambda = \pi$, when T is even, $\rho_k^2 = 0$, $k = 1, \ldots, \frac{1}{2}T-1$, $2\alpha_0^2 = 0$, and $2\alpha_{\frac{1}{2}T}^2 = 2\alpha^2$.) The expected values of the sample trigonometric coefficients whose periods do not divide the series length ($\nu \neq 2\pi k/T$, $k = 1, \ldots, [\frac{1}{2}(T-1)]$) will in general be different from 0 as will $\rho^2(\nu)$. However, if the period of the trend does not divide the length of the series, $\lambda \neq 2\pi j/T$, $j = 1, \ldots, [\frac{1}{2}(T-1)]$, then in general the expected values of all the sample trigonometric coefficients with periods dividing the length, $\nu = 2\pi k/T$, $k = 1, \ldots, [\frac{1}{2}(T-1)]$, are different from 0 as are the corresponding population amplitudes. To summarize, the population amplitudes $\rho_1^2, \ldots, \rho_{[\frac{1}{2}(T-1)]}^2$ have the special pattern that one is ρ^2 and the others are 0 only if the period of the trend is one of the special periods $T, T/2, \ldots, T/[\frac{1}{2}(T-1)]$.

The quadratic form $(0 < \nu < \pi)$

$$(29) \quad Q(\nu) = \frac{T}{2\sigma^2}[A(\nu)\; B(\nu)] \begin{bmatrix} 1 + \dfrac{\sin \nu T}{T \sin \nu} \cos \nu(T+1) & \dfrac{\sin \nu T}{T \sin \nu} \sin \nu(T+1) \\ \dfrac{\sin \nu T}{T \sin \nu} \sin \nu(T+1) & 1 - \dfrac{\sin \nu T}{T \sin \nu} \cos \nu(T+1) \end{bmatrix}^{-1} \begin{bmatrix} A(\nu) \\ B(\nu) \end{bmatrix}$$

$$= \frac{T}{2\sigma^2\left(1 - \dfrac{\sin^2 \nu T}{T^2 \sin^2 \nu}\right)} \left\{ A^2(\nu) + B^2(\nu) + \frac{\sin \nu T}{T \sin \nu}[\cos \nu(T+1)[B^2(\nu) - A^2(\nu)] - 2 \sin \nu(T+1) A(\nu) B(\nu)] \right\}$$

has a noncentral χ^2-distribution (when the y_t's have a joint normal distribution) with 2 degrees of freedom and noncentrality parameter

(30)

$$\frac{T}{2\sigma^2}[\alpha(\nu)\ \beta(\nu)]\begin{bmatrix} 1+\dfrac{\sin \nu T}{T\sin\nu}\cos\nu(T+1) & \dfrac{\sin\nu T}{T\sin\nu}\sin\nu(T+1) \\ \dfrac{\sin\nu T}{T\sin\nu}\sin\nu(T+1) & 1-\dfrac{\sin\nu T}{T\sin\nu}\cos\nu(T+1)\end{bmatrix}^{-1}\begin{bmatrix}\alpha(\nu)\\ \beta(\nu)\end{bmatrix}$$

$$= \frac{\rho^2}{2T\sigma^2\left(1-\dfrac{\sin^2\nu T}{T^2\sin^2\nu}\right)}$$

$$\left\{\left[\frac{\sin^2\frac{1}{2}(\lambda+\nu)T}{\sin^2\frac{1}{2}(\lambda+\nu)}+\frac{\sin^2\frac{1}{2}(\lambda-\nu)T}{\sin^2\frac{1}{2}(\lambda-\nu)}\right]\left[1-\frac{\sin\nu T}{T\sin\nu}\cos\left[\lambda(T+1)-2\theta\right]\right]\right.$$

$$\left.+2\frac{\sin\frac{1}{2}(\lambda+\nu)T\sin\frac{1}{2}(\lambda-\nu)T}{\sin\frac{1}{2}(\lambda+\nu)\sin\frac{1}{2}(\lambda-\nu)}\left[\cos\left[\lambda(T+1)-2\theta\right]-\frac{\sin\nu T}{T\sin\nu}\right]\right\},$$

$$\nu\neq\lambda,\ 0<\nu<\pi,$$

$$=\frac{T\rho^2}{2\sigma^2}\left\{1+\frac{\sin\lambda T}{T\sin\lambda}\cos\left[\lambda(T+1)-2\theta\right]\right\},\qquad 0<\lambda=\nu<\pi.$$

If $\nu = 2\pi j/T$, (29) reduces to $TR_j^2/(2\sigma^2)$, and (30) reduces to $T\rho_j^2/(2\sigma^2)$, $j = 1, \ldots, [\frac{1}{2}(T-1)]$. The χ^2's (central or noncentral) which are multiples of $R^2(2\pi/T), \ldots, R^2(2\pi[\frac{1}{2}(T-1)]/T)$ are independent, but a set including values $\nu \neq 2\pi j/T, j = 1, \ldots, [\frac{1}{2}(T-1)]$, will in general be dependent.

4.4.2. Decision Procedures Based on Sample Amplitudes for Periods that are Integral Divisors of the Series Length

Now let us consider using the sample intensities at $\nu = 2\pi/T, 4\pi/T, \ldots, 2\pi[\frac{1}{2}(T-1)]/T$. If the y_t's are normally distributed, $R_1^2, \ldots, R^2_{[\frac{1}{2}(T-1)]}$ are independently distributed, and the distribution of $TR_j^2/(2\sigma^2)$ is the χ^2-distribution with noncentrality parameter $T\rho_j^2/(2\sigma^2)$ given by (28).

We see that if $\lambda = 2\pi k/T$, where k is an integer between 1 and $[\frac{1}{2}(T-1)]$, then $\rho_k^2 = \rho^2$ and $\rho_j^2 = 0$ when $j \neq k$. If this is not the case [$\lambda T/(2\pi)$ not an integer], then $\rho_j^2 > 0$ for every j. That is, if the period $2\pi/\lambda$ of $f(t)$ is not one of the periods of the orthogonal functions (columns of $\boldsymbol{M}$), then the intensities corresponding to these periods are all greater than 0. In this event we may expect that the intensity for $2\pi j$ close to λT is large. We shall now show that if $2\pi k/T < \lambda < 2\pi(k+1)/T$ then either ρ_k^2 or ρ_{k+1}^2 is the largest intensity.

LEMMA 4.4.2. *If* $|K| \le 1$,

$$h(x) = \frac{1}{\sin^2 \frac{1}{2}(\lambda + x)} + \frac{1}{\sin^2 \frac{1}{2}(\lambda - x)} + 2\frac{K}{\sin \frac{1}{2}(\lambda + x) \sin \frac{1}{2}(\lambda - x)} \tag{31}$$

is monotonic increasing for $0 < x < \lambda$ *and monotonic decreasing for* $\lambda < x < \pi$.

PROOF. We can write

$$\begin{aligned} h(x) &= \left[\frac{1}{\sin \frac{1}{2}(\lambda + x)} - \frac{1}{\sin \frac{1}{2}(\lambda - x)}\right]^2 + \frac{2(K+1)}{\sin \frac{1}{2}(\lambda + x) \sin \frac{1}{2}(\lambda - x)} \\ &= \left[\frac{\sin \frac{1}{2}(\lambda - x) - \sin \frac{1}{2}(\lambda + x)}{\sin \frac{1}{2}(\lambda + x) \sin \frac{1}{2}(\lambda - x)}\right]^2 + \frac{2(K+1)}{\sin \frac{1}{2}(\lambda + x) \sin \frac{1}{2}(\lambda - x)} \\ &= \left[\frac{-4 \cos \frac{1}{2}\lambda \sin \frac{1}{2}x}{\cos x - \cos \lambda}\right]^2 + \frac{4(K+1)}{\cos x - \cos \lambda} \\ &= 4 \cos^2 \tfrac{1}{2}\lambda \left[\frac{\sin \frac{1}{2}x}{\sin^2 \frac{1}{2}\lambda - \sin^2 \frac{1}{2}x}\right]^2 + \frac{4(K+1)}{\cos x - \cos \lambda}. \end{aligned} \tag{32}$$

Each of the two terms above increases with x in $(0, \lambda)$. We can also write

$$\begin{aligned} h(x) &= \left[\frac{1}{\sin \frac{1}{2}(\lambda + x)} + \frac{1}{\sin \frac{1}{2}(\lambda - x)}\right]^2 - \frac{2(1-K)}{\sin \frac{1}{2}(\lambda + x) \sin \frac{1}{2}(\lambda - x)} \\ &= \left[\frac{4 \sin \frac{1}{2}\lambda \cos \frac{1}{2}x}{\cos x - \cos \lambda}\right]^2 - \frac{4(1-K)}{\cos x - \cos \lambda} \\ &= 4 \sin^2 \tfrac{1}{2}\lambda \left[\frac{\cos \frac{1}{2}x}{\cos^2 \frac{1}{2}x - \cos^2 \frac{1}{2}\lambda}\right]^2 + \frac{4(1-K)}{\cos \lambda - \cos x}. \end{aligned} \tag{33}$$

Each of the two terms above decreases with x in (λ, π). Q.E.D.

Since for $K = \cos [\lambda(T + 1) - 2\theta]$

$$\rho_j^2 = \frac{\rho^2}{T^2} \sin^2 \tfrac{1}{2}\lambda T \, h\left(\frac{2\pi j}{T}\right), \tag{34}$$

we see that ρ_j^2 is increasing for $j = 1, \ldots, k$ and decreasing for $j = k + 1, \ldots, [\frac{1}{2}(T - 1)]$ when $2\pi k < \lambda T < 2\pi(k + 1)$. This implies the following theorem:

THEOREM 4.4.2. *If* $2\pi k/T < \lambda < 2\pi(k + 1)/T$, *then*

$$\begin{aligned} \rho_j &< \rho_{j+1}, \qquad j = 1, \ldots, k - 1, \\ \rho_j &< \rho_{j-1}, \qquad j = k + 2, \ldots, [\tfrac{1}{2}(T - 1)], \end{aligned} \tag{35}$$

where ρ_j *is the (positive) square root of* (28).

Then either R_k^2 or R_{k+1}^2 has the distribution with the largest noncentrality parameter. In a sense the worst case for given k is when $\lambda\ [2\pi k/T < \lambda < 2\pi(k+1)/T]$ minimizes the larger of ρ_k and ρ_{k+1}. This value of λ and the minimax of ρ_k and ρ_{k+1} is, however, a complicated function of k, T, and θ.

Let us now consider ρ_k and ρ_{k+1} in terms of approximations. When $2\pi k/T < \lambda < 2\pi(k+1)/T$, the arguments of $\sin(\frac{1}{2}\lambda - k\pi/T)$ and $\sin[\frac{1}{2}\lambda - (k+1)\pi/T]$ are less than π/T in absolute value, and if T is large enough the first term of the expansion of the sine function,

$$\sin x = x - \frac{x^3}{6} + \frac{x^5}{120} - \cdots, \tag{36}$$

yields a sufficiently good approximation for $\sin x$ when $x = \frac{1}{2}\lambda - k\pi/T$ and $x = \frac{1}{2}\lambda - (k+1)\pi/T$. Let $\lambda = 2\pi(k+\varepsilon)/T\ \ (0 < \varepsilon < 1)$. Then

$$\begin{aligned} \rho_k^2 &= \frac{\rho^2}{T^2}\sin^2\pi\varepsilon \\ &\quad\times\left\{\frac{1}{\sin^2\pi\dfrac{2k+\varepsilon}{T}} + \frac{1}{\sin^2\dfrac{\pi\varepsilon}{T}} + \frac{2\cos[2\pi(T\varepsilon+\varepsilon+k)/T - 2\theta]}{\sin\pi\dfrac{2k+\varepsilon}{T}\sin\pi\dfrac{\varepsilon}{T}}\right\}, \\ \rho_{k+1}^2 &= \frac{\rho^2}{T^2}\sin^2\pi\varepsilon \\ &\quad\times\left\{\frac{1}{\sin^2\pi\dfrac{2k+1+\varepsilon}{T}} + \frac{1}{\sin^2\pi\dfrac{\varepsilon-1}{T}} + \frac{2\cos[2\pi(T\varepsilon+\varepsilon+k)/T - 2\theta]}{\sin\pi\dfrac{2k+1+\varepsilon}{T}\sin\pi\dfrac{\varepsilon-1}{T}}\right\}. \end{aligned} \tag{37}$$

Since $0 < \sin\pi\varepsilon/T < \sin[\pi(2k+\varepsilon)/T]$ and $0 < \sin[\pi(1-\varepsilon)/T] < \sin[\pi(2k+1+\varepsilon)/T]\ (k+1 < \frac{1}{2}T)$, the second term in ρ_k^2 and ρ_{k+1}^2, respectively, is larger than the first and larger than $\frac{1}{2}$ the absolute value of the third. To what extent the second term dominates depends on

$$\begin{aligned} \frac{\sin\pi\dfrac{2k+\varepsilon}{T}}{\sin\pi\dfrac{\varepsilon}{T}} &= \frac{\pi\dfrac{2k+\varepsilon}{T} - \dfrac{1}{6}\left(\pi\dfrac{2k+\varepsilon}{T}\right)^3 + \cdots}{\pi\dfrac{\varepsilon}{T} - \dfrac{1}{6}\left(\pi\dfrac{\varepsilon}{T}\right)^3 + \cdots} \\ &= \frac{2k+\varepsilon}{\varepsilon}\,\frac{1 - \pi^2\dfrac{(2k+\varepsilon)^2}{6T^2} + \cdots}{1 - \dfrac{\pi^2\varepsilon^2}{6T^2} + \cdots} \end{aligned} \tag{38}$$

in the first case, and this is roughly $2k/\varepsilon + 1$. Except when k is very small (small frequency and large period), this factor is large and the second term in ρ_k^2 dominates. If k is near $[\frac{1}{2}(T-1)]$ the first term must be examined more closely; the argument $\pi(2k+\varepsilon)/T$ is replaced by $\pi - \pi(2k+\varepsilon)/T$ and (38) is approximated by $(T-2k-\varepsilon)/\varepsilon$.

Consideration of only the second terms in ρ_k^2 and ρ_{k+1}^2 indicates that the larger of these two terms is minimized by $\varepsilon = \frac{1}{2}$, in which case, ρ_k^2 and ρ_{k+1}^2 are approximated by

$$\text{(39)} \qquad \frac{\rho^2}{T^2}\cdot\frac{\sin^2\frac{\pi}{2}}{\sin^2\frac{\pi}{2T}} = \frac{\rho^2}{T^2\left[\frac{\pi}{2T} - \frac{1}{6}\left(\frac{\pi}{2T}\right)^3 + \cdots\right]^2} = \rho^2\left(\frac{2}{\pi}\right)^2 \frac{1}{\left[1 - \frac{1}{6}\left(\frac{\pi}{2T}\right)^2 + \cdots\right]^2}.$$

If T is sufficiently large, (39) is about $\rho^2(2/\pi)^2 = \rho^2(0.6366)^2 = \rho^2 \times 0.4053$. That is, each of ρ_k^2 and ρ_{k+1}^2 is about 41% of the intensity of the harmonic term and the sum $\rho_k^2 + \rho_{k+1}^2$ about 81%. As will be seen from the next paragraph the other ρ_j^2's must then be small. [If ε is less than $\frac{1}{2}$, the second term in ρ_k^2 is approximately $\rho^2(1 - \pi^2\varepsilon^2/6)^2$.]

The total sum of squares of $\mathscr{E}y_t = f(t)$ is

$$\text{(40)} \qquad \sum_{t=1}^{T} f^2(t) = \alpha^2 \sum_{t=1}^{T} \cos^2 \lambda t + 2\alpha\beta \sum_{t=1}^{T} \cos\lambda t \sin\lambda t + \beta^2 \sum_{t=1}^{T} \sin^2 \lambda t$$
$$= \rho^2\left\{\tfrac{1}{2}T + \frac{\sin\lambda T}{2\sin\lambda}\cos[\lambda(T+1) - 2\theta]\right\}, \qquad 0 < \lambda < \pi,$$
$$= T\alpha^2, \qquad \lambda = 0, \pi.$$

This is normalized by division by $\frac{1}{2}T$. If $\lambda = 2\pi k/T$ or if $\lambda = 2\pi(k+\frac{1}{2})/T$, the second term on the right of (40) is 0 and

$$\text{(41)} \qquad \frac{2}{T}\sum_{t=1}^{T} f^2(t) = \rho^2, \qquad \lambda = \frac{2\pi k}{T}, \qquad \lambda = \frac{2\pi(k+\frac{1}{2})}{T}.$$

When $\lambda = 2\pi(k+\varepsilon)/T$, the sum is

$$\text{(42)} \qquad \frac{2}{T}\sum_{t=1}^{T} f^2(t) = \rho^2\left\{1 + \frac{\sin 2\pi\varepsilon}{T\sin 2\pi\frac{(k+\varepsilon)}{T}}\cos\left[2\pi\left(\varepsilon + \frac{k+\varepsilon}{T}\right) - 2\theta\right]\right\};$$

the second term in (42) will be small, particularly if k is large. In any event we observe that

$$\frac{2}{T}\sum_{t=1}^{T} f^2(t) = \sum_{j=1}^{[\frac{1}{2}(T-1)]} \rho_j^2 + 2\alpha_0^2 + 2\alpha_{\frac{1}{2}T}^2, \tag{43}$$

where the last term is absent if T is odd. The coefficients α_0 and $\alpha_{\frac{1}{2}T}$ are given by (23) and (24); they will usually be small (Problems 34 and 35) and contribute little to the sum of squares (43). Hence, the normalized sum of squares of the expected values of the observations is about the same as the sum of squares of the $[\frac{1}{2}(T-1)]$ intensities. The part of this latter sum not in the largest ρ_j^2 or the second largest is distributed among the other ρ_j^2's. The previous discussion indicates the largest of these ρ_j^2's accounts for at least 41% of the sum of squares.

Note that the sum of squares (40) divided by σ^2 is the noncentrality parameter (30) for $\nu = \lambda$ ($\neq 0, \pi$). The statistic $Q(\lambda)$ is a quadratic form in $A(\lambda)$ and $B(\lambda)$ and can be considered as the sum of squares of two normalized orthogonal linear combinations of $y_1, \ldots, y_T$. The sum of squares of the expected values of these two linear combinations is equal to the sum of squares of the expectations of all the observations. This shows that the noncentrality parameter of $Q(\nu)$ is maximized at $\nu = \lambda$.

The procedure involving $R_1^2, \ldots, R_q^2$ ($q = [\frac{1}{2}(T-1)]$) is to conclude H_0: $\rho^2 = 0$ if $R_j^2 \leq g\sum_{i=1}^q R_i^2$, $j = 1, \ldots, q$, where g is chosen so (92) of Section 4.3 is the specified significance level ε and to conclude H_j: $\rho^2 > 0$ and $\rho_j^2 > \rho_i^2$, $i \neq j$, $i = 1, \ldots, q$, if $R_j^2 > g\sum_{i=1}^q R_i^2$ and $R_j^2 > R_i^2$, $i \neq j$, $i = 1, \ldots, q$. The hypothesis H_j implies (Theorem 4.4.2) that λ is within $2\pi/T$ of $2\pi j/T$ and very nearly in the interval $2\pi(j - \frac{1}{2})/T$ to $2\pi(j + \frac{1}{2})/T$. In terms of $z_j = TR_j^2/(2\sigma^2)$ the regions are

$$R_0:\quad z_j \leq \frac{g}{1-g}\sum_{i \neq j} z_i, \qquad j = 1, \ldots, q, \tag{44}$$

$$R_j:\quad z_j > \frac{g}{1-g}\sum_{i \neq j} z_i, \quad z_j > z_i,\ i \neq j,\ i = 1, \ldots, q, \qquad j = 1, \ldots, q. \tag{45}$$

We shall now show that if H_j is true, the probability of concluding H_j is greater than the probability of concluding any other H_i is true, and this probability is maximum when $\lambda = 2\pi j/T$ (and hence $\rho_i^2 = 0$, $i \neq j$). Let $\tau_j^2 = T\rho_j^2/(2\sigma^2)$. Here we assume that the y_t's are normally distributed.

LEMMA 4.4.3.

$$\int_c^\infty k(z \mid \tau^2)\, dz = \int_c^\infty e^{-\frac{1}{2}\tau^2} k(z) I_0(\tau\sqrt{z})\, dz \tag{46}$$

is an increasing function of τ^2 $(0 \leq \tau^2 < \infty)$.

PROOF. If $\tau_1^2 < \tau_2^2$, $z \geq c$ is equivalent to $k(z \mid \tau_2^2)/k(z \mid \tau_1^2) > d$ for a suitable d. Then

$$\int_c^\infty k(z \mid \tau_2^2)\, dz > d \int_c^\infty k(z \mid \tau_1^2)\, dz, \tag{47}$$

(48)

$$1 - \int_c^\infty k(z \mid \tau_2^2)\, dz = \int_0^c k(z \mid \tau_2^2)\, dz < d \int_0^c k(z \mid \tau_1^2)\, dz = d\left[1 - \int_c^\infty k(z \mid \tau_1^2)\, dz\right],$$

from which the lemma follows.

LEMMA 4.4.4. *$\Pr\{R_1 \mid \tau_1^2, \tau_2^2, \ldots, \tau_q^2\}$ is an increasing function of τ_1^2 for fixed values of $\tau_2^2, \ldots, \tau_q^2$ and a decreasing function of τ_2^2 for fixed values of τ_1^2, $\tau_3^2, \ldots, \tau_q^2$.*

PROOF. Let $h(z_2, \ldots, z_q)$ be the maximum of $z_2, \ldots, z_q$ and $[g/(1-g)] \times \sum_{i=2}^q z_i$. Then

(49)

$$\Pr\{R_1 \mid \tau_1^2, \tau_2^2, \ldots, \tau_q^2\} = \int_0^\infty \cdots \int_0^\infty \prod_{i=2}^q k(z_i \mid \tau_i^2) \left\{\int_{h(z_2, \ldots, z_q)}^\infty k(z_1 \mid \tau_1^2)\, dz_1\right\} \prod_{i=2}^q dz_i.$$

The integral in the braces is an increasing function of τ_1^2 by Lemma 4.4.3. Similarly let $m(z_1, z_3, \ldots, z_q)$ be the minimum of z_1 and $[(1-g)/g]z_1 - \sum_{i=3}^q z_i$. Then

$$\begin{aligned} \Pr\{R_1 \mid \tau_1^2, \tau_2^2, \ldots, \tau_q^2\} &= \int_0^\infty \cdots \int_0^\infty \int_{\max_{i>2} z_i}^\infty \prod_{i \neq 2} k(z_i \mid \tau_i^2) \left\{\int_0^{m(z_1, z_3, \ldots, z_q)} k(z_2 \mid \tau_2^2)\, dz_2\right\} \prod_{i \neq 2} dz_i. \end{aligned} \tag{50}$$

The integral in the braces is a decreasing function of τ_2^2 by Lemma 4.4.3. Q.E.D.

THEOREM 4.4.3. *If $\tau_j^2 > \tau_k^2$, then*

$$\Pr\{R_j \mid \tau_1^2, \ldots, \tau_q^2\} > \Pr\{R_k \mid \tau_1^2, \ldots, \tau_q^2\}. \tag{51}$$

PROOF. The left-hand side of (51) is greater than the probability of R_j with τ_j^2 replaced by τ_k^2 which is equal to the probability of R_k with τ_j^2 replaced by τ_k^2 and greater than the right-hand side of (51). Q.E.D.

THEOREM 4.4.4.

$$\Pr\{R_j \mid \tau_j^2 = \tau^2, \tau_i^2 = 0, i \neq j\} > \Pr\left\{R_j \mid \tau_j^2 < \tau^2, \sum_{i=1}^q \tau_i^2 = \tau^2\right\}. \tag{52}$$

PROOF. The theorem follows from successive applications of Lemma 4.4.4 with τ_1^2 replaced by τ_j^2 and τ_2^2 replaced by τ_i^2, $i \neq j$, in turn. Q.E.D.

The probabilities $\Pr\{R_1 \mid \tau_1^2, \ldots, \tau_q^2\}, \ldots, \Pr\{R_q \mid \tau_1^2, \ldots, \tau_q^2\}$ depend on $\tau_1^2, \ldots, \tau_q^2$ and are very difficult to evaluate.

4.4.3. Use of Trigonometric Functions whose Periods are not Integral Divisors of the Series Length

It was observed that $y_1, \ldots, y_T$ can be expressed in terms of $\bar{y}, a_1, b_1, \ldots, a_{[\frac{1}{2}(T-1)]}, b_{[\frac{1}{2}(T-1)]}$, and $a_{\frac{1}{2}T}$ (if T is even). If $T = 2H + 1$,

$$y_t = \bar{y} + \sum_{j=1}^{H}\left(a_j \cos\frac{2\pi j}{T}t + b_j \sin\frac{2\pi j}{T}t\right); \tag{53}$$

if $T = 2H + 2$

$$y_t = \bar{y} + \sum_{j=1}^{H}\left(a_j \cos\frac{2\pi j}{T}t + b_j \sin\frac{2\pi j}{T}t\right) + a_{\frac{1}{2}T}(-1)^t. \tag{54}$$

The coefficients $A(\nu)$ and $B(\nu)$ can therefore also be expressed in terms of these T trigonometric coefficients. For $T = 2H + 1$

$$\begin{aligned}
A(\nu) &= \frac{2}{T}\bar{y}\sum_{t=1}^{T}\cos\nu t \\
&\quad + \frac{2}{T}\sum_{j=1}^{H}\left(a_j\sum_{t=1}^{T}\cos\frac{2\pi j}{T}t\cos\nu t + b_j\sum_{t=1}^{T}\sin\frac{2\pi j}{T}t\cos\nu t\right) \\
&= \frac{2}{T}\bar{y}\frac{\sin\frac{1}{2}\nu T}{\sin\frac{1}{2}\nu}\cos\tfrac{1}{2}\nu(T+1) \\
&\quad + \frac{1}{T}\sum_{j=1}^{H}\left\{a_j\left[\frac{\sin\left(\frac{1}{2}\nu + \frac{\pi j}{T}\right)T}{\sin\left(\frac{1}{2}\nu + \frac{\pi j}{T}\right)}\cos\left(\tfrac{1}{2}\nu + \frac{\pi j}{T}\right)(T+1)\right.\right. \\
&\quad \left.+ \frac{\sin\left(\frac{1}{2}\nu - \frac{\pi j}{T}\right)T}{\sin\left(\frac{1}{2}\nu - \frac{\pi j}{T}\right)}\cos\left(\tfrac{1}{2}\nu - \frac{\pi j}{T}\right)(T+1)\right]
\end{aligned} \tag{55}$$

$$+ b_j \left[\frac{\sin\left(\frac{1}{2}\nu + \frac{\pi j}{T}\right)T}{\sin\left(\frac{1}{2}\nu + \frac{\pi j}{T}\right)} \sin\left(\frac{1}{2}\nu + \frac{\pi j}{T}\right)(T+1) - \frac{\sin\left(\frac{1}{2}\nu - \frac{\pi j}{T}\right)T}{\sin\left(\frac{1}{2}\nu - \frac{\pi j}{T}\right)} \sin\left(\frac{1}{2}\nu - \frac{\pi j}{T}\right)(T+1) \right] \Bigg\}$$

$$= \frac{\sin \frac{1}{2}\nu T}{T} \Bigg[2\bar{y} \frac{\cos \frac{1}{2}\nu(T+1)}{\sin \frac{1}{2}\nu} + \sum_{j=1}^{H} \Bigg\{ a_j \left[\frac{\cos\left[\frac{1}{2}\nu(T+1) + \frac{\pi j}{T}\right]}{\sin\left(\frac{1}{2}\nu + \frac{\pi j}{T}\right)} + \frac{\cos\left[\frac{1}{2}\nu(T+1) - \frac{\pi j}{T}\right]}{\sin\left(\frac{1}{2}\nu - \frac{\pi j}{T}\right)} \right]$$

$$+ b_j \left[\frac{\sin\left[\frac{1}{2}\nu(T+1) + \frac{\pi j}{T}\right]}{\sin\left(\frac{1}{2}\nu + \frac{\pi j}{T}\right)} - \frac{\sin\left[\frac{1}{2}\nu(T+1) - \frac{\pi j}{T}\right]}{\sin\left(\frac{1}{2}\nu - \frac{\pi j}{T}\right)} \right] \Bigg\} \Bigg],$$

$$B(\nu) = \frac{\sin \frac{1}{2}\nu T}{T} \Bigg[2\bar{y} \frac{\sin \frac{1}{2}\nu(T+1)}{\sin \frac{1}{2}\nu} + \sum_{j=1}^{H} \Bigg\{ a_j \left[\frac{\sin\left[\frac{1}{2}\nu(T+1) + \frac{\pi j}{T}\right]}{\sin\left(\frac{1}{2}\nu + \frac{\pi j}{T}\right)} + \frac{\sin\left[\frac{1}{2}\nu(T+1) - \frac{\pi j}{T}\right]}{\sin\left(\frac{1}{2}\nu - \frac{\pi j}{T}\right)} \right] \tag{56}$$

$$+ b_j \left[- \frac{\cos\left[\frac{1}{2}\nu(T+1) + \frac{\pi j}{T}\right]}{\sin\left(\frac{1}{2}\nu + \frac{\pi j}{T}\right)} + \frac{\cos\left[\frac{1}{2}\nu(T+1) - \frac{\pi j}{T}\right]}{\sin\left(\frac{1}{2}\nu - \frac{\pi j}{T}\right)} \right] \Bigg\} \Bigg].$$

The evaluation is simplified by an orthogonal transformation of $A(\nu)$ and $B(\nu)$ and a_j and b_j, $j = 1, \ldots, H$. Let

$$C(\nu) = A(\nu)\cos\tfrac{1}{2}\nu(T+1) + B(\nu)\sin\tfrac{1}{2}\nu(T+1) \tag{57}$$

$$= \frac{\sin\frac{1}{2}\nu T}{T}\left[\frac{2\bar{y}}{\sin\frac{1}{2}\nu} + \sum_{j=1}^{H}\left\{\frac{a_j\cos\frac{\pi j}{T} + b_j\sin\frac{\pi j}{T}}{\sin\left(\frac{1}{2}\nu + \frac{\pi j}{T}\right)} + \frac{a_j\cos\frac{\pi j}{T} + b_j\sin\frac{\pi j}{T}}{\sin\left(\frac{1}{2}\nu - \frac{\pi j}{T}\right)}\right\}\right]$$

$$= \frac{\sin\frac{1}{2}\nu T}{T}\left[\frac{2\bar{y}}{\sin\frac{1}{2}\nu} + \sum_{j=1}^{H} c_j\left(\frac{1}{\sin\left(\frac{1}{2}\nu + \frac{\pi j}{T}\right)} + \frac{1}{\sin\left(\frac{1}{2}\nu - \frac{\pi j}{T}\right)}\right)\right],$$

$$D(\nu) = -A(\nu)\sin\tfrac{1}{2}\nu(T+1) + B(\nu)\cos\tfrac{1}{2}\nu(T+1) \tag{58}$$

$$= \frac{\sin\frac{1}{2}\nu T}{T}\sum_{j=1}^{H}\left\{\frac{a_j\sin\frac{\pi j}{T} - b_j\cos\frac{\pi j}{T}}{\sin\left(\frac{1}{2}\nu + \frac{\pi j}{T}\right)} + \frac{-a_j\sin\frac{\pi j}{T} + b_j\cos\frac{\pi j}{T}}{\sin\left(\frac{1}{2}\nu - \frac{\pi j}{T}\right)}\right\}$$

$$= \frac{\sin\frac{1}{2}\nu T}{T}\sum_{j=1}^{H} d_j\left\{\frac{1}{\sin\left(\frac{1}{2}\nu - \frac{\pi j}{T}\right)} - \frac{1}{\sin\left(\frac{1}{2}\nu + \frac{\pi j}{T}\right)}\right\},$$

where

$$c_j = a_j\cos\frac{\pi j}{T} + b_j\sin\frac{\pi j}{T} \tag{59}$$

$$= \frac{2}{T}\sum_{t=1}^{T} y_t\cos\frac{2\pi j}{T}(t - \tfrac{1}{2}),$$

$$d_j = -a_j\sin\frac{\pi j}{T} + b_j\cos\frac{\pi j}{T} \tag{60}$$

$$= \frac{2}{T}\sum_{t=1}^{T} y_t\sin\frac{2\pi j}{T}(t - \tfrac{1}{2}).$$

If $T = 2H + 2$, $A(\nu)$ is the right-hand side of (55) plus $2a_{\frac{1}{2}T} \sin \frac{1}{2}\nu T \cos [\frac{1}{2}\nu(T + 1) + \frac{1}{2}\pi]/[T \sin (\frac{1}{2}\nu + \frac{1}{2}\pi)]$, $B(\nu)$ is the right-hand side of (56) plus $2a_{\frac{1}{2}T} \sin \frac{1}{2}\nu T \sin [\frac{1}{2}\nu(T + 1) + \frac{1}{2}\pi]/[T \sin (\frac{1}{2}\nu + \frac{1}{2}\pi)]$, $C(\nu)$ is the right-hand side of (57), and $D(\nu)$ is the right-hand side of (58) minus $2d_{\frac{1}{2}T} \sin \frac{1}{2}\nu T/[T \sin (\frac{1}{2}\nu + \frac{1}{2}\pi)]$, where $d_{\frac{1}{2}T} = -a_{\frac{1}{2}T}$ (and $c_{\frac{1}{2}T} = b_{\frac{1}{2}T} = 0$).

The coefficients a_j's and b_j's are uncorrelated and independent in the case of normality. Similarly $\bar{y}$ and the coefficients c_j's and d_j's are uncorrelated and independent in the case of normality. Thus $C(\nu)$ and $D(\nu')$ are uncorrelated for any ν and ν', and the set $\{C(\nu)\}$ for all ν $(0 \le \nu \le \pi)$ is independent of the set $\{D(\nu)\}$ for all ν $(0 \le \nu \le \pi)$ in the case of normality. Since the transformation from $A(\nu)$, $B(\nu)$ to $C(\nu)$, $D(\nu)$ is orthogonal, $R^2(\nu) = C^2(\nu) + D^2(\nu)$. Unfortunately, the representations $C(\nu)$ and $D(\nu)$ do not prove to be very useful.

If $f(t) = \mu + \rho \cos (\lambda t - \theta)$, we may want to use deviations from the sample mean. Let

$$A^*(\nu) = \frac{2}{T} \sum_{t=1}^{T} (y_t - \bar{y}) \cos \nu t, \tag{61}$$

$$B^*(\nu) = \frac{2}{T} \sum_{t=1}^{T} (y_t - \bar{y}) \sin \nu t. \tag{62}$$

THEOREM 4.4.5. *When* $\mathscr{E} y_t = \mu + \alpha \cos \lambda t + \beta \sin \lambda t = \mu + \rho \cos (\lambda t - \theta)$,

$$\begin{aligned} \mathscr{E} A^*(\nu) = \frac{\rho}{T} \Bigg\{ & \frac{\sin \frac{1}{2}(\lambda + \nu)T}{\sin \frac{1}{2}(\lambda + \nu)} \cos [\tfrac{1}{2}(\lambda + \nu)(T + 1) - \theta] \\ & + \frac{\sin \frac{1}{2}(\lambda - \nu)T}{\sin \frac{1}{2}(\lambda - \nu)} \cos [\tfrac{1}{2}(\lambda - \nu)(T + 1) - \theta] \\ & - \frac{2}{T} \frac{\sin \frac{1}{2}\lambda T}{\sin \frac{1}{2}\lambda} \cos [\tfrac{1}{2}\lambda(T + 1) - \theta] \frac{\sin \frac{1}{2}\nu T}{\sin \frac{1}{2}\nu} \cos \tfrac{1}{2}\nu(T + 1) \Bigg\}, \\ & \qquad \nu \neq \lambda, \quad 0 < \nu < \pi, \\ = \alpha + \frac{\rho}{T} \Bigg\{ & \frac{\sin \lambda T}{\sin \lambda} \cos [\lambda(T + 1) - \theta] \\ & - \frac{2}{T} \frac{\sin^2 \frac{1}{2}\lambda T}{\sin^2 \frac{1}{2}\lambda} \cos [\tfrac{1}{2}\lambda(T + 1) - \theta] \cos \tfrac{1}{2}\lambda(T + 1) \Bigg\}, \\ & \qquad 0 < \lambda = \nu < \pi, \end{aligned} \tag{63}$$

$$(64)\quad \mathscr{E}B^*(\nu) = \frac{\rho}{T}\left\{\frac{\sin \frac{1}{2}(\lambda+\nu)T}{\sin \frac{1}{2}(\lambda+\nu)} \sin\left[\tfrac{1}{2}(\lambda+\nu)(T+1)-\theta\right]\right.$$

$$-\frac{\sin \frac{1}{2}(\lambda-\nu)T}{\sin \frac{1}{2}(\lambda-\nu)} \sin\left[\tfrac{1}{2}(\lambda-\nu)(T+1)-\theta\right]$$

$$\left.-\frac{2}{T}\frac{\sin \frac{1}{2}\lambda T}{\sin \frac{1}{2}\lambda}\cos\left[\tfrac{1}{2}\lambda(T+1)-\theta\right]\frac{\sin \frac{1}{2}\nu T}{\sin \frac{1}{2}\nu}\sin \tfrac{1}{2}\nu(T+1)\right\},$$

$$\nu \neq \lambda, \quad 0 < \nu < \pi,$$

$$= \beta + \frac{\rho}{T}\left\{\frac{\sin \lambda T}{\sin \lambda}\sin\left[\lambda(T+1)-\theta\right]\right.$$

$$\left.-\frac{2}{T}\frac{\sin^2 \frac{1}{2}\lambda T}{\sin^2 \frac{1}{2}\lambda}\cos\left[\tfrac{1}{2}\lambda(T+1)-\theta\right]\sin \tfrac{1}{2}\lambda(T+1)\right\},$$

$$0 < \lambda = \nu < \pi,$$

(65)

$$\operatorname{Var}[A^*(\nu)] = \frac{2\sigma^2}{T}\left\{1 + \frac{\sin \nu T}{T \sin \nu}\cos \nu(T+1) - 2\left[\frac{\sin \frac{1}{2}\nu T}{T \sin \frac{1}{2}\nu}\cos \tfrac{1}{2}\nu(T+1)\right]^2\right\},$$

$$0 < \nu < \pi,$$

(66)

$$\operatorname{Var}[B^*(\nu)] = \frac{2\sigma^2}{T}\left\{1 - \frac{\sin \nu T}{T \sin \nu}\cos \nu(T+1) - 2\left[\frac{\sin \frac{1}{2}\nu T}{T \sin \frac{1}{2}\nu}\sin \tfrac{1}{2}\nu(T+1)\right]^2\right\},$$

$$0 < \nu < \pi,$$

$$(67)\quad \operatorname{Cov}[A^*(\nu), B^*(\nu)] = \frac{2\sigma^2}{T}\left\{\frac{\sin \nu T}{T \sin \nu}\sin \nu(T+1)\right.$$

$$\left.-2\left[\frac{\sin \frac{1}{2}\nu T}{T \sin \frac{1}{2}\nu}\right]^2 \cos \tfrac{1}{2}\nu(T+1)\sin \tfrac{1}{2}\nu(T+1)\right\}, \qquad 0 < \nu < \pi.$$

The population squared amplitude is

$$(68)\quad \rho^{*2}(\nu) = \frac{\rho^2}{T^2}\left\{\frac{\sin^2 \frac{1}{2}(\lambda+\nu)T}{\sin^2 \frac{1}{2}(\lambda+\nu)} + \frac{\sin^2 \frac{1}{2}(\lambda-\nu)T}{\sin^2 \frac{1}{2}(\lambda-\nu)}\right.$$

$$+2\frac{\sin \frac{1}{2}(\lambda+\nu)T}{\sin \frac{1}{2}(\lambda+\nu)}\frac{\sin \frac{1}{2}(\lambda-\nu)T}{\sin \frac{1}{2}(\lambda-\nu)}\cos\left[\lambda(T+1)-2\theta\right]$$

$$-\frac{4}{T}\frac{\sin \frac{1}{2}\lambda T}{\sin \frac{1}{2}\lambda}\cos^2\left[\tfrac{1}{2}\lambda(T+1)-\theta\right]$$

$$\times \frac{\sin \frac{1}{2}\nu T}{\sin \frac{1}{2}\nu}\left[\frac{\sin \frac{1}{2}(\lambda+\nu)T}{\sin \frac{1}{2}(\lambda+\nu)} + \frac{\sin \frac{1}{2}(\lambda-\nu)T}{\sin \frac{1}{2}(\lambda-\nu)}\right]$$

$$+ \frac{4}{T^2}\frac{\sin^2 \frac{1}{2}\lambda T}{\sin^2 \frac{1}{2}\lambda}\cos^2 [\tfrac{1}{2}\lambda(T+1)-\theta]\frac{\sin^2 \frac{1}{2}\nu T}{\sin^2 \frac{1}{2}\nu}\Bigg\}, \qquad \nu \neq \lambda, \quad 0 < \nu < \pi,$$

$$= \rho^2\Bigg\{1 + \frac{\sin^2 \lambda T}{T^2 \sin^2 \lambda} + 2\frac{\sin \lambda T}{T \sin \lambda}\cos [\lambda(T+1) - 2\theta]$$

$$- \frac{4}{T^2}\frac{\sin^2 \frac{1}{2}\lambda T}{\sin^2 \frac{1}{2}\lambda}\cos^2 [\tfrac{1}{2}\lambda(T+1)-\theta]$$

$$\times\left[1 + \frac{\sin \lambda T}{T \sin \lambda} - \frac{\sin^2 \frac{1}{2}\lambda T}{T^2 \sin^2 \frac{1}{2}\lambda}\right]\Bigg\}, \qquad 0 < \lambda = \nu < \pi.$$

Note that $A^*(2\pi j/T) = A(2\pi j/T)$, $B^*(2\pi j/T) = B(2\pi j/T)$, $R^{*2}(2\pi j/T) = R^2(2\pi j/T) = R_j^2$, and $\rho^{*2}(2\pi j/T) = \rho^2(2\pi j/T) = \rho_j^2$.

A possible test of the hypothesis that $\rho = 0$ when it is assumed that $\mathscr{E}y_t = \rho \cos(\lambda t - \theta)$ for some $0 < \lambda < \pi$ and when σ^2 is unknown is to reject the null hypothesis if $\max R^2(\nu)/\sum_{t=1}^T y_t^2$ is greater than some constant. Unfortunately, we do not know the probability of this event under the null hypothesis and normality. Because of the continuity of $R^2(\nu)$ the maximum of $R^2(\nu)$, $0 \leq \nu \leq \pi$, will not differ greatly from the maximum of R_j^2, $j = 1, \ldots, [\frac{1}{2}(T-1)]$. If it is assumed that $\mathscr{E}y_t = \mu + \rho\cos(\lambda t - \theta)$, one can similarly use $\max R^{*2}(\nu)/\sum_{t=1}^T (y_t - \bar{y})^2$.

We now turn to the problem of estimating α, β, and λ when it is assumed that the trend is a nontrivial trigonometric sequence. For convenience suppose that $\mu = 0$. If the y_t's are assumed normally distributed, the logarithm of the likelihood function is

$$-\tfrac{1}{2}T \log 2\pi - \tfrac{1}{2}T \log \sigma^{*2} - \frac{1}{2\sigma^{*2}} S(\alpha^*, \beta^*, \lambda^*), \tag{69}$$

where

$$S(\alpha^*, \beta^*, \lambda^*) = \sum_{t=1}^T (y_t - \alpha^* \cos \lambda^* t - \beta^* \sin \lambda^* t)^2. \tag{70}$$

The maximum likelihood estimates of α, β, and λ are the values of α^*, β^*, and λ^* that minimize $S(\alpha^*, \beta^*, \lambda^*)$. For given λ^* the extremal values of α^* and β^* satisfy

$$\begin{aligned} \left[1 + \frac{\sin \lambda^* T}{T \sin \lambda^*}\cos \lambda^*(T+1)\right]\alpha^* + \frac{\sin \lambda^* T}{T\sin\lambda^*}\sin\lambda^*(T+1)\beta^* &= A(\lambda^*), \\ \frac{\sin \lambda^* T}{T \sin \lambda^*}\sin \lambda^*(T+1)\alpha^* + \left[1 - \frac{\sin \lambda^* T}{T \sin \lambda^*}\cos\lambda^*(T+1)\right]\beta^* &= B(\lambda^*). \end{aligned} \tag{71}$$

The solution is

$$(72)\quad \hat{\alpha} = \frac{A(\lambda^*) - \dfrac{\sin \lambda^* T}{T \sin \lambda^*}[\cos \lambda^*(T+1)A(\lambda^*) + \sin \lambda^*(T+1)B(\lambda^*)]}{1 - \dfrac{\sin^2 \lambda^* T}{T^2 \sin^2 \lambda^*}},$$

$$(73)\quad \hat{\beta} = \frac{B(\lambda^*) - \dfrac{\sin \lambda^* T}{T \sin \lambda^*}[\sin \lambda^*(T+1)A(\lambda^*) - \cos \lambda^*(T+1)B(\lambda^*)]}{1 - \dfrac{\sin^2 \lambda^* T}{T^2 \sin^2 \lambda^*}}.$$

For these values of α^* and β^* the sum of squares (70) is $\sum_{t=1}^{T} y_t^2 - \sigma^2 Q(\lambda^*)$, where $Q(\nu)$ is defined by (29). The value of λ^* that minimizes this sum of squares is the value of λ^* that maximizes $Q(\lambda^*)$. In theory this value of λ^* satisfies $dQ(\lambda^*)/d\lambda^* = 0$, but the derivative is too complicated to solve this equation explicitly. (See Problem 40.)

Let us consider some large-sample theory of these procedures.

THEOREM 4.4.6. *If* $\mathscr{E}y_t = \alpha \cos \lambda t + \beta \sin \lambda t$, $0 < \lambda < \pi$, $\text{Var}(y_t) = \sigma^2$, *and* $\text{Cov}(y_t, y_s) = 0$, $t \neq s$, *then*

$$(74)\quad \operatorname*{plim}_{T\to\infty} A(\nu) = 0, \quad \nu \neq \lambda, \quad 0 < \nu < \pi,$$
$$= \alpha, \quad \nu = \lambda,$$

$$(75)\quad \operatorname*{plim}_{T\to\infty} B(\nu) = 0, \quad \nu \neq \lambda, \quad 0 < \nu < \pi,$$
$$= \beta, \quad \nu = \lambda,$$

$$(76)\quad \operatorname*{plim}_{T\to\infty} R^2(\nu) = 0, \quad \nu \neq \lambda, \quad 0 < \nu < \pi,$$
$$= \rho^2, \quad \nu = \lambda.$$

PROOF. By Theorem 4.4.1 $\mathscr{E}A(\nu) \to 0$, $\nu \neq \lambda$, $\mathscr{E}A(\nu) \to \alpha$, $\nu = \lambda$, $\mathscr{E}B(\nu) \to 0$, $\nu \neq \lambda$, $\mathscr{E}B(\nu) \to \beta$, $\nu = \lambda$, $\text{Var}[A(\nu)] \to 0$, and $\text{Var}[B(\nu)] \to 0$. Then (74) and (75) follow by the Tchebycheff inequality. These imply (76) because $R^2(\nu)$ is a continuous function of $A(\nu)$ and $B(\nu)$. Q.E.D.

COROLLARY 4.4.2. *If* $\mathscr{E}y_t = \mu + \alpha \cos \lambda t + \beta \sin \lambda t$, $0 < \lambda < \pi$, $\text{Var}(y_t) = \sigma^2$, *and* $\text{Cov}(y_t, y_s) = 0$, $t \neq s$, *then the conclusions of Theorem* 4.4.6 *hold for* $A^*(\nu)$, $B^*(\nu)$, *and* $R^{*2}(\nu)$.

THEOREM 4.4.7. *Suppose* $y_t = \alpha \cos \lambda t + \beta \sin \lambda t + u_t$, $0 < \lambda < \pi$, $\mathscr{E}u_t = 0$, $\mathscr{E}u_t^2 = \sigma^2$, *the* u_t's *are independent, and their distributions satisfy* (3) *of Section* 2.6. *Then* $\sqrt{\tfrac{1}{2}T}[A(\lambda) - \alpha]$ *and* $\sqrt{\tfrac{1}{2}T}[B(\lambda) - \beta]$ *have a limiting normal distribution with means* 0, *variances* σ^2, *and covariance* 0. *Under these conditions* $\sqrt{\frac{1}{2}T}\,A(\nu)$ *and* $\sqrt{\frac{1}{2}T}\,B(\nu)$, $\nu \neq \lambda$, $0 < \nu < \pi$, *have the same limiting distribution, and* $TR^2(\nu)/(2\sigma^2)$ *has a limiting* χ^2*-distribution with* 2 *degrees of freedom.*

COROLLARY 4.4.3. *Suppose* $y_t = \mu + \alpha \cos \lambda t + \beta \sin \lambda t + u_t$, $0 < \lambda < \pi$, $\mathscr{E}u_t = 0$, $\mathscr{E}u_t^2 = \sigma^2$, *the* u_t*'s are independent, and their distributions satisfy* (3) *of Section* 2.6. *Then the conclusions of Theorem* 4.4.7 *hold for* $A^*(\nu)$, $B^*(\nu)$, *and* $R^{*2}(\nu)$.

Theorems 4.4.6 and 4.4.7 hold for $A(\nu)$, $B(\nu)$, and $R^2(\nu)$ for each value of ν; they can be extended to hold simultaneously for $\nu = \nu_1, \ldots, \nu_k$ for fixed k. However, we are interested in such limiting results for either the case of $\nu = 2\pi j/T$, $j = 1, \ldots, [\frac{1}{2}(T-1)]$, or the case of all ν in the range $0 < \nu < \pi$. In either case the number of points is not a fixed finite number.

Whittle (1952) stated that modified maximum likelihood estimates $\hat{\alpha}$, $\hat{\beta}$, and $\hat{\lambda}$ are consistent estimates [see also M. M. Rao (1960)], and A. M. Walker (1968) gave a rigorous proof, showing moreover that

$$\operatorname*{plim}_{T\to\infty} T(\hat{\lambda} - \lambda) = 0. \tag{77}$$

Whittle stated that these estimates are asymptotically normally distributed with a covariance matrix as given below (except for a minor error). Walker proved rigorously that $\sqrt{T}(\hat{\alpha} - \alpha)$, $\sqrt{T}(\hat{\beta} - \beta)$, and $T^{3/2}(\hat{\lambda} - \lambda)$ have a joint limiting normal distribution with means 0 and covariance matrix

$$2\sigma^2 \begin{pmatrix} 1 & 0 & \frac{1}{2}\beta \\ 0 & 1 & -\frac{1}{2}\alpha \\ \frac{1}{2}\beta & -\frac{1}{2}\alpha & \frac{1}{3}\rho^2 \end{pmatrix}^{-1} = \frac{2\sigma^2}{\rho^2} \begin{pmatrix} \alpha^2 + 4\beta^2 & -3\alpha\beta & -6\beta \\ -3\alpha\beta & 4\alpha^2 + \beta^2 & 6\alpha \\ -6\beta & 6\alpha & 12 \end{pmatrix}. \tag{78}$$

The proofs of these results are extremely lengthy and complicated. It is noteworthy that the variance of $\hat{\lambda}$ is of order $1/T^3$, instead of the usual $1/T$ and unusual $1/T^2$. Walker's assumptions are that $\lambda \neq 0, \pi$ and the u_t's are independently and identically distributed with means 0 and variances σ^2.

An essential part of Walker's proof of (77) is [A. M. Walker (1965)] that

$$A_u(\nu) = \frac{2}{T}\sum_{t=1}^{T} u_t \cos \nu t, \qquad B_u(\nu) = \frac{2}{T}\sum_{t=1}^{T} u_t \sin \nu t \tag{79}$$

converge stochastically to 0 for all ν simultaneously. Let $R_u^2(\nu) = A_u^2(\nu) + B_u^2(\nu)$. Walker (1965) proved the following:

THEOREM 4.4.8. *If the u_t's are independent with means* 0 *and variances* σ^2,

$$\operatorname*{plim}_{T\to\infty} \max_{0\le\nu\le\pi} R_u^2(\nu) = 0. \tag{80}$$

PROOF. We have

$$\mathscr{E} \max_{0\le\nu\le\pi} R_u^2(\nu) = \frac{4}{T^2}\mathscr{E} \max_{0\le\nu\le\pi} \left|\sum_{t=1}^{T} u_t e^{i\nu t}\right|^2. \tag{81}$$

However,

$$\begin{aligned}\left|\sum_{t=1}^{T} u_t e^{i\nu t}\right|^2 &= \sum_{t,t'=1}^{T} e^{i\nu(t-t')}u_t u_{t'} \\ &= \sum_{s=-(T-1)}^{T-1} e^{i\nu s}\sum_{t=1}^{T-|s|} u_t u_{t+|s|} \\ &\le \sum_{s=-(T-1)}^{T-1} \left|\sum_{t=1}^{T-|s|} u_t u_{t+|s|}\right|\end{aligned} \tag{82}$$

for every ν. The expectation is

$$\begin{aligned}\mathscr{E}\sum_{t=1}^{T} u_t^2 + 2\sum_{s=1}^{T-1}\mathscr{E}\left|\sum_{t=1}^{T-s} u_t u_{t+s}\right| &\le T\sigma^2 + 2\sum_{s=1}^{T-1}\left[\mathscr{E}\left(\sum_{t=1}^{T-s} u_t u_{t+s}\right)^2\right]^{\frac{1}{2}} \\ &= T\sigma^2 + 2\sum_{s=1}^{T-1}\left[\mathscr{E}\sum_{t,t'=1}^{T-s} u_t u_{t+s} u_{t'} u_{t'+s}\right]^{\frac{1}{2}} \\ &= T\sigma^2 + 2\sum_{s=1}^{T-1}[\sigma^4(T-s)]^{\frac{1}{2}} \\ &= \sigma^2\left\{T + 2\sum_{r=1}^{T-1} r^{\frac{1}{2}}\right\} \\ &\le \sigma^2\left\{T + 2\int_1^T r^{\frac{1}{2}}\,dr\right\} \\ &\le \sigma^2\{T + \tfrac{4}{3}T^{\frac{3}{2}}\}.\end{aligned} \tag{83}$$

Hence

$$\mathscr{E}\max_{0\le\nu\le\pi} R_u^2(\nu) \le 4\sigma^2\left\{\frac{1}{T} + \frac{4}{3}\frac{1}{\sqrt{T}}\right\}. \tag{84}$$

The theorem follows from this result and the generalized Tchebycheff inequality.

Whittle (1959) has shown under the further assumption that $\mathscr{E}|u_t|^{4+\delta} < \infty$ for some $\delta > 0$ that

$$\operatorname*{plim}_{T\to\infty} \frac{T}{\log T}\max_{0\le\nu\le\pi} R_u^2(\nu) = 4\sigma^2. \tag{85}$$

It follows from (57) of Section 4.3 that when the u_t's are normally distributed

$$\operatorname*{plim}_{T\to\infty} \frac{T}{\log T} \max_{j=1,\ldots,[\frac{1}{2}(T-1)]} R_u^2\left(\frac{2\pi j}{T}\right) = 4\sigma^2. \tag{86}$$

It is interesting that the maximum of $R_u^2(\nu)$ over all real numbers ν $(0 \leq \nu \leq \pi)$ is stochastically similar to the maximum over $\nu = 2\pi j/T, j = 1, \ldots, [\frac{1}{2}(T-1)]$.

4.5. DISCUSSION

Trend functions that are periodic are of importance in many disciplines—economics, meteorology, communications, and astronomy, among others. Frequently these periodic functions are trigonometric functions or may be expressed as linear combinations of such. In this chapter we have considered some of the problems of statistical inference about such trend functions.

The observed time series is described and analyzed in terms of the linear combinations of observed values with trigonometric functions of time as coefficients, that is, the sample trigonometric coefficients, $A(\nu)$ and $B(\nu)$. As we shall see in Chapter 8, these coefficients are used in the spectral analysis of the covariance structure of stationary processes.

The basic practical shortcoming of the procedures developed in this chapter is that they assume that the random terms are uncorrelated as well as having constant variance. In most applications, however, the effect of time enters into the structures of both the stochastic part and the systematic part. A second deficiency of the treatment in this chapter is that the methods amenable to precise mathematical treatment are limited and do not apply to all of the practical problems.

An interesting time series, particularly from a historical point of view, is the Beveridge series of wheat prices in Western and Central Europe from 1500 to 1869. The index is made up from prices in nearly 50 places in various countries. Beveridge (1921) gives this index (adjusted so the mean for 1700 to 1745 is 100) but uses for analysis a series which is derived by expressing the index for a given year as a percentage of the mean of the 31 years for which it is the center. This latter series, which is considered trend-free, is given in Table A.1.1 of Appendix A.1. Later Beveridge (1922) gave the periodogram computed by using $NR^2(2\pi k/N)/300$, where N is an integer not greater than 356 and k/N is chosen to give a large number of frequencies. The sequence of years usually started with 1545. These results are given in Table A.1.3 of Appendix A.1. Beveridge found 19 periods worthy of discussion. While this analysis is interesting, the error model does not seem appropriate because it is hard to justify the underlying idea of a cyclical trend made up of many trigonometric terms which goes on for over 300 years. As we shall see later, the model of a stationary process seems more appropriate and the spectrogram underlies the corresponding statistical treatment.

REFERENCES

Section 4.2. Whittaker and Watson (1943).

Section 4.3. T. W. Anderson (1958), Birnbaum (1959), (1961), Buijs-Ballot (1847), Cooley, Lewis, and Welch (1967), Cooley and Tukey (1965), Davis (1941), Feller (1968), Fisher (1929), (1940), Irwin (1955), Karlin and Truax (1960), Kendall (1946b), Kudo (1960), Lehmann (1957), Runge (1903), Scheffé (1970), Schuster (1898), Stevens (1939), United States Department of Agriculture (1939), G. T. Walker (1914), Whittle (1951), (1952).

Section 4.4. M. M. Rao (1960), A. M. Walker (1965), (1968), Whittle (1952), (1959).

Section 4.5. Beveridge (1921), (1922).

PROBLEMS

1. (Sec. 4.2.1.) Using $e^{i\lambda} = \cos\lambda + i\sin\lambda$, prove

(a) $\cos(a-b) = \cos a\cos b + \sin a\sin b$,

(b) $\cos(a+b) = \cos a\cos b - \sin a\sin b$,

(c) $\sin(a+b) = \sin a\cos b + \cos a\sin b$,

(d) $\sin(a-b) = \sin a\cos b - \cos a\sin b$.

2. (Sec. 4.2.1.) Verify

(a) $\cos a\cos b = \frac{1}{2}\cos(a-b) + \frac{1}{2}\cos(a+b)$,

(b) $\cos^2 a = \frac{1}{2} + \frac{1}{2}\cos(2a)$,

(c) $\sin a\sin b = \frac{1}{2}\cos(a-b) - \frac{1}{2}\cos(a+b)$,

(d) $\sin^2 a = \frac{1}{2} - \frac{1}{2}\cos(2a)$,

(e) $\sin a\cos b = \frac{1}{2}\sin(a+b) + \frac{1}{2}\sin(a-b)$.

3. (Sec. 4.2.1.) Prove (11).

4. (Sec. 4.2.2.) Write out the matrix $\boldsymbol{M}$ for $T = 4$ numerically (using $\sqrt{2}$). Graph the four functions.

5. (Sec. 4.2.2.) Write out the matrix $\boldsymbol{M}$ for $T = 6$ numerically. Graph the six functions.

6. (Sec. 4.2.2.) Let $\boldsymbol{N} = (e^{i2\pi st/T}/\sqrt{T})$ and define $\boldsymbol{M}$ by (14) and (21). Find $\boldsymbol{P}$ so $\boldsymbol{N} = \boldsymbol{MP}$. Show $\boldsymbol{N}\bar{\boldsymbol{N}}' = \boldsymbol{I}$.

7. (Sec. 4.2.3.) Give explicitly the twelve functions needed to represent a periodic sequence of period 12.

8. (Sec. 4.2.4.) Prove (36) to (39) by the method of (40).

9. (Sec. 4.2.4.) Prove (35) to (39) by the analogue of (8).

10. (Sec. 4.3.3.) Verify that (28) is the distribution of (27).

11. (Sec. 4.3.3.) Show that the transformation (29) induces the transformation of estimates described in the preceding sentence.

12. (Sec. 4.3.3.) Show that (29) is

$$y_t^* = ky_t + c_0 + \sum_{\substack{i=1 \\ i \neq j}}^{q} \left(c_i \cos \frac{2\pi k_i}{T} t + d_i \sin \frac{2\pi k_i}{T} t \right)$$

$$+ \frac{4}{T} \sin \tfrac{1}{2}\theta \sum_{s=1}^{T} \sin \left[\frac{2\pi k_j}{T} (s - t) - \tfrac{1}{2}\theta \right] ky_s + c_{\frac{1}{2}T}(-1)^t, \quad t = 1, \ldots, T.$$

13. (Sec. 4.3.3.) Show that the test $F \geq F_{2, T-p}(\varepsilon)$ is the uniformly most powerful test of H: $\rho(k_j) = 0$ among tests whose power depends only on $\rho^2(k_j)/\sigma^2$.

14. (Sec. 4.3.3.) At significance level ε indicate the t-tests for testing the null hypothesis $\alpha_0 = \alpha_0^*$, where α_0^* is a specified number, against alternatives (a) $\alpha_0 > \alpha_0^*$, (b) $\alpha_0 < \alpha_0^*$, and (c) $\alpha_0 \neq \alpha_0^*$.

15. (Sec. 4.3.3.) At significance level ε indicate the t-tests for testing the null hypothesis $\alpha_{\frac{1}{2}T} = 0$ against alternatives (a) $\alpha_{\frac{1}{2}T} > 0$, (b) $\alpha_{\frac{1}{2}T} < 0$, and (c) $\alpha_{\frac{1}{2}T} \neq 0$.

16. (Sec. 4.3.3.) Find a group of transformations for which (30) [or (36)] and $\bar{y}$ are the maximal invariants and that leaves invariant the problem for which (30) is the test criterion.

17. (Sec. 4.3.3.) Express $\alpha(kh)$ and $\beta(kh)$ in (31) as sums involving $f(t)$, $t = 1, \ldots,$ $T = hn$, and as sums involving $f(t)$, $t = 1, \ldots, n$.

18. (Sec. 4.3.3.) Prove that (30) and (36) are the same.

19. (Sec. 4.3.3.) Prove that (31) and (37) are the same.

20. (Sec. 4.3.3.) Show how the endpoints of the intervals of (42) and (43) with $F_{1,T-p}^{(\varepsilon)}$ replaced by $2F_{2,T-p}^{(\varepsilon)}$ are the slopes of the lines through the origin tangent to the circle (38).

21. (Sec. 4.3.3.) Prove that the expected value of $a(k_j) \cos \phi + b(k_j) \sin \phi$ is $\rho(k_j) \cos [\phi - \theta(k_j)]$.

22. (Sec. 4.3.3.) Find a two-sided confidence interval for $\rho(k_j) \cos [\phi - \theta(k_j)]$ with confidence $1 - \varepsilon$ when σ^2 is unknown.

23. (Sec. 4.3.3.) Show that $\sqrt{T}[a_0 - \alpha_0]$, $\sqrt{T}[a_{\frac{1}{2}T} - \alpha_{\frac{1}{2}T}]$, $\sqrt{T}[a(k_j) - \alpha(k_j)]$, $\sqrt{T}[b(k_j) - \beta(k_j)]$, $j = 1, \ldots, q$, have a joint limiting normal distribution as $T \to \infty$ if the y_t's are independently distributed with variance σ^2 and uniformly bounded moments of order $2 + \delta$ for some $\delta > 0$. [*Hint:* Use Corollary 2.6.2.]

24. (Sec. 4.3.4.) Prove the equivalence of the forms of $k(z \mid \tau^2)$ in the first two lines of (49) by use of the duplication formula for the gamma function $\sqrt{\pi}\,\Gamma(2\beta + 1) = 2^{2\beta}\Gamma(\beta + \tfrac{1}{2})\Gamma(\beta + 1)$.

25. (Sec. 4.3.4.) Prove that the noncentral χ^2-distribution with 2 degrees of freedom [with density given by (49)] is boundedly complete; that is, that

$$\int_0^\infty g(z)k(z \mid \tau^2)\, dz \equiv 0$$

identically in $\tau^2 > 0$ implies $g(z) = 0$ almost everywhere ($z \geq 0$) for every bounded function $g(z)$. [*Hint:* The above identity can be written

$$0 \equiv \sum_{\gamma=0}^{\infty} c_\gamma\, (\tau^2)^\gamma \int_0^\infty e^{-\frac{1}{2}z} g(z) z^\gamma\, dz,$$

where $c_\gamma > 0$.]

26. (Sec. 4.3.4.) Prove that the product of $q - 1$ noncentral χ^2-distributions each with 2 degrees of freedom is boundedly complete.

27. (Sec. 4.3.4.) Let x_1, x_2, x_3 ($x_1 + x_2 + x_3 = 1, x_i \geq 0$) be barycentric coordinates (distances from sides of an equilateral triangle with altitude 1). (a) Graph max $x_j \geq g$, $\frac{1}{3} \leq g \leq \frac{1}{2}$ and $\frac{1}{2} \leq g \leq 1$. (b) Graph second largest $x_j \geq g$, $0 \leq g \leq \frac{1}{3}$ and $\frac{1}{3} \leq g \leq \frac{1}{2}$.

28. (Sec. 4.3.4.) Let x_1, x_2, x_3 ($x_1 + x_2 + x_3 = 1, x_i \geq 0$) be barycentric coordinates. Graph the regions R_0, R_1, R_2, R_3, R_{12}, R_{13}, R_{23} defined by

(i) R_0: $x_i \leq g$, $\quad i = 1, 2, 3,$

(ii) R_i: $x_i > g$, $\quad x_j \leq g^* x_k$, $\quad i \neq j \neq k \neq i,$

(iii) R_{ij}: $x_i > g$, $\quad x_j > g^* x_k$, $\quad i \neq j \neq k \neq i,$

with $g^* \geq 1$ and $g \geq \frac{1}{2}$.

29. (Sec. 4.4.1.) Prove Lemma 4.4.1.

30. (Sec. 4.4.1.) Verify (23).

31. (Sec. 4.4.1.) Verify (24).

32. (Sec. 4.4.1.) Verify (30).

33. (Sec. 4.4.2.) Show that

$$2\alpha_0^2 + 2\alpha_{\frac{1}{2}T}^2 = \frac{4\rho^2 \sin^2 \frac{1}{2}\lambda T}{T^2 \sin^2 \lambda}\{1 + \cos \lambda \cos [\lambda(T + 1) - 2\theta]\}.$$

34. (Sec. 4.4.2.) Show that for $\lambda = 2\pi(k + \varepsilon)/T$

$$\alpha_0 = \frac{\rho \sin \pi\varepsilon}{T \sin \pi \dfrac{k + \varepsilon}{T}} \cos\left(\pi\varepsilon + \pi\frac{k + \varepsilon}{T} - \theta\right)$$

$$\sim \rho \frac{\varepsilon}{k + \varepsilon} \frac{1 - \dfrac{\pi^2}{6}\varepsilon^2}{1 - \dfrac{\pi^2}{6} \cdot \dfrac{(k + \varepsilon)^2}{T^2}} \cos\left(\pi\varepsilon + \pi\frac{k + \varepsilon}{T} - \theta\right).$$

35. (Sec. 4.4.2.) Show that for $\lambda = 2\pi(k + \varepsilon)/T$

$$\alpha_{\frac{1}{2}T} = -\frac{\rho \sin \pi\varepsilon}{T \cos \pi \dfrac{k+\varepsilon}{T}} \sin\left(\pi\varepsilon + \pi\frac{k+\varepsilon}{T} - \theta\right)$$

$$\sim -\frac{\rho}{T}\pi\varepsilon\left(1 - \frac{\pi^2\varepsilon^2}{6}\right)\sin\left(\pi\varepsilon + \pi\frac{k+\varepsilon}{T} - \theta\right).$$

36. (Sec. 4.4.3.) Show

$$\frac{1}{\sin\left(\frac{1}{2}\nu + \dfrac{\pi j}{T}\right)} + \frac{1}{\sin\left(\frac{1}{2}\nu - \dfrac{\pi j}{T}\right)} = 4\frac{\sin\frac{1}{2}\nu\cos\dfrac{\pi j}{T}}{\cos\dfrac{2\pi j}{T} - \cos\nu},$$

$$\frac{1}{\sin\left(\frac{1}{2}\nu - \dfrac{\pi j}{T}\right)} - \frac{1}{\sin\left(\frac{1}{2}\nu + \dfrac{\pi j}{T}\right)} = 4\frac{\cos\frac{1}{2}\nu\sin\dfrac{\pi j}{T}}{\cos\dfrac{2\pi j}{T} - \cos\nu}.$$

37. (Sec. 4.4.3.) Show for $T = 2H + 1$

$$C(\nu) = 4\frac{\sin\frac{1}{2}\nu\sin\frac{1}{2}\nu T}{T}\left\{\frac{\bar{y}}{1-\cos\nu} + \sum_{j=1}^{H}\frac{c_j\cos\dfrac{\pi j}{T}}{\cos\dfrac{2\pi j}{T} - \cos\nu}\right\},$$

$$D(\nu) = 4\frac{\cos\frac{1}{2}\nu\sin\frac{1}{2}\nu T}{T}\sum_{j=1}^{H}\frac{d_j\sin\dfrac{\pi j}{T}}{\cos\dfrac{2\pi j}{T} - \cos\nu}.$$

38. (Sec. 4.4.3.) Show that for $K = \cos[\lambda(T+1) - 2\theta]$

$$\frac{d\rho^2(\nu)}{d\nu} = -\frac{\rho^2}{T^2}\left\{-\frac{T}{2}\frac{\sin(\lambda+\nu)T}{\sin^2\frac{1}{2}(\lambda+\nu)} + \frac{\sin^2\frac{1}{2}(\lambda+\nu)T\cos\frac{1}{2}(\lambda+\nu)}{\sin^3\frac{1}{2}(\lambda+\nu)}\right.$$

$$+\frac{T}{2}\frac{\sin(\lambda-\nu)T}{\sin^2\frac{1}{2}(\lambda-\nu)} - \frac{\sin^2\frac{1}{2}(\lambda-\nu)T\cos\frac{1}{2}(\lambda-\nu)}{\sin^3\frac{1}{2}(\lambda-\nu)}$$

$$\left. + KT\frac{\sin\nu T}{\sin\frac{1}{2}(\lambda+\nu)\sin\frac{1}{2}(\lambda-\nu)} - K\frac{\sin\frac{1}{2}(\lambda+\nu)T\sin\frac{1}{2}(\lambda-\nu)T\sin\nu}{\sin^2\frac{1}{2}(\lambda+\nu)\sin^2\frac{1}{2}(\lambda-\nu)}\right\},$$

$$\nu \neq \lambda,\ 0 < \nu < \pi.$$

39. (Sec. 4.4.3.) Show that for $d\rho^2(\nu)/d\nu$ defined in Problem 38

$$\lim_{\nu \to \lambda} \frac{d\rho^2(\nu)}{d\nu} = \rho^2 \frac{T \cos \lambda T \sin \lambda - \sin \lambda T \cos \lambda}{T \sin^2 \lambda} \left\{ \frac{\sin \lambda T}{T \sin \lambda} + K \right\}.$$

40. (Sec. 4.4.3.) Obtain an expression for $dQ(\lambda^*)/d\lambda^* = 0$.

41. (Sec. 4.4.4.) Prove (86) from (57) of Section 4.3.

CHAPTER 5

Linear Stochastic Models with Finite Numbers of Parameters

5.1. INTRODUCTION

We now turn to models for the generation of time series in which the characteristic and useful properties appropriate to the time sequence are not in a deterministic mean value function but in the probability structure itself. In these cases, for example, there are not regular periodic cycles but more or less irregular fluctuations that have statistical properties of variability. These models are generally called *stochastic processes.* A process whose probability structure does not change with time is called *stationary*. (We shall treat general stationary stochastic processes more precisely and in more detail in Chapter 7.) In this book we are mainly interested in processes that are stationary or almost stationary, or such that at least the probability aspect (as distinguished from a deterministic mean value function) is roughly stationary.

One of the simplest models of this type, and perhaps the most useful, is the *autoregressive process* or the *stochastic difference equation.* A sequence of random variables $y_1, y_2, \ldots$ is said to satisfy a stochastic difference equation if there is a linear combination

$$y_t + \beta_1 y_{t-1} + \cdots + \beta_p y_{t-p} = u_t, \qquad t = p + 1, \ldots, \tag{1}$$

such that the sequence $u_{p+1}, u_{p+2}, \ldots$ consists of independently and identically distributed random variables. (Often we shall assume $\mathscr{E}u_t = 0$.) It is convenient to generalize this definition to a doubly-infinite sequence $\ldots, y_{-1}, y_0, y_1, \ldots$ resulting in a doubly-infinite sequence $\ldots, u_{-1}, u_0, u_1, \ldots$. These processes are also known as autoregressive processes.

We can also write (1) using the operator $\mathscr{P}$ (where $\mathscr{P}y_t = y_{t+1}$) as

$$(\mathscr{P}^p + \beta_1 \mathscr{P}^{p-1} + \cdots + \beta_p)y_{t-p} = u_t. \tag{2}$$

Since $\mathscr{P} = \Delta + 1$, the operator acting on y_{t-p} in (2) can also be written as a polynomial of degree p in Δ. Hence, the left-hand side of (1) can be written as a linear combination of $y_{t-p}, \Delta y_{t-p}, \ldots, \Delta^p y_{t-p}$, and (1) is called a stochastic difference equation of *degree p* when $\beta_p \neq 0$; the process is said to be an autoregressive process of *order p*. The difference operator Δ was discussed in Section 3.4.

The stochastic difference equation (1) can be thought of in another way. If u_t is distributed independently of $y_{t-1}, y_{t-2}, \ldots$, then the conditional distribution of y_t, given $y_{t-1}, y_{t-2}, \ldots$, is the distribution of $u_t - (\beta_1 y_{t-1} + \cdots + \beta_p y_{t-p})$. Thus the joint distribution of, say, $y_t, y_{t-1}, \ldots, y_1$ $(t > p)$ can be derived from the distribution of u_t and the joint distribution of $y_{t-1}, \ldots, y_1$. If we are given the distribution of u_t (identical for all t) and the joint distribution of any p successive y_t's, then we can use this procedure successively to determine the joint distribution of these and any finite number of succeeding y_t's. In the next section we shall see conditions under which this procedure can determine a process which is stationary (and for which u_t is independent of $y_{t-1}, y_{t-2}, \ldots$).

This model is useful because it can generate a wide variety of processes. The effect of a trend can easily be incorporated by adding to the left-hand side of (1) $\sum_i \gamma_i z_{it}$, where the z_{it}'s are known quantities (functions of time). We study some properties of these models in Section 5.2.

Many of the problems of statistical inference have to do with the finite number of parameters $\beta_1, \ldots, \beta_p$ and the variance of u_t. These problems include estimation and testing hypotheses. When the length of the series observed is large, these problems of inference can be treated approximately according to the least squares theory reviewed in Chapter 2. The related small sample theory of serial correlation coefficients is treated in Chapter 6.

Another simple model of a stationary stochastic process is the moving average

$$y_t = \alpha_0 v_t + \alpha_1 v_{t-1} + \cdots + \alpha_q v_{t-q}, \tag{3}$$

where the v_t's are independently and identically distributed. This model and the former can be combined by assuming the left-hand side of (1) is equal to the right-hand side of (3); that is, the disturbance u_t in (1) is the moving average defined by the right-hand side of (3). Any of these models can be further modified by assuming that what is observed is the process generated by one of these models plus an independent random "error"; that is, the observed sequence is $\{x_t\}$, where $x_t = y_t + w_t$ and $\{w_t\}$ is a sequence of independently and identically distributed random variables. Inference in all of these models is more complicated (even asymptotically) than in the case of the autoregressive process.

If all joint distributions are normal, then the joint normal distribution of the y_t's is determined by the means, variances, and covariances of these variables. For stationary processes these are determined by the mean and variance of u_t and the β's in the case of (1), the mean and variance of v_t and the α's in the case of (3), and the mean and variance of v_t and the β's and α's in the case of the combined model. Sometimes the above definitions are said to hold in the *wide sense* if the first- and second-order moments agree with the respective definitions; for example, $\{y_t\}$ satisfies a stochastic difference equation in the wide sense if the u_t's defined by (1) have $\mathscr{E}u_t = 0$, $\mathscr{E}u_t^2 = \sigma^2$, and $\mathscr{E}u_t u_s = 0$, $t \neq s$. The moving average and combined models may similarly be defined in the wide sense in terms of the v_t's.

Yule (1927) suggested the use of the autoregressive process for time series analysis and applied it to sunspot data. Gilbert Walker (1931) extended the theory and applied it to atmospheric data.

5.2. AUTOREGRESSIVE PROCESSES

5.2.1. Representation of the Time Series as an Infinite Moving Average

In the introduction to this chapter we indicated how the joint distribution of some p consecutive y_t's and the identical distribution of the (independently distributed) u_t's together with

$$\sum_{r=0}^{p} \beta_r y_{t-r} = u_t, \qquad \beta_0 = 1, \tag{1}$$

determine the joint distribution of succeeding y_t's. Now we show how any y_t is expressed as a linear combination of any earlier p successive y_r's and earlier u_r's. Under certain conditions, which we shall specify, any y_t can be written as an infinite linear combination of u_t and earlier u_r's.

A particularly simple case is the first-order equation

$$y_t = \rho y_{t-1} + u_t. \tag{2}$$

If we replace y_{t-1} in (2) by the right-hand side of (2) with t replaced by $t-1$ (that is, $y_{t-1} = \rho y_{t-2} + u_{t-1}$) we obtain

$$y_t = u_t + \rho u_{t-1} + \rho^2 y_{t-2}. \tag{3}$$

If we substitute successively, we obtain

$$y_t = u_t + \rho u_{t-1} + \cdots + \rho^s u_{t-s} + \rho^{s+1} y_{t-(s+1)}. \tag{4}$$

Thus

$$y_t - (u_t + \rho u_{t-1} + \cdots + \rho^s u_{t-s}) = \rho^{s+1} y_{t-(s+1)}. \tag{5}$$

If $\{y_t\}$ is a doubly-infinite stationary process, the difference between y_t and the linear combination of $s + 1$ u_r's is $\rho^{s+1} y_{t-(s+1)}$ and this becomes small when s increases if $|\rho| < 1$. In particular (if second-order moments exist)

$$\mathscr{E}[y_t - (u_t + \rho u_{t-1} + \cdots + \rho^s u_{t-s})]^2 = \rho^{2(s+1)} \mathscr{E} y_{t-(s+1)}^2, \tag{6}$$

which does not depend on t because of the assumed stationarity. As s increases, (6) goes to 0. Then we write

$$y_t = \sum_{r=0}^{\infty} \rho^r u_{t-r} \tag{7}$$

and say that the series on the right *converges in the mean* (or in the mean square or in quadratic mean) to y_t. (See Section 7.6.1 for further discussion of convergence in the mean.)

Now let us turn to the general case. We can write (1) as

$$y_t = u_t - \beta_1 y_{t-1} - \cdots - \beta_p y_{t-p}; \tag{8}$$

at one time unit earlier

$$y_{t-1} = u_{t-1} - \beta_1 y_{t-2} - \cdots - \beta_p y_{t-1-p}. \tag{9}$$

Substituting (9) into (8), we obtain

$$\begin{aligned} y_t &= u_t - \beta_1(u_{t-1} - \beta_1 y_{t-2} - \cdots - \beta_p y_{t-1-p}) - \beta_2 y_{t-2} - \cdots - \beta_p y_{t-p} \\ &= u_t - \beta_1 u_{t-1} - (\beta_2 - \beta_1^2) y_{t-2} - \cdots + \beta_1 \beta_p y_{t-1-p}. \end{aligned} \tag{10}$$

After following this procedure s times we arrive at the expression

$$y_t = u_t + \delta_1^* u_{t-1} + \cdots + \delta_s^* u_{t-s} + \alpha_{s1}^* y_{t-s-1} + \alpha_{s2}^* y_{t-s-2} + \cdots + \alpha_{sp}^* y_{t-s-p}. \tag{11}$$

(Each substitution leaves p consecutive y_r's on the right.) Then substitution of

$$y_{t-s-1} = u_{t-s-1} - \beta_1 y_{t-s-2} - \cdots - \beta_p y_{t-s-p-1} \tag{12}$$

into (11) yields

$$(13)\quad y_t = u_t + \delta_1^* u_{t-1} + \cdots + \delta_s^* u_{t-s} + \alpha_{s1}^* u_{t-s-1} + (\alpha_{s2}^* - \alpha_{s1}^*\beta_1) y_{t-s-2} + \cdots + (\alpha_{sp}^* - \alpha_{s1}^*\beta_{p-1}) y_{t-s-p} - \alpha_{s1}^*\beta_p y_{t-s-p-1}.$$

Thus

$$(14)\quad \begin{aligned} \delta_{s+1}^* &= \alpha_{s1}^*, \\ \alpha_{s+1,j}^* &= \alpha_{s,j+1}^* - \alpha_{s1}^*\beta_j, \qquad j = 1, \ldots, p-1, \\ \alpha_{s+1,p}^* &= -\alpha_{s1}^*\beta_p. \end{aligned}$$

These are a set of recursion relations for the coefficients. Continuation of this procedure yields

$$(15)\quad y_t = \sum_{i=0}^{\infty} \delta_i^* u_{t-i},$$

where $\delta_0^* = 1$. We shall see what the conditions are for (15) to converge in the mean. In any case (11) through (14) hold for each s, which shows that the joint distribution of some p consecutive y_t's and the subsequent u_t's determines the joint distribution of subsequent y_t's.

The procedure can be put into formal terms. Let $\mathscr{L}$ be the lag operator; that is,

$$(16)\quad \mathscr{L} y_t = y_{t-1}.$$

The stochastic difference equation (1) can be written

$$(17)\quad \sum_{r=0}^{p} \beta_r \mathscr{L}^r y_t = u_t.$$

Then formally

$$(18)\quad y_t = \left(\sum_{r=0}^{p} \beta_r \mathscr{L}^r\right)^{-1} u_t,$$

where

$$(19)\quad \left(\sum_{r=0}^{p} \beta_r \mathscr{L}^r\right)^{-1} = \sum_{r=0}^{\infty} \delta_r \mathscr{L}^r;$$

that is, the δ_r's are the coefficients in

$$(20)\quad \left(\sum_{r=0}^{p} \beta_r z^r\right)^{-1} = \sum_{r=0}^{\infty} \delta_r z^r$$

and can be determined by formal long division.

Now let us verify that $\delta_r^* = \delta_r$. We have

$$
\begin{aligned}
(21)\quad \frac{1}{1+\beta_1 z+\cdots+\beta_p z^p}
&= 1 - \frac{\beta_1 z+\cdots+\beta_p z^p}{1+\beta_1 z+\cdots+\beta_p z^p} \\
&= 1 - \beta_1 z - \frac{(\beta_2-\beta_1^2)z^2+\cdots+(\beta_p-\beta_1\beta_{p-1})z^p-\beta_1\beta_p z^{p+1}}{1+\beta_1 z+\cdots+\beta_p z^p}.
\end{aligned}
$$

It will be observed that the coefficient of z^j in the numerator of the right-hand side of (21) is the coefficient of y_{t-j} in (10), $j = 2, \ldots, p+1$. If we continue, we obtain

$$
\begin{aligned}
(22)\quad \frac{1}{1+\beta_1 z+\cdots+\beta_p z^p}
&= 1+\delta_1 z+\cdots+\delta_s z^s + \frac{\alpha_{s1}z^{s+1}+\cdots+\alpha_{sp}z^{s+p}}{1+\beta_1 z+\cdots+\beta_p z^p} \\
&= 1+\delta_1 z+\cdots+\delta_s z^s+\alpha_{s1}z^{s+1} \\
&\quad + \frac{(\alpha_{s2}-\alpha_{s1}\beta_1)z^{s+2}+\cdots+(\alpha_{sp}-\alpha_{s1}\beta_{p-1})z^{s+p}-\alpha_{s1}\beta_p z^{s+p+1}}{1+\beta_1 z+\cdots+\beta_p z^p}.
\end{aligned}
$$

Thus the δ_r's and α_{si}'s satisfy the same recursion relations as the δ_r^*'s and α_{si}^*'s as well as the same initial conditions. Hence, $\delta_r^* = \delta_r$, and $\alpha_{si}^* = \alpha_{si}$. The equation

$$(23)\qquad \sum_{r=0}^{p} \beta_r x^{p-r} = 0$$

is called the *associated polynomial equation* to (1). There are p roots, say $x_1, \ldots, x_p$. If $|x_i| < 1$, $i = 1, \ldots, p$, then the roots of

$$(24)\qquad \sum_{r=0}^{p} \beta_r z^r = 0$$

are $z_i = 1/x_i$ when $\beta_p \neq 0$ and $|z_i| > 1$. For any z such that $|z| < \min |z_i|$ the series

$$
\begin{aligned}
(25)\quad \frac{1}{\sum_{r=0}^{p}\beta_r z^r} = \frac{1}{\prod_{i=1}^{p}\left(1-\dfrac{z}{z_i}\right)} &= \prod_{i=1}^{p}\sum_{\nu=0}^{\infty}\left(\frac{z}{z_i}\right)^{\nu} \\
&= \sum_{r=0}^{\infty}\delta_r z^r
\end{aligned}
$$

converges (absolutely). Hence, we see from (22) that

$$\frac{\alpha_{s1}z + \cdots + \alpha_{sp}z^p}{1 + \beta_1 z + \cdots + \beta_p z^p} \tag{26}$$

converges to 0 for $|z| < \min |z_i|$ (in particular for $|z| = 1$). This implies $\alpha_{si} \to 0$ (as $s \to \infty$) for each i. Thus

$$\mathscr{E}\left(y_t - \sum_{r=0}^{s} \delta_r u_{t-r}\right)^2 = \mathscr{E}(\alpha_{s1}y_{t-s-1} + \cdots + \alpha_{sp}y_{t-s-p})^2 \tag{27}$$

converges to 0. We have

$$y_t = \sum_{r=0}^{\infty} \delta_r u_{t-r} \tag{28}$$

in the sense of convergence in the mean.

THEOREM 5.2.1. *If all the roots of the polynomial equation* (23) *associated with the stochastic difference equation* (1) *are less than* 1 *in absolute value,* y_t *can be written as an infinite linear combination of* $u_t, u_{t-1}, \ldots$.

COROLLARY 5.2.1. *If the roots are all less than* 1 *in absolute value,* y_t *is independent of* $u_{t+1}, u_{t+2}, \ldots$.

PROOF. y_t is a linear combination of $u_t, u_{t-1}, \ldots$ and these are independent of $u_{t+1}, u_{t+2}, \ldots$. Q.E.D.

Let us briefly look at the case in which some roots of the polynomial equation are larger than 1 in absolute value. Suppose $|x_i| > 1$, $i = 1, \ldots, q$, $|x_i| < 1$, $i = q + 1, \ldots, p$. We write the stochastic difference equation as

$$\begin{aligned} u_t &= \sum_{r=0}^{p} \beta_r \mathscr{P}^{p-r} y_{t-p} \\ &= \prod_{i=1}^{p} (\mathscr{P} - x_i) y_{t-p}. \end{aligned} \tag{29}$$

The inverse of (29) is

$$\begin{aligned} y_{t-p} &= \prod_{i=1}^{p} (\mathscr{P} - x_i)^{-1} u_t \\ &= \prod_{i=1}^{q} (\mathscr{P} - x_i)^{-1} \prod_{i=q+1}^{p} [\mathscr{P}(1 - x_i \mathscr{L})]^{-1} u_t \\ &= \prod_{i=1}^{q} \left[\left(-\frac{1}{x_i}\right)\left(1 - \frac{1}{x_i}\mathscr{P}\right)^{-1}\right] \prod_{i=q+1}^{p} [\mathscr{L}(1 - x_i \mathscr{L})^{-1}] u_t, \end{aligned} \tag{30}$$

since $\mathscr{L} = \mathscr{P}^{-1}$. Each term $(1 - x_i^{-1}\mathscr{P})^{-1}$ can be expanded in a power series in $\mathscr{P}$, and each term $(1 - x_i\mathscr{L})^{-1}$ can be expanded in a power series in $\mathscr{L}$. If $1 \le q < p$, the operator will be a power series in $\mathscr{L}(= \mathscr{P}^{-1})$ and in $\mathscr{P}(= \mathscr{L}^{-1})$; the indicated infinite series in $\ldots, u_{t-1}, u_t, u_{t+1}, \ldots$ will be doubly infinite. If $q = p$ (all roots in absolute value greater than 1), the operator is an infinite power series in $\mathscr{P}$ (only nonnegative powers); the indicated infinite series in the random variables involves $u_t, u_{t+1}, \ldots$. This discussion is in purely formal terms, but it can be justified in the manner used for the case of all roots less than 1 in absolute value. We shall see in Chapter 7 (Problem 22) that if the linear form $\prod_{i=1}^{p} (\mathscr{P} - x_i)y_{t-p}$ is replaced by the linear form (for any x_i's)

$$\prod_{i=1}^{r} (\mathscr{P} - x_i^{-1}) \prod_{i=r+1}^{p} (\mathscr{P} - x_i)y_{t-p} = u_t^* \tag{31}$$

the residuals u_t^* are also uncorrelated (but not necessarily independent).

Now consider the special case of one root equal to 1

$$y_t = y_{t-1} + u_t, \tag{32}$$

and we assume $\mathscr{E}u_t = 0$, $\mathscr{E}u_t^2 = \sigma^2$. Then

$$y_t - y_{t-s} = u_t + u_{t-1} + \cdots + u_{t-s+1}, \tag{33}$$

and

$$\mathscr{E}(y_t - y_{t-s})^2 = \mathscr{E}y_t^2 + \mathscr{E}y_{t-s}^2 - 2\mathscr{E}y_t y_{t-s} = s\sigma^2. \tag{34}$$

If the process is stationary, $\mathscr{E}y_t^2 = \mathscr{E}y_{t-s}^2$; from (34) we obtain

$$\mathscr{E}y_t y_{t-s} = \mathscr{E}y_t^2 - \tfrac{1}{2}s\sigma^2, \qquad s = 1, 2, \ldots. \tag{35}$$

This can hold for all $s > 0$ only if $\sigma^2 = 0$ and then $y_t = y_{t-s}$ with probability 1. Intuitively, we see that unless the variance of u_t is 0, (32) implies that the variance of y_t increases with t, but this fact is contrary to stationarity.

A more general case is

$$(\mathscr{P} - 1) \prod_{i=2}^{p} (\mathscr{P} - x_i)y_{t-p} = u_t, \tag{36}$$

where $|x_i| \neq 1$, $i = 2, \ldots, p$. If $\prod_{i=2}^{p}(\mathscr{P} - x_i)y_{t-p} = z_{t-1}$, then $(\mathscr{P} - 1)z_{t-1} = u_t$ and $z_t = z_{t-s}$ with probability 1, say $z_t = z$. Thus

$$\prod_{i=2}^{p} (\mathscr{P} - x_i)y_{t-p} = z, \tag{37}$$

and $y_t = \sum_{s=-\infty}^{\infty} \delta_s z$; that is, $y_t = y_{t-s}$ with probability 1. (z can be a random variable.)

THEOREM 5.2.2. *If a stationary stochastic process satisfies a stochastic difference equation for which at least one root is* 1, *all values of the process are the same with probability* 1.

From now on we shall only consider the case in which all roots of (23) are less than 1 in absolute value. Then y_t is independent of $u_{t+1}, u_{t+2}, \ldots$, and can be expressed as $\sum_{r=0}^{\infty} \delta_r u_{t-r}$. We now turn our attention to the coefficients δ_r. From (25) we see

$$(38) \qquad 1 = \left(\sum_{r=0}^{p} \beta_r z^r\right)^{-1} \sum_{s=0}^{p} \beta_s z^s = \sum_{r=0}^{\infty} \delta_r z^r \sum_{s=0}^{p} \beta_s z^s$$

$$= \sum_{r=0}^{\infty} \sum_{s=0}^{p} \beta_s \delta_r z^{s+r}$$

$$= \sum_{t=0}^{p-1} \left(\sum_{s=0}^{t} \beta_s \delta_{t-s}\right) z^t + \sum_{t=p}^{\infty} \left(\sum_{s=0}^{p} \beta_s \delta_{t-s}\right) z^t,$$

where we have replaced r by $t - s$. Since this is an identity in z (the series converging uniformly for $|z| \leq 1$), the coefficient of z^0 on the right is 1 and the coefficients of the positive powers are 0; that is,

$$(39) \qquad \begin{aligned} 1 &= \beta_0 \delta_0 = \delta_0, \\ 0 &= \beta_0 \delta_1 + \beta_1 \delta_0 = \delta_1 + \beta_1, \\ &\cdot \\ &\cdot \\ &\cdot \\ 0 &= \beta_0 \delta_{p-1} + \cdots + \beta_{p-1} \delta_0, \end{aligned}$$

$$(40) \qquad 0 = \beta_0 \delta_t + \cdots + \beta_p \delta_{t-p}, \qquad t = p, p+1, \ldots .$$

Equation (40) is the homogeneous difference equation corresponding to the nonhomogeneous difference equation (1). If the roots of the associated polynomial equation (23) are different, the general solution of the homogeneous difference equation (40) is

$$(41) \qquad \delta_r = \sum_{i=1}^{p} k_i x_i^r, \qquad r = 0, 1, \ldots .$$

If x_i is real, the coefficient k_i is real; if x_i and x_{i+1} are conjugate complex, then k_i and k_{i+1} are conjugate complex and $k_i x_i^r + k_{i+1} x_{i+1}^r$ is real, $r = 0, 1, \ldots$. In the case of multiple roots the general solution can be assembled from Problems 9 and 10.

Equations (39) give p boundary conditions. From these p equations one can determine $\delta_0, \delta_1, \ldots, \delta_{p-1}$ (by solving the first equation for δ_0, then the second for δ_1, etc.). Then (41) for $r = 0, 1, \ldots, p - 1$ are a set of p linearly independent linear equations in the p unknowns $k_1, \ldots, k_p$, which determine them uniquely. (See also Problem 4.)

If $p = 1$, $\delta_r = (-\beta_1)^r$, an exponential function of r. If $p = 2$ and x_1 and x_2 are different, $k_1 = x_1/(x_1 - x_2)$, $k_2 = -x_2/(x_1 - x_2)$, and

$$\delta_r = \frac{x_1^{r+1} - x_2^{r+1}}{x_1 - x_2}, \qquad r = 0, 1, \ldots . \tag{42}$$

If x_1 and x_2 are real, δ_r is a linear combination of two exponential functions of r. If the roots are complex, we may write them $x_1 = \alpha e^{i\theta}$ and $x_2 = \alpha e^{-i\theta}$. Then

$$k_1 = \frac{e^{i\theta}}{e^{i\theta} - e^{-i\theta}}, \qquad k_2 = -\frac{e^{-i\theta}}{e^{i\theta} - e^{-i\theta}}. \tag{43}$$

The coefficients are

$$\begin{aligned} \delta_r = k_1 x_1^r + k_2 x_2^r &= \alpha^r \frac{e^{i\theta(r+1)} - e^{-i\theta(r+1)}}{e^{i\theta} - e^{-i\theta}} \\ &= \alpha^r \frac{\sin \theta(r+1)}{\sin \theta}, \end{aligned} \tag{44}$$

a damped sine function of r. This is the same as (42) for complex conjugate roots.

5.2.2. Second-Order Moments; Covariance Function

If joint distributions are normal, they are characterized by the means, $\mathscr{E} y_t$, which we temporarily assume to be 0 for convenience, variances $\mathscr{E} y_t^2$, and covariances $\mathscr{E} y_t y_{t+s}$. If the joint distributions are not normal, these first- and second-order moments, nevertheless, give relevant and important information about the process. For example, the correlation between y_t and y_{t+s} (that is, $\mathscr{E} y_t y_{t+s}/\sqrt{\mathscr{E} y_t^2 \mathscr{E} y_{t+s}^2}$) is a measure of association between the two random variables.

If the process is stationary [in particular, if y_t can be represented by (28)], all variances are the same and covariances depend only on the difference between the two indices involved. These moments

$$\mathscr{E} y_t y_{t+s} = \sigma(s), \qquad s = \ldots, -1, 0, 1, \ldots, \tag{45}$$

constitute the *covariance function* (sometimes called correlation function). We call $\sigma(s)/\sigma(0)$ the *correlation function.*

We shall now show that the covariance function $\sigma(h)$ satisfies the homogeneous difference equation (40). Multiplying (1) by (28) with t replaced by $t - s$, we obtain

$$\sum_{r=0}^{p} \beta_r y_{t-r} y_{t-s} = \sum_{q=0}^{\infty} \delta_q u_t u_{t-s-q}. \tag{46}$$

Since $\mathscr{E} y_{t-r} y_{t-s} = \sigma(s - r)$, $\mathscr{E} u_t^2 = \sigma^2$, $\mathscr{E} u_t u_{t'} = 0$, $t \neq t'$, the expected values of the two sides of (46) satisfy for $s = 0$ and for $s > 0$

$$\sum_{r=0}^{p} \beta_r \sigma(-r) = \sigma^2, \tag{47}$$

$$\sum_{r=0}^{p} \beta_r \sigma(s - r) = 0, \qquad s = 1, 2, \ldots . \tag{48}$$

These are often called the *Yule-Walker equations*. Thus the sequence $\sigma(1 - p)$, $\sigma(2 - p), \ldots, \sigma(0), \ldots$ satisfies the homogeneous difference equation (48), and therefore

$$\sigma(h) = \sum_{i=1}^{p} c_i x_i^h, \qquad h = 1 - p, 2 - p, \ldots, \tag{49}$$

if the roots are distinct and $\beta_p \neq 0$ (so $x_i \neq 0$). There are $p - 1$ boundary conditions

$$\sigma(-h) = \sigma(h), \qquad h = 1, \ldots, p - 1. \tag{50}$$

The other boundary condition is (47) where $\sigma(-p)$ [which is not necessarily of the form (49)] is replaced by $\sigma(p)$. (This last condition simply determines a constant of proportionality.)

If $p = 1$, $\sigma(h) = (-\beta_1)^h \sigma^2/(1 - \beta_1^2)$, $h = 0, 1, \ldots$, an exponential function of h. If $p = 2$ and x_1 and x_2 are different,

$$\sigma(h) = \frac{\sigma^2}{(x_1 - x_2)(1 - x_1 x_2)} \left(\frac{x_1^{h+1}}{1 - x_1^2} - \frac{x_2^{h+1}}{1 - x_2^2} \right). \tag{51}$$

If x_1 and x_2 are real, $\sigma(h)$ is a linear combination of two exponential functions. If $p = 2$ and x_1 and x_2 are conjugate complex, $\alpha e^{\pm i\theta}$, then (51) can be written

$$\sigma(h) = \frac{\sigma^2 \alpha^h [\sin \theta(h + 1) - \alpha^2 \sin \theta(h - 1)]}{(1 - \alpha^2) \sin \theta \, [1 - 2\alpha^2 \cos 2\theta + \alpha^4]}, \tag{52}$$

a damped linear combination of sine functions of h.

Since $\sigma(h)$ is a linear combination of hth powers of the roots, all of which are less than 1 in absolute value, $|\sigma(h)|$ is bounded by an exponentially decreasing function; in fact, $|\sigma(h)| < K(\max_i |x_i|)^h$ for suitably chosen $K > 0$. The contribution of a positive root is a decreasing exponential function; the contribution of a negative root is an exponential function alternating in sign and decreasing in absolute value. The contribution of a pair of complex conjugate roots is an oscillating trigonometric function, decreasing in absolute value, whose period of oscillation depends on the argument of the complex roots.

5.2.3. Fluctuations of the Time Series

The typical time series generated by a stochastic difference equation model fluctuates up and down. Its oscillations are not regular, but tend to have an average length which depends on the difference equation. If we think of (1) as generating the series for successive values of t we see that each set of p y_t's directly affects the next. If $p = 2$, we can write the equation

$$\begin{aligned} y_t &= -\beta_1 y_{t-1} - \beta_2 y_{t-2} + u_t \\ &= -(\beta_1 + \beta_2) y_{t-1} + \beta_2 (y_{t-1} - y_{t-2}) + u_t, \end{aligned} \tag{53}$$

which indicates that the direct effect of preceding y_r's on y_t involves the value of y_{t-1} and the change between y_{t-2} and y_{t-1}. This effect will generally have a tendency to produce fluctuations.

Another way of looking at the process is in terms of the representation $y_t = \sum_{r=0}^{\infty} \delta_r u_{t-r}$. A given u_s will affect a subsequent y_{s+q} according to the coefficient δ_q. Since these coefficients oscillate, the effect on successive y_r's fluctuates, tending to produce fluctuations in the series y_t.

The fluctuating behavior is also indicated by the covariance function. Since $\sigma(s) \neq 0$ (usually), there is some statistical association between y_t and y_{t+s}, though it tends to decrease as s (> 0) increases. This association (measured by the correlation coefficient) tends to fluctuate. For example, if there is a pair of complex roots, the trigonometric functions oscillate and the association may increase with s in some intervals (of s).

5.2.4. Adding "Independent" Variables

If there are known external variables $z_{1t}, \ldots, z_{qt}$ which may affect the time series under investigation, their effect may be introduced into the model by replacing (1) by

$$\sum_{r=0}^{p} \beta_r y_{t-r} + \sum_{i=1}^{q} \gamma_i z_{it} = u_t. \tag{54}$$

In particular, a constant can be inserted (when $q = 1$ and $z_{1t} = 1$), and we can assume $\mathscr{E}u_t = 0$.

We can also write (54) as

$$u_t = \prod_{j=1}^{p} (\mathscr{P} - x_j) y_{t-p} + \sum_{i=1}^{q} \gamma_i z_{it} \tag{55}$$

$$= \prod_{j=1}^{p} (\mathscr{P} - x_j) \left[y_{t-p} + \sum_{i=1}^{q} \gamma_i \prod_{j=1}^{p} (\mathscr{P} - x_j)^{-1} z_{it} \right]$$

$$= \prod_{j=1}^{p} (\mathscr{P} - x_j) \left[y_{t-p} + \sum_{i=1}^{q} \gamma_i \sum_{s=0}^{\infty} \delta_s z_{i,t-p-s} \right]$$

$$= \sum_{r=0}^{p} \beta_r \left[y_{t-r} + \sum_{i=1}^{q} \gamma_i \sum_{s=0}^{\infty} \delta_s z_{i,t-r-s} \right].$$

In the special case of $q = 1$, $z_{1t} = 1$, (55) is

$$\sum_{r=0}^{p} \beta_r [y_{t-r} - \mu] = u_t, \tag{56}$$

for $\mu = -\gamma_1 \sum_{s=0}^{\infty} \delta_s$.

5.2.5. Prediction

The pure stochastic difference equation (1) can be written

$$y_t = -\beta_1 y_{t-1} - \cdots - \beta_p y_{t-p} + u_t, \tag{57}$$

where u_t is independent of $y_{t-1}, y_{t-2}, \ldots$. Then the conditional expectation of y_t given $y_{t-1}, y_{t-2}, \ldots$ is

$$\mathscr{E}\{y_t \mid y_{t-1}, y_{t-2}, \ldots\} = -\beta_1 y_{t-1} - \cdots - \beta_p y_{t-p}. \tag{58}$$

This conditional expectation can be used as a predictor of y_t based on the past. Its variance is

$$\mathscr{E}[-\beta_1 y_{t-1} - \cdots - \beta_p y_{t-p} - y_t]^2 = \mathscr{E}u_t^2 = \sigma^2. \tag{59}$$

The mean square error of any other predictor, say $f(y_{t-1}, y_{t-2}, \ldots)$ is

$$\begin{aligned} \mathscr{E}[f(y_{t-1}, y_{t-2}, \ldots) - y_t]^2 &= \mathscr{E}[f(y_{t-1}, y_{t-2}, \ldots) + \beta_1 y_{t-1} + \cdots + \beta_p y_{t-p} - u_t]^2 \\ &= \mathscr{E}u_t^2 + \mathscr{E}[f(y_{t-1}, y_{t-2}, \ldots) + \beta_1 y_{t-1} + \cdots + \beta_p y_{t-p}]^2. \end{aligned} \tag{60}$$

THEOREM 5.2.3. *The predictor of y_t based on $y_{t-1}, y_{t-2}, \ldots$ with minimum mean square error for the stationary stochastic process satisfying* (1) *for which the roots are less than* 1 *in absolute value is* (58).

The best predictor of y_t based on $y_{t-s-1}, y_{t-s-2}, \ldots \quad (s \geq 0)$ is

$$\mathscr{E}\{y_t \mid y_{t-s-1}, y_{t-s-2}, \ldots\} = \alpha_{s1} y_{t-s-1} + \cdots + \alpha_{sp} y_{t-s-p}, \tag{61}$$

where $\alpha_{s1} = \alpha^*_{s1}, \ldots, \alpha_{sp} = \alpha^*_{sp}$ are given in (11) and (22).

5.3. REDUCTION OF A GENERAL SCALAR EQUATION TO A FIRST-ORDER VECTOR EQUATION

As a convenience in studying the stochastic difference equation (1) in Section 5.2 we write it in the form of a first-order vector equation. Let

$$\tilde{\boldsymbol{y}}_t = \begin{bmatrix} y_t \\ y_{t-1} \\ \cdot \\ \cdot \\ \cdot \\ y_{t-p+1} \end{bmatrix}, \tag{1}$$

$$\tilde{\boldsymbol{u}}_t = \begin{bmatrix} u_t \\ 0 \\ \cdot \\ \cdot \\ \cdot \\ 0 \end{bmatrix} \qquad (p \text{ rows}), \tag{2}$$

$$\tilde{\mathbf{B}} = \begin{bmatrix} \beta_1 & \beta_2 & \beta_3 & \cdots & \beta_{p-1} & \beta_p \\ -1 & 0 & 0 & \cdots & 0 & 0 \\ 0 & -1 & 0 & \cdots & 0 & 0 \\ \cdot & \cdot & \cdot & & \cdot & \cdot \\ \cdot & \cdot & \cdot & & \cdot & \cdot \\ \cdot & \cdot & \cdot & & \cdot & \cdot \\ 0 & 0 & 0 & \cdots & -1 & 0 \end{bmatrix}. \tag{3}$$

Then the stochastic difference equation can be written as

$$\tilde{\boldsymbol{y}}_t + \tilde{\mathbf{B}} \tilde{\boldsymbol{y}}_{t-1} = \tilde{\boldsymbol{u}}_t. \tag{4}$$

The first component equation of (4) is exactly (1) of Section 5.2, and the other component equations are identities. Thus the scalar stochastic difference equation of order p is a special case of a vector stochastic difference equation of order 1.

Consider now the general stochastic difference equation of order 1 of a vector with p components

$$y_t + \mathbf{B}y_{t-1} = u_t, \tag{5}$$

where u_t is a random vector with mean $\mathscr{E}u_t = \mathbf{0}$, covariance matrix $\mathscr{E}u_t u_t' = \mathbf{\Sigma}$, and $\mathscr{E}u_t u_s' = \mathbf{0}$, $t \neq s$, and $\mathbf{B} = (\beta_{ij})$. If we substitute from

$$y_s = -\mathbf{B}y_{s-1} + u_s \tag{6}$$

repeatedly into (5) for $s = t - 1, t - 2, \ldots$, we obtain

$$y_t = \sum_{\tau=0}^{\infty} (-\mathbf{B})^\tau u_{t-\tau}. \tag{7}$$

Let $\lambda_1, \lambda_2, \ldots, \lambda_p$ be the characteristic roots of $-\mathbf{B}$, that is, the roots of

$$|\mathbf{B} + \lambda I| = 0. \tag{8}$$

If these roots are all different then there exists a matrix C such that

$$-\mathbf{B} = C\Lambda C^{-1}, \tag{9}$$

where Λ is the diagonal matrix†

$$\Lambda = \begin{bmatrix} \lambda_1 & 0 & \cdots & 0 \\ 0 & \lambda_2 & \cdots & 0 \\ \cdot & \cdot & & \cdot \\ \cdot & \cdot & & \cdot \\ \cdot & \cdot & & \cdot \\ 0 & 0 & \cdots & \lambda_p \end{bmatrix}. \tag{10}$$

† If there are p linearly independent characteristic vectors, (10) is the canonical form. The more general Jordan canonical form Λ has diagonal blocks of the form

$$\begin{bmatrix} \lambda & 1 & 0 & \cdots & 0 & 0 \\ 0 & \lambda & 1 & \cdots & 0 & 0 \\ \cdot & \cdot & \cdot & & \cdot & \cdot \\ \cdot & \cdot & \cdot & & \cdot & \cdot \\ \cdot & \cdot & \cdot & & \cdot & \cdot \\ 0 & 0 & 0 & \cdots & \lambda & 1 \\ 0 & 0 & 0 & \cdots & 0 & \lambda \end{bmatrix},$$

where λ is a root of (8). [See Halmos (1958) or Turnbull and Aitken (1952), who call this the classical canonical form.] There is one such block corresponding to each linearly independent characteristic vector. See Problems 17 and 18.

In this section and in Sections 5.5 and 5.6 we shall develop a number of results that hold when all of the characteristic roots of $\mathbf{B}$ (or $\tilde{\mathbf{B}}$) are less than 1 in absolute value. Because the calculations in terms of the general Jordan canonical form are laborious, we shall prove the results on the basis of $\mathbf{\Lambda}$ diagonal; the considerations required for the general treatment will be referred to in footnotes and problems.

From (9) we have

$$(-\mathbf{B})^\tau = \mathbf{C}\mathbf{\Lambda}^\tau\mathbf{C}^{-1}, \tag{11}$$

where†

$$\mathbf{\Lambda}^\tau = \begin{bmatrix} \lambda_1^\tau & 0 & \cdots & 0 \\ 0 & \lambda_2^\tau & \cdots & 0 \\ \cdot & \cdot & & \cdot \\ \cdot & \cdot & & \cdot \\ \cdot & \cdot & & \cdot \\ 0 & 0 & \cdots & \lambda_p^\tau \end{bmatrix}. \tag{12}$$

Then (7) can be written

$$\mathbf{y}_t = \mathbf{C}\sum_{\tau=0}^{\infty} \mathbf{\Lambda}^\tau\mathbf{C}^{-1}\mathbf{u}_{t-\tau}. \tag{13}$$

This series converges (in quadratic mean, see Section 7.6.1) if and only if $|\lambda_i| < 1$, $i = 1, \ldots, p$. For

$$\mathbf{y}_t - \sum_{\tau=0}^{s}(-\mathbf{B})^\tau\mathbf{u}_{t-\tau} = \mathbf{y}_t - \mathbf{C}\sum_{\tau=0}^{s}\mathbf{\Lambda}^\tau\mathbf{C}^{-1}\mathbf{u}_{t-\tau} = \mathbf{C}\mathbf{\Lambda}^{s+1}\mathbf{C}^{-1}\mathbf{y}_{t-(s+1)}, \tag{14}$$

and $\mathbf{C}^{-1}$ times (14) is $\mathbf{\Lambda}^{s+1}(\mathbf{C}^{-1}\mathbf{y}_{t-(s+1)})$. Under the assumption of stationarity the covariance matrix

$$\mathscr{E}\mathbf{y}_{t-(s+1)}\mathbf{y}'_{t-(s+1)} = \mathbf{F}, \tag{15}$$

say, does not depend on t or s. The covariance matrix of (14) is

$$\mathbf{C}\mathbf{\Lambda}^{s+1}\mathbf{C}^{-1}\mathbf{F}(\mathbf{C}^{-1})'\mathbf{\Lambda}^{s+1}\mathbf{C}'; \tag{16}$$

since each element of (16) is a linear combination of $(\lambda_i\lambda_j)^{s+1}$, it approaches 0 as $s \to \infty$ if each $|\lambda_i| < 1$.‡ (The "only if" statement requires that $\mathbf{y}_t$ have a nonsingular covariance matrix.)

† In the case of the general Jordan canonical form a diagonal block in $\mathbf{\Lambda}^\tau$ of order m corresponding to a root λ_i will have $\binom{\tau}{j}\lambda_i^{\tau-j}$ in each position j above the main diagonal, $j = 0, 1, \ldots, m-1$, and 0's below the main diagonal. (See Problem 17.)

‡ In the case of the general Jordan canonical form $(\lambda_i\lambda_j)^{s+1}$ may be replaced by a polynomial in s (of degree not greater than $2p-2$) times $\lambda_i^{s+1-k}\lambda_j^{s+1-l}$.

Now let us consider again (4), which is the pth order scalar equation written in vector form. We shall show that the roots of the associated polynomial equation (23) of Section 5.2 are the roots of

$$|\tilde{\mathbf{B}} + \lambda I| = 0. \tag{17}$$

The determinant

$$|\tilde{\mathbf{B}} + \lambda I| = \begin{vmatrix} \beta_1 + \lambda & \beta_2 & \beta_3 & \cdots & \beta_{p-1} & \beta_p \\ -1 & \lambda & 0 & \cdots & 0 & 0 \\ 0 & -1 & \lambda & \cdots & 0 & 0 \\ \cdot & \cdot & \cdot & & \cdot & \cdot \\ \cdot & \cdot & \cdot & & \cdot & \cdot \\ \cdot & \cdot & \cdot & & \cdot & \cdot \\ 0 & 0 & 0 & \cdots & \lambda & 0 \\ 0 & 0 & 0 & \cdots & -1 & \lambda \end{vmatrix} \tag{18}$$

can be evaluated by multiplying the ith column by λ and adding to the $(i + 1)$st column, $i = 1, \ldots, p - 1$, successively to obtain

$$\begin{vmatrix} \beta_1 + \lambda & \beta_2 + \beta_1\lambda + \lambda^2 & \cdots & \beta_p + \lambda\beta_{p-1} + \cdots + \lambda^{p-1}\beta_1 + \lambda^p \\ -1 & 0 & \cdots & 0 \\ 0 & -1 & \cdots & 0 \\ \cdot & \cdot & & \cdot \\ \cdot & \cdot & & \cdot \\ \cdot & \cdot & & \cdot \\ 0 & 0 & \cdots & 0 \end{vmatrix} \tag{19}$$

$$= \beta_p + \lambda\beta_{p-1} + \cdots + \lambda^{p-1}\beta_1 + \lambda^p,$$

which is the associated polynomial. Since all the roots of the polynomial are assumed less than 1 in absolute value, all the characteristic roots of $-\tilde{\mathbf{B}}$ are less than 1 in absolute value. It follows from this that the sum

$$\tilde{\boldsymbol{y}}_t = \sum_{r=0}^{\infty} (-\tilde{\mathbf{B}})^r \tilde{\boldsymbol{u}}_{t-r} \tag{20}$$

is well-defined [that is, that the series (20) converges in the mean].

If the roots $x_1, \ldots, x_p$ are distinct†, then we can find a matrix C such that $-\tilde{\mathbf{B}} = C\Lambda C^{-1}$, where

$$\Lambda = \begin{bmatrix} x_1 & 0 & \cdots & 0 \\ 0 & x_2 & \cdots & 0 \\ \cdot & \cdot & & \cdot \\ \cdot & \cdot & & \cdot \\ \cdot & \cdot & & \cdot \\ 0 & 0 & \cdots & x_p \end{bmatrix}. \tag{21}$$

Then (20) can be written as (13). The first component is

$$y_t = \sum_{\tau=0}^{\infty} \sum_{j=1}^{p} c_{1j} x_j^{\tau} c^{j1} u_{t-\tau}, \tag{22}$$

where‡ $(c^{ji}) = C^{-1}$. This is equivalent to (28) of Section 5.2. Thus another way of deriving (28) is to obtain (22) by finding C, whose columns are the characteristic vectors of $-\tilde{\mathbf{B}}$. (See Problem 20.)

In the case of the general vector equation, (13) can be used to find the covariance matrix of y_t, given $\mathbf{B}$ and $\mathbf{\Sigma}$, for $(\mathscr{E}y_t = 0)$

$$\begin{aligned} \mathscr{E}y_t y_t' &= \sum_{\tau,s=0}^{\infty} (-\mathbf{B})^{\tau} \mathscr{E} u_{t-\tau} u_{t-s}' (-\mathbf{B}')^{s} \\ &= \sum_{\tau=0}^{\infty} \mathbf{B}^{\tau} \mathbf{\Sigma} \mathbf{B}'^{\tau} \\ &= C \sum_{\tau=0}^{\infty} \Lambda^{\tau} C^{-1} \mathbf{\Sigma} (C^{-1})' \Lambda^{\tau} C' \\ &= F. \end{aligned} \tag{23}$$

Then§

$$\begin{aligned} f_{ij} &= \sum_{g,h,k,l=1}^{p} c_{ig} \sum_{\tau=0}^{\infty} \lambda_g^{\tau} c^{gh} \sigma_{hk} c^{lk} \lambda_l^{\tau} c_{jl} \\ &= \sum_{g,l=1}^{p} c_{ig} \frac{\sum_{h,k=1}^{p} c^{gh} \sigma_{hk} c^{lk}}{1 - \lambda_g \lambda_l} c_{jl}. \end{aligned} \tag{24}$$

† It follows from Problem 19 that if the roots are not all different, there are not p linearly independent characteristic vectors and the canonical form involves diagonal blocks as indicated in the previous footnotes.

‡ In the case of the general Jordan canonical form the sum of $c_{1j} x_j^{\tau} c^{j1}$ on j must be modified suitably.

§ In the case of the general Jordan canonical form instead of sums on τ of $\lambda_g^{\tau} \lambda_l^{\tau}$ there will be sums of powers $\lambda_g^{\tau-u} \lambda_l^{\tau-v}$ multiplied by polynomials of τ (of degrees not greater than $2p - 2$); such sums converge for $|\lambda_i| < 1$.

When C and $\boldsymbol{\Lambda}$ are defined in terms of $-\tilde{\mathbf{B}}$, (24) gives $\sigma(0) = f_{ii}$, $\sigma(1) = f_{i,i+1} = f_{i+1,i}, \ldots, \sigma(p-1) = f_{1p} = f_{p1}$.

If we multiply (5) by the transpose of (7) and by the transpose of (7) with t replaced by $t - s$, we obtain

$$\mathscr{E}y_t y_t' + \mathbf{B}\mathscr{E}y_{t-1}y_t' = \boldsymbol{\Sigma}, \tag{25}$$

$$\mathscr{E}y_t y_{t-s}' + \mathbf{B}\mathscr{E}y_{t-1}y_{t-s}' = \mathbf{0}, \qquad s = 1, 2, \ldots . \tag{26}$$

Since $\mathscr{E}y_{t-1}y_{t-s}' = \mathscr{E}y_t y_{t-(s-1)}'$, we can solve (26) successively to obtain

$$\mathscr{E}y_t y_{t-s}' = (-\mathbf{B})^s \mathscr{E}y_t y_t'. \tag{27}$$

When we substitute from (26) with $s = 1$ into (25), we obtain

$$\mathscr{E}y_t y_t' - \mathbf{B}\mathscr{E}y_t y_t' \mathbf{B}' = \boldsymbol{\Sigma}; \tag{28}$$

that is,

$$F - \mathbf{B}F\mathbf{B}' = \boldsymbol{\Sigma}. \tag{29}$$

Solving the linear equations (29) is another way to evaluate the elements of F. A numerical iterative procedure is to compute successive approximations $F_i = \boldsymbol{\Sigma} + \mathbf{B}F_{i-1}\mathbf{B}'$ starting with an initial approximation F_0, which may be $\boldsymbol{\Sigma}$. When $\mathbf{B} = \tilde{\mathbf{B}}$ and

$$\boldsymbol{\Sigma} = \begin{bmatrix} \sigma^2 & 0 & \cdots & 0 \\ 0 & 0 & \cdots & 0 \\ \cdot & \cdot & & \cdot \\ \cdot & \cdot & & \cdot \\ \cdot & \cdot & & \cdot \\ 0 & 0 & \cdots & 0 \end{bmatrix}, \tag{30}$$

this gives us a method for evaluating $\sigma(h)$.

Independent variables may be included in the first-order vector equation by letting

$$\tilde{\boldsymbol{\Gamma}} = \begin{bmatrix} \gamma_1 & \gamma_2 & \cdots & \gamma_q \\ 0 & 0 & \cdots & 0 \\ \cdot & \cdot & & \cdot \\ \cdot & \cdot & & \cdot \\ \cdot & \cdot & & \cdot \\ 0 & 0 & \cdots & 0 \end{bmatrix}, \qquad z_t = \begin{bmatrix} z_{1t} \\ z_{2t} \\ \cdot \\ \cdot \\ \cdot \\ z_{qt} \end{bmatrix}. \tag{31}$$

Then the model is

$$\tilde{y}_t + \tilde{\mathbf{B}}\tilde{y}_{t-1} + \tilde{\boldsymbol{\Gamma}}z_t = \tilde{u}_t, \tag{32}$$

and

$$\tilde{y}_t = \sum_{r=0}^{\infty} (-\tilde{\mathbf{B}})^r \tilde{u}_{t-r} - \sum_{r=0}^{\infty} (-\tilde{\mathbf{B}})^r \tilde{\mathbf{\Gamma}} z_{t-r}. \tag{33}$$

5.4. MAXIMUM LIKELIHOOD ESTIMATES IN THE CASE OF THE NORMAL DISTRIBUTION

One of the first problems of statistical inference concerning the stochastic difference equation model is the estimation of the coefficients $\beta_1, \ldots, \beta_p$ and the variance of the disturbance σ^2 based on observation of a segment of the series, say $y_1, \ldots, y_T$. If the u_t's are normally distributed and the difference equation (1) of Section 5.2 holds for all t, these coefficients and the variance (for $\mathscr{E}u_t = 0$) completely describe the distribution. We shall treat the more general problem of the estimation of the coefficients for the model

$$\sum_{r=0}^{p} \beta_r y_{t-r} + \sum_{i=1}^{q} \gamma_i z_{it} = u_t, \qquad t = 1, \ldots, T. \tag{1}$$

We shall find the maximum likelihood estimates of $\beta_1, \ldots, \beta_p, \gamma_1, \ldots, \gamma_q$, and σ^2 when the u_t's are independently and normally distributed with mean $\mathscr{E}u_t = 0$ and variance $\mathscr{E}u_t^2 = \sigma^2$. We shall modify the assumption of a stationary stochastic process (in addition to inserting the z_{it}'s) by assuming observation on y_t starts at $t = -(p-1)$ and that $y_{-(p-1)}, y_{-(p-2)}, \ldots, y_0$ are given known numbers. Then the joint distribution of $u_1, \ldots, u_T$ determines the joint distribution of $y_1, \ldots, y_T$, which are observed (as well as $z_{11}, \ldots, z_{q1}, \ldots, z_{1T}, \ldots, z_{qT}$, which are assumed as given numbers). This assumption is made for convenience in finding the maximum likelihood estimates. (In Chapter 6 we shall see the difficulties that arise when this assumption is not made.) However, any estimation procedure proposed for this model is maximum likelihood estimation or some moderate modification. In any case, for T large the effect of this assumption is small (and does not affect asymptotic theory). Mann and Wald (1943b) developed the maximum likelihood estimates as well as the asymptotic theory.

Let

$$\boldsymbol{\beta} = \begin{bmatrix} \beta_1 \\ \beta_2 \\ \cdot \\ \cdot \\ \cdot \\ \beta_p \end{bmatrix}, \qquad \boldsymbol{\gamma} = \begin{bmatrix} \gamma_1 \\ \gamma_2 \\ \cdot \\ \cdot \\ \cdot \\ \gamma_q \end{bmatrix}, \qquad \tilde{y}_t = \begin{bmatrix} y_t \\ y_{t-1} \\ \cdot \\ \cdot \\ \cdot \\ y_{t-p+1} \end{bmatrix}, \qquad z_t = \begin{bmatrix} z_{1t} \\ z_{2t} \\ \cdot \\ \cdot \\ \cdot \\ z_{qt} \end{bmatrix}. \tag{2}$$

Setting $\beta_0 = 1$, we may write (1) as

$$y_t + \boldsymbol{\beta}'\tilde{\boldsymbol{y}}_{t-1} + \boldsymbol{\gamma}'\boldsymbol{z}_t = u_t, \qquad t = 1, \ldots, T. \tag{3}$$

Since $u_1, \ldots, u_T$ are independent of $y_{-p+1}, \ldots, y_0$, the conditional probability density of $u_1, \ldots, u_T$ given $y_{-p+1}, \ldots, y_0$ is the same as the unconditional probability density and is given by

$$\frac{1}{(2\pi\sigma^2)^{\frac{1}{2}T}} \exp\left(-\frac{1}{2\sigma^2}\sum_{t=1}^{T} u_t^2\right). \tag{4}$$

When $y_{-p+1}, \ldots, y_0$ are fixed, (3) defines a one-to-one transformation of $y_1, \ldots, y_T$ into $u_1, \ldots, u_T$, and the Jacobian of this transformation is

$$\left|\frac{\partial u_t}{\partial y_s}\right| = \begin{vmatrix} 1 & 0 & 0 & \cdots & 0 \\ \beta_1 & 1 & 0 & \cdots & 0 \\ \beta_2 & \beta_1 & 1 & \cdots & 0 \\ \cdot & \cdot & \cdot & & \cdot \\ \cdot & \cdot & \cdot & & \cdot \\ \cdot & \cdot & \cdot & & \cdot \\ 0 & 0 & 0 & \cdots & 1 \end{vmatrix} = 1. \tag{5}$$

Hence the joint probability density of $y_1, \ldots, y_T$ given $y_{-p+1}, \ldots, y_0$ is

$$\frac{1}{(2\pi\sigma^2)^{\frac{1}{2}T}} \exp\left[-\frac{1}{2\sigma^2}\sum_{t=1}^{T}(y_t + \boldsymbol{\beta}'\tilde{\boldsymbol{y}}_{t-1} + \boldsymbol{\gamma}'\boldsymbol{z}_t)^2\right]. \tag{6}$$

In order to maximize this expression with respect to $\boldsymbol{\beta}$ and $\boldsymbol{\gamma}$ when it is considered the likelihood function, we must minimize the quantity

$$\sum_{t=1}^{T}(y_t + \boldsymbol{\beta}'\tilde{\boldsymbol{y}}_{t-1} + \boldsymbol{\gamma}'\boldsymbol{z}_t)^2 = \sum_{t=1}^{T}\left[y_t - (-\boldsymbol{\beta}' - \boldsymbol{\gamma}')\begin{pmatrix}\tilde{\boldsymbol{y}}_{t-1} \\ \boldsymbol{z}_t\end{pmatrix}\right]^2. \tag{7}$$

This is of the form of the usual least squares problem. The normal equations for the minimizing values $\hat{\boldsymbol{\beta}}$ and $\hat{\boldsymbol{\gamma}}$ of $\boldsymbol{\beta}$ and $\boldsymbol{\gamma}$ are then given by

$$\sum_{t=1}^{T}\begin{pmatrix}\tilde{\boldsymbol{y}}_{t-1} \\ \boldsymbol{z}_t\end{pmatrix}(\tilde{\boldsymbol{y}}_{t-1}' \; \boldsymbol{z}_t')\begin{pmatrix}-\hat{\boldsymbol{\beta}} \\ -\hat{\boldsymbol{\gamma}}\end{pmatrix} = \sum_{t=1}^{T} y_t\begin{pmatrix}\tilde{\boldsymbol{y}}_{t-1} \\ \boldsymbol{z}_t\end{pmatrix}, \tag{8}$$

which may be written

$$\begin{pmatrix}\boldsymbol{A}_{11} & \boldsymbol{A}_{12} \\ \boldsymbol{A}_{21} & \boldsymbol{A}_{22}\end{pmatrix}\begin{pmatrix}\hat{\boldsymbol{\beta}} \\ \hat{\boldsymbol{\gamma}}\end{pmatrix} = -\begin{pmatrix}\boldsymbol{a}_0 \\ \boldsymbol{d}\end{pmatrix}, \tag{9}$$

where

$$A_{11} = \sum_{t=1}^{T} \tilde{\mathbf{y}}_{t-1}\tilde{\mathbf{y}}_{t-1}', \qquad A_{12} = A_{21}' = \sum_{t=1}^{T} \tilde{\mathbf{y}}_{t-1}\mathbf{z}_t',$$

(10)

$$A_{22} = \sum_{t=1}^{T} \mathbf{z}_t\mathbf{z}_t', \qquad \mathbf{a}_0 = \sum_{t=1}^{T} y_t\tilde{\mathbf{y}}_{t-1}, \qquad \mathbf{d} = \sum_{t=1}^{T} y_t\mathbf{z}_t.$$

The maximum likelihood estimate of σ^2 is given by

$$\hat{\sigma}^2 = \frac{1}{T}\sum_{t=1}^{T}(y_t + \hat{\boldsymbol{\beta}}'\tilde{\mathbf{y}}_{t-1} + \hat{\boldsymbol{\gamma}}'\mathbf{z}_t)^2. \tag{11}$$

It is seen that the density (6) is a function of the observable variables through (7) which is

$$\sum_{t=1}^{T} y_t^2 + 2\boldsymbol{\beta}'\mathbf{a}_0 + 2\boldsymbol{\gamma}'\mathbf{d} + \boldsymbol{\beta}'A_{11}\boldsymbol{\beta} + 2\ \boldsymbol{\beta}'A_{12}\boldsymbol{\gamma} + \boldsymbol{\gamma}'A_{22}\boldsymbol{\gamma}. \tag{12}$$

Hence $\sum_{t=1}^{T} y_t^2$, $\mathbf{a}_0$, $\mathbf{d}$, A_{11}, and A_{12} form a sufficient set of statistics.

In the special case $q = 1$, $z_{1t} = 1$, let $\gamma_1 = \gamma$; the estimating equations are

$$\sum_{j=1}^{p} a_{ij}\hat{\beta}_j + T\bar{y}_{(i)}\hat{\gamma} = -a_{i0}, \qquad i = 1, \ldots, p, \tag{13}$$

$$T\sum_{j=1}^{p} \bar{y}_{(j)}\hat{\beta}_j + T\hat{\gamma} = -T\bar{y}, \tag{14}$$

where

$$a_{ij} = \sum_{t=1}^{T} y_{t-i}y_{t-j}, \qquad i, j = 1, \ldots, p, \tag{15}$$

$$a_{i0} = \sum_{t=1}^{T} y_{t-i}y_t, \qquad i = 1, \ldots, p, \tag{16}$$

$$T\bar{y}_{(i)} = \sum_{t=1}^{T} y_{t-i}, \qquad i = 0, 1, \ldots, p, \tag{17}$$

and $\bar{y} = \bar{y}_{(0)}$. If we subtract $\bar{y}_{(i)}$ times (14) from the ith equation of (13) we obtain

(18)

$$\sum_{j=1}^{p}\left[\sum_{t=1}^{T} y_{t-i}y_{t-j} - T\bar{y}_{(i)}\bar{y}_{(j)}\right]\hat{\beta}_j = -\sum_{t=1}^{T} y_{t-i}y_t + T\bar{y}_{(i)}\bar{y}, \qquad i = 1, \ldots, p,$$

or equivalently

$$(19) \quad \sum_{j=1}^{p}\sum_{t=1}^{T}(y_{t-i}-\bar{y}_{(i)})(y_{t-j}-\bar{y}_{(j)})\hat{\beta}_j = -\sum_{t=1}^{T}(y_{t-i}-\bar{y}_{(i)})(y_t-\bar{y}), \qquad i=1,\ldots,p.$$

The sums over t in (19) are the same for $i-j$ fixed except for addition and subtraction of terms at the beginning and end. We can modify these sums in various ways. For example, we can write

$$(20) \qquad \sum_{j=1}^{p} C_{i-j}^{*}b_j = -C_i^{*}, \qquad i=1,\ldots,p,$$

$$(21) \qquad \sum_{j=1}^{p} c_{i-j}^{*}b_j = -c_i^{*}, \qquad i=1,\ldots,p,$$

$$(22) \qquad \sum_{j=1}^{p} r_{i-j}^{*}b_j = -r_i^{*}, \qquad i=1,\ldots,p;$$

for convenience and for conformity with later uses we define the C_h^*'s, c_h^*'s, and r_h^*'s in terms of T observations $y_1,\ldots,y_T$ as follows:

$$(23) \quad C_h^* = C_{-h}^* = \frac{1}{T-h}\sum_{t=1}^{T-h}(y_t-\bar{y})(y_{t+h}-\bar{y}), \qquad h=0,1,\ldots,T-1,$$

$$(24) \quad c_h^* = c_{-h}^* = \frac{1}{T}\sum_{t=1}^{T-h}(y_t-\bar{y})(y_{t+h}-\bar{y}), \qquad h=0,1,\ldots,T-1,$$

$r_h^* = c_h^*/c_0^*$, $h=0,1,\ldots,T-1$, and $\bar{y}=\sum_{t=1}^{T}y_t/T$. Note that (22) is derived from (21) by dividing each c_h^* by c_0^*.

If T is relatively large, there will be little difference between the maximum likelihood equations (19) [for $y_{-p+1},\ldots,y_T$] and (20), (21), or (22) [for $y_1,\ldots,y_T$]. Each of the latter equations has more symmetry in that all elements on a given diagonal are the same and the terms on the right-hand side are $p-1$ of the same elements plus one further element.

If we write

$$(25)\qquad R^* = \begin{bmatrix} 1 & r_1^* & \cdots & r_{p-1}^* \\ r_1^* & 1 & \cdots & r_{p-2}^* \\ \cdot & \cdot & & \cdot \\ \cdot & \cdot & & \cdot \\ \cdot & \cdot & & \cdot \\ r_{p-1}^* & r_{p-2}^* & \cdots & 1 \end{bmatrix}, \qquad r^* = \begin{bmatrix} r_1^* \\ r_2^* \\ \cdot \\ \cdot \\ \cdot \\ r_p^* \end{bmatrix},$$

then (22) can be written

$$(26)\qquad R^*b = -r^*,$$

with the solution $b = -(R^*)^{-1}r^*$. These equations have the advantage that only p statistics are needed and R^* is positive definite. [The matrix R^* can be considered as proportional to the matrix whose i, jth element is $\sum_{t=-p}^{T} (y_{t+i} - \bar{y})(y_{t+j} - \bar{y})$, where $y_{t+i} - \bar{y}$ is set equal to 0 if $t + i \leq 0$ or $t + i > T$.] R^* is not only symmetric with respect to the main diagonal, but also with respect to the transverse diagonal (that is, the diagonal from the lower left-hand corner to the upper right-hand corner).

The special forms of R^* and r^* lead to a simple recursive method of calculating the coefficients for a $(p + 1)$st order equation from the solution for a pth order equation. The equations for the vector of coefficients for the $(p + 1)$st order equation, say $b(p + 1)$, are

$$(27)\qquad R_{p+1}^* b(p + 1) = -r_{p+1}^*,$$

where

$$(28)\qquad R_{p+1}^* = \begin{pmatrix} R_p^* & \tilde{r}_p^* \\ \tilde{r}_p^{*\prime} & 1 \end{pmatrix}, \qquad b(p + 1) = \begin{pmatrix} b^{(1)}(p + 1) \\ b_{p+1}(p + 1) \end{pmatrix}, \qquad r_{p+1}^* = \begin{pmatrix} r_p^* \\ r_{p+1}^* \end{pmatrix},$$

and R_p^* and r_p^* are given by (25), and $\tilde{r}_p^* = (r_p^*, r_{p-1}^*, \ldots, r_1^*)'$, which is r_p^* with its coordinates in reverse order. The partitioned equations of (27) are

$$(29)\qquad R_p^* b^{(1)}(p + 1) + \tilde{r}_p^* b_{p+1}(p + 1) = -r_p^*,$$

$$(30)\qquad \tilde{r}^{*\prime} b^{(1)}(p + 1) + b_{p+1}(p + 1) = -r_{p+1}^*.$$

Elimination of $b^{(1)}(p + 1)$ from (29) and (30) yields

$$(31)\qquad b_{p+1}(p + 1) = -\frac{r_{p+1}^* - \tilde{r}_p^{*\prime}(R_p^*)^{-1} r_p^*}{1 - \tilde{r}_p^{*\prime}(R_p^*)^{-1}\tilde{r}_p^*}$$

$$= -\frac{r_{p+1}^* + \tilde{r}_p^{*\prime} b(p)}{1 + \tilde{r}_p^{*\prime}\tilde{b}(p)},$$

where $\tilde{\boldsymbol{b}}(p) = (b_p(p), b_{p-1}(p), \ldots, b_1(p))'$, which is $\boldsymbol{b}(p)$ with its coordinates in reverse order. Substitution of (31) into (29) gives

$$\boldsymbol{b}^{(1)}(p+1) = -(\boldsymbol{R}_p^*)^{-1}\boldsymbol{r}_p^* - (\boldsymbol{R}_p^*)^{-1}\tilde{\boldsymbol{r}}_p^* b_{p+1}(p+1) \tag{32}$$

$$= \boldsymbol{b}(p) + \tilde{\boldsymbol{b}}(p) b_{p+1}(p+1).$$

In components (31) and (32) are

$$b_{p+1}(p+1) = -\frac{r_{p+1}^* + \sum_{j=1}^{p} r_{p+1-j}^* b_j(p)}{1 + \sum_{j=1}^{p} r_j^* b_j(p)}, \tag{33}$$

$$b_j(p+1) = b_j(p) + b_{p+1-j}(p) b_{p+1}(p+1), \qquad j = 1, \ldots, p. \tag{34}$$

These equations give a simple method of calculating the estimates for a given order process. The intermediate steps of the calculation give the coefficients for all lower-order processes. It will be shown in Section 5.6 that $b_{p+1}(p+1)$ is the partial correlation between y_t and y_{t+p+1} "holding $y_{t+1}, \ldots, y_{t+p}$ fixed." This method corresponds to the standard method of adding an independent variable to a regression equation; the simplified formulas for this case were given by Durbin (1960a).

We have seen that the estimation equations (9) for the coefficients of the stochastic difference equation are formally the same as least squares. In the usual regression situation the least squares estimates are the best linear unbiased estimates. The estimates here however are not linear in the y_t's. Durbin (1960b) has called the *estimating equations* linear (in the estimates) and unbiased in the sense that if the estimates are replaced by the parameters the expected values of the two sides of the normal equations are equal. Among this class the above estimating equations are best in a certain sense that the difference between right- and left-hand sides has minimum variance.

5.5. THE ASYMPTOTIC DISTRIBUTION OF THE MAXIMUM LIKELIHOOD ESTIMATES

5.5.1. Discussion of Asymptotic Properties

In this section we shall show that the maximum likelihood estimates deduced in Section 5.4 are consistent and asymptotically normally distributed as the length of period of observation increases. While the proofs are more complicated here, the conditions and results are similar to those for the least squares estimates as given in Theorem 2.6.1 and Corollaries 2.6.1 and 2.6.2. The implication of these results is that all the techniques used in the usual regression theory are applicable here for large samples.

It was shown in Section 5.3 that the pth-order scalar stochastic difference equation could be considered as a special case of the first-order p-component vector stochastic difference equation

$$y_t + \mathbf{B}y_{t-1} = u_t, \tag{1}$$

where

$$y_t = \begin{bmatrix} y_{1t} \\ y_{2t} \\ \cdot \\ \cdot \\ \cdot \\ y_{pt} \end{bmatrix}, \qquad u_t = \begin{bmatrix} u_{1t} \\ u_{2t} \\ \cdot \\ \cdot \\ \cdot \\ u_{pt} \end{bmatrix}, \tag{2}$$

$\mathbf{B}$ is a $p \times p$ matrix of coefficients, $\mathscr{E}u_t = \mathbf{0}$, $\mathscr{E}u_t u_t' = \mathbf{\Sigma}$, the u_t's are independent, and the characteristic roots of $-\mathbf{B}$ are less than 1 in absolute value. The equations defining the maximum likelihood estimates for the general-order scalar model constitute a special case of the estimating equations of the first-order vector case

$$\frac{1}{T}\sum_{t=1}^{T} y_{t-1}y_{t-1}'\hat{\mathbf{B}}' = -\frac{1}{T}\sum_{t=1}^{T} y_{t-1}y_t', \tag{3}$$

$$\hat{\mathbf{\Sigma}} = \frac{1}{T}\sum_{t=1}^{T}(y_t + \hat{\mathbf{B}}y_{t-1})(y_t + \hat{\mathbf{B}}y_{t-1})'. \tag{4}$$

$\hat{\mathbf{B}}$ defined by (3) is the maximum likelihood estimate when the u_t's are normally distributed and y_0 is assumed constant. We shall find it convenient to derive the asymptotic properties of these estimates and then specialize the results to the pth-order scalar equation.

First we shall show that the asymptotic theory (as $T \to \infty$) when (1) holds for $t = 1, 2, \ldots$ and y_0 is fixed is the same as when (1) holds for all t ($t = \ldots, -1, 0, 1, \ldots$). Then we shall show that the estimate $\hat{\mathbf{B}}$ is consistent and asymptotically normally distributed, and we shall find the variances and covariances of the limiting normal distribution and see that these can be consistently estimated. These results will be extended to models of moving averages and models of stochastic difference equations, the disturbances of which are moving averages.

In proofs we proceed on the basis that the roots of the polynomial equation associated with the scalar stochastic difference equation are different and that the matrix $-\mathbf{B}$ is diagonalizable in the vector case for convenience. However, these results hold more generally when the roots are less than 1 in absolute value and the modifications of proofs are indicated in footnotes.

5.5.2. Asymptotic Equivalence of Second-Order Sample Moments from Two Processes

Let us consider $\sum_{t=1}^T y_t y_t'$ when y_t, $t = 1, \ldots, T$, is generated by (1) with y_0 fixed and hence can be written

$$y_t = \sum_{s=0}^{t-1} (-\mathbf{B})^s u_{t-s} + (-\mathbf{B})^t y_0. \tag{5}$$

We also consider $\sum_{t=1}^T y_t^* y_t^{*\prime}$ when y_t^* satisfies (1) for $t = \ldots, -1, 0, 1, \ldots$ and hence can be written

$$y_t^* = \sum_{s=0}^{\infty} (-\mathbf{B})^s u_{t-s}. \tag{6}$$

We shall show that $\sum_{t=1}^T y_t^* y_t^{*\prime} - \sum_{t=1}^T y_t y_t'$ divided by $\sqrt{T}$ or T converges stochastically to $\mathbf{0}$.

From (5) and (6) we see that $y_t^* - y_t = (-\mathbf{B})^t (y_0^* - y_0)$, where

$$y_0^* = \sum_{s=0}^{\infty} (-\mathbf{B})^s u_{-s}. \tag{7}$$

Then

$$\sum_{t=1}^T y_t^* y_t^{*\prime} - \sum_{t=1}^T y_t y_t' = \sum_{t=1}^T (-\mathbf{B})^t (y_0^* - y_0) y_t' + \sum_{t=1}^T y_t (y_0^* - y_0)' (-\mathbf{B}')^t + \sum_{t=1}^T (-\mathbf{B})^t (y_0^* - y_0)(y_0^* - y_0)'(-\mathbf{B}')^t. \tag{8}$$

The sum of the expected values of the squares of the components of the matrix $\sum_{t=1}^T (-\mathbf{B})^t (y_0^* - y_0) y_t'$ is

$$\begin{aligned} \operatorname{tr} \mathscr{E} \sum_{t=1}^T (-\mathbf{B})^t (y_0^* - y_0) y_t' \left[\sum_{s=1}^T (-\mathbf{B})^s (y_0^* - y_0) y_s' \right]' &= \operatorname{tr} \mathscr{E} \sum_{t,s=1}^T (-\mathbf{B})^t (y_0^* - y_0) y_t' y_s (y_0^* - y_0)' (-\mathbf{B}')^s \\ &= \sum_{t,s=1}^T \operatorname{tr} [(-\mathbf{B})^t (F + y_0 y_0') (-\mathbf{B}')^s] \mathscr{E} y_t' y_s \end{aligned} \tag{9}$$

because y_0^* depending on $u_0, u_{-1}, \ldots$ is independent of $y_1, \ldots, y_T$ depending on $u_1, \ldots, u_T$ and $\mathscr{E}y_0^*y_0^{*\prime} = F$.

LEMMA 5.5.1. *If λ is greater than the maximum absolute value of the characteristic roots of $-\mathbf{B}$, then there is a constant c such that the absolute value of each component of $(-\mathbf{B})^t$ is less than $c\lambda^t$, $t = 0, 1, \ldots$.*

PROOF. We can write

$$(-\mathbf{B})^t = C\Lambda^t C^{-1}, \tag{10}$$

where Λ is the diagonal matrix with the characteristic roots of $-\mathbf{B}$ as diagonal elements and C is nonsingular. Then c can be taken† as the product of p, the maximum of the absolute values of the components of C and the maximum of the absolute values of the components of C^{-1}. Q.E.D.

Since

$$\mathscr{E}y_t y_t' = \sum_{r=0}^{t-1} (-\mathbf{B})^r \mathbf{\Sigma} (-\mathbf{B}')^r + (-\mathbf{B})^t y_0 y_0' (-\mathbf{B}')^t, \tag{11}$$

we have for λ satisfying Lemma 5.5.1 and less than 1

$$\begin{aligned}\mathscr{E}y_t'y_t = \operatorname{tr} \mathscr{E}y_t y_t' &\le \sum_{r=0}^{t-1} c^2\lambda^{2r}p^3 \max|\sigma_{ij}| + c^2\lambda^{2t}p^2 y_0'y_0 \\ &= c^2p^3 \max|\sigma_{ij}| \frac{1-\lambda^{2t}}{1-\lambda^2} + c^2p^2y_0'y_0\lambda^{2t} \\ &\le \frac{c^2p^3\max|\sigma_{ij}|}{1-\lambda^2} + c^2p^2y_0'y_0,\end{aligned} \tag{12}$$

where c is the constant from Lemma 5.5.1. Then we can bound the absolute value of the second factor on the right-hand side of (9), namely $|\mathscr{E}y_t'y_s|$, by the product of $\sqrt{\mathscr{E}y_t'y_t}$ and $\sqrt{\mathscr{E}y_s'y_s}$, each of which is bounded by the square root of the right-hand side of (12).

The other factor on the right-hand side of (9), $\operatorname{tr}[(-\mathbf{B})^t(F + y_0y_0')(-\mathbf{B}')^s]$, is bounded in absolute value by $K\lambda^{t+s}$, where K is the product of c^2, p^3, and the maximum of the absolute values of the components of $F + y_0y_0'$. Thus

$$\sum_{s,t=1}^{T} \operatorname{tr}[(-\mathbf{B})^t(F + y_0y_0')(-\mathbf{B}')^s] \le K\lambda^2\left(\frac{1-\lambda^T}{1-\lambda}\right)^2 \le \frac{K\lambda^2}{(1-\lambda)^2}. \tag{13}$$

† In the case of the general Jordan canonical form, if $|\lambda_i| < \lambda$, then a constant k_i can be found so $\binom{t}{j}\lambda_i^{t-j} < k_i\lambda^t$ and c involves $\max_i k_i$. (See Problem 17.)

Hence (9) is less than a constant independent of T. From this fact and the Tchebycheff inequality it follows that the first and second terms on the right-hand side of (8) divided by $\sqrt{T}$ converge stochastically to $\mathbf{0}$.

The third term on the right-hand side of (8) is positive semidefinite. Its expected value is

$$(14)\quad \mathscr{E}\sum_{t=1}^{T}(-\mathbf{B})^t(y_0^*-y_0)(y_0^*-y_0)'(-\mathbf{B}')^t=\sum_{t=1}^{T}(-\mathbf{B})^t(F+y_0y_0')(-\mathbf{B}')^t.$$

The trace of (14) is bounded by the product of p^3, the maximum of the absolute values of the components of $F + y_0y_0'$, and $\lambda^2c^2(1-\lambda^{2T})/(1-\lambda^2)$; hence the trace of (14) is bounded uniformly in T. Thus the sum of the diagonal elements of the third term in (8) divided by $\sqrt{T}$ converges stochastically to 0 (by a Tchebycheff inequality using the mean of a nonnegative random variable). Since the matrix is positive semidefinite, each element of the matrix converges stochastically to 0. Thus we have shown that $\sum_{t=1}^{T} y_t^*y_t^{*\prime} - \sum_{t=1}^{T} y_ty_t'$ divided by $\sqrt{T}$ converges stochastically to $\mathbf{0}$.

We now consider

$$\begin{aligned}(15)\quad \sum_{t=1}^{T}y_{t-1}^*y_t^{*\prime}-\sum_{t=1}^{T}y_{t-1}y_t' &=\sum_{t=1}^{T}y_{t-1}^*[y_{t-1}^{*\prime}(-\mathbf{B}')+u_t']-\sum_{t=1}^{T}y_{t-1}[y_{t-1}'(-\mathbf{B}')+u_t']\\ &=\left[\sum_{t=1}^{T}y_{t-1}^*y_{t-1}^{*\prime}-\sum_{t=1}^{T}y_{t-1}y_{t-1}'\right](-\mathbf{B}')+\sum_{t=1}^{T}(y_{t-1}^*-y_{t-1})u_t'.\end{aligned}$$

The sum of expected values of the squares of the elements of the second term on the right-hand side of (15) is

$$\begin{aligned}(16)\quad \operatorname{tr}\mathscr{E}\sum_{t,s=1}^{T}(-\mathbf{B})^{t-1}(y_0^*-y_0)u_t'u_s(y_0^*-y_0)'(-\mathbf{B}')^{s-1}\\ =(\operatorname{tr}\mathbf{\Sigma})\sum_{t=1}^{T}\operatorname{tr}(-\mathbf{B})^{t-1}(F+y_0y_0')(-\mathbf{B}')^{t-1},\end{aligned}$$

which is bounded by the product of $\operatorname{tr}\mathbf{\Sigma}$, p^3, the maximum of the absolute values of the elements of $F + y_0y_0'$, and $c^2(1-\lambda^{2T})/(1-\lambda^2)$ and hence is bounded uniformly in T. Thus the second term on the right-hand side of (15) divided by $\sqrt{T}$ converges stochastically to $\mathbf{0}$.

Since

$$\frac{1}{\sqrt{T}}\left[\sum_{t=1}^{T} y_{t-1}^{*}y_{t-1}^{*\prime} - \sum_{t=1}^{T} y_{t-1}y_{t-1}'\right] \tag{17}$$

$$= \frac{1}{\sqrt{T}}\left[\sum_{t=1}^{T} y_t^{*}y_t^{*\prime} - \sum_{t=1}^{T} y_t y_t' + y_0^{*}y_0^{*\prime} - y_T^{*}y_T^{*\prime} - y_0 y_0' + y_T y_T'\right],$$

and the last four terms on the right-hand side of (17) converge stochastically to $\mathbf{0}$, the left-hand side does as well. Therefore the left-hand side of (15) divided by $\sqrt{T}$ converges stochastically to $\mathbf{0}$.

THEOREM 5.5.1. *If $\{y_t\}$ is a process satisfying* (1) *for $t = 1, 2, \ldots$ and y_0 is a fixed vector, if $\{y_t^*\}$ is a process satisfying* (1) *for $t = \ldots, -1, 0, 1, \ldots$, if the u_t's are independently distributed with $\mathscr{E}u_t = \mathbf{0}$ and $\mathscr{E}u_t u_t' = \mathbf{\Sigma}$, and if $-\mathbf{B}$ has characteristic roots less than* 1 *in absolute value, then*

$$\operatorname*{plim}_{T\to\infty}\left[\frac{1}{\sqrt{T}}\sum_{t=1}^{T} y_{t-1}^{*}y_{t-1}^{*\prime} - \frac{1}{\sqrt{T}}\sum_{t=1}^{T} y_{t-1}y_{t-1}'\right] = \mathbf{0}, \tag{18}$$

$$\operatorname*{plim}_{T\to\infty}\left[\frac{1}{\sqrt{T}}\sum_{t=1}^{T} y_{t-1}^{*}y_{t}^{*\prime} - \frac{1}{\sqrt{T}}\sum_{t=1}^{T} y_{t-1}y_{t}'\right] = \mathbf{0}, \tag{19}$$

$$\operatorname*{plim}_{T\to\infty}\left[\frac{1}{\sqrt{T}}\sum_{t=1}^{T} y_{t}^{*}y_{t}^{*\prime} - \frac{1}{\sqrt{T}}\sum_{t=1}^{T} y_{t}y_{t}'\right] = \mathbf{0}. \tag{20}$$

The point of this theorem is that the asymptotic properties of $\hat{\mathbf{B}}$ and $\hat{\mathbf{\Sigma}}$ (consistency and asymptotic normal distributions) are the same for the processes $\{y_t\}$ and $\{y_t^*\}$. It will be more convenient to work with the latter (stationary) process.

5.5.3. Consistency of the Maximum Likelihood Estimates

We shall now show that the matrix of coefficients of the maximum likelihood estimates in (3) converges stochastically to a nonsingular matrix and that the matrix on the right-hand side of (3) converges stochastically to a matrix such that the resulting equation is satisfied only by the parameter matrix $\mathbf{B}$. In this section we shall now use y_t to denote the process satisfying (1) for $t = \ldots, -1, 0, 1, \ldots$. Thus

$$y_t = \sum_{s=0}^{\infty}(-\mathbf{B})^s u_{t-s}, \qquad t = \ldots, -1, 0, 1, \ldots, \tag{21}$$

and

$$\mathscr{E}y_t y_t' = \sum_{s=0}^{\infty} (-\mathbf{B})^s \mathbf{\Sigma} (-\mathbf{B}')^s = \boldsymbol{F}. \tag{22}$$

LEMMA 5.5.2. *Suppose $\boldsymbol{u}_1, \boldsymbol{u}_2, \ldots$ are independently distributed with $\mathscr{E}\boldsymbol{u}_t = \mathbf{0}$ and $\mathscr{E}\boldsymbol{u}_t\boldsymbol{u}_t' = \mathbf{\Sigma}$. If the $\boldsymbol{u}_t$'s are identically distributed or if $\mathscr{E}|u_{it}|^{2+\varepsilon} < m$, $i = 1, \ldots, p$, $t = 1, 2, \ldots$, for some $\varepsilon > 0$ and some m, then*

$$\operatorname*{plim}_{T\to\infty} \frac{1}{T} \sum_{t=1}^{T} \boldsymbol{u}_t\boldsymbol{u}_t' = \mathbf{\Sigma}. \tag{23}$$

PROOF. In the case of identically distributed random vectors the lemma follows from the law of large numbers for identically distributed random variables applied to each component equation in (23). Under the alternative conditions the i,jth element of $\boldsymbol{u}_t\boldsymbol{u}_t'$, namely, $u_{it}u_{jt}$ has expected value σ_{ij} and

$$\mathscr{E}|u_{it}u_{jt}|^{1+\frac{1}{2}\varepsilon} \le \sqrt{\mathscr{E}|u_{it}|^{2+\varepsilon}\,\mathscr{E}|u_{jt}|^{2+\varepsilon}} < m. \tag{24}$$

The result follows from a law of large numbers, known as Markov's theorem. [See Loève (1963), Section 20.1.]

LEMMA 5.5.3 *If $\boldsymbol{y}_t$ is defined by (21) with $-\mathbf{B}$ having characteristic roots less than 1 in absolute value and if the $\boldsymbol{u}_t$'s are independent with $\mathscr{E}\boldsymbol{u}_t = \mathbf{0}$ and $\mathscr{E}\boldsymbol{u}_t\boldsymbol{u}_t' = \mathbf{\Sigma}$, then*

$$\operatorname*{plim}_{T\to\infty} \frac{1}{T} \sum_{t=1}^{T} \boldsymbol{u}_t\boldsymbol{y}_{t-1}' = \mathbf{0}. \tag{25}$$

PROOF. The sum of expected values of the squares of the components of $(1/T)\sum_{t=1}^{T} \boldsymbol{u}_t\boldsymbol{y}_{t-1}'$ is

$$\begin{aligned}
\operatorname{tr}\mathscr{E}\frac{1}{T}\sum_{s=1}^{T} \boldsymbol{y}_{s-1}\boldsymbol{u}_s' \frac{1}{T}\sum_{t=1}^{T} \boldsymbol{u}_t\boldsymbol{y}_{t-1}' &= \frac{1}{T^2}\sum_{s,t=1}^{T} \mathscr{E}\operatorname{tr}\boldsymbol{y}_{s-1}\boldsymbol{u}_s'\boldsymbol{u}_t\boldsymbol{y}_{t-1}' \\
&= \frac{1}{T^2}\sum_{t=1}^{T} \mathscr{E}\boldsymbol{u}_t'\boldsymbol{u}_t\mathscr{E}\boldsymbol{y}_{t-1}'\boldsymbol{y}_{t-1} \\
&= \frac{1}{T^2}\sum_{t=1}^{T} \operatorname{tr}\mathbf{\Sigma}\operatorname{tr}\boldsymbol{F} \\
&= \frac{1}{T}\operatorname{tr}\mathbf{\Sigma}\operatorname{tr}\boldsymbol{F}
\end{aligned} \tag{26}$$

because $\mathscr{E}\boldsymbol{y}_{s-1}\boldsymbol{u}_s'\boldsymbol{u}_t\boldsymbol{y}_{t-1}' = \mathbf{0}$ if $t \neq s$ (since $\boldsymbol{u}_s$ is independent of $\boldsymbol{y}_{s-1}$ and $\boldsymbol{u}_t\boldsymbol{y}_{t-1}'$ if $s > t$ and $\boldsymbol{u}_t$ is independent of the other factors if $s < t$). The lemma follows from Tchebycheff's inequality.

Lemma 5.5.3 yields

$$\text{(27)} \qquad \operatorname*{plim}_{T\to\infty}\left[\frac{1}{T}\sum_{t=1}^{T} y_t y_{t-1}' + \mathbf{B}\frac{1}{T}\sum_{t=1}^{T} y_{t-1}y_{t-1}'\right] = \mathbf{0},$$

and Lemma 5.5.2 yields

(28)

$$\operatorname*{plim}_{T\to\infty}\left[\frac{1}{T}\sum_{t=1}^{T} y_t y_t' + \mathbf{B}\frac{1}{T}\sum_{t=1}^{T} y_{t-1}y_t' + \frac{1}{T}\sum_{t=1}^{T} y_t y_{t-1}'\mathbf{B}' + \mathbf{B}\frac{1}{T}\sum_{t=1}^{T} y_{t-1}y_{t-1}'\mathbf{B}'\right] = \mathbf{\Sigma}.$$

If we subtract from (28) the product of (27) and $\mathbf{B}'$ and the product of $\mathbf{B}$ and the transpose of (27), we obtain

$$\text{(29)} \qquad \operatorname*{plim}_{T\to\infty}\left[\frac{1}{T}\sum_{t=1}^{T} y_t y_t' - \mathbf{B}\frac{1}{T}\sum_{t=1}^{T} y_t y_t'\mathbf{B}'\right] = \mathbf{\Sigma}.$$

We have used the fact that

$$\text{(30)} \qquad \frac{1}{T}\sum_{t=1}^{T} y_t y_t' - \frac{1}{T}\sum_{t=1}^{T} y_{t-1}y_{t-1}' = \frac{1}{T}(y_T y_T' - y_0 y_0')$$

converges stochastically to $\mathbf{0}$. Since

$$\text{(31)} \qquad A - \mathbf{B}A\mathbf{B}' = \mathbf{\Sigma}$$

can be solved uniquely for A (see Problem 27), (29) determines the probability limit of $(1/T)\sum_{t=1}^{T} y_t y_t'$, which must be (22).

THEOREM 5.5.2. *If $\boldsymbol{y}_t$ is defined by (21), $t = 1, 2, \ldots,$ with $-\mathbf{B}$ having characteristic roots less than 1 in absolute value, if the $\boldsymbol{u}_t$'s are independently distributed with $\mathscr{E}\boldsymbol{u}_t = \mathbf{0}$ and $\mathscr{E}\boldsymbol{u}_t\boldsymbol{u}_t' = \mathbf{\Sigma}$, and if either the $\boldsymbol{u}_t$'s are identically distributed or $\mathscr{E}\,|u_{it}|^{2+\varepsilon} < m$, $i = 1, \ldots, p$, $t = 1, 2, \ldots,$ for some $\varepsilon > 0$ and some m, then*

$$\text{(32)} \qquad \operatorname*{plim}_{T\to\infty}\frac{1}{T}\sum_{t=1}^{T} y_t y_t' = \operatorname*{plim}_{T\to\infty}\frac{1}{T}\sum_{t=1}^{T} y_{t-1}y_{t-1}' = \boldsymbol{F},$$

where $\boldsymbol{F}$ is given by (22).

LEMMA 5.5.4. *If $\mathbf{\Sigma}$ is positive definite, $\boldsymbol{F}$ is positive definite.*

PROOF. $\boldsymbol{F}$ is positive semidefinite because it is a covariance matrix. Since $\boldsymbol{F} = \mathbf{\Sigma} + \mathbf{B}\boldsymbol{F}\mathbf{B}'$, it is the sum of a positive definite matrix and a positive semidefinite matrix and hence is positive definite. (See Problem 28.)

THEOREM 5.5.3. *Under the conditions of Theorem 5.5.2 and if $\boldsymbol{F}$ is positive definite*

$$\operatorname*{plim}_{T\to\infty} \hat{\mathbf{B}} = \mathbf{B}, \tag{33}$$

$$\operatorname*{plim}_{T\to\infty} \hat{\mathbf{\Sigma}} = \mathbf{\Sigma}. \tag{34}$$

PROOF. To prove (33) we observe that

(35)

$$\begin{aligned}\operatorname*{plim}_{T\to\infty} \hat{\mathbf{B}}' - \mathbf{B}' &= \operatorname*{plim}_{T\to\infty} \left(\frac{1}{T}\sum_{t=1}^{T} y_{t-1}y_{t-1}'\right)^{-1}\left(-\frac{1}{T}\sum_{t=1}^{T} y_{t-1}y_t' - \frac{1}{T}\sum_{t=1}^{T} y_{t-1}y_{t-1}'\mathbf{B}'\right)\\ &= -\operatorname*{plim}_{T\to\infty}\left(\frac{1}{T}\sum_{t=1}^{T} y_{t-1}y_{t-1}'\right)^{-1}\frac{1}{T}\sum_{t=1}^{T} y_{t-1}\boldsymbol{u}_t'\\ &= \mathbf{0}.\end{aligned}$$

Then (34) follows from

$$\begin{aligned}\hat{\mathbf{\Sigma}} &= \frac{1}{T}\sum_{t=1}^{T}(y_t + \hat{\mathbf{B}}y_{t-1})(y_t + \hat{\mathbf{B}}y_{t-1})'\\ &= \frac{1}{T}\sum_{t=1}^{T} y_t y_t' + \hat{\mathbf{B}}\frac{1}{T}\sum_{t=1}^{T} y_{t-1}y_t' + \frac{1}{T}\sum_{t=1}^{T} y_t y_{t-1}'\hat{\mathbf{B}}' + \hat{\mathbf{B}}\frac{1}{T}\sum_{t=1}^{T} y_{t-1}y_{t-1}'\hat{\mathbf{B}}',\end{aligned} \tag{36}$$

(28) and (33). Q.E.D.

Theorem 5.5.3 shows that the estimates of $\mathbf{B}$ and $\mathbf{\Sigma}$ are consistent under quite general conditions and in particular under more general conditions than those defining the estimates as maximum likelihood. The conditions assumed here have been selected because they are relatively simple and because they also lead to asymptotic normality. The assumption that the roots are less than 1 in absolute value is not necessary for consistency of the estimate of **B** though it is necessary for a general theorem about asymptotic normality. [See T. W. Anderson (1959).]

For consistency of the estimates for the pth-order univariate equation we need to show that $\tilde{F}$ is nonsingular.

LEMMA 5.5.5. *If σ^2 is positive, then*

$$\tilde{F} = \sum_{s=0}^{\infty} \tilde{\mathbf{B}}^s \tilde{\boldsymbol{\Sigma}} \tilde{\mathbf{B}}'^s \tag{37}$$

is positive definite, where

$$\tilde{\boldsymbol{\Sigma}} = \begin{bmatrix} \sigma^2 & 0 & 0 & \cdots & 0 \\ 0 & 0 & 0 & \cdots & 0 \\ 0 & 0 & 0 & \cdots & 0 \\ \cdot & \cdot & \cdot & & \cdot \\ \cdot & \cdot & \cdot & & \cdot \\ \cdot & \cdot & \cdot & & \cdot \\ 0 & 0 & 0 & \cdots & 0 \end{bmatrix}, \qquad \tilde{\mathbf{B}} = \begin{bmatrix} \beta_1 & \beta_2 & \beta_3 & \cdots & \beta_p \\ -1 & 0 & 0 & \cdots & 0 \\ 0 & -1 & 0 & \cdots & 0 \\ \cdot & \cdot & \cdot & & \cdot \\ \cdot & \cdot & \cdot & & \cdot \\ \cdot & \cdot & \cdot & & \cdot \\ 0 & 0 & 0 & \cdots & 0 \end{bmatrix}. \tag{38}$$

PROOF. For

$$\boldsymbol{x}'\tilde{F}\boldsymbol{x} = \sum_{s=0}^{\infty} \boldsymbol{x}'\tilde{\mathbf{B}}^s \tilde{\boldsymbol{\Sigma}} \tilde{\mathbf{B}}'^s \boldsymbol{x} \tag{39}$$

to be 0,

$$\begin{aligned} 0 &= \boldsymbol{x}'\tilde{\boldsymbol{\Sigma}}\boldsymbol{x} = x_1^2\sigma^2, \\ 0 &= \boldsymbol{x}'\tilde{\mathbf{B}}^s\tilde{\boldsymbol{\Sigma}}(\boldsymbol{x}'\tilde{\mathbf{B}}^s)', \qquad s = 1, 2, \ldots ; \end{aligned} \tag{40}$$

hence $x_1 = 0$, and similarly the first component of $\boldsymbol{x}'\tilde{\mathbf{B}}^s$ is 0, $s = 1, 2, \ldots$. But the first component of $\boldsymbol{x}'\tilde{\mathbf{B}}$ is $-x_2$ since $x_1 = 0$ and hence $x_2 = 0$. Similarly, each component of $\boldsymbol{x}$ is shown to be 0. Thus $\boldsymbol{x}'\tilde{F}\boldsymbol{x} = 0$ implies $\boldsymbol{x} = \mathbf{0}$, which proves the lemma.

THEOREM 5.5.4. *If y_t is defined by* (1) *of Section* 5.1, *for* $t = \ldots, -1, 0, 1, \ldots$, *if the roots of the associated polynomial equation are less than* 1 *in absolute value, if the u_t's are independent with $\mathscr{E}u_t = 0$ and $\mathscr{E}u_t^2 = \sigma^2 > 0$, and if either the u_t's are identically distributed or $\mathscr{E}\,|u_t|^{2+\varepsilon} < m$, $t = 1, 2, \ldots$,* *for some* $\varepsilon > 0$ *and some* m, *then*

(41) $$\operatorname*{plim}_{T\to\infty} \hat{\boldsymbol{\beta}} = \boldsymbol{\beta},$$

(42) $$\operatorname*{plim}_{T\to\infty} \hat{\sigma}^2 = \sigma^2.$$

5.5.4. Asymptotic Normality of the Estimates

The limiting distribution of

(43) $$\sqrt{T}(\hat{\mathbf{B}}' - \mathbf{B}') = -\sqrt{T}\left(\frac{1}{T}\sum_{t=1}^{T} y_{t-1}y_{t-1}'\right)^{-1}\frac{1}{T}\sum_{t=1}^{T} y_{t-1}u_t'$$

is the same as the limiting distribution of $-F^{-1}(1/\sqrt{T})\sum_{t=1}^{T} y_{t-1}u_t'$. We shall find the limiting distribution of this last matrix by showing that an arbitrary linear combination of its elements, say

(44) $$S_T = \frac{1}{\sqrt{T}}\sum_{t=1}^{T} u_t'\boldsymbol{\Phi} y_{t-1},$$

has a limiting normal distribution.

Let

(45) $$y_t^{(k)} = \sum_{s=0}^{k}(-\mathbf{B})^s u_{t-s},$$

(46) $$Z_{kT} = \frac{1}{\sqrt{T}}\sum_{t=1}^{T} u_t'\boldsymbol{\Phi} y_{t-1}^{(k)},$$

(47) $$X_{kT} = \frac{1}{\sqrt{T}}\sum_{t=1}^{T} u_t'\boldsymbol{\Phi}\sum_{s=k+1}^{\infty}(-\mathbf{B})^s u_{t-1-s}.$$

Then

(48) $$\begin{aligned}
\mathscr{E}X_{kT}^2 &= \frac{1}{T}\sum_{t,\tau=1}^{T}\sum_{s,\sigma=k+1}^{\infty}\mathscr{E}u_t'\boldsymbol{\Phi}(-\mathbf{B})^s u_{t-1-s}u_\tau'\boldsymbol{\Phi}(-\mathbf{B})^\sigma u_{\tau-1-\sigma} \\
&= \frac{1}{T}\sum_{t,\tau=1}^{T}\sum_{s,\sigma=k+1}^{\infty}\mathscr{E}\operatorname{tr}(-\mathbf{B})^s u_{t-1-s}u_{\tau-1-\sigma}'(-\mathbf{B}')^\sigma\boldsymbol{\Phi}'u_\tau u_t'\boldsymbol{\Phi} \\
&= \frac{1}{T}\sum_{t=1}^{T}\sum_{s=k+1}^{\infty}\operatorname{tr}(-\mathbf{B})^s\boldsymbol{\Sigma}(-\mathbf{B}')^s\boldsymbol{\Phi}'\boldsymbol{\Sigma}\boldsymbol{\Phi} \\
&= \operatorname{tr}\boldsymbol{\Phi}'\boldsymbol{\Sigma}\boldsymbol{\Phi}\sum_{s=k+1}^{\infty}(-\mathbf{B})^s\boldsymbol{\Sigma}(-\mathbf{B}')^s \\
&= M_k.
\end{aligned}$$

(We have used the fact that $x'x = \operatorname{tr} xx'$.) Then $M_k \to 0$ as $k \to \infty$ because $\sum_{s=0}^{\infty} (-\mathbf{B})^s \mathbf{\Sigma}(-\mathbf{B}')^s$ is convergent. For given k

$$u_t'\mathbf{\Phi} y_{t-1}^{(k)} = u_t'\mathbf{\Phi} \sum_{s=0}^{k} (-\mathbf{B})^s u_{t-1-s} \tag{49}$$

has mean 0, variance

$$\mathscr{E} \sum_{s,r=0}^{k} u_t'\mathbf{\Phi}(-\mathbf{B})^s u_{t-1-s} u_t'\mathbf{\Phi}(-\mathbf{B})^r u_{t-1-r} = \sum_{s=0}^{k} \operatorname{tr} (-\mathbf{B})^s \mathbf{\Sigma}(-\mathbf{B}')^s \mathbf{\Phi}' \mathbf{\Sigma} \mathbf{\Phi} = \sigma_k^2 \tag{50}$$

say, and covariance at t and τ $(t \neq \tau)$

$$\mathscr{E} \sum_{s,r=0}^{k} u_t'\mathbf{\Phi}(-\mathbf{B})^s u_{t-1-s} u_\tau'\mathbf{\Phi}(-\mathbf{B})^r u_{\tau-1-r} = 0, \qquad \tau \neq t. \tag{51}$$

The sequence $\{u_t'\mathbf{\Phi} y_{t-1}^{(k)}\}$ has the further property that any set of n elements, $t = t_1 < \cdots < t_n$, is distributed independently of the elements for $t < t_1 - k - 1$ and for $t > t_n + k + 1$. Application of Theorem 7.7.5 yields the following lemma:

Lemma 5.5.6. *If $y_t^{(k)}$, $t = 1, 2, \ldots,$ is defined by* (45) *and the u_t's are independently and identically distributed with $\mathscr{E} u_t = \mathbf{0}$ and $\mathscr{E} u_t u_t' = \mathbf{\Sigma}$, then the distribution of Z_{kT} defined by* (46)

$$\Pr\{Z_{kT} \leq z\} = F_{kT}(z) \tag{52}$$

has the limit on T

$$\lim_{T\to\infty} F_{kT}(z) = \Phi\left(\frac{z}{\sigma_k}\right), \tag{53}$$

where

$$\Phi(v) = \int_{-\infty}^{v} \frac{1}{\sqrt{2\pi}} e^{-\frac{1}{2}w^2}\, dw, \tag{54}$$

and σ_k is given by (50).

Then

$$\lim_{k\to\infty} \Phi\left(\frac{z}{\sigma_k}\right) = \Phi\left(\frac{z}{\sigma}\right), \tag{55}$$

where

$$\sigma^2 = \lim_{k\to\infty} \sigma_k^2 = \operatorname{tr} \mathbf{\Sigma} \mathbf{\Phi} \boldsymbol{F} \mathbf{\Phi}'. \tag{56}$$

We use Corollary 7.7.1.

LEMMA 5.5.7. *Under the conditions of Lemma* 5.5.6, *if the characteristic roots of* $-\mathbf{B}$ *are less than* 1 *in absolute value*, S_T, *defined by* (44), *has a limiting normal distribution with mean* 0 *and variance* σ^2.

Application of Theorem 7.7.7 yields the desired result. The matrix whose elements are $f_{ij}\sigma_{kl}$ will be denoted by $\boldsymbol{F} \otimes \boldsymbol{\Sigma}$, the *Kronecker product* of $\boldsymbol{F}$ and $\boldsymbol{\Sigma}$.

THEOREM 5.5.5. *Under the conditions of Lemmas* 5.5.6 *and* 5.5.7, $(1/\sqrt{T}) \times \sum_{t=1}^{T} \boldsymbol{y}_{t-1}\boldsymbol{u}_t'$ *has a limiting normal distribution with mean* $\mathbf{0}$ *and covariance matrix* $\boldsymbol{F} \otimes \boldsymbol{\Sigma}$.

THEOREM 5.5.6. *If* $\{\boldsymbol{y}_t\}$ *is a sequence of random vectors satisfying* (1) *with* $\{\boldsymbol{u}_t\}$ *independently and identically distributed with* $\mathscr{E}\boldsymbol{u}_t = \mathbf{0}$ *and* $\mathscr{E}\boldsymbol{u}_t\boldsymbol{u}_t' = \boldsymbol{\Sigma}$, *if* $-\mathbf{B}$ *has all characteristic roots less than* 1 *in absolute value, and if* $\boldsymbol{F}$ *is positive definite, then* $\sqrt{T}(\hat{\mathbf{B}}' - \mathbf{B}')$ *has a limiting normal distribution with mean* $\mathbf{0}$ *and covariance matrix* $\boldsymbol{F}^{-1} \otimes \boldsymbol{\Sigma}$.

PROOF. This follows from Theorem 5.5.5 and (43). (See Problem 29.)

THEOREM 5.5.7. *If* $\{y_t\}$ *is a sequence of random variables satisfying* (1) *of Section* 5.1 *for* $t = \ldots, -1, 0, 1, \ldots$, *if the roots of the associated polynomial equation are less than* 1 *in absolute value, and if the* u_t*'s are independently and identically distributed with mean* 0 *and variance* $\sigma^2 > 0$, *then* $\sqrt{T}(\hat{\boldsymbol{\beta}} - \boldsymbol{\beta})$ *has a limiting normal distribution with mean* $\mathbf{0}$ *and covariance matrix* $\sigma^2\tilde{\boldsymbol{F}}^{-1}$.

Theorems 5.5.6 and 5.5.7 are true if the condition on the distribution of disturbance is that moments of order $2 + \varepsilon$ are uniformly bounded instead of being identically distributed, or a Lindeberg-type condition could be used.

5.5.5. Case of Unknown Mean

We now suppose that the vector stochastic difference equation is

$$y_t + \mathbf{B}y_{t-1} + \boldsymbol{\nu} = \boldsymbol{u}_t \tag{57}$$

or equivalently

$$(y_t - \boldsymbol{\mu}) + \mathbf{B}(y_{t-1} - \boldsymbol{\mu}) = \boldsymbol{u}_t, \tag{58}$$

where

$$\boldsymbol{\nu} = -\boldsymbol{\mu} - \mathbf{B}\boldsymbol{\mu}. \tag{59}$$

Then the maximum likelihood equations for $\mathbf{B}$, $\boldsymbol{\nu}$, and $\boldsymbol{\Sigma}$ are

$$\begin{bmatrix} \sum_{t=1}^{T} y_{t-1}y'_{t-1} & \sum_{t=1}^{T} y_{t-1} \\ \sum_{t=1}^{T} y'_{t-1} & T \end{bmatrix} \begin{bmatrix} \hat{\mathbf{B}}' \\ \hat{\boldsymbol{\nu}}' \end{bmatrix} = - \begin{bmatrix} \sum_{t=1}^{T} y_{t-1}y'_t \\ \sum_{t=1}^{T} y'_t \end{bmatrix}, \tag{60}$$

$$\hat{\boldsymbol{\Sigma}} = \frac{1}{T}\sum_{t=1}^{T}(y_t + \hat{\mathbf{B}}y_{t-1} + \hat{\boldsymbol{\nu}})(y_t + \hat{\mathbf{B}}y_{t-1} + \hat{\boldsymbol{\nu}})'. \tag{61}$$

Equation (60) leads to

$$\left(\frac{1}{T}\sum_{t=1}^{T} y_{t-1}y'_{t-1} - \bar{y}_{(1)}\bar{y}'_{(1)}\right)\hat{\mathbf{B}}' = -\left(\frac{1}{T}\sum_{t=1}^{T} y_{t-1}y'_t - \bar{y}_{(1)}\bar{y}'\right), \tag{62}$$

$$\hat{\boldsymbol{\nu}} = -\bar{y} - \hat{\mathbf{B}}\bar{y}_{(1)}, \tag{63}$$

where $\bar{y} = (1/T)\sum_{t=1}^{T} y_t$ and $\bar{y}_{(1)} = (1/T)\sum_{t=1}^{T} y_{t-1}$. We want to show that the asymptotic behavior of $\hat{\mathbf{B}}$ defined by (62) is the same as that of $\hat{\mathbf{B}}$ defined by (3) when $\boldsymbol{\mu} = \mathbf{0}$. This will follow from showing that $\sqrt{T}\,\bar{y}_{(1)}\bar{y}'_{(1)}$ and $\sqrt{T}\,\bar{y}_{(1)}\bar{y}'$ converge stochastically to $\mathbf{0}$ when $\boldsymbol{\mu} = \mathbf{0}$.

First consider the case where (57) holds for $t = \ldots, -1, 0, 1, \ldots$. Then

$$y_t^* = \boldsymbol{\mu} + \sum_{s=0}^{\infty}(-\mathbf{B})^s u_{t-s}. \tag{64}$$

We see that $\mathscr{E}y_t^* = \boldsymbol{\mu}$. The covariance matrix of y_t^* and $y_{t'}^*$ for $t \leq t'$ is

$$\begin{aligned} \mathscr{E}(y_t^* - \boldsymbol{\mu})(y_{t'}^* - \boldsymbol{\mu})' &= \sum_{s,s'=0}^{\infty}(-\mathbf{B})^s \mathscr{E}u_{t-s}u'_{t'-s'}(-\mathbf{B}')^{s'} \\ &= \sum_{s=0}^{\infty}(-\mathbf{B})^s\boldsymbol{\Sigma}(-\mathbf{B}')^{s+t'-t} \\ &= F(-\mathbf{B}')^{t'-t}. \end{aligned} \tag{65}$$

Then

$$\begin{aligned} T\mathscr{E}(\bar{y}^* - \boldsymbol{\mu})(\bar{y}^* - \boldsymbol{\mu})' &= \frac{1}{T}\mathscr{E}\sum_{t,t'=1}^{T}(y_t^* - \boldsymbol{\mu})(y_{t'}^* - \boldsymbol{\mu})' \\ &= F + \frac{1}{T}\sum_{s=1}^{T-1}(T-s)F(-\mathbf{B}')^s + \frac{1}{T}\sum_{s=1}^{T-1}(T-s)(-\mathbf{B})^s F, \end{aligned} \tag{66}$$

which converges to

$$\text{(67)}\quad \lim_{T\to\infty} T\mathscr{E}(\bar{\boldsymbol{y}}^* - \boldsymbol{\mu})(\bar{\boldsymbol{y}}^* - \boldsymbol{\mu})' = \boldsymbol{F} + \boldsymbol{F}\sum_{s=1}^{\infty}(-\mathbf{B}')^s + \sum_{s=1}^{\infty}(-\mathbf{B})^s\boldsymbol{F}$$
$$= \boldsymbol{F} + \boldsymbol{F}(-\mathbf{B}')(\boldsymbol{I}+\mathbf{B}')^{-1} + (\boldsymbol{I}+\mathbf{B})^{-1}(-\mathbf{B})\boldsymbol{F}$$
$$= \boldsymbol{F}(\boldsymbol{I}+\mathbf{B}')^{-1} + (\boldsymbol{I}+\mathbf{B})^{-1}\boldsymbol{F} - \boldsymbol{F}.$$

By the Tchebycheff inequality $\sqrt{T}(\bar{\boldsymbol{y}}^* - \boldsymbol{\mu})(\bar{\boldsymbol{y}}^* - \boldsymbol{\mu})'$ converges stochastically to $\mathbf{0}$.

In the case where (57) holds for $t = 1, 2, \ldots$ and $\boldsymbol{y}_0$ is a fixed vector

$$\text{(68)}\qquad \boldsymbol{y}_t = \boldsymbol{\mu} + (-\mathbf{B})^t(\boldsymbol{y}_0 - \boldsymbol{\mu}) + \sum_{s=0}^{t-1}(-\mathbf{B})^s\boldsymbol{u}_{t-s}.$$

The difference between $\boldsymbol{y}_t^*$ defined by (64) and $\boldsymbol{y}_t$ defined by (68) is

$$\text{(69)}\qquad (-\mathbf{B})^t[\boldsymbol{y}_0^* - (\boldsymbol{y}_0 - \boldsymbol{\mu})],$$

where $\boldsymbol{y}_0^*$ is given by (7), and the difference between $\sqrt{T}\bar{\boldsymbol{y}}^*$ and $\sqrt{T}\bar{\boldsymbol{y}}$ defined by the two is

$$\text{(70)}\qquad \frac{1}{\sqrt{T}}\sum_{t=1}^{T}(-\mathbf{B})^t[\boldsymbol{y}_0^* - (\boldsymbol{y}_0 - \boldsymbol{\mu})];$$

the sum of expected values of the squares of the components of (70) converges to 0.

It follows from these facts that $\sqrt{T}\bar{\boldsymbol{y}}_{(1)}\bar{\boldsymbol{y}}_{(1)}'$ and $\sqrt{T}\bar{\boldsymbol{y}}_{(1)}\bar{\boldsymbol{y}}'$ converge stochastically to $\mathbf{0}$ if $\boldsymbol{\mu}$ is $\mathbf{0}$ in the second case as well. Thus for both models $\hat{\mathbf{B}}$ defined by (62) has the asymptotic properties of $\hat{\mathbf{B}}$ defined by (3).

THEOREM 5.5.8. *If $\boldsymbol{y}_t$ is defined by* (64) *or* (68), *where* $-\mathbf{B}$ *has characteristic roots less than* 1 *in absolute value, and if the $\boldsymbol{u}_t$'s are independently and identically distributed with $\mathscr{E}\boldsymbol{u}_t = \mathbf{0}$ and $\mathscr{E}\boldsymbol{u}_t\boldsymbol{u}_t' = \boldsymbol{\Sigma}$, then $\sqrt{(T}\bar{\boldsymbol{y}} - \boldsymbol{\mu})$ has a limiting normal distribution with mean* $\mathbf{0}$ *and covariance matrix* (67).

PROOF. Let $S_T = \sqrt{T}\boldsymbol{\phi}'(\bar{\boldsymbol{y}} - \boldsymbol{\mu})$; let $\boldsymbol{y}_t^{(k)}$ be (45) plus $\boldsymbol{\mu}$, $Z_{kT} = (1/\sqrt{T}) \times \sum_{t=1}^{T}\boldsymbol{\phi}'(\boldsymbol{y}_t^{(k)} - \boldsymbol{\mu})$, and $X_{kT} = S_T - Z_{kT}$. Then if $\boldsymbol{y}_t$ is defined by (64)

$$\text{(71)}\quad \mathscr{E}X_{kT}^2 = \boldsymbol{\phi}'(-\mathbf{B})^{k+1}[\boldsymbol{F}(\boldsymbol{I}+\mathbf{B}')^{-1} + (\boldsymbol{I}+\mathbf{B})^{-1}\boldsymbol{F} - \boldsymbol{F}](-\mathbf{B}')^{k+1}\boldsymbol{\phi} = M_k,$$

which converges to 0. By Theorem 7.7.5 we find that Z_{kT} has a limiting normal distribution as $T \to \infty$ with mean 0 and a variance depending on k which converges as $k \to \infty$. It follows by Corollary 7.7.1 that S_T has a limiting normal distribution, and hence $\sqrt{T}(\bar{\boldsymbol{y}} - \boldsymbol{\mu})$ has a limiting normal distribution. Use of

(70) and another application of Corollary 7.7.1 yields the theorem for $\mathbf{y}_t$ defined by (68).

The scalar stochastic difference equation of order p can be written in vector form with $\tilde{\boldsymbol{\nu}}$ having only the first component different from 0, namely ν. In fact, equations (60) for the model (57) with $\mathbf{B}$ replaced by $\tilde{\mathbf{B}}$, $\mathbf{y}_t$ by $\tilde{\mathbf{y}}_t$, and $\mathbf{u}_t$ by $\tilde{\mathbf{u}}_t$ forces the last $p - 1$ components of $\hat{\tilde{\boldsymbol{\nu}}}$ to be 0. Then (59) implies all of the components of $\tilde{\boldsymbol{\mu}}$ are $\mu = \nu/(1 + \sum_{j=1}^{p} \beta_j)$. The above results apply in this case; $\sqrt{T}(\bar{\tilde{\mathbf{y}}} - \tilde{\boldsymbol{\mu}})$ has the limiting distribution of $\sqrt{T}(\bar{y} - \mu)\boldsymbol{\varepsilon}$, where $\boldsymbol{\varepsilon} = (1, 1, \ldots, 1)'$.

5.5.6. Case of Fixed Variates

The maximum likelihood estimates were obtained when the model was

$$\sum_{s=0}^{p} \beta_s y_{t-s} + \sum_{j=1}^{q} \gamma_j z_{jt} = u_t. \tag{72}$$

This can be considered as a special case of the p-component vector model

$$\mathbf{y}_t + \mathbf{B}\mathbf{y}_{t-1} + \boldsymbol{\Gamma}\mathbf{z}_t = \mathbf{u}_t, \tag{73}$$

where $\boldsymbol{\Gamma}$ is a $p \times q$ matrix of constants and $\mathbf{z}_t = (z_{1t}, \ldots, z_{qt})'$. The maximum likelihood estimates of $\mathbf{B}$, $\boldsymbol{\Gamma}$, and $\boldsymbol{\Sigma}$ are defined by

$$\begin{bmatrix} \frac{1}{T}\sum_{t=1}^{T} \mathbf{y}_{t-1}\mathbf{y}_{t-1}' & \frac{1}{T}\sum_{t=1}^{T} \mathbf{y}_{t-1}\mathbf{z}_t' \\ \frac{1}{T}\sum_{t=1}^{T} \mathbf{z}_t\mathbf{y}_{t-1}' & \frac{1}{T}\sum_{t=1}^{T} \mathbf{z}_t\mathbf{z}_t' \end{bmatrix} \begin{bmatrix} \hat{\mathbf{B}}' \\ \hat{\boldsymbol{\Gamma}}' \end{bmatrix} = - \begin{bmatrix} \frac{1}{T}\sum_{t=1}^{T} \mathbf{y}_{t-1}\mathbf{y}_t' \\ \frac{1}{T}\sum_{t=1}^{T} \mathbf{z}_t\mathbf{y}_t' \end{bmatrix}, \tag{74}$$

$$\hat{\boldsymbol{\Sigma}} = \frac{1}{T}\sum_{t=1}^{T} (\mathbf{y}_t + \hat{\mathbf{B}}\mathbf{y}_{t-1} + \hat{\boldsymbol{\Gamma}}\mathbf{z}_t)(\mathbf{y}_t + \hat{\mathbf{B}}\mathbf{y}_{t-1} + \hat{\boldsymbol{\Gamma}}\mathbf{z}_t)'. \tag{75}$$

The asymptotic theory developed for $\hat{\mathbf{B}}$ and $\hat{\boldsymbol{\Sigma}}$ defined by (3) and (4) can be extended to $\hat{\mathbf{B}}$, $\hat{\boldsymbol{\Gamma}}$, and $\hat{\boldsymbol{\Sigma}}$ defined here. The details, however, are extremely laborious. Instead of giving them in full we shall only sketch the developments. [See T. W. Anderson and Rubin (1950) and Koopmans, Rubin, and Leipnik (1950).]

If (73) holds for $t = 1, 2, \ldots$ and y_0 is fixed, then

$$y_t = \sum_{s=0}^{t-1} (-\mathbf{B})^s u_{t-s} + w_t + (-\mathbf{B})^t y_0, \tag{76}$$

where

$$w_t = -\sum_{s=0}^{t-1} (-\mathbf{B})^s \mathbf{\Gamma} z_{t-s}. \tag{77}$$

If (73) holds for $t = \ldots, -1, 0, 1, \ldots$ and $z_t = \mathbf{0}$ for $t = 0, -1, \ldots$, then the solution is

$$y_t^* = \sum_{s=0}^{\infty} (-\mathbf{B})^s u_{t-s} + w_t. \tag{78}$$

The difference $y_t^* - y_t$ is $(-\mathbf{B})^t(y_0^* - y_0)$, where y_0^* is given by (7).

THEOREM 5.5.9. *If y_t is defined by (73) for $t = 1, 2, \ldots$ with y_0 a fixed vector and y_t^* is defined by (73) for $t = \ldots, -1, 0, 1, \ldots$, where the u_t's are independently distributed with $\mathscr{E} u_t = \mathbf{0}$ and $\mathscr{E} u_t u_t' = \mathbf{\Sigma}$, if $-\mathbf{B}$ has characteristic roots less than 1 in absolute value, and if $z_t' z_t < N$, $t = 1, 2, \ldots$, for some N, and $z_t = \mathbf{0}$, $t = 0, -1, \ldots$, then* (18), (19), (20) *hold and*

$$\operatorname*{plim}_{T \to \infty} \left[\frac{1}{\sqrt{T}} \sum_{t=1}^{T} y_{t-1}^* z_t' - \frac{1}{\sqrt{T}} \sum_{t=1}^{T} y_{t-1} z_t' \right] = \mathbf{0}, \tag{79}$$

$$\operatorname*{plim}_{T \to \infty} \left[\frac{1}{\sqrt{T}} \sum_{t=1}^{T} z_t y_t^{*\prime} - \frac{1}{\sqrt{T}} \sum_{t=1}^{T} z_t y_t' \right] = \mathbf{0}. \tag{80}$$

PROOF. We have

$$\begin{aligned} (81)\quad & \sum_{t=1}^{T} y_t^* y_t^{*\prime} - \sum_{t=1}^{T} y_t y_t' \\ &= \sum_{t=1}^{T} (-\mathbf{B})^t (y_0^* - y_0) \left[\sum_{s=0}^{t-1} (-\mathbf{B})^s u_{t-s} + (-\mathbf{B})^t y_0 \right]' + \sum_{t=1}^{T} (-\mathbf{B})^t (y_0^* - y_0) w_t' \\ &\quad + \sum_{t=1}^{T} \left[\sum_{s=0}^{t-1} (-\mathbf{B})^s u_{t-s} + (-\mathbf{B})^t y_0 \right] (y_0^* - y_0)' (-\mathbf{B}')^t \\ &\quad + \sum_{t=1}^{T} w_t (y_0^* - y_0)' (-\mathbf{B}')^t + \sum_{t=1}^{T} (-\mathbf{B})^t (y_0^* - y_0)(y_0^* - y_0)' (-\mathbf{B}')^t. \end{aligned}$$

The first, third, and fifth terms divided by $\sqrt{T}$ converge stochastically to $\mathbf{0}$ by the proof of Theorem 5.5.1. The sum of expected values of squares of components of the second term is

$$(82)\quad \operatorname{tr} \mathscr{E} \sum_{t,t'=1}^{T} (-\mathbf{B})^t(y_0^* - y_0)w_t'w_{t'}(y_0^* - y_0)'(-\mathbf{B}')^{t'}$$
$$= \sum_{t,t'=1}^{T} w_t'w_{t'} \operatorname{tr} (-\mathbf{B})^t(F + y_0y_0')(-\mathbf{B}')^{t'}.$$

Since $|w_t'w_{t'}| \le \sqrt{w_t'w_t}\sqrt{w_{t'}'w_{t'}}$,

$$(83)\qquad w_t'w_t = \sum_{s,s'=0}^{t-1} z_{t-s}'\mathbf{\Gamma}'(-\mathbf{B}')^s(-\mathbf{B})^{s'}\mathbf{\Gamma} z_{t-s'}$$

and each element of z_t is less than $\sqrt{N}$, (82) is uniformly bounded (by Lemma 5.5.1). Hence, (81) divided by $\sqrt{T}$ converges stochastically to $\mathbf{0}$.

Now consider

$$(84)\qquad \sum_{t=1}^{T} y_{t-1}^* z_t' - \sum_{t=1}^{T} y_{t-1}z_t' = \sum_{t=1}^{T} (-\mathbf{B})^{t-1}(y_0^* - y_0)z_t'.$$

The sum of expected values of the squares of the components of (84) is bounded; the argument is that given above with w_t replaced by z_t. This proves (79).

Then

$$(85)\quad \sum_{t=1}^{T} y_{t-1}^* y_t^{*\prime} - \sum_{t=1}^{T} y_{t-1}y_t' = -\left[\sum_{t=1}^{T} y_{t-1}^* y_{t-1}^{*\prime} - \sum_{t=1}^{T} y_{t-1}y_{t-1}'\right]\mathbf{B}'$$
$$- \left[\sum_{t=1}^{T} y_{t-1}^* z_t' - \sum_{t=1}^{T} y_{t-1}z_t'\right]\mathbf{\Gamma}' + \sum_{t=1}^{T} (y_{t-1}^* - y_{t-1})u_t',$$

and the sum of expected values of the squares of the components of the last matrix is uniformly bounded. (See the proof of Theorem 5.5.1.)

Finally

$$(86)\qquad \sum_{t=1}^{T} y_t^* z_t' - \sum_{t=1}^{T} y_t z_t' = -\mathbf{B}\left[\sum_{t=1}^{T} y_{t-1}^* z_t' - \sum_{t=1}^{T} y_{t-1}z_t'\right].$$

Thus, the theorem is proved.

As in Section 5.5.2 the asymptotic properties of $\hat{\mathbf{B}}$, $\hat{\mathbf{\Gamma}}$ and $\hat{\mathbf{\Sigma}}$ (under appropriate conditions) are the same for y_t defined by (73) for $t = 1, 2, \ldots$ with y_0 a fixed vector as for y_t^* defined by (73) for $t = \ldots, -1, 0, 1, \ldots$ or equivalently by (78). In the sequel we shall for convenience treat only the latter case and shall use the notation y_t to denote (78).

LEMMA 5.5.8. *If $\mathbf{y}_t$ is defined by* (78), *where the $\mathbf{u}_t$'s are independently distributed with $\mathscr{E}\mathbf{u}_t = \mathbf{0}$ and $\mathscr{E}\mathbf{u}_t\mathbf{u}_t' = \mathbf{\Sigma}$, if $-\mathbf{B}$ has characteristic roots less than* 1 *in absolute value, and if $z_t'z_t < N$, $t = 1, 2, \ldots,$ for some N, then*

(87) $$\operatorname*{plim}_{T\to\infty} \frac{1}{T}\sum_{t=1}^{T} \mathbf{u}_t\mathbf{y}_{t-1}' = \mathbf{0},$$

(88) $$\operatorname*{plim}_{T\to\infty} \frac{1}{T}\sum_{t=1}^{T} \mathbf{u}_t\mathbf{z}_t' = \mathbf{0}.$$

PROOF. The sum of the expected values of the squares of the components of (87) is

(89) $$\begin{aligned}\frac{1}{T^2}\operatorname{tr}\sum_{t,t'=1}^{T}\mathscr{E}\mathbf{y}_{t-1}\mathbf{u}_t'\mathbf{u}_{t'}\mathbf{y}_{t'-1}' &= \frac{1}{T^2}\operatorname{tr}\mathbf{\Sigma}\operatorname{tr}\sum_{t=1}^{T}\mathscr{E}\mathbf{y}_{t-1}\mathbf{y}_{t-1}'\\ &= \frac{1}{T}\operatorname{tr}\mathbf{\Sigma}\left(\operatorname{tr}\boldsymbol{F} + \frac{1}{T}\sum_{t=1}^{T}\mathbf{w}_{t-1}'\mathbf{w}_{t-1}\right),\end{aligned}$$

which converges to 0. The sum of expected values of the squares of the components of (88) is

(90) $$\begin{aligned}\frac{1}{T^2}\operatorname{tr}\sum_{t,t'=1}^{T}\mathscr{E}\mathbf{u}_t\mathbf{z}_t'\mathbf{z}_{t'}\mathbf{u}_{t'}' &= \frac{1}{T^2}\operatorname{tr}\mathbf{\Sigma}\sum_{t=1}^{T}\mathbf{z}_t'\mathbf{z}_t\\ &< \frac{1}{T}\operatorname{tr}\mathbf{\Sigma}N,\end{aligned}$$

which converges to 0. Q.E.D.

Lemma 5.5.8 yields

(91) $$\operatorname*{plim}_{T\to\infty}\left[\frac{1}{T}\sum_{t=1}^{T}\mathbf{y}_t\mathbf{y}_{t-1}' + \mathbf{B}\frac{1}{T}\sum_{t=1}^{T}\mathbf{y}_{t-1}\mathbf{y}_{t-1}' + \mathbf{\Gamma}\frac{1}{T}\sum_{t=1}^{T}\mathbf{z}_t\mathbf{y}_{t-1}'\right] = \mathbf{0},$$

(92) $$\operatorname*{plim}_{T\to\infty}\left[\frac{1}{T}\sum_{t=1}^{T}\mathbf{y}_t\mathbf{z}_t' + \mathbf{B}\frac{1}{T}\sum_{t=1}^{T}\mathbf{y}_{t-1}\mathbf{z}_t' + \mathbf{\Gamma}\frac{1}{T}\sum_{t=1}^{T}\mathbf{z}_t\mathbf{z}_t'\right] = \mathbf{0}.$$

Lemma 5.5.2, (91), and (92) imply

(93) $$\operatorname*{plim}_{T\to\infty}\left[\frac{1}{T}\sum_{t=1}^{T}\mathbf{y}_t\mathbf{y}_t' + \mathbf{B}\frac{1}{T}\sum_{t=1}^{T}\mathbf{y}_{t-1}\mathbf{y}_t' + \mathbf{\Gamma}\frac{1}{T}\sum_{t=1}^{T}\mathbf{z}_t\mathbf{y}_t'\right] = \mathbf{\Sigma}.$$

Now we assume the existence of the following limits:

$$\lim_{T\to\infty} \frac{1}{T}\sum_{t=1}^{T} z_t z_t' = M, \tag{94}$$

$$\lim_{T\to\infty} \frac{1}{T}\sum_{t=1}^{T} w_{t-1} z_t' = L. \tag{95}$$

Then

$$\mathscr{E}\frac{1}{T}\sum_{t=1}^{T} y_{t-1} z_t' = \frac{1}{T}\sum_{t=1}^{T} w_{t-1} z_t' \to L, \tag{96}$$

and

$$\begin{aligned}\operatorname{tr}\mathscr{E}&\left[\frac{1}{T}\sum_{t=1}^{T}(y_{t-1}-w_{t-1})z_t'\right]\left[\frac{1}{T}\sum_{t'=1}^{T}(y_{t'-1}-w_{t'-1})z_{t'}'\right]' \\ &= \frac{1}{T^2}\sum_{t,t'=1}^{T} z_t' z_{t'} \operatorname{tr}\mathscr{E}(y_{t-1}-w_{t-1})(y_{t'-1}-w_{t'-1})' \\ &< N\frac{1}{T}\operatorname{tr}\left[F(I+\mathbf{B}')^{-1}+(I+\mathbf{B})^{-1}F-F\right],\end{aligned} \tag{97}$$

which converges to 0. Thus

$$\operatorname*{plim}_{T\to\infty} \frac{1}{T}\sum_{t=1}^{T} y_{t-1} z_t' = L. \tag{98}$$

From (92) and (98) we deduce

$$\operatorname*{plim}_{T\to\infty} \frac{1}{T}\sum_{t=1}^{T} y_t z_t' = -\mathbf{B}L - \boldsymbol{\Gamma}M; \tag{99}$$

from (91) and (93) and the fact that (30) converges stochastically to $\mathbf{0}$ we have

$$\begin{aligned}\boldsymbol{\Sigma} &= \operatorname*{plim}_{T\to\infty}\left(\frac{1}{T}\sum_{t=1}^{T} y_t y_t' - \mathbf{B}\frac{1}{T}\sum_{t=1}^{T} y_{t-1}y_{t-1}'\mathbf{B}' - \mathbf{B}\frac{1}{T}\sum_{t=1}^{T} y_{t-1}z_t'\boldsymbol{\Gamma}' + \boldsymbol{\Gamma}\frac{1}{T}\sum_{t=1}^{T} z_t y_t'\right) \\ &= \operatorname*{plim}_{T\to\infty}\left(\frac{1}{T}\sum_{t=1}^{T} y_t y_t' - \mathbf{B}\frac{1}{T}\sum_{t=1}^{T} y_t y_t'\mathbf{B}'\right) - \mathbf{B}L\boldsymbol{\Gamma}' - \boldsymbol{\Gamma}L'\mathbf{B}' - \boldsymbol{\Gamma}M\boldsymbol{\Gamma}'.\end{aligned} \tag{100}$$

This shows that $(1/T)\sum_{t=1}^{T} y_t y_t'$ has a probability limit which must be

$$\lim_{T\to\infty}\mathscr{E}\frac{1}{T}\sum_{t=1}^{T} y_t y_t' = F + \lim_{T\to\infty}\frac{1}{T}\sum_{t=1}^{T} w_t w_t' = F + H, \tag{101}$$

where we assume the existence of the limit

$$H = \lim_{T\to\infty} \frac{1}{T}\sum_{t=1}^{T} w_t w_t'. \tag{102}$$

THEOREM 5.5.10. *If the characteristic roots of* $-\mathbf{B}$ *are less than* 1 *in absolute value, if the limits* (94), (95), *and* (102) *exist, if* $z_t'z_t < N$, $t = 1, 2, \ldots,$ *for some* N, *if either the* $\mathbf{u}_t$*'s are identically distributed or* $\mathscr{E}\,|u_{it}|^{2+\varepsilon} < m$, $i = 1, \ldots, p$, $t = 1, 2, \ldots,$ *for some* $\varepsilon > 0$ *and some* m, *and if the* $\mathbf{u}_t$*'s are independently distributed with* $\mathscr{E}\mathbf{u}_t = \mathbf{0}$ *and* $\mathscr{E}\mathbf{u}_t\mathbf{u}_t' = \mathbf{\Sigma}$, *then*

$$\operatorname*{plim}_{T\to\infty}\begin{bmatrix} \frac{1}{T}\sum_{t=1}^{T} y_{t-1}y_{t-1}' & \frac{1}{T}\sum_{t=1}^{T} y_{t-1}z_t' \\ \frac{1}{T}\sum_{t=1}^{T} z_t y_{t-1}' & \frac{1}{T}\sum_{t=1}^{T} z_t z_t' \end{bmatrix} = \begin{pmatrix} F + H & L \\ L' & M \end{pmatrix}. \tag{103}$$

To prove consistency of $\hat{\mathbf{B}}$ and $\hat{\mathbf{\Gamma}}$ we want (103) to be nonsingular.

LEMMA 5.5.9. *If* F *and* M *are positive definite,* (103) *is positive definite.*

PROOF. Since

$$\begin{aligned} (x'\, y')\begin{pmatrix} H & L \\ L' & M \end{pmatrix}\begin{pmatrix} x \\ y \end{pmatrix} &= \lim_{T\to\infty}\frac{1}{T}\sum_{t=1}^{T}(x'\, y')\begin{pmatrix} w_{t-1} \\ z_t \end{pmatrix}(w_{t-1}'\, z_t')\begin{pmatrix} x \\ y \end{pmatrix} \\ &= \lim_{T\to\infty}\frac{1}{T}\sum_{t=1}^{T}[x'w_{t-1} + y'z_t]^2 \geq 0, \end{aligned} \tag{104}$$

the matrix of the quadratic form is positive semidefinite. The quadratic form

$$(x'\, y')\begin{pmatrix} F + H & L \\ L' & M \end{pmatrix}\begin{pmatrix} x \\ y \end{pmatrix} = x'Fx + (x'\, y')\begin{pmatrix} H & L \\ L' & M \end{pmatrix}\begin{pmatrix} x \\ y \end{pmatrix} \tag{105}$$

is the sum of two nonnegative quadratic forms and hence can be 0 only if both forms are 0. Since F is positive definite, $x'Fx = 0$ implies $x = \mathbf{0}$. Then for (105) to be 0, $y'My = 0$, which implies $y = \mathbf{0}$. Q.E.D.

THEOREM 5.5.11. *Under the conditions of Theorem* 5.5.10 *if* F *is positive definite, and if* M *defined by* (94) *is nonsingular, then*

$$\operatorname*{plim}_{T\to\infty}\hat{\mathbf{B}} = \mathbf{B}, \tag{106}$$

$$\operatorname*{plim}_{T\to\infty}\hat{\mathbf{\Gamma}} = \mathbf{\Gamma}, \tag{107}$$

$$\operatorname*{plim}_{T\to\infty}\hat{\mathbf{\Sigma}} = \mathbf{\Sigma}. \tag{108}$$

We now turn to the asymptotic normality of $(1/\sqrt{T})\sum_{t=1}^T y_{t-1}u_t'$ and $(1/\sqrt{T})\sum_{t=1}^T z_t u_t'$. The means are $\mathbf{0}$. Evaluations similar to those of (89) and (90) show

$$\lim_{T\to\infty}\frac{1}{T}\mathscr{E}\sum_{t=1}^T y_{t-1}u_t' \otimes \sum_{t=1}^T u_t y_{t-1}' = (F+H)\otimes\Sigma, \tag{109}$$

$$\lim_{T\to\infty}\frac{1}{T}\mathscr{E}\sum_{t=1}^T y_{t-1}u_t' \otimes \sum_{t=1}^T u_t z_t' = L\otimes\Sigma, \tag{110}$$

$$\lim_{T\to\infty}\frac{1}{T}\mathscr{E}\sum_{t=1}^T z_t u_t' \otimes \sum_{t=1}^T u_t z_t' = M\otimes\Sigma. \tag{111}$$

Now let

$$S_T = \frac{1}{\sqrt{T}}\sum_{t=1}^T u_t'(\Phi y_{t-1} + \Theta z_t), \tag{112}$$

$$Z_{kT} = \frac{1}{\sqrt{T}}\sum_{t=1}^T u_t'(\Phi y_{t-1}^{(k)} + \Theta z_t), \tag{113}$$

with $y_t^{(k)}$ and X_{kT} as before. Here the terms in the sum Z_{kT} are not identically distributed even if the u_t's are identically distributed. If the fourth-order moments of the u_t's are uniformly bounded, it is easy to show that the fourth-order moments of the summands in $\sqrt{T}Z_{kT}$ are uniformly bounded. It is more difficult to show (by use of Hölder's inequality) that if $\mathscr{E}|u_{it}|^{2+\varepsilon}$ is uniformly bounded, then the moments of the summands of order $2+\varepsilon$ are uniformly bounded.

THEOREM 5.5.12. *If y_t is defined by* (76) *or* (78), *where the u_t's are independently distributed with $\mathscr{E}u_t = \mathbf{0}$ and $\mathscr{E}u_t u_t' = \Sigma$, if the characteristic roots of $-\mathbf{B}$ are less than* 1 *in absolute value, if $z_t'z_t < N$, $t = 1, 2, \ldots,$ for some N, if the limits* (94), (95), *and* (102) *exist, and if $\mathscr{E}u_{it}^4 < N^*$, $i = 1, \ldots, p, t = 1, 2, \ldots,$ for some N^*, then*

$$\begin{bmatrix} \dfrac{1}{\sqrt{T}}\sum_{t=1}^T y_{t-1}u_t' \\ \dfrac{1}{\sqrt{T}}\sum_{t=1}^T z_t u_t' \end{bmatrix} \tag{114}$$

has a limiting normal distribution with mean $\mathbf{0}$ *and covariance matrix*

$$\begin{pmatrix} \boldsymbol{F}+\boldsymbol{H} & \boldsymbol{L} \\ \boldsymbol{L}' & \boldsymbol{M} \end{pmatrix} \otimes \boldsymbol{\Sigma}. \tag{115}$$

THEOREM 5.5.13. *If the conditions of Theorem* 5.5.12 *hold and if* $\boldsymbol{F} = \sum_{s=0}^{\infty} \mathbf{B}^s \boldsymbol{\Sigma} \mathbf{B}'^s$ *and* $\boldsymbol{M} = \lim_{T\to\infty} (1/T) \sum_{t=1}^{T} z_t z_t'$ *are positive definite, then*

$$\sqrt{T}(\hat{\mathbf{B}}' - \mathbf{B}'), \qquad \sqrt{T}(\hat{\boldsymbol{\Gamma}}' - \boldsymbol{\Gamma}') \tag{116}$$

have limiting normal distribution with mean $\mathbf{0}$ *and covariance matrix*

$$\begin{pmatrix} \boldsymbol{F}+\boldsymbol{H} & \boldsymbol{L} \\ \boldsymbol{L}' & \boldsymbol{M} \end{pmatrix}^{-1} \otimes \boldsymbol{\Sigma}. \tag{117}$$

The nonsingularity of $\boldsymbol{F}$ and $\boldsymbol{M}$ is assumed so that the inverse matrix in (117) exists. If $\boldsymbol{\Sigma}$ is singular, the limiting normal distribution is singular. If z_t is constant and scalar, then this theorem follows if the $\boldsymbol{u}_t$'s are identically distributed with covariance matrix $\boldsymbol{\Sigma}$, with no condition on higher-order moments because then the $\boldsymbol{u}_t'(\boldsymbol{\Phi} \boldsymbol{y}_{t-1}^{(k)} + \boldsymbol{\theta} z_t)$ form a stationary finitely-dependent process. In general the conditions for z_t and $\boldsymbol{u}_t$ can be weakened to the extent that the $\boldsymbol{u}_t'(\boldsymbol{\Phi} \boldsymbol{y}_{t-1}^{(k)} + \boldsymbol{\Theta} z_t)$ satisfy a Lindeberg-type condition. The condition $z_t' z_t < N$ can be weakened to $\sum_{t=1}^{\infty} \lambda^t |z_{jt}| < \infty$, $j = 1, \ldots, q$.

THEOREM 5.5.14. *If* y_t *is defined by* (1) *of Section* 5.4, *where the* u_t*'s are independently distributed with* $\mathscr{E} u_t = 0$ *and* $\mathscr{E} u_t^2 = \sigma^2 > 0$, *if the roots of the associated polynomial equation are less than* 1 *in absolute value, if* $z_t' z_t < N$, $t = 1, 2, \ldots,$ *for some* N, *if* $\mathscr{E} u_t^4 < m$, $t = 1, 2, \ldots,$ *for some* m, *if* $\boldsymbol{M}$ *defined by* (94) *is positive definite, and if the limits* (95) *and* (102) *exist, where*

$$\tilde{\boldsymbol{w}}_t = -\sum_{s=0}^{t-1} (-\tilde{\mathbf{B}})^s \begin{bmatrix} \boldsymbol{\gamma}' z_{t-s} \\ 0 \\ \cdot \\ \cdot \\ \cdot \\ 0 \end{bmatrix}, \tag{118}$$

then $\sqrt{T}(\hat{\boldsymbol{\beta}} - \boldsymbol{\beta})$, $\sqrt{T}(\hat{\boldsymbol{\gamma}} - \boldsymbol{\gamma})$ *have a limiting normal distribution with mean* $\mathbf{0}$ *and covariance matrix*

$$\sigma^2 \begin{pmatrix} \tilde{F} + \tilde{H} & \tilde{L} \\ \tilde{L}' & M \end{pmatrix}^{-1}. \tag{119}$$

5.6. STATISTICAL INFERENCE ABOUT AUTOREGRESSIVE MODELS BASED ON LARGE-SAMPLE THEORY

5.6.1. Relation to Regression Methods

In Section 5.4 we showed that the maximum likelihood estimates of $\boldsymbol{\beta} = (\beta_1, \ldots, \beta_p)'$, $\boldsymbol{\gamma} = (\gamma_1, \ldots, \gamma_q)'$, and σ^2 are based on minimizing

$$\sum_{t=1}^{T} \left(y_t + \sum_{i=1}^{p} b_i y_{t-i} + \sum_{j=1}^{q} c_j z_{jt} \right)^2, \tag{1}$$

and hence these estimates are formally least squares estimates. This implies that all computational procedures of ordinary regression analysis are available.

In Section 5.5 we showed that under appropriate conditions the estimates $\hat{\boldsymbol{\beta}}$ and $\hat{\boldsymbol{\gamma}}$ had asymptotically a normal distribution just as in the case of ordinary regression. Thus for large samples (that is, large T) we can treat them as normally distributed and the procedures of ordinary regression analysis can be used. The approximating multivariate normal distribution of $\sqrt{T}(\hat{\boldsymbol{\beta}} - \boldsymbol{\beta})$ and $\sqrt{T}(\hat{\boldsymbol{\gamma}} - \boldsymbol{\gamma})$ has covariance matrix

$$\sigma^2 \begin{pmatrix} F + H & L \\ L' & M \end{pmatrix}^{-1}, \tag{2}$$

which in turn is approximated by

$$s^2 \begin{bmatrix} A_T & L_T \\ L_T' & M_T \end{bmatrix}^{-1}, \tag{3}$$

where

(4) $$s^2 = \frac{1}{T-p-q}\sum_{t=1}^{T}(y_t + \hat{\boldsymbol{\beta}}'\tilde{\boldsymbol{y}}_{t-1} + \hat{\boldsymbol{\gamma}}'\boldsymbol{z}_t)^2 = \frac{T\hat{\sigma}^2}{T-p-q},$$

(5) $$\boldsymbol{A}_T = \frac{1}{T}\sum_{t=1}^{T}\tilde{\boldsymbol{y}}_{t-1}\tilde{\boldsymbol{y}}_{t-1}'$$

$$= \begin{bmatrix} \frac{1}{T}\sum_{t=1}^{T} y_{t-1}^2 & \frac{1}{T}\sum_{t=1}^{T} y_{t-1}y_{t-2} & \cdots & \frac{1}{T}\sum_{t=1}^{T} y_{t-1}y_{t-p} \\ \frac{1}{T}\sum_{t=1}^{T} y_{t-2}y_{t-1} & \frac{1}{T}\sum_{t=1}^{T} y_{t-2}^2 & \cdots & \frac{1}{T}\sum_{t=1}^{T} y_{t-2}y_{t-p} \\ \vdots & \vdots & & \vdots \\ \frac{1}{T}\sum_{t=1}^{T} y_{t-p}y_{t-1} & \frac{1}{T}\sum_{t=1}^{T} y_{t-p}y_{t-2} & \cdots & \frac{1}{T}\sum_{t=1}^{T} y_{t-p}^2 \end{bmatrix},$$

(6) $$\boldsymbol{L}_T = \frac{1}{T}\sum_{t=1}^{T}\tilde{\boldsymbol{y}}_{t-1}\boldsymbol{z}_t'$$

$$= \begin{bmatrix} \frac{1}{T}\sum_{t=1}^{T} y_{t-1}z_{1t} & \frac{1}{T}\sum_{t=1}^{T} y_{t-1}z_{2t} & \cdots & \frac{1}{T}\sum_{t=1}^{T} y_{t-1}z_{qt} \\ \frac{1}{T}\sum_{t=1}^{T} y_{t-2}z_{1t} & \frac{1}{T}\sum_{t=1}^{T} y_{t-2}z_{2t} & \cdots & \frac{1}{T}\sum_{t=1}^{T} y_{t-2}z_{qt} \\ \vdots & \vdots & & \vdots \\ \frac{1}{T}\sum_{t=1}^{T} y_{t-p}z_{1t} & \frac{1}{T}\sum_{t=1}^{T} y_{t-p}z_{2t} & \cdots & \frac{1}{T}\sum_{t=1}^{T} y_{t-p}z_{qt} \end{bmatrix},$$

(7) $$M_T = \frac{1}{T}\sum_{t=1}^{T} z_t z_t'$$

$$= \begin{bmatrix} \frac{1}{T}\sum_{t=1}^{T} z_{1t}^2 & \frac{1}{T}\sum_{t=1}^{T} z_{1t}z_{2t} & \cdots & \frac{1}{T}\sum_{t=1}^{T} z_{1t}z_{qt} \\ \frac{1}{T}\sum_{t=1}^{T} z_{2t}z_{1t} & \frac{1}{T}\sum_{t=1}^{T} z_{2t}^2 & \cdots & \frac{1}{T}\sum_{t=1}^{T} z_{2t}z_{qt} \\ \vdots & \vdots & & \vdots \\ \frac{1}{T}\sum_{t=1}^{T} z_{qt}z_{1t} & \frac{1}{T}\sum_{t=1}^{T} z_{qt}z_{2t} & \cdots & \frac{1}{T}\sum_{t=1}^{T} z_{qt}^2 \end{bmatrix}.$$

Usually, there will be a constant which is included by letting $z_{1t} = 1$. Then when $\hat{\gamma}_1$ is eliminated all the sums of squares and cross-products are in terms of deviations from the corresponding means. (We have dropped $\sim$ from $\boldsymbol{F}$, $\boldsymbol{H}$, and $\boldsymbol{L}$ because in Section 5.6 the general vector case is not treated.)

As remarked at the end of Section 5.4, the terms on any diagonal of $(1/T)\sum_{t=1}^{T} \tilde{y}_{t-1}\tilde{y}_{t-1}'$ differ only with respect to the terms at the beginning and end of the series. As far as the large-sample theory goes, these end-effects are negligible and the same term could be used in each position on a given diagonal. Moreover, the divisor T could be modified. This suggests

(8) $$\frac{1}{T+p-s}\sum_{t=-p+s+1}^{T} y_t y_{t-s}, \qquad s = 0, 1, \ldots, p,$$

on the diagonal s above or below the main diagonal. If the mean $\bar{y}^* = \sum_{t=-(p-1)}^{T} y_t/(T+p)$ is subtracted, the sum is

(9) $$\frac{1}{T+p-s}\sum_{t=-p+s+1}^{T} (y_t - \bar{y}^*)(y_{t-s} - \bar{y}^*).$$

Other modifications are possible such as $\sum_{t=-p+s+1}^{T} y_t y_{t-s}/(T+p-s) - \bar{y}^{*2}$. Note that we have available observations from $-p+1$ to T.

5.6.2. Tests of Hypotheses and Confidence Regions

From the usual regression theory we verify that each of

$$(10) \qquad \sqrt{T}\frac{\hat{\beta}_i - \beta_i}{s\sqrt{\bar{a}^{ii}}}, \qquad \sqrt{T}\frac{\hat{\gamma}_j - \gamma_j}{s\sqrt{\bar{m}^{jj}}},$$

where $s^2\bar{a}^{ii}$ and $s^2\bar{m}^{jj}$ are the ith diagonal element and $(p+j)$th diagonal element, respectively, of the matrix in (3), has approximately a normal distribution with mean 0 and variance 1. This enables us to test a hypothesis about β_i or γ_j or set up a confidence interval for β_i or γ_j in the usual way.

Suppose we are interested in a set of coefficients, say, $\beta_{i_1}, \ldots, \beta_{i_g}, \gamma_{j_1}, \ldots, \gamma_{j_h}$. We can use the fact that

$$(11) \qquad \frac{1}{s^2}\left\{\sum_{u,v=1}^{g}(\hat{\beta}_{i_u} - \beta_{i_u})(\hat{\beta}_{i_v} - \beta_{i_v})\bar{a}_{i_u i_v} + 2\sum_{u=1}^{g}\sum_{v=1}^{h}(\hat{\beta}_{i_u} - \beta_{i_u})(\hat{\gamma}_{j_v} - \gamma_{j_v})\bar{l}_{i_u j_v} + \sum_{u,v=1}^{h}(\hat{\gamma}_{j_u} - \gamma_{j_u})(\hat{\gamma}_{j_v} - \gamma_{j_v})\bar{m}_{j_u j_v}\right\},$$

where $\bar{a}_{i_u i_v}$, $\bar{l}_{i_u j_v}$, $\bar{m}_{j_u j_v}$ form the matrix inverse to the matrix consisting of rows and columns $i_1, \ldots, i_g$, $j_1 + p, \ldots, j_h + p$ of $\begin{pmatrix} A_T & L_T \\ L_T' & M_T \end{pmatrix}^{-1}$, has an approximate χ^2-distribution with $g + h$ degrees of freedom. This leads to tests of hypotheses and confidence regions. (The F-distribution with $g + h$ and $T - (p + q)$ degrees of freedom could also be used.)

5.6.3. Testing the Order of an Autoregressive Process

The order of a stochastic difference equation is determined by the highest lag which has a nonzero coefficient; as we have written the equation the order is p $(\beta_p \neq 0)$. Apart from the effect of any "independent variables" z_{it}, at most p lagged variables, $y_{t-1}, \ldots, y_{t-p}$, directly affect y_t; only these need be used in prediction. It is of interest, therefore, to determine the order of an autoregressive model. We shall now study this problem when the difference equation is

$$(12) \qquad y_t + \beta_1 y_{t-1} + \cdots + \beta_p y_{t-p} + \gamma = u_t.$$

In this case the estimating equations for $\beta_1, \ldots, \beta_p$ are given by

$$(13) \qquad \sum_{j=1}^{p} a_{ij}^* \hat{\beta}_j = -a_{i0}^*, \qquad i = 1, \ldots, p,$$

where

$$a_{ij}^* = \sum_{t=1}^{T} y_{t-i}y_{t-j} - T\bar{y}_{(i)}\bar{y}_{(j)}, \qquad i, j = 1, \ldots, p, \tag{14}$$

$$a_{i0}^* = \sum_{t=1}^{T} y_{t-i}y_t - T\bar{y}_{(i)}\bar{y}, \qquad i = 1, \ldots, p, \tag{15}$$

and $\bar{y} = \bar{y}_{(0)}$ and $\bar{y}_{(i)}$ is given by (17) of Section 5.4. [The equations (13) are (18) of Section 5.4.]

First let us consider deciding whether the order is $p - 1$ or p; that is, deciding whether $\beta_p = 0$. We formulate this as a problem of testing the null hypothesis H: $\beta_p = 0$ against alternatives $\beta_p \neq 0$. The likelihood ratio test is the analogue of the two-sided t-test in the regression problem; we reject the null hypothesis if

$$\frac{|\hat{\beta}_p|}{s\sqrt{a^{*pp}}} > t(\varepsilon), \tag{16}$$

where $(a^{*ij}) = (a_{ij}^*)^{-1}$ and $t(\varepsilon)$ is the (two-sided) ε-significance point of the standardized normal distribution. For sufficiently large T, this test has approximate significance level ε.

Suppose the null hypothesis is not true. In general

$$\operatorname*{plim}_{T\to\infty} \left(\frac{1}{T} a_{ij}^*\right) = F \tag{17}$$

given by (37) of Section 5.5, $\text{plim}_{T\to\infty} s = \sigma$, and $\text{plim}_{T\to\infty} \hat{\beta}_p = \beta_p$. Then roughly the test statistic behaves like $\sqrt{T}\,|\beta_p|/(\sigma\sqrt{f^{pp}})$. This argument can be made rigorous to prove that the test is consistent; that is, for a given alternative $\beta_p \neq 0$, the probability of rejecting the null hypothesis approaches 1 as T increases.

A more interesting study of the power of the test can be based on considering a sequence of alternative hypotheses β_{pT} (with σ^2, $\beta_1, \ldots, \beta_{p-1}$ fixed and u_t normally distributed, say) such that

$$\lim_{T\to\infty} \sqrt{T}\,\beta_{pT} = \phi \neq 0. \tag{18}$$

Then $\hat{\beta}_p/(s\sqrt{a^{*pp}})$ has a limiting normal distribution with variance 1 and mean $\phi/(\sigma\sqrt{f^{pp}})$. To make the argument rigorous (and some later arguments rigorous) we could prove that $\sqrt{T}\,(\hat{\mathbf{B}} - \mathbf{B}_T)$ has a limiting normal distribution when $\lim_{T\to\infty} \sqrt{T}\,(\mathbf{B} - \mathbf{B}_T) = \mathbf{\Phi}$ for suitable $\mathbf{B}$, with mean matrix $\mathbf{\Phi}$ and covariances corresponding to $\mathbf{B}$. (Here $\hat{\mathbf{B}}$, $\mathbf{B}$, and $\mathbf{B}_T$ refer to the Markov vector case.)

An interesting point is the following:

LEMMA 5.6.1. *If* $\beta_p = 0$, *then*

$$\sigma^2 f^{pp} = 1. \tag{19}$$

PROOF. We have (Problem 8 of Chapter 2)

$$\frac{1}{f^{pp}} = f_{pp} - (f_{p1} \cdots f_{p,p-1}) \begin{bmatrix} f_{11} & \cdots & f_{1,p-1} \\ \cdot & & \cdot \\ \cdot & & \cdot \\ \cdot & & \cdot \\ f_{p-1,1} & \cdots & f_{p-1,p-1} \end{bmatrix}^{-1} \begin{bmatrix} f_{1p} \\ \cdot \\ \cdot \\ \cdot \\ f_{p-1,p} \end{bmatrix}. \tag{20}$$

Since in the stationary case

$$\begin{aligned} f_{ij} &= \mathscr{E}(y_{t-i} - \mathscr{E}y_{t-i})(y_{t-j} - \mathscr{E}y_{t-j}) \\ &= \mathscr{E}(y_s - \mathscr{E}y_s)(y_{s-j+i} - \mathscr{E}y_{s-j+i}) \\ &= \sigma(i-j), \qquad i,j = 1, \ldots, p, \end{aligned} \tag{21}$$

corresponds to the difference equation of order $p - 1$ when $\beta_p = 0$, the coefficients satisfy [by (48) of Section 5.2]

$$\begin{bmatrix} f_{11} & \cdots & f_{1,p-1} \\ \cdot & & \cdot \\ \cdot & & \cdot \\ \cdot & & \cdot \\ f_{p-1,1} & \cdots & f_{p-1,p-1} \end{bmatrix} \begin{bmatrix} \beta_{p-1} \\ \cdot \\ \cdot \\ \cdot \\ \beta_1 \end{bmatrix} = - \begin{bmatrix} f_{1p} \\ \cdot \\ \cdot \\ \cdot \\ f_{p-1,p} \end{bmatrix}. \tag{22}$$

Then (20) becomes [by (47) of Section 5.2]

$$\begin{aligned} \frac{1}{f^{pp}} &= \sigma(0) + \sigma(p-1)\beta_{p-1} + \cdots + \sigma(1)\beta_1 \\ &= \sigma^2. \end{aligned} \tag{23}$$

This proves the lemma, which in turn implies the following theorem:

THEOREM 5.6.1. *If* $\beta_p = 0$, *then under the conditions of Theorem* 5.5.7 $\sqrt{T}\,\hat{\beta}_p$ *has a limiting normal distribution with mean* 0 *and variance* 1.

Now consider testing the null hypothesis that the order of a stochastic difference equation is m against alternative hypotheses that the order is p ($> m$). Let $A^* = (a^*_{ij})$ be partitioned into m and $p - m$ rows and columns and $\hat{\boldsymbol{\beta}}$ correspondingly

$$A^* = \begin{pmatrix} A^*_{11} & A^*_{12} \\ A^*_{21} & A^*_{22} \end{pmatrix}, \qquad \hat{\boldsymbol{\beta}} = \begin{pmatrix} \hat{\boldsymbol{\beta}}^{(1)} \\ \hat{\boldsymbol{\beta}}^{(2)} \end{pmatrix}. \tag{24}$$

Then the likelihood ratio criterion for testing the hypothesis that $\beta_{m+1} = \cdots = \beta_p = 0$ is a monotonic function of

$$\frac{\hat{\boldsymbol{\beta}}^{(2)\prime}(A_{22}^* - A_{21}^*(A_{11}^*)^{-1}A_{12}^*)\hat{\boldsymbol{\beta}}^{(2)}}{s^2}, \tag{25}$$

which has a limiting χ^2-distribution with $p - m$ degrees of freedom when the null hypothesis is true. Analogy with the usual regression procedures suggests that the power of this test depends mainly on the function of the parameters measuring the excess in the mean square error in prediction by using m terms instead of p. [See Section 2.5; this was pointed out by A. M. Walker (1952).]

Another approach to testing the hypothesis that $\beta_{m+1} = \cdots = \beta_p = 0$ has been given by Quenouille (1947) and extended by Bartlett and Diananda (1950) and others. For simplification in presentation we assume $\gamma = 0$, and therefore shall use A in place of A^*. If $\{y_t\}$ is a stationary autoregressive process of arbitrary order, say p, with coefficients $\beta_1, \ldots, \beta_m, 0, \ldots, 0$, then

$$\begin{aligned} g_i &= \frac{1}{\sqrt{T}} \sum_{t=1}^{T} y_{t-i} u_t \\ &= \frac{1}{\sqrt{T}} \left[\sum_{t=1}^{T} y_t y_{t-i} + \beta_1 \sum_{t=1}^{T} y_{t-1} y_{t-i} + \cdots + \beta_m \sum_{t=1}^{T} y_{t-m} y_{t-i} \right] \\ &= \frac{1}{\sqrt{T}} \sum_{l=0}^{m} \beta_l a_{li}, \qquad i = 1, \ldots, p, \end{aligned} \tag{26}$$

have means 0 and covariance matrix $\sigma^2 F = \sigma^2[\sigma(i-j)]$, which is of order p. It follows from Theorem 5.5.5 that $g_1, \ldots, g_p$ have a limiting joint normal distribution. The limiting distribution of a given number of g_i's is the same as the distribution of that number of σy_t's when the σy_t's are normally distributed. Then

$$\begin{aligned} h_j &= \sum_{k=0}^{m} \beta_k g_{j-k} \\ &= \frac{1}{\sqrt{T}} \sum_{k=0}^{m} \beta_k \sum_{t=1}^{T} y_{t-j+k} u_t \\ &= \frac{1}{\sqrt{T}} \sum_{k,l=0}^{m} \beta_k \beta_l \sum_{t=1}^{T} y_{t-j+k} y_{t-l} \\ &= \frac{1}{\sqrt{T}} \sum_{k,l=0}^{m} \beta_k \beta_l a_{l,j-k}, \qquad j = m+1, \ldots, p, \end{aligned} \tag{27}$$

have a limiting joint normal distribution. For any T the joint distribution has means 0 and covariances $\mathscr{E}h_j h_{j'}$, based on

$$\mathscr{E}h_j g_i = \sum_{k=0}^{m} \beta_k \mathscr{E} g_{j-k} g_i \tag{28}$$

$$= \sigma^2 \sum_{k=0}^{m} \beta_k \sigma(j-k-i)$$

$$= 0, \qquad j > i,$$

$$= \sigma^4, \qquad j = i,$$

by (47) and (48) of Section 5.2. Thus

$$\mathscr{E}h_j h_{j'} = \sum_{k'=0}^{m} \beta_{k'} \mathscr{E} h_j g_{j'-k'} \tag{29}$$

$$= 0, \qquad j > j',$$

$$= \sigma^4, \qquad j = j'.$$

That is, when the null hypothesis is true, $h_{m+1}/\sigma^2, h_{m+2}/\sigma^2, \ldots, h_p/\sigma^2$ have a joint distribution with means 0, variances 1, and covariances 0; their limiting distribution is normal.

We could use the h_j's to test the hypothesis that $\beta_{m+1} = \cdots = \beta_p = 0$ if σ^2 and $\beta_1, \ldots, \beta_m$ are known because then

$$\frac{1}{\sigma^4} \sum_{j=m+1}^{p} h_j^2 \tag{30}$$

has a limiting χ^2-distribution with $p - m$ degrees of freedom. To make this procedure more efficient A. M. Walker (1952) has suggested adding another statistic so the criterion has a limiting χ^2-distribution with p degrees of freedom.

To test the hypothesis $\beta_{m+1} = \cdots = \beta_p = 0$ when the other parameters are unknown it has been suggested to use (30) with the unknown parameters replaced by estimates. Let

$$\hat{h}_j = \sum_{k=0}^{m} \hat{\beta}_k \hat{g}_{j-k} \tag{31}$$

$$= \frac{1}{\sqrt{T}} \sum_{k,l=0}^{m} \hat{\beta}_k \hat{\beta}_l a_{l,j-k}, \qquad j = m+1, \ldots, p,$$

where $\hat{\beta}_0 = 1, \hat{\beta}_1, \ldots, \hat{\beta}_m$ are the maximum likelihood estimates given in Section 5.4 (with p replaced by m). We propose the criterion $\sum_{j=m+1}^{p} \hat{h}_j^2/s^4$, and now verify that it is asymptotically equivalent to (30) when the null hypothesis is true. Since $\hat{\beta}_k$ is a consistent estimate of β_k and $\text{plim}_{T\to\infty} a_{ij}/T = \sigma(i-j)$,

$$\text{(32)} \qquad \operatorname*{plim}_{T\to\infty} \sum_{k=0}^{m} \hat{\beta}_k \frac{1}{T} a_{l,j-k} = \sum_{k=0}^{m} \beta_k \sigma(j-k-l) = 0, \qquad j > l,$$

$$\text{(33)} \qquad \operatorname*{plim}_{T\to\infty} \sum_{l=0}^{m} \hat{\beta}_l \frac{1}{T} a_{l,j-k} = \sum_{l=0}^{m} \beta_l \sigma(j-k-l) = 0, \qquad j > k.$$

Hence

$$\text{(34)} \quad \operatorname*{plim}_{T\to\infty} (\hat{h}_j - h_j)$$

$$= \operatorname*{plim}_{T\to\infty} \sqrt{T}\left[\sum_{k,l=0}^{m} \hat{\beta}_k\hat{\beta}_l \frac{1}{T} a_{l,j-k} - \sum_{k,l=0}^{m} \beta_k\beta_l \frac{1}{T} a_{l,j-k}\right]$$

$$= \operatorname*{plim}_{T\to\infty} \sum_{k,l=0}^{m} \sqrt{T}(\hat{\beta}_k - \beta_k)\hat{\beta}_l \frac{1}{T} a_{l,j-k} + \operatorname*{plim}_{T\to\infty} \sum_{k,l=0}^{m} \beta_k \sqrt{T}(\hat{\beta}_l - \beta_l) \frac{1}{T} a_{l,j-k}$$

$$= 0, \qquad j = m+1, \ldots, p.$$

Since s^2 (based on the mth-order stochastic difference equation) is a consistent estimate of σ^2 when the null hypothesis is true, our verification is complete.

THEOREM 5.6.2. *If the roots of the polynomial equation associated with the* mth-*order stochastic difference equation are less than* 1 *in absolute value,*

$$\text{(35)} \qquad \frac{\sum_{j=m+1}^{p} \hat{h}_j^2}{s_m^4},$$

where $\hat{h}_j$ *is defined by* (31) *and* s_m^2 *is the estimate of* σ^2 *based on the* mth-*order stochastic difference equation, has a limiting* χ^2-*distribution with* $p - m$ *degrees of freedom.*

Now we shall relate these two approaches to testing the order of the stochastic difference equation.

LEMMA 5.6.2. *When* $p = m + 1$

$$\text{(36)} \qquad \sqrt{T}\,\hat{\beta}_{m+1}(m+1) = -\frac{\hat{h}_{m+1}}{\tilde{s}_m^2},$$

where $\hat{\beta}_{m+1}(m+1)$ *is the estimate of* β_{m+1} *on the assumption of an* $(m+1)$st-*order process and where* $\tilde{s}_m^2$ *is an estimate of* σ^2 *based on* (45) *below.*

PROOF. Let the matrix of sums of squares and cross-products with elements

$$a_{ij} = \sum_{t=1}^{T} y_{t-i}y_{t-j}, \qquad i, j = 0, 1, \ldots, m+1, \tag{37}$$

be partitioned

$$A = \begin{pmatrix} a_{00} & A_{01} & a_{0,m+1} \\ A_{10} & A_{11} & A_{1,m+1} \\ a_{m+1,0} & A_{m+1,1} & a_{m+1,m+1} \end{pmatrix}. \tag{38}$$

Then the estimate of $\boldsymbol{\beta}(m) = [\beta_1(m), \ldots, \beta_m(m)]'$ based on a difference equation of order m is

$$\hat{\boldsymbol{\beta}}(m) = -A_{11}^{-1}A_{10}. \tag{39}$$

Use of these coefficients in the definition of $\hat{h}_{m+1}$ yields

$$\begin{aligned} \sqrt{T}\,\hat{h}_{m+1} &= a_{0,m+1} + \sum_{k=1}^{m} \hat{\beta}_k(m)a_{0,m+1-k} + \sum_{l=1}^{m} \hat{\beta}_l(m)a_{l,m+1} + \sum_{k,l=1}^{m} \hat{\beta}_k(m)\hat{\beta}_l(m)a_{l,m+1-k} \\ &= a_{0,m+1} + \sum_{j=1}^{m} \hat{\beta}_{m+1-j}(m)a_{0j} + A_{m+1,1}\hat{\boldsymbol{\beta}}(m) + \sum_{j,l=1}^{m} \hat{\beta}_l(m)\hat{\beta}_{m+1-j}(m)a_{lj} \\ &= a_{0,m+1} - A_{m+1,1}A_{11}^{-1}A_{10}. \end{aligned} \tag{40}$$

The equations for the estimates of $\beta_1, \ldots, \beta_{m+1}$ based on a difference equation of order $m+1$ are

$$\begin{pmatrix} A_{11} & A_{1,m+1} \\ A_{m+1,1} & a_{m+1,m+1} \end{pmatrix} \begin{pmatrix} \hat{\boldsymbol{\beta}}^{(1)}(m+1) \\ \hat{\beta}_{m+1}(m+1) \end{pmatrix} = -\begin{pmatrix} A_{10} \\ a_{m+1,0} \end{pmatrix}, \tag{41}$$

that is,

$$A_{11}\hat{\boldsymbol{\beta}}^{(1)}(m+1) + A_{1,m+1}\hat{\beta}_{m+1}(m+1) = -A_{10}, \tag{42}$$

$$A_{m+1,1}\hat{\boldsymbol{\beta}}^{(1)}(m+1) + a_{m+1,m+1}\hat{\beta}_{m+1}(m+1) = -a_{m+1,0}. \tag{43}$$

[Note that these correspond to (29) and (30) of Section 5.4.] Multiplication of (42) by $A_{m+1,1}A_{11}^{-1}$ and subtraction from (43) yields

$$(a_{m+1,m+1} - A_{m+1,1}A_{11}^{-1}A_{1,m+1})\hat{\beta}_{m+1}(m+1) = -(a_{m+1,0} - A_{m+1,1}A_{11}^{-1}A_{10}). \tag{44}$$

The right-hand side is $-\sqrt{T}\,\hat{h}_{m+1}$. The coefficient of $\hat{\beta}_{m+1}(m+1)$ on the left-hand side is T times

$$\frac{a_{m+1,m+1} - A_{m+1,1}A_{11}^{-1}A_{1,m+1}}{T} = \tilde{s}_m^2 \tag{45}$$

which is asymptotically equivalent to $\hat{\sigma}_m^2$ (based on an equation of order m); the corresponding sums in the matrices defining $\tilde{s}_m^2$ and $\hat{\sigma}_m^2$ differ only with respect to end terms. This proves the lemma.

LEMMA 5.6.3. *If* $\beta_{m+1} = \cdots = \beta_p = 0$ *and* $\gamma = 0$,

$$\operatorname*{plim}_{T\to\infty}\left[\frac{\hat{\boldsymbol{\beta}}^{(2)\prime}(p)(A_{22} - A_{21}A_{11}^{-1}A_{12})\hat{\boldsymbol{\beta}}^{(2)}(p)}{s^2} - T\sum_{j=m+1}^{p}\hat{\beta}_j^2(j)\right] = 0. \tag{46}$$

PROOF. The proof of Lemma 5.6.2 shows that $\hat{\beta}_j(j)$ is the ratio of the right-hand side of the normal equations to the coefficient of $\hat{\beta}_j$ after the first $j-1$ $\hat{\beta}_i$'s have been eliminated from the normal equations. This corresponds to $b_j^* = c_j^*/a_{jj}^*$ in (20) of Section 2.3. In this case the coefficient of $\hat{\beta}_j$ is $T\tilde{s}_{j-1}^2$ as defined in (45) for $j = m+1$. Thus as shown in Section 2.3 the numerator of the left-hand term in (46) is

$$\sum_{j=m+1}^{p} T\tilde{s}_{j-1}^2\hat{\beta}_j^2(j). \tag{47}$$

The lemma follows because

$$\operatorname*{plim}_{T\to\infty}\frac{\tilde{s}_{j-1}^2}{s^2} = 1. \tag{48}$$

Q.E.D.

THEOREM 5.6.3. *If* $\beta_{m+1} = \cdots = \beta_p = 0$ *and* $\gamma = 0$, *the differences between* (25), (30), (35), *and*

$$T\sum_{j=m+1}^{p}\hat{\beta}_j^2(j) \tag{49}$$

have probability limits 0.

PROOF. These probability limits follow from Lemmas 5.6.2 and 5.6.3, (48), and $\operatorname{plim}_{T\to\infty} s^2 = \sigma^2$.

Each criterion has a limiting χ^2-distribution with $p-m$ degrees of freedom under the null hypothesis $\beta_{m+1} = \cdots = \beta_p = 0$ if $\gamma = 0$; if $\gamma \neq 0$, A is replaced by A^*.

A variant on $\sum_{j=m+1}^{p} h_j^2/\sigma^4$ is obtained by replacing h_j by

$$\begin{aligned}\sum_{k=0}^{m}\beta_k g_{j+k} &= \frac{1}{\sqrt{T}}\sum_{k=0}^{m}\beta_k\sum_{t=1}^{T} y_{t-j-k}u_t \\ &= \frac{1}{\sqrt{T}}\sum_{t=1}^{T} u_{t-j}u_t, \qquad j = m+1, \ldots, p.\end{aligned} \tag{50}$$

These variables have means 0, variances σ^4, are uncorrelated, and have a limiting joint normal distribution. However, when $\beta_1, \ldots, \beta_m$ are replaced by estimates in the definition, the limiting distribution is altered.

A. M. Walker (1952) has considered the limiting power of these tests against alternatives approaching the null hypothesis.

The asymptotic theory is not altered by changing end terms in the sums. In particular, when $\gamma \neq 0$, we can replace $\boldsymbol{A}^*$ by

$$\boldsymbol{C}^* = \begin{bmatrix} c_0^* & c_1^* & \cdots & c_{p-1}^* & c_p^* \\ c_1^* & c_0^* & \cdots & c_{p-2}^* & c_{p-1}^* \\ \cdot & \cdot & & \cdot & \cdot \\ \cdot & \cdot & & \cdot & \cdot \\ \cdot & \cdot & & \cdot & \cdot \\ c_{p-1}^* & c_{p-2}^* & \cdots & c_0^* & c_1^* \\ c_p^* & c_{p-1}^* & \cdots & c_1^* & c_0^* \end{bmatrix}, \tag{51}$$

where $c_0^*, \ldots, c_p^*$ are defined by (24) of Section 5.4 when the observations are $y_1, \ldots, y_T$. In turn, the c_j^* can be replaced by $r_j^* = c_j^*/c_0^*$, $j = 0, 1, \ldots, p$. Note that (44) of this section corresponds to (31) of Section 5.4 (with p and m interchanged).

Let us partition $\boldsymbol{C}^*$ into

$$\boldsymbol{C}^* = \begin{bmatrix} c_0^* & \boldsymbol{c}_{p-1}^{*\prime} & c_p^* \\ \boldsymbol{c}_{p-1}^* & \boldsymbol{C}_{p-1}^* & \tilde{\boldsymbol{c}}_{p-1}^* \\ c_p^* & \tilde{\boldsymbol{c}}_{p-1}^{*\prime} & c_0^* \end{bmatrix}. \tag{52}$$

Then the partial correlation between y_t and y_{t+p} "holding $y_{t+1}, \ldots, y_{t+p-1}$ fixed" is

$$\frac{c_p^* - \tilde{\boldsymbol{c}}_{p-1}^{*\prime}(\boldsymbol{C}_{p-1}^*)^{-1}\boldsymbol{c}_{p-1}^*}{c_0^* - \tilde{\boldsymbol{c}}_{p-1}^{*\prime}(\boldsymbol{C}_{p-1}^*)^{-1}\tilde{\boldsymbol{c}}_{p-1}^*}. \tag{53}$$

The normal equations for the estimate of $\boldsymbol{\beta}(p)$ based on c_j^* as an estimate of $\sigma(j)$, $j = 0, 1, \ldots, p$, are

$$\begin{pmatrix} \boldsymbol{C}_{p-1}^* & \tilde{\boldsymbol{c}}_{p-1}^* \\ \tilde{\boldsymbol{c}}_{p-1}^{*\prime} & c_0^* \end{pmatrix} \begin{pmatrix} \boldsymbol{b}^{(1)}(p) \\ b_p(p) \end{pmatrix} = -\begin{pmatrix} \boldsymbol{c}_{p-1}^* \\ c_p^* \end{pmatrix}. \tag{54}$$

The solution of (54) for $-b_p(p)$ is (53). Thus, except possibly for end effects, the estimate of β_p under the assumption that the order is p is the partial correlation between the y_t and y_{t+p} taking into account $y_{t+1}, \ldots, y_{t+p-1}$.

Another question is that of choosing the suitable order of the stochastic difference equation. The analogy with ordinary regression (Section 3.2) suggests that if the investigator can specify a maximum order p and a minimum order m he can test $\beta_p = 0$, then $\beta_{p-1} = 0$, etc. We shall treat this question in Chapter 6 in terms of a modified model that permits the use of an exact theory similar to that of Section 3.2.

For large T the normal theory can be used. The maximum order p can be taken quite large, say $T/10$ or $T/4$, if ε_i, $i = p, p-1, \ldots, m+1$, is taken very small, such as 10^{-3}. Since the series is presumably studied because it is known that the observations are not independent, the minimum order m might be taken positive but small, say 2. To test $\beta_p = 0$ on a large-sample basis one needs only $\hat{\beta}_p(p)$. In fact, to carry out the multiple-decision procedure one only needs $\hat{\beta}_j(j)$ for $j = p$ and smaller j until the hypothesis $\beta_j = 0$ is rejected. The calculation of these on the basis of $r_1^*, \ldots, r_p^*$ was indicated in Section 5.4. Alternatively the forward solution of $\boldsymbol{R}_p^* \boldsymbol{b}(p) = -\boldsymbol{r}_p^*$ can be carried out to obtain the statistics needed.

5.7. THE MOVING AVERAGE MODEL

5.7.1. The Model

Let

$$y_t = \sum_{j=0}^{q} \alpha_j v_{t-j}, \tag{1}$$

where $\alpha_0 = 1$ and $\{v_t\}$ is a sequence of independent random variables with $\mathscr{E} v_t = 0$ and $\mathscr{E} v_t^2 = \sigma^2$. Then $\mathscr{E} y_t = 0$ and

$$\begin{aligned} \mathscr{E} y_t y_s = \sigma(t-s) &= \sigma^2 \sum_{j=0}^{q-|t-s|} \alpha_j \alpha_{j+|t-s|}, && |t-s| \leq q, \\ &= 0, && |t-s| > q. \end{aligned} \tag{2}$$

These first-order and second-order moments, of course, only depend on the v_t's being uncorrelated. We can also write (1) as

$$\begin{aligned} y_t &= \sum_{j=0}^{q} \alpha_j \mathscr{L}^j v_t \\ &= \mathscr{L}^q M(\mathscr{L}^{-1}) v_t, \end{aligned} \tag{3}$$

where $\mathscr{L}$ is the lag operator ($\mathscr{L}v_t = v_{t-1}$),

$$M(z) = \sum_{j=0}^{q} \alpha_j z^{q-j}, \tag{4}$$

and $\mathscr{L}^q M(\mathscr{L}^{-1})$ is the linear operator formed by replacing z in $z^q M(z^{-1}) = \sum_{j=0}^{q} \alpha_j z^j$.

The *covariance generating function* of any stationary process is $\sum_{h=-\infty}^{\infty} \sigma(h) z^h$; the coefficient of z^k is $\sigma(k)$, $k = 0, \pm 1, \ldots$. In the case of the moving average (1) with covariances (2) the covariance generating function is

$$\sum_{h=-q}^{q} \sigma(h) z^h = \sigma^2 M(z) M(z^{-1}). \tag{5}$$

[Note that this equation, and hence (2), can be obtained from Lemma 3.4.1.] Given the coefficients $\alpha_0 = 1, \alpha_1, \ldots, \alpha_q$, the covariances can be calculated by forming the product on the right-hand side of (5) and identifying the coefficients of powers of z. Since the α_j's are real, the roots $z_1, \ldots, z_q$ of

$$M(z) = 0 \tag{6}$$

are real or complex conjugate in pairs. When $\alpha_q \neq 0$ [and hence $\sigma(q) \neq 0$], the roots of

$$\sum_{g=0}^{2q} \sigma(g - q) z^g = 0 \tag{7}$$

are $z_1, \ldots, z_q, 1/z_1, \ldots, 1/z_q$.

Now suppose we are given an arbitrary set of covariances of a stationary process $\sigma(0), \sigma(1), \ldots, \sigma(q) \neq 0$ with $\sigma(q+1) = \cdots = 0$. Then if x_j is a root of (7), $1/x_j$ also is a root because $\sigma(h) = \sigma(-h)$. The multiplicity of a root less than 1 in absolute value must be the same as the multiplicity of its reciprocal (because if the derivative of any order vanishes at x_j it must vanish at $1/x_j$). Thus the roots different from 1 in absolute value can be paired $(x_j, 1/x_j)$. Moreover, any root of absolute value 1 must have even multiplicity because $\sum_{g=-q}^{q} \sigma(g) z^g$ is real and nonnegative for $|z| = 1$, say $z = e^{i\lambda}$, for then the function is a spectral density (Section 7.3). (See Problem 35.) The $2q$ roots of (7) can be numbered and divided into two sets $(x_1, \ldots, x_q)$ and $(x_{q+1}, \ldots, x_{2q})$ such that if $|x_i| < 1$ then $x_{q+i} = 1/x_i$ and if $|x_i| = 1$ then $x_{q+i} = x_i$, $i = 1, \ldots, q$.

Since the coefficients in (7) are real, the roots are real or occur in conjugate pairs. Then

$$(8) \qquad M^*(z) = \prod_{j=1}^{q} (z - x_j) = \sum_{j=0}^{q} \alpha_j^* z^{q-j}$$

has real coefficients and (5) holds for $M(z)$ replaced by $M^*(z)$. Thus the covariance sequence $\{\sigma(h)\}$ with $\sigma(q+1) = \cdots = 0$ can be generated by a finite moving average process (1) with coefficients $\alpha_0^* = 1, \alpha_1^*, \ldots, \alpha_q^*$, and $\operatorname{Var}(v_t) = \sigma^2$.

Given the covariances $\sigma(0), \sigma(1), \ldots, \sigma(q)$ or the variance $\sigma(0)$ and correlations $\rho_h = \sigma(h)/\sigma(0)$, $h = 1, \ldots, q$, the coefficients of the moving average process can be obtained by solving (7) for $x_1, \ldots, x_{2q}$, composing $M^*(z)$ and finding the coefficients of the powers of z.

If the roots of (6) are less than 1 in absolute value,

$$(9) \qquad \left(\sum_{j=0}^{q} \alpha_j x^j\right)^{-1} = \prod_{j=1}^{q} (1 - z_j x)^{-1} = \prod_{j=1}^{q} \sum_{\nu=0}^{\infty} (z_j x)^{\nu} = \sum_{r=0}^{\infty} \gamma_r x^r$$

converges for $|x| < 1/\max_j |z_j|$. We can write

$$(10) \qquad v_t = \left(\sum_{j=0}^{q} \alpha_j \mathscr{L}^j\right)^{-1} y_t = \sum_{r=0}^{\infty} \gamma_r \mathscr{L}^r y_t = \sum_{r=0}^{\infty} \gamma_r y_{t-r}.$$

If the roots of (6) are less than 1 in absolute value, (10) converges in mean square; the argument is similar to that used to show that (28) in Section 5.2 converges in the mean. [See also (34), (35), and (36) of Section 7.5.] In fact,

$\gamma_r = \sum_{i=1}^{q} k_i z_i^r$ for suitable constants $k_1, \ldots, k_q$ if the roots are all different. [See (39), (40), and (41) of Section 5.2.] Note that the condition excludes the simple average $\frac{1}{2}(v_t + v_{t-1})$ and the difference $\Delta v_t = v_{t+1} - v_t$.

5.7.2. Estimation of Parameters

If the v_t's are normally distributed, the y_t's are normally distributed with means 0 and covariances (2). We now consider estimating the $q + 1$ parameters on the basis of T observations, $y_1, \ldots, y_T$. Unfortunately, although the covariance matrix has a simple form, the inverse matrix is not simple. In fact, a minimal sufficient set of statistics has T components, and the maximum likelihood equations are complicated and cannot be solved directly. (See Problems 4 and 5 of Chapter 6.)

The approach followed by A. M. Walker (1961) was to apply the method of maximum likelihood to the approximate normal distribution of some n sample correlations. As is shown in Theorem 5.7.1, $\sqrt{T}(r_1 - \rho_1), \ldots, \sqrt{T}(r_n - \rho_n)$ have a limiting normal distribution with means 0 and covariances $w_{gh} = w_{gh}(\rho_1, \ldots, \rho_q)$, which are functions of the correlations $\rho_h = \sigma(h)/\sigma(0)$, $h = 1, \ldots, q$. The sample correlations are $r_h = c_h/c_0$, where

$$c_h = c_{-h} = \frac{1}{T}\sum_{t=1}^{T-h} y_t y_{t+h}, \qquad h = 0, 1, \ldots, T-1. \tag{11}$$

Here n ($n \geq q$, $n \leq T - 1$) is fixed. Actually, the maximization of the approximate normal distribution of the sample correlations is only to motivate the estimation equations obtained; the properties of the estimating procedure do not depend on this approach.

Let $\boldsymbol{\rho} = (\rho_1, \ldots, \rho_q)'$. The logarithm of the approximate likelihood function of $\boldsymbol{r} = (r_1, \ldots, r_n)'$ is

$$\begin{aligned} \log L = {} & -\tfrac{1}{2}n \log 2\pi - \tfrac{1}{2}\log|\boldsymbol{W}| \\ & -\tfrac{1}{2}T(\boldsymbol{r}^{(1)\prime} - \boldsymbol{\rho}' \quad \boldsymbol{r}^{(2)\prime})\begin{pmatrix} \boldsymbol{W}^{11} & \boldsymbol{W}^{12} \\ \boldsymbol{W}^{21} & \boldsymbol{W}^{22} \end{pmatrix}\begin{pmatrix} \boldsymbol{r}^{(1)} - \boldsymbol{\rho} \\ \boldsymbol{r}^{(2)} \end{pmatrix}, \end{aligned} \tag{12}$$

where

$$\boldsymbol{r} = \begin{pmatrix} \boldsymbol{r}^{(1)} \\ \boldsymbol{r}^{(2)} \end{pmatrix}, \qquad \boldsymbol{W}^{-1} = \begin{pmatrix} \boldsymbol{W}^{11} & \boldsymbol{W}^{12} \\ \boldsymbol{W}^{21} & \boldsymbol{W}^{22} \end{pmatrix} \tag{13}$$

are partitioned into q and $n - q$ rows and columns. Then the vector of partial derivatives is

$$(14)\quad \frac{\partial \log L}{\partial \boldsymbol{\rho}} = -\frac{1}{2}\frac{1}{|W|}\frac{\partial |W|}{\partial \boldsymbol{\rho}} - \tfrac{1}{2}T(\boldsymbol{r}^{(1)\prime} - \boldsymbol{\rho}' \quad \boldsymbol{r}^{(2)\prime})$$

$$\times \frac{\partial}{\partial \boldsymbol{\rho}}\begin{pmatrix} W^{11} & W^{12} \\ W^{21} & W^{22} \end{pmatrix}\begin{pmatrix} \boldsymbol{r}^{(1)} - \boldsymbol{\rho} \\ \boldsymbol{r}^{(2)} \end{pmatrix} + TW^{11}(\boldsymbol{r}^{(1)} - \boldsymbol{\rho}) + TW^{12}\boldsymbol{r}^{(2)}.$$

We set this vector of derivatives equal to $\mathbf{0}$. Since $\sqrt{T}(\boldsymbol{r}^{(1)} - \boldsymbol{\rho})$ and $\sqrt{T}\boldsymbol{r}^{(2)}$ have a limiting normal distribution, we normalize the resulting equations by dividing by $\sqrt{T}$. Then the first two terms are of order $1/\sqrt{T}$ and converge stochastically to $\mathbf{0}$. The resulting equations are

$$(15)\quad \begin{aligned} \hat{\boldsymbol{\rho}} &= \boldsymbol{r}^{(1)} + (W^{11})^{-1}W^{12}\boldsymbol{r}^{(2)} \\ &= \boldsymbol{r}^{(1)} - W_{12}W_{22}^{-1}\boldsymbol{r}^{(2)} \\ &= \boldsymbol{r}^{(1)} - W_{12}(\boldsymbol{\rho})W_{22}^{-1}(\boldsymbol{\rho})\boldsymbol{r}^{(2)}. \end{aligned}$$

(See Problem 8 of Chapter 2.) An estimating procedure is to use $\boldsymbol{r}^{(1)}$ as a consistent estimate of $\boldsymbol{\rho}$, calculate $W_{12}(\boldsymbol{r}^{(1)})W_{22}^{-1}(\boldsymbol{r}^{(1)})$ and estimate $\boldsymbol{\rho}$ by

$$(16)\quad \hat{\boldsymbol{\rho}} = \boldsymbol{r}^{(1)} - W_{12}(\boldsymbol{r}^{(1)})W_{22}^{-1}(\boldsymbol{r}^{(1)})\boldsymbol{r}^{(2)}.$$

Then $\sqrt{T}(\hat{\boldsymbol{\rho}} - \boldsymbol{\rho})$ calculated according to (16) has the same limiting distribution as that of $\sqrt{T}(\hat{\boldsymbol{\rho}} - \boldsymbol{\rho})$ calculated according to (15), which is normal with mean $\mathbf{0}$ and covariance matrix

$$(17)\quad W_{11} - W_{12}W_{22}^{-1}W_{21}.$$

One can iterate the above procedure. Let $\hat{\boldsymbol{\rho}}^{(0)}$ be an initial consistent estimate; then iterate by

$$(18)\quad \hat{\boldsymbol{\rho}}^{(j)} = \boldsymbol{r}^{(1)} - W_{12}(\hat{\boldsymbol{\rho}}^{(j-1)})W_{22}^{-1}(\hat{\boldsymbol{\rho}}^{(j-1)})\boldsymbol{r}^{(2)}.$$

Walker gives the following example, using a series of 100 observations that Durbin (1959) generated by the model with $q = 1$ and $\alpha_1 = \frac{1}{2}$. The first five sample correlations were 0.35005, -0.06174, -0.08007, -0.14116, and -0.15629. The matrix W is taken from the covariance matrix of the first five correlations

$$(19)\qquad \begin{bmatrix} 1-3\rho^2+4\rho^4 & 2\rho(1-\rho^2) & \rho^2 & 0 & 0 \\ 2\rho(1-\rho^2) & 1+2\rho^2 & 2\rho & \rho^2 & 0 \\ \rho^2 & 2\rho & 1+2\rho^2 & 2\rho & \rho^2 \\ 0 & \rho^2 & 2\rho & 1+2\rho^2 & 2\rho \\ 0 & 0 & \rho^2 & 2\rho & 1+2\rho^2 \end{bmatrix},$$

where $\rho = \rho_1$. For $n = 2$ and $q = 1$ (15) is

$$(20)\qquad \hat{\rho} = r_1 - \frac{2\rho(1-\rho^2)}{1+2\rho^2} r_2.$$

We take $\hat{\rho}^{(0)} = r_1$; then $\hat{\rho}^{(1)} = 0.38051$ and $\hat{\rho}^{(2)} = 0.38121$ ($\rho = 0.4$). The estimated standard deviation of the asymptotic distribution is

$$(21)\qquad \frac{1}{\sqrt{T}}\sqrt{w_{11}(\hat{\rho}) - w_{12}^2(\hat{\rho})/w_{22}(\hat{\rho})} = 0.0565.$$

It will be observed that (17) is a monotonically nonincreasing function of n in the sense that for every $\boldsymbol{x}$ the quadratic form $\boldsymbol{x}'(\boldsymbol{W}_{11} - \boldsymbol{W}_{12}\boldsymbol{W}_{22}^{-1}\boldsymbol{W}_{21})\boldsymbol{x}$ is nonincreasing with n. More precisely, $\boldsymbol{x}'[\boldsymbol{W}_{11}^{(n)} - \boldsymbol{W}_{12}^{(n)}(\boldsymbol{W}_{22}^{(n)})^{-1}\boldsymbol{W}_{21}^{(n)}]\boldsymbol{x}$ is the variance of the limiting normal distribution of $\sqrt{T}\boldsymbol{x}'(\hat{\boldsymbol{\rho}} - \boldsymbol{\rho})$ when $r_1, \ldots, r_n$ are used; this asymptotic variance is monotonically nonincreasing in n for every $\boldsymbol{x}$. This fact suggests choosing n relatively large (as long as T is large enough for the asymptotic normality to hold reasonably well).

The estimates $\hat{\boldsymbol{\rho}}$ of $\boldsymbol{\rho}$ and c_0 of $\sigma(0)$ can be used to calculate estimates of $\alpha_1, \ldots, \alpha_q$, and σ^2. Then $\sqrt{T}(\hat{\alpha}_1 - \alpha_1), \ldots, \sqrt{T}(\hat{\alpha}_q - \alpha_q)$ have a limiting normal distribution.

Another approach was made by Durbin (1959). He suggested that the infinite sum (10) could be approximated by a finite sum

$$(22)\qquad v_t^* = \sum_{r=0}^{n} \gamma_r y_{t-r},$$

where n is fairly large. (Note $\gamma_0 = 1$.) The sequence $\{v_t^*\}$ is a finite moving average process. [In fact (22) corresponds to (11) of Section 5.2 with suitable changes of notation such as u_r's replaced by y_r's and y_r's replaced by v_r's and with the resulting linear combination of $v_{t-n-1}, \ldots, v_{t-n-q}$ transposed to the left.] v_t^* is v_t plus a linear combination of $v_{t-n-1}, \ldots, v_{t-n-q}$; the coefficients in this linear combination are small when n is large. The coefficients are linear combinations of $\gamma_{n+1}, \ldots, \gamma_{n+q}$ as defined by (9) or (10), and these coefficients are linear combinations of $(n+1)$st to $(n+q)$th powers of the roots of $M(z) = 0$, which are assumed less than 1 in absolute value. [See (14) or (22) and (26) of Section 5.2.] Hence, the v_t^*'s are nearly uncorrelated if n is large, and (22) is roughly a stochastic difference equation. This suggests estimating $\boldsymbol{\gamma} = (\gamma_1, \ldots, \gamma_n)'$ by

$$(23) \qquad \begin{bmatrix} c_0 & c_1 & \cdots & c_{n-1} \\ c_1 & c_0 & \cdots & c_{n-2} \\ \cdot & \cdot & & \cdot \\ \cdot & \cdot & & \cdot \\ \cdot & \cdot & & \cdot \\ c_{n-1} & c_{n-2} & \cdots & c_0 \end{bmatrix} \begin{bmatrix} \hat{\gamma}_1 \\ \hat{\gamma}_2 \\ \cdot \\ \cdot \\ \cdot \\ \hat{\gamma}_n \end{bmatrix} = - \begin{bmatrix} c_1 \\ c_2 \\ \cdot \\ \cdot \\ \cdot \\ c_n \end{bmatrix}.$$

The equation is written equivalently in terms of the correlations $r_0 = 1, r_1, \ldots, r_n$. Theorem 5.7.1 asserts that $\sqrt{T}(r_1 - \rho_1), \ldots, \sqrt{T}(r_n - \rho_n)$ have a limiting normal distribution and $\text{plim}_{T\to\infty} r_i = \rho_i$, $i = 1, \ldots, n$. Therefore, $\sqrt{T}(\hat{\boldsymbol{\gamma}} - \boldsymbol{\gamma}^*)$ has a limiting normal distribution, where $\boldsymbol{\gamma}^* = -\boldsymbol{\Sigma}^{-1}\boldsymbol{\sigma}$, $\boldsymbol{\sigma} = [\sigma(1), \ldots, \sigma(n)]'$ and

$$(24) \qquad \boldsymbol{\Sigma} = \begin{bmatrix} \sigma(0) & \sigma(1) & \cdots & \sigma(n-1) \\ \sigma(1) & \sigma(0) & \cdots & \sigma(n-2) \\ \cdot & \cdot & & \cdot \\ \cdot & \cdot & & \cdot \\ \cdot & \cdot & & \cdot \\ \sigma(n-1) & \sigma(n-2) & \cdots & \sigma(0) \end{bmatrix},$$

the $\sigma(h)$'s given by (2). $\boldsymbol{\gamma}^*(= -\boldsymbol{\Sigma}^{-1}\boldsymbol{\sigma})$ is an approximation to $\boldsymbol{\gamma}$ [defined by the coefficients $\gamma_1, \ldots, \gamma_n$ in (9)]. Note that $\text{plim}_{T\to\infty} \hat{\boldsymbol{\gamma}} = \boldsymbol{\gamma}^*$. The resemblance of (22) to a stochastic difference equation suggests that the limiting normal distribution has a covariance matrix approximately σ^2 times the inverse of $\mathbf{\Sigma}$. Then the approximate density has in the exponent the quadratic form which is $-T/(2\sigma^2)$ times

$$
(25)\qquad
\begin{aligned}
(\hat{\boldsymbol{\gamma}} - \boldsymbol{\gamma}^*)'\mathbf{\Sigma}(\hat{\boldsymbol{\gamma}} - \boldsymbol{\gamma}^*) &= \hat{\boldsymbol{\gamma}}'\mathbf{\Sigma}\hat{\boldsymbol{\gamma}} - 2\hat{\boldsymbol{\gamma}}'\mathbf{\Sigma}\boldsymbol{\gamma}^* + \boldsymbol{\gamma}^{*\prime}\mathbf{\Sigma}\boldsymbol{\gamma}^* \\
&= \hat{\boldsymbol{\gamma}}'\mathbf{\Sigma}\hat{\boldsymbol{\gamma}} + 2\hat{\boldsymbol{\gamma}}'\boldsymbol{\sigma} - \boldsymbol{\gamma}^{*\prime}\boldsymbol{\sigma} \\
&= (1 \quad \hat{\boldsymbol{\gamma}}')\begin{pmatrix} \sigma^*(0) - \sigma^2 & \boldsymbol{\sigma}' \\ \boldsymbol{\sigma} & \mathbf{\Sigma} \end{pmatrix}\begin{pmatrix} 1 \\ \hat{\boldsymbol{\gamma}} \end{pmatrix},
\end{aligned}
$$

where $\sigma^*(0) = \sigma^2 - \boldsymbol{\gamma}^{*\prime}\boldsymbol{\sigma} = \sigma^2 - \sum_{r=1}^n \gamma_r^*\sigma(r)$. If $\mathbf{\Sigma}$ and $\boldsymbol{\sigma}$ corresponded to an autoregressive process of order n, then $\sigma^*(0)$ would be the variance of the process. [In Section 5.2 (47) can be written as $\sigma_a^2 = \sum_{r=0}^n \beta_{ar}\sigma_a(r)$, and (48) can be written as $\boldsymbol{\beta}_a = -\mathbf{\Sigma}_a^{-1}\,\boldsymbol{\sigma}_a$, where the subscript a denotes the autoregressive process studied in that section and n replaces p.] We shall replace $\sigma^*(0)$ in (25) by $\sigma(0)$. Then the approximate density has in the exponent the quadratic form $-\frac{1}{2}$ times

(26)

$$
\begin{bmatrix} \hat{\gamma}_0 \\ \hat{\gamma}_1 \\ \hat{\gamma}_2 \\ \vdots \\ \hat{\gamma}_n \end{bmatrix}'
\begin{bmatrix}
\sum_{j=1}^q \alpha_j^2 & \alpha_1 + \sum_{j=1}^{q-1}\alpha_j\alpha_{j+1} & \alpha_2 + \sum_{j=1}^{q-2}\alpha_j\alpha_{j+2} & \cdots & 0 \\
\alpha_1 + \sum_{j=1}^{q-1}\alpha_j\alpha_{j+1} & 1 + \sum_{j=1}^q \alpha_j^2 & \alpha_1 + \sum_{j=1}^{q-1}\alpha_j\alpha_{j+1} & \cdots & 0 \\
\alpha_2 + \sum_{j=1}^{q-2}\alpha_j\alpha_{j+2} & \alpha_1 + \sum_{j=1}^{q-1}\alpha_j\alpha_{j+1} & 1 + \sum_{j=1}^q \alpha_j^2 & \cdots & 0 \\
\vdots & \vdots & \vdots & & \vdots \\
0 & 0 & 0 & \cdots & 1 + \sum_{j=1}^q \alpha_j^2
\end{bmatrix}
\begin{bmatrix} \hat{\gamma}_0 \\ \hat{\gamma}_1 \\ \hat{\gamma}_2 \\ \vdots \\ \hat{\gamma}_n \end{bmatrix}
$$

$$
\begin{aligned}
= \sum_{j=1}^q \alpha_j^2 &+ \left(1 + \sum_{j=1}^q \alpha_j^2\right)\sum_{u=1}^n \hat{\gamma}_u^2 + 2\left(\alpha_1 + \sum_{j=1}^{q-1}\alpha_j\alpha_{j+1}\right)\sum_{u=0}^{n-1}\hat{\gamma}_u\hat{\gamma}_{u+1} \\
&+ 2\left(\alpha_2 + \sum_{j=1}^{q-2}\alpha_j\alpha_{j+2}\right)\sum_{u=0}^{n-2}\hat{\gamma}_u\hat{\gamma}_{u+2} + \cdots + 2\alpha_q\sum_{u=0}^{n-q}\hat{\gamma}_u\hat{\gamma}_{u+q},
\end{aligned}
$$

when $\hat{\gamma}_0 = 1$. Setting the derivatives of (26) with respect to α_i, $i = 1, \ldots, q$, equal to 0 we obtain

(27)

$$\begin{bmatrix} \sum_{u=0}^{n} \hat{\gamma}_u^2 & \sum_{u=0}^{n-1} \hat{\gamma}_u\hat{\gamma}_{u+1} & \sum_{u=0}^{n-2} \hat{\gamma}_u\hat{\gamma}_{u+2} & \cdots & \sum_{u=0}^{n-q+1} \hat{\gamma}_u\hat{\gamma}_{u+q-1} \\ \sum_{u=0}^{n-1} \hat{\gamma}_u\hat{\gamma}_{u+1} & \sum_{u=0}^{n} \hat{\gamma}_u^2 & \sum_{u=0}^{n-1} \hat{\gamma}_u\hat{\gamma}_{u+1} & \cdots & \sum_{u=0}^{n-q+2} \hat{\gamma}_u\hat{\gamma}_{u+q-2} \\ \sum_{u=0}^{n-2} \hat{\gamma}_u\hat{\gamma}_{u+2} & \sum_{u=0}^{n-1} \hat{\gamma}_u\hat{\gamma}_{u+1} & \sum_{u=0}^{n} \hat{\gamma}_u^{\,2} & \cdots & \sum_{u=0}^{n-q+3} \hat{\gamma}_u\hat{\gamma}_{u+q-3} \\ \vdots & \vdots & \vdots & & \vdots \\ \sum_{u=0}^{n-q+1} \hat{\gamma}_u\hat{\gamma}_{u+q-1} & \sum_{u=0}^{n-q+2} \hat{\gamma}_u\hat{\gamma}_{u+q-2} & \sum_{u=0}^{n-q+3} \hat{\gamma}_u\hat{\gamma}_{u+q-3} & \cdots & \sum_{u=0}^{n} \hat{\gamma}_u^2 \end{bmatrix} \begin{bmatrix} \hat{\alpha}_1 \\ \hat{\alpha}_2 \\ \hat{\alpha}_3 \\ \vdots \\ \hat{\alpha}_q \end{bmatrix} = - \begin{bmatrix} \sum_{u=0}^{n-1} \hat{\gamma}_u\hat{\gamma}_{u+1} \\ \sum_{u=0}^{n-2} \hat{\gamma}_u\hat{\gamma}_{u+2} \\ \sum_{u=0}^{n-3} \hat{\gamma}_u\hat{\gamma}_{u+3} \\ \vdots \\ \sum_{u=0}^{n-q} \hat{\gamma}_u\hat{\gamma}_{u+q} \end{bmatrix}.$$

These equations are the equations that one would obtain for q coefficients of a stochastic difference equation when the observations were $1, \hat{\gamma}_1, \ldots, \hat{\gamma}_n$. The solution $\hat{\boldsymbol{\alpha}} = (\hat{\alpha}_1, \ldots, \hat{\alpha}_q)'$ is a consistent estimate of the vector $\boldsymbol{\alpha}^*$, which is the solution of (27) when $\hat{\boldsymbol{\gamma}}$ is replaced by $\boldsymbol{\gamma}^*$; the vector $\boldsymbol{\alpha}^*$ is approximately $\boldsymbol{\alpha}$.

Then as $T \to \infty$, $\sqrt{T}(\hat{\alpha}_1 - \alpha_1^*), \ldots, \sqrt{T}(\hat{\alpha}_q - \alpha_q^*)$ have a limiting normal distribution with means 0 for fixed n. The values $\alpha_1^*, \ldots, \alpha_q^*$ to which $\hat{\alpha}_1, \ldots, \hat{\alpha}_q$ converge stochastically may differ from $\alpha_1, \ldots, \alpha_q$, but presumably by small amounts if n is sufficiently large.

5.7.3. A Central Limit Theorem for Moving Average Processes

To justify the asymptotic normality of the estimates we need to prove that the sample correlations have a limiting normal distribution. Although the second-order moments of the sample covariances depend on the fourth-order moments of the stochastic process, the second-order moments of the asymptotic distribution of the correlation coefficients depend only on the second-order moments of the process, and the asymptotic normality only depends on the variance being finite.

THEOREM 5.7.1. *Let y_t, $t = 1, 2, \ldots$, be defined by* (1), *where the v_t's are independently identically distributed with $\mathscr{E}v_t = 0$ and $\mathscr{E}v_t^2 = \sigma^2$. Then $\sqrt{T}(r_1 - \rho_1), \ldots, \sqrt{T}(r_n - \rho_n)$ have a limiting normal distribution with means* 0 *and covariances*

(28)

$$w_{hh'} = \sum_{g=-(q+n)}^{q+n} (\rho_{g-h}\rho_{g-h'} + \rho_{g+h}\rho_{g-h'} - 2\rho_h\rho_g\rho_{g+h'} - 2\rho_{h'}\rho_g\rho_{g+h} + 2\rho_h\rho_{h'}\rho_g^2).$$

PROOF. Consider for $0 < h \leq q$

$$\begin{aligned}
(29)\quad \sqrt{T}(c_h - \rho_h c_0) &= \frac{1}{\sqrt{T}}\sum_{t=1}^{T-h} y_t y_{t+h} - \rho_h \frac{1}{\sqrt{T}}\sum_{t=1}^{T} y_t^2 \\
&= \frac{1}{\sqrt{T}}\left[\sum_{t=1}^{T-h}\sum_{i,j=0}^{q} \alpha_i\alpha_j v_{t-i}v_{t+h-j} - \rho_h \sum_{t=1}^{T}\sum_{i,j=0}^{q} \alpha_i\alpha_j v_{t-i}v_{t-j}\right] \\
&= \frac{1}{\sqrt{T}}\left[\sum_{t=1}^{T-h}\left(\sum_{i=0}^{q-h} \alpha_i\alpha_{i+h}v_{t-i}^2 + \sum_{\substack{i,j=0 \\ i\neq j-h}}^{q} \alpha_i\alpha_j v_{t-i}v_{t+h-j}\right)\right. \\
&\qquad \left. - \rho_h \sum_{t=1}^{T}\left(\sum_{i=0}^{q} \alpha_i^2 v_{t-i}^2 + \sum_{\substack{i,j=0 \\ i\neq j}}^{q} \alpha_i\alpha_j v_{t-i}v_{t-j}\right)\right].
\end{aligned}$$

The first sum on the right-hand side of (29) is

$$(30)\qquad \frac{1}{\sqrt{T}}\sum_{i=0}^{q-h} \alpha_i\alpha_{i+h}\sum_{t=1}^{T-q} v_t^2$$

except for at most $2q^2$ terms. Similarly the third sum on the right-hand side of (29) is (30) except for at most $2q^2$ terms, because $\rho_h = \sum_{i=0}^{q-h} \alpha_i\alpha_{i+h}/\sum_{i=0}^{q} \alpha_i^2$. Hence, the limiting distribution of $\sqrt{T}(c_h - \rho_h c_0)$ is the limiting distribution of

$$(31)\quad \frac{1}{\sqrt{T}}\left[\sum_{t=1}^{T-h} \sum_{\substack{i,j=0\\ i\neq j-h}}^{q} \alpha_i\alpha_j v_{t-i}v_{t+h-j} - \rho_h \sum_{t=1}^{T} \sum_{\substack{i,j=0\\ i\neq j}}^{q} \alpha_i\alpha_j v_{t-i}v_{t-j}\right]$$

$$= \frac{1}{\sqrt{T}}\left[\sum_{t=1}^{T-h} \sum_{i=0}^{q} \sum_{\substack{k=-h\\ k\neq i}}^{q-h} \alpha_i\alpha_{k+h} v_{t-i}v_{t-k} - \rho_h \sum_{t=1}^{T} \sum_{i=0}^{q} \sum_{\substack{k=0\\ k\neq i}}^{q} \alpha_i\alpha_k v_{t-i}v_{t-k}\right].$$

Except for at most $4q^3$ terms (31) is

$$(32)\quad \frac{1}{\sqrt{T}}\left[\sum_{s=1}^{T-q} \sum_{i=0}^{q} \sum_{\substack{k=-h\\ k\neq i}}^{q-h} \alpha_i\alpha_{k+h} v_s v_{s+i-k} - \rho_h \sum_{s=1}^{T-q} \sum_{i=0}^{q} \sum_{\substack{k=0\\ k\neq i}}^{q} \alpha_i\alpha_k v_s v_{s+i-k}\right]$$

$$= \frac{1}{\sqrt{T}} \sum_{s=1}^{T-q} \sum_{i=0}^{q} \sum_{\substack{k=-h\\ k\neq i}}^{q} (\alpha_i'\alpha_{k+h}' - \rho_h\alpha_i'\alpha_k') v_s v_{s+i-k},$$

where $\alpha_i' = 0$ for $i < 0$ and for $i > q$ and $\alpha_i' = \alpha_i$ for $i = 0, 1, \ldots, q$. Then (32) is

$$(33)\quad \frac{1}{\sqrt{T}} \sum_{s=1}^{T-q} \sum_{\substack{g=-q\\ g\neq 0}}^{q+h} \sum_{i=0}^{q} (\alpha_i'\alpha_{i+h-g}' - \rho_h\alpha_i'\alpha_{i-g}') v_s v_{s+g}$$

$$= \frac{1}{\sqrt{T}} \sum_{s=1}^{T-q} \left[\sum_{g=1}^{q+h} \sum_{i=0}^{q} (\alpha_i'\alpha_{i+h-g}' - \rho_h\alpha_i'\alpha_{i-g}') v_s v_{s+g} + \sum_{f=1}^{q} \sum_{i=0}^{q} (\alpha_i'\alpha_{i+h+f}' - \rho_h\alpha_i'\alpha_{i+f}') v_s v_{s-f}\right],$$

which differs by less than $2(q+1)^3$ terms from

$$(34)\quad \frac{1}{\sqrt{T}} \sum_{s=1}^{T-q} \sum_{g=1}^{q+h} \sum_{i=0}^{q} (\alpha_i'\alpha_{i+h-g}' + \alpha_i'\alpha_{i+h+g}' - 2\rho_h\alpha_i'\alpha_{i-g}') v_s v_{s+g}.$$

This is of the form $\sum_{s=1}^{T-q} (\sum_{g=1}^{n} k_g v_s v_{s+g})$. The terms $\sum_{g=1}^{n} k_g v_s v_{s+g}$ have means 0, variances

$$(35)\quad \mathscr{E} \sum_{g,g'=1}^{n} k_g k_{g'} v_s v_{s+g} v_s v_{s+g'} = \sigma^4 \sum_{g=1}^{n} k_g^2,$$

and correlations 0; the sequence constitutes a $(q+h)$-dependent stationary stochastic process. The asymptotic normality then follows from Theorem 7.7.5.

For $q < h \leq n$ $\rho_h = 0$ and

$$(36) \qquad \sqrt{T}\, c_h = \frac{1}{\sqrt{T}} \sum_{t=1}^{T-h} y_t y_{t+h}$$

$$= \frac{1}{\sqrt{T}} \sum_{t=1}^{T-h} \sum_{i,j=0}^{q} \alpha_i \alpha_j v_{t-i} v_{t+h-j},$$

which differs by less than $2q(q+1)^2$ terms from

$$(37) \qquad \frac{1}{\sqrt{T}} \sum_{s=1}^{T-h} \sum_{i,j=0}^{q} \alpha_i \alpha_j v_s v_{s+h+i-j} = \frac{1}{\sqrt{T}} \sum_{s=1}^{T-h} \sum_{g=-q+h}^{q+h} \sum_{i=0}^{q} \alpha_i' \alpha_{i+h-g}' v_s v_{s+g}.$$

Theorem 7.7.5 then implies that (37) has a limiting normal distribution. Note that the coefficient of $v_t v_{t+j} + v_t v_{t-j}$ in (37) for $j \neq 0$ is the same as the coefficient of $v_t v_{t+j}$ in (34) since $\alpha_j' = 0$ for $j > q$ and $\rho_h = 0$.

Since

$$(38) \qquad \sum_{i=0}^{q} \alpha_i' \alpha_{i+k}' = \left(\sum_{j=0}^{q} \alpha_j^2 \right) \rho_k,$$

the covariance between (34) and (34) with h replaced by h', $0 < h \leq h'$, is

$$(39) \qquad \sigma^4 \left(\sum_{i=0}^{q} \alpha_i^2 \right)^2 \sum_{g=1}^{q+h'} (\rho_{h-g} + \rho_{g+h} - 2\rho_h \rho_g)(\rho_{h'-g} + \rho_{g+h'} - 2\rho_{h'} \rho_g)$$

$$= \tfrac{1}{2} \sigma^4 \left(\sum_{i=0}^{q} \alpha_i^2 \right)^2 \sum_{g=-(q+h')}^{q+h'} (\rho_{h-g} + \rho_{g+h} - 2\rho_h \rho_g)(\rho_{h'-g} + \rho_{g+h'} - 2\rho_{h'} \rho_g)$$

$$= \sigma^2(0) w_{hh'}$$

as defined in (28). [If $\mathscr{E} v_t^4 < \infty$, (39) is $\lim_{T \to \infty} T \mathscr{E}(c_h - \rho_h c_0)(c_{h'} - \rho_{h'} c_0)$.] We have

$$(40) \qquad c_0 = \frac{1}{T} \sum_{t=1}^{T} y_t^2$$

$$= \sum_{i,j=0}^{q} \alpha_i \alpha_j \frac{1}{T} \sum_{t=1}^{T} v_{t-i} v_{t-j}$$

$$= \sum_{i=0}^{q} \alpha_i^2 \frac{1}{T} \sum_{t=1}^{T} v_{t-i}^2 + \sum_{\substack{i,j=0 \\ i \neq j}}^{q} \alpha_i \alpha_j \frac{1}{T} \sum_{t=1}^{T} v_{t-i} v_{t-j}.$$

Each term $(1/T) \sum_{t=1}^{T} v_{t-i}^2$ converges stochastically to σ^2 by the law of large numbers, and each term $(1/T) \sum_{t=1}^{T} v_{t-i} v_{t-j}$, $i \neq j$, converges stochastically to 0 because

$$\mathscr{E}\left(\frac{1}{T}\sum_{t=1}^{T} v_{t-i}v_{t-j}\right)^2 = \frac{1}{T^2}\sum_{t,s=1}^{T} \mathscr{E} v_{t-i}v_{t-j}v_{s-i}v_{s-j} \tag{41}$$

$$= \frac{1}{T}\sigma^4, \qquad i \neq j,$$

converges to 0. Thus c_0 converges stochastically to $\sigma^2 \sum_{i=0}^{q} \alpha_i^2 = \sigma(0)$. Thus the limiting distribution of

$$\sqrt{T}\,(r_h - \rho_h) = \frac{\sqrt{T}\,(c_h - \rho_h c_0)}{c_0} \tag{42}$$

is $N(0, w_{hh})$. The theorem is proved by considering an arbitrary linear combination $\sum_{h=1}^{n} k_h \sqrt{T}\,(r_h - \rho_h)$.

5.8. AUTOREGRESSIVE PROCESSES WITH MOVING AVERAGE RESIDUALS

5.8.1. The Model

Let $\{y_t\}$ satisfy

$$\sum_{s=0}^{p} \beta_s y_{t-s} = \sum_{j=0}^{q} \alpha_j v_{t-j}, \tag{1}$$

where $\beta_0 = \alpha_0 = 1$ and $\{v_t\}$ is a sequence of independent random variables with $\mathscr{E} v_t = 0$ and $\mathscr{E} v_t^2 = \sigma^2$. To avoid trivialities we assume $\beta_p \neq 0$ and $\alpha_q \neq 0$. Then $\mathscr{E} y_t = 0$. We can write (1) as

$$\sum_{s=0}^{p} \beta_s \mathscr{L}^s y_t = \sum_{j=0}^{q} \alpha_j \mathscr{L}^j v_t. \tag{2}$$

Let the roots of

$$\sum_{s=0}^{p} \beta_s x^{p-s} = 0 \tag{3}$$

be $x_1, \ldots, x_p$ and the roots of

$$\sum_{j=0}^{q} \alpha_j z^{q-j} = 0 \tag{4}$$

be $z_1, \ldots, z_q$. Then (2) can be written

$$\prod_{i=1}^{p} (1 - x_i \mathscr{L}) y_t = \prod_{j=1}^{q} (1 - z_j \mathscr{L}) v_t. \tag{5}$$

The relationship between $\{y_t\}$ and $\{v_t\}$ is not affected by deleting common factors from both sides of (5); thus there is no loss of generality in assuming that (3) and (4) have no roots in common.

In order to be able to express y_t as a linear combination of $v_t, v_{t-1}, \ldots$ (when $\sigma^2 > 0$) the roots of (3) must be less than 1 in absolute value. Then

$$y_t = \prod_{i=1}^{p}(1 - x_i\mathscr{L})^{-1} \prod_{j=1}^{q}(1 - z_j\mathscr{L})v_t = \sum_{s=0}^{\infty} \delta_s \sum_{j=0}^{q} \alpha_j v_{t-s-j} = \sum_{r=0}^{\infty} \gamma_r v_{t-r}. \tag{6}$$

The coefficients $\{\delta_s\}$ are those obtained in Section 5.2.1; the coefficients $\{\gamma_r\}$ are obtained by

$$\begin{aligned} \gamma_0 &= \delta_0\alpha_0 = 1, \\ \gamma_1 &= \delta_0\alpha_1 + \delta_1\alpha_0 = \alpha_1 + \delta_1, \\ &\cdot \\ &\cdot \\ &\cdot \\ \gamma_{q-1} &= \delta_0\alpha_{q-1} + \delta_1\alpha_{q-2} + \cdots + \delta_{q-1}\alpha_0, \\ \gamma_r &= \delta_{r-q}\alpha_q + \delta_{r-q+1}\alpha_{q-1} + \cdots + \delta_r\alpha_0, \qquad r = q, q+1, \ldots. \end{aligned} \tag{7}$$

If we multiply (1) by y_{t-h}, $h = 0, 1, \ldots$, and take expected values, we obtain

$$\sum_{s=0}^{p} \beta_s \sigma(h - s) = \sum_{j=0}^{q}\sum_{r=0}^{\infty} \alpha_j \gamma_r \mathscr{E} v_{t-j} v_{t-h-r} = \sigma^2 \sum_{j=0}^{q}\sum_{r=0}^{\infty} \alpha_j \gamma_r \delta_{j,r+h}, \qquad h = 0, 1, \ldots, \tag{8}$$

where $\delta_{j,r+h}$ is the Kronecker delta. In particular we have

$$\sum_{s=0}^{p} \beta_s \sigma(h - s) = 0, \qquad h = q + 1, \ldots. \tag{9}$$

Thus the sequence $\{\sigma(r)\}$ satisfies the homogeneous difference equation for $r = q + 1 - p, \ldots$. [It will be shown in Section 5.8.2 that these equations and $\beta_0 = 1$ determine $\beta_1, \ldots, \beta_p$ if $\{\sigma(r)\}$ is known.]

For some purposes it is convenient to consider the parameters of the process to be $\beta_1, \ldots, \beta_p, \sigma(0), \rho_1, \ldots, \rho_q$, where $\rho_h = \sigma(h)/\sigma(0)$, instead of $\beta_1, \ldots, \beta_p, \alpha_1, \ldots, \alpha_q, \sigma^2$. If we let (1) be u_t, the expected value of $u_t u_{t+h}$ can be written in terms of the left-hand side of (1) as

$$\mathscr{E} u_t u_{t+h} = \sum_{s,r=0}^{p} \beta_s \beta_r \mathscr{E} y_{t-s} y_{t+h-r} = \sum_{s,r=0}^{p} \beta_s \beta_r \sigma(h + s - r). \tag{10}$$

In terms of the right-hand side of (1) the moment is

$$\begin{aligned} \mathscr{E} u_t u_{t+h} &= \sigma^2 \sum_{j=0}^{q-|h|} \alpha_j \alpha_{j+|h|}, \qquad & h &= 0, \pm 1, \ldots, \pm q, \\ &= 0, & h &= \pm(q+1), \ldots . \end{aligned} \tag{11}$$

The covariance generating function of $\{u_t\}$ can be written in terms of (10) and (11) as

$$\sigma(0) \sum_{h=-q}^{q} \sum_{s,r=0}^{p} \beta_s \beta_r \rho_{h+s-r} z^h = \sigma^2 M(z) M(z^{-1}), \tag{12}$$

where $M(z)$ is given by (4) of Section 5.7. The use of (9) permits us to write (12) as

$$\sigma(0) \sum_{s,r=0}^{p} \beta_s \beta_r \sum_{h=-(q-r)}^{q-s} \rho_{h+s-r} z^h = \sigma^2 M(z) M(z^{-1}). \tag{13}$$

The left-hand side of (13) only involves $\beta_1, \ldots, \beta_p, \sigma(0), \rho_1, \ldots, \rho_q$, and the right-hand side only involves $\sigma^2, \alpha_1, \ldots, \alpha_q$; thus the latter set of parameters can be determined from the former. The left-hand side of (13) multiplied by z^q is a polynomial of degree $2q$ and the roots occur in pairs as in Section 5.7.1. One set of roots determines $M(z)$ and hence $\alpha_1, \ldots, \alpha_q$. [See Doob (1944).]

5.8.2. Estimation of Parameters

The observed variables, the y_t's, are normally distributed if the v_t's are normally distributed. As in the case of the simple moving average studied in Section 5.7, the inverse of the covariance matrix of a finite set $y_1, \ldots, y_T$ is not simple, the minimal sufficient set of statistics is T in number, and the maximum likelihood equations are complicated and cannot be solved directly. However, any finite set of sample correlations has an asymptotic normal distribution (as stated in Theorem 8.4.6). A. M. Walker (1962) has extended the estimation method used for the simple moving average to the present case, namely, maximizing the approximate likelihood function of the first n sample serial correlations.

If the v_t's are independently and identically distributed with $\mathscr{E}v_t = 0$ and $\mathscr{E}v_t^2 = \sigma^2$, then the limiting distribution of $\sqrt{T}(r_1 - \rho_1), \ldots, \sqrt{T}(r_n - \rho_n)$ is multivariate normal with means 0 and covariances

$$(14)\quad w_{gh} = \sum_{r=-\infty}^{\infty} (\rho_{r+g}\rho_{r+h} + \rho_{r-g}\rho_{r+h} - 2\rho_h\rho_r\rho_{r+g} - 2\rho_g\rho_r\rho_{r+h} + 2\rho_g\rho_h\rho_r^2);$$

Theorem 8.4.6 will generalize this. [If the roots of (3) are less than 1 in absolute value, the conditions of Theorem 8.4.6 are satisfied; see (41) of Section 5.2 and (6) and (7).] In developing an estimation procedure based on this limiting distribution it is convenient to use the variables

$$(15)\quad \begin{aligned} x_j &= r_j, & j &= 1, \ldots, q, \\ x_j &= \sum_{s=0}^{p} \beta_s r_{j-s}, & j &= q+1, \ldots, q+p, \\ x_j &= \sum_{s,t=0}^{p} \beta_s \beta_t r_{j-s-t}, & j &= q+p+1, \ldots, n, \end{aligned}$$

where $r_0 = 1$ and $r_{-j} = r_j$, $j = 1, 2, \ldots$. Since

$$(16)\quad \sum_{s=0}^{p} \beta_s \rho_{h-s} = 0, \qquad h = q+1, \ldots,$$

from (9), $\sqrt{T}(x_1 - \rho_1), \ldots, \sqrt{T}(x_q - \rho_q), \sqrt{T}\,x_{q+1}, \ldots, \sqrt{T}\,x_n$ have a limiting normal distribution with means 0 and covariances ϕ_{ij}, $i, j = 1, \ldots, n$, which can be deduced from (w_{ij}). Let $\boldsymbol{\rho} = (\rho_1, \ldots, \rho_q)'$, $\boldsymbol{x}^{(1)} = (x_1, \ldots, x_q)'$, $\boldsymbol{x}^{(2)} = (x_{q+1}, \ldots, x_{q+p})'$, $\boldsymbol{x}^{(3)} = (x_{q+p+1}, \ldots, x_n)'$, $\boldsymbol{x} = (\boldsymbol{x}^{(1)\prime}, \boldsymbol{x}^{(2)\prime}, \boldsymbol{x}^{(3)\prime})'$, and let the covariance matrix $\boldsymbol{\Phi} = (\phi_{ij})$ and its inverse be partitioned into q, p, and $n - (q+p)$ rows and columns

$$(17)\quad \boldsymbol{\Phi} = \begin{pmatrix} \boldsymbol{\Phi}_{11} & \boldsymbol{\Phi}_{12} & \boldsymbol{\Phi}_{13} \\ \boldsymbol{\Phi}_{21} & \boldsymbol{\Phi}_{22} & \boldsymbol{\Phi}_{23} \\ \boldsymbol{\Phi}_{31} & \boldsymbol{\Phi}_{32} & \boldsymbol{\Phi}_{33} \end{pmatrix}, \qquad \boldsymbol{\Phi}^{-1} = \begin{pmatrix} \boldsymbol{\Phi}^{11} & \boldsymbol{\Phi}^{12} & \boldsymbol{\Phi}^{13} \\ \boldsymbol{\Phi}^{21} & \boldsymbol{\Phi}^{22} & \boldsymbol{\Phi}^{23} \\ \boldsymbol{\Phi}^{31} & \boldsymbol{\Phi}^{32} & \boldsymbol{\Phi}^{33} \end{pmatrix}.$$

Then the logarithm of the approximate likelihood function of $\boldsymbol{x}$ is

$$(18)\quad \log L = -\tfrac{1}{2}n \log 2\pi - \tfrac{1}{2}\log|\boldsymbol{\Phi}| \\ -\tfrac{1}{2}T(\boldsymbol{x}^{(1)\prime} - \boldsymbol{\rho}' \quad \boldsymbol{x}^{(2)\prime} \quad \boldsymbol{x}^{(3)\prime}) \begin{pmatrix} \boldsymbol{\Phi}^{11} & \boldsymbol{\Phi}^{12} & \boldsymbol{\Phi}^{13} \\ \boldsymbol{\Phi}^{21} & \boldsymbol{\Phi}^{22} & \boldsymbol{\Phi}^{23} \\ \boldsymbol{\Phi}^{31} & \boldsymbol{\Phi}^{32} & \boldsymbol{\Phi}^{33} \end{pmatrix} \begin{pmatrix} \boldsymbol{x}^{(1)} - \boldsymbol{\rho} \\ \boldsymbol{x}^{(2)} \\ \boldsymbol{x}^{(3)} \end{pmatrix}.$$

We consider as estimates a set of values of $\beta_1, \ldots, \beta_p, \rho_1, \ldots, \rho_q$ that satisfy the derivative equations $\partial \log L/\partial\beta_s = 0$, $s = 1, \ldots, p$, and $\partial \log L/\partial\rho_h = 0$, $h = 1, \ldots, q$. In these derivative equations the terms involving the partial derivatives of the ϕ_{ij}'s (or ϕ^{ij}'s) with respect to the β_s's and ρ_h's are of lower order in T than the other terms. If we ignore these terms and multiply by suitable constants, the resulting equations are

$$\mathbf{\Phi}^{11}(\boldsymbol{x}^{(1)} - \boldsymbol{\rho}) + \mathbf{\Phi}^{12}\boldsymbol{x}^{(2)} + \mathbf{\Phi}^{13}\boldsymbol{x}^{(3)} = \mathbf{0}, \tag{19}$$

$$\frac{\partial \boldsymbol{x}^{(2)\prime}}{\partial \boldsymbol{\beta}}[\mathbf{\Phi}^{21}(\boldsymbol{x}^{(1)} - \boldsymbol{\rho}) + \mathbf{\Phi}^{22}\boldsymbol{x}^{(2)} + \mathbf{\Phi}^{23}\boldsymbol{x}^{(3)}] + \frac{\partial \boldsymbol{x}^{(3)\prime}}{\partial \boldsymbol{\beta}}[\mathbf{\Phi}^{31}(\boldsymbol{x}^{(1)} - \boldsymbol{\rho}) + \mathbf{\Phi}^{32}\boldsymbol{x}^{(2)} + \mathbf{\Phi}^{33}\boldsymbol{x}^{(3)}] = \mathbf{0}, \tag{20}$$

where

$$\frac{\partial \boldsymbol{x}^{(2)\prime}}{\partial \boldsymbol{\beta}} = \left(\frac{\partial x_j}{\partial \beta_s}\right), \qquad j = q+1, \ldots, q+p,\ s = 1, \ldots, p, \tag{21}$$

$$\frac{\partial \boldsymbol{x}^{(3)\prime}}{\partial \boldsymbol{\beta}} = \left(\frac{\partial x_j}{\partial \beta_s}\right), \qquad j = q+p+1, \ldots, n,\ s = 1, \ldots, p. \tag{22}$$

To simplify (20) we note that

$$\frac{\partial x_j}{\partial \beta_s} = r_{j-s}, \qquad j = q+1, \ldots, q+p,\ s = 1, \ldots, p, \tag{23}$$

$$\frac{\partial x_j}{\partial \beta_s} = 2\sum_{t=0}^{p} \beta_t r_{j-s-t}, \qquad j = q+p+1, \ldots, n,\ s = 1, \ldots, p. \tag{24}$$

Since $\sqrt{T}\,(r_{j-s} - \rho_{j-s})$, $j = q+1, \ldots, q+p$, $s = 1, \ldots, p$, have a limiting distribution, $\partial x_j/\partial\beta_s$ can be replaced by ρ_{j-s} in (20) for these values of j and s. Because of (16), $\sqrt{T}\sum_{t=0}^{p} \beta_t r_{j-s-t}$, $j = q+p+1, \ldots, n$, $s = 1, \ldots, p$, have a limiting distribution and $\partial x_j/\partial\beta_s$ can be replaced by 0 in (20) for these values of j and s. Then (20) is asymptotically equivalent to

$$\mathbf{R}[\mathbf{\Phi}^{21}(\boldsymbol{x}^{(1)} - \boldsymbol{\rho}) + \mathbf{\Phi}^{22}\boldsymbol{x}^{(2)} + \mathbf{\Phi}^{23}\boldsymbol{x}^{(3)}] = \mathbf{0}, \tag{25}$$

where

$$\mathbf{R} = (\rho_{j-s}) = \begin{pmatrix} \rho_q & \rho_{q+1} & \cdots & \rho_{q+p-1} \\ \rho_{q-1} & \rho_q & \cdots & \rho_{q+p-2} \\ \cdot & \cdot & & \cdot \\ \cdot & \cdot & & \cdot \\ \cdot & \cdot & & \cdot \\ \rho_{q-p+1} & \rho_{q-p+2} & \cdots & \rho_q \end{pmatrix}. \tag{26}$$

We now want to argue that $\mathbf{R}$ is nonsingular. If $\mathbf{R}$ were singular, there would be a vector of constants $\boldsymbol{c} = (c_1, \ldots, c_p)'$ such that

$$\sum_{s=1}^{p} c_s \rho_{j-s} = 0, \qquad j = q+1, \ldots, q+p. \tag{27}$$

From (16) we have $\rho_h = -\sum_{t=1}^{p} \beta_t \rho_{h-t}$, $h = q+1, \ldots$. Application of this to (27) extends the range of validity of (27) to $j = q+1, \ldots$. This fact would imply that the process would satisfy a model (1) with $\beta_0, \ldots, \beta_p$ replaced by $c_1, \ldots, c_p$ and values of $\alpha_1, \ldots, \alpha_q$ determined by $c_1, \ldots, c_p$ and $\rho_1, \ldots, \rho_q$ in (13). However, this would contradict the fact that $\beta_p \neq 0$ and $p+1$ constants are needed on the left-hand side of (1).

Since $\mathbf{R}$ is nonsingular, (19) and (25) can be written

$$\begin{pmatrix} \boldsymbol{\Phi}^{11} & \boldsymbol{\Phi}^{12} & \boldsymbol{\Phi}^{13} \\ \boldsymbol{\Phi}^{21} & \boldsymbol{\Phi}^{22} & \boldsymbol{\Phi}^{23} \end{pmatrix} \begin{pmatrix} \boldsymbol{x}^{(1)} - \boldsymbol{\rho} \\ \boldsymbol{x}^{(2)} \\ \boldsymbol{x}^{(3)} \end{pmatrix} = \mathbf{0}. \tag{28}$$

Then

$$\begin{aligned} \begin{pmatrix} \boldsymbol{x}^{(1)} - \boldsymbol{\rho} \\ \boldsymbol{x}^{(2)} \end{pmatrix} &= -\begin{pmatrix} \boldsymbol{\Phi}^{11} & \boldsymbol{\Phi}^{12} \\ \boldsymbol{\Phi}^{21} & \boldsymbol{\Phi}^{22} \end{pmatrix}^{-1} \begin{pmatrix} \boldsymbol{\Phi}^{13} \\ \boldsymbol{\Phi}^{23} \end{pmatrix} \boldsymbol{x}^{(3)} \\ &= \begin{pmatrix} \boldsymbol{\Phi}_{13} \\ \boldsymbol{\Phi}_{23} \end{pmatrix} \boldsymbol{\Phi}_{33}^{-1} \boldsymbol{x}^{(3)}. \end{aligned} \tag{29}$$

(See Problem 8 of Chapter 2, for example.) The right-hand side of (29) is the regression of $\boldsymbol{x}^{(1)} - \boldsymbol{\rho}$ and $\boldsymbol{x}^{(2)}$ on $\boldsymbol{x}^{(3)}$ based on the covariance matrix of the asymptotic distribution. Then

$$\hat{\boldsymbol{\rho}} = \boldsymbol{x}^{(1)} - \boldsymbol{\Phi}_{13} \boldsymbol{\Phi}_{33}^{-1} \boldsymbol{x}^{(3)} \tag{30}$$

consists of the (asymptotic) residuals of the correlations $r_1, \ldots, r_q$ on $\sum_{s,t=0}^{p} \beta_s \beta_t r_{j-s-t}$, $j = q+p+1, \ldots, n$. The other set of equations in (29) states that the (asymptotic) residuals of $\sum_{s=0}^{p} \beta_s r_{j-s}$, $j = q+1, \ldots, q+p$, on $\sum_{s,t=0}^{p} \beta_s \beta_t r_{j-s-t}$, $j = q+p+1, \ldots, n$, are 0. These equations then can be written

$$\sum_{s=0}^{p} \hat{\beta}_s r_{j-s}(\boldsymbol{x}^{(3)}) = 0, \qquad j = q+1, \ldots, q+p, \tag{31}$$

where $r_{j-s}(\boldsymbol{x}^{(3)})$ denotes the residual of r_{j-s} from its (asymptotic) regression on $\boldsymbol{x}^{(3)}$.

The equations are nonlinear because $\boldsymbol{\Phi}$ and $\boldsymbol{x}^{(3)}$ depend on the unknown parameters. However, (30) and (31) can be solved iteratively by computing $\boldsymbol{\Phi}$

and $x^{(3)}$ on the basis of initial estimates of $\boldsymbol{\rho}$ and the β_t's and solving (30) and (31) for revised estimates. Initial consistent estimates can be found from $\hat{\rho}_i^0 = r_i$, $i = 1, \ldots, q$, and $\sum_{s=0}^{p} \hat{\beta}_s^0 r_{j-r} = 0$, $j = q + 1, \ldots, q + p$.

The estimates defined by (30) and (31) are consistent and asymptotically normally distributed. In fact

$$(32) \qquad \sqrt{T}\,(\hat{\boldsymbol{\rho}} - \boldsymbol{\rho}) = \sqrt{T}\,(x^{(1)} - \boldsymbol{\rho}) - \boldsymbol{\Phi}_{13}\boldsymbol{\Phi}_{33}^{-1}\sqrt{T}\,x^{(3)}$$

has a limiting normal distribution with mean $\mathbf{0}$ covariance matrix $\boldsymbol{\Phi}_{11} - \boldsymbol{\Phi}_{13}\boldsymbol{\Phi}_{33}^{-1}\boldsymbol{\Phi}_{31}$. The equations (31) are

$$(33) \qquad \sum_{s=1}^{p} \hat{\beta}_s r_{j-s}(x^{(3)}) = -r_j(x^{(3)}), \qquad j = q + 1, \ldots, q + p,$$

since $\hat{\beta}_0 = \beta_0 = 1$. The set $\sqrt{T}\,[r_i(x^{(3)}) - \rho_i]$, $i = 1, \ldots, q + p$, has a limiting normal distribution. Then (33) is equivalent to

$$(34) \qquad \sum_{s=1}^{p} r_{j-s}(x^{(3)})\sqrt{T}\,(\hat{\beta}_s - \beta_s) = -\sqrt{T}\sum_{s=0}^{p} r_{j-s}(x^{(3)})\beta_s, \qquad j = q + 1, \ldots, q + p.$$

By virtue of (16) the right-hand side of (34) can be written as

$$(35) \qquad -\sum_{s=0}^{p} \sqrt{T}\,[r_{j-s}(x^{(3)}) - \rho_{j-s}]\beta_s, \qquad j = q + 1, \ldots, q + p,$$

which have a limiting normal distribution with means 0 and covariance matrix $\boldsymbol{\Phi}_{22} - \boldsymbol{\Phi}_{23}\boldsymbol{\Phi}_{33}^{-1}\boldsymbol{\Phi}_{32}$. Since $[r_{j-s}(x^{(3)})]$ converges stochastically to $\mathbf{R}'$, (34) shows that $\sqrt{T}(\hat{\boldsymbol{\beta}} - \boldsymbol{\beta})$, where $\hat{\boldsymbol{\beta}} = (\hat{\beta}_1, \ldots, \hat{\beta}_p)'$ and $\boldsymbol{\beta} = (\beta_1, \ldots, \beta_p)'$, has the limiting distribution of

$$(36) \qquad (\mathbf{R}')^{-1}\sqrt{T}\,(x^{(2)} - \boldsymbol{\Phi}_{23}\boldsymbol{\Phi}_{33}^{-1}x^{(3)})$$

which is normal with mean $\mathbf{0}$ and covariance matrix

$$(37) \qquad (\mathbf{R}')^{-1}(\boldsymbol{\Phi}_{22} - \boldsymbol{\Phi}_{23}\boldsymbol{\Phi}_{33}^{-1}\boldsymbol{\Phi}_{32})\mathbf{R}^{-1}.$$

Then $\sqrt{T}(\hat{\boldsymbol{\rho}} - \boldsymbol{\rho})$ and $\sqrt{T}(\hat{\boldsymbol{\beta}} - \boldsymbol{\beta})$ have a limiting joint normal distribution and the covariance between the two vectors in this limiting distribution is

$$(38) \qquad (\boldsymbol{\Phi}_{12} - \boldsymbol{\Phi}_{13}\boldsymbol{\Phi}_{33}^{-1}\boldsymbol{\Phi}_{32})\mathbf{R}^{-1}.$$

Durbin's procedure for the moving average model can be modified [Durbin (1960b)]. Suppose one starts with some initial estimates of $\beta_1, \ldots, \beta_p$, say $\hat{\beta}_1^{(0)}, \ldots, \hat{\beta}_p^{(0)}$ (as suggested above for iteration); let $u_t^{(0)} = \sum_{s=0}^{p} \hat{\beta}_s^{(0)} y_{t-s}$.

Then use Durbin's method described in Section 5.7.2 to estimate the coefficients in the moving average model

$$(39)\qquad u_t^{(0)} = \sum_{j=0}^{q} \alpha_j^{(0)} v_{t-j},$$

where the $u_t^{(0)}$'s are taken as the observed variables. Call the resulting estimates $\hat{\alpha}_1^{(0)}, \ldots, \hat{\alpha}_q^{(0)}$.

Let $\{x_t\}$ be a stochastic process such that

$$(40)\qquad \sum_{s=0}^{p} \beta_s x_{t-s} = v_t, \qquad t = \ldots, -1, 0, 1, \ldots.$$

Then the stochastic process defined by

$$(41)\qquad y_t^* = x_t + \alpha_1 x_{t-1} + \cdots + \alpha_q x_{t-q}, \qquad t = \ldots, -1, 0, 1, \ldots,$$

satisfies (1) for $t = \ldots, -1, 0, 1, \ldots$. If $\{x_t\}$ is given for $t = \ldots, -1, 0$, and $\{y_t^*\}$ is given for $t = 1, 2, \ldots$, then $\{x_t\}$ could be continued by solving (41) recursively for $x_1, x_2, \ldots$. This suggests using $\hat{\alpha}_1^{(0)}, \ldots, \hat{\alpha}_q^{(0)}$ and arbitrary $x_0^{(0)}, \ldots, x_{-q+1}^{(0)}$ to define

$$(42)\qquad x_t^{(0)} = y_t - (\hat{\alpha}_1^{(0)} x_{t-1}^{(0)} + \cdots + \hat{\alpha}_q^{(0)} x_{t-q}^{(0)}), \qquad t = 1, \ldots, T.$$

Then the β_r's are estimated on the basis of $\{x_t^{(0)}\}$ satisfying (41). This leads to estimates $\hat{\beta}_1^{(1)}, \ldots, \hat{\beta}_p^{(1)}$. Durbin proposes to alternate back and forth between these procedures, calculating $\hat{\alpha}_1^{(1)}, \ldots, \hat{\alpha}_q^{(1)}, \hat{\beta}_1^{(2)}, \ldots, \hat{\beta}_p^{(2)}$, etc. It is hoped that for large T (and large n) the iteration converges.

5.9. SOME EXAMPLES

Wold (1965) has generated artificial time series by using second-order difference equations with u_t's taken as independent random normal deviates. Three of these series are tabulated in Appendix A.2 and the three series are graphed. The coefficients were $\beta_1 = -\gamma$, $\beta_2 = \gamma^2$ and the roots of the associated polynomial equations were $\gamma e^{\pm i2\pi/6}$ with $\gamma = 0.25$, 0.7, and 0.9, respectively. In each case the variance of u_t was taken so the variance of y_t was 1. There can be observed in the graphs a tendency towards oscillation with periods of length roughly 6; the period is more pronounced with the larger amplitude γ.

The first three correlations ($r_h = C_h/C_0$) of the three tabulated series for $T = 200$ and the corresponding population correlations $[\rho_h = \sigma(h)/\sigma(0)]$ are given in Table 5.1.

TABLE 5.1

SAMPLE AND POPULATION CORRELATION COEFFICIENTS FOR THREE AUTOREGRESSIVE PROCESSES ($T = 200$)

γ	r_1	r_2	r_3	ρ_1	ρ_2	ρ_3
0.25	0.2473	0.1120	0.0492	0.23529	−0.00368	−0.015625
0.70	0.5113	−0.0473	−0.3001	0.46980	−0.16114	−0.34300
0.90	0.5011	−0.3858	−0.7726	0.49724	−0.36249	−0.72900

Estimates of β_1 and β_2 under the assumption that $p = 2$ and estimates of β_1, β_2, and β_3 under the assumption that $p = 3$ were obtained from the equations

$$(1) \qquad \begin{pmatrix} 1 & r_1 \\ r_1 & 1 \end{pmatrix} \begin{pmatrix} b_1 \\ b_2 \end{pmatrix} = -\begin{pmatrix} r_1 \\ r_2 \end{pmatrix}, \qquad \begin{pmatrix} 1 & r_1 & r_2 \\ r_1 & 1 & r_1 \\ r_2 & r_1 & 1 \end{pmatrix} \begin{pmatrix} b_1 \\ b_2 \\ b_3 \end{pmatrix} = -\begin{pmatrix} r_1 \\ r_2 \\ r_3 \end{pmatrix},$$

respectively. The results are given in Table 5.2.

The test of the null hypothesis $\beta_3 = 0$ can be made on the basis of $\sqrt{T}\, b_3$; the three values are −0.1458, 1.208, and −0.7706. Under the null hypothesis the values of $\sqrt{T}\, b_3$ should be approximately distributed according to $N(0, 1)$. In none of the three cases would the null hypothesis be rejected at any reasonable level of significance such as 1%, 5%, or 10% for two-sided tests.

Yule (1927) proposed the autoregressive process as a statistical model in preference to the model of an error superimposed on trigonometric trends such as considered in Chapter 4. He applied it to Wolfer's sunspot numbers, which are annual measures of sunspot activity, from 1749 to 1924. [The data have been expanded and given in more detail by Waldmeier (1961).] The figures used by Yule are tabulated in Table A.3.1 of Appendix A.3. Similar data analyzed by

TABLE 5.2

ESTIMATES OF COEFFICIENTS OF AUTOREGRESSIVE PROCESSES

γ	β_1	β_2	b_1	b_2	b_1	b_2	b_3
0.25	−0.25	0.0625	−0.2339	−0.0542	−0.2334	−0.0518	−0.0103
0.70	−0.70	0.49	−0.7250	0.4180	−0.6893	0.3561	0.0854
0.90	−0.90	0.81	−0.9273	0.8505	−0.9736	0.9010	−0.0545

TABLE 5.3

CORRELATIONS FOR SUNSPOT NUMBERS

Lag	Correlation
1	0.811180
2	0.433998
3	0.031574
4	−0.264463
5	−0.404119

Craddock (1967) are graphed below. Roughly, the series looks somewhat similar to the artificial series of Wold.

The first five correlations as given by Yule are recorded in Table 5.3.

For a second-order process Yule obtains the estimates $b_1 = -1.34254$, $b_2 = 0.65504$, $\hat{\nu} = -13.854$, $\hat{\sigma} = 15.41$. The roots of the associated polynomial equation are $0.67127 \pm 0.45215i = 0.80935e^{\pm i33.963°}$. The tendency of the fitted model is to have a period of $360°/33.963° = 10.600$ (years). This is rather less than the usually accepted period of a little over 11 years. [Schuster (1906) arrived at 11.125 years.]

From the data in Table 5.3 Yule worked out the estimates of β_p on the basis of a pth-order process (the partial correlation coefficient) for $p = 1, 2, 3, 4, 5$. These are given in Table 5.4. It will be seen that $\sqrt{T}\, b_p(p)$, which is approximately normally distributed if $\beta_p = \beta_{p+1} = \cdots = 0$, are relatively small numerically for $p = 3, 4, 5$.

Another study of the sunspot data was made by Craddock (1967) on the basis of annual mean numbers from 1700 to 1965. A graph of the record is

TABLE 5.4

LAST COEFFICIENTS

p	$b_p(p)$	$\sqrt{T}\, b_p(p)$
1	−0.811180	−10.761
2	0.655040	8.690
3	0.101043	1.340
4	−0.013531	−0.180
5	0.050001	0.663

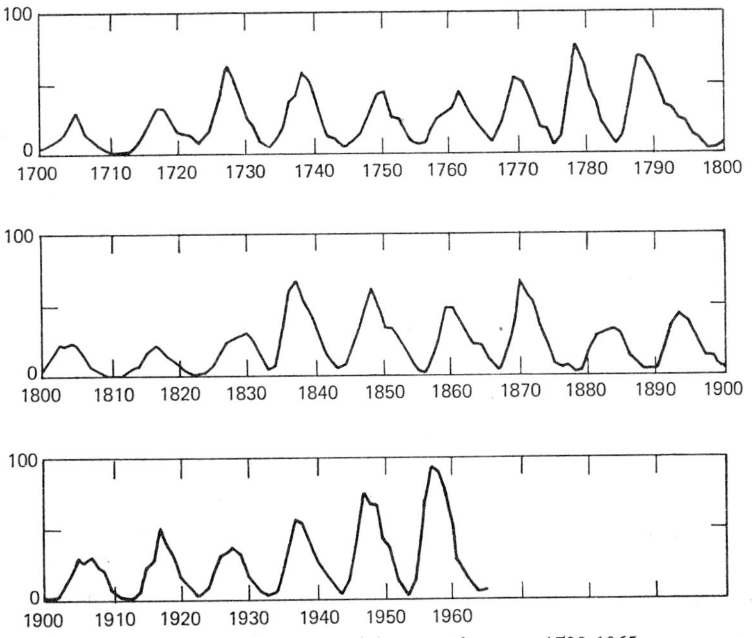

Figure 5.1. Sunspot activity over the years 1700-1965.

given in Figure 5.1. The correlations are given in Figure 5.2. Craddock fitted alternative autoregressive models for values of p from 1 to 30. In Figure 5.3 is plotted the ratio $\hat{\sigma}_p^2/\hat{\sigma}_0^2$ (in terms of percentages), where $\hat{\sigma}_p^2$ is the estimate of variance based on a pth-order model (with a constant included). The difference $T(\hat{\sigma}_{p-1}^2 - \hat{\sigma}_p^2)$ is proportional to $\hat{\beta}_p^2/a^{*pp}$, which is used in testing the hypothesis $\beta_p = 0$. [See (16) of Section 5.6.] In fact the t^2-statistic is

$$(T-p)\frac{\hat{\sigma}_{p-1}^2 - \hat{\sigma}_p^2}{\hat{\sigma}_p^2} = (T-p)\left(\frac{\hat{\sigma}_{p-1}^2}{\hat{\sigma}_p^2} - 1\right). \tag{2}$$

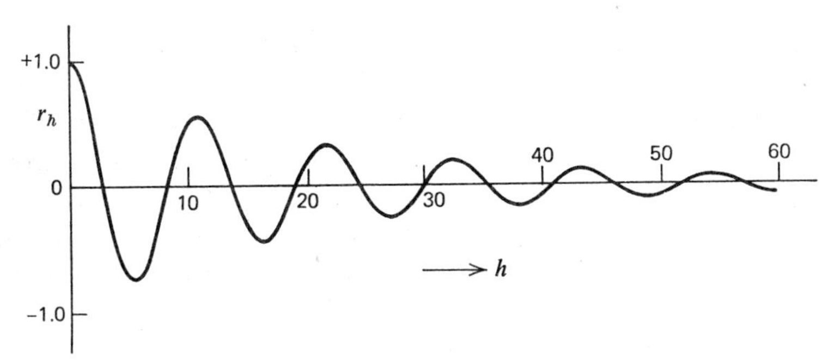

Figure 5.2. Correlation coefficients for sunspot numbers by Craddock.

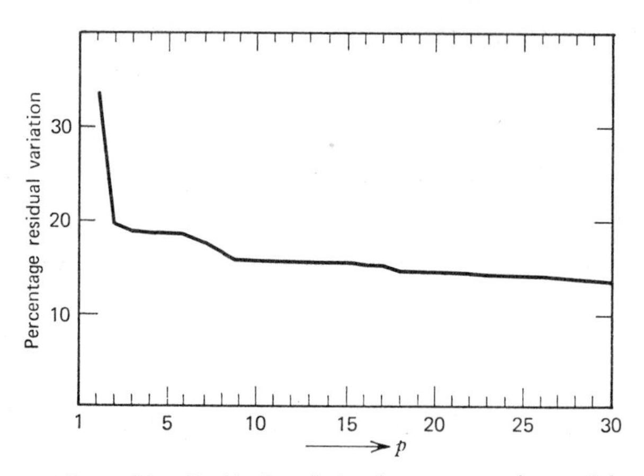

Figure 5.3. Residual variation in autoregressive models.

The data indicate clearly that β_1 and β_2 should be taken different from 0. The set $\beta_3, \ldots, \beta_9$ are doubtfully 0, $\beta_{10}, \ldots, \beta_{18}$ are more likely to be taken 0, and $\beta_{19}, \ldots, \beta_{30}$ can be concluded to be 0. In formal terms of the multiple decision procedure described at the end of Section 5.6.3 we take $p = 30$ and $m = 0$. The coefficients $\sqrt{T}\,\hat{\beta}_j$, $j = 9, \ldots, 30$, are roughly 1.2, except for $\sqrt{T}\,\hat{\beta}_{18}$ of about 2.2 and $\sqrt{T}\,\hat{\beta}_9$ of about 3.8. If the sequence of ε_i, $i = 9, \ldots, 30$, is such that $0.0002 < \varepsilon_i < 0.02$, then the order would be decided as 9; if the sequence of ε_i, $i = 18, \ldots, 20$, is such that $0.03 < \varepsilon_i < 0.1$, the order would be decided as 18. ($\varepsilon_1 = \cdots = \varepsilon_{30} = 0.0035$ corresponds to $p_0 = 0.9$.)

Whittle (1954) has studied the sunspot phenomenon on the basis of semi-annual data from 1886 to 1945 and found an autoregressive model using lags of 1 and 22 (that is, 6 months and 11 years). Many other statistical investigations have been made.

Schaerf (1964) has also studied Wolfer's series as well as Whittle's. She fits an autoregressive model with lags 1, 2, and 9.

As another example, consider the Beveridge trend-free wheat price index tabulated in Appendix A.1 together with the correlations. The estimated coefficients of a pth-order autoregressive process are given in Table 5.5 for $p = 2, 3, \ldots, 8$. The penultimate column gives $\sqrt{T}\,b_p(p)$, which can be used to test the null hypothesis $\beta_p = 0$ against the alternative $\beta_p \neq 0$ under the assumption that the order is not greater than p. It will be observed that the null hypothesis $\beta_2 = 0$ is rejected (at any reasonable significance level); the null hypothesis $\beta_p = 0$ is accepted for $p = 3$, 4, or 5 (at any reasonable significance level); and the null hypothesis $\beta_p = 0$ is rejected for $p = 6$, 7, or 8 at the 2.5% significance level and accepted for $p = 6$ or 7 at the 1% level. If one assumes

TABLE 5.5

ESTIMATED COEFFICIENTS OF AUTOREGRESSIVE PROCESSES FOR THE BEVERIDGE TREND-FREE WHEAT PRICE INDEX

p	$b_1(p)$	$b_2(p)$	$b_3(p)$	$b_4(p)$	$b_5(p)$	$b_6(p)$	$b_7(p)$	$b_8(p)$	$\sqrt{370}\, b_p(p)$	$\dfrac{\hat{\sigma}_p^2}{\hat{\sigma}_0^2}$
2	−0.7368	0.3110							5.984	0.6179
3	−0.7489	0.3397	−0.0388						−0.746	0.6170
4	−0.7503	0.3521	−0.0662	0.0367					0.706	0.6162
5	−0.7488	0.3494	−0.0516	0.0055	−0.0415				−0.798	0.6151
6	−0.7435	0.3501	−0.0582	0.0504	−0.0546	0.1284			2.470	0.6050
7	−0.7285	0.3437	−0.0524	0.0436	−0.0138	0.0417	0.1165		2.241	0.5968
8	−0.7123	0.3495	−0.0543	0.0496	−0.0211	0.0895	0.0153	0.1390	2.674	0.5853

that the order is not greater than p for $p = 6$, 7, or 8 and uses the multiple-decision procedure suggested earlier one would decide on the value of p if ε_p is at least 0.025. On the other hand, if one used as a multiple-decision procedure testing hypotheses $\beta_j = 0$ in sequence, $j = 2, 3, \ldots$, until one was accepted and then deciding on the order of the last rejected hypothesis, he would settle on the second-order process [as did Sargan (1953)]. However, Table 5.5 suggests that an order greater than 8 might be suitable.

5.10 DISCUSSION

In this chapter we have discussed statistical models for stationary processes that depend on a finite number of parameters in a form that may be said to have linear structure: linear in the coefficients, in the observed variables, and in the hypothetical random variables. In some cases the model corresponds to a theory of the generation of the data and the coefficients have substantive meaning; in other cases the model is an approximation that is adequate for many purposes.

The autoregressive model has several advantages over the moving average model and over the autoregressive process with moving average residuals, though in certain instances the latter models may give a good description of the formation of the observed time series. The estimates of the coefficients of the autoregressive process are readily calculated, and inference based on large samples is easy to carry out because it corresponds to ordinary least squares. In many cases the coefficients have a direct interpretation, and the linear functions of lagged variables can be used for prediction.

As shall be seen later, any of these models can be used for approximating and estimating the spectral density.

REFERENCES

Section 5.1. Gilbert Walker (1931), Yule (1927).
Section 5.3. Halmos (1958), Turnbull and Aitken (1952).
Section 5.4. Durbin (1960a), (1960b), Mann and Wald (1943b).
Section 5.5. T. W. Anderson (1959), T. W. Anderson and Rubin (1950), Koopmans, Rubin, and Leipnik (1950), Loève (1963).
Section 5.6. Bartlett and Diananda (1950), Quenouille (1947), A. M. Walker (1952).
Section 5.7. Durbin (1959), A. M. Walker (1961).
Section 5.8. Doob (1944), Durbin (1960b), A. M. Walker (1962).
Section 5.9. Craddock (1967), Sargan (1953), Schaerf (1964), Schuster (1906). Waldmeier (1961), Whittle (1954), Wold (1965), Yule (1927).
Problems. Haavelmo (1947), Kuznets (1954).

PROBLEMS

1. (Sec. 5.1.) Find the coefficients $\alpha_1, \ldots, \alpha_p$ in (2) written as

$$(\Delta^p + \alpha_1\Delta^{p-1} + \cdots + \alpha_p)y_{t-p} = u_t.$$

2. (Sec. 5.2.1.) Carry through in detail (30) in the case of $q = 1, p = 2$. Express u_t^* defined by (31) in terms of u_t for $r = 1$. Verify that u_t^* are uncorrelated.

3. (Sec. 5.2.1.) If $\{y_t\}$ is a stationary process and satisfies

$$y_t + y_{t-1} = u_t,$$

with $\mathscr{E}u_t = 0$, $\mathscr{E}u_t^2 = \sigma^2$, and $\mathscr{E}u_tu_s = 0$, $t \neq s$, prove $y_t = (-1)^s y_{t-s}$ with probability 1.

4. (Sec. 5.2.1.) Prove that a linear difference equation of order p

$$\beta_0(t)z_t + \beta_1(t)z_{t-1} + \cdots + \beta_p(t)z_{t-p} = \alpha(t),$$

$\beta_0(t) \neq 0$, $\beta_p(t) \neq 0$, over a set S of consecutive integer values of t has one and only one solution z_t for which values at p consecutive t-values are arbitrarily prescribed. [*Hint:* Prove by induction on t, showing that the values of a solution z_t at any p consecutive integers determine its value at the next integer.]

5. (Sec. 5.2.1.) Show that the tth power of any root x_i of the polynomial equation (23) associated with

$$\beta_0 w_t + \beta_1 w_{t-1} + \cdots + \beta_p w_{t-p} = 0$$

is a solution to the difference equation ($\beta_0 \neq 0$, $\beta_p \neq 0$). [*Hint:* Substitute $w_t = x_i^t$.]

6. (Sec. 5.2.1.) Show that

$$(\mathscr{P} - a)P_q(t)a^t = P_{q-1}(t)a^{t+1},$$

where $P_q(t)$ and $P_{q-1}(t)$ are polynomials of degree q and $q - 1$, respectively.

7. (Sec. 5.2.1.) Show that

$$(\mathscr{P} - a)^m P_{m-1}(t)a^t = 0.$$

8. (Sec. 5.2.1.) Show that $c_0 a^t + c_1 t a^{t-1} + \cdots + c_{m-1} t^{m-1} a^{t-m+1}$ can be written $P_{m-1}(t)a^t$ (for $a \neq 0$).

9. (Sec. 5.2.1.) Prove that if a is a real root of multiplicity m of the polynomial equation (23) associated with the homogenous difference equation of Problem 5, then $w_t = P_{m-1}(t)a^t$ is a solution to the difference equation.

10. (Sec. 5.2.1.) Prove that if $\alpha e^{i\theta}$ and $\alpha e^{-i\theta}$ are a pair of conjugate complex roots of multiplicity m of the polynomial equation (23) associated with the homogeneous difference equation of Problem 5, then

$$w_t = P_{m-1}(t)\alpha^t e^{it\theta} + \bar{P}_{m-1}(t)\alpha^t e^{-it\theta}$$

is a real solution to the difference equation, where $P_{m-1}(t)$ and $\bar{P}_{m-1}(t)$ are polynomials whose corresponding coefficients are conjugate complex numbers.

11. (Sec. 5.2.1.) Prove that if the roots of (23) are distinct, every solution of the equation of Problem 5 is $w_t = \sum_{i=1}^{p} k_i x_i^t$; w_t real implies k_i is real if x_i is real and k_i and k_{i+1} are conjugate complex if x_i and x_{i+1} are conjugate complex. [*Hint:* Use Problem 5 to show the above is a solution and use Problem 4 to show it is unique for specified values of w_t for p consecutive values of t.]

12. (Sec. 5.2.1.) In a second-order difference equation model is $\beta_1 = 0$ a necessary and sufficient condition that y_t and y_{t-1} are independent?

13. (Sec. 5.2.2.) Verify from (51) that

$$\sigma(0) = \frac{\sigma^2}{(1 + \beta_2)^2 - \beta_1^2} \cdot \frac{1 + \beta_2}{1 - \beta_2},$$

$$\sigma(1) = \frac{\sigma^2}{(1 + \beta_2)^2 - \beta_1^2} \cdot \frac{-\beta_1}{1 - \beta_2}.$$

14. (Sec. 5.2.2.) Verify the results of Problem 13 from (42) and

$$\sigma(0) = \mathscr{E}\left(\sum_{r=0}^{\infty} \delta_r u_{t-r}\right)^2 = \sigma^2 \sum_{r=0}^{\infty} \delta_r^2,$$

$$\sigma(1) = \mathscr{E} \sum_{r=0}^{\infty} \delta_r u_{t-r} \sum_{s=0}^{\infty} \delta_s u_{t-1-s} = \sigma^2 \sum_{r=0}^{\infty} \delta_r \delta_{r+1}.$$

15. (Sec. 5.2.2.) Graph δ_r and $\sigma(r)$ for $\sigma^2 = 1$ and (a) $\beta_1 = 0.9$, (b) $\beta_1 = -0.9$, (c) $\beta_1 = 0.1$, and (d) $\beta_1 = -0.1$.

16. (Sec. 5.2.2.) Graph δ_r and $\sigma(r)$ for $\sigma^2 = 1$ and (a) $\beta_1 = 0.25$, $\beta_2 = 0.0625$, (b) $\beta_1 = 0.7$, $\beta_2 = 0.49$, and (c) $\beta_1 = 0.9$, $\beta_2 = 0.81$.

17. (Sec. 5.3.) Find $\mathbf{\Lambda}^\tau$ when

$$\mathbf{\Lambda} = \begin{bmatrix} \lambda & 1 & 0 & \cdots & 0 & 0 \\ 0 & \lambda & 1 & \cdots & 0 & 0 \\ 0 & 0 & \lambda & \cdots & 0 & 0 \\ \cdot & \cdot & \cdot & & \cdot & \cdot \\ \cdot & \cdot & \cdot & & \cdot & \cdot \\ \cdot & \cdot & \cdot & & \cdot & \cdot \\ 0 & 0 & 0 & \cdots & \lambda & 1 \\ 0 & 0 & 0 & \cdots & 0 & \lambda \end{bmatrix}.$$

18. (Sec. 5.3.) Prove that $\mathbf{\Lambda}$ of Problem 17 has the characteristic root λ of multiplicity equal to the order of the matrix and that there is only one linearly independent characteristic vector.

19. (Sec. 5.3.) Prove that the one and only characteristic vector of $-\tilde{\mathbf{B}}$ (except for multiplication by a constant) corresponding to a root x_i of any multiplicity is $(x_i^{p-1}, x_i^{p-2}, \ldots, 1)'$. Hence, prove that the number of linearly independent characteristic vectors of $-\tilde{\mathbf{B}}$ is equal to the number of distinct roots. Argue from this that if the roots are not all distinct then $-\tilde{\mathbf{B}}$ cannot be written as $\boldsymbol{C}\mathbf{\Lambda}\boldsymbol{C}^{-1}$ with $\mathbf{\Lambda}$ diagonal.

20. (Sec. 5.3.) Verify that (28) of Section 5.2 is identical to (22) of Section 5.3.

21. (Sec. 5.4.) Find conditions under which

$$\frac{1}{T}\sum_{t=1}^{T}\frac{y_t y_{t-1}}{y_{t-1}^2} = \frac{1}{T}\sum_{t=1}^{T}\frac{y_t}{y_{t-1}}$$

is an unbiased estimate of $-\beta_1$ when $p = 1$ (and $q = 0$).

22. (Sec. 5.4.) The estimates in (b) of Problem 18 in Chapter 3 are $\hat{\gamma}_0 = 72.31$, $\hat{\gamma}_1 = 2.206$, and $\hat{\gamma}_2 = -0.1109$.

(a) Find the (maximum likelihood) estimates of the coefficients β_1 and β_2 and the variance σ^2 in the model

$$\tilde{y}_t + \beta_1 \tilde{y}_{t-1} + \beta_2 \tilde{y}_{t-2} = u_t,$$

where u_t has mean 0 and variance σ^2,

$$\tilde{y}_t = y_t - \gamma_0 \phi_{0T}(t) - \gamma_1 \phi_{1T}(t),$$

and the 1897 price is y_{-1}, the 1898 price is y_0, etc.

(b) Find estimates of the roots x_1 and x_2 of the associated polynomial equation (23) of Section 5.2, estimates of coefficients δ_r in (42) or (44) of Section 5.2, and estimates of $\sigma(r)$ in (51) or (52) of Section 5.2.

(c) Show that the estimates of β_1 and β_2 in (a) are approximately the maximum likelihood estimates in

$$y_t + \beta_1 y_{t-1} + \beta_2 y_{t-2} + \alpha_0 + \alpha_1 t = u_t.$$

23. (Sec. 5.4.) (a) Let estimates of β_1, β_2 be defined by

$$\begin{pmatrix} 1 & r_1^* \\ r_1^* & 1 \end{pmatrix} \begin{pmatrix} b_1 \\ b_2 \end{pmatrix} = -\begin{pmatrix} r_1^* \\ r_2^* \end{pmatrix}.$$

Show that

$$b_2 = -\frac{r_2^* - r_1^{*2}}{1 - r_1^{*2}}.$$

(b) Let the estimates of β_1, β_2, β_3 be defined by

$$\begin{pmatrix} 1 & r_1^* & r_2^* \\ r_1^* & 1 & r_1^* \\ r_2^* & r_1^* & 1 \end{pmatrix} \begin{pmatrix} b_1 \\ b_2 \\ b_3 \end{pmatrix} = -\begin{pmatrix} r_1^* \\ r_2^* \\ r_3^* \end{pmatrix}.$$

Show that

$$b_3 = -\frac{r_3^* - r_1^{*2} r_3^* + r_1^{*3} - 2r_1^* r_2^* + r_1^* r_2^{*2}}{1 - 2r_1^{*2} + 2r_1^{*2} r_2^* - r_2^{*2}}.$$

24. (Sec. 5.5.1.) Write out (3) in detail when $y_t = \tilde{y}_t$, and verify that (3) defines the same estimates as (8) of Section 5.4 with z_{-t} and $\hat{\gamma}$ deleted.

25. (Sec. 5.5.1.) Show that (3) and (4) define the maximum likelihood estimates of $\mathbf{B}$ and $\mathbf{\Sigma}$ defined by (1) and the fact that $\boldsymbol{u}_1, \ldots, \boldsymbol{u}_T$ are independently distributed, each according to $N(\mathbf{0}, \mathbf{\Sigma})$, and $\boldsymbol{y}_0$ is a given vector.

26. (Sec. 5.5.2.) Show that the conditions that $\boldsymbol{A}$ is positive semidefinite and $\operatorname{tr} \boldsymbol{A} = 0$ imply that $\boldsymbol{A} = \mathbf{0}$.

27. (Sec. 5.5.3.) Prove that (31) can be solved uniquely for $\boldsymbol{A}$. [*Hint:* Transform (31) to

$$\boldsymbol{A}^* - \mathbf{\Lambda} \boldsymbol{A}^* \mathbf{\Lambda}' = \mathbf{\Sigma}^*,$$

where $\boldsymbol{A}^* = \boldsymbol{C}^{-1} \boldsymbol{A} (\boldsymbol{C}')^{-1}$ and $\mathbf{\Sigma}^* = \boldsymbol{C}^{-1} \mathbf{\Sigma} (\boldsymbol{C}')^{-1}$.]

28. (Sec. 5.5.3.) Show that if $\boldsymbol{A} = \boldsymbol{B} + \boldsymbol{C}$, where $\boldsymbol{B}$ is positive definite and $\boldsymbol{C}$ is positive semidefinite, then $\boldsymbol{A}$ is positive definite. [*Hint:* $\boldsymbol{x}'\boldsymbol{A}\boldsymbol{x} = \boldsymbol{x}'\boldsymbol{B}\boldsymbol{x} + \boldsymbol{x}'\boldsymbol{C}\boldsymbol{x}$ and $\boldsymbol{A}$ is positive definite if and only if $\boldsymbol{x}'\boldsymbol{A}\boldsymbol{x} \geq 0$ and $\boldsymbol{x}'\boldsymbol{A}\boldsymbol{x} = 0$ implies $\boldsymbol{x} = \mathbf{0}$.]

29. (Sec. 5.5.4.) Verify that if the random matrix $\boldsymbol{W}$ has covariance matrix $\boldsymbol{A} \otimes \mathbf{\Sigma}$, then $\boldsymbol{A}^{-1}\boldsymbol{W}$ has covariance matrix $\boldsymbol{A}^{-1} \otimes \mathbf{\Sigma}$.

30. (Sec. 5.5.5.) Show that (60) and (61) define the maximum likelihood estimates of $\mathbf{B}$, $\boldsymbol{\nu}$, and $\mathbf{\Sigma}$ defined by (57) and the facts that $\boldsymbol{u}_1, \ldots, \boldsymbol{u}_T$ are independently distributed, each according to $N(\mathbf{0}, \mathbf{\Sigma})$, and $\boldsymbol{y}_0$ is a given vector.

31. (Sec. 5.5.5.) Evaluate (67) for $p = 2$ and $\mathbf{B} = \tilde{\mathbf{B}}$.

32. (Sec. 5.5.5.) Show that the sum of expected values of the squares of the components of (70) converges to 0.

33. (Sec. 5.6.3.) Verify that (50) has mean 0 and variance σ^4 and that the covariance between (50) and (50) with j replaced by j' $(\neq j)$ is 0.

34. (Sec. 5.7.1.) For $q = 1$ show that $|\sigma(1)| \leq \frac{1}{2}\sigma(0)$.

35. (Sec. 5.7.1.) Prove that any root of (7) of absolute value 1 is of even multiplicity. [*Hint:* For $z = e^{i\lambda}$

$$\sum_{g=-q}^{q} \sigma(g)z^g = \sum_{g=-q}^{q} \sigma(g) \cos \lambda g = f(\lambda),$$

say, is a spectral density and hence is nonnegative. (See Section 7.3.) Show that if $\lambda = \nu$ is a root of $f(\lambda)$, it is of even multiplicity by showing that the highest order of the derivative of $f(\lambda)$ vanishing at $\lambda = \nu$ must be odd.]

36. (Sec. 5.7.1.) Prove that (10) converges in the mean.

37. (Sec. 5.7.1.) For $q = 1$ show that $|\mathbf{\Sigma}_T| = \sigma^{2T}(1 - \alpha_1^{2(T+1)})(1 - \alpha_1^2)^{-1}$, where $\mathbf{\Sigma}_T = [\sigma(t - s)]$ is of order T. [*Hint:* See Section 6.7.7.]

38. (Sec. 5.7.1.) For $q = 1$ show

$$\lim_{T\to\infty} \sigma_T^{ts} = \frac{(-\alpha_1)^{|t-s|}(1 - \alpha_1^{2\min(t,s)})}{\sigma^2(1 - \alpha_1^2)},$$

where (σ_T^{ts}) is the inverse of the Tth-order matrix $[\sigma(t - s)] = \mathbf{\Sigma}_T$.

39. (Sec. 5.7.3.) For $q = 1$ and $\rho_1 = \rho$ show that for positive indices

$$w_{11} = 1 - 3\rho^2 + 4\rho^4, \qquad w_{jj} = 1 + 2\rho^2, \qquad j = 2, 3, \ldots,$$
$$w_{12} = 2\rho - 2\rho^3, \qquad w_{j,j+1} = 2\rho, \qquad j = 2, 3, \ldots,$$
$$w_{j,j+2} = \rho^2, j = 1, 2, \ldots, \qquad w_{ij} = 0, \qquad |i - j| > 2.$$

40. (Sec. 5.9.) The following table gives Deflated Aggregate Disposable Income for the United States, 1919 – 1941, in billions of dollars:

1919	44.32	1927	59.26	1935	57.47
1920	45.45	1928	61.58	1936	65.87
1921	44.08	1929	65.04	1937	67.39
1922	47.67	1930	59.18	1938	62.34
1923	54.10	1931	54.91	1939	68.09
1924	54.65	1932	46.72	1940	72.77
1925	56.28	1933	48.12	1941	84.29
1926	58.00	1934	53.25		

The data were obtained from Haavelmo (1947) for 1922-1941 and from Kuznets [(1954), pp. 151-153] for 1919, 1920, and 1921.

(a) Graph income against year.

(b) Find the (maximum likelihood) estimates of the coefficients β_1, β_2, γ_1, γ_2 and the standard deviation σ in the model

$$y_t + \beta_1 y_{t-1} + \beta_2 y_{t-2} + \gamma_1 + \gamma_2 t = u_t, \qquad t = 1, \ldots, 21,$$

where u_t has mean 0 and variance σ^2, y_t is income for year $t + 1920$, $t = -1, 0, 1, \ldots,$ 21, and y_{-1} and y_0 are treated as "given." The matrix of sums of squares and cross-products in the order 1, t, y_{t-1}, y_{t-2}, y_t is

$$\begin{bmatrix} 21 & 231 & 1{,}202 & 1{,}174 & 1{,}241 \\ 231 & 3{,}311 & 13{,}905 & 13{,}551 & 14{,}473 \\ 1{,}202 & 13{,}905 & 70{,}112 & 68{,}247 & 72{,}365 \\ 1{,}174 & 13{,}551 & 68{,}247 & 66{,}781 & 70{,}313 \\ 1{,}241 & 14{,}473 & 72{,}365 & 70{,}313 & 75{,}151 \end{bmatrix}.$$

(c) Find the (maximum likelihood) estimates of μ and δ (the coefficients of the trend) when the model is written

$$y_t - (\mu + \delta t) + \beta_1\{y_{t-1} - [\mu + \delta(t-1)]\} + \beta_2\{y_{t-2} - [\mu + \delta(t-2)]\} = u_t.$$

(d) Find the roots of the polynomial equation associated with the estimated coefficients of the second-order stochastic difference equation, and give the angle and modulus of these (complex) roots.

CHAPTER 6

Serial Correlation

6.1. INTRODUCTION

One of the first problems of time series analysis is to decide whether an observed time series is from a process of independent random variables. A simple alternative to independence is a process in which successive observations are correlated. For example, the process may be generated by a first-order stochastic difference equation. In such cases it may be appropriate to test the null hypothesis of independence by some kind of first-order serial correlation.

Given a series of observations $y_1, \ldots, y_T$ from a process with means known to be 0, one definition of a first-order serial correlation coefficient is

$$r_1 = \frac{\sum_{t=2}^{T} y_t y_{t-1}}{\sum_{t=1}^{T} y_t^2}. \tag{1}$$

In this chapter we shall consider various definitions of serial correlation. When the means are known to be 0, these definitions are modifications of (1); when the means are not known, y_t in (1) is replaced by the deviation from the sample mean (or from a fitted trend). Many modifications consist in changes in the so-called end terms, that is, changes in terms involving y_1, y_2, y_{T-1}, y_T; other modifications may involve multiplication by a factor such as $T/(T-1)$.

A serial correlation is a measurement of serial dependence in a sequence of observations that is similar to the measurement of dependence between two sets of observations $x_1, \ldots, x_N$ and $z_1, \ldots, z_N$ as furnished by the usual product-moment correlation coefficient (when the means are assumed 0)

$$(2) \qquad r = \frac{\sum_{i=1}^{N} x_i z_i}{\sqrt{\sum_{i=1}^{N} x_i^2}\sqrt{\sum_{i=1}^{N} z_i^2}}.$$

In the case of the serial correlation x_i is replaced by y_t and z_i is replaced by y_{t-1} in the numerator. Since both $\sum_{t=2}^{T} y_t^2$ and $\sum_{t=2}^{T} y_{t-1}^2 = \sum_{t=1}^{T-1} y_t^2$ estimate $(T-1)\sigma^2$ in the case of a stationary process (with mean 0), these may be replaced by $\sum_{t=1}^{T} y_t^2$ [possibly multiplied by $(T-1)/T$]. This gives (1) as an analogue of (2). An alternative term for serial correlation is *autocorrelation.*

Higher-order serial correlations are defined similarly. One definition of a jth-order serial correlation is

$$(3) \qquad r_j = \frac{\sum_{t=j+1}^{T} y_t y_{t-j}}{\sum_{t=1}^{T} y_t^2};$$

this measures the serial dependence between observations j time units apart. Such a correlation coefficient might be used to test the null hypothesis of independence against the alternative that there is dependence between observations j time units apart.

In most situations, however, if there is dependence between observations i time units apart, it would be assumed that there is dependence between observations that are $1, 2, \ldots,$ and $i-1$ time units apart. The ith-order serial correlation would customarily be used to test the null hypothesis of $(i-1)$st-order dependence against alternatives of ith-order dependence. For example, a process generated by an ith-order difference equation would be said to exhibit ith-order dependence. The ith-order serial correlation coefficient would be used in conjunction with the serial correlation coefficients of lower order.

A more general problem is the determination of the order of dependence when it is assumed that the order is at least m (≥ 0) and not greater than p $(>m)$. Such a multiple-decision procedure will be based on the serial correlation coefficients up to order p.

Tests of independence, tests of order of dependence, and determination of dependence were considered in Chapter 5 within the context of the stochastic difference equation model (roughly, the stationary Gaussian process of finite order). The properties of procedures and the distributions of estimates and test criteria were developed on a large-sample basis. In this chapter optimal procedures and exact distributions are derived for models which are moderate modifications of the stochastic difference equation model. (For instance, the estimate $-\hat{\beta}_1$ in the first-order difference equation is a type of first-order serial correlation coefficient.)

The general types of models will be developed from the stochastic difference equation model (Section 6.2). For these exponential models optimal tests of dependence (one-sided or unbiased) will be derived (Section 6.3). Multiple-decision procedures will be developed from these tests (Section 6.4) in a fashion similar to the development of procedures for choosing the degree of polynomial trend (in Section 3.2.2). Some explicit models will be defined for known means (Section 6.5) and for unknown means, including trends (Section 6.6); these models imply definitions of serial correlation coefficients, and the treatment includes the circular serial correlation coefficient and the coefficient based on the mean square successive difference. Some exact distributions of serial correlations are obtained (Section 6.7), as well as approximate distributions (Section 6.8). Some joint distributions are also found (Section 6.9). Serial correlations as estimates will be considered later (Chapter 8).

6.2. TYPES OF MODELS

In Chapter 5 we studied time series that satisfied a stochastic difference equation of order p,

$$y_t + \beta_1 y_{t-1} + \cdots + \beta_p y_{t-p} = u_t, \tag{1}$$

where $\{u_t\}$ is a sequence of independent random variables with means 0 and variances σ^2. If the distribution of $y_1, \ldots, y_p$ is given and u_t is independent of $y_{t-1}, y_{t-2}, \ldots, y_1$, $t = p + 1, \ldots, T$, then the distributions of $u_{p+1}, \ldots, u_T$ determine the joint distribution of $y_1, \ldots, y_T$, which are the observed variables. If the u_t's are identically distributed, the joint distribution of $y_1, \ldots, y_p$ can be taken so that the y_t's form part of a stationary process when the roots of the associated polynomial equation are less than 1 in absolute value. [It was shown in Section 5.2.1 that a necessary and sufficient condition that a stochastic process $\ldots, y_{-1}, y_0, y_1, \ldots$ satisfying a stochastic difference equation (1) be stationary such that y_t is independent of $u_{t+1}, u_{t+2}, \ldots$ is that the roots of the associated polynomial equation be less than 1 in absolute value.] If σ^2 is the only unknown parameter in the distribution of u_t, then $\beta_1, \ldots, \beta_p$, and σ^2 constitute the parameters of the process.

We shall now consider time series with joint normal distributions and with means 0; cases of nonzero means will be taken up later in this chapter. Models with normal distributions are called *Gaussian*. Here the joint distribution of $y_1, \ldots, y_p$ is determined by its covariance matrix, $\mathbf{\Sigma}_{pp}$, say. Let

$$\mathbf{\Sigma}_{pp}^{-1} = (\sigma_{(p)}^{ts}). \tag{2}$$

Then the joint density of $y_1, \ldots, y_p$ is

$$\frac{1}{(2\pi)^{\frac{1}{2}p}\,|\mathbf{\Sigma}_{pp}|^{\frac{1}{2}}} \exp\left(-\frac{1}{2}\sum_{t,s=1}^{p}\sigma_{(p)}^{ts}y_t y_s\right), \tag{3}$$

and the joint density of $u_{p+1}, \ldots, u_T$ is

$$\frac{1}{(2\pi)^{\frac{1}{2}(T-p)}(\sigma^2)^{\frac{1}{2}(T-p)}} \exp\left(-\frac{1}{2\sigma^2}\sum_{t=p+1}^{T}u_t^2\right). \tag{4}$$

Thus the joint density of $y_1, \ldots, y_T$ is

$$\frac{1}{(2\pi)^{\frac{1}{2}T}\,|\mathbf{\Sigma}_{pp}|^{\frac{1}{2}}\,(\sigma^2)^{\frac{1}{2}(T-p)}} \exp\left\{-\frac{1}{2}\left[\sum_{t,s=1}^{p}\sigma_{(p)}^{ts}y_t y_s + \frac{1}{\sigma^2}\sum_{t=p+1}^{T}(y_t + \beta_1 y_{t-1} + \cdots + \beta_p y_{t-p})^2\right]\right\}. \tag{5}$$

The exponent in (5) is $-\frac{1}{2}$ times

$$\begin{aligned}
&\sum_{t,s=1}^{p}\sigma_{(p)}^{ts}y_t y_s + \frac{1}{\sigma^2}\sum_{t=p+1}^{T}[y_t^2 + \beta_1^2 y_{t-1}^2 + \cdots + \beta_p^2 y_{t-p}^2 + 2\beta_1 y_t y_{t-1} \\
&\qquad + 2\beta_1\beta_2 y_{t-1}y_{t-2} + \cdots + 2\beta_{p-1}\beta_p y_{t-p+1}y_{t-p} + \cdots + 2\beta_p y_t y_{t-p}] \\
&= \frac{1}{\sigma^2}\left[(1 + \beta_1^2 + \cdots + \beta_p^2)\sum_{t=p+1}^{T-p} y_t^2 + 2(\beta_1 + \beta_1\beta_2 + \cdots + \beta_{p-1}\beta_p)\sum_{t=p+1}^{T-p+1} y_t y_{t-1}\right. \\
&\qquad \left. + \ldots + 2\beta_p \sum_{t=p+1}^{T} y_t y_{t-p}\right] + \sum_{t,s=1}^{p} a_{ts}y_t y_s + \sum_{t,s=1}^{p} b_{ts}y_{T+1-t}y_{T+1-s}.
\end{aligned} \tag{6}$$

(We assume $T \geq 2p$.) Let (6) be $\boldsymbol{y}'\mathbf{\Sigma}^{-1}\boldsymbol{y}$, where $\boldsymbol{y} = (y_1, \ldots, y_T)'$.

If $\mathbf{\Sigma}_{pp}$ is chosen to obtain a stationary process, the fact that

$$\sigma(s - t) = \sigma[(T - s + 1) - (T - t + 1)] \tag{7}$$

implies that the element in row t and column s of $\mathbf{\Sigma}^{-1}$ is the same as the element in row $(T - t + 1)$ and column $(T - s + 1)$ of $\mathbf{\Sigma}^{-1}$ and therefore that the quadratic form in $y_1, \ldots, y_p$ is the same as the quadratic form in $y_T, \ldots, y_{T+1-p}$; that is, $a_{ts} = b_{ts}$, $t, s = 1, \ldots, p$. The coefficients of the form in $y_T, \ldots, y_{T+1-p}$ can be obtained from the left-hand side of (6).

It may be noted that the quadratic form (6) has no term $y_t y_s$ for which $|t - s| > p$. Except for the end terms, the quadratic form is a linear combination of the sums $\sum_t y_t y_{t-j}$, $j = 0, 1, \ldots, p$, namely,

$$(8) \quad \frac{1 + \beta_1^2 + \cdots + \beta_p^2}{\sigma^2} \sum_t y_t^2 + 2\frac{\beta_1 + \beta_1\beta_2 + \cdots + \beta_{p-1}\beta_p}{\sigma^2} \sum_t y_t y_{t-1} + \cdots + 2\frac{\beta_p}{\sigma^2} \sum_t y_t y_{t-p}.$$

The $p + 1$ coefficients of the quadratic forms form a transformation of the $p + 1$ parameters $\sigma^2, \beta_1, \ldots, \beta_p$. Hypotheses and multiple-decision problems can be formulated in terms of these coefficients. In particular, $2\beta_p/\sigma^2$ is the coefficient of $\sum_t y_t y_{t-p}$.

The end terms in (6) involve coefficients that are other functions of $\sigma^2, \beta_1, \ldots, \beta_p$. This fact implies that a sufficient set of statistics includes some of $y_t y_s$ and $y_{T+1-t} y_{T+1-s}$, $t, s = 1, \ldots, p$, as well as the $p + 1$ sums $\sum_t y_t y_{t-j}$, $j = 0, 1, \ldots, p$. In turn, the fact that the number of statistics in the minimal sufficient set is a good deal greater than the number of parameters makes it impossible to formulate inference problems in a way that leads to relatively simple optimum procedures.

If T is large compared to p, the sums $\sum_t y_t y_{t-j}$ will tend to dominate the other terms. This feature suggests replacing the density (5) with densities that have these same sums but end terms modified to permit families of $p + 1$ sufficient statistics.

We shall suppose that the density of $y_1, \ldots, y_T$ is given by

$$(9) \quad f(y_1, \ldots, y_T \mid \gamma_0, \ldots, \gamma_p) = K \exp\left[-\tfrac{1}{2}(\gamma_0 Q_0 + \cdots + \gamma_p Q_p)\right],$$

where Q_j is a quadratic form in $\mathbf{y} = (y_1, \ldots, y_T)'$,

$$(10) \quad Q_j = \mathbf{y}' \mathbf{A}_j \mathbf{y},$$

$\mathbf{A}_j$ is a given symmetric matrix, $j = 0, 1, \ldots, p$, with $\mathbf{A}_0$ positive definite, and $\gamma_0, \ldots, \gamma_p$ are parameters with γ_0 positive such that

$$(11) \quad Q = \gamma_0 Q_0 + \cdots + \gamma_p Q_p = \mathbf{y}'(\gamma_0 \mathbf{A}_0 + \cdots + \gamma_p \mathbf{A}_p)\mathbf{y}$$

is positive definite. The density (9) is then multivariate normal; the constant $K = K(\gamma_0, \ldots, \gamma_p)$ depends on the values of $\gamma_0, \ldots, \gamma_p$ and is given by

$$K(\gamma_0, \ldots, \gamma_p) = (2\pi)^{-\frac{1}{2}T} \, |\gamma_0 A_0 + \cdots + \gamma_p A_p|^{\frac{1}{2}}. \tag{12}$$

Usually A_0 will be taken to be the identity matrix I.

The relationship between the parameters in (9) and those in (5) is

$$\begin{aligned}
\gamma_0 &= \frac{1 + \beta_1^2 + \cdots + \beta_p^2}{\sigma^2}, \\
\gamma_1 &= 2\,\frac{\beta_1 + \beta_1\beta_2 + \cdots + \beta_{p-1}\beta_p}{\sigma^2}, \\
&\;\;\vdots \\
\gamma_{p-1} &= 2\,\frac{\beta_{p-1} + \beta_1\beta_p}{\sigma^2}, \\
\gamma_p &= 2\,\frac{\beta_p}{\sigma^2}.
\end{aligned} \tag{13}$$

Note that $\gamma_{m+1} = \cdots = \gamma_p = 0$ if and only if $\beta_{m+1} = \cdots = \beta_p = 0$. Usually the quadratic form Q_j is $\sum_{t=j+1}^{T} y_t y_{t-j}$, plus a quadratic form in $y_1, \ldots, y_j$, plus a similar quadratic form in $y_{T-j+1}, \ldots, y_T$. These $p + 1$ quadratic forms are chosen conveniently (for example, with common characteristic vectors). Three examples of systems of quadratic forms are given in Section 6.5. The basic theorems of this chapter apply to some other problems, such as those involving components of variance.

In terms of this model, we consider tests of the hypotheses that $\gamma_{m+1} = \cdots = \gamma_p = 0$ for specified m and p. We are particularly interested in the case $m = p - 1$; the case $p = 1$ is most important and has been considered most in the literature. We find the uniformly most powerful test against alternatives $\gamma_p < 0$ (equivalently $\beta_p < 0$) and the uniformly most powerful unbiased test. The actual use of these tests is limited by the available theory and tabulation of the test statistics.

The second type of problem we treat is a multiple-decision procedure for choosing the largest index i for which $\gamma_i \neq 0$ for $i = m, \ldots, p$; this is the order of dependence. We assign the probability of overstating the order, given the true order i for $i = m, \ldots, p - 1$, and ask for the procedure that is uniformly most powerful against positive dependence ($\gamma_i < 0$) or uniformly most powerful unbiased. Such procedures are built up from the tests of hypotheses in a manner similar to that of Section 3.2.2. [See T. W. Anderson (1963).]

6.3. UNIFORMLY MOST POWERFUL TESTS OF A GIVEN ORDER

6.3.1. Uniformly Most Powerful Tests Against Alternatives of Positive Dependence

The basic model for the observations $y_1, \ldots, y_T$ is the multivariate normal density (9) of Section 6.2. In this section we shall be interested in testing a hypothesis about γ_i, $i = 1, \ldots, p$, when we assume $\gamma_{i+1} = \cdots = \gamma_p = 0$. In that case the density is

(1)
$$f_i(y_1, \ldots, y_T \mid \gamma_0, \ldots, \gamma_i) = K_i(\gamma_0, \ldots, \gamma_i) \exp\left[-\tfrac{1}{2}(\gamma_0 Q_0 + \cdots + \gamma_i Q_i)\right],$$

where

(2)
$$K_i(\gamma_0, \ldots, \gamma_i) = K(\gamma_0, \ldots, \gamma_i, 0, \ldots, 0) = (2\pi)^{-\frac{1}{2}T} \left|\gamma_0 \boldsymbol{A}_0 + \cdots + \gamma_i \boldsymbol{A}_i\right|^{\frac{1}{2}},$$

and $\gamma_0 Q_0 + \cdots + \gamma_i Q_i$ is positive definite. Then the null hypothesis that $\gamma_i = 0$ is the hypothesis that the order of dependence is less than i (on the assumption that the order is not greater than i). When $i = 1$, the null hypothesis $\gamma_1 = 0$ is the hypothesis of independence (when $\boldsymbol{A}_0$ is diagonal as, for instance, when $\boldsymbol{A}_0 = \boldsymbol{I}$). In this subsection we shall consider testing the hypothesis that γ_i is a specific value against one-sided alternatives. In the case of $i = 1$ and the null hypothesis $\gamma_1 = 0$ $(\beta_1 = 0)$ a natural alternative is $\gamma_1 < 0$ (that is, $\beta_1 < 0$), which corresponds to positive population correlation between successive observations. In the case of $i > 1$, the alternative to the null hypothesis of $\gamma_i = 0$ $(\beta_i = 0)$ is not necessarily $\gamma_i < 0$ $(\beta_i < 0)$ because $-\beta_i$ represents the partial correlation between y_t and y_{t+i} given the correlations of lower order. Since the optimum tests against alternatives $\gamma_i > 0$ are constructed in a manner similar to the optimum tests against alternatives $\gamma_i < 0$ we shall restrict the actual treatment to the one-sided alternatives corresponding to positive dependence ($\gamma_i < 0$).

The problems of statistical inference are relatively simple for an exponential family of densities such as (1). One feature is that the minimal sufficient set of statistics is of the same dimensionality as the set of parameters.

THEOREM 6.3.1. *$Q_0, \ldots, Q_i$ form a sufficient set of statistics for $\gamma_0, \ldots, \gamma_i$ when $y_1, \ldots, y_T$ have the density $f_i(y_1, \ldots, y_T \mid \gamma_0, \ldots, \gamma_i)$ given by* (1).

The exponential family is of the classical Koopman–Darmois form, which obviously satisfies the factorization criterion that the density factor into a nonnegative (measurable) function of the parameters and sufficient statistics and a nonnegative (measurable) function of the observations that does not depend on the parameters. [See Lehmann (1959), p. 49, or Kendall and Stuart (1961), p. 22, for example.] In this case the first function can be taken to be the density, and the second function can be taken to be 1. The implication of the theorem is that inference can be based on $Q_0, \ldots, Q_i$; for instance, the significance level and power function of any test based on $y_1, \ldots, y_T$ can be duplicated by a (possibly randomized) test based on $Q_0, \ldots, Q_i$.

THEOREM 6.3.2. *The family of distributions of $Q_0, \ldots, Q_i$ based on densities $f_i(y_1, \ldots, y_T \mid \gamma_0, \ldots, \gamma_i)$ is complete as $\gamma_0, \ldots, \gamma_i$ range over the set of values for which $\gamma_0 Q_0 + \cdots + \gamma_i Q_i$ is positive definite.*

The property of completeness is also a property of the exponential family. [See Lehmann (1959), p. 132, or Kendall and Stuart (1961), p. 190, for example.] A proof is outlined in Problem 7.

Theorem 6.3.2 implies that a test of the null hypothesis that γ_i is a given value, say $\gamma_i^{(1)}$, at assigned level of significance ε_i and determined by a similar region has Neyman structure; that is, if S_i is a set in the space of $Q_0, \ldots, Q_i$ such that

$$\Pr\{S_i \mid \gamma_0, \gamma_1, \ldots, \gamma_{i-1}, \gamma_i^{(1)}, 0, \ldots, 0\} = \varepsilon_i, \tag{3}$$

then

$$\Pr\{S_i \mid Q_0, \ldots, Q_{i-1}; \gamma_0, \ldots, \gamma_{i-1}, \gamma_i^{(1)}, 0, \ldots, 0\} = \varepsilon_i \tag{4}$$

almost everywhere in the set of possible $Q_0, \ldots, Q_{i-1}$. [See Lehmann (1959), Section 4.3, or Kendall and Stuart (1961), Section 23.19.] A proof is outlined in Problem 9. Note that since $Q_0, \ldots, Q_{i-1}$ are sufficient for $\gamma_0, \ldots, \gamma_{i-1}$, $\Pr\{S_i \mid Q_0, \ldots, Q_{i-1}; \gamma_0, \ldots, \gamma_{i-1}, \gamma_i^{(1)}, 0, \ldots, 0\}$ does not depend on $\gamma_0, \ldots, \gamma_{i-1}$.

Let $h_i(Q_i \mid Q_0, \ldots, Q_{i-1}; \gamma_0, \ldots, \gamma_i)$ be the conditional density of Q_i, given $Q_0, \ldots, Q_{i-1}$, when $\gamma_{i+1} = \cdots = \gamma_p = 0$. We are interested in this conditional density on the set V_{i-1} of possible values of $Q_0, \ldots, Q_{i-1}$ (that is, the set for which the marginal density of $Q_0, \ldots, Q_{i-1}$ is positive). Let $h_i(Q_0, \ldots, Q_i \mid \gamma_0, \ldots, \gamma_i)$ be the density of $Q_0, \ldots, Q_i$ when $\gamma_{i+1} = \cdots = \gamma_p = 0$. This density is

$$\begin{aligned} h_i(Q_0, \ldots, Q_i \mid \gamma_0, \ldots, \gamma_i) \\ = K_i(\gamma_0, \ldots, \gamma_i) \exp\left[-\tfrac{1}{2}(\gamma_0 Q_0 + \cdots + \gamma_i Q_i)\right] k_i(Q_0, \ldots, Q_i). \end{aligned} \tag{5}$$

The density (5) is obtained from the density of $y_1, \ldots, y_T$ by transforming $y_1, \ldots, y_T$ to $Q_0, \ldots, Q_i$ and some $T - i - 1$ of the y_t's and integrating these y_t's over a range determined by $Q_0, \ldots, Q_i$; this operation yields $k_i(Q_0, \ldots, Q_i)$. The set V_i is the set of $Q_0, \ldots, Q_i$ for which $k_i(Q_0, \ldots, Q_i) > 0$. Then the marginal density of $Q_0, \ldots, Q_{i-1}$ is

$$(6) \quad K_i(\gamma_0, \ldots, \gamma_i) \exp\left[-\tfrac{1}{2}(\gamma_0 Q_0 + \cdots + \gamma_{i-1}Q_{i-1})\right] m_i(Q_0, \ldots, Q_{i-1}; \gamma_i),$$

where

$$(7) \qquad m_i(Q_0, \ldots, Q_{i-1}; \gamma_i) = \int_{-\infty}^{\infty} e^{-\frac{1}{2}\gamma_i Q_i} k_i(Q_0, \ldots, Q_i)\, dQ_i.$$

[Note that $m_i(Q_0, \ldots, Q_{i-1}; \gamma_i) > 0$ for every value of γ_i.] Thus the conditional density of Q_i, given $Q_0, \ldots, Q_{i-1}$ in V_{i-1}, is

$$(8) \qquad \begin{aligned} h_i(Q_i \mid Q_0, \ldots, Q_{i-1}; \gamma_0, \ldots, \gamma_i) &= e^{-\frac{1}{2}\gamma_i Q_i} \frac{k_i(Q_0, \ldots, Q_i)}{m_i(Q_0, \ldots, Q_{i-1}; \gamma_i)} \\ &= h_i(Q_i \mid Q_0, \ldots, Q_{i-1}; \gamma_i), \end{aligned}$$

say. The conditional density of Q_i, given $Q_0, \ldots, Q_{i-1}$, is in an exponential family with parameter γ_i; for fixed $Q_0, \ldots, Q_{i-1}$, $k_i(Q_0, \ldots, Q_i)$ is a function of Q_i and $m_i(Q_0, \ldots, Q_{i-1}; \gamma_i)$ is a function of γ_i and is the normalizing constant. It will be understood that the conditional density $h_i(Q_i \mid Q_0, \ldots, Q_{i-1}; \gamma_i)$ will be written only for $Q_0, \ldots, Q_{i-1}$ in V_{i-1}.

Since a test of the hypothesis $\gamma_i = \gamma_i^{(1)}$ has Neyman structure, we can apply the Neyman-Pearson Fundamental Lemma (Problem 10) to the conditional distribution of Q_i. The best test of the hypothesis that $\gamma_i = \gamma_i^{(1)}$ against the alternative $\gamma_i = \gamma_i^{(2)}$ has a rejection region

(9)

$$k < \frac{h_i(Q_i \mid Q_0, \ldots, Q_{i-1}; \gamma_i^{(2)})}{h_i(Q_i \mid Q_0, \ldots, Q_{i-1}; \gamma_i^{(1)})} = \exp\left[\tfrac{1}{2}(\gamma_i^{(1)} - \gamma_i^{(2)})Q_i\right] \frac{m_i(Q_0, \ldots, Q_{i-1}; \gamma_i^{(1)})}{m_i(Q_0, \ldots, Q_{i-1}; \gamma_i^{(2)})},$$

where k, depending on $Q_0, \ldots, Q_{i-1}$, is determined so that the conditional probability of the inequality is the desired significance level when $\gamma_i = \gamma_i^{(1)}$. If $\gamma_i^{(1)} > \gamma_i^{(2)}$, then the rejection region (9) can be written

$$(10) \qquad Q_i > c_i(Q_0, \ldots, Q_{i-1}; \gamma_i^{(1)}, \gamma_i^{(2)}).$$

In particular, the test of $\gamma_i = 0$ against an alternative $\gamma_i = \gamma_i^{(2)} < 0$ is of this form.

THEOREM 6.3.3. *The best similar test of the null hypothesis $\gamma_i = \gamma_i^{(1)}$ against the alternative $\gamma_i = \gamma_i^{(2)} < \gamma_i^{(1)}$ at significance level ε_i is given by* (10), *where*

$c_i(Q_0, \ldots, Q_{i-1}; \gamma^{(1)}, \gamma_i^{(2)})$ *is determined so the probability of* (10) *according to the density* (8) *is* ε_i *when* $\gamma_i = \gamma_i^{(1)}$.

Since $c_i(Q_0, \ldots, Q_{i-1}; \gamma_i^{(1)}, \gamma_i^{(2)})$ is determined from the conditional density (8) with $\gamma_i = \gamma_i^{(1)}$, it does not depend on the value of $\gamma_i^{(2)}$, and we shall write it as $c_i(Q_0, \ldots, Q_{i-1}; \gamma_i^{(1)})$. Thus the region (10) does not depend on $\gamma_i^{(2)}$; it gives the best test against any alternative $\gamma_i^{(2)} < \gamma_i^{(1)}$.

THEOREM 6.3.4. *The uniformly most powerful similar test of the null hypothesis* $\gamma_i = \gamma_i^{(1)}$ *against alternatives* $\gamma_i < \gamma_i^{(1)}$ *at significance level* ε_i *has the rejection region*

$$Q_i > c_i(Q_0, \ldots, Q_{i-1}; \gamma_i^{(1)}), \tag{11}$$

where $c_i(Q_0, \ldots, Q_{i-1}; \gamma_i^{(1)})$ *is determined so the probability of* (11) *according to the density* (8) *is* ε_i *when* $\gamma_i = \gamma_i^{(1)}$.

We are particularly interested in the case $\gamma_i^{(1)} = 0$. Then the conditional density of Q_i under the null hypothesis is

$$h_i(Q_i \mid Q_0, \ldots, Q_{i-1}; 0) = \frac{k_i(Q_0, \ldots, Q_i)}{k_{i-1}(Q_0, \ldots, Q_{i-1})}. \tag{12}$$

COROLLARY 6.3.1. *The uniformly most powerful similar test of the null hypothesis* $\gamma_i = 0$ *against alternatives* $\gamma_i < 0$ *at significance level* ε_i *has the rejection region*

$$Q_i > c_i(Q_0, \ldots, Q_{i-1}), \tag{13}$$

where $c_i(Q_0, \ldots, Q_{i-1})$ *is determined so the integral of* (12) *over* (13) *is* ε_i.

When we come to consider specific models we shall see that usually the rejection region (13) is

$$r_i > C_i(r_1, \ldots, r_{i-1}), \tag{14}$$

where $r_1 = Q_1/Q_0, \ldots, r_i = Q_i/Q_0$ are serial correlation coefficients. In particular, when $i = 1$, a test of the null hypothesis of independence against alternatives of positive dependence has the rejection region $r_1 > C_1$.

6.3.2. Uniformly Most Powerful Unbiased Tests

In testing that the order of dependence is less than i (that is, testing $\gamma_i = 0$), we may be interested in alternatives of dependence of order i (that is, $\gamma_i \neq 0$). Against two-sided alternatives no uniformly most powerful test exists. For example, the best similar test of $\gamma_i = \gamma_i^{(1)}$ against an alternative $\gamma_i = \gamma_i^{(2)} < \gamma_i^{(1)}$

at significance level ε_i is (11), while the best similar test against an alternative $\gamma_i = \gamma_i^{(2)} > \gamma_i^{(1)}$ has a rejection region

$$Q_i < c_i'(Q_0, \ldots, Q_{i-1}; \gamma_i^{(1)}), \tag{15}$$

where $c_i'(Q_0, \ldots, Q_{i-1}; \gamma_i^{(1)})$ is determined so the probability of (15) according to the density (8) is ε_i when $\gamma_i = \gamma_i^{(1)}$. When we are interested in two-sided alternatives, we shall require that the test be unbiased, that is, that the power function have its minimum at $\gamma_i = \gamma_i^{(1)}$. Among the class of unbiased tests at a given significance level we shall seek a uniformly most powerful test. Such a test exists, because the family of densities is exponential, and its acceptance region is an interval. We shall sketch the derivation of the uniformly most powerful unbiased test; a complete and rigorous proof is given in Lehmann (1959), Chapter 4, Section 4.

For given $Q_0, \ldots, Q_{i-1}$ let R be a rejection region with significance level ε_i; that is,

$$\int_R h_i(Q_i \mid Q_0, \ldots, Q_{i-1}; \gamma_i^{(1)})\, dQ_i = \varepsilon_i. \tag{16}$$

Let the power function be $\beta(\gamma_i)$; that is, $\beta(\gamma_i^{(1)}) = \varepsilon_i$, where

$$\beta(\gamma_i) = \int_R \frac{k_i(Q_0, \ldots, Q_i)}{m_i(Q_0, \ldots, Q_{i-1}; \gamma_i)} e^{-\frac{1}{2}\gamma_i Q_i}\, dQ_i. \tag{17}$$

For the test to be locally unbiased we want the first derivative of the power function to be zero at $\gamma_i = \gamma_i^{(1)}$. Interchanging the order of integration and differentiation gives us

(18)

$$\begin{aligned}
0 &= \frac{d}{d\gamma_i}\beta(\gamma_i)\bigg|_{\gamma_i=\gamma_i^{(1)}} \\
&= -\frac{1}{2}\int_R Q_i \frac{k_i(Q_0, \ldots, Q_i)}{m_i(Q_0, \ldots, Q_{i-1}; \gamma_i^{(1)})} \exp(-\tfrac{1}{2}\gamma_i^{(1)} Q_i)\, dQ_i \\
&\quad + \frac{d}{d\gamma_i}\left[\frac{1}{m_i(Q_0, \ldots, Q_{i-1}; \gamma_i)}\right]\bigg|_{\gamma_i=\gamma_i^{(1)}} \int_R k_i(Q_0, \ldots, Q_i) \exp(-\tfrac{1}{2}\gamma_i^{(1)} Q_i)\, dQ_i \\
&= -\frac{1}{2}\int_R Q_i h_i(Q_i \mid Q_0, \ldots, Q_{i-1}; \gamma_i^{(1)})\, dQ_i \\
&\quad + \varepsilon_i m_i(Q_0, \ldots, Q_{i-1}; \gamma_i^{(1)}) \frac{d}{d\gamma_i}\left[\frac{1}{m_i(Q_0, \ldots, Q_{i-1}; \gamma_i)}\right]\bigg|_{\gamma_i=\gamma_i^{(1)}}.
\end{aligned}$$

If we similarly evaluate the derivative of

$$1 = \int_{-\infty}^{\infty} \frac{k_i(Q_0, \ldots, Q_i)}{m_i(Q_0, \ldots, Q_{i-1}; \gamma_i)} e^{-\frac{1}{2}\gamma_i Q_i} \, dQ_i \tag{19}$$

at $\gamma_i = \gamma_i^{(1)}$ and compare with (18) we find

(20)

$$\int_R Q_i h_i(Q_i \mid Q_0, \ldots, Q_{i-1}; \gamma_i^{(1)}) \, dQ_i = \varepsilon_i \int_{-\infty}^{\infty} Q_i h_i(Q_i \mid Q_0, \ldots, Q_{i-1}; \gamma_i^{(1)}) \, dQ_i.$$

Thus (20) is the condition of local unbiasedness.

We now want to find the region R satisfying (16) and (20) that maximizes the corresponding $\beta(\gamma_i)$ at some particular value of γ_i, say $\gamma_i = \gamma_i^{(2)} > \gamma_i^{(1)}$. By a generalized Neyman-Pearson Fundamental Lemma (Problem 12) that region will be defined by

$$h_i(Q_i \mid Q_0, \ldots, Q_{i-1}; \gamma_i^{(2)}) > a h_i(Q_i \mid Q_0, \ldots, Q_{i-1}; \gamma_i^{(1)}) + b Q_i h_i(Q_i \mid Q_0, \ldots, Q_{i-1}; \gamma_i^{(1)}) \tag{21}$$

if there are functions of $Q_0, \ldots, Q_{i-1}$ a and b such that the region satisfies (16) and (20). The inequality (21) is equivalent to $g(Q_i) > 0$, where

$$g(Q_i) = \exp\left[-\tfrac{1}{2}(\gamma_i^{(2)} - \gamma_i^{(1)})Q_i\right] - (A + BQ_i), \tag{22}$$

and

$$A = a \frac{m_i(Q_0, \ldots, Q_{i-1}; \gamma_i^{(2)})}{m_i(Q_0, \ldots, Q_{i-1}; \gamma_i^{(1)})}, \qquad B = b \frac{m_i(Q_0, \ldots, Q_{i-1}; \gamma_i^{(2)})}{m_i(Q_0, \ldots, Q_{i-1}; \gamma_i^{(1)})}. \tag{23}$$

Then

$$\frac{d}{dQ_i} g(Q_i) = -\tfrac{1}{2}(\gamma_i^{(2)} - \gamma_i^{(1)}) \exp\left[-\tfrac{1}{2}(\gamma_i^{(2)} - \gamma_i^{(1)})Q_i\right] - B. \tag{24}$$

If $B > 0$ (and $\gamma_i^{(2)} > \gamma_i^{(1)}$), the derivative is negative and (21) defines a half-interval $(-\infty, c)$; but such a region cannot satisfy (16) and (20). (See Problem 13.) Thus B should be negative. Then the derivative is a continuous increasing function with one 0; $g(Q_i)$ has a minimum at this point; if the minimum is negative (which will be the case if A is sufficiently large), (21) will define two half-intervals $(-\infty, c)$ and (d, ∞). The acceptance region will be the interval (c, d). Note that the values of c and d will be defined by (16) and (20), which are independent of $\gamma_i^{(2)}$. In a similar fashion, when $\gamma_i^{(2)} < \gamma_i^{(1)}$, we argue that B should be positive and that then the acceptance region is an interval satisfying

(16) and (20) independently of $\gamma_i^{(2)}$. Thus we obtain the uniformly most powerful unbiased test.

THEOREM 6.3.5. *The uniformly most powerful unbiased test of the null hypothesis* $\gamma_i = \gamma_i^{(1)}$ *against alternatives* $\gamma_i \neq \gamma_i^{(1)}$ *at significance level* ε_i *has the rejection region*

$$Q_i > c_{Ui}(Q_0, \ldots, Q_{i-1}; \gamma_i^{(1)}), \qquad Q_i < c_{Li}(Q_0, \ldots, Q_{i-1}; \gamma_i^{(1)}), \tag{25}$$

where $c_{Ui}(Q_0, \ldots, Q_{i-1};\ \gamma_i^{(1)})$ *and* $c_{Li}(Q_0, \ldots, Q_{i-1};\ \gamma_i^{(1)})$ *are determined so that*

$$\int_{c_{Li}(Q_0,\ldots,Q_{i-1};\ \gamma_i^{(1)})}^{c_{Ui}(Q_0,\ldots,Q_{i-1};\ \gamma_i^{(1)})} h_i(Q_i \mid Q_0, \ldots, Q_{i-1}; \gamma_i^{(1)})\, dQ_i = 1 - \varepsilon_i \tag{26}$$

and

$$\int_{c_{Li}(Q_0,\ldots,Q_{i-1};\gamma_i^{(1)})}^{c_{Ui}(Q_0,\ldots,Q_{i-1};\gamma_i^{(1)})} Q_i h_i(Q_i \mid Q_0, \ldots, Q_{i-1}; \gamma_i^{(1)})\, dQ_i = (1 - \varepsilon_i)\int_{-\infty}^{\infty} Q_i h_i(Q_i \mid Q_0, \ldots, Q_{i-1}; \gamma_i^{(1)})\, dQ_i. \tag{27}$$

COROLLARY 6.3.2. *The uniformly most powerful unbiased test of the null hypothesis* $\gamma_i = 0$ *against alternatives* $\gamma_i \neq 0$ *at significance level* ε_i *has the rejection region*

$$Q_i > c_{Ui}(Q_0, \ldots, Q_{i-1}), \qquad Q_i < c_{Li}(Q_0, \ldots, Q_{i-1}), \tag{28}$$

where $c_{Ui}(Q_0, \ldots, Q_{i-1})$ *and* $c_{Li}(Q_0, \ldots, Q_{i-1})$ *are determined by*

$$\int_{c_{Li}(Q_0,\ldots,Q_{i-1})}^{c_{Ui}(Q_0,\ldots,Q_{i-1})} \frac{k_i(Q_0, \ldots, Q_i)}{k_{i-1}(Q_0, \ldots, Q_{i-1})}\, dQ_i = 1 - \varepsilon_i \tag{29}$$

and

$$\int_{c_{Li}(Q_0,\ldots,Q_{i-1})}^{c_{Ui}(Q_0,\ldots,Q_{i-1})} Q_i \frac{k_i(Q_0, \ldots, Q_i)}{k_{i-1}(Q_0, \ldots, Q_{i-1})}\, dQ_i = (1 - \varepsilon_i)\int_{-\infty}^{\infty} Q_i \frac{k_i(Q_0, \ldots, Q_i)}{k_{i-1}(Q_0, \ldots, Q_{i-1})}\, dQ_i. \tag{30}$$

If the conditional density of Q_i, given $Q_0, \ldots, Q_{i-1}$ for $\gamma_i = 0$, is symmetric, then (30) is automatically satisfied for

$$c_{Li}(Q_0, \ldots, Q_{i-1}) = -c_{Ui}(Q_0, \ldots, Q_{i-1}), \tag{31}$$

since both sides are zero, and (29) can be written

$$\int_{cU_i(Q_0,\ldots,Q_{i-1})}^{\infty} \frac{k_i(Q_0,\ldots,Q_i)}{k_{i-1}(Q_0,\ldots,Q_{i-1})}\,dQ_i = \tfrac{1}{2}\varepsilon_i. \tag{32}$$

The uniformly most powerful one-sided similar tests can also be termed uniformly most powerful one-sided unbiased tests, for unbiasedness of a test implies similarity of the rejection region.

It should be noted that the conditional distribution of Q_i, given $Q_0, \ldots, Q_{i-1}$, does not depend on the values of $\gamma_0, \ldots, \gamma_{i-1}$. Hence these parameters can be assigned convenient values, such as $\gamma_0 = 1$, $\gamma_1 = \cdots = \gamma_{i-1} = 0$.

Usually Q_0 will be $\sum_{t=1}^{T} y_t^2$. The ratio

$$r_j = \frac{Q_j}{Q_0} \tag{33}$$

can be considered as a *serial correlation coefficient*. When $\gamma_1 = \cdots = \gamma_i = 0$, the joint distribution of $r_1, \ldots, r_i$ is independent of $\gamma_0\,(> 0)$ and Q_0. (See Theorem 6.7.2.) The definitions of the optimum tests (13) and (28) can be given in terms of $r_i = Q_i/Q_0$ conditional on $r_1, \ldots, r_{i-1}$.

A partial serial correlation of order i, given the serial correlations of order $1, \ldots, i-1$, can be defined in the usual way. It has been suggested that this partial serial correlation coefficient be used to test the hypothesis that $\gamma_i = 0$. Since this partial correlation coefficient is a linear function of r_i (with coefficients depending on $r_1, \ldots, r_{i-1}$), using it conditional on $r_1, \ldots, r_{i-1}$ is equivalent to using r_i conditional on $r_1, \ldots, r_{i-1}$. (See Section 6.9.2.)

6.3.3. Some Cases Where Uniformly Most Powerful Tests Do Not Exist

There are a number of problems of serial correlation whose solutions cannot be obtained from Theorems 6.3.4 and 6.3.5. If we wish to test the null hypothesis

$$\gamma_1 = \cdots = \gamma_p = 0 \tag{34}$$

against an alternative $\gamma_1 = \gamma_1^0, \ldots, \gamma_p = \gamma_p^0$, where $\gamma_1^0, \ldots, \gamma_p^0$ are specified values, the most powerful similar test may have the rejection region

$$\sum_{j=1}^{p} \gamma_j^0 Q_j < c(Q_0), \tag{35}$$

where $c(Q_0)$ is a constant determined so the conditional probability of (35) is ε under the null hypothesis. If the alternative values of $\gamma_1, \ldots, \gamma_p$ are given by

$$\gamma_j = k\gamma_j^*, \qquad j = 1, \ldots, p, \tag{36}$$

where the γ_j^*'s are fixed, then (35) with γ_j^0 replaced by γ_j^* is the best region for all alternatives $k > 0$. If the alternative values do not lie on a line (36) in the γ

space, the region (35) is not the same for all admissible alternatives $(\gamma_1^0, \ldots, \gamma_p^0)$. Hence there does not exist a uniformly most powerful test.

Consider the stationary stochastic process defined by

$$y_t + \beta y_{t-1} = u_t, \qquad t = \ldots, -1, 0, 1, \ldots, \tag{37}$$

where $|\beta| < 1$ and the distribution of u_t is normal with mean zero and variance σ^2 and is independent of $y_{t-1}, y_{t-2}, \ldots$ and $u_{t-1}, u_{t-2}, \ldots$. Then the marginal distribution of a sample $y_1, \ldots, y_T$ is given by (1) with

$$A_0 = I,$$

$$A_1 = \frac{1}{2}\begin{bmatrix} 0 & 1 & 0 & \cdots & 0 & 0 \\ 1 & 0 & 1 & \cdots & 0 & 0 \\ 0 & 1 & 0 & \cdots & 0 & 0 \\ \cdot & \cdot & \cdot & & \cdot & \cdot \\ \cdot & \cdot & \cdot & & \cdot & \cdot \\ \cdot & \cdot & \cdot & & \cdot & \cdot \\ 0 & 0 & 0 & \cdots & 0 & 1 \\ 0 & 0 & 0 & \cdots & 1 & 0 \end{bmatrix}, \tag{38}$$

$$A_2 = \begin{bmatrix} 0 & 0 & 0 & \cdots & 0 & 0 \\ 0 & 1 & 0 & \cdots & 0 & 0 \\ 0 & 0 & 1 & \cdots & 0 & 0 \\ \cdot & \cdot & \cdot & & \cdot & \cdot \\ \cdot & \cdot & \cdot & & \cdot & \cdot \\ \cdot & \cdot & \cdot & & \cdot & \cdot \\ 0 & 0 & 0 & \cdots & 1 & 0 \\ 0 & 0 & 0 & \cdots & 0 & 0 \end{bmatrix}.$$

The set of alternatives for $\gamma_1 = 2\beta/\sigma^2$ and $\gamma_2 = \beta^2/\sigma^2$ does not lie on a straight line. Hence, no uniformly most powerful test exists, even for one-sided alternatives.

If $|\beta|$ is small (that is, near 0) this case is similar to the model of Section 6.5.4; if β is small (that is, near -1) this case is similar to the model of Section 6.5.3. Thus one might conjecture that one of the two relevant correlations would be a good statistic to use in testing $\beta = 0$ in this population (when the mean is known). Positive dependence corresponds to $\beta < 0$.

Suppose (37) defines the distribution for $t = 1, \ldots, T$ with the condition

$y_0 \equiv 0$. Then A_0 and A_1 are the same as in the previous case and

$$A_2 = \begin{bmatrix} 1 & 0 & 0 & \cdots & 0 & 0 \\ 0 & 1 & 0 & \cdots & 0 & 0 \\ & 0 & 1 & \cdots & 0 & 0 \\ \vdots & \vdots & \vdots & & \vdots & \vdots \\ 0 & 0 & 0 & \cdots & 1 & 0 \\ 0 & 0 & 0 & \cdots & 0 & 0 \end{bmatrix}. \tag{39}$$

The set of alternatives for $\gamma_1 = 2\beta/\sigma^2$ and $\gamma_2 = \beta^2/\sigma^2$ does not lie on a straight line. Hence no uniformly most powerful test exists for the hypotheses $\beta = 0$.

A model defined by a second-order difference equation

$$y_t + \beta_1 y_{t-1} + \beta_2 y_{t-2} = u_t, \qquad t = 1, \ldots, T, \tag{40}$$

with the circular definition $y_0 \equiv y_T$, $y_{-1} \equiv y_{T-1}$ leads to a density (1) with

$$A_0 = I,$$

$$A_1 = \tfrac{1}{2}\begin{bmatrix} 0 & 1 & 0 & \cdots & 0 & 1 \\ 1 & 0 & 1 & \cdots & 0 & 0 \\ 0 & 1 & 0 & \cdots & 0 & 0 \\ \vdots & \vdots & \vdots & & \vdots & \vdots \\ 0 & 0 & 0 & \cdots & 0 & 1 \\ 1 & 0 & 0 & \cdots & 1 & 0 \end{bmatrix}, \tag{41}$$

$$A_2 = \tfrac{1}{2}\begin{bmatrix} 0 & 0 & 1 & 0 & \cdots & 1 & 0 \\ 0 & 0 & 0 & 1 & \cdots & 0 & 1 \\ 1 & 0 & 0 & 0 & \cdots & 0 & 0 \\ 0 & 1 & 0 & 0 & \cdots & 0 & 0 \\ \vdots & \vdots & \vdots & \vdots & & \vdots & \vdots \\ 1 & 0 & 0 & 0 & \cdots & 0 & 0 \\ 0 & 1 & 0 & 0 & \cdots & 0 & 0 \end{bmatrix}.$$

If the alternative hypotheses leave β_1 and β_2 free, there is no uniformly most powerful test of the hypothesis $\beta_1 = \beta_2 = 0$.

The theory of this section was developed in T. W. Anderson (1948), (1963).

6.4. CHOICE OF THE ORDER OF DEPENDENCE AS A MULTIPLE-DECISION PROBLEM

We now treat the question of determining the order of dependence on the basis of the observations $y_1, \ldots, y_T$. We suppose that the investigator can specify some minimum order of dependence m and some maximum order of dependence p. In many cases m will be 0 (indicating independence), but in other cases the investigator may have reason to expect the order to be at least some positive m. Because the observations are not farther apart in time than $T - 1$, it is clear that the maximum order cannot be greater than $T - 1$ but that p must be a good deal less than this if the procedure is to have good properties corresponding to power.

The investigator has a multiple-decision problem, namely, choosing the order $m, m + 1, \ldots, p - 1$, or p. This problem is similar to the problem of choosing the degree of polynomial to fit as trend, which was treated in Section 3.2.2. Although the distributions here are not the same as in Section 3.2.2, the general structure of the problem and its solution will be the same. We formalize the investigator's problem by saying that he wants to decide to which of the following mutually exclusive sets the parameter point $(\gamma_{m+1}, \ldots, \gamma_p)$ belongs:

$$
\begin{aligned}
&H_p:\ \gamma_p \neq 0,\\
&H_{p-1}:\ \gamma_p = 0, \qquad \gamma_{p-1} \neq 0,\\
&\qquad \vdots\\
&H_{m+1}:\ \gamma_p = \cdots = \gamma_{m+2} = 0, \qquad \gamma_{m+1} \neq 0,\\
&H_m:\ \gamma_p = \cdots = \gamma_{m+1} = 0.
\end{aligned} \tag{1}
$$

The set H_i implies that the dependence is of order i. An alternative formulation is that he wishes to decide which (if any) of the following null hypotheses are true:

$$
\begin{aligned}
&H_p^*:\ \gamma_p = 0,\\
&H_{p,p-1}^*:\ \gamma_p = \gamma_{p-1} = 0,\\
&\qquad \vdots\\
&H_{p,p-1,\ldots,m+1}^*:\ \gamma_p = \cdots = \gamma_{m+1} = 0.
\end{aligned} \tag{2}
$$

If any hypothesis of (2) is true, the preceding hypotheses must be true, and if any hypothesis is false the succeeding ones must be false; that is,

$$H^*_{p,p-1,\ldots,m+1} \subset \cdots \subset H^*_p. \tag{3}$$

The two families of sets are related by

$$\begin{aligned} H^*_p &= H_{p-1} \cup \cdots \cup H_m, \\ H^*_{p,p-1} &= H_{p-2} \cup \cdots \cup H_m, \\ &\cdot \\ &\cdot \\ &\cdot \\ H^*_{p,p-1,\ldots,m+1} &= H_m. \end{aligned} \tag{4}$$

We suppose that the investigator wants to control directly the probabilities of errors of saying coefficients are not zero when they are, or correspondingly of choosing a higher order than necessary. We assume that the investigator assigns a significance level to each null hypothesis:

$$\begin{aligned} p_p &= \Pr\{\text{reject } H^*_p \mid H^*_p\}, \\ p_p + p_{p-1} &= \Pr\{\text{reject } H^*_{p,p-1} \mid H^*_{p,p-1}\}, \\ &\cdot \\ &\cdot \\ &\cdot \\ p_p + \cdots + p_{m+1} &= \Pr\{\text{reject } H^*_{p,p-1,\ldots,m+1} \mid H^*_{p,p-1,\ldots,m+1}\} \\ &= \Pr\{\text{reject } H_m \mid H_m\}, \end{aligned} \tag{5}$$

where $p_i \geq 0$ and $p_p + \cdots + p_{m+1} \leq 1$. Since one null hypothesis includes the next (that is, each subsequent null hypothesis is more stringent), the probabilities of rejection are taken monotonically nondecreasing (that is, the probability of rejecting a more restricted null hypothesis when it is true is not less than that of rejecting a less restricted one when it is true). In terms of the mutually exclusive categories the specification is

$$\begin{aligned} p_p &= \Pr\{\text{accept } H_p \mid H_{p-1} \cup \cdots \cup H_m\}, \\ p_{p-1} &= \Pr\{\text{accept } H_{p-1} \mid H_{p-2} \cup \cdots \cup H_m\}, \\ &\cdot \\ &\cdot \\ &\cdot \\ p_{m+1} &= \Pr\{\text{accept } H_{m+1} \mid H_m\}. \end{aligned} \tag{6}$$

Let $p_m = 1 - p_p - \cdots - p_{m+1} = \Pr\{\text{accept } H_m \mid H_m\}$. In another way of putting the formulation (6), the investigator specifies the probability of deciding on order of dependence i when the order of dependence needed is less than i, $i = m + 1, \ldots, p$.

A statistical procedure for this multiple-decision problem consists of a set of $p - m + 1$ (mutually exclusive and exhaustive) regions in the space of $Q_0, \ldots, Q_p$ (or in the original space of $y_1, \ldots, y_T$), say $R_m, R_{m+1}, \ldots, R_p$. If the sample point falls in R_i, then H_i is accepted. The assignment of significance levels implies that these regions are similar regions in the sense that when $\gamma_i = \cdots = \gamma_p = 0$ the probabilities of the sample point falling in $R_i, \ldots, R_p$ are $p_i, \ldots, p_p$, respectively (independent of $\gamma_0, \ldots, \gamma_{i-1}$); that is, when the order of the dependence needed is less than i, the probability of making an error by saying the order is i does not depend on what that (lower-order) dependence is.

We shall consider first alternatives that the dependence is positive. This corresponds to $\gamma_i < 0$. (Later we shall consider unbiased procedures when $\gamma_i \neq 0$.)

Subject to the foregoing requirements, we ask for the best regions in the sense that we want to maximize the probability of R_i when H_i is true, $i = m + 1, \ldots, p$. We are trying to maximize simultaneously the probabilities of $p - m$ different regions (each for all negative values of the corresponding parameter). We shall show that under these conditions the maximized probabilities of one region are not affected by the choice of another. This fact permits us to optimize $R_{m+1}, \ldots, R_p$ simultaneously.

The requirements (5) or equivalently (6) imply that each set R_i, or equivalently $R_i \cup R_{i+1} \cup \cdots \cup R_p$, the region of rejection of $H^*_{p,p-1,\ldots,i}$, has Neyman structure (because of the properties of sufficiency and completeness); that is

$$\Pr\{R_i \mid Q_0, \ldots, Q_{i-1};\ \gamma_i = \cdots = \gamma_p = 0\} = p_i, \qquad i = m + 1, \ldots, p, \tag{7}$$

almost everywhere in the set of possible $Q_0, \ldots, Q_{i-1}$. We make use of this fact to show that the choice of $R_p, \ldots, R_{i+1}$ [subject to (6)] does not affect the choice of R_i in the sense that the probability of R_i (a function of γ_i), when $\gamma_{i+1} = \cdots = \gamma_p = 0$, does not depend on how the $R_p, \ldots, R_{i+1}$ are chosen. Note that we are interested in γ_i when $\gamma_{i+1} = \cdots = \gamma_p = 0$, and, if one of $\gamma_{i+1}, \ldots, \gamma_p$ is not 0 (that is, the order is greater than i), we are not interested in γ_i and may assume $\gamma_i \neq 0$.

LEMMA 6.4.1. *Let S_{i+1} be a set in the space of $Q_0, \ldots, Q_p$ such that*

$$\Pr\{S_{i+1} \mid \gamma_{i+1} = \cdots = \gamma_p = 0\} = p_p + \cdots + p_{i+1}, \tag{8}$$

and let T_i *be a set defined by* $Q_0, \ldots, Q_i$. *Then*

(9) $$\Pr\{\bar{S}_{i+1} \cap T_i \mid \gamma_{i+1} = \cdots = \gamma_p = 0\} = (1 - p_p - \cdots - p_{i+1}) \Pr\{T_i \mid \gamma_{i+1} = \cdots = \gamma_p = 0\},$$

where $\bar{S}_{i+1}$ *is the complement of* S_{i+1}.

The point of the lemma is that however $R_p, \ldots, R_{i+1}$ are chosen, subject to (6), which implies (8) for $S_{i+1} = R_p \cup \cdots \cup R_{i+1}$, the probability of a region R_i defined as an intersection $\bar{S}_{i+1} \cap T_i$ depends only on T_i and not on S_{i+1} (when $H^*_{p,p-1,\ldots,i+1}$ is true). The proof is given in Section 3.2.2 in terms of regression coefficients, as is the proof of Lemma 6.4.2. The proofs, however, depend only on the properties of sufficiency and completeness; the terminology and notation of Section 3.2.2 can be translated into that of this section.

Let T_i^* be the region

(10) $$Q_i > c_i(Q_0, \ldots, Q_{i-1}),$$

where $c_i(Q_0, \ldots, Q_{i-1})$ is determined so the conditional probability of (10) given $Q_0, \ldots, Q_{i-1}$ when $\gamma_i = 0$ is ε_i and $\varepsilon_i = p_i/(1 - p_p - \cdots - p_{i+1})$.

LEMMA 6.4.2. *Given* S_{i+1} *satisfying* (8) *and any* R_i *disjoint with* S_{i+1} *such that*

(11) $$\Pr\{R_i \mid \gamma_i = \gamma_{i+1} = \cdots = \gamma_p = 0\} = p_i,$$

then

(12) $$\Pr\{\bar{S}_{i+1} \cap T_i^* \mid \gamma_{i+1} = \cdots = \gamma_p = 0\} \geq \Pr\{R_i \mid \gamma_{i+1} = \cdots = \gamma_p = 0\}$$

for $\gamma_i < 0$.

The implication of the two lemmas is that, whatever $R_p, \ldots, R_{i+1}$ are, the best choice of R_i is the part of T_i^* disjoint from $R_p, \ldots, R_{i+1}$ and for such a choice

(13) $$\Pr\{R_i \mid \gamma_{i+1} = \cdots = \gamma_p = 0\} = (1 - p_p - \cdots - p_{i+1}) \Pr\{T_i^* \mid \gamma_{i+1} = \cdots = \gamma_p = 0\}.$$

This does not depend on the choice of $R_p, \ldots, R_{i+1}$.

THEOREM 6.4.1. *Let* $R_m, R_{m+1}, \ldots, R_p$ *be* $p - m + 1$ *disjoint regions in the sample space such that* $R_m \cup R_{m+1} \cup \cdots \cup R_p$ *is the entire space and such that*

(14) $$\Pr\{R_i \mid \gamma_i = \gamma_{i+1} = \cdots = \gamma_p = 0\} = p_i, \qquad i = m+1, \ldots, p,$$

where $p_{m+1} + \cdots + p_p \leq 1$. *Then for every value of* $\gamma_i < 0$ $\Pr\{R_i \mid \gamma_i, \gamma_{i+1} = \cdots = \gamma_p = 0\}$ *is maximized by* R_i *defined by* (10) *for* $\varepsilon_i = p_i/(1 - p_p - \cdots - p_{i+1})$ *complementary to* $R_p \cup \cdots \cup R_{i+1}$, $i = m+1, \ldots, p$.

The optimum procedure, therefore, is

$$
\begin{aligned}
R_p &= T_p^*, \\
R_{p-1} &= \bar{T}_p^* \cap T_{p-1}^*, \\
&\;\;\vdots \\
R_i &= \bar{T}_p^* \cap \bar{T}_{p-1}^* \cap \cdots \cap \bar{T}_{i+1}^* \cap T_i^*, \qquad (15)\\
&\;\;\vdots \\
R_{m+1} &= \bar{T}_p^* \cap \bar{T}_{p-1}^* \cap \cdots \cap \bar{T}_{m+2}^* \cap T_{m+1}^*, \\
R_m &= \bar{T}_p^* \cap \bar{T}_{p-1}^* \cap \cdots \cap \bar{T}_{m+2}^* \cap \bar{T}_{m+1}^*.
\end{aligned}
$$

What the procedure amounts to is to test $\gamma_p = 0$, $\gamma_{p-1} = 0, \ldots$ in turn until either one rejects such a hypothesis, say rejects $\gamma_i = 0$, and hence decides H_i, or one accepts all such hypotheses and eventually H_m. Thus the procedure is a sequence of hypothesis tests; this result is a consequence of the requirement that the probability of deciding that the order of dependence is less than a given integer when that is the case should not depend on what the actual dependence is.

Now let us consider procedures based on two-sided tests. We impose conditions of unbiasedness.

THEOREM 6.4.2. *Let $R_m, R_{m+1}, \ldots, R_p$ be $p - m + 1$ disjoint regions in the sample space such that $R_m \cup R_{m+1} \cup \cdots \cup R_p$ is the entire space and such that*

$$
\begin{aligned}
(16) \quad \Pr\{R_i \cup \cdots \cup R_p \mid \gamma_i = \gamma_{i+1} = \cdots = \gamma_p = 0\} &\le p_i + \cdots + p_p \\
&\le \Pr\{R_i \cup \cdots \cup R_p \mid \gamma_i, \gamma_{i+1} = \cdots = \gamma_p = 0\}, \\
& \qquad i = m+1, \ldots, p,
\end{aligned}
$$

where $p_{m+1} + \cdots + p_p \le 1$. Then for every γ_i $\Pr\{R_i \mid \gamma_i, \gamma_{i+1} = \cdots = \gamma_p = 0\}$ is maximized by R_i defined by Corollary 6.3.2 for $\varepsilon_i = p_i/(1 - p_p - \cdots - p_{i+1})$ complementary to $R_{i+1} \cup \cdots \cup R_p$, $i = m+1, \ldots, p$.

We have not stated here how $p_m, p_{m+1}, \ldots, p_p$ are to be chosen. If ε_i is fixed, say ε, then $p_p = \varepsilon$, $p_{p-1} = \varepsilon(1-\varepsilon), \ldots, p_i = \varepsilon(1-\varepsilon)^{p-i}, \ldots, p_{m+1} = \varepsilon(1-\varepsilon)^{p-m-1}$, $p_m = (1-\varepsilon)^{p-m}$. In general, we want to balance the desirability of not overestimating the order with sensitivity to nonzero coefficients. An alternative is to take $p_p < p_{p+1} < \cdots < p_{m+1}$; if the p_i's for i large are taken very small, one can afford to let the maximum order p be fairly large.

The developments of this section lead to a test of the hypothesis

$$H: \quad \gamma_{m+1} = \cdots = \gamma_p = 0. \tag{17}$$

The procedure in Theorem 6.4.1 or 6.4.2 can be interpreted as a test at significance level $p_{m+1} + \cdots + p_p$ in which the hypothesis (17) is rejected if the sample point falls in any region but R_m.

The procedure of Theorem 6.4.1 consists of one-sided tests against one-sided alternatives of positive dependence. There is an analogous procedure of one-sided tests of the form (15) of Section 6.3 when negative dependence ($\gamma_i > 0$) is the alternative to independence. Moreover, composite procedures can be constructed when the alternative to $\gamma_i = 0$ is specified as $\gamma_i < 0$, $\gamma_i > 0$, or $\gamma_i \neq 0$, $i = m + 1, \ldots, p$.

The comments that were made in Section 3.2.2 on this type of multiple-decision procedure for determining the degree of polynomial trend apply here as well. The theoretical advantage of this procedure is that the investigator controls the probability of error of saying the order of dependence is i when it is actually lower. A disadvantage is that the maximum order of dependence p must be specified in advance; this may be difficult to do, but it is essential. A practical disadvantage is that $p + 1$ quadratic forms must be calculated. The disadvantage of having to specify p in advance might be mitigated by specifying p fairly large, but taking p_i very small for i near p; that is, make the probabilities small of deciding on a large order of dependence when that is not the case.

An alternative multiple-decision procedure is to run significance tests in the opposite direction. If it is assumed that the order of dependence is at least m, the investigator may test the hypothesis $\gamma_{m+1} = 0$ at significance level ε_{m+1}. If the hypothesis is accepted, the investigator decides the order of dependence is m; if the hypothesis is rejected, he then tests the hypothesis $\gamma_{m+2} = 0$. The procedure may continue with such a sequence of significance tests. If a hypothesis $\gamma_i = 0$ is accepted, the decision is made that the order of dependence is $i - 1$; if the hypothesis $\gamma_i = 0$ is rejected, the hypothesis $\gamma_{i+1} = 0$ is tested. If successive hypotheses are rejected until the hypothesis $\gamma_p = 0$ is rejected for p specified a priori, the order of dependence is said to be p. The advantage of this procedure is that not all $p + 1$ quadratic forms may need to be computed, because the sequence of tests may stop before p is reached. Because of this fact the specification of p is not as important.

It was pointed out in Section 3.2.2 that the corresponding procedure for determining the degree of polynomial trend had the disadvantage that if the coefficient of degree i was large it would tend to swell the estimates of error variance used to test $\gamma_{m+1}, \ldots, \gamma_{i-1}$ and hence would tend to increase the probability of accepting one of the hypotheses $\gamma_{m+1} = 0, \ldots, \gamma_{i-1} = 0$; that is, the effect would be to increase the probability of deciding the degree is lower than i. In fact, the larger γ_i is, the greater is this probability of error (holding the other γ_j's fixed). It is more difficult to analyze the case here of testing order of dependence. It is possible that if the dependence one time unit apart is small, but the dependence two time units apart is large, the probability will not be large of rejecting $\gamma_1 = 0$ and getting to decide the order of dependence is at least two.

This procedure seems to require that $\gamma_{m+1}, \ldots, \gamma_{i-1}$ be substantial so that none of them is accepted as zero if $\gamma_i \neq 0$ for some particular i. For a situation where the γ_i's fall off monotonically (or perhaps sharply) this procedure may be desirable. However, no theoretically optimum properties have been proved.

Thus far in this chapter we have assumed that $\mathscr{E}y_1 = \cdots = \mathscr{E}y_T = 0$; if the means are different from 0 but known, then the differences between the variates and their means can be treated under the above assumption. Knowledge of the means, however, is not a very realistic supposition. We shall take up the cases of unknown means in Section 6.6 after we discuss more specific models in Section 6.5. [See T. W. Anderson (1963).]

6.5. MODELS: SYSTEMS OF QUADRATIC FORMS

6.5.1. General Discussion

The densities of $y_1, \ldots, y_T$, the observed time series, which are studied here, are normal with exponent $-\frac{1}{2}\sum_{j=0}^{p}\gamma_j Q_j$, where $Q_j = \boldsymbol{y}'\boldsymbol{A}_j\boldsymbol{y}$ is a quadratic form in $\boldsymbol{y} = (y_1, \ldots, y_T)'$. The serial correlation coefficients obtained from these densities are $r_j = Q_j/Q_0$. In this section we shall suggest several different systems of matrices $\boldsymbol{A}_0, \boldsymbol{A}_1, \ldots, \boldsymbol{A}_p$; in each system Q_j is approximately $\sum_t y_t y_{t-j}$ as suggested in Section 6.2. In each system we can use $r_1, \ldots, r_p$ to test hypotheses about the order of dependence as developed in Section 6.3 and to choose the order of dependence as described in Section 6.4.

First we discuss sets of matrices in general to prepare for the analysis of the systems of matrices. The characteristic roots $\lambda_1, \ldots, \lambda_T$ and corresponding characteristic vectors $\boldsymbol{v}_1, \ldots, \boldsymbol{v}_T$ of a $T \times T$ matrix $\boldsymbol{A}$ satisfy

$$\boldsymbol{A}\boldsymbol{v}_s = \lambda_s \boldsymbol{v}_s, \qquad s = 1, \ldots, T. \tag{1}$$

Since (1) can be written $(\boldsymbol{A} - \lambda_s\boldsymbol{I})\boldsymbol{v}_s = \boldsymbol{0}$ and $\boldsymbol{v}_s \neq \boldsymbol{0}$, the matrix $\boldsymbol{A} - \lambda_s\boldsymbol{I}$ must be singular; this is equivalent to saying that λ_s is a root of

$$|\boldsymbol{A} - \lambda\boldsymbol{I}| = 0. \tag{2}$$

If A is symmetric (which we assume for quadratic forms $y'Ay$), the roots are real and so are the components of the characteristic vectors. If the roots are distinct, the vectors are orthogonal. If m roots are equal, say λ, there are m linearly independent solutions of $(A - \lambda I)v = 0$; any such set may serve as the characteristic vectors corresponding to this multiple root. It will be convenient to take a set of vectors that are mutually orthogonal; they are necessarily orthogonal to the other characteristic vectors. If each characteristic vector is normalized so the sum of squares of the components is 1,

$$v_s'v_t = \delta_{st}, \qquad s, t = 1, \ldots, T, \tag{3}$$

where $\delta_{ss} = 1$ and $\delta_{st} = 0$ for $s \neq t$.

Let

$$\Lambda = \begin{bmatrix} \lambda_1 & 0 & \cdots & 0 \\ 0 & \lambda_2 & \cdots & 0 \\ \cdot & \cdot & & \cdot \\ \cdot & \cdot & & \cdot \\ \cdot & \cdot & & \cdot \\ 0 & 0 & \cdots & \lambda_T \end{bmatrix}, \qquad V = (v_1, \ldots, v_T). \tag{4}$$

Then (1) and (3) can be summarized as

$$AV = V\Lambda, \tag{5}$$

$$V'V = I. \tag{6}$$

If we multiply (5) on the left by V' we obtain

$$V'AV = \Lambda; \tag{7}$$

that is, the orthogonal matrix V diagonalizes the symmetric matrix A. A quadratic form $y'Ay$ is transformed to diagonal form $x'\Lambda x$ by the linear transformation $y = Vx$. We shall use these transformations to simplify the problems of finding distributions of quadratic forms and serial correlation coefficients.

We are interested in systems of matrices, $A_0, A_1, \ldots, A_p$. Usually $A_0 = I$. Any nonzero vector is a characteristic vector of I, and any orthogonal matrix V diagonalizes I as $V'IV = I$. It is convenient if the same matrix V diagonalizes all the A_j; that is, if

$$V'A_jV = \Lambda_j, \qquad j = 1, \ldots, p. \tag{8}$$

This will occur if every matrix A_j has $v_1, \ldots, v_T$ as its characteristic vectors. The following theorem will be useful:

THEOREM 6.5.1. *If $\lambda_1, \ldots, \lambda_T$ and $v_1, \ldots, v_T$ are characteristic roots and vectors of a matrix A and $P(A)$ is a polynomial in A, then $P(\lambda_1), \ldots, P(\lambda_T)$ and $v_1, \ldots, v_T$ are characteristic roots and vectors of $P(A)$.*

We use the following lemmas to prove this theorem:

LEMMA 6.5.1. *If $\lambda_1, \ldots, \lambda_T$ and $v_1, \ldots, v_T$ are characteristic roots and vectors of A, then $\lambda_1^g, \ldots, \lambda_T^g$ and $v_1, \ldots, v_T$ are characteristic roots and vectors of A^g, $g = 0, 1, \ldots$.*

PROOF OF LEMMA. We prove the lemma by induction. It obviously holds for $g = 1$. Suppose it holds for $g = G - 1$. Then

$$\begin{aligned} A^G v_s &= A(A^{G-1}v_s) = A\lambda_s^{G-1}v_s \\ &= \lambda_s^{G-1}(Av_s) = \lambda_s^{G-1}\lambda_s v_s \\ &= \lambda_s^G v_s, \qquad s = 1, \ldots, T. \end{aligned} \tag{9}$$

LEMMA 6.5.2. *If $\lambda_{h1}, \ldots, \lambda_{hT}$ and $v_1, \ldots, v_T$ are characteristic roots and vectors of A_h, $h = 0, \ldots, H$, then $\sum_{h=0}^H a_h\lambda_{h1}, \ldots, \sum_{h=0}^H a_h\lambda_{hT}$ and $v_1, \ldots, v_T$ are characteristic roots and vectors of $\sum_{h=0}^H a_h A_h$.*

PROOF OF LEMMA. We have

$$\sum_{h=0}^H a_h A_h v_s = \sum_{h=0}^H a_h \lambda_{hs} v_s, \qquad s = 1, \ldots, T. \tag{10}$$

PROOF OF THEOREM. Let

$$P(A) = a_0 I + a_1 A + \cdots + a_H A^H. \tag{11}$$

Then the theorem follows from the two lemmas.

6.5.2. The Circular Model

One serial correlation that has simple mathematical properties and has been studied considerably is the circular serial correlation coefficient. It is based on a circular probability model.

Let $\boldsymbol{B}$ be the circulant

$$(12) \qquad \boldsymbol{B} = \begin{bmatrix} 0 & 1 & 0 & 0 & \cdots & 0 & 0 \\ 0 & 0 & 1 & 0 & \cdots & 0 & 0 \\ 0 & 0 & 0 & 1 & \cdots & 0 & 0 \\ 0 & 0 & 0 & 0 & \cdots & 0 & 0 \\ \cdot & \cdot & \cdot & \cdot & & \cdot & \\ \cdot & \cdot & \cdot & \cdot & & \cdot & \cdot \\ \cdot & \cdot & \cdot & \cdot & & \cdot & \cdot \\ 0 & 0 & 0 & 0 & \cdots & 0 & 1 \\ 1 & 0 & 0 & 0 & \cdots & 0 & 0 \end{bmatrix}.$$

This matrix is orthogonal, $\boldsymbol{B}^{-1} = \boldsymbol{B}'$. Let

$$(13) \qquad \boldsymbol{A}_1 = \tfrac{1}{2}(\boldsymbol{B} + \boldsymbol{B}') = \tfrac{1}{2}\begin{bmatrix} 0 & 1 & 0 & 0 & \cdots & 0 & 1 \\ 1 & 0 & 1 & 0 & \cdots & 0 & 0 \\ 0 & 1 & 0 & 1 & \cdots & 0 & 0 \\ 0 & 0 & 1 & 0 & \cdots & 0 & 0 \\ \cdot & \cdot & \cdot & \cdot & & \cdot & \cdot \\ \cdot & \cdot & \cdot & \cdot & & \cdot & \cdot \\ \cdot & \cdot & \cdot & \cdot & & \cdot & \cdot \\ 0 & 0 & 0 & 0 & \cdots & 0 & 1 \\ 1 & 0 & 0 & 0 & \cdots & 1 & 0 \end{bmatrix}.$$

Then

$$(14) \qquad \begin{aligned} Q_1 = \boldsymbol{y}'\boldsymbol{A}_1\boldsymbol{y} &= \frac{1}{2}\left(\sum_{t=1}^{T-1} y_t y_{t+1} + \sum_{t=2}^{T} y_t y_{t-1} + y_1 y_T + y_T y_1\right) \\ &= \sum_{t=1}^{T} y_t y_{t-1}, \end{aligned}$$

where $y_0 \equiv y_T$. This amounts to $\sum_{t=2}^{T} y_t y_{t-1}$ plus the term $y_1 y_T$. In effect the last term is included for mathematical convenience. We define

$$(15) \qquad \boldsymbol{A}_j = \tfrac{1}{2}[\boldsymbol{B}^j + (\boldsymbol{B}')^j] = \tfrac{1}{2}(\boldsymbol{B}^j + \boldsymbol{B}^{-j}), \qquad j < \tfrac{1}{2}T.$$

This matrix has $\frac{1}{2}$'s on the diagonals j elements above and j elements below the main diagonal, corresponding to $\sum_t y_t y_{t-j}$, which is in the exponent of the density of a stationary Gaussian process of order at least j. A_j also has $\frac{1}{2}$'s on the diagonals $T-j$ elements above and $T-j$ elements below the main diagonal. We have

$$
\begin{aligned}
(16) \qquad A_1^{2r} &= (\tfrac{1}{2})^{2r}[B + B^{-1}]^{2r} \\
&= (\tfrac{1}{2})^{2r}\left[B^{2r} + 2rB^{2r-2} + \binom{2r}{2}B^{2r-4} + \cdots + 2rB^{-(2r-2)} + B^{-2r}\right] \\
&= (\tfrac{1}{2})^{2r-1}\left[A_{2r} + 2rA_{2r-2} + \cdots + \binom{2r}{r-1}A_2 + \tfrac{1}{2}\binom{2r}{r}I\right],
\end{aligned}
$$

$$
(17) \qquad A_1^{2r+1} = (\tfrac{1}{2})^{2r}\left[A_{2r+1} + (2r+1)A_{2r-1} + \cdots + \binom{2r+1}{r}A_1\right].
$$

Thus A_j can be written as a polynomial in A_1:

$$
(18) \qquad
\begin{aligned}
A_2 &= 2A_1^2 - I, \\
A_3 &= 4A_1^3 - 3A_1, \\
A_4 &= 8A_1^4 - 8A_1^2 + I, \\
A_5 &= 16A_1^5 - 20A_1^3 + 5A_1, \\
A_6 &= 32A_1^6 - 48A_1^4 + 18A_1^2 - I.
\end{aligned}
$$

THEOREM 6.5.2. *The characteristic roots of* A_1 *are*

$$
(19) \qquad \cos 0 = 1,\ \cos\frac{2\pi}{T},\ \cos\frac{2\pi}{T},\ \cos\frac{4\pi}{T},\ \cos\frac{4\pi}{T},\ \ldots,\ \cos\frac{\pi(T-1)}{T},\ \cos\frac{\pi(T-1)}{T}, \qquad T \text{ odd},
$$

$$
(20) \qquad \cos 0 = 1,\ \cos\frac{2\pi}{T},\ \cos\frac{2\pi}{T},\ \cos\frac{4\pi}{T},\ \cos\frac{4\pi}{T},\ \ldots,\ \cos\frac{\pi(T-2)}{T},\ \cos\frac{\pi(T-2)}{T},\ \cos\pi = -1, \qquad T \text{ even};
$$

the characteristic vector corresponding to the root 1 *is* $(1, 1, \ldots, 1)'$, *the characteristic vectors corresponding to the roots* $\cos 2\pi s/T$ $(s \neq 0, \frac{1}{2}T)$ *are*

$$\begin{bmatrix} \cos \frac{2\pi s}{T} \\ \cos \frac{4\pi s}{T} \\ \cdot \\ \cdot \\ \cdot \\ 1 \end{bmatrix}, \qquad \begin{bmatrix} \sin \frac{2\pi s}{T} \\ \sin \frac{4\pi s}{T} \\ \cdot \\ \cdot \\ \cdot \\ 0 \end{bmatrix}, \tag{21}$$

and the characteristic vector corresponding to the root -1 (*if* T *is even*) *is* $(-1, 1, -1, \ldots, 1)'$.

PROOF. The equation defining the characteristic roots and vectors of $\boldsymbol{B}$,

$$\boldsymbol{Bx} = \nu \boldsymbol{x}, \tag{22}$$

is written in components as

$$x_{t+1} = \nu x_t, \qquad t = 1, \ldots, T-1, \tag{23}$$

$$x_1 = \nu x_T. \tag{24}$$

Thus (23) implies

$$x_t = \nu^{t-1} x_1, \qquad t = 2, \ldots, T, \tag{25}$$

and (24) then implies

$$x_1 = \nu^T x_1, \tag{26}$$

$$\nu^T = 1. \tag{27}$$

Thus the roots are the Tth roots of 1, namely $1, e^{i2\pi/T}, e^{i4\pi/T}, \ldots, e^{i2\pi(T-1)/T}$. The characteristic vector corresponding to the root $e^{i2\pi s/T}$, $s = 0, 1, \ldots, T-1$, has as its tth component $e^{i2\pi ts/T}$ if the first component is $e^{i2\pi s/T}$. Let this characteristic vector be $\boldsymbol{u}_s$. Then

$$\begin{aligned} \boldsymbol{A}_1 \boldsymbol{u}_s &= \tfrac{1}{2}(\boldsymbol{B} + \boldsymbol{B}^{-1})\boldsymbol{u}_s = \tfrac{1}{2}\boldsymbol{B}\boldsymbol{u}_s + \tfrac{1}{2}\boldsymbol{B}^{-1}\boldsymbol{u}_s \\ &= \tfrac{1}{2}e^{i2\pi s/T}\boldsymbol{u}_s + \tfrac{1}{2}e^{-i2\pi s/T}\boldsymbol{u}_s \\ &= \cos \frac{2\pi}{T} s \, \boldsymbol{u}_s. \end{aligned} \tag{28}$$

If $s \neq 0, \frac{1}{2}T$, then $\cos(2\pi s/T) = \cos[2\pi(T-s)/T]$ is a double root. Corresponding to this double root, we take as characteristic vectors $\frac{1}{2}(\boldsymbol{u}_s + \boldsymbol{u}_{T-s})$ and $\dfrac{1}{2i}(\boldsymbol{u}_s - \boldsymbol{u}_{T-s})$, which have as tth components

$$\tfrac{1}{2}(e^{i2\pi ts/T} + e^{i2\pi t(T-s)/T}) = \cos\frac{2\pi ts}{T}, \tag{29}$$

$$\frac{1}{2i}(e^{i2\pi ts/T} - e^{i2\pi t(T-s)/T}) = \sin\frac{2\pi ts}{T}, \tag{30}$$

respectively. This proves the theorem.

THEOREM 6.5.3. *The characteristic roots of* $\boldsymbol{A}_j$ *are*

$$\cos 0 = 1, \cos\frac{2\pi j}{T}, \cos\frac{2\pi j}{T}, \cos\frac{4\pi j}{T}, \cos\frac{4\pi j}{T}, \ldots, \cos\frac{\pi j(T-1)}{T}, \cos\frac{\pi j(T-1)}{T}, \quad T \textit{ odd}, \tag{31}$$

$$\cos 0 = 1, \cos\frac{2\pi j}{T}, \cos\frac{2\pi j}{T}, \cos\frac{4\pi j}{T}, \cos\frac{4\pi j}{T}, \ldots, \cos\frac{\pi j(T-2)}{T}, \cos\frac{\pi j(T-2)}{T}, \cos \pi j = (-1)^j, \quad T \textit{ even}; \tag{32}$$

the characteristic vector corresponding to the root 1 *is* $(1, 1, \ldots, 1)'$, *the characteristic vectors corresponding to the roots* $\cos 2\pi js/T$ $(s \neq 0, \frac{1}{2}T)$ *are* (21), *and the characteristic vector corresponding to the root* $(-1)^j$ (*if T is even*) *is* $(-1, 1, -1, \ldots, 1)'$.

PROOF. Since $\boldsymbol{A}_j$ is a polynomial in $\boldsymbol{A}_1$, it has the same characteristic vectors as $\boldsymbol{A}_1$ and the characteristic roots are the polynomials in the roots of $\boldsymbol{A}_1$ (by Theorem 6.5.1). Also, the characteristic vectors and roots of a power of $\boldsymbol{B}$ are the vectors and powers of the roots of $\boldsymbol{B}$. Thus

$$\begin{aligned} \boldsymbol{A}_j\boldsymbol{u}_s &= \tfrac{1}{2}(\boldsymbol{B}^j + \boldsymbol{B}^{-j})\boldsymbol{u}_s \\ &= \tfrac{1}{2}(e^{i2\pi js/T} + e^{-i2\pi js/T})\boldsymbol{u}_s \\ &= \cos\frac{2\pi j}{T} s\, \boldsymbol{u}_s. \end{aligned} \tag{33}$$

It will be observed that the characteristic vectors are the sequences of trigonometric functions that were used in the cyclic trends of Chapter 4. When these characteristic vectors are normalized, they form the orthogonal matrices of Section 4.2.2, namely

(34)

$$M = \sqrt{\frac{2}{T}} \begin{bmatrix} \frac{1}{\sqrt{2}} & \cos\frac{2\pi}{T} & \sin\frac{2\pi}{T} & \cos\frac{4\pi}{T} & \cdots & \sin\frac{\pi(T-2)}{T} & -\frac{1}{\sqrt{2}} \\ \frac{1}{\sqrt{2}} & \cos\frac{4\pi}{T} & \sin\frac{4\pi}{T} & \cos\frac{8\pi}{T} & \cdots & \sin\frac{2\pi(T-2)}{T} & \frac{1}{\sqrt{2}} \\ \vdots & \vdots & \vdots & \vdots & & \vdots & \vdots \\ \frac{1}{\sqrt{2}} & 1 & 0 & 1 & \cdots & 0 & \frac{1}{\sqrt{2}} \end{bmatrix}$$

if T is even and

$$(35) \qquad M = \sqrt{\frac{2}{T}} \begin{bmatrix} \frac{1}{\sqrt{2}} & \cos\frac{2\pi}{T} & \sin\frac{2\pi}{T} & \cos\frac{4\pi}{T} & \cdots & \sin\frac{\pi(T-1)}{T} \\ \frac{1}{\sqrt{2}} & \cos\frac{4\pi}{T} & \sin\frac{4\pi}{T} & \cos\frac{8\pi}{T} & \cdots & \sin\frac{2\pi(T-1)}{T} \\ \vdots & \vdots & \vdots & \vdots & & \vdots \\ \frac{1}{\sqrt{2}} & 1 & 0 & 1 & \cdots & 0 \end{bmatrix}$$

if T is odd.

The same orthogonal matrix diagonalizes all of the quadratic forms. If we transform the vector $(y_1, \ldots, y_T)'$ by this matrix to $(z_1, \ldots, z_T)'$, the jth quadratic form is

$$(36) \qquad Q_j = \sum_{s=1}^{T} \cos\frac{2\pi j}{T} s\, z_s^2, \qquad j = 0, 1, \ldots, p.$$

(Note $\cos 2\pi js/T$ is the same for $s = 0$ and $s = T$.) If the y_t's are independently normally distributed with means 0 and variances σ^2 ($\gamma_0 = 1/\sigma^2$, $\gamma_1 = \cdots = \gamma_p = 0$), then the z_t's are similarly distributed.

The jth-order circular serial correlation coefficient is

$$r_j = \frac{Q_j}{Q_0} = \frac{\sum_{t=1}^{T} y_t y_{t-j}}{\sum_{t=1}^{T} y_t^2} \tag{37}$$

$$= \frac{\sum_{s=1}^{T} \cos \frac{2\pi j}{T} s\, z_s^2}{\sum_{s=1}^{T} z_s^2},$$

where $y_{t-j} = y_{t-j+T}$, $t = 1, \ldots, j$, $j = 1, \ldots, p$. It is called "circular" because the quadratic form Q_j is made up by taking products of all y_t's j units apart when $y_1, \ldots, y_T$ are equally spaced on the circumference of a circle.

The weights $\cos 2\pi s/T$ can be visualized by placing T equidistant points on the unit circle with one point at $(1, 0)$ and dropping perpendiculars to the horizontal axis. The weights are the x-coordinates of these points.

The advantages of the circular model are that the roots occur in pairs for the first-order coefficient, except for possibly one or two roots, and all the matrices have the same characteristic vectors; it will be seen in Section 6.7 that when the roots occur in pairs except for possibly one or two roots the distribution of r_1 or r_1^* can be given in closed form. There is a practical disadvantage in adding terms such as $y_1 y_T$, for such a term does not reflect serial dependence, and if there is a trend (which has been ignored) such a term may be numerically large and will tend to distort the measurement of dependence.

6.5.3. A Model Based on Successive Differences

Here we shall develop another system of matrices $A_0, A_1, \ldots, A_p$. In Section 3.4 we considered using sums of squares of differences to estimate the variance of a time series. If $y_1, \ldots, y_T$ have means 0, variances σ^2, and are uncorrelated,

$$V_r = \frac{\sum_{t=1}^{T-r} (\Delta^r y_t)^2}{(T - r)\binom{2r}{r}} \tag{38}$$

is an unbiased estimate of σ^2. If the means are possibly not 0 (but the y_t's are uncorrelated), V_r/V_q $(q > r)$ can be used to test the null hypothesis that the trend $f(t) = \mathscr{E}y_t$ satisfies $\Delta^r f(t) = 0$ under the assumption that $\Delta^q f(t) = 0$. If the means are 0 and the y_t's are possibly correlated, then the variate difference will reflect the correlation. In particular, for $r = 1$ the expected value of $2(T-1)V_1$ is

$$\mathscr{E}\sum_{t=1}^{T-1}(\Delta y_t)^2 = \mathscr{E}\sum_{t=1}^{T-1}(y_{t+1}^2 - 2y_{t+1}y_t + y_t^2) \tag{39}$$
$$= \mathscr{E}y_1^2 + 2\sum_{t=2}^{T-1}\mathscr{E}y_t^2 + \mathscr{E}y_T^2 - 2\sum_{t=1}^{T-1}\mathscr{E}y_{t+1}y_t.$$

If the successive pairs of y_t's are positively correlated, V_1 will tend to underestimate the variance.

A statistic suggested for testing serial correlation is the ratio of the *mean-square successive difference* to the sum of squares. If the means are known to be 0, this statistic is

$$\frac{\sum_{t=2}^{T}(y_t - y_{t-1})^2}{\sum_{t=1}^{T} y_t^2} = \frac{y_1^2 + 2\sum_{t=2}^{T} y_t^2 + y_T^2 - 2\sum_{t=2}^{T} y_t y_{t-1}}{\sum_{t=1}^{T} y_t^2} \tag{40}$$
$$= 2\left[1 - \frac{\sum_{t=2}^{T} y_t y_{t-1} + \frac{1}{2}y_1^2 + \frac{1}{2}y_T^2}{\sum_{t=1}^{T} y_t^2}\right].$$

As an alternative to the circular quadratic forms, (40) suggests defining $Q_0 = \sum_{t=1}^{T} y_t^2$ and Q_1 as the numerator in the fraction in the last term of (40). The matrix of this quadratic form is

$$A_1 = \frac{1}{2}\begin{pmatrix} 1 & 1 & 0 & 0 & \cdots & 0 \\ 1 & 0 & 1 & 0 & \cdots & 0 \\ 0 & 1 & 0 & 1 & \cdots & 0 \\ 0 & 0 & 1 & 0 & \cdots & 0 \\ \cdot & \cdot & \cdot & \cdot & & \cdot \\ \cdot & \cdot & \cdot & \cdot & & \cdot \\ \cdot & \cdot & \cdot & \cdot & & \cdot \\ 0 & 0 & 0 & 0 & \cdots & 1 \end{pmatrix}. \tag{41}$$

This matrix differs from the matrix of the quadratic form $\sum_{t=2}^{T} y_t y_{t-1}$ by having $\frac{1}{2}$'s in the upper left-hand corner and in the lower right-hand corner. We can generate the subsequent matrices in this model by the polynomials in (18) [implied by (16) and (17)] in this A_1. These are the polynomials which we used in the circular case. The first are

$$A_2 = 2A_1^2 - I = \frac{1}{2}\begin{bmatrix} 0 & 1 & 1 & 0 & \cdots & 0 \\ 1 & 0 & 0 & 1 & \cdots & 0 \\ 1 & 0 & 0 & 0 & \cdots & 0 \\ 0 & 1 & 0 & 0 & \cdots & 0 \\ \cdot & \cdot & \cdot & \cdot & & \cdot \\ \cdot & \cdot & \cdot & \cdot & & \cdot \\ \cdot & \cdot & \cdot & \cdot & & \cdot \\ 0 & 0 & 0 & 0 & \cdots & 0 \end{bmatrix}, \tag{42}$$

$$A_3 = 4A_1^3 - 3A_1 = \frac{1}{2}\begin{bmatrix} 0 & 0 & 1 & 1 & 0 & \cdots & 0 \\ 0 & 1 & 0 & 0 & 1 & \cdots & 0 \\ 1 & 0 & 0 & 0 & 0 & \cdots & 0 \\ 1 & 0 & 0 & 0 & 0 & \cdots & 0 \\ 0 & 1 & 0 & 0 & 0 & \cdots & 0 \\ \cdot & \cdot & \cdot & \cdot & \cdot & & \cdot \\ \cdot & \cdot & \cdot & \cdot & \cdot & & \cdot \\ \cdot & \cdot & \cdot & \cdot & \cdot & & \cdot \\ 0 & 0 & 0 & 0 & 0 & \cdots & 0 \end{bmatrix}. \tag{43}$$

It will be noted that A_j differs from the matrix of the desired form $\sum_t y_t y_{t-j}$ only by insertion of j $\frac{1}{2}$'s in the upper left-hand corner and in the lower right-hand corner.

Defining A_j, $j = 2, \ldots, p$, as polynomials in A_1 assures that the characteristic vectors of A_j are the same as the characteristic vectors of A_1. The use of the same polynomials in this system as were used in the system of circular serial correlation coefficients implies that A_j here will not differ very much from the matrix with $\frac{1}{2}$'s j elements off the main diagonal because A_1 here differs from $\frac{1}{2}(B + B')$ only in corner elements, and hence A_j differs from $\frac{1}{2}[B^j + (B')^j]$ by only a small number of elements when j is small.

A characteristic vector x with components $x_1, \ldots, x_T$ corresponding to a characteristic root λ satisfying $A_1 x = \lambda x$ satisfies

$$\tfrac{1}{2}(x_1 + x_2) = \lambda x_1, \tag{44}$$

$$\tfrac{1}{2}(x_{t-1} + x_{t+1}) = \lambda x_t, \qquad t = 2, \ldots, T-1, \tag{45}$$

$$\tfrac{1}{2}(x_{T-1} + x_T) = \lambda x_T. \tag{46}$$

Equations (45) may be written

$$x_{t+1} - 2\lambda x_t + x_{t-1} = 0, \qquad t = 2, \ldots, T-1, \tag{47}$$

which is a second-order difference equation. The solution is

$$x_t = c_1 \xi_1^t + c_2 \xi_2^t, \tag{48}$$

where ξ_1 and ξ_2 are the roots of the associated polynomial equation

$$x^2 - 2\lambda x + 1 = 0. \tag{49}$$

The roots $\lambda \pm \sqrt{\lambda^2 - 1}$ are distinct unless $\lambda = 1$ or $\lambda = -1$. Since $\xi_1 \xi_2 = 1$ and $\xi_1 + \xi_2 = 2\lambda$, we can write (48) as

$$x_t = c_1 \xi^t + c_2 \xi^{-t}, \tag{50}$$

and $2\lambda = \xi + \xi^{-1}$. Then (44) can be written as

$$\begin{aligned} 0 &= x_2 + (1 - 2\lambda)x_1 \\ &= c_1\xi^2 + c_2\xi^{-2} + (1 - \xi - \xi^{-1})(c_1\xi + c_2\xi^{-1}) \\ &= c_1(1 - \xi^{-1})\xi + c_2(1 - \xi)\xi^{-1} \\ &= (\xi - 1)(c_1 - c_2\xi^{-1}). \end{aligned} \tag{51}$$

Thus $c_2 = c_1\xi$ and we can take $c_1 = \xi^{-\frac{1}{2}}$ and $c_2 = \xi^{\frac{1}{2}}$. Then

$$x_t = \xi^{t-\frac{1}{2}} + \xi^{-(t-\frac{1}{2})}. \tag{52}$$

Equation (46) is

$$\begin{aligned} 0 &= (2\lambda - 1)x_T - x_{T-1} \\ &= (\xi + \xi^{-1} - 1)(\xi^{T-\frac{1}{2}} + \xi^{-(T-\frac{1}{2})}) - (\xi^{T-\frac{3}{2}} + \xi^{-(T-\frac{3}{2})}) \\ &= \xi^{T+\frac{1}{2}} + \xi^{-(T+\frac{1}{2})} - \xi^{T-\frac{1}{2}} - \xi^{-(T-\frac{1}{2})} \\ &= (\xi^T - \xi^{-T})(\xi^{\frac{1}{2}} - \xi^{-\frac{1}{2}}). \end{aligned} \tag{53}$$

Thus either $\xi = 1$ or $\xi^{2T} = 1$. The roots of $\xi^{2T} = 1$ are $\xi = e^{i2\pi s/(2T)} = e^{is\pi/T}$, $s = 0, 1, \ldots, 2T-1$. Since $e^{i2\pi s/(2T)} = e^{-i2\pi(2T-s)/(2T)}$, there are $T+1$ different values of $\lambda = \frac{1}{2}(\xi + \xi^{-1})$ for $\xi = e^{i2\pi s/(2T)}$, $s = 0, 1, \ldots, T$.

However, the value of $\lambda = -1$ corresponding to $\xi = -1$ for $s = T$ is inadmissible because $x_t \equiv 0$. The remaining λ's are characteristic roots of A_1 and since there are T characteristic roots, these must constitute all of them. It will be convenient to multiply $e^{i(2t-1)\pi s/(2T)} + e^{-i(2t-1)\pi s/(2T)}$ by $\frac{1}{2}$ to obtain the tth component of the sth characteristic vector.

THEOREM 6.5.4. *The characteristic roots of* A_1 *given by* (41) *are* $\cos \pi s/T$, $s = 0, 1, \ldots, T-1$, *and the corresponding characteristic vectors are*

$$
(54) \qquad \begin{bmatrix} \cos \dfrac{\pi s}{2T} \\ \cos \dfrac{3\pi s}{2T} \\ \cdot \\ \cdot \\ \cdot \\ \cos \dfrac{(2T-1)\pi s}{2T} \end{bmatrix}.
$$

COROLLARY 6.5.1. *The characteristic roots of* A_j *implied by* (16) *and* (17) *and based on* A_1 *given by* (41) *are* $\cos \pi js/T$, $s = 0, 1, \ldots, T-1$, *and the corresponding characteristic vectors are* (54).

PROOF. A_j is defined as a polynomial in A_1, say $P_j(A_1)$, and the characteristic roots of A_j are $P_j(\lambda_s)$, where λ_s are the characteristic roots of A_1, $s = 0, 1, \ldots, T-1$, by Theorem 6.5.1. The proof of Theorem 6.5.3, where $A_1 = \frac{1}{2}(B + B')$, implies $P_j(\cos \nu) = \cos j\nu$. Corollary 6.5.1 now follows from Theorem 6.5.4.

COROLLARY 6.5.2. *The orthogonal matrix that diagonalizes* A_j *of Corollary* 6.5.1 *is*

$$
(55) \qquad \sqrt{\frac{2}{T}} \begin{bmatrix} \dfrac{1}{\sqrt{2}} & \cos \dfrac{\pi}{2T} & \cos \dfrac{2\pi}{2T} & \cdots & \cos \dfrac{\pi(T-1)}{2T} \\ \dfrac{1}{\sqrt{2}} & \cos \dfrac{3\pi}{2T} & \cos \dfrac{6\pi}{2T} & \cdots & \cos \dfrac{3\pi(T-1)}{2T} \\ \cdot & \cdot & \cdot & & \cdot \\ \cdot & \cdot & \cdot & & \cdot \\ \cdot & \cdot & \cdot & & \cdot \\ \dfrac{1}{\sqrt{2}} & \cos \dfrac{(2T-1)\pi}{2T} & \cos \dfrac{(2T-1)2\pi}{2T} & \cdots & \cos \dfrac{(2T-1)\pi(T-1)}{2T} \end{bmatrix}.
$$

All of the matrices are diagonalized by the same orthogonal matrix. In terms of the transformed variables, the jth quadratic form is

$$Q_j = \sum_{s=1}^{T} \cos \frac{\pi j}{T}(s-1)\, z_s^2, \qquad j = 0, 1, \ldots, p. \tag{56}$$

The matrices developed in this section are closely related to the matrices used in the variate-difference method (Section 3.4). Let

$$S_r = \sum_{t=1}^{T-r} (\Delta^r y_t)^2 = \sum_{s,t=1}^{T} c_{st}^{(r)} y_s y_t, \tag{57}$$

and $C_r = (c_{st}^{(r)})$. Then C_r is approximately C_1^r; in fact, C_r and C_1^r differ only in the elements for s, $t \leq r$ and $(T - s + 1)$, $(T - t + 1) \leq r$. The matrices C_1, C_2, C_3, and C_4 were given in (43), (44), (45), and (46) of Section 3.4 (denoted by A_1, A_2, A_3, and A_4). We have

$$C_1^2 = \begin{bmatrix} 2 & -3 & 1 & 0 & \cdots & 0 \\ -3 & 6 & -4 & 1 & \cdots & 0 \\ 1 & -4 & 6 & -4 & \cdots & 0 \\ 0 & 1 & -4 & 6 & \cdots & 0 \\ \cdot & \cdot & \cdot & \cdot & & \cdot \\ \cdot & \cdot & \cdot & \cdot & & \cdot \\ \cdot & \cdot & \cdot & \cdot & & \cdot \\ 0 & 0 & 0 & 0 & \cdots & 2 \end{bmatrix}, \tag{58}$$

$$C_1^3 = \begin{bmatrix} 5 & -9 & 5 & -1 & 0 & \cdots & 0 \\ -9 & 19 & -15 & 6 & -1 & \cdots & 0 \\ 5 & -15 & 20 & -15 & 6 & \cdots & 0 \\ -1 & 6 & -15 & 20 & -15 & \cdots & 0 \\ 0 & -1 & 6 & -15 & 20 & \cdots & 0 \\ \cdot & \cdot & \cdot & \cdot & \cdot & & \cdot \\ \cdot & \cdot & \cdot & \cdot & \cdot & & \cdot \\ \cdot & \cdot & \cdot & \cdot & \cdot & & \cdot \\ 0 & 0 & 0 & 0 & 0 & \cdots & 5 \end{bmatrix}, \tag{59}$$

(60)
$$C_1^4 = \begin{bmatrix} 14 & -28 & 20 & -7 & 1 & 0 & \cdots & 0 \\ -28 & 62 & -55 & 28 & -8 & 1 & \cdots & 0 \\ 20 & -55 & 70 & -56 & 28 & -8 & \cdots & 0 \\ -7 & 28 & -56 & 70 & -56 & 28 & \cdots & 0 \\ 1 & -8 & 28 & -56 & 70 & -56 & \cdots & 0 \\ 0 & 1 & -8 & 28 & -56 & 70 & \cdots & 0 \\ \cdot & \cdot & \cdot & \cdot & \cdot & \cdot & & \cdot \\ \cdot & \cdot & \cdot & \cdot & \cdot & \cdot & & \cdot \\ \cdot & \cdot & \cdot & \cdot & \cdot & \cdot & & \cdot \\ 0 & 0 & 0 & 0 & 0 & 0 & \cdots & 14 \end{bmatrix}.$$

Since $C_1 = 2(I - A_1)$, $C_1^r = 2^r(I - A_1)^r$ has (54) as its sth characteristic vector corresponding to the characteristic root $2^r(1 - \cos \pi s/T)^r$. Since A_j is a polynomial in A_1 of degree j, C_1^r is a linear combination of $A_0 = I$, $A_1, \ldots, A_r$; in fact

(61)
$$C_1^r = \binom{2r}{r} I + 2 \sum_{j=1}^{r} (-1)^j \binom{2r}{r+j} A_j.$$

6.5.4. Another Model

It is appealing to take Q_1 as $\sum_{t=2}^{T} y_t y_{t-1}$, since the latter is the desired sum of lagged products. The matrix of this quadratic form is

(62)
$$A_1 = \tfrac{1}{2} \begin{bmatrix} 0 & 1 & 0 & \cdots & 0 \\ 1 & 0 & 1 & \cdots & 0 \\ 0 & 1 & 0 & \cdots & 0 \\ \cdot & \cdot & \cdot & & \cdot \\ \cdot & \cdot & \cdot & & \cdot \\ \cdot & \cdot & \cdot & & \cdot \\ 0 & 0 & 0 & \cdots & 0 \end{bmatrix}.$$

Another system of matrices can be generated by the polynomials (18) [implied by (16) and (17)] using $\boldsymbol{A}_1$ defined in (62). These are

$$
\text{(63)} \qquad \boldsymbol{A}_2 = 2\boldsymbol{A}_1^2 - \boldsymbol{I} = \frac{1}{2}\begin{bmatrix} -1 & 0 & 1 & 0 & \cdots & 0 \\ 0 & 0 & 0 & 1 & \cdots & 0 \\ 1 & 0 & 0 & 0 & \cdots & 0 \\ 0 & 1 & 0 & 0 & \cdots & 0 \\ \cdot & \cdot & \cdot & \cdot & & \cdot \\ \cdot & \cdot & \cdot & \cdot & & \cdot \\ \cdot & \cdot & \cdot & \cdot & & \cdot \\ 0 & 0 & 0 & 0 & \cdots & -1 \end{bmatrix},
$$

$$
\text{(64)} \qquad \boldsymbol{A}_3 = 4\boldsymbol{A}_1^3 - 3\boldsymbol{A}_1 = \frac{1}{2}\begin{bmatrix} 0 & -1 & 0 & 1 & 0 & \cdots & 0 \\ -1 & 0 & 0 & 0 & 1 & \cdots & 0 \\ 0 & 0 & 0 & 0 & 0 & \cdots & 0 \\ 1 & 0 & 0 & 0 & 0 & \cdots & 0 \\ 0 & 1 & 0 & 0 & 0 & \cdots & 0 \\ \cdot & \cdot & \cdot & \cdot & \cdot & & \cdot \\ \cdot & \cdot & \cdot & \cdot & \cdot & & \cdot \\ \cdot & \cdot & \cdot & \cdot & \cdot & & \cdot \\ 0 & 0 & 0 & 0 & 0 & \cdots & 0 \end{bmatrix}.
$$

It will be observed that $\boldsymbol{A}_j$ here differs from the desired form $\sum_t y_t y_{t-j}$ only by $j - 1$ $\frac{1}{2}$'s in the upper left-hand and lower right-hand corners of the matrix.

A characteristic vector $\boldsymbol{x}$ with components $x_1, \ldots, x_T$ corresponding to a root λ satisfies the second-order difference equation (45) and hence is of the form (50) $x_t = c_1\xi^t + c_2\xi^{-t}$ with $2\lambda = \xi + \xi^{-1}$. The equation $x_2 = 2\lambda x_1$ is

$$
\text{(65)} \qquad c_1\xi^2 + c_2\xi^{-2} = (\xi + \xi^{-1})(c_1\xi + c_2\xi^{-1}),
$$

which implies $c_1 + c_2 = 0$. We can take $c_1 = 1$ and $c_2 = -1$. Then $x_t = \xi^t - \xi^{-t}$. The equation $x_{T-1} = 2\lambda x_T$ is

$$\xi^{T-1} - \xi^{-(T-1)} = (\xi + \xi^{-1})(\xi^T - \xi^{-T}), \tag{66}$$

which implies $\xi^{T+1} - \xi^{-(T+1)} = 0$; that is, $\xi^{2(T+1)} = 1$. The roots of this equation are the $2(T+1)$st roots of 1, namely, $e^{i2\pi s/[2(T+1)]}$, $s = 0, 1, \ldots, 2T+1$. ($\xi = 1$ and $\xi = -1$ are inadmissible values because then $x_t = 0$, $t = 1, \ldots, T$, which is trivial.) The sth characteristic root is

$$\begin{aligned}\lambda &= \tfrac{1}{2}(\xi + \xi^{-1}) \\ &= \tfrac{1}{2}(e^{i2\pi s/[2(T+1)]} + e^{-i2\pi s/[2(T+1)]}) \\ &= \cos \frac{\pi s}{T+1}, \qquad s = 1, \ldots, T,\end{aligned} \tag{67}$$

and the tth component of the corresponding characteristic vector is taken as

$$\begin{aligned}\frac{1}{2i} x_t &= \frac{1}{2i}(e^{i2\pi st/[2(T+1)]} - e^{-i2\pi st/[2(T+1)]}) \\ &= \sin \frac{\pi st}{T+1}.\end{aligned} \tag{68}$$

THEOREM 6.5.5. *The characteristic roots of A_1 given by* (62) *are* $\cos \pi s/(T+1)$, $s = 1, \ldots, T$, *and the corresponding characteristic vectors are*

$$\begin{pmatrix} \sin \dfrac{\pi s}{T+1} \\ \sin \dfrac{2\pi s}{T+1} \\ \cdot \\ \cdot \\ \cdot \\ \sin \dfrac{T\pi s}{T+1} \end{pmatrix}. \tag{69}$$

COROLLARY 6.5.3. *The characteristic roots of A_j implied by* (16) *and* (17) *and based on A_1 given by* (62) *are* $\cos \pi js/(T+1)$, $s = 1, \ldots, T$, *and the corresponding characteristic vectors are* (69).

COROLLARY 6.5.4. *The orthogonal matrix that diagonalizes A_j of Corollary 6.5.3 is*

$$\sqrt{\frac{2}{T+1}}\begin{pmatrix} \sin\frac{\pi}{T+1} & \sin\frac{2\pi}{T+1} & \cdots & \sin\frac{T\pi}{T+1} \\ \sin\frac{2\pi}{T+1} & \sin\frac{4\pi}{T+1} & \cdots & \sin\frac{2T\pi}{T+1} \\ \cdot & \cdot & & \cdot \\ \cdot & \cdot & & \cdot \\ \cdot & \cdot & & \cdot \\ \sin\frac{T\pi}{T+1} & \sin\frac{2T\pi}{T+1} & \cdots & \sin\frac{T^2\pi}{T+1} \end{pmatrix}. \tag{70}$$

The result of transformation by the above orthogonal matrix is the quadratic form

$$Q_j = \sum_{s=1}^{T} \cos\frac{\pi j}{T+1} s\, z_s^2, \qquad j = 0, 1, \ldots, p. \tag{71}$$

It may be observed that the roots here for $T = T^*$ correspond to the roots in the preceding case for $T = T^* + 1$ except for the omission of the root 1 (corresponding to the characteristic vector with all components equal). However, the characteristic vectors are different. The orthogonal matrix (70) is symmetric.

6.5.5. Models with Double Roots

In Section 6.7 we shall study the distributions of the serial correlations. In the case of independence ($\gamma_1 = \cdots = \gamma_p = 0$) we shall find that the distribution of a serial correlation can be found explicitly in a relatively simple form if the characteristic roots of the numerator quadratic form occur in pairs with possibly one single root. For example, this is the case in the circular model for T odd and $j = 1$. In the other two types of models the roots do not occur in pairs when $j = 1$, and the distribution of r_1 cannot be written in closed form for arbitrary values of T.

It has been suggested by Watson and Durbin (1951) that matrices $\boldsymbol{A}_1$ be chosen so that the roots occur in pairs. This can be done by taking $\boldsymbol{A}_1$ as

$$\boldsymbol{A}_1 = \begin{pmatrix} \boldsymbol{A}_1^* & \boldsymbol{0} \\ \boldsymbol{0} & \boldsymbol{A}_1^* \end{pmatrix} \tag{72}$$

in case T is even and

$$\boldsymbol{A}_1 = \begin{pmatrix} \boldsymbol{A}_1^* & \boldsymbol{0} & \boldsymbol{0} \\ \boldsymbol{0} & 0 & \boldsymbol{0} \\ \boldsymbol{0} & \boldsymbol{0} & \boldsymbol{A}_1^* \end{pmatrix} \tag{73}$$

in case T is odd; $\boldsymbol{A}_1^*$ is a matrix with all roots different and in (73) the center 0 is a scalar. In (72) $\boldsymbol{A}_1$ will have all roots in pairs. In (73) all roots of $\boldsymbol{A}_1$ will occur in pairs except for the root 0; it will occur as a single root if 0 is not a root of $\boldsymbol{A}_1^*$, and it will occur as a triple root if 0 is a root of $\boldsymbol{A}_1^*$. As an example, if $\boldsymbol{A}_1^*$ is taken from Section 6.5.4, then

$$Q_1 = \sum_{t=2}^{\frac{1}{2}T} y_t y_{t-1} + \sum_{t=\frac{1}{2}T+2}^{T} y_t y_{t-1}, \qquad T \text{ even}, \tag{74}$$

$$Q_1 = \sum_{t=2}^{\frac{1}{2}(T-1)} y_t y_{t-1} + \sum_{t=\frac{1}{2}(T+1)+2}^{T} y_t y_{t-1}, \qquad T \text{ odd}. \tag{75}$$

If $\boldsymbol{A}_j$ is the polynomial $P_j(\boldsymbol{A}_1)$, then

$$\boldsymbol{A}_j = P_j(\boldsymbol{A}_1) = \begin{pmatrix} P_j(\boldsymbol{A}_1^*) & \boldsymbol{0} \\ \boldsymbol{0} & P_j(\boldsymbol{A}_1^*) \end{pmatrix}, \tag{76}$$

or

$$\boldsymbol{A}_j = P_j(\boldsymbol{A}_1) = \begin{pmatrix} P_j(\boldsymbol{A}_1^*) & \boldsymbol{0} & \boldsymbol{0} \\ \boldsymbol{0} & P_j(0) & \boldsymbol{0} \\ \boldsymbol{0} & \boldsymbol{0} & P_j(\boldsymbol{A}_1^*) \end{pmatrix}. \tag{77}$$

6.6. CASES OF UNKNOWN MEANS

6.6.1. Constant Mean

In most cases of statistical interest the means $\mathscr{E}y_t$ are constant, say μ, but unknown. The modification of the density

$$Ke^{-\frac{1}{2}(\gamma_0 Q_0 + \cdots + \gamma_p Q_p)}, \tag{1}$$

where $Q_j = y'A_jy$, to take this into account is to replace the vector y by $y - \mu\varepsilon$, where ε is the vector with each component 1. Then the exponent in the normal density is $-\frac{1}{2}$ times

$$(2)\quad \gamma_0(y-\mu\varepsilon)'A_0(y-\mu\varepsilon)+\cdots+\gamma_p(y-\mu\varepsilon)'A_p(y-\mu\varepsilon)$$
$$=\gamma_0Q_0+\cdots+\gamma_pQ_p-2\mu(\gamma_0\varepsilon'A_0y+\cdots+\gamma_p\varepsilon'A_py)$$
$$+\mu^2\varepsilon'(\gamma_0A_0+\cdots+\gamma_pA_p)\varepsilon.$$

A sufficient set of statistics for the parameters $\gamma_0, \ldots, \gamma_p$, and μ is $Q_0, \ldots, Q_p$ and $\varepsilon'A_0y, \ldots, \varepsilon'A_py$. If ε is a characteristic vector of A_j associated with the characteristic root λ_j, $j = 0, 1, \ldots, p$, then (2) is

$$(3)\quad \gamma_0Q_0+\cdots+\gamma_pQ_p-2\mu(\gamma_0\varepsilon'A_0+\cdots+\gamma_p\varepsilon'A_p)y$$
$$+\mu^2(\gamma_0\varepsilon'A_0\varepsilon+\cdots+\gamma_p\varepsilon'A_p\varepsilon)$$
$$=\gamma_0Q_0+\cdots+\gamma_pQ_p-2\mu(\gamma_0\lambda_0+\cdots+\gamma_p\lambda_p)\varepsilon'y+\mu^2(\gamma_0\lambda_0+\cdots+\gamma_p\lambda_p)T,$$

and $Q_0, \ldots, Q_p$, and $\bar{y} = \varepsilon'y/T$ form a sufficient set of statistics for the parameters $\gamma_0, \ldots, \gamma_p$, and μ. An equivalent sufficient set of statistics is $Q_0^*, \ldots, Q_p^*$, and $\bar{y}$, where

$$(4)\quad Q_j^* = (y-\bar{y}\varepsilon)'A_j(y-\bar{y}\varepsilon) = y'A_jy - T\lambda_j\bar{y}^2.$$

If ε is not a characteristic vector of A_j, $j = 0, 1, \ldots, p$, then a minimal sufficient set of statistics is $Q_0, \ldots, Q_p$ and a linearly independent subset (not necessarily proper) of $\varepsilon'A_0y, \ldots, \varepsilon'A_py$.

As an example, in the circular case

$$(5)\quad Q_1^* = \sum_{t=2}^{T} y_ty_{t-1} + y_Ty_1 - T\bar{y}^2.$$

In the model based on successive differences

$$(6)\quad Q_1^* = \sum_{t=2}^{T} y_ty_{t-1} + \tfrac{1}{2}y_1^2 + \tfrac{1}{2}y_T^2 - T\bar{y}^2.$$

In the third model considered, ε is not a characteristic vector of A_1, and hence Q_1^* cannot be written as the right-hand side of (4). In the cases of models based on using A_1^* in the partitioned matrix, when T is even, ε (of order T) will be a characteristic vector of A_1 if ε (of order $\frac{1}{2}T$) is a characteristic vector of A_1^*; when T is odd, ε (of order T) will be a characteristic vector of A_1 if ε [of order $\frac{1}{2}(T-1)$] is a characteristic vector of A_1 corresponding to a root of 0. (Note the $\frac{1}{2}(T+1)$st component of $A_1\varepsilon$ must be 0.)

In the remainder of Section 6.6.1 we shall assume that $\boldsymbol{\varepsilon}$ is a characteristic vector of $\boldsymbol{A}_j$ corresponding to the root λ_j, $j = 0, 1, \ldots, p$. Then $\boldsymbol{\varepsilon}$ is a characteristic vector of $\sum_{j=0}^{p} \gamma_j \boldsymbol{A}_j$ corresponding to the root $\sum_{j=0}^{p} \gamma_j \lambda_j$. The least squares estimate $\bar{y}$ of μ [which minimizes $\sum_{t=1}^{T} (y_t - m)^2$] is the Markov estimate (see Section 2.4); that is, $\bar{y}$ is the best linear unbiased estimate of μ.

THEOREM 6.6.1. *The mean $\bar{y}$ and the vector of residuals $\boldsymbol{y} - \bar{y}\boldsymbol{\varepsilon}$ are independently distributed if $\boldsymbol{\varepsilon}$ is a characteristic vector of $\boldsymbol{A}_j$, $j = 0, 1, \ldots, p$.*

PROOF. Since $\bar{y}$ and the components of $\boldsymbol{y} - \bar{y}\boldsymbol{\varepsilon}$ consist of linear combinations of the components of $\boldsymbol{y}$ which has a normal distribution, $\bar{y}$ and $\boldsymbol{y} - \bar{y}\boldsymbol{\varepsilon}$ have a joint (singular) normal distribution. The covariance matrix $\boldsymbol{\Sigma}$ of $\boldsymbol{y}$ is found from $\boldsymbol{\Sigma}^{-1} = \sum_{j=0}^{p} \gamma_j \boldsymbol{A}_j$. Because

$$
\begin{aligned}
(7) \quad \mathscr{E}\bar{y}(\boldsymbol{y} - \bar{y}\boldsymbol{\varepsilon}) &= \mathscr{E}[(\boldsymbol{y} - \mu\boldsymbol{\varepsilon}) - (\bar{y} - \mu)\boldsymbol{\varepsilon}](\bar{y} - \mu) \\
&= \mathscr{E}\left[\frac{1}{T}(\boldsymbol{y} - \mu\boldsymbol{\varepsilon})(\boldsymbol{y} - \mu\boldsymbol{\varepsilon})'\boldsymbol{\varepsilon} - \boldsymbol{\varepsilon}\frac{1}{T^2}\boldsymbol{\varepsilon}'(\boldsymbol{y} - \mu\boldsymbol{\varepsilon})(\boldsymbol{y} - \mu\boldsymbol{\varepsilon})'\boldsymbol{\varepsilon}\right] \\
&= \frac{1}{T}\boldsymbol{\Sigma}\boldsymbol{\varepsilon} - \boldsymbol{\varepsilon}\frac{1}{T^2}\boldsymbol{\varepsilon}'\boldsymbol{\Sigma}\boldsymbol{\varepsilon} \\
&= \frac{1}{T}\left(\sum_{j=0}^{p}\gamma_j\lambda_j\right)^{-1}\boldsymbol{\varepsilon} - \boldsymbol{\varepsilon}\frac{1}{T^2}\left(\sum_{j=0}^{p}\gamma_j\lambda_j\right)^{-1}T \\
&= \boldsymbol{0},
\end{aligned}
$$

$\bar{y}$ is uncorrelated with the components of $\boldsymbol{y} - \bar{y}\boldsymbol{\varepsilon}$ and hence is independent of $\boldsymbol{y} - \bar{y}\boldsymbol{\varepsilon}$.

COROLLARY 6.6.1. *The mean $\bar{y}$ is distributed independently of the quadratic forms in the residuals, $Q_0^*, \ldots, Q_p^*$.*

The development of uniformly most powerful tests in Section 6.3 and multiple-decision procedures in Section 6.4 can be extended to these models in which $\mathscr{E}y_t = \mu$ and $\boldsymbol{\varepsilon}$ is a characteristic vector of $\boldsymbol{A}_j$, $j = 0, 1, \ldots, p$. For in this case $Q_0^*, \ldots, Q_i^*$, and $\bar{y}$ form a sufficient set of statistics for $\gamma_0, \ldots, \gamma_i$ and μ when $\gamma_{i+1} = \cdots = \gamma_p = 0$, $i = 0, 1, \ldots, p$. The rejection region S_i^* (in the space of $Q_0^*, \ldots, Q_i^*, \bar{y}$) of a similar test of the hypothesis $\gamma_i = \gamma_i^{(1)}$ at significance level ε_i (when $\gamma_{i+1} = \cdots = \gamma_p = 0$) has Neyman structure; that is, it satisfies

$$
(8) \qquad \Pr\{S_i^* \mid Q_0^*, \ldots, Q_{i-1}^*, \bar{y}; \gamma_0, \ldots, \gamma_{i-1}, \gamma_i^{(1)}, 0, \ldots, 0, \mu\} = \varepsilon_i
$$

almost everywhere in the set of possible $Q_0^*, \ldots, Q_{i-1}^*$, and $\bar{y}$. Since $Q_0^*, \ldots, Q_i^*$ are statistically independent of $\bar{y}$, the conditional probability of Q_i^* falling in S_i^* given the values of $Q_0^*, \ldots, Q_{i-1}^*$, and $\bar{y}$ does not depend on the given value of $\bar{y}$, and since $Q_0^*, \ldots, Q_i^*$ are functions of the residuals $\mathbf{y} - \bar{y}\boldsymbol{\varepsilon}$, the distribution of which does not depend on μ, the probability in (8) does not depend on μ. Hence we can write (8) as

$$\Pr\{S_i^* \mid Q_0^*, \ldots, Q_{i-1}^*; \gamma_0, \ldots, \gamma_{i-1}, \gamma_i^{(1)}, 0, \ldots, 0\} = \varepsilon_i. \tag{9}$$

The joint density of $Q_0^*, \ldots, Q_i^*$ (when $\gamma_{i+1} = \cdots = \gamma_p = 0$) can be written

$$\begin{aligned} &h_i^*(Q_0^*, \ldots, Q_i^* \mid \gamma_0, \ldots, \gamma_i) \\ &\quad = K_i^*(\gamma_0, \ldots, \gamma_i) \exp\left[-\tfrac{1}{2}(\gamma_0 Q_0^* + \cdots + \gamma_i Q_i^*)\right] k_i^*(Q_0^*, \ldots, Q_i^*). \end{aligned} \tag{10}$$

The marginal density of $Q_0^*, \ldots, Q_{i-1}^*$ is

$$K_i^*(\gamma_0, \ldots, \gamma_i) \exp\left[-\tfrac{1}{2}(\gamma_0 Q_0^* + \cdots + \gamma_{i-1} Q_{i-1}^*)\right] m_i^*(Q_0^*, \ldots, Q_{i-1}^*; \gamma_i), \tag{11}$$

where

$$m_i^*(Q_0^*, \ldots, Q_{i-1}^*; \gamma_i) = \int_{-\infty}^{\infty} e^{-\frac{1}{2}\gamma_i Q_i^*} k_i^*(Q_0^*, \ldots, Q_i^*)\, dQ_i^*. \tag{12}$$

The conditional density of Q_i^*, given $Q_0^*, \ldots, Q_{i-1}^*$, is

$$h_i^*(Q_i^* \mid Q_0^*, \ldots, Q_{i-1}^*; \gamma_i) = \exp\left[-\tfrac{1}{2}\gamma_i Q_i^*\right] \frac{k_i^*(Q_0^*, \ldots, Q_i^*)}{m_i^*(Q_0^*, \ldots, Q_{i-1}^*; \gamma_i)}. \tag{13}$$

The conditional density will only be written for the set of $Q_0^*, \ldots, Q_{i-1}^*$ for which $k_{i-1}^*(Q_0^*, \ldots, Q_{i-1}^*) > 0$, and hence for which $m_i^*(Q_0^*, \ldots, Q_{i-1}^*; \gamma_i) > 0$. The Neyman-Pearson Fundamental Lemma can then be applied.

THEOREM 6.6.2. *The best similar test of the null hypothesis* $\gamma_i = \gamma_i^{(1)}$ *against the alternative* $\gamma_i = \gamma_i^{(2)} < \gamma_i^{(1)}$ *at significance level* ε_i *is given by*

$$Q_i^* > c_i^*(Q_0^*, \ldots, Q_{i-1}^*; \gamma_i^{(1)}), \tag{14}$$

where $c_i^*(Q_0^*, \ldots, Q_{i-1}^*; \gamma_i^{(1)})$ *is determined so the probability of* (14) *according to the density* (13) *is* ε_i *when* $\gamma_i = \gamma_i^{(1)}$.

THEOREM 6.6.3. *The uniformly most powerful similar test of the null hypothesis* $\gamma_i = \gamma_i^{(1)}$ *against alternatives* $\gamma_i < \gamma_i^{(1)}$ *at significance level* ε_i *has rejection region* (14), *where* $c_i^*(Q_0^*, \ldots, Q_{i-1}^*; \gamma_i^{(1)})$ *is determined so the probability of* (14) *according to the density* (13) *is* ε_i *when* $\gamma_i = \gamma_i^{(1)}$.

COROLLARY 6.6.2. *The uniformly most powerful similar test of the null hypothesis* $\gamma_i = 0$ *against alternatives* $\gamma_i < 0$ *at significance level* ε_i *has rejection region*

$$Q_i^* > c_i^*(Q_0^*, \ldots, Q_{i-1}^*), \tag{15}$$

where $c_i^*(Q_0^*, \ldots, Q_{i-1}^*)$ *is determined so the integral of*

$$h_i^*(Q_i^* \mid Q_0^*, \ldots, Q_{i-1}^*; 0) = \frac{k_i^*(Q_0^*, \ldots, Q_i^*)}{k_{i-1}^*(Q_0^*, \ldots, Q_{i-1}^*)} \tag{16}$$

over (15) *is* ε_i.

The best test against an alternative $\gamma_i = \gamma_i^{(2)} > \gamma_i^{(1)}$ has the rejection region

$$Q_i^* < c_i^{*\prime}(Q_0^*, \ldots, Q_{i-1}^*; \gamma_i^{(1)}). \tag{17}$$

THEOREM 6.6.4. *The uniformly most powerful unbiased test of the null hypothesis* $\gamma_i = \gamma_i^{(1)}$ *against alternatives* $\gamma_i \neq \gamma_i^{(1)}$ *at significance level* ε_i *has the rejection region*

$$\begin{aligned} Q_i^* &> c_{Ui}^*(Q_0^*, \ldots, Q_{i-1}^*; \gamma_i^{(1)}), \\ Q_i^* &< c_{Li}^*(Q_0^*, \ldots, Q_{i-1}^*; \gamma_i^{(1)}), \end{aligned} \tag{18}$$

where $c_{Ui}^*(Q_0^*, \ldots, Q_{i-1}^*;\ \gamma_i^{(1)})$ *and* $c_{Li}^*(Q_0^*, \ldots, Q_{i-1}^*;\ \gamma_i^{(1)})$ *are determined so that*

$$\int_{c_{Li}^*(Q_0^*, \ldots, Q_{i-1}^*; \gamma_i^{(1)})}^{c_{Ui}^*(Q_0^*, \ldots, Q_{i-1}^*; \gamma_i^{(1)})} h_i^*(Q_i^* \mid Q_0^*, \ldots, Q_{i-1}^*; \gamma_i^{(1)})\, dQ_i^* = 1 - \varepsilon_i \tag{19}$$

and

$$\begin{aligned} \int_{c_{Li}^*(Q_0^*, \ldots, Q_{i-1}^*; \gamma_i^{(1)})}^{c_{Ui}^*(Q_0^*, \ldots, Q_{i-1}^*; \gamma_i^{(1)})} & Q_i^* h_i^*(Q_i^* \mid Q_0^*, \ldots, Q_{i-1}^*; \gamma_i^{(1)})\, dQ_i^* \\ &= (1 - \varepsilon_i) \int_{-\infty}^{\infty} Q_i^* h_i^*(Q_i^* \mid Q_0^*, \ldots, Q_{i-1}^*; \gamma_i^{(1)})\, dQ_i^*. \end{aligned} \tag{20}$$

COROLLARY 6.6.3. *The uniformly most powerful unbiased test of the null hypothesis* $\gamma_i = 0$ *against alternatives* $\gamma_i \neq 0$ *at significance level* ε_i *has the rejection region*

$$\begin{aligned} Q_i^* &> c_{Ui}^*(Q_0^*, \ldots, Q_{i-1}^*), \\ Q_i^* &< c_{Li}^*(Q_0^*, \ldots, Q_{i-1}^*), \end{aligned} \tag{21}$$

where $c_{Ui}^*(Q_0^*, \ldots, Q_{i-1}^*)$ *and* $c_{Li}^*(Q_0^*, \ldots, Q_{i-1}^*)$ *are determined by*

$$\int_{c_{Li}^*(Q_0^* \ldots, Q_{i-1}^*)}^{c_{Ui}^*(Q_0^*, \ldots, Q_{i-1}^*)} \frac{k_i^*(Q_0^*, \ldots, Q_i^*)}{k_{i-1}^*(Q_0^*, \ldots, Q_{i-1}^*)}\, dQ_i^* = 1 - \varepsilon_i \tag{22}$$

and

$$\begin{aligned} \int_{c_{Li}^*(Q_0^*, \ldots, Q_{i-1}^*)}^{c_{Ui}^*(Q_0^*, \ldots, Q_{i-1}^*)} & Q_i^* \frac{k_i^*(Q_0^*, \ldots, Q_i^*)}{k_{i-1}^*(Q_0^*, \ldots, Q_{i-1}^*)}\, dQ_i^* \\ &= (1 - \varepsilon_i) \int_{-\infty}^{\infty} Q_i^* \frac{k_i^*(Q_0^*, \ldots, Q_i^*)}{k_{i-1}^*(Q_0^*, \ldots, Q_{i-1}^*)}\, dQ_i^*. \end{aligned} \tag{23}$$

If the conditional density of Q_i^*, given $Q_0^*, \ldots, Q_{i-1}^*$ for $\gamma_i = 0$, is symmetric, then (23) is satisfied for

$$c_{Li}^*(Q_0^*, \ldots, Q_{i-1}^*) = -c_{Ui}^*(Q_0^*, \ldots, Q_{i-1}^*), \tag{24}$$

since both sides are zero, and (22) can be written

$$\int_{c_{Ui}^*(Q_0^*, \ldots, Q_{i-1}^*)}^{\infty} \frac{k_i^*(Q_0^*, \ldots, Q_i^*)}{k_{i-1}^*(Q_0^*, \ldots, Q_{i-1}^*)}\, dQ_i^* = \tfrac{1}{2}\varepsilon_i. \tag{25}$$

It should be noted that the conditional distribution of Q_i^*, given $Q_0^*, \ldots, Q_{i-1}^*$, does not depend on the values of $\gamma_0, \ldots, \gamma_{i-1}$. Hence these parameters can be assigned convenient values, such as $\gamma_0 = 1$, $\gamma_1 = \cdots = \gamma_{i-1} = 0$.

Usually Q_0^* will be $\sum_{t=1}^T (y_t - \bar{y})^2 = \sum_{t=1}^T y_t^2 - T\bar{y}^2$. The ratio (for $\lambda_j = 1$)

$$r_j^* = \frac{Q_j^*}{Q_0^*} = \frac{\sum_{s,t=1}^T a_{st}^{(j)}(y_s - \bar{y})(y_t - \bar{y})}{\sum_{t=1}^T (y_t - \bar{y})^2} = \frac{\sum_{s,t=1}^T a_{st}^{(j)} y_s y_t - T\bar{y}^2}{\sum_{t=1}^T y_t^2 - T\bar{y}^2} \tag{26}$$

can be considered as a *serial correlation coefficient*. When $\gamma_0 > 0$ and $\gamma_1 = \cdots = \gamma_i = 0$, the joint distribution of $r_1^*, \ldots, r_i^*$ is independent of γ_0 and Q_0^*. (See Theorem 6.7.2.) The definitions of the optimum tests (15) and (21) can then be given in terms of r_i^* conditional on $r_1^*, \ldots, r_{i-1}^*$.

The theorems of Section 6.4 on optimum multiple-decision procedures can similarly be phrased in terms of $Q_0^*, \ldots, Q_p^*$ when $\mathscr{E} y_t = \mu$ and $\boldsymbol{\varepsilon}$ is a characteristic vector of $\boldsymbol{A}_0, \ldots, \boldsymbol{A}_p$.

6.6.2. Certain Regression Functions

More general mean value functions can be included in the treatment of nonzero means. Suppose

$$\mathscr{E}\boldsymbol{y} = \sum_{\nu=1}^m \alpha_\nu \boldsymbol{v}_{s_\nu}, \tag{27}$$

where $\boldsymbol{v}_1, \ldots, \boldsymbol{v}_T$ are characteristic vectors,

$$\boldsymbol{A}_j \boldsymbol{v}_t = \lambda_{jt} \boldsymbol{v}_t, \qquad t = 1, \ldots, T, \quad j = 0, 1, \ldots, p, \tag{28}$$

and $s_1, \ldots, s_m$ is a subset of $1, \ldots, T$. The vectors are orthogonal: $\boldsymbol{v}_t'\boldsymbol{v}_s = \delta_{ts}$. Then the exponent in the normal density is $-\frac{1}{2}$ times

$$(29)\quad \sum_{j=0}^{p} \gamma_j\left(\boldsymbol{y} - \sum_{\nu=1}^{m} \alpha_\nu \boldsymbol{v}_{s_\nu}\right)' \boldsymbol{A}_j\left(\boldsymbol{y} - \sum_{\nu=1}^{m} \alpha_\nu \boldsymbol{v}_{s_\nu}\right)$$

$$= \sum_{j=0}^{p} \gamma_j \boldsymbol{y}'\boldsymbol{A}_j\boldsymbol{y} - 2\sum_{j=0}^{p}\sum_{\nu=1}^{m} \gamma_j \lambda_{js_\nu}\alpha_\nu \boldsymbol{y}'\boldsymbol{v}_{s_\nu} + \sum_{j=0}^{p}\sum_{\nu=1}^{m} \gamma_j\lambda_{js_\nu}\alpha_\nu^2 \boldsymbol{v}_{s_\nu}'\boldsymbol{v}_{s_\nu}.$$

Thus $Q_0, \ldots, Q_p$ and $\boldsymbol{y}'\boldsymbol{v}_{s_1}, \ldots, \boldsymbol{y}'\boldsymbol{v}_{s_m}$ form a sufficient set of statistics for the parameters $\gamma_0, \ldots, \gamma_p$ and $\alpha_1, \ldots, \alpha_m$. An equivalent sufficient set of statistics is

$$(30)\quad \left(\boldsymbol{y} - \sum_{\nu=1}^{m} \hat{\alpha}_\nu \boldsymbol{v}_{s_\nu}\right)' \boldsymbol{A}_j\left(\boldsymbol{y} - \sum_{\nu=1}^{m} \hat{\alpha}_\nu \boldsymbol{v}_{s_\nu}\right) = \boldsymbol{y}'\boldsymbol{A}_j\boldsymbol{y} - \sum_{\nu=1}^{m} \lambda_{js_\nu}\hat{\alpha}_\nu^2 \boldsymbol{v}_{s_\nu}'\boldsymbol{v}_{s_\nu}, \qquad j = 0, 1, \ldots, p,$$

and

$$(31)\quad \hat{\alpha}_\nu = \frac{\boldsymbol{v}_{s_\nu}'\boldsymbol{y}}{\boldsymbol{v}_{s_\nu}'\boldsymbol{v}_{s_\nu}}, \qquad \nu = 1, \ldots, m.$$

Note that the least squares estimates (31) of α_ν, $\nu = 1, \ldots, m$, obtained by minimizing $(\boldsymbol{y} - \sum_{\nu=1}^m \alpha_\nu \boldsymbol{v}_{s_\nu})'(\boldsymbol{y} - \sum_{\nu=1}^m \alpha_\nu \boldsymbol{v}_{s_\nu})$, are also the Markov estimates, that is, the best linear unbiased estimates, because the $\boldsymbol{v}_{s_\nu}$ are characteristic vectors of $\sum_{j=0}^p \gamma_j \boldsymbol{A}_j$. (See Section 2.4.)

As an example, consider $\boldsymbol{A}_0 = \boldsymbol{I}$, $\boldsymbol{A}_1, \ldots, \boldsymbol{A}_p$ of the circular model of Section 6.5.2. The characteristic roots of $\boldsymbol{A}_j$ are $\cos 2\pi sj/T$, $s = 1, \ldots, T$; and the characteristic vectors have tth components $\cos 2\pi st/T$ and $\sin 2\pi st/T$, $s = 1, \ldots, \frac{1}{2}(T-2)$ if T is even or $\frac{1}{2}(T-1)$ if T is odd; $(1, 1, \ldots, 1)'$ corresponds to the root $1 = \cos 2\pi Tj/T$, and if T is even $(-1, 1, \ldots, 1)'$ corresponds to the root $(-1)^j = \cos 2\pi \frac{1}{2}Tj/T$. These characteristic vectors are the sequences that were used in the periodic trends studied in Chapter 4. Suppose that

$$(32)\quad \mathscr{E} y_t = \mu_t = \alpha_0 + \sum_{h=1}^{q}\left(\alpha(\nu_h)\cos\frac{2\pi\nu_h}{T}t + \beta(\nu_h)\sin\frac{2\pi\nu_h}{T}t\right),$$

where $\nu_1, \ldots, \nu_q$ form a subset of the integers $1, \ldots, \frac{1}{2}(T-2)$ if T is even or $\frac{1}{2}(T-1)$ if T is odd, or suppose that

$$(33)\quad \mathscr{E} y_t = \mu_t = \alpha_0 + \sum_{h=1}^{q}\left(\alpha(\nu_h)\cos\frac{2\pi\nu_h}{T}t + \beta(\nu_h)\sin\frac{2\pi\nu_h}{T}t\right) + \alpha_{\frac{1}{2}T}(-1)^t$$

if T is even. The index m in (27) corresponds to $2q + 1$ for (32) and $2q + 2$ for (33). Then the least squares estimates of the coefficients are given by

$$a(\nu_h) = \frac{2}{T}\sum_{t=1}^{T} y_t \cos\frac{2\pi\nu_h}{T}t, \tag{34}$$
$$b(\nu_h) = \frac{2}{T}\sum_{t=1}^{T} y_t \sin\frac{2\pi\nu_h}{T}t, \qquad \nu_h \neq 0, \tfrac{1}{2}T,$$

$$a_0 = \bar{y}, \tag{35}$$

and

$$a_{\frac{1}{2}T} = \frac{1}{T}\sum_{t=1}^{T} y_t(-1)^t, \tag{36}$$

if T is even. Let

$$m_t = a_0 + \sum_{h=1}^{q}\left(a(\nu_h)\cos\frac{2\pi\nu_h}{T}t + b(\nu_h)\sin\frac{2\pi\nu_h}{T}t\right) \tag{37}$$

or

$$m_t = a_0 + \sum_{h=1}^{q}\left(a(\nu_h)\cos\frac{2\pi\nu_h}{T}t + b(\nu_h)\sin\frac{2\pi\nu_h}{T}t\right) + a_{\frac{1}{2}T}(-1)^t \tag{38}$$

if T is even. The jth-order circular serial correlation coefficient using residuals from the fitted periodic trend is

$$r_j^* = \frac{\sum_{t=1}^{T}(y_t - m_t)(y_{t-j} - m_{t-j})}{\sum_{t=1}^{T}(y_t - m_t)^2} \tag{39}$$
$$= \frac{\sum_{t=1}^{T} y_t y_{t-j} - T\bar{y}^2 - \frac{1}{2}T\sum_{h=1}^{q}\cos\frac{2\pi\nu_h j}{T}\,[a^2(\nu_h) + b^2(\nu_h)]}{\sum_{t=1}^{T} y_t^2 - T\bar{y}^2 - \frac{1}{2}T\sum_{h=1}^{q}[a^2(\nu_h) + b^2(\nu_h)]},$$

or

$$r_j^* = \frac{\sum_{t=1}^{T} y_t y_{t-j} - T[\bar{y}^2 + (-1)^j a_{\frac{1}{2}T}^2] - \frac{1}{2}T\sum_{h=1}^{q}\cos\frac{2\pi\nu_h j}{T}\,[a^2(\nu_h) + b^2(\nu_h)]}{\sum_{t=1}^{T} y_t^2 - T[\bar{y}^2 + a_{\frac{1}{2}T}^2] - \frac{1}{2}T\sum_{h=1}^{q}[a^2(\nu_h) + b^2(\nu_h)]} \tag{40}$$

if T is even. [See R. L. Anderson and T. W. Anderson (1950).]

Best tests and multiple-decision procedures can be phrased for the cases considered in this subsection.

Example 4.1 in Chapter 4 gave monthly receipts of butter at five markets over three years. The fitted trend m_t based on twelve trigonometric functions (including the constant) was given in Table 4.1. The circular serial correlation of first order based on the residuals is

$$(41)\quad r_1^* = \frac{(2.4)(-0.6) + (-0.6)(-4.6) + \cdots + (1.6)(2.5) + (2.5)(2.4)}{(2.4)^2 + (-0.6)^2 + \cdots + (1.6)^2 + (2.5)^2}$$

$$= \frac{232.18}{474.51} = 0.489.$$

The computation in (41) is carried out in rounded-off residuals for the sake of easier reading.

6.7. DISTRIBUTIONS OF SERIAL CORRELATION COEFFICIENTS

6.7.1. The General Problem

The various serial correlation coefficients discussed in earlier sections are ratios of quadratic forms, say $s_j = P_j/P_0$, $j = 1, \ldots, p$, where $P_j = \boldsymbol{y}'\boldsymbol{B}_j\boldsymbol{y}$, $j = 0, 1, \ldots, p$. We are interested in their distributions when $\boldsymbol{y}$ has a multivariate normal distribution with covariance matrix $\boldsymbol{\Sigma}$. The general discussion will be carried out in broad terms because of applications to other problems. In the applications in this chapter the density has in the exponent the matrix $\boldsymbol{\Sigma}^{-1} = \sum_{j=0}^{p} \gamma_j \boldsymbol{A}_j$. In the simplest cases $\boldsymbol{B}_j = \boldsymbol{A}_j$; in some other cases

$$(1)\qquad \boldsymbol{B}_j = \left(\boldsymbol{I} - \frac{1}{T}\,\boldsymbol{\varepsilon}\boldsymbol{\varepsilon}'\right)\boldsymbol{A}_j\left(\boldsymbol{I} - \frac{1}{T}\,\boldsymbol{\varepsilon}\boldsymbol{\varepsilon}'\right),$$

where $\boldsymbol{\varepsilon} = (1, 1, \ldots, 1)'$, and then

$$(2)\qquad P_j = (\boldsymbol{y} - \bar{y}\boldsymbol{\varepsilon})'\boldsymbol{A}_j(\boldsymbol{y} - \bar{y}\boldsymbol{\varepsilon}).$$

In still other cases $\boldsymbol{B}_j$ is such that P_j is a quadratic form in the residuals of $\boldsymbol{y}$ from a more general regression function; that is,

$$(3)\qquad \boldsymbol{B}_j = (\boldsymbol{I} - \boldsymbol{V}^*(\boldsymbol{V}^{*\prime}\boldsymbol{V}^*)^{-1}\boldsymbol{V}^{*\prime})\boldsymbol{A}_j(\boldsymbol{I} - \boldsymbol{V}^*(\boldsymbol{V}^{*\prime}\boldsymbol{V}^*)^{-1}\boldsymbol{V}^{*\prime}),$$

$$(4)\qquad P_j = (\boldsymbol{y} - \boldsymbol{V}^*(\boldsymbol{V}^{*\prime}\boldsymbol{V}^*)^{-1}\boldsymbol{V}^{*\prime}\boldsymbol{y})'\boldsymbol{A}_j(\boldsymbol{y} - \boldsymbol{V}^*(\boldsymbol{V}^{*\prime}\boldsymbol{V}^*)^{-1}\boldsymbol{V}^{*\prime}\boldsymbol{y}),$$

where V^* is a matrix with T rows and rank equal to the number of columns. When $B_j = A_j$ we shall write Q_j for P_j and r_j for s_j. When B_j is given by (1) we shall write Q_j^* for P_j and r_j^* for s_j.

THEOREM 6.7.1. *The joint distribution of $P_0, \ldots, P_p$ given by* (4) *does not depend on $\boldsymbol{\beta}$ when y is distributed according to $N(V^*\boldsymbol{\beta}, \boldsymbol{\Sigma})$.*

PROOF. Let $y = u + V^*\boldsymbol{\beta}$; then u is distributed according to $N(\mathbf{0}, \boldsymbol{\Sigma})$ and

$$y - V^*(V^{*\prime}V^*)^{-1}V^{*\prime}y = (I - V^*(V^{*\prime}V^*)^{-1}V^{*\prime})(u + V^*\boldsymbol{\beta}) = (I - V^*(V^{*\prime}V^*)^{-1}V^{*\prime})u, \tag{5}$$

$$P_j = (u - V^*(V^{*\prime}V^*)^{-1}V^{*\prime}u)'A_j(u - V^*(V^{*\prime}V^*)^{-1}V^{*\prime}u). \tag{6}$$

Q.E.D.

If $\boldsymbol{\Sigma}^{-1} = \sum_{j=0}^{i} \gamma_j A_j$ and $B_j = A_j$ ($P_j = Q_j$ and $s_j = r_j$), then the conditional density of Q_i, given $Q_0, \ldots, Q_{i-1}$, is

$$h_i(Q_i \mid Q_0, \ldots, Q_{i-1}; \gamma_i) = e^{-\frac{1}{2}\gamma_i Q_i} \frac{k_i(Q_0, \ldots, Q_i)}{m_i(Q_0, \ldots, Q_{i-1}; \gamma_i)}, \tag{7}$$

where

$$m_i(Q_0, \ldots, Q_{i-1}; \gamma_i) = \int_{-\infty}^{\infty} e^{-\frac{1}{2}\gamma_i Q_i} k_i(Q_0, \ldots, Q_i)\, dQ_i \tag{8}$$

(assumed to be positive); thus it does not depend on $\gamma_0, \ldots, \gamma_{i-1}$. Equivalently, the conditional distribution of r_i, given Q_0 and $r_1, \ldots, r_{i-1}$, does not depend on $\gamma_0, \ldots, \gamma_{i-1}$. It is this conditional distribution of r_i that will be used in testing the null hypothesis that the order of dependence is less than i (that is, $\gamma_i = 0$). It will be convenient to obtain the conditional distribution on the simple assignment of parameter values of $\gamma_0 = 1$, $\gamma_1 = \cdots = \gamma_{i-1} = 0$. This conditional distribution when $\gamma_i = 0$ does not depend on Q_0 (because $r_1, \ldots, r_i$ are invariant with respect to scale transformations $y_t \rightarrow cy_t$).

The following lemma and theorem will be useful.

LEMMA 6.7.1. *The conditional density of $\tilde{Q}_i$, given $\tilde{Q}_0 = a_0, \ldots, \tilde{Q}_{i-1} = a_{i-1}$, where $\tilde{Q}_0 = c^2Q_0, \ldots, \tilde{Q}_i = c^2Q_i$, is the conditional density of Q_i, given $Q_0 = a_0, \ldots, Q_{i-1} = a_{i-1}$, with γ_i replaced by γ_i/c^2, namely $h_i(\tilde{Q}_i \mid \tilde{Q}_0, \ldots, \tilde{Q}_{i-1};\ \gamma_i/c^2)$.*

PROOF. The quadratic form $\tilde{Q}_j = c^2y'A_jy = (cy)'A_j(cy)$ can be written $\tilde{Q}_j = x'A_jx$, $j = 0, \ldots, i$, where $x = cy$ has the density with exponent equal to $-\frac{1}{2}$ times $\sum_{j=0}^{i} \gamma_j(c^{-1}x)'A_j(c^{-1}x) = \sum_{j=0}^{i} (\gamma_j/c^2)\tilde{Q}_j$. Thus $\tilde{Q}_0, \ldots, \tilde{Q}_i$ have the joint density of $Q_0, \ldots, Q_i$ with γ_j replaced by γ_j/c^2, $j = 0, \ldots, i$; the lemma follows.

THEOREM 6.7.2. *The conditional density of* $r_i = Q_i/Q_0$, *given* Q_0 *and* $r_j = Q_j/Q_0$, $j = 1, \ldots, i-1$, *is*

$$h_i(r_i \mid 1, r_1, \ldots, r_{i-1};\ \gamma_i Q_0). \tag{9}$$

If $\gamma_i = 0$, *this conditional density does not depend on* Q_0.

PROOF. The conditional density of r_i, given $Q_0 = b_0$ and $r_j = b_j$, $j = 1, \ldots, i-1$, is that of $c^2 Q_i$ for $c^2 = 1/Q_0$, given $c^2 Q_0 = 1$ and $r_j = c^2 Q_j = b_j$ (with γ_i replaced by $\gamma_i/c^2 = \gamma_i Q_0$), by Lemma 6.7.1. This gives the first part of the theorem. If $\gamma_i = 0$, it will be seen that the conditional density (9) does not depend on Q_0 by substitution of the arguments of (9) into the definitions (7) and (8).

If $\gamma_1 = \cdots = \gamma_i = 0$, then Theorem 6.7.2 implies that the joint conditional density of $r_1, \ldots, r_i$ given Q_0 [the product of i conditional densities of the form of (9)] does not depend on Q_0. Thus, the conditional density of r_i given Q_0 does not depend on Q_0.

It should be noted that Lemma 6.7.1 and Theorem 6.7.2 hold similarly for Q_i^* and r_i^*.

6.7.2. Characteristic Functions

One approach to finding the distributions of $P_0, \ldots, P_p$ or $s_1, \ldots, s_p$ is to use the characteristic function of $P_0, \ldots, P_p$.

THEOREM 6.7.3. *If* $\boldsymbol{y}$ *is distributed according to* $N(\boldsymbol{0}, \boldsymbol{\Sigma})$, *the characteristic function of* $P_0, \ldots, P_p$ *is*

$$\mathscr{E} e^{i(t_0 P_0 + \cdots + t_p P_p)} = \sqrt{\frac{|\boldsymbol{\Sigma}^{-1}|}{|\boldsymbol{\Sigma}^{-1} - 2i(t_0 \boldsymbol{B}_0 + \cdots + t_p \boldsymbol{B}_p)|}} = \frac{1}{\sqrt{|\boldsymbol{I} - 2i(t_0 \boldsymbol{B}_0 + \cdots + t_p \boldsymbol{B}_p)\boldsymbol{\Sigma}|}}. \tag{10}$$

PROOF. For purely imaginary $t_0, \ldots, t_p$ sufficiently small, (10) can be obtained from the fact that the integral of $e^{-\frac{1}{2}\boldsymbol{y}'\boldsymbol{A}\boldsymbol{y}}$ is $|\boldsymbol{A}|^{-\frac{1}{2}} (2\pi)^{\frac{1}{2}T}$; to justify (10) for real $t_0, \ldots, t_p$ we reduce the multivariate integral to a product of univariate integrals.

For fixed real $t_0, \ldots, t_p$ let $\boldsymbol{C}$ be a matrix such that

$$\boldsymbol{C}' \boldsymbol{\Sigma}^{-1} \boldsymbol{C} = \boldsymbol{I}, \tag{11}$$

$$\boldsymbol{C}'(t_0 \boldsymbol{B}_0 + \cdots + t_p \boldsymbol{B}_p)\boldsymbol{C} = \boldsymbol{D}, \tag{12}$$

where $\boldsymbol{D}$ is diagonal. (If $\boldsymbol{A}$ is positive definite and $\boldsymbol{B}$ is symmetric, there is a matrix $\boldsymbol{C}$ such that $\boldsymbol{C}'\boldsymbol{A}\boldsymbol{C} = \boldsymbol{I}$ and $\boldsymbol{C}'\boldsymbol{B}\boldsymbol{C}$ is diagonal. See Problem 30.) Then (when $\boldsymbol{y} = \boldsymbol{C}\boldsymbol{z}$)

(13)

$$
\begin{aligned}
\mathscr{E} e^{i(t_0 P_0 + \cdots + t_p P_p)} &= \int_{-\infty}^{\infty} \cdots \int_{-\infty}^{\infty} \frac{1}{(2\pi)^{\frac{1}{2}T} |\boldsymbol{\Sigma}|^{\frac{1}{2}}} \\
&\quad \times \exp\left[-\tfrac{1}{2}\boldsymbol{y}'\boldsymbol{\Sigma}^{-1}\boldsymbol{y} + i(t_0 \boldsymbol{y}'\boldsymbol{B}_0\boldsymbol{y} + \cdots + t_p \boldsymbol{y}'\boldsymbol{B}_p\boldsymbol{y})\right] dy_1 \cdots dy_T \\
&= \int_{-\infty}^{\infty} \cdots \int_{-\infty}^{\infty} \frac{1}{(2\pi)^{\frac{1}{2}T} |\boldsymbol{\Sigma}|^{\frac{1}{2}}} \\
&\quad \times \exp\left\{-\tfrac{1}{2}\boldsymbol{y}'[\boldsymbol{\Sigma}^{-1} - 2i(t_0\boldsymbol{B}_0 + \cdots + t_p\boldsymbol{B}_p)]\boldsymbol{y}\right\} dy_1 \cdots dy_T \\
&= \int_{-\infty}^{\infty} \cdots \int_{-\infty}^{\infty} \frac{1}{(2\pi)^{\frac{1}{2}T}} \exp\left[-\tfrac{1}{2}\boldsymbol{z}'(\boldsymbol{I} - 2i\boldsymbol{D})\boldsymbol{z}\right] dz_1 \cdots dz_T \\
&= \int_{-\infty}^{\infty} \cdots \int_{-\infty}^{\infty} \frac{1}{(2\pi)^{\frac{1}{2}T}} \exp\left[-\frac{1}{2}\sum_{s=1}^{T} z_s^2 + i\sum_{s=1}^{T} d_{ss} z_s^2\right] dz_1 \cdots dz_T \\
&= \prod_{s=1}^{T} \int_{-\infty}^{\infty} \frac{1}{(2\pi)^{\frac{1}{2}}} \exp\left[-\tfrac{1}{2}z_s^2 + i d_{ss} z_s^2\right] dz_s \\
&= \prod_{s=1}^{T} \frac{1}{\sqrt{1 - 2i d_{ss}}},
\end{aligned}
$$

because $1/\sqrt{1 - 2it}$ is the characteristic function of z_s^2 (χ^2 with 1 degree of freedom) when z_s is distributed according to $N(0, 1)$. Using (11) and (12) again, we derive from (13)

$$
\begin{aligned}
\text{(14)} \qquad \mathscr{E} e^{i(t_0 P_0 + \cdots + t_p P_p)} &= \frac{1}{\sqrt{\prod_{s=1}^{T} (1 - 2i d_{ss})}} \\
&= \frac{1}{\sqrt{|\boldsymbol{I} - 2i\boldsymbol{D}|}} \\
&= \frac{1}{\sqrt{|\boldsymbol{C}'|\,|\boldsymbol{\Sigma}^{-1} - 2i(t_0\boldsymbol{B}_0 + \cdots + t_p\boldsymbol{B}_p)|\,|\boldsymbol{C}|}},
\end{aligned}
$$

which is equal to (10). Q.E.D.

COROLLARY 6.7.1. *If $\boldsymbol{y}$ is distributed with density* $K \exp\left[-\frac{1}{2}\sum_{j=0}^{p} \gamma_j Q_j\right]$, *the characteristic function of* $Q_0 = \boldsymbol{y}'\boldsymbol{A}_0\boldsymbol{y}, \ldots, Q_p = \boldsymbol{y}'\boldsymbol{A}_p\boldsymbol{y}$ *is*

$$
\text{(15)} \qquad \mathscr{E} e^{i(t_0 Q_0 + \cdots + t_p Q_p)} = \sqrt{\frac{|\gamma_0\boldsymbol{A}_0 + \cdots + \gamma_p\boldsymbol{A}_p|}{|(\gamma_0 - 2it_0)\boldsymbol{A}_0 + \cdots + (\gamma_p - 2it_p)\boldsymbol{A}_p|}}.
$$

In principle, the characteristic function can be inverted to find the joint distribution of $P_0, P_1, \ldots, P_p$, though in practice it cannot be carried out in all cases. It can be used to obtain the joint distribution of $s_1, \ldots, s_p$ by use of

$$(16)\quad \Pr\{s_1 \le w_1, \ldots, s_p \le w_p\} = \Pr\left\{\frac{P_1}{P_0} \le w_1, \ldots, \frac{P_p}{P_0} \le w_p\right\}$$
$$= \Pr\{P_1 - w_1P_0 \le 0, \ldots, P_p - w_pP_0 \le 0\}.$$

The characteristic function of $P_1 - w_1P_0, \ldots, P_p - w_pP_0$ is obtained from Theorem 6.7.3 by letting $t_0 = -w_1t_1 - \cdots - w_pt_p$.

6.7.3. Canonical Forms

As we have observed in Section 6.5, if $\boldsymbol{v}_t$ is the tth normalized characteristic vector of $\boldsymbol{B}$ and ϕ_t the corresponding characteristic root, then

$$(17)\qquad \boldsymbol{V}'\boldsymbol{B}\boldsymbol{V} = \boldsymbol{\Phi}, \qquad \boldsymbol{V}'\boldsymbol{V} = \boldsymbol{I},$$

where $\boldsymbol{\Phi}$ is a diagonal matrix with $\phi_1, \ldots, \phi_T$ as diagonal elements and $\boldsymbol{V} = (\boldsymbol{v}_1, \ldots, \boldsymbol{v}_T)$. (If the orthogonality conditions $\boldsymbol{v}_t'\boldsymbol{v}_s = 0$, $t \ne s$, are not necessarily satisfied, linear combinations of the $\boldsymbol{v}_t$'s corresponding to common roots can be defined as the characteristic vectors so as to satisfy them.) If the columns of $\boldsymbol{V}$ are the characteristic vectors of $\boldsymbol{B}_0, \ldots, \boldsymbol{B}_p$ with characteristic roots $\phi_{01}, \ldots, \phi_{0T}$; $\phi_{11}, \ldots, \phi_{1T}; \ldots; \phi_{p1}, \ldots, \phi_{pT}$, respectively, then

$$(18)\qquad \boldsymbol{V}'\boldsymbol{B}_j\boldsymbol{V} = \boldsymbol{\Phi}_j, \qquad \boldsymbol{V}'\boldsymbol{V} = \boldsymbol{I}, \qquad j = 0, 1, \ldots, p,$$

where $\boldsymbol{\Phi}_j$ is diagonal with $\phi_{j1}, \ldots, \phi_{jT}$ as its diagonal elements. Then when $\boldsymbol{B}_0 = \boldsymbol{I}$ the quadratic forms can be written

$$(19)\qquad P_0 = \sum_{t=1}^{T} z_t^2, \qquad P_j = \sum_{t=1}^{T} \phi_{jt} z_t^2, \qquad j = 1, \ldots, p,$$

where $\boldsymbol{y} = \boldsymbol{V}\boldsymbol{z}$. A necessary and sufficient condition for $\boldsymbol{B}_1, \ldots, \boldsymbol{B}_p$ to have a common set of characteristic vectors is

$$(20)\qquad \boldsymbol{B}_j\boldsymbol{B}_k = \boldsymbol{B}_k\boldsymbol{B}_j, \qquad j, k = 1, \ldots, p.$$

(See Problem 31.)

We are particularly interested in the case where $\boldsymbol{\Sigma}^{-1} = \sum_{j=0}^{p} \gamma_j \boldsymbol{A}_j$, $\boldsymbol{A}_0 = \boldsymbol{I}$, and

$$(21)\qquad \boldsymbol{B}_j = [\boldsymbol{I} - \boldsymbol{V}^*\boldsymbol{V}^{*\prime}]\boldsymbol{A}_j[\boldsymbol{I} - \boldsymbol{V}^*\boldsymbol{V}^{*\prime}], \qquad j = 0, 1, \ldots, p,$$

where V^* is a matrix whose columns are columns of V, namely the characteristic vectors of A_j corresponding to roots which are the diagonal elements of the diagonal matrix Λ_j^*, $j = 0, 1, \ldots, p$. Suppose that the roots and vectors are numbered in such a way that V^* consists of the last m columns of V and Λ_j^* is the lower right-hand corner of Λ_j. It is assumed that $V'A_jV = \Lambda_j$, where Λ_j is diagonal and $V'V = I$. Then

$$V'V^* = \begin{pmatrix} 0 \\ I \end{pmatrix}, \tag{22}$$

$$V'(I - V^*V^{*\prime}) = V' - \begin{pmatrix} 0 \\ V^{*\prime} \end{pmatrix}, \tag{23}$$

$$\begin{aligned} V'B_jV &= V'A_jV - \begin{pmatrix} 0 \\ V^{*\prime} \end{pmatrix} A_jV - V'A_j(0 \quad V^*) + \begin{pmatrix} 0 \\ V^{*\prime} \end{pmatrix} A_j(0 \quad V^*) \\ &= \Lambda_j - \begin{pmatrix} 0 & 0 \\ 0 & \Lambda_j^* \end{pmatrix}. \end{aligned} \tag{24}$$

Thus (for $y = Vz$)

$$\begin{aligned} P_j &= y'B_jy = z'V'B_jVz \\ &= z'\Lambda_jz - z'\begin{pmatrix} 0 & 0 \\ 0 & \Lambda_j^* \end{pmatrix} z \\ &= \sum_{t=1}^{T} \lambda_{jt} z_t^2 - \sum_{t=T-m+1}^{T} \lambda_{jt} z_t^2 \\ &= \sum_{t=1}^{T-m} \lambda_{jt} z_t^2, \end{aligned} \tag{25}$$

$$P_0 = \sum_{t=1}^{T-m} z_t^2. \tag{26}$$

If the distribution of y is $N(V^*\boldsymbol{\beta}, \Sigma)$, where $\Sigma^{-1} = \sum_{j=0}^{p} \gamma_j A_j$, then the density of $z = V^{-1}y$ is

$$\begin{aligned} &(2\pi)^{-\frac{1}{2}T} \prod_{t=1}^{T} \left(\sum_{j=0}^{p} \gamma_j \lambda_{jt} \right)^{\frac{1}{2}} \\ &\quad \times \exp\left\{ -\frac{1}{2} \left[\sum_{t=1}^{T-m} \left(\sum_{j=0}^{p} \gamma_j \lambda_{jt} \right) z_t^2 + \sum_{t=T-m+1}^{T} \left(\sum_{j=0}^{p} \gamma_j \lambda_{jt} \right) (z_t - \beta_t)^2 \right] \right\}, \end{aligned} \tag{27}$$

where the components of $\boldsymbol{\beta}$ have been numbered from $T - m + 1$ to T.

In the case of the circular model $\boldsymbol{\varepsilon} = (1, \ldots, 1)'$ is a characteristic vector of A_j corresponding to the characteristic root 1 (the Tth root), $j = 0, 1, \ldots, p$. The canonical form of Q_j^* with residuals from the sample mean is

$$(28) \qquad Q_j^* = \sum_{t=1}^{T-1} \cos \frac{2\pi j}{T} t \, z_t^2, \qquad T = 2, 3, \ldots .$$

In the case of the mean square successive difference model $\boldsymbol{\varepsilon}$ is also a characteristic vector of $\boldsymbol{A}_j$ corresponding to the root 1 and

$$(29) \qquad Q_j^* = \sum_{t=1}^{T-1} \cos \frac{\pi j}{T} t \, z_t^2, \qquad T = 2, 3, \ldots .$$

In the case of the third model $\boldsymbol{\varepsilon}$ is not a characteristic vector of $\boldsymbol{A}_j$, $j = 1, \ldots, p$, and a reduction of Q_j^* to a weighted sum of $T - 1$ squares is not possible.

In general we may have $p + 1$ quadratic forms $P_0, \ldots, P_p$ defining p serial correlations $s_1 = P_1/P_0, \ldots, s_p = P_p/P_0$ and a covariance matrix $\boldsymbol{\Sigma}$. It is always possible to find a matrix $\boldsymbol{C}$ such that $\boldsymbol{C}'\boldsymbol{\Sigma}^{-1}\boldsymbol{C} = \boldsymbol{I}$ [equivalently $\boldsymbol{C}^{-1}\boldsymbol{\Sigma}(\boldsymbol{C}')^{-1} = \boldsymbol{I}$] and $\boldsymbol{C}'\boldsymbol{B}_j\boldsymbol{C}$ is diagonal for some j. Unless special relations hold among these matrices, however, only one $\boldsymbol{B}_j$ can be diagonalized by some one matrix $\boldsymbol{C}$. Thus if $\boldsymbol{\Sigma}$ is transformed to $\boldsymbol{I}$ and $s_j = \boldsymbol{x}'(\boldsymbol{C}'\boldsymbol{B}_j\boldsymbol{C})\boldsymbol{x}/[\boldsymbol{x}'(\boldsymbol{C}'\boldsymbol{B}_0\boldsymbol{C})\boldsymbol{x}]$, then either the numerator or denominator quadratic form may be a weighted sum of squares, but not necessarily both.

The following is a corollary to Theorem 6.7.3:

COROLLARY 6.7.2. *If $\boldsymbol{y}$ is distributed according to $N(\boldsymbol{V}^*\boldsymbol{\beta}, \boldsymbol{\Sigma})$, where $\boldsymbol{\Sigma}^{-1} = \sum_{j=0}^{p} \gamma_j \boldsymbol{A}_j$ and the columns of $\boldsymbol{V}^*$ are characteristic vectors of $\boldsymbol{A}_j$ corresponding to characteristic roots λ_{jt}, $t = T - m + 1, \ldots, T$, $j = 0, 1, \ldots, p$, the characteristic function of $P_j^* = (\boldsymbol{y} - \boldsymbol{V}^*\boldsymbol{b})'\boldsymbol{A}_j(\boldsymbol{y} - \boldsymbol{V}^*\boldsymbol{b})$, $j = 0, 1, \ldots, p$, where $\boldsymbol{b}$ is the least squares estimate of $\boldsymbol{\beta}$, is*

$$(30) \qquad \mathscr{E} e^{i(t_0 P_0^* + \cdots + t_p P_p^*)} = \prod_{s=1}^{T-m} \sqrt{\frac{\sum_{j=0}^{p} \gamma_j \lambda_{js}}{\sum_{j=0}^{p} (\gamma_j - 2it_j)\lambda_{js}}} .$$

6.7.4. Distributions of Serial Correlation Coefficients with Paired Roots Under Independence

To carry out the procedures derived in Sections 6.3 and 6.4 we need the conditional distribution of r_i, given Q_0 and $r_1, \ldots, r_{i-1}$. In particular, to determine a test of the null hypothesis $\gamma_1 = 0$ (given $\gamma_j = 0, j > 1$) at a specified significance level we want the distribution of r_1 when $\gamma_1 = 0$; this distribution does not depend on γ_0. We shall therefore consider the (marginal) distribution of a single serial correlation coefficient when $\gamma_1 = \cdots = \gamma_p = 0$. Since $A_0 = I$, the observations are independent. The distribution of a serial correlation coefficient can be expressed simply when the characteristic roots occur in pairs. Inasmuch as the distribution may be valid for other serial correlations, we shall give a general treatment and hence shall drop the index. Later we shall discuss the conditional distribution of r_i, given $r_1, \ldots, r_{i-1}$.

Suppose $T - m = 2H$, and that there are H different roots

$$\lambda_{jh} = \lambda_{j,T-m+1-h} = \nu_h, \qquad h = 1, \ldots, H. \tag{31}$$

For instance, in the case of the first-order circular serial correlation with residuals from the mean, $m = 1$, T is odd, and $\nu_h = \cos 2\pi h/T$. Let

$$x_h = z_h^2 + z_{T-m+1-h}^2. \tag{32}$$

Then

$$r = \frac{\sum_{h=1}^{H} \nu_h x_h}{\sum_{h=1}^{H} x_h}. \tag{33}$$

When $\gamma_j = 0$, $j > 0$, the z_t's are independently distributed with variance 1 ($\gamma_0 = 1$). Then $x_1, \ldots, x_H$ are independently, identically distributed, each according to a χ^2-distribution with 2 degrees of freedom. The density of x_h is $\frac{1}{2}e^{-\frac{1}{2}x_h}$, and the joint density of $x_1, \ldots, x_H$ is

$$(\tfrac{1}{2})^H \exp\left(-\tfrac{1}{2}\sum_{h=1}^{H} x_h\right), \qquad x_h \geq 0, \quad h = 1, \ldots, H, \tag{34}$$

and 0 elsewhere. The surfaces of constant density are hyperplanes in the positive orthant

$$\sum_{h=1}^{H} x_h = c, \qquad x_h \geq 0, \quad h = 1, \ldots, H, \tag{35}$$

for $c > 0$. Thus the conditional distribution of $x_1, \ldots, x_H$ given (35) is uniform on that $(H - 1)$-dimensional regular tetrahedron. (If $H = 3$, the distribution is uniform on the equilateral triangle.)

The conditional probability is

$$\text{(36)}\qquad \Pr\left\{r \geq R \,\middle|\, \sum_{h=1}^{H} x_h = c\right\} = \Pr\left\{\frac{\sum_{h=1}^{H} \nu_h x_h}{\sum_{h=1}^{H} x_h} \geq R \,\middle|\, \sum_{h=1}^{H} x_h = c\right\}$$

$$= \Pr\left\{\sum_{h=1}^{H} \nu_h \frac{x_h}{c} \geq R \,\middle|\, \sum_{h=1}^{H} \frac{x_h}{c} = 1\right\},$$

where $x_1/c, \ldots, x_H/c$ are conditionally uniformly distributed on the $(H-1)$-dimensional tetrahedron $\sum_{h=1}^{H} (x_h/c) = 1$ and $x_h/c \geq 0$, $h = 1, \ldots, H$. (This is a special case of Theorem 6.7.2.) Thus the conditional probability (36) is independent of c; for convenience we can take $c = 1$. This shows that r is distributed independently of $\sum_{h=1}^{H} x_h$.

We shall, therefore, find the probability

$$\text{(37)}\qquad \Pr\{r \geq R\} = \Pr\left\{\sum_{h=1}^{H} \nu_h x_h \geq R \,\middle|\, \sum_{h=1}^{H} x_h = 1\right\},$$

where $x_1, \ldots, x_H$ are uniformly distributed on

$$\text{(38)}\qquad \sum_{h=1}^{H} x_h = 1, \qquad x_h \geq 0, \qquad h = 1, \ldots, H.$$

The probability (37) is the $(H-1)$-dimensional volume of the intersection of (38) and

$$\text{(39)}\qquad \sum_{h=1}^{H} \nu_h x_h \geq R$$

relative to the $(H-1)$-dimensional volume of the regular tetrahedron (38).

The vertices of the regular tetrahedron (38) are $X_1, \ldots, X_H$, where all of the coordinates of X_j are 0, except the jth which is 1. The hyperplane $\sum_{h=1}^{H} \nu_h x_h = R$ intersects an edge $X_i X_j$ ($x_i + x_j = 1$, $x_k = 0$, $k \neq i, k \neq j, i \neq j$) or its extension beyond the positive orthant in the point W_{ij} with coordinates

$$\text{(40)}\qquad x_i = \frac{R - \nu_j}{\nu_i - \nu_j}, \qquad x_j = \frac{\nu_i - R}{\nu_i - \nu_j}, \qquad x_k = 0,$$

$$k \neq i, \quad k \neq j, \quad i \neq j.$$

(Note that $W_{ij} = W_{ji}$.) It will be convenient to number the roots so

$$\text{(41)}\qquad \nu_H < \nu_{H-1} < \cdots < \nu_2 < \nu_1.$$

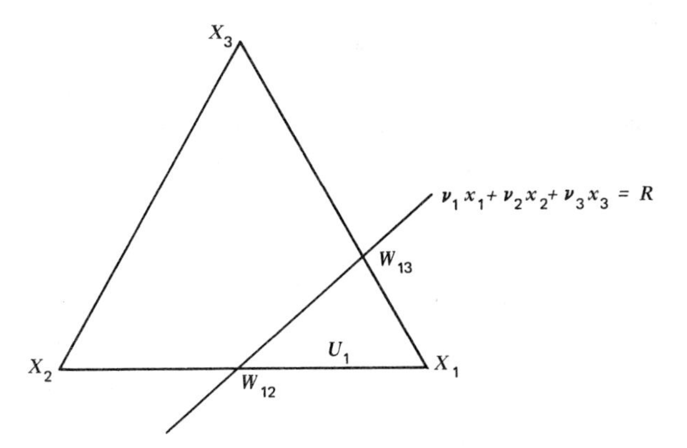

Figure 6.1. Regions when $\nu_2 \leq R \leq \nu_1$.

Note that $\nu_H \leq \sum_{h=1}^{H} \nu_h x_h \leq \nu_1$. If $\nu_2 \leq R \leq \nu_1$, the hyperplane intersects each edge $X_1 X_j$ between the vertices; that is, the coordinates x_1 and x_j of W_{1j} (40) are nonnegative. Then the region $r \geq R$ is a tetrahedron with vertices $X_1, W_{12}, \ldots, W_{1H}$. We shall show that for each value of R the set $r \geq R$ can be expressed in terms of tetrahedrons. The probability (37) will then be expressed in terms of volumes of tetrahedrons. See Figures 6.1 and 6.2.

LEMMA 6.7.2. *Let T be an $(H - 1)$-dimensional tetrahedron with $V_1, V_2, V_3, \ldots, V_H$ as vertices, and let T^* be the $(H - 1)$-dimensional tetrahedron with $V_1, V_2^*, V_3, \ldots, V_H$ as vertices, where V_2^* is on the line $V_1 V_2$, and the length of the edge $V_1 V_2^*$ is k_2^* times the length of the edge $V_1 V_2$. Then the $(H - 1)$-dimensional volume of T^* is k_2^* times the $(H - 1)$-dimensional volume of T.*

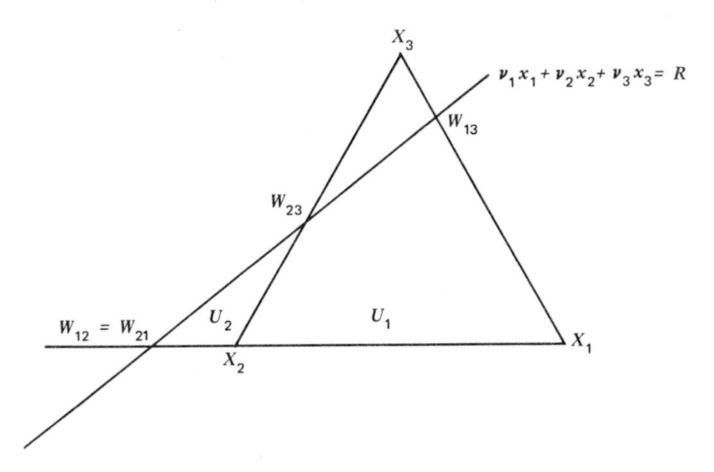

Figure 6.2. Regions when $\nu_3 \leq R \leq \nu_2$.

PROOF. Let $y_1, \ldots, y_{H-2}$ be arbitrary (rectangular) coordinates in an $(H-2)$-dimensional hyperplane orthogonal to V_1V_2; let $h(y_1, \ldots, y_{H-2})$ be the length (possibly 0) of the intersection of T with the line through $(y_1, \ldots, y_{H-2})$ parallel to V_1V_2 and let $h^*(y_1, \ldots, y_{H-2})$ be the length (possibly 0) of the intersection of T^* with this line. Then $h^*(y_1, \ldots, y_{H-2}) = k_2^* h(y_1, \ldots, y_{H-2})$ and the volume of T^* is

$$\text{Vol}(T^*) = \int_{-\infty}^{\infty} \cdots \int_{-\infty}^{\infty} h^*(y_1, \ldots, y_{H-2})\, dy_1 \cdots dy_{H-2} \tag{42}$$
$$= k_2^* \int_{-\infty}^{\infty} \cdots \int_{-\infty}^{\infty} h(y_1, \ldots, y_{H-2})\, dy_1 \cdots dy_{H-2}$$
$$= k_2^* \,\text{Vol}(T).$$

Q.E.D.

LEMMA 6.7.3. *Let T be an $(H-1)$-dimensional tetrahedron with $V_1, V_2, \ldots, V_H$ as vertices, and let T^* be the $(H-1)$-dimensional tetrahedron with $V_1, V_2^*, \ldots, V_H^*$ as vertices, where V_j^* is on the line V_1V_j, and the length of the edge $V_1V_j^*$ is k_j^* times the length of the edge V_1V_j, $j = 2, \ldots, H$. Then the $(H-1)$-dimensional volume of T^* is $k_2^* \cdots k_H^*$ times the $(H-1)$-dimensional volume of T.*

PROOF. This lemma follows from Lemma 6.7.2 by induction.

Let S be the set which is the intersection of (38) and (39). We now require $\sum_{h=1}^H x_h = 1$ and assume $2 \le m \le H-1$. If $\nu_{m+1} \le R \le \nu_m$, $X_1, \ldots, X_m$ are in S and $X_{m+1}, \ldots, X_H$ are outside of S. (Strictly speaking X_{m+1} is in S if $R = \nu_{m+1}$.) Let the sets $U_1, \ldots, U_m$ be defined as follows:

$$\begin{aligned}
&U_1: && \sum_{h=1}^H \nu_h x_h \ge R, && x_2 \ge 0, \ldots, x_H \ge 0,\\
&U_2: && \sum_{h=1}^H \nu_h x_h \ge R, && x_1 < 0,\, x_3 \ge 0, \ldots, x_H \ge 0,\\
&U_3: && \sum_{h=1}^H \nu_h x_h \ge R, && x_1 < 0,\, x_2 < 0,\, x_4 \ge 0, \ldots, x_H \ge 0,\\
& && \vdots \\
&U_{m-1}: && \sum_{h=1}^H \nu_h x_h \ge R, && x_1 < 0, \ldots, x_{m-2} < 0,\, x_m \ge 0, \ldots, x_H \ge 0,\\
&U_m: && \sum_{h=1}^H \nu_h x_h \ge R, && x_1 < 0, \ldots, x_{m-1} < 0,\, x_{m+1} \ge 0, \ldots, x_H \ge 0.
\end{aligned} \tag{43}$$

Each set is a tetrahedron; the vertices are as follows:

(44)

$$
\begin{aligned}
&U_1: \quad X_1, W_{12}, W_{13}, \ldots, W_{1H},\\
&U_2: \quad W_{21}, X_2, W_{23}, \ldots, W_{2H},\\
&U_3: \quad W_{31}, W_{32}, X_3, \ldots, W_{3H},\\
&\quad \vdots\\
&U_{m-1}: \quad W_{m-1,1}, W_{m-1,2}, W_{m-1,3}, \ldots, W_{m-1,m-2}, X_{m-1}, W_{m-1,m}, \ldots, W_{m-1,H},\\
&U_m: \quad W_{m1}, W_{m2}, W_{m3}, \ldots, W_{m,m-1}, X_m, W_{m,m+1}, \ldots, W_{mH}.
\end{aligned}
$$

Since $x_j \geq 0$ in U_{j-1} and $x_j < 0$ in U_{j+i}, $i \geq 1$,

$$U_{j-1} \cap U_{j+i} = 0, \qquad i = 1, \ldots, m - j,\ j = 2, \ldots, m - 1\ (\geq 2), \tag{45}$$

the empty set. Also

$$U_j \subset U_{j-1} \cup U_{j+1}, \qquad j = 2, \ldots, m - 1\ (\geq 2). \tag{46}$$

Since $x_1 < 0, \ldots, x_{m-1} < 0$ and $x_{m+1} \geq 0, \ldots, x_H \geq 0$ in U_m, $\sum_{h=1}^H \nu_h x_h \geq R \geq \nu_{m+1}$ and $\sum_{h=1}^H x_h = 1$ imply

$$\sum_{h=m}^{H} (\nu_h - \nu_{m+1}) x_h \geq \sum_{h=1}^{m-1} (\nu_{m+1} - \nu_h) x_h > 0, \tag{47}$$

and, hence, $x_m > 0$ in U_m; thus

$$U_m \subset U_{m-1}. \tag{48}$$

LEMMA 6.7.4.

$$S = U_1 \cup U_3 \cup \cdots \cup U_m - (U_2 \cup U_4 \cup \cdots \cup U_{m-1}), \qquad m \text{ odd}, \tag{49}$$

$$S = U_1 \cup U_3 \cup \cdots \cup U_{m-1} - (U_2 \cup U_4 \cup \cdots \cup U_m), \qquad m \text{ even}. \tag{50}$$

PROOF. Relations (46) and (48) indicate that the set in parentheses on the right-hand side of (49) or (50) is contained in the set outside parentheses, and hence the indicated subtraction is possible. Since $S \subset U_1$ and $S \cap U_j = 0$, $j > 1$, S is contained in the right-hand side of (49) or (50). On the other hand any point in the set on the right-hand side is in S, for a point in the right-hand side must be in some one set U_{2j-1} (because $U_1, U_3, \ldots$ are disjoint) and it must not be in any U_{2j}; because of (46) and (48) such a point must be in the part of U_1 disjoint to U_2 and hence it is in S. This proves the lemma.

LEMMA 6.7.5. *The $(H-1)$-dimensional volume of U_i relative to* (38) *is*

$$(-1)^{i-1}\prod_{\substack{j=1\\j\neq i}}^{H}\frac{\nu_i - R}{\nu_i - \nu_j}. \tag{51}$$

PROOF. The coordinates of the vertex W_{ij} $(j \neq i)$ of U_i are (40). The length of the edge X_iW_{ij} relative to the length of the edge X_iX_j is $-(\nu_i - R)/(\nu_i - \nu_j)$ for $j < i$ and is $(\nu_i - R)/(\nu_i - \nu_j)$ for $j > i$. Lemma 6.7.5 then follows from Lemma 6.7.3.

THEOREM 6.7.4. $\Pr\{r \geq R\} = 1$ *for* $R \leq \nu_H$,

$$\Pr\{r \geq R\} = \sum_{i=1}^{m}\prod_{\substack{j=1\\j\neq i}}^{H}\frac{\nu_i - R}{\nu_i - \nu_j}, \qquad \nu_{m+1} \leq R \leq \nu_m, \quad m = 1, \ldots, H-1, \tag{52}$$

$\Pr\{r \geq R\} = 0$ *for* $\nu_1 \leq R$.

PROOF. The theorem follows from Lemmas 6.7.4 and 6.7.5 and (45).

COROLLARY 6.7.3. *The density of r is*

$$(H-1)\sum_{i=1}^{m}\frac{(\nu_i - r)^{H-2}}{\prod\limits_{\substack{j=1\\j\neq i}}^{H}(\nu_i - \nu_j)}, \qquad \nu_{m+1} \leq r \leq \nu_m, \quad m = 1, \ldots, H-1, \tag{53}$$

and is 0 *outside* (ν_H, ν_1).

The distribution of the serial correlation when there is one single root in addition to double roots can be derived from (52). It will be convenient to derive a more general result from which the distribution may be obtained. We let $\mu_1, \ldots, \mu_M, \mu_{M+1}$ be arbitrary weights.

THEOREM 6.7.5. *Let*

$$s = \frac{\sum_{j=1}^{M}\mu_j z_j^2}{\sum_{j=1}^{M} z_j^2}, \qquad t = \frac{\sum_{j=1}^{M}\mu_j z_j^2 + \mu_{M+1}\sum_{j=M+1}^{M+L} z_j^2}{\sum_{j=1}^{M+L} z_j^2}, \tag{54}$$

where $z_1, \ldots, z_{M+L}$ *are independently distributed, each according to* $N(0, 1)$. *Then*

$$\Pr\{t \geq T\} = \int_0^\infty \Pr\{s \geq T + (T - \mu_{M+1})u\}\frac{\Gamma[\frac{1}{2}(M+L)]}{\Gamma(\frac{1}{2}M)\Gamma(\frac{1}{2}L)}\frac{u^{\frac{1}{2}L-1}}{(1+u)^{\frac{1}{2}(M+L)}}\,du. \tag{55}$$

PROOF.

$$\begin{aligned}
\text{(56)}\qquad \Pr\{t \geq T\} &= \Pr\left\{\sum_{j=1}^{M} \mu_j z_j^2 + \mu_{M+1} \sum_{j=M+1}^{M+L} z_j^2 \geq T \sum_{j=1}^{M+L} z_j^2\right\} \\
&= \Pr\left\{\sum_{j=1}^{M} \mu_j z_j^2 \geq T \sum_{j=1}^{M} z_j^2 + (T - \mu_{M+1}) \sum_{j=M+1}^{M+L} z_j^2\right\} \\
&= \Pr\left\{\frac{\sum_{j=1}^{M} \mu_j z_j^2}{\sum_{j=1}^{M} z_j^2} \geq T + (T - \mu_{M+1}) \frac{\sum_{j=M+1}^{M+L} z_j^2}{\sum_{j=1}^{M} z_j^2}\right\} \\
&= \Pr\{s \geq T + (T - \mu_{M+1})u\},
\end{aligned}$$

where $u = \chi_L^2/\chi_M^2$ has the distribution of $LF_{L,M}/M$. The last step in (56) is a result of the fact that s is distributed independently of $\sum_{j=1}^{M} z_j^2$.

COROLLARY 6.7.4. *If* $L = 1$ *and* $M = 2H$,

$$\text{(57)}\qquad \Pr\{t \geq T\} = \int_0^\infty \Pr\{s \geq T + (T - \mu_{2H+1})u\} \frac{\Gamma(H + \frac{1}{2})u^{-\frac{1}{2}}}{\Gamma(H)\Gamma(\frac{1}{2})(1 + u)^{H+\frac{1}{2}}}\, du.$$

To evaluate (57) when s is r of (52) we use a special case of the following lemma:

LEMMA 6.7.6.

$$\text{(58)}\qquad \int_0^{c/d} (c - du)^{\frac{1}{2}M-1} \frac{\Gamma[\frac{1}{2}(M + L)]}{\Gamma(\frac{1}{2}M)\Gamma(\frac{1}{2}L)} u^{\frac{1}{2}L-1}(1 + u)^{-\frac{1}{2}(M+L)}\, du = c^{\frac{1}{2}(M+L)-1}(c + d)^{-\frac{1}{2}L}.$$

PROOF. Upon the substitution

$$\text{(59)}\qquad w = \frac{c + d}{c} \frac{u}{1 + u}$$

into the left-hand side of (58) we obtain

$$\text{(60)}\qquad \frac{\Gamma[\frac{1}{2}(M + L)]}{\Gamma(\frac{1}{2}M)\Gamma(\frac{1}{2}L)} c^{\frac{1}{2}M-1}\left(\frac{c}{c + d}\right)^{\frac{1}{2}L} \int_0^1 (1 - w)^{\frac{1}{2}M-1} w^{\frac{1}{2}L-1}\, dw,$$

which gives the right-hand side of (58),because the integral is the Beta function $B(\frac{1}{2}M, \frac{1}{2}L)$.

COROLLARY 6.7.5.

$$\text{(61)}\qquad \int_0^{c/d} (c - du)^{H-1} \frac{\Gamma(H + \frac{1}{2})}{\Gamma(H)\Gamma(\frac{1}{2})} u^{-\frac{1}{2}}(1 + u)^{-(H+\frac{1}{2})}\, du = c^{H-\frac{1}{2}}(c + d)^{-\frac{1}{2}}.$$

THEOREM 6.7.6. *Let*

$$r = \frac{\sum_{h=1}^{H} \nu_h(z_{2h-1}^2 + z_{2h}^2) + \nu_{H+1} z_{2H+1}^2}{\sum_{h=1}^{2H+1} z_h^2}, \tag{62}$$

where $z_1, \ldots, z_{2H+1}$ *are independently distributed, each according to* $N(0, 1)$, *and* $\nu_{H+1} < \nu_H < \cdots < \nu_1$. *Then*

$$\Pr\{r \geq R\} = \sum_{i=1}^{n} \frac{(\nu_i - R)^{H-\frac{1}{2}}}{\sqrt{\nu_i - \nu_{H+1}} \prod_{\substack{j=1 \\ j \neq i}}^{H} (\nu_i - \nu_j)}, \qquad \begin{array}{l} \nu_{n+1} \leq R \leq \nu_n, \\ n = 1, \ldots, H. \end{array} \tag{63}$$

PROOF. In the right-hand side of (52) we shall replace R by $R + (R - \nu_{H+1})u$. If $\nu_{n+1} \leq R \leq \nu_n$, then $\nu_{n+1} \leq R + (R - \nu_{H+1})u$, because $\nu_{H+1} \leq R$ and $u \geq 0$. For $\nu_{m+1} \leq R + (R - \nu_{H+1})u \leq \nu_m$, $m = 1, \ldots, n - 1$, we have

$$\frac{\nu_{m+1} - R}{R - \nu_{H+1}} \leq u \leq \frac{\nu_m - R}{R - \nu_{H+1}}, \tag{64}$$

and for $\nu_{n+1} \leq R + (R - \nu_{H+1})u \leq \nu_n$ we have $0 \leq u \leq (\nu_n - R)/(R - \nu_{H+1})$. Hence, from (57) we have $(n = 1, \ldots, H - 1)$

(65) $\Pr\{r \geq R\}$

$$= \sum_{m=1}^{n-1} \int_{\frac{\nu_{m+1}-R}{R-\nu_{H+1}}}^{\frac{\nu_m-R}{R-\nu_{H+1}}} \sum_{i=1}^{m} \frac{[\nu_i - R - (R - \nu_{H+1})u]^{H-1}}{\prod_{\substack{j=1 \\ j \neq i}}^{H} (\nu_i - \nu_j)} \frac{\Gamma(H + \frac{1}{2})}{\Gamma(H)\Gamma(\frac{1}{2})} \frac{u^{-\frac{1}{2}}}{(1 + u)^{H+\frac{1}{2}}} du$$

$$+ \int_0^{\frac{\nu_n-R}{R-\nu_{H+1}}} \sum_{i=1}^{n} \frac{[\nu_i - R - (R - \nu_{H+1})u]^{H-1}}{\prod_{\substack{j=1 \\ j \neq i}}^{H} (\nu_i - \nu_j)} \frac{\Gamma(H + \frac{1}{2})}{\Gamma(H)\Gamma(\frac{1}{2})} \frac{u^{-\frac{1}{2}}}{(1 + u)^{H+\frac{1}{2}}} du$$

$$= \sum_{i=1}^{n} \int_0^{\frac{\nu_i-R}{R-\nu_{H+1}}} \frac{[\nu_i - R - (R - \nu_{H+1})u]^{H-1}}{\prod_{\substack{j=1 \\ j \neq i}}^{H} (\nu_i - \nu_j)} \frac{\Gamma(H + \frac{1}{2})}{\Gamma(H)\Gamma(\frac{1}{2})} \frac{u^{-\frac{1}{2}}}{(1 + u)^{H+\frac{1}{2}}} du$$

by interchange of the order of summation on m and i. This gives (63) for $n = 1, \ldots, H-1$. To prove (63) for $n = H$ we must integrate over values of u for which $R + (R - \nu_{H+1})u \le \nu_H$. To justify the analytic expressions in (65) we need to verify that (52) with $m = H$ holds for $R \le \nu_H$ and corresponds to $\Pr\{r \ge R\} = 1$ for $R \le \nu_H$. This is a special case of the following lemma:

LEMMA 6.7.7.

$$\sum_{i=1}^{H} \frac{(\nu_i - x)^{H-1}}{\prod_{\substack{j=1\\ j\ne i}}^{H} (\nu_i - \nu_j)} \equiv 1. \tag{66}$$

PROOF. If we put $R = \nu_H$ in (52) we obtain $\Pr\{r \ge \nu_H\} = 1$. If we restate this when H is replaced by $H + 1$ and $\nu_1, \ldots, \nu_H$ are replaced by $\nu_1, \ldots, \nu_H$, x, we obtain

$$1 = \sum_{i=1}^{H} \frac{(\nu_i - x)^H}{(\nu_i - x)\prod_{\substack{j=1\\ j\ne i}}^{H} (\nu_i - \nu_j)}, \tag{67}$$

which is (66). Since this holds for all $x \le \nu_H$ and the left-hand side of (66) is a polynomial in x, the equality in (66) is an identity and holds for all x.

It might be noted that $\Pr\{r < R\} = 1 - \Pr\{r \ge R\}$ can be derived from Lemma 6.7.7 and Theorem 6.7.4 and has an algebraic form similar to that of (52). (See Problem 35.) It might also be noted that Theorem 6.7.5 can be used to prove Theorem 6.7.4 by induction; in fact, R. L. Anderson (1942) found (52) by a method of induction (applied to the joint distribution of r and Q_0). (See Problem 36.)

Another method of obtaining Theorem 6.7.4 is to invert the characteristic function. [See Koopmans (1942).] For

$$\Pr\{r \ge R\} = \Pr\left\{\sum_{j=1}^{H} (\nu_j - R)x_j \ge 0\right\} \tag{68}$$

and

$$\mathscr{E} \exp\left[it\sum_{j=1}^{H} (\nu_j - R)x_j\right] = \frac{1}{\prod_{j=1}^{H} [1 - 2it(\nu_j - R)]}. \tag{69}$$

Because the roots λ_t appear in pairs, namely ν_j, which are different, the characteristic function (69) has simple poles, and the method of residues can be used to invert the characteristic function. (See Problem 37.)

6.7.5. Distributions of Circular Serial Correlation Coefficients

The characteristic roots of A_1 in the circular model are $\cos 2\pi t/T$, $t = 1, \ldots, T$. The roots occur in pairs except for $\cos 2\pi T/T = 1$ and $\cos 2\pi \frac{1}{2}T/T = -1$ in case T is even. Since $\boldsymbol{\varepsilon}$ is a characteristic vector of A_1 corresponding to the characteristic root 1, the weights for deviations from the sample mean are

$$\nu_j = \cos\frac{2\pi j}{T}, \qquad j = 1, \ldots, \tfrac{1}{2}(T-1) = H, \quad T \text{ odd}, \tag{70}$$

$$\nu_j = \cos\frac{2\pi j}{T}, \qquad j = 1, \ldots, \tfrac{1}{2}T - 1 = H, \quad \nu_{H+1} = -1, \quad T \text{ even}. \tag{71}$$

THEOREM 6.7.7. *If $y_1, \ldots, y_T$ are independently normally distributed, each according to $N(\mu, \sigma^2)$, and*

$$r_1^* = \frac{\sum_{t=1}^{T}(y_t - \bar{y})(y_{t-1} - \bar{y})}{\sum_{t=1}^{T}(y_t - \bar{y})^2} = \frac{\sum_{t=1}^{T} y_t y_{t-1} - T\bar{y}^2}{\sum_{t=1}^{T} y_t^2 - T\bar{y}^2}, \qquad y_0 \equiv y_T, \tag{72}$$

then

$$\Pr\{r_1^* \geq R\} = \sum_{i=1}^{m} \frac{(\nu_i - R)^{\frac{1}{2}(T-3)}}{\alpha_i}, \qquad \nu_{m+1} \leq R \leq \nu_m, \tag{73}$$

where $\nu_i = \cos 2\pi i/T$ and

$$\alpha_i = \prod_{\substack{j=1\\ j\neq i}}^{\frac{1}{2}(T-1)} (\nu_i - \nu_j), \qquad T = 3, 5, \ldots, \tag{74}$$

$$\alpha_i = \sqrt{\nu_i + 1} \prod_{\substack{j=1\\ j\neq i}}^{\frac{1}{2}T-1} (\nu_i - \nu_j), \qquad T = 4, 6, \ldots, \tag{75}$$

and $m = 1, \ldots, \frac{1}{2}(T-3)$ if T is odd and $m = 1, \ldots, \frac{1}{2}T - 1$ if T is even.

R. L. Anderson (1942) has given this distribution and prepared a table of the values of R for which the cumulative distribution is 0.01, 0.05, 0.95, and 0.99. These are given in Table 6.1.

The distribution of $\sum_{t=1}^{T} y_t y_{t-1}/\sum_{t=1}^{T} y_t^2$ can be deduced from Theorem 6.7.6 if T is odd, because the roots occur in pairs except for $\cos 2\pi T/T = 1$.

TABLE 6.1

CUMULATIVE DISTRIBUTION OF THE CIRCULAR SERIAL CORRELATION WITH DEVIATIONS FROM THE SAMPLE MEAN†

T	0.01	0.05	0.95	0.99
5	−0.798	−0.753	0.253	0.297
6	−0.863	−0.708	0.345	0.447
7	−0.799	−0.674	0.370	0.510
8	−0.764	−0.625	0.371	0.531
9	−0.737	−0.593	0.366	0.533
10	−0.705	−0.564	0.360	0.525
11	−0.679	−0.539	0.353	0.515
12	−0.655	−0.516	0.348	0.505
13	−0.634	−0.497	0.341	0.495
14	−0.615	−0.479	0.335	0.485
15	−0.597	−0.462	0.328	0.475
20	−0.524	−0.399	0.299	0.432
25	−0.473	−0.356	0.276	0.398
30	−0.433	−0.324	0.257	0.370
(35)	−0.401	−0.300	0.242	0.347
(40)	−0.376	−0.279	0.229	0.329
45	−0.356	−0.262	0.218	0.313
(50)	−0.339	−0.248	0.208	0.301
(55)	−0.324	−0.236	0.199	0.289
(60)	−0.310	−0.225	0.191	0.278
(65)	−0.298	−0.216	0.184	0.268
(70)	−0.287	−0.207	0.178	0.259
75	−0.276	−0.201	0.174	0.250

† Parentheses indicate a graphically interpolated value. Some values are corrections furnished by R. L. Anderson to Dixon (1944).

THEOREM 6.7.8. *If $y_1, \ldots, y_T$ are independently distributed, each according to $N(0, \sigma^2)$, T is odd, and*

$$r_1 = \frac{\sum_{t=1}^{T} y_t y_{t-1}}{\sum_{t=1}^{T} y_t^2}, \qquad y_0 \equiv y_T, \tag{76}$$

then

$$\Pr\{r_1 \le R^*\} = \sum_{i=m}^{\frac{1}{2}(T-1)} \frac{(R^* - \nu_i^*)^{\frac{1}{2}(T-2)}}{\alpha_i^*}, \qquad \nu_m^* \le R^* \le \nu_{m-1}^*, \quad m = 1, \ldots, \tfrac{1}{2}(T-1), \tag{77}$$

where $\nu^* = \cos 2\pi i/T$, $i = 0, \ldots, \frac{1}{2}(T-1)$, *and*

$$\alpha_i^* = \sqrt{1-\nu_i^*}\prod_{\substack{j=1\\ j\neq i}}^{\frac{1}{2}(T-1)}(\nu_j^* - \nu_i^*). \tag{78}$$

PROOF. The theorem follows from Theorem 6.7.6 by replacing ν_i by $-\nu_{H+1-i}^*$, ν_{H+1} by $-\nu_0^* = -1$, R by $-R^*$, and H by $\frac{1}{2}(T-1)$ [R.L. Anderson (1941)].

6.7.6. Other Distributions of Serial Correlation Coefficients

In the case of the first-order serial correlation coefficient based on the mean square successive difference with residuals from the sample mean, namely,

$$r_1^* = \frac{\sum_{t=2}^{T}(y_t-\bar{y})(y_{t-1}-\bar{y}) + \frac{1}{2}(y_1-\bar{y})^2 + \frac{1}{2}(y_T-\bar{y})^2}{\sum_{t=1}^{T}(y_t-\bar{y})^2}, \tag{79}$$

the roots are $\cos \pi t/T$, $t = 1, \ldots, T-1$, $T = 2, 3, \ldots$; the roots are all different. Let

$$s = \frac{\sum_{t=1}^{T-1}\lambda_t z_t^2}{\sum_{t=1}^{T-1} z_t^2}, \tag{80}$$

where $\lambda_{T-1} \leq \lambda_{T-2} \leq \cdots \leq \lambda_1$ and the λ_t's are arbitrary. Von Neumann (1941) has shown that if $T-1 = 2H$ and $f(s)$ is the density of s then

$$\begin{aligned}\frac{d^{H-1}}{ds^{H-1}}f(s) &= 0, \qquad \lambda_{2h+1} < s < \lambda_{2h}, \quad h = 1, \ldots, H-1,\\ &= \frac{(-1)^{H-h}(H-1)!}{\pi\sqrt{-\prod_{t=1}^{2H}(s-\lambda_t)}}, \qquad \lambda_{2h} < s < \lambda_{2h-1}, \quad h = 1, \ldots, H,\end{aligned} \tag{81}$$

for such h that the inequalities may be satisfied. (For instance, if $\lambda_1 = \lambda_2 > \lambda_3 = \lambda_4 > \cdots > \lambda_{2H-1} = \lambda_{2H}$, then $\lambda_{2h} < s < \lambda_{2h-1}$ is impossible.) In an interval $\lambda_{2h+1} < s < \lambda_{2h}$, the density is a polynomial of degree at most $H-2$. (In the case of double roots, all different, this result agrees with the results of Section 6.7.4.)

It was shown by von Neumann (1941) that if the λ_t's are all different, $d^k f(s)/ds^k$ are continuous, $\lambda_{2H} \le s \le \lambda_1$, $k = 0, 1, \ldots, H-2$ (even at $s = \lambda_t$, $t = 1, \ldots, 2H$). In principle, (81) can be integrated to obtain the density $f(s)$. The constants of integration are determined by the fact that derivatives of order less than $H-1$ in adjacent intervals are equal at the common endpoints (and are 0 at $s = \lambda_{2H}$ and $s = \lambda_1$). In intervals $\lambda_{2h} < s < \lambda_{2h-1}$ the denominator of the $(H-1)$st derivative of the density is the square root of a polynomial in s of degree $2H$. If $H = 1$ $(T = 3)$, the density is

$$(82) \qquad f(s) = \frac{1}{\pi\sqrt{\left(\frac{\lambda_1 - \lambda_2}{2}\right)^2 - \left(s - \frac{\lambda_1 + \lambda_2}{2}\right)^2}}, \qquad \lambda_2 < s < \lambda_1,$$

$$= 0, \qquad \text{otherwise};$$

the cumulative distribution function is $F(s) = 0$ for $s \le \lambda_2$,

$$(83) \qquad F(s) = \int_{\lambda_2}^{s} f(t)\, dt$$

$$= \tfrac{1}{2} + \frac{1}{\pi} \arcsin \frac{2s - (\lambda_1 + \lambda_2)}{\lambda_1 - \lambda_2}, \qquad \lambda_2 < s < \lambda_1,$$

and $F(s) = 1$ for $s \ge \lambda_1$. For $H > 1$ elliptic, hyperelliptic and more complicated integrals are involved.

The distribution for $T-1$ odd can be obtained from the distribution for $T-1$ even with the help of Corollary 6.7.4.

In the case of the serial correlation (79) $\lambda_t = \cos \pi t/T$, $t = 1, \ldots, T-1$. The distribution of r_1^* is symmetric because $\lambda_{T-t} = -\lambda_t$, $t = 1, \ldots, T-1$. The roots are paired (positive and negative values) except when T is even and then $\lambda_{\frac{1}{2}T} = 0$. The range of values of r_1^* is $(-\cos \pi/T, \cos \pi/T)$.

6.7.7. Moment Generating Functions and Moments

If $r_1^* = Q_1^*/Q_0^*$,

$$(84) \qquad \mathscr{E} r_1^{*h} = \mathscr{E} Q_1^{*h} Q_0^{*-h}.$$

The moments of Q_1^* and Q_0^* can be found from the characteristic function or moment generating function of Q_0^* and Q_1^* (the moments of positive order by differentiation and the moments of negative order by integration). Under the null hypothesis of independence $(\gamma_1 = 0)$, however, the deduction of the moments is accomplished effectively by virtue of the fact that r_1^* and Q_0^* are independent (Theorem 6.7.2).

LEMMA 6.7.8. *If* $r_1^* = Q_1^*/Q_0^*$ *and* $\gamma_1 = 0$,

$$\mathscr{E} r_1^{*h} Q_0^{*g} = \mathscr{E} r_1^{*h} \mathscr{E} Q_0^{*g}. \tag{85}$$

THEOREM 6.7.9. *If* $r_1^* = Q_1^*/Q_0^*$ *and* $\gamma_1 = 0$,

$$\mathscr{E} r_1^{*h} = \frac{\mathscr{E} Q_1^{*h}}{\mathscr{E} Q_0^{*h}}. \tag{86}$$

PROOF. This follows from Lemma 6.7.8 by letting $g = h$.

If $r_1 = Q_1/Q_0$ and $\gamma_1 = 0$, then similarly $\mathscr{E} r_1^h = \mathscr{E} Q_1^h / \mathscr{E} Q_0^h$.

Usually Q_0^* has the canonical form $\sum_{t=1}^{T-1} z_t^2$, and Q_0^* has the χ^2-distribution with $T - 1$ degrees of freedom. Similarly Q_0 has the canonical form $\sum_{t=1}^{T} z_t^2$, and Q_0 has the χ^2-distribution with T degrees of freedom. The moments are

$$\begin{aligned} \mathscr{E} Q_0^{*h} &= 2^h \frac{\Gamma[\frac{1}{2}(T-1)+h]}{\Gamma[\frac{1}{2}(T-1)]}, & -\tfrac{1}{2}(T-1) &< h, \\ \mathscr{E} Q_0^h &= 2^h \frac{\Gamma(\frac{1}{2}T+h)}{\Gamma(\frac{1}{2}T)}, & -\tfrac{1}{2}T &< h. \end{aligned} \tag{87}$$

To find $\mathscr{E} Q_1^h$ it is convenient to use the moment generating function. If $Q_1 = \boldsymbol{y}' \boldsymbol{A}_1 \boldsymbol{y}$, where $\boldsymbol{A}_1$ is an arbitrary symmetric matrix with characteristic roots $\lambda_1, \ldots, \lambda_T$, and $\boldsymbol{A}_0 = \boldsymbol{I}$, then from Corollaries 6.7.1 and 6.7.2

$$\mathscr{E} e^{\theta Q_1} = \frac{1}{\sqrt{|\boldsymbol{I} - 2\theta \boldsymbol{A}_1|}} = \frac{1}{\sqrt{\prod_{t=1}^{T} (1 - 2\theta\lambda_t)}} \tag{88}$$

for θ sufficiently small in absolute value. (See Problem 29.) We shall evaluate (88) in the first three specific models defined in Section 6.5.

LEMMA 6.7.9. *The determinant* $D_T = |\boldsymbol{I} - 2\theta \boldsymbol{A}_1|$, *where* $\boldsymbol{A}_1$ *has* $\frac{1}{2}$'s *in the diagonals immediately above and below the main diagonal and* 0's *elsewhere, is*

$$D_T = \frac{1}{\sqrt{1-4\theta^2}} \left[\left(\frac{1+\sqrt{1-4\theta^2}}{2} \right)^{T+1} - \left(\frac{1-\sqrt{1-4\theta^2}}{2} \right)^{T+1} \right], \tag{89}$$

$$T = 1, 2, \ldots.$$

PROOF. Expansion of D_T by elements of the first row gives

$$
(90) \qquad D_T = \begin{vmatrix} 1 & -\theta & 0 & 0 & \cdots & 0 & 0 \\ -\theta & 1 & -\theta & 0 & \cdots & 0 & 0 \\ 0 & -\theta & 1 & -\theta & \cdots & 0 & 0 \\ 0 & 0 & -\theta & 1 & \cdots & 0 & 0 \\ \cdot & \cdot & \cdot & \cdot & & \cdot & \cdot \\ \cdot & \cdot & \cdot & \cdot & & \cdot & \cdot \\ \cdot & \cdot & \cdot & \cdot & & \cdot & \cdot \\ 0 & 0 & 0 & 0 & \cdots & 1 & -\theta \\ 0 & 0 & 0 & 0 & \cdots & -\theta & 1 \end{vmatrix}
$$

$$
= D_{T-1} + \theta \begin{vmatrix} -\theta & -\theta & 0 & \cdots & 0 & 0 \\ 0 & 1 & -\theta & \cdots & 0 & 0 \\ 0 & -\theta & 1 & \cdots & 0 & 0 \\ \cdot & \cdot & \cdot & & \cdot & \cdot \\ \cdot & \cdot & \cdot & & \cdot & \cdot \\ \cdot & \cdot & \cdot & & \cdot & \cdot \\ 0 & 0 & 0 & \cdots & 1 & -\theta \\ 0 & 0 & 0 & \cdots & -\theta & 1 \end{vmatrix}
$$

$$
= D_{T-1} - \theta^2 D_{T-2}, \qquad T = 3, 4, \ldots,
$$

where $D_1 = 1$. Thus D_T satisfies the difference equation

$$
(91) \qquad D_T - D_{T-1} + \theta^2 D_{T-2} = 0,
$$

which has the associated polynomial equation

$$
(92) \qquad x^2 - x + \theta^2 = 0,
$$

whose roots are $\frac{1}{2}(1 \pm \sqrt{1 - 4\theta^2})$. Thus the solution for $\theta^2 \neq 1/4$ is

$$
(93) \quad D_T = c_1\left(\frac{1 + \sqrt{1 - 4\theta^2}}{2}\right)^T + c_2\left(\frac{1 - \sqrt{1 - 4\theta^2}}{2}\right)^T, \qquad T = 1, 2, \ldots .
$$

The constants c_1 and c_2 can be determined by $D_1 = 1$ and $D_2 = 1 - \theta^2$ for $\theta \neq 0$. This proves the lemma, which is needed for the model of Section 6.5.4.

LEMMA 6.7.10. *The determinant* $\Delta_T = |\boldsymbol{I} - 2\theta \boldsymbol{A}_1|$, *where* $\boldsymbol{A}_1$ *is the matrix defining the first-order circular serial correlation coefficient, is*

$$(94) \qquad \Delta_T = \left(\frac{1 + \sqrt{1 - 4\theta^2}}{2}\right)^T + \left(\frac{1 - \sqrt{1 - 4\theta^2}}{2}\right)^T - 2\theta^T$$

$$= \left[\left(\frac{\sqrt{1 + 2\theta} + \sqrt{1 - 2\theta}}{2}\right)^T - \left(\frac{\sqrt{1 + 2\theta} - \sqrt{1 - 2\theta}}{2}\right)^T\right]^2,$$

$$T = 2, 3, \ldots .$$

PROOF. Expansion of Δ_T by elements of the first row and then by elements of the first column of cofactors gives

$$(95) \qquad \Delta_T = \begin{vmatrix} 1 & -\theta & 0 & 0 & \cdots & 0 & 0 & -\theta \\ -\theta & 1 & -\theta & 0 & \cdots & 0 & 0 & 0 \\ 0 & -\theta & 1 & -\theta & \cdots & 0 & 0 & 0 \\ 0 & 0 & -\theta & 1 & \cdots & 0 & 0 & 0 \\ \cdot & \cdot & \cdot & \cdot & & \cdot & \cdot & \cdot \\ \cdot & \cdot & \cdot & \cdot & & \cdot & \cdot & \cdot \\ \cdot & \cdot & \cdot & \cdot & & \cdot & \cdot & \cdot \\ 0 & 0 & 0 & 0 & \cdots & 1 & -\theta & 0 \\ 0 & 0 & 0 & 0 & \cdots & -\theta & 1 & -\theta \\ -\theta & 0 & 0 & 0 & \cdots & 0 & -\theta & 1 \end{vmatrix}$$

$$= D_{T-1} + \theta \begin{vmatrix} -\theta & -\theta & 0 & \cdots & 0 & 0 & 0 \\ 0 & 1 & -\theta & \cdots & 0 & 0 & 0 \\ 0 & -\theta & 1 & \cdots & 0 & 0 & 0 \\ \cdot & \cdot & \cdot & & \cdot & \cdot & \cdot \\ \cdot & \cdot & \cdot & & \cdot & \cdot & \cdot \\ \cdot & \cdot & \cdot & & \cdot & \cdot & \cdot \\ 0 & 0 & 0 & \cdots & 1 & -\theta & 0 \\ 0 & 0 & 0 & \cdots & -\theta & 1 & -\theta \\ -\theta & 0 & 0 & \cdots & 0 & -\theta & 1 \end{vmatrix}$$

$$
+ (-1)^T \theta \begin{vmatrix}
-\theta & 1 & -\theta & 0 & \cdots & 0 & 0 \\
0 & -\theta & 1 & -\theta & \cdots & 0 & 0 \\
0 & 0 & -\theta & 1 & \cdots & 0 & 0 \\
\cdot & \cdot & \cdot & \cdot & & \cdot & \cdot \\
\cdot & \cdot & \cdot & \cdot & & \cdot & \cdot \\
\cdot & \cdot & \cdot & \cdot & & \cdot & \cdot \\
0 & 0 & 0 & 0 & \cdots & 1 & -\theta \\
0 & 0 & 0 & 0 & \cdots & -\theta & 1 \\
-\theta & 0 & 0 & 0 & \cdots & 0 & -\theta
\end{vmatrix}
$$

$$
\begin{aligned}
&= D_{T-1} - \theta^2 D_{T-2} + (-1)^{T-1}\theta^2(-\theta)^{T-2} \\
&\quad - (-1)^T\theta^2(-\theta)^{T-2} + (-1)^T\theta^2(-1)^{T-1}D_{T-2} \\
&= D_{T-1} - 2\theta^2 D_{T-2} - 2\theta^T, \qquad T = 3, 4, \ldots .
\end{aligned}
$$

Substitution from (89) gives (94). For $T = 2$ $a_{12} = a_{21} = 1$, and, hence, $\Delta_2 = 1 - 4\theta^2$.

LEMMA 6.7.11. *The determinant $C_T = |\boldsymbol{I} - 2\theta \boldsymbol{A}_1|$, where $\boldsymbol{A}_1$ is the matrix based on successive differences, is*

$$
C_T = \frac{1 - 2\theta}{\sqrt{1 - 4\theta^2}}\left[\left(\frac{1 + \sqrt{1 - 4\theta^2}}{2}\right)^T - \left(\frac{1 - \sqrt{1 - 4\theta^2}}{2}\right)^T\right], \qquad T = 2, 3, \ldots . \tag{96}
$$

PROOF. Expansion of C_T (by elements of the first row and next by elements of the last row of one cofactor and the first column of the other) gives

$$
C_T = \begin{vmatrix}
1-\theta & -\theta & 0 & 0 & \cdots & 0 & 0 & 0 \\
-\theta & 1 & -\theta & 0 & \cdots & 0 & 0 & 0 \\
0 & -\theta & 1 & -\theta & \cdots & 0 & 0 & 0 \\
0 & 0 & -\theta & 1 & \cdots & 0 & 0 & 0 \\
\cdot & \cdot & \cdot & \cdot & & \cdot & \cdot & \cdot \\
\cdot & \cdot & \cdot & \cdot & & \cdot & \cdot & \cdot \\
\cdot & \cdot & \cdot & \cdot & & \cdot & \cdot & \cdot \\
0 & 0 & 0 & 0 & \cdots & 1 & -\theta & 0 \\
0 & 0 & 0 & 0 & \cdots & -\theta & 1 & -\theta \\
0 & 0 & 0 & 0 & \cdots & 0 & -\theta & 1-\theta
\end{vmatrix} \tag{97}
$$

$$= (1 - \theta)\begin{vmatrix} 1 & -\theta & 0 & \cdots & 0 & 0 & 0 \\ -\theta & 1 & -\theta & \cdots & 0 & 0 & 0 \\ 0 & -\theta & 1 & \cdots & 0 & 0 & 0 \\ \cdot & \cdot & \cdot & & \cdot & \cdot & \cdot \\ \cdot & \cdot & \cdot & & \cdot & \cdot & \cdot \\ \cdot & \cdot & \cdot & & \cdot & \cdot & \cdot \\ 0 & 0 & 0 & \cdots & 1 & -\theta & 0 \\ 0 & 0 & 0 & \cdots & -\theta & 1 & -\theta \\ 0 & 0 & 0 & \cdots & 0 & -\theta & 1 - \theta \end{vmatrix}$$

$$+ \theta\begin{vmatrix} -\theta & -\theta & 0 & \cdots & 0 & 0 & 0 \\ 0 & 1 & -\theta & \cdots & 0 & 0 & 0 \\ 0 & -\theta & 1 & \cdots & 0 & 0 & 0 \\ \cdot & \cdot & \cdot & & \cdot & \cdot & \cdot \\ \cdot & \cdot & \cdot & & \cdot & \cdot & \cdot \\ \cdot & \cdot & \cdot & & \cdot & \cdot & \cdot \\ 0 & 0 & 0 & \cdots & 1 & -\theta & 0 \\ 0 & 0 & 0 & \cdots & -\theta & 1 & -\theta \\ 0 & 0 & 0 & \cdots & 0 & -\theta & 1 - \theta \end{vmatrix}$$

$$= (1 - \theta)^2 D_{T-2} + (1 - \theta)\theta\begin{vmatrix} 1 & -\theta & 0 & \cdots & 0 & 0 \\ -\theta & 1 & -\theta & \cdots & 0 & 0 \\ 0 & -\theta & 1 & \cdots & 0 & 0 \\ \cdot & \cdot & \cdot & & \cdot & \cdot \\ \cdot & \cdot & \cdot & & \cdot & \cdot \\ \cdot & \cdot & \cdot & & \cdot & \cdot \\ 0 & 0 & 0 & \cdots & 1 & 0 \\ 0 & 0 & 0 & \cdots & -\theta & -\theta \end{vmatrix}$$

$$
-\theta^2 \begin{vmatrix} 1 & -\theta & \cdots & 0 & 0 & 0 \\ -\theta & 1 & \cdots & 0 & 0 & 0 \\ \cdot & \cdot & & \cdot & \cdot & \cdot \\ \cdot & \cdot & & \cdot & \cdot & \cdot \\ \cdot & \cdot & & \cdot & \cdot & \cdot \\ 0 & 0 & \cdots & 1 & -\theta & 0 \\ 0 & 0 & \cdots & -\theta & 1 & -\theta \\ 0 & 0 & \cdots & 0 & -\theta & 1-\theta \end{vmatrix}
$$

$$
= (1-\theta)^2 D_{T-2} - 2(1-\theta)\theta^2 D_{T-3}
$$

$$
-\theta^3 \begin{vmatrix} 1 & -\theta & \cdots & 0 & 0 \\ -\theta & 1 & \cdots & 0 & 0 \\ \cdot & \cdot & & \cdot & \cdot \\ \cdot & \cdot & & \cdot & \cdot \\ \cdot & \cdot & & \cdot & \cdot \\ 0 & 0 & \cdots & 1 & 0 \\ 0 & 0 & \cdots & -\theta & -\theta \end{vmatrix}
$$

$$
= (1-\theta)^2 D_{T-2} - 2(1-\theta)\theta^2 D_{T-3} + \theta^4 D_{T-4}, \quad T = 5, 6, \ldots .
$$

Substitution from (89) gives

$$
\begin{aligned}
(98) \quad C_T = \frac{1-2\theta}{2\sqrt{1-4\theta^2}} \Bigg\{ & [(1-3\theta^2) + (1-\theta^2)\sqrt{1-4\theta^2}] \left(\frac{1+\sqrt{1-4\theta^2}}{2} \right)^{T-3} \\
& - [(1-3\theta^2) - (1-\theta^2)\sqrt{1-4\theta^2}] \left(\frac{1-\sqrt{1-4\theta^2}}{2} \right)^{T-3} \Bigg\}.
\end{aligned}
$$

Rearrangement of terms gives (96) for $T = 5, 6, \ldots$. Direct calculation verifies (96) for $T = 2, 3, 4$.

Another form of the determinant, obtained by expanding the two Tth powers in (96) according to the binomial theorem, is

$$
\begin{aligned}
(99) \qquad C_T &= \frac{1}{2^T} \frac{1-2\theta}{\sqrt{1-4\theta^2}} \sum_{\nu=0}^{T} \binom{T}{\nu} (1-4\theta^2)^{\frac{1}{2}\nu} [1-(-1)^\nu] \\
&= \frac{1-2\theta}{2^{T-1}} \sum_{\alpha=0}^{H} \binom{T}{2\alpha+1} (1-4\theta^2)^\alpha,
\end{aligned}
$$

where $H = \frac{1}{2}(T-1)$ if T is odd and $H = \frac{1}{2}T - 1$ if T is even.

LEMMA 6.7.12. *If* $Q_1^* = (y - \bar{y}\varepsilon)' A_1 (y - \bar{y}\varepsilon)$, $A_0 = I$, ε *is a characteristic vector of* A_1 *corresponding to a characteristic root* 1, $\gamma_0 = 1$, *and* $\gamma_1 = 0$, *then*

$$\mathscr{E} e^{\theta Q_1^*} = \sqrt{\frac{1 - 2\theta}{|I - 2\theta A_1|}} \tag{100}$$

for θ *sufficiently small in absolute value.*

PROOF. This lemma is a special case of Corollary 6.7.2.

THEOREM 6.7.10. *The moment generating function of* Q_1^* *for the circular serial correlation coefficient when* $\gamma_0 = 1$ *and* $\gamma_1 = 0$ *is*

$$\mathscr{E} e^{\theta Q_1^*} = \frac{\sqrt{1 - 2\theta}}{\left(\dfrac{\sqrt{1 + 2\theta} + \sqrt{1 - 2\theta}}{2}\right)^T - \left(\dfrac{\sqrt{1 + 2\theta} - \sqrt{1 - 2\theta}}{2}\right)^T}, \qquad T = 2, 3, \ldots, \tag{101}$$

for θ *sufficiently small in absolute value.*

THEOREM 6.7.11. *The moment generating function of* Q_1^* *based on successive differences when* $\gamma_0 = 1$ *and* $\gamma_1 = 0$ *is*

$$\mathscr{E} e^{\theta Q_1^*} = \sqrt{\frac{\sqrt{1 - 4\theta^2}}{\left(\dfrac{1 + \sqrt{1 - 4\theta^2}}{2}\right)^T - \left(\dfrac{1 - \sqrt{1 - 4\theta^2}}{2}\right)^T}}, \qquad T = 2, 3, \ldots, \tag{102}$$

for θ *sufficiently small in absolute value.*

It will be observed that (102) is the same as $1/\sqrt{D_{T-1}}$, which is the moment generating function of $Q_1 = \sum_{t=2}^{T-1} y_t y_{t-1}$ defined in Section 6.5.4 for $T - 1$ observations.

The moments of the circular Q_1 can be found from the moment generating function

$$\begin{aligned}
\frac{1}{\sqrt{\Delta_T}} \tag{103}
&= \frac{1}{\left(\dfrac{1 + \sqrt{1 - 4\theta^2}}{2}\right)^{\frac{1}{2}T} - \left(\dfrac{\sqrt{1 + 2\theta} - \sqrt{1 - 2\theta}}{2}\right)^T} \\
&= \left(\frac{1 + \sqrt{1 - 4\theta^2}}{2}\right)^{-\frac{1}{2}T} \left[1 - \left(\frac{\sqrt{1 + 2\theta} - \sqrt{1 - 2\theta}}{\sqrt{1 + 2\theta} + \sqrt{1 - 2\theta}}\right)^T\right]^{-1} \\
&= \left(\frac{1 + \sqrt{1 - 4\theta^2}}{2}\right)^{-\frac{1}{2}T} \left[1 + \left(\frac{\sqrt{1 + 2\theta} - \sqrt{1 - 2\theta}}{\sqrt{1 + 2\theta} + \sqrt{1 - 2\theta}}\right)^T + \cdots\right], \qquad T = 2, 3, \ldots,
\end{aligned}$$

for θ sufficiently small in absolute value. Laplace (1829) has shown that if u is the root $1 + \sqrt{1 - c}$ of $v^2 - 2v + c = 0$, then

$$u^{-n} = \frac{1}{2^n} + \frac{n}{2^n}\left\{\frac{c}{4} + \frac{n+3}{2!}\left(\frac{c}{4}\right)^2 + \frac{(n+4)(n+5)}{3!}\left(\frac{c}{4}\right)^3 + \cdots + \frac{\Gamma(n+2k)}{k!\,\Gamma(n+k+1)}\left(\frac{c}{4}\right)^k + \cdots\right\}. \tag{104}$$

The derivation of this expansion is not straightforward and will not be given here. If we let $c = 4\theta^2$ and $n = \frac{1}{2}T$, we have

$$\left(\frac{1 + \sqrt{1 - 4\theta^2}}{2}\right)^{-\frac{1}{2}T} = 1 + \tfrac{1}{2}T\left\{\theta^2 + \frac{\frac{1}{2}T + 3}{2!}\theta^4 + \frac{(\frac{1}{2}T+4)(\frac{1}{2}T+5)}{3!}\theta^6 + \cdots + \frac{\Gamma(\frac{1}{2}T + 2k)}{k!\,\Gamma(\frac{1}{2}T + k + 1)}\theta^{2k} + \cdots\right\}. \tag{105}$$

Since

$$\begin{aligned}\frac{\sqrt{1+2\theta} - \sqrt{1-2\theta}}{\sqrt{1+2\theta} + \sqrt{1-2\theta}} &= \frac{\sqrt{1+2\theta} - \sqrt{1-2\theta}}{\sqrt{1+2\theta} + \sqrt{1-2\theta}} \cdot \frac{\sqrt{1+2\theta} + \sqrt{1-2\theta}}{\sqrt{1+2\theta} + \sqrt{1-2\theta}} \\ &= \frac{2\theta}{(1 + \sqrt{1 - 4\theta^2}} \\ &= \theta\left(\frac{1 + \sqrt{1-4\theta^2}}{2}\right)^{-1},\end{aligned} \tag{106}$$

the second term in the square brackets in (103) is a power series starting with θ^T and higher powers. Thus the square brackets contain $1 +$ powers of θ at least T. The moments of Q_1 to order $T - 1$ are

$$\mathscr{E}Q_1^n = \frac{d^n}{d\theta^n}\frac{1}{\sqrt{\Delta_T}}\bigg|_{\theta=0} = \frac{d^n}{d\theta^n}\left(\frac{1 + \sqrt{1 - 4\theta^2}}{2}\right)^{-\frac{1}{2}T}\bigg|_{\theta=0}, \qquad n = 0, 1, \ldots, T-1. \tag{107}$$

(See Problem 51.) From (105) and (107) we obtain

$$\begin{aligned}\mathscr{E}Q_1^{2k-1} &= 0, && 1 \le k \le \tfrac{1}{2}T, \quad T = 2, 3, \ldots, \\ \mathscr{E}Q_1^{2k} &= \frac{T}{2}\frac{\Gamma(\frac{1}{2}T + 2k)(2k)!}{\Gamma(\frac{1}{2}T + k + 1)k!}, && 0 \le k \le \tfrac{1}{2}(T-1), \quad T = 2, 3, \ldots.\end{aligned} \tag{108}$$

From this result and (87) we have

$$\mathscr{E}r_1^{2k-1} = 0, \qquad 1 \le k \le \tfrac{1}{2}T, \quad T = 2, 3, \ldots,$$

(109)
$$\mathscr{E}r_1^{2k} = \frac{\Gamma(\frac{1}{2}T + 1)\Gamma(k + \frac{1}{2})}{\Gamma(\frac{1}{2}T + k + 1)\Gamma(\frac{1}{2})}, \quad 0 \le k \le \tfrac{1}{2}(T - 1), \quad T = 2, 3, \ldots.$$

The duplication formula for the Gamma function,

(110)
$$\Gamma(2k + 1) = \frac{2^{2k}\Gamma(k + \frac{1}{2})\Gamma(k + 1)}{\sqrt{\pi}},$$

has been used.

The moments of r_1^*, the circular serial correlation in terms of residuals from the mean, can be obtained from the moments of Q_1^* and Q_0^*; the moments of Q_1^* can be found from the moment generating function given in Theorem 6.7.10. Moments up to order $T - 1$ can be obtained from the approximate moment generating function

(111)
$$\left(\frac{1 + \sqrt{1 - 4\theta^2}}{2}\right)^{-\frac{1}{2}T} \sqrt{1 - 2\theta}, \qquad T = 2, 3, \ldots,$$

since the first $T - 1$ derivatives of (111) and (101) are equal at $\theta = 0$.

LEMMA 6.7.13. *Let*

(112)
$$\phi(T) = \left(\frac{1 + \sqrt{1 - 4\theta^2}}{2}\right)^{-\frac{1}{2}T}, \quad T = \ldots, -1, 0, 1, \ldots,$$

and

(113)
$$\phi^*(T) = \left(\frac{1 + \sqrt{1 - 4\theta^2}}{2}\right)^{-\frac{1}{2}T} \sqrt{1 - 2\theta}, \quad T = \cdots, -1, 0, 1, \ldots.$$

Then for real θ, $-\frac{1}{2} \le \theta \le \frac{1}{2}$,

(114)
$$\phi^*(T) = \phi(T - 1) - \theta\phi(T + 1), \quad T = \ldots, -1, 0, 1, \ldots.$$

PROOF. We have

(115)
$$\phi(T - 1) - \theta\phi(T + 1) = \left(\frac{1 + \sqrt{1 - 4\theta^2}}{2}\right)^{-\frac{1}{2}(T+1)} \left(\frac{1 - 2\theta + \sqrt{1 - 4\theta^2}}{2}\right)$$

$$= \left(\frac{1 + \sqrt{1 - 4\theta^2}}{2}\right)^{-\frac{1}{2}T} \sqrt{1 - 2\theta} \;\; \frac{\sqrt{1 - 2\theta} + \sqrt{1 + 2\theta}}{\sqrt{2}\,(1 + \sqrt{1 - 4\theta^2})^{\frac{1}{2}}}$$

$$= \phi^*(T) \frac{\sqrt{1 - 2\theta} + \sqrt{1 + 2\theta}}{\sqrt{2}\,(1 + \sqrt{1 - 4\theta^2})^{\frac{1}{2}}},$$

which is $\phi^*(T)$ for real θ, $-\frac{1}{2} \le \theta \le \frac{1}{2}$. Q.E.D.

It follows from the lemma and the fact that $\phi(T-1)$ and $\phi(T+1)$ are expanded in even powers of θ that

$$\begin{aligned}(116) \qquad \mathscr{E}[Q_1^*(T)]^{2k-1} &= \left.\frac{d^{2k-1}\phi^*(T)}{d\theta^{2k-1}}\right|_{\theta=0} \\ &= -\left.\frac{d^{2k-1}\theta\phi(T+1)}{d\theta^{2k-1}}\right|_{\theta=0}, \\ & \qquad 1 \le k \le \tfrac{1}{2}T, \ T = 2, 3, \ldots,\end{aligned}$$

which is $-(2k-1)\mathscr{E}Q_1^{2k-2}(T+1)$, and

$$\begin{aligned}(117) \qquad \mathscr{E}[Q_1^*(T)]^{2k} &= \left.\frac{d^{2k}\phi^*(T)}{d\theta^{2k}}\right|_{\theta=0} \\ &= \left.\frac{d^{2k}\phi(T-1)}{d\theta^{2k}}\right|_{\theta=0}, \\ & \qquad 0 \le k \le \tfrac{1}{2}(T-1), \quad T = 2, 3, \ldots,\end{aligned}$$

which is $\mathscr{E}Q_1^{2k}(T-1)$ for $0 \le k \le \frac{1}{2}(T-2)$, $T = 3, 4, \ldots$, where $Q_1^*(T)$ and $Q_1(T)$ denote Q_1^* and Q_1 based on T observations, respectively; the limits for k and T are limits for the validity of the right-hand sides as determined earlier. Thus

$$(118) \qquad \mathscr{E}Q_1^{*2k-1} = -\frac{T+1}{2} \cdot \frac{\Gamma(\frac{1}{2}T - \frac{3}{2} + 2k)(2k-1)!}{\Gamma(\frac{1}{2}T + \frac{1}{2} + k)(k-1)!}, \qquad 1 \le k \le \tfrac{1}{2}T, \quad T = 2, 3, \ldots,$$

$$(119) \qquad \mathscr{E}Q_1^{*2k} = \frac{T-1}{2} \cdot \frac{\Gamma(\frac{1}{2}T - \frac{1}{2} + 2k)(2k)!}{\Gamma(\frac{1}{2}T + \frac{1}{2} + k)k!}, \qquad 0 \le k \le \tfrac{1}{2}(T-1), \quad T = 2, 3, \ldots.$$

Then

$$\begin{aligned}(120) \quad \mathscr{E}r_1^{*2k-1} &= -\frac{T+1}{T-1} \cdot \frac{\Gamma(\frac{1}{2}T + \frac{1}{2})\Gamma(k + \frac{1}{2})}{\Gamma(\frac{1}{2}T + \frac{1}{2} + k)\Gamma(\frac{1}{2})}, 1 \le k \le \tfrac{1}{2}T, \quad T = 2, 3, \ldots, \\ &= -\frac{T+1}{T-1}\mathscr{E}r_1^{*2k}, \qquad 1 \le k \le \tfrac{1}{2}(T-1), \quad T = 2, 3, \ldots,\end{aligned}$$

$$(121) \quad \mathscr{E}r_1^{*2k} = \frac{\Gamma(\frac{1}{2}T + \frac{1}{2})\Gamma(k + \frac{1}{2})}{\Gamma(\frac{1}{2}T + \frac{1}{2} + k)\Gamma(\frac{1}{2})}, \quad 0 \le k \le \tfrac{1}{2}(T-1), \quad T = 2, 3, \ldots.$$

Note that $\mathscr{E}r_1^{*2k}$ is the same as $\mathscr{E}r_1^{2k}$ for $T-1$ observations.

Another way of finding the moments is to note that $Q_1 = Q_1^* + z_T^2$ and Q_1^* and z_T are independent. Then

$$\mathscr{E}Q_1^h = \mathscr{E}(Q_1^* + z_T^2)^h = \sum_{j=0}^{h} \binom{h}{j} \mathscr{E}(z_T^2)^j \mathscr{E}Q_1^{*h-j} = \sum_{j=0}^{h} \binom{h}{j} \frac{\Gamma(j+\frac{1}{2})}{\Gamma(\frac{1}{2})} 2^j \mathscr{E}Q_1^{*h-j}. \tag{122}$$

Consider now the general quadratic form Q with characteristic roots $\lambda_1, \ldots, \lambda_M$. Then when $\gamma_0 = 1$ and $\gamma_1 = 0$

$$\begin{aligned} \mathscr{E}e^{\theta Q} &= \prod_{t=1}^{M} (1 - 2\theta\lambda_t)^{-\frac{1}{2}} \\ &= \exp\left\{-\frac{1}{2}\sum_{t=1}^{M} \log(1 - 2\theta\lambda_t)\right\} \\ &= \exp\left\{\frac{1}{2}\sum_{t=1}^{M}\left[2\theta\lambda_t + \frac{(2\theta\lambda_t)^2}{2} + \frac{(2\theta\lambda_t)^3}{3} + \cdots\right]\right\} \\ &= \exp\left\{\frac{1}{2}\left[2\theta\sum_{t=1}^{M}\lambda_t + \frac{(2\theta)^2}{2}\sum_{t=1}^{M}\lambda_t^2 + \frac{(2\theta)^3}{3}\sum_{t=1}^{M}\lambda_t^3 + \cdots\right]\right\}. \end{aligned} \tag{123}$$

The kth cumulant is $k!$ times

$$\nu_k = \frac{2^{k-1}}{k}\sum_{t=1}^{M}\lambda_t^k; \tag{124}$$

then

$$\mathscr{E}e^{\theta Q} = \exp(\nu_1\theta + \nu_2\theta^2 + \nu_3\theta^3 + \cdots) = 1 + \psi_1\theta + \psi_2\theta^2 + \psi_3\theta^3 + \cdots, \tag{125}$$

where

$$\begin{aligned} \psi_1 &= \nu_1, \\ \psi_2 &= \nu_2 + \tfrac{1}{2}\nu_1^2, \\ \psi_3 &= \nu_3 + \nu_1\nu_2 + \tfrac{1}{6}\nu_1^3, \\ \psi_4 &= \nu_4 + \tfrac{1}{2}\nu_2^2 + \nu_1\nu_3 + \tfrac{1}{2}\nu_1^2\nu_2 + \tfrac{1}{24}\nu_1^4. \end{aligned} \tag{126}$$

More of the ψ_k's can be computed from the previous identity. Then

$$\mathscr{E} Q^h = h! \, \psi_h. \tag{127}$$

If $\lambda_t = -\lambda_{M+1-t}$ (as for Q_1^* in the successive difference case), $\nu_1 = \nu_3 = \cdots = 0$, and $\psi_1 = \psi_3 = \cdots = 0$; then

$$\begin{aligned} \psi_2 &= \nu_2, \\ \psi_4 &= \nu_4 + \tfrac{1}{2}\nu_2^2, \\ \psi_6 &= \nu_6 + \nu_2\nu_4 + \tfrac{1}{6}\nu_2^3, \\ \psi_8 &= \nu_8 + \tfrac{1}{2}\nu_4^2 + \nu_2\nu_6 + \tfrac{1}{2}\nu_2^2\nu_4 + \tfrac{1}{24}\nu_2^4. \end{aligned} \tag{128}$$

If $M = T - 1$ and $\lambda_t = \cos \pi t/T$, then for k even

$$\begin{aligned} \nu_k &= \frac{2^{k-1}}{k} \sum_{t=1}^{T-1} \cos^k \frac{\pi t}{T} \\ &= \frac{1}{2k} \sum_{t=1}^{T-1} (e^{i\pi t/T} + e^{-i\pi t/T})^k \\ &= \frac{1}{2k} \sum_{j=0}^{k} \binom{k}{j} \left[\sum_{t=0}^{T-1} e^{i2\pi t(j-\frac{1}{2}k)/T} - 1 \right]. \end{aligned} \tag{129}$$

The sum inside the brackets is T if $j - \frac{1}{2}k$ is divisible by T (including $j - \frac{1}{2}k = 0$) and is 0 otherwise. Then

$$\nu_k = \frac{T}{2k} {\sum_j}' \binom{k}{j} - \frac{2^{k-1}}{k}, \tag{130}$$

where $\sum_j'$ indicates the sum over the values of j, $j = 0, 1, \ldots, k$, for which $j - \frac{1}{2}k$ is divisible by T. These values are $j = \frac{1}{2}k \pm hT$, $h = 0, 1, 2, \ldots$ $(0 \leq hT \leq \frac{1}{2}k)$. Since

$$\binom{k}{\frac{1}{2}k + hT} = \binom{k}{\frac{1}{2}k - hT}, \tag{131}$$

we have $\nu_{2l-1} = 0$ and

$$\nu_{2l} = \frac{1}{4l} \left\{ T \left[\binom{2l}{l} + 2 \sum_{h=1}^{[l/T]} \binom{2l}{l - hT} \right] - 2^{2l} \right\}. \tag{132}$$

The number of terms in the sum is 0 if $l < T$, is 1 if $T \leq l < 2T$, is 2 if $2T \leq l < 3T$, etc. In particular, we have

(133)
$$\nu_{2l} = \frac{1}{4l}\left\{\binom{2l}{l}T - 2^{2l}\right\}, \qquad l < T,$$

(134)
$$\nu_2 = \tfrac{1}{2}(T-2), \qquad T = 2, 3, \ldots,$$
$$\nu_4 = \frac{3T-8}{4}, \qquad T = 3, 4, \ldots,$$
$$\nu_6 = \frac{5T-16}{3}, \qquad T = 4, 5, \ldots,$$
$$\nu_8 = \frac{35T-128}{8}, \qquad T = 5, 6, \ldots,$$

(135)
$$\psi_2 = \tfrac{1}{2}(T-2), \qquad T = 2, 3, \ldots,$$
$$\psi_4 = \frac{T^2 + 2T - 12}{8}, \qquad T = 3, 4, \ldots,$$
$$\psi_6 = \frac{T^3 + 12T^2 + 8T - 168}{48}, \qquad T = 4, 5, \ldots,$$
$$\psi_8 = \frac{T^4 + 28T^3 + 212T^2 - 64T - 3696}{384}, \qquad T = 5, 6, \ldots.$$

As noted before, $\psi_k = 0$ for k odd.

The moments of r_1^* based on the successive differences are

(136)
$$\mathscr{E}r_1^{*2} = \frac{T-2}{(T-1)(T+1)} = \frac{T-2}{T^2-1}, \qquad T = 2, 3, \ldots,$$
$$\mathscr{E}r_1^{*4} = \frac{3(T^2+2T-12)}{(T-1)(T+1)(T+3)(T+5)}, \qquad T = 3, 4, \ldots,$$
$$\mathscr{E}r_1^{*6} = \frac{15(T^3+12T^2+8T-168)}{(T-1)(T+1)(T+3)(T+5)(T+7)(T+9)}, \qquad T = 4, 5, \ldots,$$
$$\mathscr{E}r_1^{*8} = \frac{105(T^4+28T^3+212T^2-64T-3696)}{(T-1)(T+1)(T+3)(T+5)(T+7)(T+9)(T+11)(T+13)},$$
$$T = 5, 6, \ldots.$$

The expansion of the moment generating function in (123) was given by von Neumann (1941). It can be used for other serial correlation coefficients. For example, in the case of the first-order circular serial correlation coefficient in terms of residuals from the sample mean the coefficients in the exponent of the moment generating function are

$$\begin{aligned}
\nu_k &= \frac{2^{k-1}}{k} \sum_{t=1}^{T-1} \cos^k \frac{2\pi}{T} t \\
&= \frac{1}{2k} \sum_{t=1}^{T-1} (e^{i2\pi t/T} + e^{-i2\pi t/T})^k \\
&= \frac{1}{2k} \sum_{t=1}^{T-1} \sum_{j=0}^{k} \binom{k}{j} e^{i2\pi t(2j-k)/T} \\
&= \frac{1}{2k} \sum_{j=0}^{k} \binom{k}{j} \left[\sum_{t=0}^{T-1} e^{i2\pi t(2j-k)/T} - 1 \right].
\end{aligned} \tag{137}$$

The sum inside the brackets is T if $2j - k$, $j = 0, 1, \ldots, k$, is divisible by T (including $2j - k = 0$) and is 0 otherwise. Then

$$\nu_k = \frac{T}{2k} \sum_j{}' \binom{k}{j} - \frac{2^{k-1}}{k}, \tag{138}$$

where $\sum'$ indicates the sum over the values of j, $j = 0, 1, \ldots, k$, for which $2j - k$ is divisible by T. These values are obtained as follows. Suppose k is even, say $k = 2l$. Then the sum is over values of j such that $2(j - l)$ is divisible by T. If T is even, say $T = 2H$, then the sum is over values of j such that $j - l$ is divisible by H; these values are $j = l \pm hH = \frac{1}{2}k \pm \frac{1}{2}hT$, $h = 0, 1, 2, \ldots$ $(0 \leq hT \leq k)$. If T is odd, then the sum is over values of j such that $j - l$ is divisible by T; these values are $j = l \pm hT, h = 0, 1, 2, \ldots$ $(0 \leq hT \leq l = \frac{1}{2}k)$. Thus

$$\begin{aligned}
\nu_{2l} &= \frac{1}{4l}\left\{T\left[\binom{2l}{l} + 2\sum_{h=1}^{[2l/T]} \binom{2l}{l - \frac{1}{2}hT}\right] - 2^{2l}\right\}, \quad T = 2, 4, \ldots, \\
&= \frac{1}{4l}\left\{T\left[\binom{2l}{l} + 2\sum_{h=1}^{[l/T]} \binom{2l}{l - hT}\right] - 2^{2l}\right\}, \quad T = 3, 5, \ldots .
\end{aligned} \tag{139}$$

Suppose k is odd, say $k = 2l - 1$. Then the sum is over values of j such that $2j - 2l + 1 = 2(j - l) + 1$ is divisible by T. If T is even, there are no such values of j. If T is odd, we have $2(j - l) + 1 = \pm hT$, and hence h must be odd, say $h = 2g - 1$ $(g = 1, 2, \ldots)$; thus $2(j - l) + 1 = \pm(2g - 1)T$ and

$$j = l \pm (gT - \tfrac{1}{2}T) - \tfrac{1}{2}. \tag{140}$$

If $T > 2l - 1$, there is no g possible and, hence, no j. We have

$$\begin{aligned} \nu_{2l-1} &= -\frac{2^{2l-1}}{2(2l-1)}, & T &= 2, 4, \ldots, \\ &= \frac{1}{2(2l-1)}\left\{2T \sum_{g=1}^{[(l-\frac{1}{2})/T+\frac{1}{2}]} \binom{2l-1}{l - \frac{1}{2} - (g - \frac{1}{2})T} - 2^{2l-1}\right\}, & T &= 3, 5, \ldots. \end{aligned} \tag{141}$$

In particular, we have

$$\nu_{2l-1} = -\frac{2^{2l-1}}{2(2l-1)}, \quad 1 \le l < \tfrac{1}{2}(T+1), \quad T = 2, 3, \ldots, \tag{142}$$

$$\begin{aligned} \nu_{2l} = \frac{T}{2(2l)}\binom{2l}{l} - \frac{2^{2l}}{2(2l)}, \quad & 1 \le l < T, & T &= 3, 5, \ldots, \\ & 1 \le l < \tfrac{1}{2}T, & T &= 2, 4, \ldots, \end{aligned} \tag{143}$$

$$\begin{aligned} \nu_1 &= -1, & T &= 2, 3, \ldots, \\ \nu_2 &= \tfrac{1}{2}(T-2), & T &= 3, 4, \ldots, \\ \nu_3 &= -\tfrac{4}{3}, & T &= 4, 5, \ldots, \\ \nu_4 &= \frac{3T-8}{4}, & T &= 5, 6, \ldots, \end{aligned} \tag{144}$$

$$\begin{aligned} \psi_1 &= -1, & T &= 2, 3, \ldots, \\ \psi_2 &= \tfrac{1}{2}(T-1), & T &= 3, 4, \ldots, \\ \psi_3 &= -\tfrac{1}{2}(T+1), & T &= 4, 5, \ldots, \\ \psi_4 &= \frac{(T-1)(T+5)}{8}, & T &= 5, 6, \ldots. \end{aligned} \tag{145}$$

The first few moments of the circular r_1^* are

(146)
$$\begin{aligned} \mathscr{E} r_1^* &= -\frac{1}{T-1}, && T = 2, 3, \dots, \\ \mathscr{E} r_1^{*2} &= \frac{1}{T+1}, && T = 3, 4, \dots, \\ \mathscr{E} r_1^{*3} &= -\frac{3}{(T-1)(T+3)}, && T = 4, 5, \dots, \\ \mathscr{E} r_1^{*4} &= \frac{3}{(T+1)(T+3)}, && T = 5, 6, \dots . \end{aligned}$$

These, of course, agree with (120) and (121).

The expansion of von Neumann can be related to the use of part of the generating function to get the first few moments. In the case of the serial correlation coefficient r_1^* based on successive differences we use the expression (133) for ν_k for $k < 2T$ and k even, say ν_k^*. We approximate $\log \mathscr{E} e^{\theta Q}$ by

(147)
$$\begin{aligned} \sum_{j=1}^{\infty} \theta^{2j} \nu_{2j}^* &= \sum_{j=1}^{\infty} \frac{\theta^{2j}}{4j} \left\{ \binom{2j}{j} T - 2^{2j} \right\} \\ &= T \sum_{j=1}^{\infty} \frac{\theta^{2j}}{4j} \frac{(2j)!}{(j!)^2} - \sum_{j=1}^{\infty} \frac{(4\theta^2)^j}{4j} \\ &= T \sum_{j=1}^{\infty} \frac{\theta^{2j}}{4j} \frac{\Gamma(2j+1)}{j!\,\Gamma(j+1)} + \tfrac{1}{4} \log(1 - 4\theta^2). \end{aligned}$$

The first sum above can be written (by use of the duplication formula for the Gamma function)

(148)
$$T \sum_{j=1}^{\infty} \frac{(4\theta^2)^j \Gamma(j + \frac{1}{2})}{4j \cdot j!\, \Gamma(\frac{1}{2})}.$$

From

(149)
$$\sum_{h=0}^{\infty} \frac{x^h \Gamma(h + \frac{1}{2})}{h!\, \Gamma(\frac{1}{2})} = (1 - x)^{-\frac{1}{2}},$$

we find

(150)
$$\begin{aligned} \sum_{h=1}^{\infty} \frac{x^{h-1} \Gamma(h + \frac{1}{2})}{h!\, \Gamma(\frac{1}{2})} &= \frac{1}{x} \left[(1 - x)^{-\frac{1}{2}} - 1 \right] \\ &= \frac{1 - \sqrt{1 - x}}{x \sqrt{1 - x}}. \end{aligned}$$

This leads to

$$\sum_{h=1}^{\infty} \frac{x^h \Gamma(h + \frac{1}{2})}{h \cdot h!\, \Gamma(\frac{1}{2})} = -2 \log \left(\frac{1 + \sqrt{1 - x}}{2} \right); \tag{151}$$

this is verified by differentiating both sides of (151) and checking that the two sides are equal for $x = 0$. Putting these results together, we obtain the approximation

(152)

$$\exp\left(\sum_{j=1}^{\infty} \theta^{2j} v_{2j}^* \right) = \exp\left\{ \frac{T}{4}\left[-2 \log \left(\frac{1 + \sqrt{1 - 4\theta^2}}{2} \right) \right] + \tfrac{1}{4} \log (1 - 4\theta^2) \right\}$$

$$= \frac{(1 - 4\theta^2)^{\frac{1}{4}}}{\left(\dfrac{1 + \sqrt{1 - 4\theta^2}}{2} \right)^{\frac{1}{2}T}}.$$

This is equivalent to the approximate moment generating function obtained by deleting $([1 - \sqrt{1 - 4\theta^2}]/2)^T$ from (102).

6.8. APPROXIMATE DISTRIBUTIONS OF SERIAL CORRELATION COEFFICIENTS

6.8.1. Approximate Distributions of the Circular Serial Correlation Coefficients

Although the distribution of the circular serial correlation coefficient in terms of residuals from the sample mean is known and is fairly simple in the case of independent normal observations and although it has been tabulated to some extent (Table 6.1), it is convenient to be able to use approximate distributions for determining significance points and appraising the variability of this r_1^*. We shall also consider approximate distributions for the circular serial correlation coefficient r_1 in terms of observations measured from their known mean; these approximations are simple because the approximate distributions are symmetric.

Most of the approximations are based on fitting a Beta-type density

$$f(x) = \frac{\Gamma(p + q)}{\Gamma(p)\Gamma(q)} \frac{(x - a)^{p-1}(b - x)^{q-1}}{(b - a)^{p+q-1}}, \qquad a \le x \le b, \tag{1}$$

and $f(x) = 0$ otherwise, with $p > 0$ and $q > 0$. The hth moment of this density is

$$\int_a^b x^h f(x)\,dx = \frac{\Gamma(p+q)}{\Gamma(p)\Gamma(q)} \int_a^b \frac{(x-a)^{p-1}(b-x)^{q-1}x^h}{(b-a)^{p+q-1}}\,dx \tag{2}$$

$$= \frac{\Gamma(p+q)}{\Gamma(p)\Gamma(q)} \int_0^1 z^{p-1}(1-z)^{q-1}[a+(b-a)z]^h\,dz$$

$$= \frac{\Gamma(p+q)}{\Gamma(p)\Gamma(q)} \sum_{j=0}^{h} \binom{h}{j} (b-a)^j a^{h-j} \int_0^1 z^{p+j-1}(1-z)^{q-1}\,dz$$

$$= \frac{\Gamma(p+q)}{\Gamma(p)} \sum_{j=0}^{h} \binom{h}{j} (b-a)^j a^{h-j} \frac{\Gamma(p+j)}{\Gamma(p+q+j)}.$$

For example,

$$\mathscr{E}X = a + (b-a)\frac{p}{p+q}, \tag{3}$$

$$\mathscr{E}X^2 = a^2 + 2a(b-a)\frac{p}{p+q} + (b-a)^2\frac{p(p+1)}{(p+q)(p+q+1)}, \tag{4}$$

$$\mathscr{E}X^3 = a^3 + 3a^2(b-a)\frac{p}{p+q} + 3a(b-a)^2\frac{p(p+1)}{(p+q)(p+q+1)} + (b-a)^3\frac{p(p+1)(p+2)}{(p+q)(p+q+1)(p+q+2)}, \tag{5}$$

$$\mathscr{E}X^4 = a^4 + 4a^3(b-a)\frac{p}{p+q} + 6a^2(b-a)^2\frac{p(p+1)}{(p+q)(p+q+1)} + 4a(b-a)^3\frac{p(p+1)(p+2)}{(p+q)(p+q+1)(p+q+2)} + (b-a)^4\frac{p(p+1)(p+2)(p+3)}{(p+q)(p+q+1)(p+q+2)(p+q+3)}. \tag{6}$$

If $a = -b$ and $q = p$, the density is symmetric,

$$f(x) = \frac{\Gamma(2p)}{\Gamma^2(p)} \frac{(b+x)^{p-1}(b-x)^{p-1}}{(2b)^{2p-1}} \tag{7}$$

$$= \frac{\Gamma(2p)}{\Gamma^2(p)} \frac{(b^2-x^2)^{p-1}}{(2b)^{2p-1}}, \qquad -b \le x \le b.$$

The odd moments are 0, and the even moments are

$$\int_{-b}^{b} x^{2k} f(x)\,dx = \frac{\Gamma(2p)}{\Gamma^2(p)(2b)^{2p-1}} \int_{-b}^{b} (b^2 - x^2)^{p-1} x^{2k}\,dx \tag{8}$$

$$= \frac{\Gamma(2p) b^{2k}}{\Gamma^2(p) 2^{2p-1}} \int_0^1 (1-y)^{p-1} y^{k-\frac{1}{2}}\,dy$$

$$= b^{2k} \frac{\Gamma(2p)\Gamma(k+\frac{1}{2})}{2^{2p-1}\Gamma(p)\Gamma(p+k+\frac{1}{2})}, \qquad k = 0, 1, \ldots .$$

Since (8) is 1 for $k = 0$, we have the duplication formula for the Gamma function

$$\Gamma(2p) = \frac{2^{2p-1}\Gamma(p)\Gamma(p+\frac{1}{2})}{\Gamma(\frac{1}{2})}. \tag{9}$$

The density (7) and moments (8) can be written respectively

$$f(x) = \frac{\Gamma(p+\frac{1}{2})}{\Gamma(\frac{1}{2})\Gamma(p)} \frac{(b^2 - x^2)^{p-1}}{b^{2p-1}}, \qquad -b \le x \le b, \tag{10}$$

$$\mathscr{E} X^{2k} = b^{2k} \frac{\Gamma(p+\frac{1}{2})\Gamma(k+\frac{1}{2})}{\Gamma(\frac{1}{2})\Gamma(p+k+\frac{1}{2})}. \tag{11}$$

Examples of the moments are $\mathscr{E} X = \mathscr{E} X^3 = 0$, and

$$\mathscr{E} X^2 = \frac{b^2}{2p+1}, \tag{12}$$

$$\mathscr{E} X^4 = \frac{3b^4}{(2p+1)(2p+3)}. \tag{13}$$

The ordinary Pearson correlation coefficient r (using residuals from the sample means) based on N pairs of independent observations from a bivariate normal distribution with 0 correlation has such a density, namely

$$\frac{\Gamma[\frac{1}{2}(N-1)]}{\Gamma(\frac{1}{2})\Gamma[\frac{1}{2}(N-2)]} (1 - r^2)^{\frac{1}{2}(N-4)}, \qquad -1 \le r \le 1, \tag{14}$$

with $2k$th moment

$$\mathscr{E} r^{2k} = \frac{\Gamma[\frac{1}{2}(N-1)]\Gamma(k+\frac{1}{2})}{\Gamma(\frac{1}{2})\Gamma[\frac{1}{2}(N-1)+k]}. \tag{15}$$

The odd moments of r_1 are 0; in fact, the distribution of r_1 is symmetric if T is even. The low-order even moments of r_1 are

$$\mathscr{E} r_1^{2k} = \frac{\Gamma(\frac{1}{2}T + 1)\Gamma(k + \frac{1}{2})}{\Gamma(\frac{1}{2}T + k + 1)\Gamma(\frac{1}{2})}, \qquad 0 \le 2k \le T - 1, \quad T = 2, 3, \ldots . \tag{16}$$

We approximate the density of r_1 by the density (14) of the Pearson correlation with $N = T + 3$. The moments of the exact and approximate distributions agree to order $T - 1$.

One way of arriving at these moments and distribution is to approximate the moment generating function of Q_1,

$$\begin{aligned} \mathscr{E} e^{\theta Q_1} &= \prod_{t=1}^{T} \left(1 - 2\theta \cos \frac{2\pi t}{T}\right)^{-\frac{1}{2}} \\ &= \exp\left[-\frac{1}{2}\sum_{t=1}^{T} \log\left(1 - 2\theta\cos\frac{2\pi t}{T}\right)\right]. \end{aligned} \tag{17}$$

The exponent in (17) is

$$-\frac{T}{4\pi}\sum_{t=1}^{T} \log\left(1 - 2\theta\cos\frac{2\pi t}{T}\right)\frac{2\pi}{T}, \tag{18}$$

which is an approximating sum for the integral

$$-\frac{T}{4\pi}\int_0^{2\pi} \log(1 - 2\theta\cos u)\, du = -\frac{T}{2}\log\left(\frac{1 + \sqrt{1 - 4\theta^2}}{2}\right). \tag{19}$$

(See Problem 69 for an outline of the verification of this equality.) This gives as an approximation to the moment generating function

$$\exp\{-\tfrac{1}{2}T \log[(1 + \sqrt{1 - 4\theta^2})/2]\} = \left(\frac{1 + \sqrt{1 - 4\theta^2}}{2}\right)^{-\frac{1}{2}T}, \tag{20}$$

which was given in Section 6.7.7. Koopmans (1942) used this idea to find an expression for the approximate density, and Rubin (1945) showed it is the same as (14) with N replaced by $T + 3$.

We now turn to r_1^*, the circular serial correlation coefficient in terms of residuals from the mean. Since the even moments of r_1^* are the same as those of r_1 with one less observation we could use the distribution of Pearson's r with $N = T + 2$; this approximation has the same even moments as the exact distribution to order $T - 1$, but the odd moments will not agree.

Another way to use the Beta-type distribution is to set $b = -a = 1$ and find p and q in (1) so that (3) is $\mathscr{E}r_1^* = -1/(T-1)$ and (4) is $\mathscr{E}r_1^{*2} = 1/(T+1)$, $T = 3, 4, \ldots$. ($r_1^* \equiv -1$ for $T = 2$.) The resulting values of p and q are

$$p = \frac{(T-1)(T-2)}{2(T-3)}, \qquad q = \frac{T(T-1)}{2(T-3)}. \tag{21}$$

Dixon (1944) has compared this approximation with the exact distribution. (See Table 6.2.)

TABLE 6.2

EXACT AND APPROXIMATE CUMULATIVE DISTRIBUTIONS OF THE CIRCULAR SERIAL CORRELATION WITH DEVIATIONS FROM THE SAMPLE MEAN†

	0.01				0.05			
T	Exact	Beta $p \neq q$	r	Normal	Exact	Beta $p \neq q$	r	Normal
5	−0.798	−0.858	−0.989	−1.000	−0.753	−0.742	−0.821	−0.781
10	−0.705	−0.702	−0.734	−0.763	−0.564	−0.562	−0.576	−0.572
15	−0.597	−0.596	−0.609	−0.629	−0.462	−0.461	−0.467	−0.466
20	−0.524	−0.524	−0.532	−0.545	−0.399	−0.399	−0.402	−0.401
25	−0.473	−0.473	−0.477	−0.487	−0.356	−0.356	−0.357	−0.357
30	−0.433	−0.433	−0.436	−0.444	−0.324	−0.324	−0.324	−0.324
45	−0.356	−0.356	−0.357	−0.362	−0.262	−0.262	−0.262	−0.262
75	−0.276	−0.276	−0.277	−0.278	−0.201	−0.201	−0.201	−0.201

	0.95				0.99			
T	Exact	Beta $p \neq q$	r	Normal	Exact	Beta $p \neq q$	r	Normal
5	0.253	0.317	0.421	0.281	0.297	0.527	0.589	0.501
10	0.360	0.362	0.376	0.350	0.525	0.533	0.534	0.541
15	0.328	0.329	0.333	0.323	0.475	0.477	0.476	0.486
20	0.299	0.299	0.302	0.296	0.432	0.433	0.432	0.440
25	0.276	0.276	0.277	0.274	0.398	0.398	0.397	0.404
30	0.257	0.257	0.258	0.255	0.370	0.371	0.370	0.375
45	0.218	0.218	0.218	0.217	0.313	0.313	0.313	0.316
75	0.174	0.174	0.175	0.174	0.250	0.250	0.250	0.251

† The column labelled Beta $p \neq q$ was calculated by Dixon (1944) for the Beta distribution fitted by (21). The column labelled r is computed by treating $r_1^* + 1/T$ as a Pearson correlation with $N = T + 3$; see (28).

Another use of the Beta-type distribution takes into account the range of the circular serial correlation coefficient. Equating (3) and (4) to $\mathscr{E}r_1^*$ and $\mathscr{E}r_1^{*2}$, respectively, we obtain

$$p + q = \frac{T+1}{T^2 - 3T}[-(T-1)a - 1][(T-1)b + 1] - 1, \tag{22}$$

$$p = \frac{-(T-1)a - 1}{(T-1)(b-a)}(p+q), \tag{23}$$

$$q = \frac{(T-1)b + 1}{(T-1)(b-a)}(p+q). \tag{24}$$

If T is even, the range is $-1 \leq r_1^* \leq \cos 2\pi/T$; we take $a = -1$ and $b = \cos 2\pi/T$. If T is odd, the range is $\cos \pi(T-1)/T \leq r_1^* \leq \cos 2\pi/T$; we take $a = -\cos \pi/T$ and $b = \cos 2\pi/T$.

Another way of using the Beta-type distribution has been suggested by Hannan (1960), namely, that $r_1^* + 1/T$ is approximately distributed as Pearson's r based on $T + 3$ observations. If we take

$$a = -1 - \frac{1}{T-1} = -\frac{T}{T-1}, \tag{25}$$

$$b = 1 - \frac{1}{T-1} = \frac{T-2}{T-1}, \tag{26}$$

then

$$p = q = \frac{1}{2}\left(T + 1 + \frac{5T+1}{T^2 - 3T}\right). \tag{27}$$

We can round the values off to

$$a = -1 - \frac{1}{T}, \qquad b = 1 - \frac{1}{T}, \qquad p = q = \frac{T+1}{2} \tag{28}$$

and obtain the rule that r_1^* based on T observations is treated like $r - 1/T$, where r is based on $T + 3$ observations.

The limiting distribution of

$$\left(r_1^* + \frac{1}{T-1}\right)\frac{(T-1)\sqrt{T+1}}{\sqrt{T^2 - 3T}} \tag{29}$$

is $N(0, 1)$. (Asymptotically, r_1^* is equivalent to $-\hat{\beta}_1$ of Chapter 5.) This suggests using Fisher's z-transformation

$$z = \tfrac{1}{2}\log\frac{1 + (r_1^* + 1/T)}{1 - (r_1^* + 1/T)}; \tag{30}$$

then $\sqrt{T}\,z$ is approximately distributed according to $N(0, 1)$. (The preceding approximation could possibly be improved by replacing $r_1^* + 1/T$ by $f(T)[r_1^* + 1/T]$, where $f(T)$ is an easily calculated function, slightly greater than 1.)

A numerical comparison of some of the approximations is given in Table 6.2. It will be observed that the Beta approximation is accurate for T at least equal to 15 or 20. The use of r is slightly less accurate; for most purposes it is accurate for $T \geq 30$.

6.8.2. Approximate Distributions of the Serial Correlation Based on Successive Differences

The exact distribution of r_1^* based on successive differences was given explicitly by Hart and von Neumann (1942) for $T = 3$, 4, and 5. The integrals required for the distributions for $T = 6$ and 7 were evaluated numerically to obtain tables. For $T = 8, 9, \ldots$ the density was approximated by a few terms of the series

$$\sum_{j=0}^{\infty} a_j \left(\cos^2 \frac{\pi}{T} - r_1^{*2}\right)^{\frac{1}{2}T-2+j}. \tag{31}$$

This series was chosen because r_1^* is symmetrically distributed, has range $(-\cos \pi/T, \cos \pi/T)$, and has a density with order of vanishing $\frac{1}{2}T - 2$ at $\pm\cos \pi/T$. Four terms were used ($j = 0, 1, 2, 3$); the coefficients were chosen to duplicate the first three even moments of r_1^*. For $T = 7$ the approximation compares well with the exact distribution. Over the range of values of the cdf from 0.00678 to 0.07020 the error is less than 1.7%. The eighth moments of the exact and approximate distributions are 0.00413 and 0.00412, respectively; the tenth moments are 0.00202 and 0.00201. The significance points are given in Table 6.3.

We can consider approximating the density of r_1^* by a symmetric Beta-type distribution with $b = -a = \cos \pi/T$. Then $\mathscr{E} r_1^{*2} = \mathscr{E} X^2$ for

$$p = q = \frac{1}{2}\left[\frac{T^2 - 1}{T - 2} \cdot \cos^2 \frac{\pi}{T} - 1\right]. \tag{32}$$

If we approximate $\cos \pi/T$ by $1 - \frac{1}{2}(\pi/T)^2$ and $\cos^2 \pi/T$ by $1 - (\pi/T)^2$ and this by $1 - 9/T^2$, we find

$$\begin{aligned} p = q &= \frac{1}{2}\left[\frac{T^2 - 1}{T - 2} \cdot \frac{T^2 - 9}{T^2} - 1\right] \\ &= \tfrac{1}{2}(T + 1) - \frac{3(2T^2 - 3)}{2(T^3 - 2T^2)}. \end{aligned} \tag{33}$$

TABLE 6.3

ONE-SIDED SIGNIFICANCE POINTS FOR THE SERIAL CORRELATION COEFFICIENT BASED ON SUCCESSIVE DIFFERENCES†

	Significance Levels						
T	.05	.01	.001	T	.05	.01	.001
4	0.610	0.687	0.705	32	0.282	0.391	0.504
5	0.590	0.731	0.792	33	0.278	0.386	0.497
				34	0.274	0.381	0.491
6	0.555	0.719	0.818	35	0.271	0.376	0.485
7	0.532	0.693	0.815				
8	0.509	0.669	0.798	36	0.267	0.371	0.479
9	0.488	0.646	0.779	37	0.264	0.366	0.474
10	0.469	0.624	0.759	38	0.260	0.362	0.468
				39	0.257	0.358	0.463
11	0.452	0.604	0.740	40	0.254	0.353	0.458
12	0.436	0.586	0.722				
13	0.422	0.569	0.705	41	0.251	0.349	0.452
14	0.409	0.553	0.689	42	0.248	0.345	0.448
15	0.397	0.539	0.673	43	0.245	0.341	0.443
				44	0.242	0.338	0.438
16	0.386	0.525	0.659	45	0.240	0.334	0.434
17	0.376	0.513	0.645				
18	0.367	0.501	0.632	46	0.237	0.331	0.430
19	0.358	0.490	0.619	47	0.235	0.327	0.426
20	0.350	0.480	0.607	48	0.232	0.324	0.422
				49	0.230	0.321	0.418
21	0.343	0.470	0.596	50	0.228	0.319	0.415
22	0.335	0.461	0.586				
23	0.329	0.452	0.576	51	0.226	0.316	0.411
24	0.322	0.444	0.567	52	0.224	0.313	0.408
25	0.316	0.436	0.558	53	0.222	0.310	0.404
				54	0.220	0.308	0.401
26	0.311	0.429	0.549	55	0.218	0.305	0.398
27	0.305	0.422	0.541				
28	0.300	0.415	0.533	56	0.216	0.303	0.395
29	0.295	0.409	0.525	57	0.215	0.300	0.392
30	0.291	0.402	0.518	58	0.213	0.298	0.389
				59	0.211	0.295	0.386
31	0.287	0.397	0.511	60	0.209	0.293	0.383

† Modification by T. W. Anderson (1948) of a table by Hart (1942), calculated from exact distribution for $T = 4, 5, 6, 7$ and approximate distribution for $T \geq 8$.

Young (1941) considered fitting a symmetric Beta-type distribution with the same second and fourth moments as r_1^*. Then

$$p = q = \tfrac{1}{2}(T + 1) - \frac{18}{T^3 - 13T + 24}, \tag{34}$$

$$a^2 = b^2 = 1 - \frac{3T}{T^3 - 13T + 24}, \tag{35}$$

$$b \sim 1 - \frac{3T}{2(T^3 - 13T + 24)}. \tag{36}$$

It will be observed that b corresponds to $\cos \pi/T \sim 1 - 9/(2T^2)$. The tables given by Young based on b in (35) for the .05 and .01 one-sided significance points agree with those given in Table 6.3 to 3 decimal places (with one exception) for $T \geq 8$. The rules can be approximated by saying that $r_1^*/[1 - 9/(2T^2)]$ is treated as the Pearson r for $N = T + 3$ observations; that is, a significance point of r is multiplied by $1 - 9/(2T^2)$ to obtain the significance point for r_1^*. This gives values for .05 and .01 that differ from those of Table 6.3 only in the third decimal place for $T \geq 15$. Replacing $1 - 9/(2T^2)$ by $\cos \pi/T$ makes little difference.

Another approximation is to take $b = -a = 1$ and

$$p = q = \tfrac{1}{2}(T + 1) + \frac{3}{2(T - 2)}. \tag{37}$$

This leads to the rule that r_1^* is treated as the Pearson r corresponding to $N = T + 3$ observations. In Table 6.4 some comparisons are made.

TABLE 6.4

APPROXIMATE ONE-SIDED SIGNIFICANCE POINTS OF THE SERIAL CORRELATION COEFFICIENT BASED ON SUCCESSIVE DIFFERENCES

	Significance Levels					
	.05			.01		
T	Hart	r	br	Hart	r	br
10	0.469	0.476	0.455	0.624	0.634	0.605
15	0.397	0.400	0.392	0.539	0.542	0.532
20	0.350	0.352	0.348	0.480	0.482	0.477
25	0.316	0.317	0.315	0.436	0.437	0.434
30	0.291	0.291	0.290	0.402	0.403	0.401
45	0.240	0.240	0.239	0.334	0.334	0.334

This serial correlation coefficient is also asymptotically normally distributed. It will be observed that the use of r here is about as good as the use of $r - 1/T$ in the circular case.

The entries labelled Hart are taken from Table 6.3. The entries in the column labelled r were computed by treating r_1^* as a Pearson correlation with $N = T + 3$. The entries in the column labelled br are the entries for the preceding column (before rounding) multiplied by $b = 1 - 9/(2T^2)$.

As an example, consider the data in Table 6.5, which gives the ratios of total Republican to total Democratic votes for candidates for the House of Representatives, 1920–1954.

TABLE 6.5

RATIO OF TOTAL REPUBLICAN TO TOTAL DEMOCRATIC VOTES, CANDIDATES FOR HOUSE OF REPRESENTATIVES, 1920–1954†

Year	Ratio	Year	Ratio
1920	1.65	1938	0.97
1922	1.16	1940	0.89
1924	1.38	1942	1.10
1926	1.41	1944	0.93
1928	1.33	1946	1.21
1930	1.18	1948	0.88
1932	0.76	1950	1.00
1934	0.78	1952	1.00
1936	0.71	1954	0.90

† *Statistical Abstract of the United States* [United States Bureau of the Census (1955)], Table 390, p. 330.

The relevant statistics are

$$\bar{y} = 1.06\tfrac{8}{9}, \tag{38}$$

$$Q_0^* = 1.1029\tfrac{7}{9}, \tag{39}$$

$$Q_1^* = 0.67362\tfrac{7}{9}, \tag{40}$$

$$r_1^* = 0.6107. \tag{41}$$

The observed r_1^* is significantly different from 0 at the .01 significance level.

6.9. JOINT AND CONDITIONAL DISTRIBUTIONS OF SERIAL CORRELATION COEFFICIENTS

6.9.1. Joint Distributions

The joint distribution of p serial correlation coefficients $r_1^*, \ldots, r_p^*$ can be determined, in principle, from the characteristic function. As in the case of a single coefficient, the distribution can be found explicitly when the roots of the matrices occur in pairs, except for possibly one root.

Let

$$r_j^* = \frac{(y - \bar{y}\varepsilon)' A_j (y - \bar{y}\varepsilon)}{(y - \bar{y}\varepsilon)'(y - \bar{y}\varepsilon)}, \qquad j = 1, \ldots, p, \tag{1}$$

$$\bar{y} = \frac{y'\varepsilon}{\varepsilon'\varepsilon}, \tag{2}$$

where $\varepsilon = (1, 1, \ldots, 1)'$. Let the characteristic roots of A_j be $\lambda_{j1}, \ldots, \lambda_{jT}$. Suppose $A_1, \ldots, A_p$ have the same set of characteristic vectors, and suppose ε is a characteristic vector of A_j corresponding to the root $\lambda_{jT} = 1, j = 1, \ldots, p$. If y is distributed according to $N(\mu\varepsilon, \sigma^2 I)$, then $r_1^*, \ldots, r_p^*$ are distributed as

$$r_j^* = \frac{\sum_{t=1}^{T-1} \lambda_{jt} z_t^2}{\sum_{t=1}^{T-1} z_t^2}, \qquad j = 1, \ldots, p, \tag{3}$$

where $z_1, \ldots, z_{T-1}$ are independently distributed, each according to $N(0, 1)$. Suppose T is odd, say $T = 2H + 1$, and the roots occur in pairs. Let

$$\lambda_{j,2h-1} = \lambda_{j,2h} = \nu_{jh}, \qquad h = 1, \ldots, H, \tag{4}$$

and let

$$z_{2h-1}^2 + z_{2h}^2 = x_h, \qquad h = 1, \ldots, H. \tag{5}$$

Then $x_1, \ldots, x_H$ are independently distributed as χ^2 with 2 degrees of freedom. Conditional on $\sum_{h=1}^H x_h = 1$, $x_1, \ldots, x_H$ are uniformly distributed on the part of the hyperplane in the positive orthant

$$\sum_{h=1}^{H} x_h = 1, \qquad x_h \geq 0, \qquad h = 1, \ldots, H. \tag{6}$$

Then the distribution of r_j^* is that of

$$r_j^* = \sum_{h=1}^{H} \nu_{jh} x_h, \qquad j = 1, \ldots, p, \tag{7}$$

and

$$\Pr\{r_1^* \geq R_1^*, \ldots, r_p^* \geq R_p^*\} \tag{8}$$

is the $(H - 1)$-dimensional volume of the intersection of (6) and

$$\sum_{h=1}^{H} \nu_{jh} x_h \geq R_j^*, \qquad j = 1, \ldots, p, \tag{9}$$

relative to the $(H - 1)$-dimensional volume of (6). This is a geometric problem similar to that of finding $\Pr\{r_1^* \geq R_1^*\}$. The set of possible values of $(r_1^*, \ldots, r_p^*)$ is difficult to find explicitly. It includes the points $(\nu_{1h}, \ldots, \nu_{ph})$ corresponding to $x_h = 1$, $x_j = 0$, $j \neq h$, for $h = 1, \ldots, H$; it also includes all linear combinations of these with nonnegative weights. Thus the set is the least convex body enclosing these points and $(0, \ldots, 0)$.

The distribution (8) has been given by Quenouille (1949a) and Watson (1956).

6.9.2. Conditional Distributions

As pointed out in Section 6.3, to test $\gamma_i = 0$ when $\gamma_{i+1} = \cdots = \gamma_p = 0$ and μ is known, we use Q_i, given $Q_0, \ldots, Q_{i-1}$. Under the null hypothesis this is equivalent to using r_i, given $r_1, \ldots, r_{i-1}$. Thus the conditional distribution of r_i, given $r_1, \ldots, r_{i-1}$, when $\gamma_i = 0$, is to be used to test this hypothesis. When deviations from the mean are used, we want the conditional distribution of r_i^*, given $r_1^*, \ldots, r_{i-1}^*$.

If T is odd and the roots (other than 1) occur in corresponding pairs, $r_1^*, \ldots, r_i^*$ can be represented as (7). The conditional probability

$$\Pr\{r_i^* \geq R_i^* \mid r_1^* = R_1^*, \ldots, r_{i-1}^* = R_{i-1}^*\} \tag{10}$$

is the $(H - i)$-dimensional volume of the intersection of (6) and

$$\sum_{h=1}^{H} \nu_{ih} x_h \geq R_i^*, \tag{11}$$

$$\sum_{h=1}^{H} \nu_{jh} x_h = R_j^*, \qquad j = 1, \ldots, i - 1, \tag{12}$$

relative to the $(H - i)$-dimensional volume of the intersection of (6) and (12). Again this is a straightforward, but complicated, geometric problem.

Approximate distributions can be found. Let r'_i be the ith partial circular serial correlation coefficient; that is, let r'_i be the partial correlation between y_t and y_{t+i} "holding $y_{t+1}, \ldots, y_{t+i-1}$ fixed." The r'_i is calculated from $r_1^*, \ldots, r_i^*$ as a partial correlation. [See T. W. Anderson (1958), Section 4.3.1, for example.] Specifically, let

$$\begin{bmatrix} 1 & r_1^* & r_2^* & \cdots & r_{i-1}^* & r_i^* \\ r_1^* & 1 & r_1^* & \cdots & r_{i-2}^* & r_{i-1}^* \\ r_2^* & r_1^* & 1 & \cdots & r_{i-3}^* & r_{i-2}^* \\ \cdot & \cdot & \cdot & & \cdot & \cdot \\ \cdot & \cdot & \cdot & & \cdot & \cdot \\ \cdot & \cdot & \cdot & & \cdot & \cdot \\ r_{i-1}^* & r_{i-2}^* & r_{i-3}^* & \cdots & 1 & r_1^* \\ r_i^* & r_{i-1}^* & r_{i-2}^* & \cdots & r_1^* & 1 \end{bmatrix} = \begin{bmatrix} 1 & \boldsymbol{r}^{*\prime} & r_i^* \\ \boldsymbol{r}^* & \boldsymbol{R}^* & \tilde{\boldsymbol{r}}^* \\ r_i^* & \tilde{\boldsymbol{r}}^{*\prime} & 1 \end{bmatrix}; \tag{13}$$

then

$$r'_i = \frac{r_i^* - \tilde{\boldsymbol{r}}^{*\prime}(\boldsymbol{R}^*)^{-1}\boldsymbol{r}^*}{\sqrt{1 - \tilde{\boldsymbol{r}}^{*\prime}(\boldsymbol{R}^*)^{-1}\tilde{\boldsymbol{r}}^*}\,\sqrt{1 - \boldsymbol{r}^{*\prime}(\boldsymbol{R}^*)^{-1}\boldsymbol{r}^*}} = \frac{r_i^* - \tilde{\boldsymbol{r}}^{*\prime}(\boldsymbol{R}^*)^{-1}\boldsymbol{r}^*}{1 - \tilde{\boldsymbol{r}}^{*\prime}(\boldsymbol{R}^*)^{-1}\tilde{\boldsymbol{r}}^*}. \tag{14}$$

In particular,

$$r'_1 = r_1^*, \tag{15}$$

$$r'_2 = \frac{r_2^* - r_1^{*2}}{1 - r_1^{*2}}. \tag{16}$$

Daniels (1956) [and Jenkins (1956) for r'_2, r'_3, and r'_4] has found that the joint distribution of $r'_1, \ldots, r'_i$ can be approximated by a joint distribution in which the partial serial correlations are independent, with the density of r'_{2j} being

$$\begin{aligned} &\left\{\frac{\Gamma(\frac{1}{2})\Gamma[\frac{1}{2}(T-1)]}{\Gamma(\frac{1}{2}T)} + \frac{\Gamma(\frac{3}{2})\Gamma[\frac{1}{2}(T-1)]}{\Gamma(\frac{1}{2}T+1)}\right\}^{-1}(1 - r'_{2j})^2(1 - r'^2_{2j})^{\frac{1}{2}(T-3)} \\ &\quad = \frac{\Gamma(T+1)}{2^T\Gamma[\frac{1}{2}(T-1)]\Gamma[\frac{1}{2}(T+3)]}(1 + r'_{2j})^{\frac{1}{2}(T-3)}(1 - r'_{2j})^{\frac{1}{2}(T+1)}, \end{aligned} \tag{17}$$

and the density of r'_{2j-1} being

$$\begin{aligned} &\frac{\Gamma(\frac{1}{2}T)}{\Gamma(\frac{1}{2})\Gamma[\frac{1}{2}(T-1)]}(1 - r'_{2j-1})(1 - r'^2_{2j-1})^{\frac{1}{2}(T-3)} \\ &\quad = \frac{\Gamma(T)}{2^{T-1}\Gamma[\frac{1}{2}(T-1)]\Gamma[\frac{1}{2}(T+1)]}(1 + r'_{2j-1})^{\frac{1}{2}(T-3)}(1 - r'_{2j-1})^{\frac{1}{2}(T-1)}. \end{aligned} \tag{18}$$

Hannan (1960) suggests treating $r'_{2j} + 2/T$ and $r'_{2j-1} + 1/T$ as Pearson correlations based on $T + 2$ observations. [See also Quenouille (1949b).]

6.10. DISTRIBUTIONS WHEN THE OBSERVATIONS ARE NOT INDEPENDENT

For the models we have been studying, the distributions of serial correlations when the observations are not independent are easily derived from the corresponding distributions when the observations are independent (as given in Sections 6.7.5 and 6.7.6, for instance). [See Leipnik (1947) and Madow (1945).]

We consider the density of $y_1, \ldots, y_T$ as

$$(1)\quad K_i(\gamma_0, \ldots, \gamma_i) \exp[-\tfrac{1}{2}(\mathbf{y} - \mu\boldsymbol{\varepsilon})'(\gamma_0 \mathbf{A}_0 + \gamma_1 \mathbf{A}_1 + \cdots + \gamma_i \mathbf{A}_i)(\mathbf{y} - \mu\boldsymbol{\varepsilon})] = K_i(\gamma_0, \ldots, \gamma_i) \times \exp\{-\tfrac{1}{2}[\gamma_0 Q_0^* + \cdots + \gamma_i Q_i^* + T(\gamma_0\lambda_0 + \gamma_1\lambda_1 + \cdots + \gamma_i\lambda_i)(\bar{y} - \mu)^2]\},$$

where $\boldsymbol{\varepsilon} = (1, 1, \ldots, 1)'$ is a characteristic vector of $\mathbf{A}_j$ corresponding to the characteristic root λ_j,

$$(2)\quad Q_j^* = (\mathbf{y} - \bar{y}\boldsymbol{\varepsilon})'\mathbf{A}_j(\mathbf{y} - \bar{y}\boldsymbol{\varepsilon}), \qquad j = 0, 1, \ldots, i,$$

$\gamma_0, \ldots, \gamma_i$ are numbers such that $\gamma_0 \mathbf{A}_0 + \gamma_1 \mathbf{A}_1 + \cdots + \gamma_i \mathbf{A}_i$ is positive definite, $\gamma_0 > 0$, and

$$(3)\quad K_i(\gamma_0, \ldots, \gamma_i) = |\gamma_0 \mathbf{A}_0 + \gamma_1 \mathbf{A}_1 + \cdots + \gamma_i \mathbf{A}_i|^{\frac{1}{2}}/(2\pi)^{\frac{1}{2}T}.$$

The density of $Q_0^*, \ldots, Q_i^*$ is

$$(4)\quad h_i^*(Q_0^*, \ldots, Q_i^* \mid \gamma_0, \ldots, \gamma_i) = K_i^*(\gamma_0, \ldots, \gamma_i) \exp[-\tfrac{1}{2}(\gamma_0 Q_0^* + \cdots + \gamma_i Q_i^*)] k_i^*(Q_0^*, \ldots, Q_i^*)$$

over the set of possible $Q_0^*, \ldots, Q_i^*$ and 0 elsewhere, where

$$(5)\quad K_i^*(\gamma_0, \ldots, \gamma_i) = \frac{\sqrt{2\pi}\, K_i(\gamma_0, \ldots, \gamma_i)}{\sqrt{\gamma_0\lambda_0 + \gamma_1\lambda_1 + \cdots + \gamma_i\lambda_i}}$$

[and $k_i^*(Q_0^*, \ldots, Q_i^*)$ is normalized correspondingly]. The joint density of $Q_0^*, r_1^* = Q_1^*/Q_0^*, \ldots, r_i^* = Q_i^*/Q_0^*$ is

$$(6)\quad h_i^*(Q_0^*, Q_0^* r_1^*, \ldots, Q_0^* r_i^* \mid \gamma_0, \gamma_0\rho_1, \ldots, \gamma_0\rho_i) = K_i^*(\gamma_0, \gamma_0\rho_1, \ldots, \gamma_0\rho_i) \times \exp\left[-\tfrac{1}{2}\gamma_0 Q_0^*\left(1 + \sum_{j=1}^{i} \rho_j r_j^*\right)\right] Q_0^{*i} k_i^*(Q_0^*, Q_0^* r_1^*, \ldots, Q_0^* r_i^*)$$

over the set of possible $Q_0^*, r_1^*, \ldots, r_i^*$ and 0 elsewhere; here $\rho_1 = \gamma_1/\gamma_0, \ldots, \rho_i = \gamma_i/\gamma_0$. The density of $r_1^*, \ldots, r_i^*$ is

$$(7)\quad g_i^*(r_1^*, \ldots, r_i^* \mid \rho_1, \ldots, \rho_i)$$
$$= K_i^*(\gamma_0, \gamma_0\rho_1, \ldots, \gamma_0\rho_i)$$
$$\times \int \exp\left[-\tfrac{1}{2}\gamma_0 Q_0^*\left(1 + \sum_{j=1}^{i} \rho_j r_j^*\right)\right] Q_0^{*i} k_i^*(Q_0^*, Q_0^* r_1^*, \ldots, Q_0^* r_i^*)\, dQ_0^*,$$

where the integration is over values of Q_0^* compatible with $r_1^*, \ldots, r_i^*$. [We shall show that this range is $(0, \infty)$ and that g_i^* does not depend on γ_0.] If $\rho_1 = \cdots = \rho_i = 0$, then (7) is equivalent to

$$(8)\quad g_i^*(r_1^*, \ldots, r_i^* \mid 0, \ldots, 0)\, \gamma_0^{-\frac{1}{2}(T-1)} \lambda_0^{\frac{1}{2}} |A_0|^{-\frac{1}{2}} (2\pi)^{\frac{1}{2}(T-1)}$$
$$= \int \exp(-\tfrac{1}{2}\gamma_0 Q_0^*)\, Q_0^{*i} k_i^*(Q_0^*, Q_0^* r_1^*, \ldots, Q_0^* r_i^*)\, dQ_0^*,$$

which is an identity in γ_0. Since $r_1^*, \ldots, r_i^*$ are independent of Q_0^* in this case, the range of Q_0^* does not depend on $r_1^*, \ldots, r_i^*$ and hence is $(0, \infty)$. Special cases of the density $g_i^*(r_1^*, \ldots, r_i^* \mid 0, \ldots, 0)$ were discussed in Sections 6.7 and 6.9. Since (8) is an identity in γ_0, we can find the integral in (7) by replacing γ_0 in (8) by $\gamma_0(1 + \sum_{j=1}^{i} \rho_j r_j^*)$; since $\rho_1, \ldots, \rho_i$ are values such that $A_0 + \rho_1 A_1 + \cdots + \rho_i A_i$ is positive definite, $1 + \rho_1 r_1^* + \cdots + \rho_i r_i^* > 0$. Thus

$$(9)\quad g_i^*(r_1^*, \ldots, r_i^* \mid \rho_1, \ldots, \rho_i)$$
$$= \lambda_0^{\frac{1}{2}} |A_0|^{-\frac{1}{2}} (2\pi)^{\frac{1}{2}(T-1)} K_i^*(\gamma_0, \gamma_0\rho_1, \ldots, \gamma_0\rho_i) \gamma_0^{-\frac{1}{2}(T-1)}$$
$$\times \left(1 + \sum_{j=1}^{i} \rho_j r_j^*\right)^{-\frac{1}{2}(T-1)} g_i^*(r_1^*, \ldots, r_i^* \mid 0, \ldots, 0)$$
$$= \lambda_0^{\frac{1}{2}} |A_0|^{-\frac{1}{2}} (2\pi)^{\frac{1}{2}(T-1)} K_i^*(1, \rho_1, \ldots, \rho_i)$$
$$\times \left(1 + \sum_{j=1}^{i} \rho_j r_j^*\right)^{-\frac{1}{2}(T-1)} g_i^*(r_1^*, \ldots, r_i^* \mid 0, \ldots, 0),$$

since

$$(10)\quad (2\pi)^{\frac{1}{2}(T-1)} K_i^*(\gamma_0, \gamma_0\rho_1, \ldots, \gamma_0\rho_i) = \frac{|\gamma_0 A_0 + \gamma_0\rho_1 A_1 + \cdots + \gamma_0\rho_i A_i|^{\frac{1}{2}}}{\sqrt{\gamma_0\lambda_0 + \gamma_0\rho_1\lambda_1 + \cdots + \gamma_0\rho_i\lambda_i}}$$
$$= \gamma_0^{\frac{1}{2}(T-1)} \frac{|A_0 + \rho_1 A_1 + \cdots + \rho_i A_i|^{\frac{1}{2}}}{\sqrt{\lambda_0 + \rho_1\lambda_1 + \cdots + \rho_i\lambda_i}}.$$

Moments can be found from (9) by expanding $(1 + \sum_{j=1}^{i} \rho_j r_j^*)^{-\frac{1}{2}(T-1)}$ into a series in $(\sum_{j=1}^{i} \rho_j r_j^*)$ and integrating with respect to $r_1^*, \ldots, r_i^*$ for small ρ_j's.

Another approach to the distribution of the serial correlation coefficients involves the canonical form. Let

$$r^* = \frac{\sum_{t=1}^{T-1} \lambda_t z_t^2}{\sum_{t=1}^{T-1} z_t^2}, \tag{11}$$

where $z_1, \ldots, z_{T-1}$ are independently normally distributed with means 0 and variances $1/(\gamma_0 + \gamma_1\lambda_1), \ldots, 1/(\gamma_0 + \gamma_1\lambda_{T-1})$, respectively. Then

$$u_t = \sqrt{\gamma_0 + \gamma_1\lambda_t}\; z_t \tag{12}$$

is distributed according to $N(0, 1)$. We write

$$\begin{aligned} \Pr\{r^* \geq R^*\} &= \Pr\left\{\sum_{t=1}^{T-1} \frac{\lambda_t u_t^2}{\gamma_0 + \gamma_1\lambda_t} \geq R^* \sum_{t=1}^{T-1} \frac{u_t^2}{\gamma_0 + \gamma_1\lambda_t}\right\} \\ &= \Pr\left\{\sum_{t=1}^{T-1} \frac{\lambda_t - R^*}{\gamma_0 + \gamma_1\lambda_t} u_t^2 \geq 0\right\} \\ &= \Pr\left\{\sum_{t=1}^{T-1} \frac{\lambda_t - R^*}{1 + \rho\lambda_t} u_t^2 \geq 0\right\}, \end{aligned} \tag{13}$$

where $\rho = \gamma_1/\gamma_0$ for the case of $r^* = r_1^*$. The distributions of Section 6.7 may be used with $\lambda_t - R$ replaced by $(\lambda_t - R^*)/(1 + \rho\lambda_t)$. For example, if $\lambda_h = \lambda_{T-h} = \nu_h$, $h = 1, \ldots, H = \frac{1}{2}(T - 1)$, (52) may be rewritten

$$\Pr\{r^* \geq R^*\} = \sum_{i=1}^{m} \left(\frac{\nu_i - R^*}{1 + \rho R^*}\right)^{H-1} \prod_{\substack{j=1 \\ j \neq i}}^{H} \frac{1 + \rho\nu_j}{\nu_i - \nu_j}, \tag{14}$$

$$\nu_{m+1} \leq R^* \leq \nu_m, \quad m = 1, \ldots, H - 1.$$

6.11. SOME MAXIMUM LIKELIHOOD ESTIMATES

6.11.1. The First-Order Stationary Process

The first-order stationary Gaussian process satisfies

$$y_t + \beta y_{t-1} = u_t, \qquad t = \ldots, -1, 0, 1, \ldots, \tag{1}$$

where the u_t's are independently distributed, each according to $N(0, \sigma^2)$ and $|\beta| < 1$. Then $\mathscr{E}y_t = 0$, $\mathscr{E}y_t^2 = \sigma^2/(1 - \beta^2)$ and the marginal distribution of y_t is $N[0, \sigma^2/(1 - \beta^2)]$. The joint density of $y_1, \ldots, y_T$ is $N(\mathbf{0}, \boldsymbol{\Sigma})$, where the covariance matrix $\mathbf{\Sigma}$ has elements $\sigma_{st} = \sigma^2(-\beta)^{|s-t|}/(1-\beta^2)$, $s, t = 1, \ldots, T$. The density is a special case of (5) of Section 6.2 and was considered in Section 6.3.3.

Let us find the maximum likelihood estimate of β in this model. [Koopmans (1942) treated this problem.] Let

$$P_0 = \sum_{t=2}^{T-1} y_t^2, \tag{2}$$

$$P_0' = y_1^2 + y_T^2, \tag{3}$$

$$P_1 = \sum_{t=2}^{T} y_t y_{t-1}. \tag{4}$$

Then the logarithm of the likelihood function can be written as

$$\log L = \tfrac{1}{2}\log(1-\beta^2) - \tfrac{1}{2}T\log 2\pi - \tfrac{1}{2}T\log\sigma^2 - \tfrac{1}{2}[P_0' + 2\beta P_1 + (1+\beta^2)P_0]/\sigma^2. \tag{5}$$

The derivatives with respect to β and σ^2 are

$$\frac{\partial \log L}{\partial \beta} = -\frac{\beta}{1-\beta^2} - \frac{P_1}{\sigma^2} - \frac{\beta P_0}{\sigma^2}, \tag{6}$$

$$\frac{\partial \log L}{\partial \sigma^2} = -\frac{T}{2}\cdot\frac{1}{\sigma^2} + \frac{P_0' + 2\beta P_1 + (1+\beta^2)P_0}{2\sigma^4}. \tag{7}$$

Setting these equal to zero, we obtain

$$\hat{\sigma}^2 = \frac{P_0' + 2\hat{\beta}P_1 + (1+\hat{\beta}^2)P_0}{T}, \tag{8}$$

and

$$(1-\hat{\beta}^2)(P_1 + \hat{\beta}P_0) + \hat{\beta}\left\{\frac{P_0' + 2\hat{\beta}P_1 + (1+\hat{\beta}^2)P_0}{T}\right\} = 0; \tag{9}$$

that is,

$$\hat{\beta}^3\frac{T-1}{T}P_0 + \hat{\beta}^2\frac{T-2}{T}P_1 - \hat{\beta}\left(\frac{T+1}{T}P_0 + \frac{1}{T}P_0'\right) - P_1 = 0. \tag{10}$$

The left-hand side of (10) is positive for $\hat{\beta} = -1$, is negative for $\hat{\beta} = 1$, and is $-P_1$ for $\hat{\beta} = 0$; hence, there is one root of 0 if $P_1 = 0$, one root between -1 and 0 if P_1 is positive, and one root between 0 and 1 if P_1 is negative. There is one root less than -1 and one root greater than 1. As $T \to \infty$, this equation approaches (in probability)

$$\hat{\beta}^3 + \frac{P_1}{P_0}\hat{\beta}^2 - \hat{\beta} - \frac{P_1}{P_0} = 0. \tag{11}$$

The roots are $\hat{\beta} = \pm 1$, and $-P_1/P_0$. A closer approximation to the relevant solution of (10) is $\hat{\beta} = -(P_1/P_0)(1 - 1/T)$. (See Problem 78.)

This suggests the use of $-P_1/P_0$ or $-P_1/(P_0' + P_0)$ as an estimate of β. It will be observed that the cubic (11) does not have a simple solution, for example, a ratio of polynomials in P_0, P_1, and P_0'. Numerically, of course, the solution can be approximated arbitrarily closely.

6.11.2. The Circular Model

Let

$$y_t + \beta y_{t-1} = u_t, \qquad t = 1, \ldots, T, \tag{12}$$

where $y_0 \equiv y_T$ and $u_1, \ldots, u_T$ are independently distributed, each according to $N(0, \sigma^2)$. The joint density function of $y_1, \ldots, y_T \equiv y_0$ is

$$\frac{1 - (-\beta)^T}{(\sqrt{2\pi\sigma^2})^T} \exp\left[-\frac{1}{2\sigma^2}\sum_{t=1}^{T}(y_t + \beta y_{t-1})^2\right], \tag{13}$$

since the Jacobian of (12) in going from the u_t's to the y_t's is $1 - (-\beta)^T$. The matrix of the quadratic form in (13) is $-\frac{1}{2}$ times $\gamma_0 A_0 + \gamma_1 A_1$ as given in Section 6.5.2 with $\gamma_0 = (1 + \beta^2)/\sigma^2$ and $\gamma_1 = 2\beta/\sigma^2$. The derivatives of the logarithm of the likelihood function are

$$\frac{\partial \log L}{\partial \beta} = \frac{T(-\beta)^{T-1}}{1 - (-\beta)^T} - \frac{1}{\sigma^2}\sum_{t=1}^{T}(y_t + \beta y_{t-1})y_{t-1}, \tag{14}$$

$$\frac{\partial \log L}{\partial \sigma^2} = -\frac{T}{2}\cdot\frac{1}{\sigma^2} + \frac{1}{2\sigma^4}\sum_{t=1}^{T}(y_t + \beta y_{t-1})^2. \tag{15}$$

Setting these equal to zero, we obtain the maximum likelihood estimates from the equations

$$\hat{\sigma}^2 = \frac{1}{T}\left[(1 + \hat{\beta}^2)\sum_{t=1}^{T} y_t^2 + 2\hat{\beta}\sum_{t=1}^{T} y_t y_{t-1}\right], \tag{16}$$

$$(-\hat{\beta})^T \sum_{t=1}^{T} y_t y_{t-1} - (-\hat{\beta})^{T-1}\sum_{t=1}^{T} y_t^2 + \hat{\beta}\sum_{t=1}^{T} y_t^2 + \sum_{t=1}^{T} y_t y_{t-1} = 0. \tag{17}$$

If we denote by r the circular serial correlation coefficient,

$$r = \frac{\sum_{t=1}^{T} y_t y_{t-1}}{\sum_{t=1}^{T} y_t^2}, \tag{18}$$

the equations are

$$\hat{\sigma}^2 = \frac{1}{T}\sum_{t=1}^{T} y_t^2[1 + \hat{\beta}^2 + 2\hat{\beta}r], \tag{19}$$

$$r(-\hat{\beta})^T - (-\hat{\beta})^{T-1} + \hat{\beta} + r = 0. \tag{20}$$

If $r > 0$, the real roots of (20) are $\hat{\beta} = -a$ and $\hat{\beta} = -1/a$ for some $a > 0$ if T is even and $-a$, $-1/a$, and 1 if T is odd; if $r = 0$, the real roots are 0 if T is even and 0 and 1 if T is odd; if $r < 0$, the real roots are a and $1/a$ for $a > 0$ if T is even and a, $1/a$, and 1 if T is odd and $-(1\text{-}2/T) < r$, but only 1 if $-\cos \pi/T < r \leq -(1\text{-}2/T)$. Since the same process is defined for reciprocal values of β (and adjusted σ^2) we take the root between -1 and 1. (The root 1 does not give the absolute maximum of L.) It can be shown that asymptotically $\hat{\beta} = -r$ is a root of the second equation (because $r^T \to 0$ and $\hat{\beta}^T \to 0$ in probability). (See Problem 85.) It can also be shown that the likelihood ratio criterion for testing $\beta = 0$ is

$$\frac{[1 + \hat{\beta}^2 + 2\hat{\beta}r]^{\frac{1}{2}T}}{1 - (-\hat{\beta})^T}. \tag{21}$$

The hypothesis is rejected if (21) is less than a suitably chosen constant.

It may be noted that each root of (20) is a function of r, but not a rational function of r. The method based on maximum likelihood (using $\hat{\beta}$) may be approximated by the approach given in previous sections (using r). [Dixon (1944) treated these problems.]

We can write (12) as $(\boldsymbol{I} + \beta\boldsymbol{B}')\boldsymbol{y} = \boldsymbol{u}$, which is equivalent to

$$\boldsymbol{y} = (\boldsymbol{I} + \beta\boldsymbol{B}')^{-1}\boldsymbol{u}. \tag{22}$$

Since $\boldsymbol{B}^T = \boldsymbol{I}$, we have

$$(\boldsymbol{I} + \beta\boldsymbol{B}')^{-1} = \sum_{s=0}^{T-1}(-\beta)^s\boldsymbol{B}'^s + (-\beta)^T(\boldsymbol{I} + \beta\boldsymbol{B}')^{-1} \tag{23}$$

and, equivalently,

$$(\boldsymbol{I} + \beta\boldsymbol{B}')^{-1} = \frac{1}{1 - (-\beta)^T}\sum_{s=0}^{T-1}(-\beta)^s\boldsymbol{B}'^s. \tag{24}$$

Then the covariance matrix of $\boldsymbol{y}$ is

$$\begin{aligned} \mathscr{E}\boldsymbol{y}\boldsymbol{y}' &= (\boldsymbol{I} + \beta\boldsymbol{B}')^{-1}\mathscr{E}\boldsymbol{u}\boldsymbol{u}'(\boldsymbol{I} + \beta\boldsymbol{B})^{-1} \\ &= \frac{\sigma^2}{[1 - (-\beta)^T]^2}\sum_{s,t=0}^{T-1}(-\beta)^{s+t}\boldsymbol{B}^{t-s} \\ &= \frac{\sigma^2}{[1 - (-\beta)^T]^2}\sum_{r=-(T-1)}^{T-1}\sum_{s\in S_r}(-\beta)^{2s+r}\boldsymbol{B}^r, \end{aligned} \tag{25}$$

where $S_r = \{0, 1, \ldots, T-1-r\}$ for $r \geq 0$ and $S_r = \{-r, -r+1, \ldots T-1\}$ for $r \leq 0$, since $\boldsymbol{B}'^q = \boldsymbol{B}^{-q}$. Thus (since $\boldsymbol{B}^{-q} = \boldsymbol{B}^{T-q}$)

(26)

$$\begin{aligned}
\mathscr{E}\boldsymbol{y}\boldsymbol{y}' &= \frac{\sigma^2}{[1-(-\beta)^T]^2}\left\{\sum_{s=0}^{T-1}(-\beta)^{2s}\boldsymbol{I} + \sum_{q=1}^{T-1}\left[\sum_{s=0}^{T-1-q}(-\beta)^{2s+q}\boldsymbol{B}^q + \sum_{s=q}^{T-1}(-\beta)^{2s-q}\boldsymbol{B}^{-q}\right]\right\} \\
&= \frac{\sigma^2}{[1-(-\beta)^T]^2(1-\beta^2)} \\
&\quad \times \left\{(1-\beta^{2T})\boldsymbol{I} + \sum_{q=1}^{T-1}[(-\beta)^q - (-\beta)^{2T-q}](\boldsymbol{B}^q + \boldsymbol{B}^{-q})\right\} \\
&= \frac{\sigma^2}{[1-(-\beta)^T]^2(1-\beta^2)} \\
&\quad \times \left\{(1-\beta^{2T})\boldsymbol{I} + \sum_{q=1}^{T-1}(-\beta)^q[1 + (-\beta)^{T-2q} - (-\beta)^T - (-\beta)^{2T-2q}]\boldsymbol{B}^q\right\} \\
&= \frac{\sigma^2}{[1-(-\beta)^T](1-\beta^2)}\left\{[1+(-\beta)^T]\boldsymbol{I} + \sum_{q=1}^{T-1}[(-\beta)^q + (-\beta)^{T-q}]\boldsymbol{B}^q\right\}.
\end{aligned}$$

The covariance between y_t and y_{t+h} is

$$\sigma(h) = \mathscr{E}y_t y_{t+h} = \frac{(-\beta)^h + (-\beta)^{T-h}}{[1-(-\beta)^T](1-\beta^2)}\,\sigma^2, \qquad h = 0, 1, \ldots, T-1. \tag{27}$$

The correlation, $[(-\beta)^h + (-\beta)^{T-h}]/[1 + (-\beta)^T]$, is made up of two terms, one depending on the time between t and $t+h$ and the other between t and $t+h-T$. The matrix (26) is the inverse of $[(1+\beta^2)\boldsymbol{I} + \beta(\boldsymbol{B} + \boldsymbol{B}')]/\sigma^2$.

6.12. DISCUSSION

The first-order serial correlation coefficients give methods for testing the null hypothesis that $y_1, \ldots, y_T$ are independently distributed against certain alternatives that they are serially correlated; under the assumption of normality such tests can be carried out at specified significance levels. To test higher order dependence partial serial correlations may be used.

In Chapter 10 we shall discuss tests of independence and tests of order of dependence when $\mathscr{E}y_t = f(t)$ is some nontrivial trend.

REFERENCES

Section 6.2. T. W. Anderson (1963).

Section 6.3. T. W. Anderson (1948), (1963), Kendall and Stuart (1961), Lehmann (1959).

Section 6.4. T. W. Anderson (1963).

Section 6.5. T. W. Anderson (1963), Watson and Durbin (1951).

Section 6.6. R. L. Anderson and T. W. Anderson (1950).

Section 6.7. R. L. Anderson (1941), (1942), Dixon (1944), Koopmans (1942), Laplace (1829), von Neumann (1941).

Section 6.8. T. W. Anderson (1948), Dixon (1944), Hannan (1960), Hart (1942), Hart and von Neumann (1942), Koopmans (1942), Rubin (1945), United States Bureau of the Census (1955), Young (1941).

Section 6.9. T. W. Anderson (1958), Daniels (1956), Hannan (1960), Jenkins (1956), Quenouille (1949a), (1949b), Watson (1956).

Section 6.10. Leipnik (1947), Madow (1945).

Section 6.11. Dixon (1944), Koopmans (1942).

PROBLEMS

1. (Sec. 6.2.) Show that for the sequence $y_1, \ldots, y_T$ from a stationary Gaussian process of order p the following functions of the observations form a sufficient set of statistics:

$$\sum_{t=p+1}^{T-p} y_t^2, \quad y_1^2 + y_T^2, \quad y_2^2 + y_{T-1}^2, \ldots, \quad y_p^2 + y_{T-p+1}^2,$$

$$\sum_{t=p+1}^{T-p+1} y_t y_{t-1}, \quad y_1 y_2 + y_{T-1} y_T, \ldots, \quad y_{p-1} y_p + y_{T-p+2} y_{T-p+1},$$

$$\vdots$$

$$\sum_{t=p+1}^{T} y_t y_{t-p}.$$

2. (Sec. 6.2.) Let $y_1, \ldots, y_T$ be a sequence with a covariance matrix $\mathbf{\Sigma}_T = [\sigma(i-j)]$ from a stationary process generated by the moving average $y_t = v_t + \alpha_1 v_{t-1}$, where $\mathscr{E}v_t = 0$, $\mathscr{E}v_t^2 = \sigma^2$, and $\mathscr{E}v_t v_s = 0$, $t \neq s$. Then $\sigma(0) = \sigma^2(1 + \alpha_1^2)$ and $\sigma(1) = \sigma^2\alpha_1 = \sigma(0)\rho$, say, and $\sigma(h) = 0$, $h > 1$. Show

$$\boldsymbol{\Sigma}_3^{-1} = \frac{1}{(1-2\rho^2)\sigma(0)}\begin{bmatrix} 1-\rho^2 & -\rho & \rho^2 \\ -\rho & 1 & -\rho \\ \rho^2 & -\rho & 1-\rho^2 \end{bmatrix},$$

$$\boldsymbol{\Sigma}_4^{-1} = \frac{1}{(1-3\rho^2+\rho^4)\sigma(0)}\begin{bmatrix} 1-2\rho^2 & -\rho(1-\rho^2) & \rho^2 & -\rho^3 \\ -\rho(1-\rho^2) & 1-\rho^2 & -\rho & \rho^2 \\ \rho^2 & -\rho & 1-\rho^2 & -\rho(1-\rho^2) \\ -\rho^3 & \rho^2 & -\rho(1-\rho^2) & 1-2\rho^2 \end{bmatrix}.$$

3. (Sec. 6.2.) Show that $\boldsymbol{\Sigma}_T^{-1}$ defined in Problem 2 is the product of $\sigma^{T-1}(0)/|\boldsymbol{\Sigma}_T|$ and a matrix whose elements are polynomials in ρ. Show that all powers of ρ from 0 to $T-1$ appear in these polynomials and that there are T linearly independent polynomials. [*Hint:* An element of the inverse is the ratio of the corresponding cofactor and the determinant.]

4. (Sec. 6.2.) Assume that $y_1, \ldots, y_T$ are T successive values from a Gaussian process defined as in Problem 2. Show that the dimensionality of the minimal sufficient set of statistics for $\sigma(0)$ and ρ is T. [*Hint:* Use the result of Problem 3 to show that the exponent in the density is a factor times $\sum_{j=0}^{T-1} \rho^j P_j$, where the P_j's are linearly independent quadratic forms. If $Q_0, \ldots, Q_S$ $(S \leq T-1)$ is another sufficient set of statistics, the exponent is a factor times $\sum_{j=0}^{S} f_j(\rho) R_j$, the R_j's are functions of the Q_j's, and the two exponents are identical.]

5. (Sec. 6.2.) Let $\boldsymbol{\Sigma}_T$ be defined as in Problem 2 and

$$a_{ij} = [-\rho\sigma(0)]^{j-i}\, |\boldsymbol{\Sigma}_{i-1}| \cdot |\boldsymbol{\Sigma}_{T-j}| / |\boldsymbol{\Sigma}_T|, \qquad i \leq j.$$

Show $A = [a_{ij}] = \boldsymbol{\Sigma}_T^{-1}$. [*Hint:* Compute the cofactor of the i, jth element by the Laplacian expansion using the first $i-1$ columns and then the last $T-j$ columns. An alternative method is to apply Lemma 6.7.9 to show that $|\boldsymbol{\Sigma}_i|$ satisfies a second-order difference equation and to obtain a solution for $|\boldsymbol{\Sigma}_i|$. Then verify $\boldsymbol{\Sigma}_T A = \boldsymbol{I}$. Verification of the diagonal terms (except the first and last) uses the solution for $|\boldsymbol{\Sigma}_i|$; verification of the off-diagonal terms uses the second-order difference equation directly.]

6. (Sec. 6.2.) Show that the solution of (13) for $p = 1$ is $\sigma^2 = 2(\gamma_0 \pm \sqrt{\gamma_0^2 - \gamma_1^2}\,)/\gamma_1^2$ and $\beta_1 = (\gamma_0 \pm \sqrt{\gamma_0^2 - \gamma_1^2}\,)/\gamma_1$.

7. (Sec. 6.3.1.) Prove Theorem 6.3.2, namely, that

$$\text{(i)} \qquad \int_{-\infty}^{\infty} \cdots \int_{-\infty}^{\infty} h_i(Q_0, \ldots, Q_i \mid \gamma_0, \ldots, \gamma_i) g(Q_0, \ldots, Q_i)\, dQ_0 \cdots dQ_i = 0$$

identically in $\gamma_0, \ldots, \gamma_i$ (for which $\sum_{j=0} \gamma_j Q_j$ is positive definite) implies

$$\text{(ii)} \qquad g(Q_0, \ldots, Q_i) = 0$$

almost everywhere relative to the densities $h_i(Q_0, \ldots, Q_i \mid \gamma_0, \ldots, \gamma_i)$. [*Hint:* Let $g(Q_0, \ldots, Q_i) = g^+(Q_0, \ldots, Q_i) - g^-(Q_0, \ldots, Q_i)$, where g^+ and g^- are non-negative. Then show

$$\text{(iii)} \quad \int_{-\infty}^{\infty} \cdots \int_{-\infty}^{\infty} \exp\left[-\tfrac{1}{2}\sum_{j=0}^{i} \gamma_j Q_j\right] g^+(Q_0, \ldots, Q_i) k_i(Q_0, \ldots, Q_i)\, dQ_0 \cdots dQ_i$$

$$\equiv \int_{-\infty}^{\infty} \cdots \int_{-\infty}^{\infty} \exp\left[-\tfrac{1}{2}\sum_{j=0}^{i} \gamma_j Q_j\right] g^-(Q_0, \ldots, Q_i) k_i(Q_0, \ldots, Q_i)\, dQ_0 \cdots dQ_i.$$

Deduce (ii) from (iii) by the theory of characteristic functions.]

8. (Sec. 6.3.1.) Show that (4) implies (3).

9. (Sec. 6.3.1.) Show that (3) implies (4) almost everywhere. [*Hint:* Use the property of completeness as proved in Problem 7 with

$$g(Q_0, \ldots, Q_{i-1}) = \frac{\Pr\{S_i \mid Q_i, \ldots, Q_{i-1}\}}{\varepsilon_i} - 1.]$$

10. (Sec. 6.3.1.) *The Neyman-Pearson Fundamental Lemma.* Let $f_0(x)$ and $f_1(x)$ be density functions on some Euclidean space, and let R be any (measurable) set such that

$$\text{(i)} \quad \int_R f_0(x)\, dx = \varepsilon,$$

where dx indicates Lebesgue integration in the space and $0 \leq \varepsilon \leq 1$. Prove that if

$$\text{(ii)} \quad R^* = \{x \mid f_1(x) > k f_0(x)\}$$

for some k and satisfies (i), then

$$\text{(iii)} \quad \int_{R^*} f_1(x)\, dx \geq \int_R f_1(x)\, dx.$$

[*Hint:* The contribution of the integral over $R^* \cap R$ is the same to both sides of (iii). Compare the integrals of $f_1(x)$ over $R^* \cap \bar{R}$ and $\bar{R}^* \cap R$ to the integrals of $f_0(x)$.]

11. (Sec. 6.3.1.) Let

$$y_{ij} = a_i + e_{ij}, \qquad i = 1, \ldots, M,\ j = 1, \ldots, N,$$

where the a_i's and e_{ij}'s are independently distributed, the distribution of a_i is $N(0, \sigma_a^2)$, and the distribution of e_{ij} is $N(0, \sigma_e^2)$. Give the joint density of the y_{ij}'s in the form of (1). Find the uniformly most powerful similar test of the hypothesis $\sigma_a^2 = 0$ against the alternatives $\sigma_a^2 > 0$ at significance level α. [*Hint:* The covariance matrix of $y_{11}, \ldots, y_{1N}, y_{21}, \ldots, y_{MN}$ is $\sigma_e^2 \boldsymbol{I}_{MN} + \sigma_a^2 \boldsymbol{I}_M \otimes \boldsymbol{\varepsilon\varepsilon}'$, where $\boldsymbol{\varepsilon} = (1, 1, \ldots, 1)'$ is of N components and $\otimes$ denotes the Kronecker product, and the inverse is of the form $b\boldsymbol{I}_{MN} + c\boldsymbol{I}_M \otimes \boldsymbol{\varepsilon\varepsilon}'$.]

12. (Sec. 6.3.2.) *A generalized Neyman-Pearson Fundamental Lemma.* Let $f_0(x)$ and $f_2(x)$ be density functions on some Euclidean space and $f_1(x)$ another function, and let R be any (measurable) set such that

$$\int_R f_0(x)\,dx = \varepsilon, \tag{i}$$

$$\int_R f_1(x)\,dx = c, \tag{ii}$$

where dx indicates Lebesgue integration, $0 \le \varepsilon \le 1$, and c is a constant. Prove that if

$$R^* = \{x \mid f_2(x) > k_0 f_0(x) + k_1 f_1(x)\} \tag{iii}$$

for some k_0 and k_1 and satisfies (i) and (ii), then

$$\int_{R^*} f_2(x)\,dx \ge \int_R f_2(x)\,dx. \tag{iv}$$

[*Hint:* The contribution of the integral over $R^* \cap R$ is the same to both sides of (iv). Compare the integrals of $f_2(x)$ over $R^* \cap \bar{R}$ and $\bar{R}^* \cap R$ to the integrals of $k_0 f_0(x) + k_1 f_1(x)$.]

13. (Sec. 6.3.2.) Prove that $R = (-\infty, c)$ cannot satisfy (16) and (20). [*Hint:* Show that (16) and (20) are then equivalent to (16) and

$$\int_{-\infty}^{c} (Q_i - \mu) h_i(Q_i \mid Q_0, \ldots, Q_{i-1};\ \gamma_i^{(1)})\,dQ_i = 0, \tag{i}$$

where

$$\mu = \int_{-\infty}^{\infty} Q_i h_i(Q_i \mid Q_0, \ldots, Q_{i-1};\ \gamma_i^{(1)})\,dQ_i.] \tag{ii}$$

14. (Sec. 6.3.2.) Let $\sigma(h) = \sigma(0)\rho_h$ for a stationary process. Show that the partial correlation between y_t and y_{t+2} "holding y_{t+1} fixed" is

$$\frac{\rho_2 - \rho_1^2}{1 - \rho_1^2}.$$

Show that the partial correlation between y_t and y_{t+3} "holding y_{t+1} and y_{t+2} fixed" is

$$\frac{\rho_3 - \rho_1^2\rho_3 + \rho_1^3 - 2\rho_1\rho_2 + \rho_1\rho_2^2}{1 - 2\rho_1^2 + 2\rho_1^2\rho_2 - \rho_2^2}.$$

15. (Sec. 6.5.2.) Prove that $\boldsymbol{B}^j$ has all elements 0 except j above the main diagonal and $T - j$ below the main diagonal, which are 1, $j = 0, 1, \ldots, T-1$.

16. (Sec. 6.5.2.) Prove that $\boldsymbol{B}^T = \boldsymbol{I}$.

17. (Sec. 6.5.2.) Prove Theorem 6.5.2 by the method used to prove Theorem 6.5.4.

18. (Sec. 6.5.3.) Using A_1 given by (41), find A_4, A_5, and A_6 according to (18).

19. (Sec. 6.5.3.) Verify that (55) is orthogonal by using sums evaluated in Chapter 4.

20. (Sec. 6.5.3.) Prove C_r and C_1^r differ only in the elements for $s, t \leq r$ and $(T - s + 1), (T - t + 1) \leq r$.

21. (Sec. 6.5.3.) Write out C_1^5 explicitly.

22. (Sec. 6.5.3.) Prove (61). [*Hint:* $C_1 = 2(I - A_1)$ and A_j is a polynomial in A_1 of degree j.]

23. (Sec. 6.5.4.) Prove Corollary 6.5.4.

24. (Sec. 6.6.1.) Let x be distributed according to $N(\mu\varepsilon, \Sigma)$. Show that an unbiased linear estimate $m = \gamma' x$ of μ must satisfy $\gamma'\varepsilon = 1$. Show that for such an estimate to be independent of the residuals $x - m\varepsilon$, it must be Markov; that is,

$$\gamma = \frac{1}{\varepsilon'\Sigma^{-1}\varepsilon}\,\Sigma^{-1}\varepsilon.$$

25. (Sec. 6.6.1.) Show that when x is distributed according to $N(\mu\varepsilon, \Sigma)$ $\bar{x}$ and $x - \bar{x}\varepsilon$ are independent if and only if ε is a characteristic vector of Σ.

26. (Sec. 6.6.1.) State Theorems 6.4.1 and 6.4.2 when $\mathscr{E}y_t = \mu$ and ε is a characteristic vector of $A_0, \ldots, A_p$.

27. (Sec. 6.7.1.) Let y for $T = 4$ have the density

$$\text{(i)} \qquad K(\gamma_0, \gamma_1)e^{-\frac{1}{2}(\gamma_0 Q_0 + \gamma_1 Q_1)},$$

where $Q_0 = \sum_{t=1}^4 y_t^2$ and $Q_1 = y_1^2 + y_2^2 - (y_3^2 + y_4^2)$, $K(\gamma_0, \gamma_1) = (\gamma_0^2 - \gamma_1^2)/(2\pi)^2$ and $|\gamma_1| < \gamma_0$. (a) Show that the joint density of Q_0 and Q_1 is

$$\text{(ii)} \qquad \frac{\gamma_0^2 - \gamma_1^2}{8}\,e^{-\frac{1}{2}(\gamma_0 Q_0 + \gamma_1 Q_1)}, \qquad 0 \leq |Q_1| \leq Q_0.$$

[*Hint:* $(\gamma_0 + \gamma_1)(y_1^2 + y_2^2)$ and $(\gamma_0 - \gamma_1)(y_3^2 + y_4^2)$ are independently distributed as χ^2 with 2 degrees of freedom.] (b) Show that the joint density of Q_0 and $r = Q_1/Q_0$ is

$$\text{(iii)} \qquad \frac{\gamma_0^2 - \gamma_1^2}{8}\,e^{-\frac{1}{2}(\gamma_0 + \gamma_1 r)Q_0}\,Q_0, \qquad |r| \leq 1.$$

(c) Show that the marginal density of Q_0 when $\gamma_1 \neq 0$ is

$$\text{(iv)} \qquad \frac{\gamma_0^2 - \gamma_1^2}{4\gamma_1}\left[e^{-\frac{1}{2}(\gamma_0 - \gamma_1)Q_0} - e^{-\frac{1}{2}(\gamma_0 + \gamma_1)Q_0}\right], \qquad Q_0 \geq 0.$$

(d) Find the conditional density of r given Q_0. (e) Find the marginal density of r.

28. (Sec. 6.7.1.) Let x have density $f(x, \theta)$, where $\theta \in \Omega$. Suppose Q is a sufficient statistic for θ, and suppose there is a group of transformations g on x with a corresponding group $\bar{g}$ on Q. Suppose that there is some value q_0 of Q such that for every value q_1 of Q there is a transformation g such that $\bar{g}q_1 = q_0$. Then every statistic r that is invariant with respect to the group of transformations is distributed independently of Q and θ.

29. (Sec. 6.7.2.) Prove (10) for purely imaginary $t_0, \ldots, t_p$ by using the fact that

$$\int_{-\infty}^{\infty} \cdots \int_{-\infty}^{\infty} e^{-\frac{1}{2}x'Bx}\, dx_1 \cdots dx_T = \frac{(2\pi)^{\frac{1}{2}T}}{\sqrt{|B|}} .$$

30. (Sec. 6.7.2.) Show that if A is positive definite and B is symmetric there exists a nonsingular matrix C such that

$$C'AC = I, \qquad C'BC = D,$$

where D is diagonal and the diagonal elements of D are the roots of $|B - dA| = 0$. Show that the columns of C are characteristic vectors of $A^{-1}B$. [*Hint:* Let E be a matrix so $E'AE = I$; take $C = E\Theta$, where Θ is an orthogonal matrix that diagonalizes $E'BE$.]

31. (Sec. 6.7.3.) Show that $A_jA_k = A_kA_j$, $j, k = 1, \ldots, p$, is a necessary and sufficient condition for the existence of an orthogonal matrix Z such that $Z'A_jZ = \Lambda_j$, diagonal, $j = 1, \ldots, p$, where $A_1, \ldots, A_p$ are symmetric matrices. [*Hint:* $Z'A_jZ = \Lambda_j$ is equivalent to $A_j = Z\Lambda_jZ'$. To prove sufficiency, assume $A_1A_k = A_kA_1$ and use $A_1 = Z\Lambda Z'$.]

32. (Sec. 6.7.3.) Let A_0 be an arbitrary positive definite matrix and let $A_1, \ldots, A_p$ be symmetric matrices. Show that if there exists a nonsingular matrix C such that its columns are the characteristic vectors of $A_0^{-1}A_1, \ldots, A_0^{-1}A_p$ and are properly normalized, then

$$C'A_0C = I, \qquad C'A_jC = \Lambda_j, \qquad j = 1, \ldots, p,$$

where Λ_j is diagonal with the roots of $|A_j - \lambda A_0| = 0$ as diagonal elements. Prove that a necessary and sufficient condition for the above is that

$$A_jA_0^{-1}A_k = A_kA_0^{-1}A_j, \qquad j, k = 1, \ldots, p.$$

[*Hint:* See Problems 30 and 31.]

33. (Sec. 6.7.4.) Prove $h^*(y_1, \ldots, y_{H-2}) = k_2^* h(y_1, \ldots, y_{H-2})$. [*Hint:* Let Z_1, Z_2, and Z_2^* be the points of intersection of the line parallel to V_1V_2 through $(y_1, \ldots, y_{H-2})$ and the faces $V_1V_3 \cdots V_H$, $V_2V_3 \cdots V_H$, and $V_2^*V_3 \cdots V_H$, respectively, and let W be the point of intersection of $V_3V_4 \cdots V_H$ and the plane through V_1V_2 and the line. Consider the triangles V_1V_2W, $V_1V_2^*W$, Z_1Z_2W, and $Z_1Z_2^*W$.]

34. (Sec. 6.7.4.) Indicate U_1, U_2, and U_3 for $R < \nu_3$ (in barycentric coordinates with $H = 3$).

35. (Sec. 6.7.4.) Show that Theorem 6.7.4 can be stated

$$\Pr\{r < R\} = \sum_{i=m+1}^{H} \frac{(\nu_i - R)^{H-1}}{\prod\limits_{\substack{j=1\\ j\neq i}}^{H} (\nu_i - \nu_j)}, \quad \nu_{m+1} \le R \le \nu_m, \qquad m = 1, \ldots, H-1.$$

[*Hint:* Use Lemma 6.7.7.]

36. (Sec. 6.7.4.) Prove Theorem 6.7.4 by induction by applying Theorem 6.7.5. [*Hint:* Assume (52) for $H = H^* - 1$ and use Theorem 6.7.5 with $M = 2(H^* - 1)$ and $L = 2$ to verify (52) for $H = H^*$.]

37. (Sec. 6.7.4.) Prove Theorem 6.7.4 by inverting (69). [*Hint:* Show that $u = \sum_{j=1}^{H} (\nu_j - R)x_j$ has the density for $u \geq 0$.

$$f(u) = \tfrac{1}{2} \sum_{j=1}^{m} e^{-\frac{1}{2}u/(\nu_j - R)} \frac{(\nu_j - R)^{H-2}}{\prod_{\substack{k=1 \\ k \neq j}}^{H} (\nu_j - \nu_k)}, \quad \nu_{m+1} \leq R \leq \nu_m, \qquad m = 1, \ldots, H-1,$$

by integrating (69) times $e^{-itu}/(2\pi)$ along a closed contour containing the poles in the lower half of the complex plane. Then integrate the density from 0 to ∞; see (68).]

38. (Sec. 6.7.5.) Show that in the circular case

(a) $r_1^* = -1$ for $T = 2$,

(b) $r_1^* = -\frac{1}{2}$ for $T = 3$,

(c) $-2r_1^*/(1 + r_1^*)$ has the F-distribution with 1 and 2 degrees of freedom (under the null hypothesis) for $T = 4$,

(d) $(c_1 - r_1^*)/(r_1^* - c_2)$ has the F-distribution with 2 and 2 degrees of freedom (under the null hypothesis), where $c_1 = \cos 2\pi/5$ and $c_2 = \cos 4\pi/5$, for $T = 5$.

39. (Sec. 6.7.5.) Show that in the circular case under the null hypothesis

(a) $(1 + r_1)/(1 - r_1)$ has the F-distribution with 1 and 1 degrees of freedom for $T = 2$,

(b) $(1 + 2r_1)/(1 - r_1)$ has the F-distribution with 1 and 2 degrees of freedom for $T = 3$,

(c) r_1 is distributed as

$$\frac{F_{1,1} - 1}{F_{1,1} + 1} \Big/ (F_{2,2} + 1),$$

where $F_{1,1}$ and $F_{2,2}$ are independently distributed according to F-distributions with 1 and 1 degrees of freedom and 2 and 2 degrees of freedom, respectively, for $T = 4$.

40. (Sec. 6.7.5.) A section of Wolfer's series of sunspot numbers (1858-1869) is the following: 55, 94, 96, 77, 59, 44, 47, 30, 16, 7, 37, 74. Using the circular definition, calculate Q_0^*, Q_1^*, and r_1^*. Test for significance at the 1% level against alternatives of first-order positive correlation.

41. (Sec. 6.7.6.) Show that if u, v, and w are independently distributed according to χ^2-distributions with l, m, and n degrees of freedom, respectively, then $(\alpha u + \beta v + \gamma w)/(u + v + w)$ is distributed as

$$\left(\frac{\alpha x + \beta}{x + 1} + \gamma y\right) \Big/ (1 + y),$$

where x and y are independently distributed as $lF_{l,m}/m$ and $nF_{n,l+m}/(l + m)$, respectively.

42. (Sec. 6.7.6.) Find the density of the serial correlation coefficient r_1^* based on the mean square successive difference for $T = 3$ by deriving the joint density of $r_1^* = (\lambda_1 u_1 + \lambda_2 u_2)/(u_1 + u_2)$ and $v = u_1 + u_2$ when u_1 and u_2 are independently distributed as χ^2 with 1 degree of freedom.

43. (Sec. 6.7.6.) Show that the distribution of r_2^* based on successive differences is the same as the distribution of the circular r_1^*. [*Hint:* Show that the two sets of roots are the same.]

44. (Sec. 6.7.7.) Show that if $\gamma_1 \neq 0$ the moment generating function of Q_1 is obtained from the moment generating function of Q_1 when $\gamma_1 = 0$ by replacing -2θ by $\gamma_1 - 2\theta$ and multiplying by $K_1(\gamma_0, \gamma_1)/K_0(\gamma_0)$.

45. (Sec. 6.7.7.) Show that D_T in Lemma 6.7.9 is

$$D_3 = 1 - 2\theta^2,$$

$$D_4 = 1 - 3\theta^2 + \theta^4.$$

46. (Sec. 6.7.7.) Show that Δ_T in Lemma 6.7.10 is

$$\Delta_3 = 1 - 3\theta^2 - 2\theta^3 = (1 - 2\theta)(1 + \theta)^2,$$

$$\Delta_4 = 1 - 4\theta^2 = (1 - 2\theta)(1 + 2\theta).$$

47. (Sec. 6.7.7.) Show that C_T in Lemma 6.7.11 is

$$C_2 = 1 - 2\theta,$$

$$C_3 = 1 - 2\theta - \theta^2 + 2\theta^3 = (1 - 2\theta)(1 - \theta^2),$$

$$C_4 = 1 - 2\theta - 2\theta^2 + 4\theta^3 = (1 - 2\theta)(1 - 2\theta^2),$$

$$C_5 = 1 - 2\theta - 3\theta^2 + 6\theta^3 + \theta^4 - 2\theta^5 = (1 - 2\theta)(1 - 3\theta^2 + \theta^4).$$

48. (Sec. 6.7.7.) Show that if $\gamma_1 \neq 0$ and $\boldsymbol{\varepsilon} = (1, \ldots, 1)'$ is a characteristic vector of A_0 and A_1 corresponding to the root 1, the moment generating function of Q_1^* is obtained from the moment generating function of Q_1^* when $\gamma_1 = 0$ by replacing -2θ by $\gamma_1 - 2\theta$ and multiplying by $K_1(\gamma_0, \gamma_1)/K_0(\gamma_0)$ and by $\sqrt{\gamma_0/(\gamma_0 + \gamma_1)}$.

49. (Sec. 6.7.7.) Show that

$$\frac{d^n}{d\theta^n} F(\theta)G(\theta) = \sum_{j=0}^{n} \binom{n}{j} F^{(j)}(\theta)G^{(n-j)}(\theta),$$

where $F^{(j)}(\theta) = d^j F(\theta)/d\theta^j$ and $G^{(j)}(\theta) = d^j G(\theta)/d\theta^j$. [*Hint:* Prove by induction (in analogy to the binomial theorem).]

50. (Sec. 6.7.7.) Show that if $f(0) = 0$, and $f^{(n)}(0)$ exists and is finite, $n = 1, \ldots, H - 1$, then

$$\left.\frac{d^n f^H(\theta)}{d\theta^n}\right|_{\theta=0} = 0, \qquad n = 0, 1, \ldots, H - 1.$$

[*Hint:* Use the previous problem with $F(\theta) = f^{H-1}(\theta)$, $G(\theta) = f(\theta)$ for a proof by induction.]

51. (Sec. 6.7.7.) Show that if $G^{(n)}(0) = 0$, $n = 0, 1, \ldots, T-1$, and if $\phi(\theta) = F(\theta)[1 + G(\theta)]$, then

$$\phi^{(n)}(0) = F^{(n)}(0), \qquad n = 0, 1, \ldots, T-1.$$

[*Hint:* Use Problem 49.]

52. (Sec. 6.7.7.) Use Problems 49, 50, and 51 to prove (107).

53. (Sec. 6.7.7.) Show that

$$h(\theta) = \left(\frac{1 - \sqrt{1 - 4\theta^2}}{1 + \sqrt{1 - 4\theta^2}}\right)^{\frac{1}{2}}$$

does not have a derivative at $\theta = 0$,

54. (Sec. 6.7.7.) Show that the moment generating function for the circular definition of Q_1 for $T = 4$ is

$$\Delta_4^{-\frac{1}{2}} = (1 - 4\theta^2)^{-\frac{1}{2}} = 1 + 2\theta^2 + 6\theta^4 + \cdots.$$

55. (Sec. 6.7.7.) Show

$$\left(\frac{1 + \sqrt{1 - 4\theta^2}}{2}\right)^{-2} = 1 + 2\theta^2 + 5\theta^4 + \cdots.$$

56. (Sec. 6.7.7.) Verify (114) for $T = 3$ by direct computation.

57. (Sec. 6.7.7.) (a) Show

$$\phi(2) = 1 + \theta^2 + \cdots,$$

$$\phi(4) = 1 + \cdots,$$

$$\phi^*(3) = 1 - \theta + \theta^2 - \cdots.$$

(b) Verify (116) and (117) for $T = 3$.

58. (Sec. 6.7.7.) Deduce $\mathscr{E}Q_1^{*h}$ from $Q_1 = Q_1^* + z_T^2$ for $h = 1, 2, 3, 4$ in the circular case.

59. (Sec. 6.7.7.) Express ψ_5, ψ_6, ψ_7, and ψ_8 in terms of $\nu_1, \ldots, \nu_8$ without assuming $\nu_{2j-1} = 0$.

60. (Sec. 6.7.7.) In the case of Q_1^* based on the mean square successive difference show from (129) that $\nu_k = 0$, $k = 1, 2, \ldots$, for $T = 1$ and 2, that $\nu_{2l} = 1/(2l)$, $l = 1, 2, \ldots$, for $T = 3$, and that $\nu_{2l} = 2^{l-1}/l$, $l = 1, 2, \ldots$, for $T = 4$.

61. (Sec. 6.7.7.) In the case of Q_1^* based on the mean square successive difference show that $\psi_k = 0$, $k = 1, 2, \ldots$, for $T = 1$ and 2, that $\psi_6 = 5/16$ and $\psi_8 = 35/128$ for $T = 3$, and $\psi_8 = 35/8$ for $T = 4$. Show that $\mathscr{E}r_1^{*6} = 5/1024$ and $\mathscr{E}r_1^{*8} = 35/32{,}768$ for $T = 3$, and $\mathscr{E}r_1^{*8} = 112/21{,}879$ for $T = 4$.

62. (Sec. 6.7.7.) Find ν_{10}, ψ_{10}, $\mathscr{E}Q_1^{*10}$, and $\mathscr{E}r_1^{*10}$ for the coefficient based on successive differences.

63. (Sec. 6.7.7.) In the case of Q_1^* for the circular serial correlation coefficient show that for $k = 1, 2, \ldots$ $\nu_k = 0$ for $T = 1$, $\nu_k = (-1)^k 2^{k-1}/k$ for $T = 2$, $\nu_k = (-1)^k/k$ for $T = 3$, and $\nu_k = (-1)^k 2^{k-1}/k$ for $T = 4$. [Note that the respective moments for $T = 2$ and $T = 4$ are the same.]

64. (Sec. 6.7.7.) In the case of Q_1^* for the circular serial correlation coefficient show that $\psi_k = 0$, $k = 1, 2, \ldots,$ for $T = 1$, $\psi_2 = 3/2$, $\psi_3 = -5/2$, and $\psi_4 = 35/8$ for $T = 2$, $\psi_3 = -1$ and $\psi_4 = 1$ for $T = 3$, and $\psi_4 = 35/8$ for $T = 4$. Show that $\mathscr{E} r_1^{*2} = 1$, $\mathscr{E} r_1^{*3} = -1$, and $\mathscr{E} r_1^{*4} = 1$ for $T = 2$, $\mathscr{E} r_1^{*3} = -1/8$ and $\mathscr{E} r_1^{*4} = 1/16$ for $T = 3$, and $\mathscr{E} r_1^{*4} = 1/9$ for $T = 4$.

65. (Sec. 6.7.7.) Find ν_5, ψ_5, $\mathscr{E} Q_1^{*5}$, and $\mathscr{E} r_1^{*5}$ for the circular coefficient.

66. (Sec. 6.7.7.) Let $\lambda_t^* = \cos \pi t/T^*$, $t = 1, \ldots, T^* - 1$, be the roots involved in r_1^* based on successive differences, and let $\lambda_t = \cos 2\pi t/T$, $t = 1, \ldots, T - 1$, be the roots involved in the circular r_1^*.

(a) Show that if $T = 2T^*$

$$\sum_{t=1}^{T-1} \lambda_t^k = 2 \sum_{t=1}^{T^*-1} \lambda_t^{*k} + (-1)^k, \qquad k = 0, 1, \ldots.$$

(b) Show that if $T = T^*$ is odd

$$\sum_{t=1}^{T-1} \lambda_t^{2l} = \sum_{t=1}^{T^*-1} \lambda_t^{*2l}, \qquad l = 0, 1, \ldots.$$

67. (Sec. 6.7.7.) Show that

$$-\log(1 - x) = \sum_{j=1}^{\infty} \frac{x^j}{j}$$

by integrating

$$(1 - x)^{-1} = \sum_{j=1}^{\infty} x^{j-1}.$$

68. (Sec. 6.7.7.) Let $x_1, y_1, \ldots, x_N, y_N$ be independently and normally distributed with means 0 and variances 1. Let $r = A/\sqrt{BC}$, where $A = \sum_{i=1}^N x_i y_i$, $B = \sum_{i=1}^N x_i^2$, and $C = \sum_{i=1}^N y_i^2$. (a) Show that r is distributed independently of B and C. (b) Show that B and C are independently distributed according to χ^2-distributions with N degrees of freedom. (c) Show that the moment generating function of A is

$$\text{(i)} \qquad \mathscr{E} e^{\theta A} = (1 - \theta^2)^{-\frac{1}{2}N}$$

$$= \sum_{h=0}^{\infty} \frac{\theta^{2h} \Gamma(\frac{1}{2}N + h)}{h!\,\Gamma(\frac{1}{2}N)}.$$

(d) Show that

(ii) $$\mathscr{E}A^{2k-1} = 0, \qquad k = 1, 2, \ldots,$$

(iii) $$\mathscr{E}A^{2k} = \frac{(2k)!\Gamma(\tfrac{1}{2}N + k)}{k!\Gamma(\tfrac{1}{2}N)} = \frac{2^{2k}\Gamma(k + \tfrac{1}{2})\Gamma(\tfrac{1}{2}N + k)}{\Gamma(\tfrac{1}{2})\Gamma(\tfrac{1}{2}N)}, \qquad k = 0, 1, \ldots,$$

(iv) $$\mathscr{E}r^{2k-1} = 0, \qquad k = 1, 2, \ldots,$$

(v) $$\mathscr{E}r^{2k} = \frac{\Gamma(k + \tfrac{1}{2})\Gamma(\tfrac{1}{2}N)}{\Gamma(\tfrac{1}{2})\Gamma(\tfrac{1}{2}N + k)}, \qquad k = 0, 1, \ldots .$$

69. (Sec. 6.8.1.) Show

$$\int_0^{2\pi} \log (a + b \cos x)\, dx = 2 \int_0^{\pi} \log (a + b \cos x)\, dx$$

$$= 2\pi \log \tfrac{1}{2}(a + \sqrt{a^2 - b^2}), \qquad a > b \geq 0.$$

[*Hint:* Differentiate with respect to a and verify the second equality. The subsequent integral can be evaluated by letting $\cos x = \cos^2 \tfrac{1}{2}x - \sin^2 \tfrac{1}{2}x$ and then letting $y = (a - b) \tan \tfrac{1}{2}x/\sqrt{a^2 - b^2}$.]

70. (Sec. 6.8.1.) Prove

$$\frac{2\pi}{T} \sum_{t=1}^{T} (e^{i2\pi t/T} + e^{-i2\pi t/T})^k = \int_0^{2\pi} (e^{iu} + e^{-iu})^k\, du, \qquad k = 0, 1, \ldots, T - 1.$$

71. (Sec. 6.8.1.) Show that up to order $T - 1$ the semi-invariants of the circular Q_1 are the same as those of the random variable with moment generating function (20) by comparing the expansions of (18) and (19) in power series in θ. [*Hint:* Use Problem 70.]

72. (Sec. 6.8.1.) Show that if X is a random variable with mean 0, variance σ^2, and lies in (a, b) with probability 1, then $\sigma^2 < -ab$. [*Hint:* Show

$$\int_a^b (x - a)(b - x) f(x)\, dx > 0,$$

where $f(x)$ is the density of X.]

73. (Sec. 6.8.1.) Show that the odd moments of r_1^* to order $T - 2$ and the even moments up to order $T - 1$ are duplicated by

$$\frac{\Gamma(\tfrac{1}{2}T + \tfrac{1}{2})}{\Gamma(\tfrac{1}{2}T)\Gamma(\tfrac{1}{2})} (1 - r_1^{*2})^{\frac{1}{2}(T-2)} \left(1 - \frac{T + 1}{T - 1} r_1^*\right), \qquad -1 \leq r_1^* \leq 1.$$

[Note that the above is not a density because it is negative for $1 - 2/(T + 1) < r_1^* \leq 1$.]

74. (Sec. 6.9.2.) Show

$$r_3' = \frac{r_3^* - r_1^{*2}r_3^* + r_1^{*3} - 2r_1^*r_2^* + r_1^*r_2^{*2}}{1 - 2r_1^{*2} + 2r_1^{*2}r_2^* - r_2^{*2}}.$$

75. (Sec. 6.10.) Show that the derivative of 1 minus (14) with respect to R^* is (9) for $g_1^*(r_1^* \mid 0)$ given by (53) of Section 6.7 with r and r_1^* replaced by R^* and ρ_1 replaced by ρ.

76. (Sec. 6.11.1.) Show that the matrix in the quadratic form in the exponent of the distribution is $-\frac{1}{2}\mathbf{\Sigma}^{-1} = -1/(2\sigma^2)\boldsymbol{T}'\boldsymbol{T}$, where

$$\boldsymbol{T} = \begin{bmatrix} \sqrt{1-\beta^2} & 0 & 0 & \cdots & 0 & 0 \\ \beta & 1 & 0 & \cdots & 0 & 0 \\ 0 & \beta & 1 & \cdots & 0 & 0 \\ \cdot & \cdot & \cdot & & \cdot & \cdot \\ \cdot & \cdot & \cdot & & \cdot & \cdot \\ \cdot & \cdot & \cdot & & \cdot & \cdot \\ 0 & 0 & 0 & \cdots & 1 & 0 \\ 0 & 0 & 0 & \cdots & \beta & 1 \end{bmatrix}.$$

77. (Sec. 6.11.1.) Verify that the cubic (10) does not have a solution that is rational in polynomials of P_0', P_1, and P_0.

78. (Sec. 6.11.1.) Show that

$$\hat{\beta} = -\left(1 - \frac{1}{T}\right)\frac{P_1}{P_0}$$

is a solution of (10) to order $1/T$. [*Hint:* Let $\hat{\beta} = -P_1/P_0 + v/T$ and solve for v from terms of appropriate degree in $1/T$; note P_0'/P_0 is of order $1/T$ in probability.]

79. (Sec. 6.11.1.) Find equations for the maximum likelihood estimates when $\mathscr{E}y_t = \mu$.

80. (Sec. 6.11.1.) In the case of the first-order stationary Gaussian process [$\mathscr{E}y_t = \mu$, $\mathscr{E}(y_t - \mu)(y_s - \mu) = \sigma(0)\rho^{|t-s|}$] show that the maximum likelihood estimate of μ when ρ $(0 < \rho < 1)$ is known is

$$\hat{\mu} = \frac{y_1 + (1-\rho)\sum_{t=2}^{T-1} y_t + y_T}{(T-2)(1-\rho) + 2},$$

and verify that its variance is

$$\operatorname{Var} \hat{\mu} = \sigma(0)\frac{1+\rho}{(T-2)(1-\rho)+2}.$$

[*Hint:* Use the results of Section 2.4.]

81. (Sec. 6.11.1.) Find the variance of $\bar{y}$ for the first-order stationary Gaussian process and compare it with the variance of $\hat{\mu}$.

82. (Sec. 6.11.1.) In the first-order stationary Gaussian process let $\rho = e^{-k/T}$ $(0 < k)$. Show the limiting variances of $\hat{\mu}$ and $\bar{y}$ as $T \to \infty$ are

$$\lim_{T\to\infty} \operatorname{Var} \hat{\mu} = \sigma(0)\frac{2}{2+k},$$

$$\lim_{T\to\infty} \operatorname{Var} \bar{y} = \sigma(0)\, 2\, \frac{e^{-k} - 1 + k}{k^2}.$$

From these expressions verify

$$\lim_{T\to\infty} \operatorname{Var} \hat{\mu} < \lim_{T\to\infty} \operatorname{Var} \bar{y} < \sigma(0).$$

83. (Sec. 6.11.2.) Verify that the Jacobian of (12) is $1 - (-\beta)^T$.

84. (Sec. 6.11.2.) Is there a solution $-\hat{\beta}$ of (20) that is a monotonically increasing function of r?

85. (Sec. 6.11.2.) Let $-\hat{\beta}_T$ be the root of (20) that lies between -1 and 1 and let $r = r_T$. Show that $T^k(r_T + \hat{\beta}_T) \to 0$ in probability as $T \to \infty$ for arbitrary $k > 0$. [*Hint:* Show that for T sufficiently large the probability is arbitrarily large that $|r_T + \hat{\beta}_T| < \varepsilon$ and $a < -\hat{\beta}_T < b$ for suitable ε, a, and b with $-1 < a < b < 1$. Then show that $T^k(-\hat{\beta}_T)^{T-1} \to 0$ in probability.]

86. (Sec. 6.11.2.) Verify that (21) is the likelihood ratio criterion for testing the null hypothesis $\beta = 0$.

87. (Sec. 6.11.2.) Does the likelihood ratio test correspond to rejecting the hypothesis $\beta = 0$ if $r > r_1$ and if $r < r_2$?

88. (Sec. 6.11.2.) Find equations for the maximum likelihood estimates when $\mathscr{E} y_t = \mu$.

89. (Sec. 6.11.2.) Find the maximum likelihood equations for $\hat{\sigma}^2$ and $\hat{\beta}$ in the model based on successive differences, that is, the model of Section 6.5.3 with $\gamma_0 = (1 + \beta^2)/\sigma^2$ and $\gamma_1 = 2\beta/\sigma^2$. [*Hint:* Use Lemma 6.7.11 with $\theta = -\beta/(1 + \beta^2)$ to evaluate the determinant in the normal distribution.]

90. (Sec. 6.11.2.) Find equations for the maximum likelihood estimates of σ^2, β, and μ in the model based on successive differences when $\gamma_0 = (1 + \beta^2)/\sigma^2$, $\gamma_1 = 2\beta/\sigma^2$, and $\mathscr{E} y_t = \mu$.

CHAPTER 7

Stationary Stochastic Processes

7.1. INTRODUCTION

The sequence of T observations constituting the time series for statistical analysis may often be considered as a sampling at T consecutive equally-spaced time points of a much longer sequence of random variables. The objective of statistical inference may be the probability structure of the longer sequence. It is frequently convenient to treat this longer sequence as infinite, going indefinitely into the future and possibly going indefinitely into the past. Such a sequence of random variables, $y_1, y_2, \ldots$, or $\ldots, y_{-1}, y_0, y_1, \ldots$, is known as a *stochastic process with a discrete time parameter.*

Although the probability model for such a sequence of random variables can be quite arbitrary, it is useful to make a distinction between a stochastic process and a set of random variables in that the stochastic process takes the idea of time into account. Roughly speaking, in a stochastic process the variables close together in time behave more similarly than those far apart in time. Moreover, the model usually has some simplifications in the structure; then a finite set of observations has implications for the infinite sequences.

One simplifying property is that of *stationarity*. The idea is that the behavior of a set of random variables at one time is probabilistically the same as the behavior of a set at another time. This notion was discussed in connection with processes generated by stochastic difference equations and moving averages.

In Chapter 5 we were concerned with processes defined in terms of a finite number of parameters. Now we shall consider processes in which the number of parameters may be infinite. For example, if the process is Gaussian (normal) and stationary, there is one mean, one variance, and an infinite number of covariances. We shall be interested in what information about these can be gleaned from a finite number of observations.

A stochastic process as an infinite sequence of random variables involves a probability measure on an infinite-dimensional space. Some interesting questions, particularly about limits of functions of sequences, are questions about probabilities of sets in the infinite-dimensional space. Since we are interested in inferences from a finite number of observations (though possibly a large number), we do not need to use probabilities of such sets. In this chapter the mathematical theory of stochastic processes is discussed, but not developed thoroughly and rigorously, since it is not required for our purposes. The purpose of the discussion is to develop the ideas.

A stochastic process $y(t)$ of a continuous time parameter t can be defined for $0 \leq t < \infty$ or $-\infty < t < \infty$. Such a process can be treated in terms of a probability measure on a space of functions $y(t)$. A sample from such a process could consist of observations on the process at a finite number of times, or it could consist of a continuous observation on the process over an interval of time. For instance, if the process serves as a theoretical model of the temperature observable at some location, the sample might be a sequence of consecutive hourly readings or it might be a graph of a continuous reading. Often a stochastic process of a discrete time parameter may be thought of as a sampling at equally-spaced time points of a stochastic process of a continuous time parameter.

7.2. DEFINITIONS AND DISCUSSION OF STATIONARY STOCHASTIC PROCESSES

7.2.1. Definitions

A stochastic process with a discrete time parameter consists of an infinite sequence of random variables, $y_1, y_2, \ldots$, or $\ldots, y_{-1}, y_0, y_1, \ldots$. An observation on a process, often called a *realization*, is a point in the corresponding infinite-dimensional space. The process involves a probability measure defined on this infinite-dimensional space. Probability is assigned to certain sets, called measurable sets. This class of sets includes the complement of every set in the class, the union of a denumerable number of sets in the class, and the intersection of a denumerable number of such sets; the probability measure is defined on these sets with the property that the measure of the union of disjoint sets is the sum of the probabilities of the sets.

We are particularly interested in probabilities associated with a finite number of the random variables. These probabilities include the cumulative distribution function $\Pr\{y_1 \leq b_1, \ldots, y_T \leq b_T\}$ and probabilities derived from it.

Therefore, we want the class of measurable sets to include the sets $\{y_1 \le b_1, \ldots, y_T \le b_T\}$; these sets and other sets in the class which are defined in terms of a finite number of coordinates, $y_{t_1}, \ldots, y_{t_n}$, are called *cylinder sets*. Thus the probability measure of the stochastic process implies the probabilities of (measurable) cylinder sets, including the cumulative distribution functions.

Conversely, if cumulative distribution functions (cdf's) for all finite sets of t's are given and are consistent, then a probability measure can be defined on the infinite-dimensional space so that the probability measure of cylinder sets agrees with the probability as computed from the given cdf's; this extension is unique. That the family of cdf's is *consistent* means that (i) any marginal (joint) cdf derived from a given cdf corresponds to the given marginal cdf, and that (ii) the cdf's for the arguments written in different orders agree. More precisely, let $F(a_1, \ldots, a_n;\ t_1, \ldots, t_n)$ denote a cdf at $a_1, \ldots, a_n$; suppose this is given for every real n-tuplet $a_1, \ldots, a_n$, every $t_1, \ldots, t_n$ (from $\ldots, -1, 0, 1, \ldots$ or $1, 2, \ldots$), and $n = 1, 2, \ldots$. The requirements are

$$(1)\quad F(a_1, \ldots, a_m, \infty, \ldots, \infty;\ t_1, \ldots, t_m, t_{m+1}, \ldots, t_n) = F(a_1, \ldots, a_m;\ t_1, \ldots, t_m), \quad m = 1, \ldots, n,$$

$$(2)\quad F(a_1, \ldots, a_n;\ t_1, \ldots, t_n) = F(a_{i_1}, \ldots, a_{i_n};\ t_{i_1}, \ldots, t_{i_n}),$$

where $(i_1, \ldots, i_n)$ is any permutation of $(1, \ldots, n)$. Then $F(a_1, \ldots, a_n;\ t_1, \ldots, t_n)$ may be interpreted as the cdf of $y_{t_1}, \ldots, y_{t_n}$, that is, $\Pr\{y_{t_1} \le a_1, \ldots, y_{t_n} \le a_n\}$ for all a_i's, t_i's, and n. This theorem was given by Kolmogorov (1933); we shall not prove it here.

As an example, we can define the cdf's as the cdf's of multivariate normal distributions, specifying the means and covariances,

$$(3)\quad \mathscr{E} y_t = m(t),$$

$$(4)\quad \mathscr{E}[y_t - m(t)][y_s - m(s)] = \operatorname{Cov}(y_t, y_s) = \sigma(t, s).$$

The sequence $m(t)$ is arbitrary. The double sequence $\sigma(t, s)$ must satisfy the usual properties of covariances; namely, $\sigma(t, s) = \sigma(s, t)$ for every pair s, t and any matrix $[\sigma(t_i, t_j)]$, $i, j = 1, \ldots, n$, is positive semidefinite. This specification determines the process. Such a process is called *Gaussian*.

A stochastic process with a discrete time parameter is said to be *stationary* (or *stationary in the strict sense*) if the distribution of $y_{t_1}, y_{t_2}, \ldots, y_{t_n}$ is the same as the distribution of $y_{t_1+t}, y_{t_2+t}, \ldots, y_{t_n+t}$ for every finite set of integers $\{t_1, t_2, \ldots, t_n\}$ and for every integer t. This definition is equivalent to requiring that the probability measure for the sequence $\{y_s\}$ be the same as that of $\{y_{s+t}\}$ for every integer t.

If the first-order moments exist, stationarity implies that

$$\mathscr{E}y_s = \mathscr{E}y_{s+t}, \qquad s, t = \ldots, -1, 0, 1, \ldots, \tag{5}$$

or

$$m(s) = m(s + t) = m, \tag{6}$$

say, for all s and t. Since (y_{t_1}, y_{t_2}) has the same distribution as (y_{t_1+t}, y_{t_2+t}), existence of the second-order moments and stationarity imply

$$\sigma(t_1, t_2) = \sigma(t_1 + t, t_2 + t); \tag{7}$$

setting $t = -t_2$ gives

$$\sigma(t_1, t_2) = \sigma(t_1 - t_2, 0) = \sigma(t_1 - t_2), \tag{8}$$

say. In the normal case properties (5) and (7) determine that the stochastic process is stationary.

A stochastic process is said to be *stationary in the wide sense* or *weakly stationary* or *stationary to second order* if the mean function $m(t)$ and the covariance function $\sigma(t, s)$ defined as in (3) and (4) exist and satisfy the relations (5) and (7); that is, the mean is a constant, independent of time, and the covariance of any two variables depends only on their distance apart in time. Obviously, any process which is stationary in the strict sense and has finite variance is also stationary in the wide sense. (In the normal case discussed above stationarity in the strict sense and in the wide sense are equivalent.)

7.2.2. Examples of Stationary Stochastic Processes with a Discrete Time Parameter

We now give some examples of stationary stochastic processes with a discrete time parameter.

Example 7.1. Let the y_t's be independently and identically distributed with

$$\mathscr{E}y_t = m, \qquad \operatorname{Var} y_t = \sigma^2; \tag{9}$$

then

$$\begin{aligned} \sigma(t, s) &= \sigma^2, & t &= s, \\ &= 0, & t &\neq s. \end{aligned} \tag{10}$$

This process is stationary in the strict sense, but if the requirement of identical distribution is dropped [while (9) and (10) are retained], the resulting process is still stationary in the wide sense.

Example 7.2. Let all the y_t's be identically equal to a random variable y. Then, if the first two moments of y exist,

$$\mathscr{E} y_t = \mathscr{E} y = m, \tag{11}$$

$$\sigma(t, s) = \operatorname{Var} y = \sigma^2. \tag{12}$$

This process is stationary in the strict sense.

Example 7.3. Define the sequence $\{y_t\}$ as follows:

$$y_t = \sum_{j=1}^{q} (A_j \cos \lambda_j t + B_j \sin \lambda_j t), \qquad t = \ldots, -1, 0, 1, \ldots, \tag{13}$$

where the λ_j's are constants and $A_1, \ldots, A_q, B_1, \ldots, B_q$ are random variables such that

$$\mathscr{E} A_j = \mathscr{E} B_j = 0, \qquad j = 1, \ldots, q, \tag{14}$$

$$\mathscr{E} A_j^2 = \mathscr{E} B_j^2 = \sigma_j^2, \qquad j = 1, \ldots, q, \tag{15}$$

$$\mathscr{E} A_i A_j = \mathscr{E} B_i B_j = 0, \quad i \neq j, \qquad i, j = 1, \ldots, q, \tag{16}$$

$$\mathscr{E} A_i B_j = 0, \qquad i, j = 1, \ldots, q. \tag{17}$$

Then

$$\mathscr{E} y_t = 0, \tag{18}$$

$$\begin{aligned} \mathscr{E} y_t y_s &= \sum_{i,j=1}^{q} \mathscr{E} (A_i \cos \lambda_i t + B_i \sin \lambda_i t)(A_j \cos \lambda_j s + B_j \sin \lambda_j s) \\ &= \sum_{j=1}^{q} [\mathscr{E} A_j^2 \cos \lambda_j t \cos \lambda_j s + \mathscr{E} B_j^2 \sin \lambda_j t \sin \lambda_j s] \\ &= \sum_{j=1}^{q} \sigma_j^2 [\cos \lambda_j t \cos \lambda_j s + \sin \lambda_j t \sin \lambda_j s] \\ &= \sum_{j=1}^{q} \sigma_j^2 \cos \lambda_j (t - s). \end{aligned} \tag{19}$$

Hence, the process is stationary in the wide sense. If the A_j's and B_j's are normally distributed, the y_t's will also be normally distributed since they are linear combinations of the A_j's and B_j's. Then the process will be stationary in the strict sense.

This process may also be represented by

$$y_t = \sum_{j=1}^{q} R_j \cos(\lambda_j t - \theta_j), \qquad t = \ldots, -1, 0, 1, \ldots, \tag{20}$$

where (as in Figure 7.1) $R_j \geq 0$,

$$R_j^2 = A_j^2 + B_j^2, \qquad j = 1, \ldots, q, \tag{21}$$

$$\tan \theta_j = B_j / A_j, \qquad j = 1, \ldots, q, \tag{22}$$

and $0 < \theta_j < \pi$ if $B_j > 0$ and $\pi < \theta_j < 2\pi$ if $B_j < 0$. If A_j and B_j are normally distributed, then R_j^2 is proportional to a chi-square variable with 2 degrees of freedom and θ_j is uniformly distributed between 0 and 2π (by the symmetry of the normal distribution) and is independent of R_j^2. The stationarity of the process may be seen from this definition by noting that

$$y_{t+s} = \sum_{j=1}^{q} R_j \cos[\lambda_j t - (\theta_j - \lambda_j s)], \tag{23}$$

where $(\theta_j - \lambda_j s)$ is distributed uniformly over an interval of length 2π; thus y_t and y_{t+s} have the same distribution. If the A_j's and B_j's are not normally distributed, $\{y_t\}$ is not necessarily stationary in the strict sense.

This example is important because in a sense every weakly stationary stochastic process with finite variance can be approximated by a linear combination, such as the right-hand side of (13).

Example 7.4. Moving average. Let $\ldots, v_{-1}, v_0, v_1, \ldots$ be a sequence of independently and identically distributed random variables, and let $\alpha_0, \alpha_1, \ldots, \alpha_q$ be $q + 1$ numbers. Then the moving average

$$y_t = \alpha_0 v_t + \alpha_1 v_{t-1} + \cdots + \alpha_q v_{t-q}, \qquad t = \ldots, -1, 0, 1, \ldots, \tag{24}$$

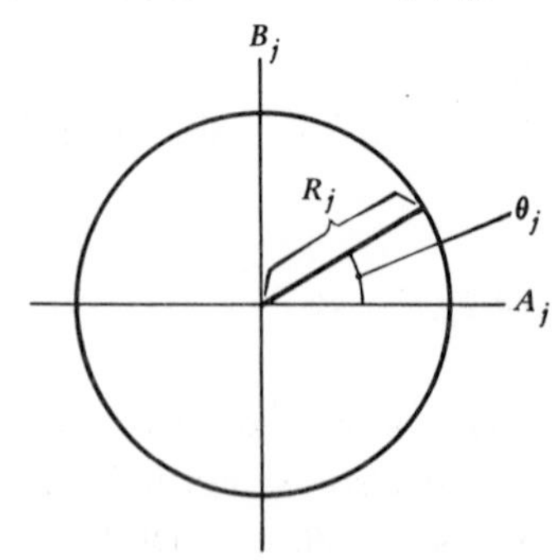

Figure 7.1. Coordinates of trigonometric functions.

is a stationary stochastic process. If $\mathscr{E}v_t = \nu$ and $\operatorname{Var} v_t = \sigma^2$, then

$$\mathscr{E}y_t = \nu(\alpha_0 + \alpha_1 + \cdots + \alpha_q), \tag{25}$$

$$\begin{aligned} \operatorname{Cov}(y_t, y_{t+s}) &= \sigma^2(\alpha_0\alpha_s + \cdots + \alpha_{q-s}\alpha_q), \qquad & s &= 0, \ldots, q, \\ &= 0, & s &= q+1, \ldots, \end{aligned} \tag{26}$$

and $\{y_t\}$ is stationary in the wide sense. For $\{y_t\}$ defined by (24) to be stationary in the wide sense the only conditions necessary on the v_t's is that they have the same mean and variance and be mutually uncorrelated.

We can also define infinite moving averages. The infinite sum

$$\sum_{s=0}^{\infty} \alpha_s v_{t-s} \tag{27}$$

means the random variable, say y_t, if it exists, such that

$$\lim_{n \to \infty} \mathscr{E}\left(y_t - \sum_{s=0}^{n} \alpha_s v_{t-s}\right)^2 = 0. \tag{28}$$

If

$$\sum_{s=0}^{\infty} \alpha_s^2 < \infty, \tag{29}$$

such a random variable exists (Corollary 7.6.1). Then (27) is said to converge in the mean or in quadratic mean. The doubly infinite sum

$$\sum_{s=-\infty}^{\infty} \alpha_s v_{t-s} \tag{30}$$

is defined similarly as the random variable y_t such that

$$\lim_{n, m \to \infty} \mathscr{E}\left(y_t - \sum_{s=-m}^{n} \alpha_s v_{t-s}\right)^2 = 0; \tag{31}$$

a sufficient condition is

$$\sum_{s=-\infty}^{\infty} \alpha_s^2 < \infty. \tag{32}$$

The conditions (29) and (32) are sufficient if the v_s's are uncorrelated with common mean and variance, but are not necessarily independent.

Example 7.5. Autoregressive process. In Chapter 5 we considered a process satisfying a stochastic difference equation

$$\sum_{r=0}^{p} \beta_r y_{t-r} = u_t, \qquad t = \ldots, -1, 0, 1, \ldots, \tag{33}$$

where the u_t's are independently and identically distributed with mean 0. If the roots of the associated polynomial equation

$$\sum_{r=0}^{p} \beta_r x^{p-r} = 0 \tag{34}$$

are less than 1 in absolute value, (33) can be inverted to yield

$$y_t = \sum_{s=0}^{\infty} \delta_s u_{t-s}, \qquad t = \ldots, -1, 0, 1, \ldots, \tag{35}$$

where the right-hand side converges in the mean. Thus $\{y_t\}$ is stationary in the strict sense. If the u_t's are uncorrelated and have common mean and variance (but are not necessarily independently and identically distributed), the process is stationary in the wide sense. It was shown in Section 5.2 that the covariance sequence satisfies the pure difference equation

$$\sum_{r=0}^{p} \beta_r \sigma(s-r) = 0, \qquad s = 1, 2, \ldots. \tag{36}$$

If the roots of the associated polynomial equation (34) are distinct, say $x_1, \ldots, x_p$, then

$$\sigma(h) = \sum_{j=1}^{p} c_j x_j^h, \qquad h = 1-p, 2-p, \ldots, -1, 0, 1, \ldots, \tag{37}$$

and the constants $c_1, \ldots, c_p$ must satisfy

$$\sigma(-h) = \sigma(h), \qquad h = 1, \ldots, p-1, \tag{38}$$

$(\beta_p \neq 0)$ and

$$\sum_{r=0}^{p} \beta_r \sigma(r) = \sigma^2, \tag{39}$$

where $\mathscr{E} u_t^2 = \sigma^2$.

Example 7.6. Autoregressive process with moving average residual. The two previous models can be combined; that is, we can take a moving average as the disturbance u_t of the stochastic difference equation. The model is

$$\sum_{r=0}^{p} \beta_r y_{t-r} = \sum_{j=0}^{q} \alpha_j v_{t-j}, \qquad t = \ldots, -1, 0, 1, \ldots, \tag{40}$$

where the v_t's are independently and identically distributed. Then $\{y_t\}$ is stationary in the strict sense. If the v_t's are uncorrelated and have common mean and variance, then $\{y_t\}$ is stationary in the wide sense.

Example 7.7. Autoregressive process with superimposed error. Let y_t satisfy the stochastic difference equation (33) and define

$$z_t = y_t + w_t, \tag{41}$$

where the w_t's are independently identically distributed, independently of the y_t's. If $\{y_t\}$ is stationary in the strict sense, the process $\{z_t\}$ is stationary in the strict sense. If $\{y_t\}$ is stationary in the wide sense and the w_t's are only uncorrelated with common mean and variance, $\{z_t\}$ is stationary in the wide sense.

Example 7.8. An extreme example of a process that is stationary in the wide sense but not in the strict sense is

$$y_t = \cos tw, \qquad t = 1, 2, \ldots, \tag{42}$$

where w is uniformly distributed in the interval $(0, 2\pi)$. Then $\mathscr{E} y_t = 0$, $\mathscr{E} y_t^2 = \frac{1}{2}$, and $\mathscr{E} y_t y_s = 0$, $t \neq s$. These random variables are uncorrelated although dependent functionally and statistically.

Example 7.9. Another process stationary in the wide sense related to Example 7.3 is

$$y_t = \sum_{j=1}^{q} \sigma_j \sqrt{2} \cos(\lambda_j t - v_j), \tag{43}$$

where $v_1, \ldots, v_q$ are independently uniformly distributed in the interval $(0, 2\pi)$. Then $\mathscr{E} y_t = 0$ and

$$\begin{aligned}
\mathscr{E} y_t y_s &= \sum_{i,j=1}^{q} 2\sigma_i \sigma_j \mathscr{E} \cos(\lambda_i t - v_i) \cos(\lambda_j s - v_j) \\
&= 2 \sum_{j=1}^{q} \sigma_j^2 \int_0^{2\pi} \cos(\lambda_j t - v_j) \cos(\lambda_j s - v_j)\, dv_j/(2\pi) \\
&= 2 \sum_{j=1}^{q} \sigma_j^2 \int_0^{2\pi} (\cos \lambda_j t \cos v_j + \sin \lambda_j t \sin v_j) \\
&\qquad \times (\cos \lambda_j s \cos v_j + \sin \lambda_j s \sin v_j)\, dv_j/(2\pi) \\
&= \sum_{j=1}^{q} \sigma_j^2 (\cos \lambda_j t \cos \lambda_j s + \sin \lambda_j t \sin \lambda_j s) \\
&= \sum_{j=1}^{q} \sigma_j^2 \cos \lambda_j (t - s),
\end{aligned} \tag{44}$$

which is (19).

7.2.3. Stochastic Process with a Continuous Time Parameter

A stochastic process $y(t)$ with a continuous time parameter, t being real-valued in some finite or infinite interval, such as $0 \leq t < \infty$ or $-\infty < t < \infty$, is based on a probability measure on a space of functions of t. The probability measure defines probabilities of cylinder sets, in particular, cdf's of every finite set of variables, $y(t_1), \ldots, y(t_n)$, where $t_1, \ldots, t_n$ are any n real numbers in the range of t. Conversely, a family of consistently defined cdf's $F(a_1, \ldots, a_n; t_1, \ldots, t_n)$ for every real n-tuplet $t_1, \ldots, t_n$ in the range of t and every n can be extended uniquely to a probability measure on the function space.

Stationarity in the strict sense and stationarity in the wide sense are defined in a manner similar to that for processes with a discrete time parameter.

7.3. THE SPECTRAL DISTRIBUTION FUNCTION AND THE SPECTRAL DENSITY

A stochastic process with discrete time parameter which is stationary in the wide sense defines a sequence of covariances, $\sigma(0), \sigma(1), \ldots$. The Fourier transform of the sequence,

$$f(\lambda) = \frac{1}{2\pi}\sigma(0) + \frac{1}{\pi}\sum_{h=1}^{\infty}\sigma(h)\cos\lambda h \tag{1}$$

$$= \frac{1}{2\pi}\sum_{h=-\infty}^{\infty}\sigma(h)\cos\lambda h, \qquad -\pi \leq \lambda \leq \pi,$$

when the series converges, is of special interest. If

$$\sigma(0) + 2\sum_{h=1}^{\infty}|\sigma(h)| = \sum_{h=-\infty}^{\infty}|\sigma(h)| < \infty, \tag{2}$$

(1) converges uniformly; multiplication of (1) by $\cos \lambda k$ and integration gives

$$\sigma(k) = \int_{-\pi}^{\pi}\cos\lambda k\, f(\lambda)\, d\lambda. \tag{3}$$

(See Section 4.2.4 for a review of Fourier series.) Thus the information in the covariance sequence is the same as the information in the function $f(\lambda)$. The function $f(\lambda)$ (when it exists) is called the *spectral density*. Note that $f(\lambda)$ is even; that is, $f(\lambda) = f(-\lambda)$. It is sometimes convenient to treat $g(\lambda) = 2f(\lambda)$,

$0 \le \lambda \le \pi$; that is,

$$g(\lambda) = \frac{1}{\pi}\,\sigma(0) + \frac{2}{\pi}\sum_{h=1}^{\infty} \sigma(h)\cos \lambda h \tag{4}$$

$$= \frac{1}{\pi}\sum_{h=-\infty}^{\infty} \sigma(h)\cos \lambda h, \qquad 0 \le \lambda \le \pi.$$

We may also call $g(\lambda)$ a spectral density. Then

$$\sigma(k) = \int_0^{\pi} \cos \lambda k\, g(\lambda)\, d\lambda. \tag{5}$$

We define the *spectral distribution function* [when $f(\lambda)$ is defined] as

$$F(\lambda) = \int_{-\pi}^{\lambda} f(\nu)\, d\nu. \tag{6}$$

When (2) holds, we can integrate (1) term by term to obtain

$$F(\lambda) = \frac{\sigma(0)}{2\pi}(\lambda + \pi) + \frac{1}{\pi}\sum_{h=1}^{\infty} \frac{\sigma(h)\sin \lambda h}{h}. \tag{7}$$

Then (3) can be written

$$\sigma(k) = \int_{-\pi}^{\pi} \cos \lambda k\, dF(\lambda). \tag{8}$$

We shall show that for any covariance sequence $\{\sigma(k)\}$ there exists a function $F(\lambda)$ such that (8) is true and $F(\lambda)$ is monotonically nondecreasing. This latter property corresponds to $f(\lambda) \ge 0$ [when $F(\lambda)$ is absolutely continuous].

As in Section 4.4.1, we define the sample periodogram

$$R^2(\lambda) = A^2(\lambda) + B^2(\lambda), \qquad -\pi \le \lambda \le \pi, \tag{9}$$

where

$$A(\lambda) = \frac{2}{T}\sum_{t=1}^{T} y_t \cos \lambda t, \qquad -\pi \le \lambda \le \pi, \tag{10}$$

$$B(\lambda) = \frac{2}{T}\sum_{t=1}^{T} y_t \sin \lambda t, \qquad -\pi \le \lambda \le \pi. \tag{11}$$

Here $y_1, \ldots, y_T$ are T successive variables from a stochastic process that is stationary in the wide sense and has mean 0. Let

$$
\begin{aligned}
(12)\qquad I_T(\lambda) &= \frac{T}{8\pi} R^2(\lambda) \\
&= \frac{1}{2\pi T}\left[\sum_{s,t=1}^{T} y_s y_t \cos \lambda s \cos \lambda t + \sum_{s,t=1}^{T} y_s y_t \sin \lambda s \sin \lambda t\right] \\
&= \frac{1}{2\pi T}\sum_{s,t=1}^{T} y_s y_t \cos \lambda(t-s) \\
&= \frac{1}{2\pi T}\sum_{s,t=1}^{T} y_s y_t e^{i\lambda(t-s)} \\
&= \frac{1}{2\pi T}\left|\sum_{t=1}^{T} y_t e^{i\lambda t}\right|^2.
\end{aligned}
$$

Then

$$
\begin{aligned}
(13)\qquad \mathscr{E} I_T(\lambda) &= \frac{1}{2\pi T}\sum_{s,t=1}^{T} \sigma(t-s)\cos \lambda(t-s) \\
&= \frac{1}{2\pi}\sum_{h=-(T-1)}^{T-1}\left(1-\frac{|h|}{T}\right)\sigma(h)\cos \lambda h \\
&= f_T(\lambda),
\end{aligned}
$$

say, since $T - |h|$ is the number of pairs $s, t,\ s, t = 1, \ldots, T$, such that $t - s = h$, $h = -(T-1), \ldots, T-1$. Since $R^2(\lambda) \geq 0$,

$$
(14)\qquad f_T(\lambda) \geq 0;
$$

also

$$
(15)\qquad
\begin{aligned}
\sigma(k)\left(1-\frac{|k|}{T}\right) &= \int_{-\pi}^{\pi} \cos \lambda k\, f_T(\lambda)\, d\lambda, \qquad k = -(T-1), \ldots, T-1, \\
0 &= \int_{-\pi}^{\pi} \cos \lambda k\, f_T(\lambda)\, d\lambda, \qquad k = \pm T, \pm(T+1), \ldots .
\end{aligned}
$$

Then

$$
\begin{aligned}
(16)\qquad F_T(\lambda) &= \int_{-\pi}^{\lambda} f_T(\nu)\, d\nu \\
&= \frac{\sigma(0)}{2\pi}(\lambda+\pi) + \frac{1}{\pi}\sum_{h=1}^{T-1}\left(1-\frac{h}{T}\right)\frac{\sigma(h)\sin \lambda h}{h}
\end{aligned}
$$

is monotonically nondecreasing, $F_T(-\pi) = 0$, and $F_T(\pi) = \sigma(0)$. By the weak compactness theorem [Loève (1963), Section 11.2], there is a sequence of values $T_1, T_2, \ldots$ such that the subsequence $F_{T_j}(\lambda)$ converges. The resulting function $F(\lambda)$, say, is monotonically nondecreasing, $F(-\pi) = 0$, and $F(\pi) = \sigma(0)$. The subsequence converges at every continuity point of $F(\lambda)$, and (8) holds as the limit (over the subsequence) of (15) [Helly-Bray Lemma, Loève (1963), Section 11.3]. Then (7) holds at all continuity points of $F(\lambda)$, implying uniqueness at those points. Define $F(\lambda)$ as continuous on the right except at $\lambda = -\pi$:

$$F(\lambda+) = F(\lambda), \qquad -\pi < \lambda < \pi. \tag{17}$$

THEOREM 7.3.1. *For any covariance sequence $\{\sigma(h)\}$ there exists a uniquely defined monotonically nondecreasing function $F(\lambda)$, $-\pi \leq \lambda \leq \pi$, with symmetric increments, with $F(-\pi) = 0$, and continuous on the right $(-\pi < \lambda < \pi)$ such that*

$$\sigma(k) = \int_{-\pi}^{\pi} \cos \lambda k \, dF(\lambda), \qquad k = 0, \pm 1, \ldots. \tag{18}$$

We can let

$$G(\lambda) = 2F(\lambda) - F(0) - F(0-), \qquad 0 < \lambda \leq \pi, \tag{19}$$

and $G(0) = 0$. Then G is continuous on the right except at $\lambda = 0$.

It may be instructive to consider the spectral distribution functions of some of the examples discussed in Section 7.2.2.

Example 7.1. Let the y_t's be independent with $\operatorname{Var} y_t = \sigma^2$. Then the covariance sequence

$$\sigma(0) = \sigma^2, \qquad \sigma(h) = 0, \qquad h = \pm 1, \pm 2, \ldots, \tag{20}$$

corresponds to

$$F(\lambda) = \frac{\sigma^2}{2\pi}(\lambda + \pi), \qquad f(\lambda) = \frac{\sigma^2}{2\pi}, \qquad -\pi \leq \lambda \leq \pi. \tag{21}$$

We verify the inverse relationship

$$\sigma(0) = \frac{\sigma^2}{2\pi} \int_{-\pi}^{\pi} d\lambda = \sigma^2, \tag{22}$$

$$\sigma(h) = \frac{\sigma^2}{2\pi} \int_{-\pi}^{\pi} \cos \lambda h \, d\lambda = 0, \qquad h = \pm 1, \pm 2, \ldots. \tag{23}$$

Example 7.2. Let $y_t \equiv y$ with $\operatorname{Var} y = \sigma^2$. Then the covariance sequence

$$\sigma(h) = \sigma^2, \qquad h = 0, \pm 1, \ldots, \tag{24}$$

corresponds to

$$F(\lambda) = 0, \qquad -\pi \le \lambda < 0, \tag{25}$$
$$= \sigma^2, \qquad 0 \le \lambda \le \pi.$$

We verify the inverse relationship

$$\begin{aligned} \sigma(h) &= \int_{-\pi}^{\pi} \cos \lambda h \, dF(\lambda) \\ &= \sigma^2 \cos 0 \\ &= \sigma^2, \qquad h = 0, \pm 1, \ldots . \end{aligned} \tag{26}$$

Example 7.3. Let

$$y_t = \sum_{j=1}^{q} (A_j \cos \lambda_j t + B_j \sin \lambda_j t), \qquad t = \ldots, -1, 0, 1, \ldots, \tag{27}$$

where $\mathscr{E} A_j = \mathscr{E} B_j = 0$, $\mathscr{E} A_j^2 = \mathscr{E} B_j^2 = \sigma_j^2$, and the A_j's and B_j's are mutually uncorrelated. Suppose the λ_j's are ordered so $0 \le \lambda_1 < \lambda_2 < \cdots < \lambda_q \le \pi$. We shall determine $F(\lambda)$ so that

$$\sigma(h) = \sum_{j=1}^{q} \sigma_j^2 \cos \lambda_j h, \qquad h = \ldots, -1, 0, 1, \ldots . \tag{28}$$

This is accomplished by letting $F(\lambda)$ be a step function with discontinuities of size $\frac{1}{2}\sigma_j^2$ at $\lambda = \pm\lambda_j$, $j = 1, \ldots, q$ ($\lambda_1 \ne 0$, or σ_1^2 at $\lambda = \lambda_1 = 0$), as in Figure 7.2. In this case

$$\sigma(0) = F(\pi) = \sum_{j=1}^{q} \sigma_j^2. \tag{29}$$

An interpretation of the process is that a realization of y_t is the sum of q cosine functions and q sine functions if $\lambda_1 > 0$. The weights of the jth cosine and sine functions are a pair of uncorrelated drawings from a population with variance σ_j^2. The randomness enters the process through these weights.

If the A_j's and B_j's are normally distributed, the process can be represented

$$y_t = \sum_{j=1}^{q} \sigma_j \sqrt{U_j} \cos (\lambda_j t - \theta_j), \qquad t = \ldots, -1, 0, 1, \ldots, \tag{30}$$

where $U_1, \ldots, U_q$ are independent χ^2-variables, each with 2 degrees of freedom, and $\theta_1, \ldots, \theta_q$ are independently distributed, each uniformly between 0 and 2π. A realization is the sum of q cosine functions of specified frequencies. The phase of each is uniformly distributed; the amplitudes are identically distributed except for a constant of proportionality σ_j.

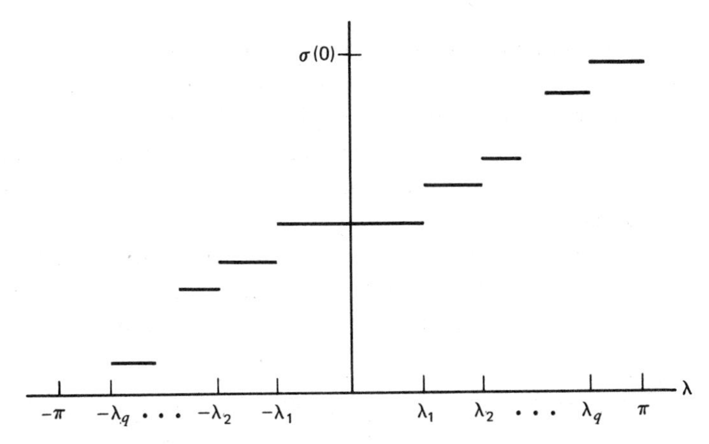

Figure 7.2. A discrete spectral distribution function.

For any stationary stochastic process we can construct a process of the type given by (27) having spectral distribution and covariance functions approximating those of the original process by choosing $\sigma_1^2, \sigma_2^2, \ldots, \sigma_q^2$ to make the step function spectral distribution of the approximating process close to the $F(\lambda)$ of the original process. By taking q large enough, the approximation can be made arbitrarily good. In Section 7.4 we shall make this point clearer by showing that a process $\{y_t\}$ can be expressed as a sum of two integrals.

In addition to being a mathematical convenience, the approximation of stochastic processes by sums like (30) or integrals permits the interpretation of a process as a sum or integral of cosine functions with random phases and random amplitudes. Often the phases are of no interest, but the amplitudes are important. The sum of the mean squared amplitudes of the frequencies in a given interval is the increase of the spectral distribution function in that interval.

In many substantive areas the spectral representation is simpler and more natural and informative than the sequence of observed values. This is true for many natural phenomena, such as sound, light, and electricity, where wave forms are cosine functions. For instance, a pure sound tone is described by a cosine function where frequency corresponds to the pitch and amplitude to the energy in the tone; the natural receptor, the ear, responds to the pitch and amplitude. The spectrum of light displays the energy in the different frequencies. This spectral approach is important in communications engineering, where the spectral function indicates the energy in waves of various frequencies. In each of these examples, the cosine function is the basic description of the simplest situation and provides the basis for analysis of complex situations. It should be noted that this spectral analysis is in terms of the trigonometric functions; if the "natural" description of some periodic phenomenon were in terms of some other basic and simple periodic functions (such as saw-toothed functions or step functions), this spectral analysis might not be appropriate. The cosine and sine functions arise as solutions to second-order differential or difference equations; where such equations describe behavior, the trigonometric functions may be appropriate.

The significance of the sample periodogram and spectral density (when it exists) can be given in another way. The computed sample covariance between y_t and $z_t = \cos(\lambda t - \theta)$ (without correcting for the mean) is

$$\sum_{t=1}^{T} y_t \cos(\lambda t - \theta) = \cos\theta \sum_{t=1}^{T} y_t \cos\lambda t + \sin\theta \sum_{t=1}^{T} y_t \sin\lambda t \tag{31}$$

$$= \frac{T}{2}[\cos\theta\, A(\lambda) + \sin\theta\, B(\lambda)];$$

the maximum with respect to θ is

$$\max_{\theta} \sum_{t=1}^{T} y_t \cos(\lambda t - \theta) = \frac{T}{2} R(\lambda). \tag{32}$$

Thus for a given observed time series $R(\lambda)$ is proportional to the covariance between the observed time series and the cosine function of frequency $\lambda/(2\pi)$ maximized with respect to phase. Except to the extent that the sum of squares of $\cos(\lambda t - \theta)$ depends on θ, $R(\lambda)$ is proportional to the corresponding correlation; roughly $R(\lambda)$ is proportional to the multiple correlation coefficient between $\{y_t\}$ and $\{(\cos\lambda t, \sin\lambda t)\}$. In this sense $R(\lambda)$ measures the similarity between the observed time series and a trigonometric function of frequency $\lambda/(2\pi)$. (See Section 4.3.2.) The spectral density can be given the corresponding interpretation as a population average.

The spectral function $F(\lambda)$ is the same as a cumulative distribution on the interval $[-\pi, \pi]$, except for the fact that $F(\pi) = \sigma(0)$ is not necessarily 1. The spectral distribution function of a stationary stochastic process may be decomposed as

$$F(\lambda) = F_1(\lambda) + F_2(\lambda) + F_3(\lambda), \tag{33}$$

where $F_1(\lambda)$, $F_2(\lambda)$, and $F_3(\lambda)$ are each nondecreasing; $F_1(\lambda)$ is a pure step function, $F_2(\lambda)$ is absolutely continuous, that is,

$$F_2(\lambda) = \int_{-\pi}^{\lambda} F_2'(\mu)\, d\mu, \tag{34}$$

and $F_3(\lambda)$ is a singular function, that is, $F_3(\lambda)$ is continuous and may be increasing although $F_3'(\lambda) = 0$ almost everywhere.

A process $y(t)$ of a continuous time parameter t, $-\infty < t < \infty$, stationary in the wide sense, with mean 0, say, and covariance function

$$\mathscr{E} y(t)y(s) = \sigma(t - s) \tag{35}$$

has a spectral distribution function $F(\lambda)$, $-\infty < \lambda < \infty$, which is monotonically nondecreasing with symmetric increments and bounded. Then

$$\sigma(t) = \int_{-\infty}^{\infty} \cos \lambda t \, dF(\lambda) = \int_{-\infty}^{\infty} e^{i\lambda t} \, dF(\lambda). \tag{36}$$

If we let $G(\lambda) = 2F(\lambda) - F(0) - F(0-)$, $0 < \lambda < \infty$, and $G(0) = 0$, then (36) can be written

$$\sigma(t) = \int_0^{\infty} \cos \lambda t \, dG(\lambda). \tag{37}$$

Except for normalization, $F(\lambda)$ is a cumulative distribution function and $\sigma(t)$ is its characteristic function.

Often a sequence of observations y_t made at equally-spaced time points is a sequence of observations on a process with a continuous parameter, for example, temperature measured every hour and water level measured every day. Thus $y_t = y(kt)$, $t = \ldots, -1, 0, 1, \ldots$, where k is the interval of time. The covariance function of the process with a discrete parameter is

$$\begin{aligned} \mathscr{E} y_t y_{t+h} &= \mathscr{E} y(tk)y(tk + hk) \\ &= \sigma(hk) \\ &= \int_{-\infty}^{\infty} \cos \lambda hk \, dF(\lambda) \\ &= \sum_{j=-\infty}^{\infty} \int_{(2j-1)\pi/k}^{(2j+1)\pi/k} \cos \lambda hk \, dF(\lambda) \\ &= \sum_{j=-\infty}^{\infty} \int_{-\pi}^{\pi} \cos (\nu + 2\pi j)h \, dF\left(\frac{\nu + 2\pi j}{k}\right) \\ &= \int_{-\pi}^{\pi} \cos \nu h \left\{ \sum_{j=-\infty}^{\infty} dF\left(\frac{\nu + 2\pi j}{k}\right)\right\} \end{aligned} \tag{38}$$

by the substitution $\lambda = (\nu + 2\pi j)/k$ and the periodicity of $\cos(\nu + 2\pi j)h$. Suppose $F(\lambda)$ is absolutely continuous with density $f(\lambda)$. Then

$$(39)\quad \begin{aligned} f_k(\nu) &= \frac{1}{k}\sum_{j=-\infty}^{\infty} f\left(\frac{\nu + 2\pi j}{k}\right) \\ &= \frac{1}{k} f\left(\frac{\nu}{k}\right) + \frac{1}{k}\sum_{j=1}^{\infty}\left[f\left(\frac{\nu + 2\pi j}{k}\right) + f\left(\frac{\nu - 2\pi j}{k}\right)\right] \\ &= \frac{1}{k} f\left(\frac{\nu}{k}\right) + \frac{1}{k}\sum_{j=1}^{\infty}\left[f\left(\frac{2\pi j + \nu}{k}\right) + f\left(\frac{2\pi j - \nu}{k}\right)\right], \qquad -\pi \le \nu \le \pi, \end{aligned}$$

is the spectral density of the process $\{y_t\}$. The weight given frequency $\nu/(2\pi)$ in the process with discrete time parameter is the sum of the weights given frequencies $\nu/(2\pi k)$, $(2\pi - \nu)/(2\pi k) = 1/k - \nu/(2\pi k)$, $(2\pi + \nu)/(2\pi k) = 1/k + \nu/(2\pi k), \ldots$. For example, suppose temperature is measured hourly; $k = 1$ if continuous time t is measured in units of an hour. The period of 4 hours corresponds to frequency $\frac{1}{4}$ and $\nu = \pi/2$. From the continuous process we obtain contributions from frequencies $\frac{1}{4}$, $1 - \frac{1}{4} = \frac{3}{4}$, $1 + \frac{1}{4} = \frac{5}{4}$, $2 - \frac{1}{4} = \frac{7}{4}$, $2 + \frac{1}{4} = \frac{9}{4}, \ldots$, that is, periods 4, 4/3, 4/5, 4/7, 4/9, ... This effect is called *aliasing* or *folding the spectrum*, and $1/(2k)$ is called the *folding* or *Nyquist* frequency.

If (2) holds, the series (1) for $f(\lambda)$ converges absolutely and uniformly, and hence $f(\lambda)$ is continuous. There are various conditions on $f(\lambda)$ such that (2) holds; that is, such that the sum of the absolute values of the Fourier coefficients converges. A sufficient condition is that

$$(40)\qquad |f(\lambda_1) - f(\lambda_2)| \le K\,|\lambda_1 - \lambda_2|^{\alpha}$$

for some $K > 0$ and α with $\frac{1}{2} < \alpha < 1$ [Hobson (1907), Section 359].

Since $f(\lambda)$ and $\cos \lambda h$ are even and $\sin \lambda h$ is odd, (1) and (3) can be written

$$(41)\qquad f(\lambda) = \frac{1}{2\pi}\sum_{h=-\infty}^{\infty} \sigma(h) e^{i\lambda h},$$

$$(42)\qquad \sigma(k) = \int_{-\pi}^{\pi} e^{i\lambda k} f(\lambda)\, d\lambda.$$

The *normalized spectral density* is $\bar{f}(\lambda) = f(\lambda)/\sigma(0)$, which is (1) or (41) with $\sigma(h)$ replaced by $\rho_h = \sigma(h)/\sigma(0)$. The Fourier integrals of $\bar{f}(\lambda)$ give ρ_h; in particular, the integral of $\bar{f}(\lambda)$ over $[-\pi, \pi]$ is $\rho_0 = 1$.

Now let us extend the spectral theory to complex-valued random variables. Let $y_t = u_t + iv_t$, where $\{u_t, v_t\}$ is a real bivariate stochastic process, stationary in the wide sense, with $\mathscr{E}u_t = \mathscr{E}v_t = 0$ and

$$\begin{aligned}(43)\qquad \sigma(h) &= \mathscr{E}y_{t+h}\bar{y}_t = \mathscr{E}(u_{t+h} + iv_{t+h})(u_t - iv_t)\\ &= \mathscr{E}u_{t+h}u_t + \mathscr{E}v_{t+h}v_t + i(\mathscr{E}v_{t+h}u_t - \mathscr{E}u_{t+h}v_t),\\ &\qquad h = \ldots, -1, 0, 1, \ldots .\end{aligned}$$

Note that $\{\sigma(h)\}$ is a sequence of complex-valued numbers with the property that $\sigma(-h) = \overline{\sigma(h)}$, where $\overline{\sigma(h)}$ denotes the complex conjugate of $\sigma(h)$; if $\mathscr{E}u_{t+h}u_t = \mathscr{E}v_{t+h}v_t$ and $\mathscr{E}u_{t+h}v_t = -\mathscr{E}v_{t+h}u_t$, then $\{\sigma(h)\}$ determines all the covariances of $\{u_t, v_t\}$.

We define $I_T(\lambda)$ as

$$\begin{aligned}(44)\qquad I_T(\lambda) &= \frac{1}{2\pi T}\left|\sum_{t=1}^{T} y_t e^{-i\lambda t}\right|^2\\ &= \frac{1}{2\pi T}\sum_{t,s=1}^{T} y_t\bar{y}_s e^{-i\lambda(t-s)}.\end{aligned}$$

Then

$$\begin{aligned}(45)\qquad \mathscr{E}I_T(\lambda) &= \frac{1}{2\pi T}\sum_{t,s=1}^{T}\sigma(t-s)e^{-i\lambda(t-s)}\\ &= \frac{1}{2\pi}\sum_{h=-(T-1)}^{T-1}\left(1 - \frac{|h|}{T}\right)\sigma(h)e^{-i\lambda h}\\ &= f_T(\lambda),\end{aligned}$$

which is nonnegative by (44). [$f_T(\lambda)$ is usually not equal to $f_T(-\lambda)$.] Then

$$\begin{aligned}(46)\qquad \sigma(k)\left(1 - \frac{|k|}{T}\right) &= \int_{-\pi}^{\pi} e^{i\lambda k} f_T(\lambda)\, d\lambda, \qquad k = -(T-1), \ldots, T-1,\\ 0 &= \int_{-\pi}^{\pi} e^{i\lambda k} f_T(\lambda)\, d\lambda, \qquad k = \pm T, \pm(T+1), \ldots .\end{aligned}$$

Then $F_T(\lambda) = \int_{-\pi}^{\lambda} f_T(\nu)\, d\nu$ corresponding to (16) is monotonically nondecreasing with $F_T(-\pi) = 0$ and $F_T(\pi) = \sigma(0)$. The same argument leads to the limit of a subsequence $F(\lambda)$, which is monotonically nondecreasing, continuous on the right, with $F(-\pi) = 0$ and $F(\pi) = \sigma(0)$, and

$$(47)\qquad \sigma(k) = \int_{-\pi}^{\pi} e^{i\lambda k}\, dF(\lambda).$$

[For the complex-valued process (41) must be written with $e^{i\lambda h}$ replaced by $e^{-i\lambda h}$.]

Now consider a vector stochastic process $\{y_t\}$ with p complex components, stationary in the wide sense. Let

$$\Sigma(h) = \mathscr{E} y_{t+h} y_t^*, \tag{48}$$

where y_t^* denotes the transpose of $\bar{y}_t$, whose respective components are the complex conjugates of the components of y_t. Note that

$$\bar{\Sigma}(h) = \mathscr{E} \bar{y}_{t+h} y_t' = (\mathscr{E} y_t y_{t+h}^*)' = \Sigma'(-h). \tag{49}$$

For any fixed vector c the random sequence $\{c'y_t\}$ is a complex-valued stochastic process that is stationary in the wide sense with $\mathscr{E} c'y_t = 0$ and covariance sequence

$$\mathscr{E} c' y_{t+h} y_t^* \bar{c} = c' \Sigma(h) \bar{c} = \int_{-\pi}^{\pi} e^{i\lambda h}\, dF_c(\lambda), \tag{50}$$

where $F_c(\lambda)$ is real and monotonically nondecreasing with $F_c(-\pi) = 0$ and $F_c(\pi) = c'\Sigma(0)\bar{c}$ and continuous on the right. For c with 1 in the jth position and 0's elsewhere we have

$$\sigma_{jj}(h) = \int_{-\pi}^{\pi} e^{i\lambda h}\, dF_{jj}(\lambda), \tag{51}$$

where $F_c(\lambda)$ has been written as $F_{jj}(\lambda)$ for such c, $j = 1, \ldots, p$. For c with 1 in the jth position and 1 in the kth and 0's elsewhere we have

$$\sigma_{jj}(h) + \sigma_{kk}(h) + \sigma_{jk}(h) + \sigma_{kj}(h) = \int_{-\pi}^{\pi} e^{i\lambda h}\, dF_{\mathrm{I}}(\lambda); \tag{52}$$

for c with 1 in the jth position and i in the kth position and 0's elsewhere we have

$$\sigma_{jj}(h) + \sigma_{kk}(h) - i\sigma_{jk}(h) + i\sigma_{kj}(h) = \int_{-\pi}^{\pi} e^{i\lambda h}\, dF_{\mathrm{II}}(\lambda). \tag{53}$$

Then

$$\sigma_{jk}(h) = \int_{-\pi}^{\pi} e^{i\lambda h}\, dF_{jk}(\lambda), \qquad j \neq k, \tag{54}$$

where

$$F_{jk}(\lambda) = \tfrac{1}{2}[F_{\mathrm{I}}(\lambda) - F_{jj}(\lambda) - F_{kk}(\lambda)] + \tfrac{1}{2}i\,[F_{\mathrm{II}}(\lambda) - F_{jj}(\lambda) - F_{kk}(\lambda)], \qquad j \neq k. \tag{55}$$

Note that $F_{jk}(\lambda)$, $j \neq k$, need not be real. We have

$$\sigma_{jk}(h) = \int_{-\pi}^{\pi} e^{i\lambda h}\, dF_{jk}(\lambda), \qquad j, k = 1, \ldots, p, \tag{56}$$

which is written

$$\boldsymbol{\Sigma}(h) = \int_{-\pi}^{\pi} e^{i\lambda h}\, d\boldsymbol{F}(\lambda). \tag{57}$$

Since (50) determines $F_c(\lambda)$ uniquely

$$F_c(\lambda) = \boldsymbol{c}'\boldsymbol{F}(\lambda)\bar{\boldsymbol{c}}\,. \tag{58}$$

Since $F_c(\lambda)$ is real and monotonically nondecreasing

$$\boldsymbol{c}'[\boldsymbol{F}(\lambda_2) - \boldsymbol{F}(\lambda_1)]\bar{\boldsymbol{c}} \geq 0, \qquad \lambda_1 < \lambda_2. \tag{59}$$

Thus $\boldsymbol{F}(\lambda)$ and increments in $\boldsymbol{F}(\lambda)$ are Hermitian $[\boldsymbol{F}(\lambda) = \boldsymbol{F}^*(\lambda)]$ and positive semidefinite.

Now suppose $\{\boldsymbol{y}_t\}$ is real. Then $\boldsymbol{\Sigma}(h)$ is real; the imaginary part of

$$\begin{aligned}\boldsymbol{\Sigma}(h) &= \int_{-\pi}^{\pi} [\cos \lambda h + i \sin \lambda h][d\mathscr{R}\boldsymbol{F}(\lambda) + id\mathscr{I}\boldsymbol{F}(\lambda)] \\ &= \int_{-\pi}^{\pi} \{\cos \lambda h\, d\mathscr{R}\boldsymbol{F}(\lambda) - \sin \lambda h\, d\mathscr{I}\boldsymbol{F}(\lambda) \\ &\qquad + i[\cos \lambda h\, d\mathscr{I}\boldsymbol{F}(\lambda) + \sin \lambda h\, d\mathscr{R}\boldsymbol{F}(\lambda)]\}\end{aligned} \tag{60}$$

must be $\boldsymbol{0}$ for every h. If $\boldsymbol{F}(\lambda)$ is absolutely continuous with density $\boldsymbol{f}(\lambda)$, the real part of $f_{jk}(\lambda)$ is called the *cospectral* density and the imaginary part the *quadrature spectral* density for $j \neq k$. Then

$$\mathscr{R}f_{jk}(-\lambda) = \mathscr{R}f_{jk}(\lambda), \qquad j \neq k, \tag{61}$$

$$\mathscr{I}f_{jk}(-\lambda) = -\mathscr{I}f_{jk}(\lambda), \qquad j \neq k, \tag{62}$$

and $f_{jj}(\lambda)$ is real and symmetric. In terms of the spectral density (60) can be written

$$\sigma_{jj}(h) = \int_{-\pi}^{\pi} \cos \lambda h\, f_{jj}(\lambda)\, d\lambda = 2\int_{0}^{\pi} \cos \lambda h\, f_{jj}(\lambda)\, d\lambda, \tag{63}$$

$$\begin{aligned}\sigma_{jk}(h) &= \int_{-\pi}^{\pi} [\cos \lambda h\, \mathscr{R}f_{jk}(\lambda) - \sin \lambda h\, \mathscr{I}f_{jk}(\lambda)]\, d\lambda \\ &= 2\int_{0}^{\pi} [\cos \lambda h\, \mathscr{R}f_{jk}(\lambda) - \sin \lambda h\, \mathscr{I}f_{jk}(\lambda)]\, d\lambda, \qquad j \neq k.\end{aligned} \tag{64}$$

In terms of the spectral distribution function (61) and (62) correspond to $\boldsymbol{F}(\lambda_2) - \boldsymbol{F}(\lambda_1) = \overline{\boldsymbol{F}(-\lambda_1) - \boldsymbol{F}(-\lambda_2)}$ for $0 < \lambda_1 < \lambda_2$.

7.4. THE SPECTRAL REPRESENTATION OF A STATIONARY STOCHASTIC PROCESS

7.4.1. Stochastic Integrals

The spectral representation of a stochastic process is a representation of the process $\{y_t\}$ in terms of integrals of stochastic processes of a continuous parameter. These integrals of stochastic processes correspond to the integrals of spectral distribution functions in the spectral representation of the covariance sequence.

The purpose of presenting the spectral representation of the stochastic process is to give an interpretation of the spectral distribution function in terms of the variances of random amplitudes of trigonometric functions composing the process. Since no further *mathematical* developments will be based on this spectral representation, we shall not develop it completely rigorously, but shall discuss it in such a way as to make the statistical ideas clear.

We shall be interested in stochastic processes $C(\lambda)$, $0 \leq \lambda \leq \pi$. In particular, the definition of the stochastic process specifies joint distributions of finite collections $C(\lambda_1), \ldots, C(\lambda_n)$. We suppose the process to be a *process of uncorrelated increments;* that is,

$$\mathscr{E}[C(\lambda_2) - C(\lambda_1)][C(\lambda_4) - C(\lambda_3)] = 0, \qquad 0 \leq \lambda_1 < \lambda_2 \leq \lambda_3 < \lambda_4 \leq \pi, \tag{1}$$

where $\mathscr{E}C(\lambda) = 0, 0 \leq \lambda \leq \pi$. The differences of $C(\lambda)$ over two nonoverlapping intervals are uncorrelated. For (1) to have meaning we assume

$$\mathscr{E}C^2(\lambda) = H(\lambda) < \infty, \qquad 0 \leq \lambda \leq \pi. \tag{2}$$

In keeping with (1), we require $C(0)$ to be uncorrelated with $C(\lambda_2) - C(\lambda_1)$, $0 \leq \lambda_1 < \lambda_2 \leq \pi$. For $\lambda_1 < \lambda_2$, (1) implies

$$\begin{aligned} H(\lambda_2) &= \mathscr{E}C^2(\lambda_2) \\ &= \mathscr{E}\{C(\lambda_1) + [C(\lambda_2) - C(\lambda_1)]\}^2 \\ &= H(\lambda_1) + \mathscr{E}[C(\lambda_2) - C(\lambda_1)]^2. \end{aligned} \tag{3}$$

Thus

$$H(\lambda_2) \geq H(\lambda_1), \qquad \lambda_2 > \lambda_1; \tag{4}$$

$H(\lambda)$ has to be monotonically nondecreasing. Moreover, for $\lambda_1 < \lambda_2$,

$$\begin{aligned} \mathscr{E}C(\lambda_1)C(\lambda_2) &= \mathscr{E}C(\lambda_1)\{C(\lambda_1) + [C(\lambda_2) - C(\lambda_1)]\} \\ &= \mathscr{E}C^2(\lambda_1) \\ &= H(\lambda_1), \qquad \lambda_1 < \lambda_2. \end{aligned} \tag{5}$$

If the increments of $C(\lambda)$ over every finite set of nonoverlapping intervals are mutually independent, the process is said to be a *process of independent increments*. In that case $C(\lambda)$ for fixed λ is a sum of independent random variables.

We now undertake to define

$$\int_0^\pi h(\lambda)\, dC(\lambda) \tag{6}$$

in analogy to the Riemann-Stieltjes integral

$$\int_0^\pi h(\lambda)\, dH(\lambda), \tag{7}$$

where $h(\lambda)$ is continuous in the interval $[0, \pi]$ and $H(\lambda)$ is monotonically nondecreasing. Suppose for the moment that $H(\lambda)$ is continuous at 0 and π. Then (7) is approximated by

$$S = \sum_{j=1}^{n} h(\lambda_j')[H(\lambda_j) - H(\lambda_{j-1})], \tag{8}$$

where

$$\begin{gathered} 0 = \lambda_0 < \lambda_1 < \cdots < \lambda_{n-1} < \lambda_n = \pi, \\ \lambda_0 \le \lambda_1' \le \lambda_1, \ldots, \lambda_{n-1} \le \lambda_n' \le \lambda_n. \end{gathered} \tag{9}$$

We want to argue that if the partition is dense the sum (8) does not depend much on which partition is used.

Since $h(\lambda)$ is continuous in the closed interval $[0, \pi]$, it is uniformly continuous. Given an $\varepsilon > 0$, there exists a δ such that

$$|h(\nu) - h(\mu)| < \varepsilon, \qquad |\nu - \mu| < \delta. \tag{10}$$

LEMMA 7.4.1. *If S is defined by* (8) *and R by*

$$R = \sum_{j=1}^{m} h(\nu_j')[H(\nu_j) - H(\nu_{j-1})], \tag{11}$$

where $\nu_0, \nu_1, \ldots, \nu_m, \nu_1', \ldots, \nu_m'$ is another partition of $[0, \pi]$, *and if*

$$\max_{j=1,\ldots,n} (\lambda_j - \lambda_{j-1}) < \tfrac{1}{2}\delta, \qquad \max_{j=1,\ldots,m} (\nu_j - \nu_{j-1}) < \tfrac{1}{2}\delta, \tag{12}$$

for δ chosen so (10) *holds for given ε, then*

$$|S - R| < \varepsilon[H(\pi) - H(0)]. \tag{13}$$

PROOF. Let $0 = \mu_0 < \mu_1 < \cdots < \mu_l = \pi$ be the distinct values of $\lambda_0, \ldots, \lambda_n, \nu_0, \ldots, \nu_m$. Let λ_k'' be λ_j' if $(\mu_{k-1}, \mu_k) \subset (\lambda_{j-1}, \lambda_j)$, and let ν_k'' be ν_j' if

$(\mu_{k-1}, \mu_k) \subset (\nu_{j-1}, \nu_j)$. Then

$$(14)\quad |S - R| = \left| \sum_{k=1}^{l} h(\lambda_k'')[H(\mu_k) - H(\mu_{k-1})] - \sum_{k=1}^{l} h(\nu_k'')[H(\mu_k) - H(\mu_{k-1})] \right|$$

$$\leq \sum_{k=1}^{l} |h(\lambda_k'') - h(\nu_k'')|\, [H(\mu_k) - H(\mu_{k-1})]$$

$$\leq \varepsilon[H(\pi) - H(0)]$$

because $|\lambda_k'' - \nu_k''| < \delta$. Q.E.D.

If we take a sequence of partitions $\{\lambda_j^{(n)}, \lambda_j'^{(n)}\}$ such that $\max_j [\lambda_j^{(n)} - \lambda_{j-1}^{(n)}] \to 0$, the corresponding sums $S^{(n)}$ converge and the limit is the integral (7). By Lemma 7.4.1 the partial sums $R^{(m)}$ defined by any other sequence of partitions $\{\nu_j^{(m)}, \nu_j'^{(m)}\}$ converge to the same value.

We shall define the stochastic integral (6) as a limit in the mean of approximating sums

$$(15)\qquad S = \sum_{j=1}^{n} h(\lambda_j')[C(\lambda_j) - C(\lambda_{j-1})].$$

Note that the sum is a linear combination of uncorrelated random variables. The expected value of any approximating sum (15) is 0, and the variance is

$$(16)\qquad \mathscr{E}\left\{\sum_{j=1}^{n} h(\lambda_j')[C(\lambda_j) - C(\lambda_{j-1})]\right\}^2 = \sum_{j=1}^{n} h^2(\lambda_j')[H(\lambda_j) - H(\lambda_{j-1})].$$

We shall show that the difference between S and

$$(17)\qquad R = \sum_{j=1}^{m} h(\nu_j')[C(\nu_j) - C(\nu_{j-1})]$$

is small in the sense that $\mathscr{E}(S - R)^2$ is small.

Lemma 7.4.2. *If S and R are defined by* (15) *and* (17), *respectively, if $H(\lambda)$ is the covariance function of $C(\lambda)$, and if* (12) *is satisfied for δ chosen so* (10) *holds for given ε, then*

$$(18)\qquad \mathscr{E}(S - R)^2 < \varepsilon^2[H(\pi) - H(0)].$$

Proof. Let μ_k, λ_k'', and ν_k'', $k = 1, \ldots, l$, be defined as before. Then

$$(19)\qquad \mathscr{E}(S - R)^2 = \mathscr{E}\left\{\sum_{k=1}^{l} [h(\lambda_k'') - h(\nu_k'')][C(\lambda_k) - C(\lambda_{k-1})]\right\}^2$$

$$= \sum_{k=1}^{l} [h(\lambda_k'') - h(\nu_k'')]^2[H(\lambda_k) - H(\lambda_{k-1})]$$

$$< \varepsilon^2[H(\pi) - H(0)].$$

Q.E.D.

A sequence of sums $S^{(n)}$ defined by a sequence of partitions $\{\lambda_j^{(n)}, \lambda_j'^{(n)}\}$ such that $\max_j [\lambda_j^{(n)} - \lambda_{j-1}^{(n)}] \to 0$ converges in the mean to a random variable, which is termed the integral (6) (Theorem 7.6.1). Lemma 7.4.2 implies that any other sequence of sums $R^{(m)}$ converges in the mean to the same random variable.

Since $\mathscr{E}C(\lambda) = 0$, the expectation of the integral (6) is 0. From (16) we obtain

$$\mathscr{E}\left\{\int_0^\pi h(\lambda)\, dC(\lambda)\right\}^2 = \lim_{n\to\infty} \sum_{j=1}^{n} h^2(\lambda_j'^{(n)})[H(\lambda_j^{(n)}) - H(\lambda_{j-1}^{(n)})] = \int_0^\pi h^2(\lambda)\, dH(\lambda). \tag{20}$$

If $k(\lambda)$ is another continuous function in $[0, \pi]$, then

$$\mathscr{E}\int_0^\pi h(\lambda)\, dC(\lambda) \int_0^\pi k(\lambda)\, dC(\lambda) = \int_0^\pi h(\lambda)k(\lambda)\, dH(\lambda). \tag{21}$$

These follow from Theorem 7.6.4.

We have assumed that $H(\lambda)$ is continuous at 0 and π. If it is not, the integral (6) will also include $h(0)[C(0+) - C(0)]$ and/or $h(\pi)[C(\pi) - C(\pi-)]$.

For a fuller treatment of stochastic integrals the reader is referred to Doob (1953), Section 2 of Chapter IX.

7.4.2. The Spectral Representation

Let $C(\lambda)$ and $S(\lambda)$ be uncorrelated stochastic processes of uncorrelated increments with continuous parameter λ, $0 \le \lambda \le \pi$. Suppose $\mathscr{E}C(\lambda) = \mathscr{E}S(\lambda) = 0$ and

$$\mathscr{E}[C(\lambda_2) - C(\lambda_1)]^2 = \mathscr{E}[S(\lambda_2) - S(\lambda_1)]^2 = G(\lambda_2) - G(\lambda_1), \qquad 0 \le \lambda_1 < \lambda_2 \le \pi, \tag{22}$$

for some monotonically nondecreasing function $G(\lambda)$. We define the process $\{y_t\}$ by

$$y_t = \int_0^\pi \cos \lambda t\, dC(\lambda) + \int_0^\pi \sin \lambda t\, dS(\lambda), \qquad t = \ldots, -1, 0, 1, \ldots . \tag{23}$$

According to the discussion in Section 7.4.1 $\mathscr{E}y_t = 0$ and

$$\mathscr{E}y_t y_s = \int_0^\pi \cos \lambda t \cos \lambda s\, dG(\lambda) + \int_0^\pi \sin \lambda t \sin \lambda s\, dG(\lambda) = \int_0^\pi \cos \lambda(t-s)\, dG(\lambda). \tag{24}$$

Thus $G(\lambda)$ is the spectral distribution function (on $[0, \pi]$) of the process defined by (23).

Conversely, given a stochastic process $\{y_t^*\}$ which is stationary in the wide sense with spectral distribution function $G^*(\lambda)$ (on $[0, \pi]$), we can construct a pair of uncorrelated processes of uncorrelated increments $C^*(\lambda)$ and $S^*(\lambda)$ such that $\{y_t^*\}$ can be written as an integral (23) and

$$\mathscr{E}[C^*(\lambda_2) - C^*(\lambda_1)]^2 = \mathscr{E}[S^*(\lambda_2) - S^*(\lambda_1)]^2 = G^*(\lambda_2) - G^*(\lambda_1), \quad 0 \leq \lambda_1 < \lambda_2 \leq \pi. \tag{25}$$

These processes are essentially the limits of the integrals of the cosine and sine transforms $\sum_t y_t^* \cos \lambda t$ and $\sum_t y_t^* \sin \lambda t$. On the basis of observations $y_{-T}^*, \ldots, y_0^*, \ldots, y_T^*$, let

$$c_T(\nu) = \frac{1}{\pi} \sum_{t=-T}^{T} y_t^* \cos \nu t, \qquad 0 \leq \nu \leq \pi, \tag{26}$$

$$s_T(\nu) = \frac{1}{\pi} \sum_{t=-T}^{T} y_t^* \sin \nu t, \qquad 0 \leq \nu \leq \pi, \tag{27}$$

$$C_T(\lambda) = \int_0^\lambda c_T(\nu)\, d\nu = \sum_{t=-T}^{T} \alpha_{\lambda,t} y_t^*, \tag{28}$$

$$S_T(\lambda) = \int_0^\lambda s_T(\nu)\, d\nu = \sum_{t=-T}^{T} \beta_{\lambda,t} y_t^*, \tag{29}$$

where

$$\begin{aligned} \alpha_{\lambda,t} &= \frac{1}{\pi} \int_0^\lambda \cos \nu t \, d\nu \\ &= \frac{\sin \lambda t}{\pi t}, \qquad t = \pm 1, \pm 2, \ldots, \\ &= \frac{\lambda}{\pi}, \qquad t = 0, \end{aligned} \tag{30}$$

$$\begin{aligned} \beta_{\lambda,t} &= \frac{1}{\pi} \int_0^\lambda \sin \nu t \, d\nu \\ &= \frac{1 - \cos \lambda t}{\pi t}, \qquad t = \pm 1, \pm 2, \ldots, \\ &= 0, \qquad t = 0. \end{aligned} \tag{31}$$

These coefficients are the Fourier coefficients of the step functions

$$\begin{aligned} \alpha_\lambda(\nu) &= 1, & -\lambda < \nu < \lambda, \\ &= 0, & \lambda \le |\nu|, \end{aligned} \tag{32}$$

$$\begin{aligned} \beta_\lambda(\nu) &= -1, & -\lambda < \nu < 0, \\ &= 1, & 0 < \nu < \lambda, \\ &= 0, & \lambda \le |\nu|. \end{aligned} \tag{33}$$

These Fourier series

$$\alpha_\lambda^*(\nu) = \sum_{t=-\infty}^{\infty} \alpha_{\lambda,t} \cos \nu t, \tag{34}$$

$$\beta_\lambda^*(\nu) = \sum_{t=-\infty}^{\infty} \beta_{\lambda,t} \sin \nu t \tag{35}$$

converge pointwise except at $\nu = \pm\lambda$ and converge in mean square. We define

$$C^*(\lambda) = \sum_{t=-\infty}^{\infty} \alpha_{\lambda,t} y_t^*, \tag{36}$$

$$S^*(\lambda) = \sum_{t=-\infty}^{\infty} \beta_{\lambda,t} y_t^* \tag{37}$$

as the limits in the mean of $C_T(\lambda)$ and $S_T(\lambda)$, respectively. These random variables exist because the limits of $\mathscr{E} C_T(\lambda) C_{T'}(\lambda)$ and $\mathscr{E} S_T(\lambda) S_{T'}(\lambda)$ exist and are finite (Corollary 7.6.1).

We calculate

$$\begin{aligned} \mathscr{E} C^*(\lambda) C^*(\mu) &= \mathscr{E} \sum_{s,t=-\infty}^{\infty} \alpha_{\lambda,s}\alpha_{\mu,t} y_s^* y_t^* \\ &= \sum_{s,t=-\infty}^{\infty} \alpha_{\lambda,s}\alpha_{\mu,t} \int_0^\pi \cos \nu(s-t)\, dG^*(\nu) \\ &= \int_0^\pi \sum_{s,t=-\infty}^{\infty} \alpha_{\lambda,s}\alpha_{\mu,t}(\cos \nu s \cos \nu t + \sin \nu s \sin \nu t)\, dG^*(\nu) \\ &= \int_0^\pi \left(\sum_{s=-\infty}^{\infty} \alpha_{\lambda,s} \cos \nu s \sum_{t=-\infty}^{\infty} \alpha_{\mu,t} \cos \nu t + \sum_{s=-\infty}^{\infty} \alpha_{\lambda,s} \sin \nu s \sum_{t=-\infty}^{\infty} \alpha_{\mu,t} \sin \nu t \right) dG^*(\nu) \\ &= \int_0^\pi \alpha_\lambda(\nu)\alpha_\mu(\nu)\, dG^*(\nu) \\ &= \min\,[G^*(\lambda), G^*(\mu)], \end{aligned} \tag{38}$$

(39) $\mathscr{E}S^*(\lambda)S^*(\mu)$

$$= \int_0^\pi \left(\sum_{s=-\infty}^{\infty} \beta_{\lambda,s} \cos \nu s \sum_{t=-\infty}^{\infty} \beta_{\mu,t} \cos \nu t + \sum_{s=-\infty}^{\infty} \beta_{\lambda,s} \sin \nu s \sum_{t=-\infty}^{\infty} \beta_{\mu,t} \sin \nu t \right) dG^*(\nu)$$

$$= \int_0^\pi \beta_\lambda(\nu)\beta_\mu(\nu)\, dG^*(\nu)$$

$$= \min\,[G^*(\lambda), G^*(\mu)],$$

(40) $\mathscr{E}C^*(\lambda)S^*(\mu)$

$$= \int_0^\pi \left(\sum_{s=-\infty}^{\infty} \alpha_{\lambda,s} \cos \nu s \sum_{t=-\infty}^{\infty} \beta_{\mu,t} \cos \nu t + \sum_{s=-\infty}^{\infty} \alpha_{\lambda,s} \sin \nu s \sum_{t=-\infty}^{\infty} \beta_{\mu,t} \sin \nu t \right) dG^*(\nu)$$

$$= 0,$$

because

$$\sum_{t=-\infty}^{\infty} \alpha_{\mu,t} \sin \nu t = 0, \tag{41}$$

$$\sum_{t=-\infty}^{\infty} \beta_{\mu,t} \cos \nu t = 0, \tag{42}$$

by virtue of $\alpha_{\mu,t} = \alpha_{\mu,-t}$, $\sin \nu t = -\sin(-\nu t)$, $\beta_{\mu,t} = -\beta_{\mu,-t}$, and $\cos \nu t = \cos(-\nu t)$. Since the Fourier series for $\alpha_\lambda(\nu)$ and $\beta_\lambda(\nu)$ converge and are continuous except at $-\lambda$, 0, λ, the above argument holds if 0, λ, and μ are continuity points of $G^*(\nu)$. To make the argument rigorous we need to carry out the limits in the mean in detail. The calculation of the variances and covariances is justified by Theorem 7.6.4. [See Doob (1953), Section 4, Chapter X.]

The preceding discussion shows how processes $C^*(\lambda)$ and $S^*(\lambda)$ can be constructed for a given process $\{y_t^*\}$ that is stationary in the wide sense such that the process can be represented in the spectral manner. The processes $C^*(\lambda)$ and $S^*(\lambda)$ so defined are essentially unique.

The purpose of this development is to show that any stationary (wide sense) stochastic process can be considered as a weighted sum or integral of trigonometric functions of time with the weights being random variables. The effect of these weights on the average is determined by the variances of the random variables.

7.5. LINEAR OPERATIONS ON STATIONARY PROCESSES

7.5.1. Covariance and Spectral Functions of Processes Obtained by Linear Operations on Stationary Processes

Let $\{y_t\}$ be a stochastic process with mean $\mathscr{E}y_t = 0$, covariance function $\mathscr{E}y_t y_s = \sigma(s-t)$, and spectral distribution function $F(\lambda)$. (The process $\{y_t\}$

is thus stationary in the wide sense.) A new process $\{z_t\}$ may be obtained by a linear operation on $\{y_t\}$,

$$z_t = \sum_r c_r y_{t-r}, \qquad t = \ldots, -1, 0, 1, \ldots, \tag{1}$$

where $\{c_r\}$ is a sequence of constants. If the series in (1) is infinite, we shall define z_t as the limit in the mean (assuming it exists). The operation is sometimes called a *linear filter*.

The mean of the derived process is $\mathscr{E} z_t = 0$. The covariance function is

$$\begin{aligned} \mathscr{E} z_t z_s &= \mathscr{E} \sum_{r,q} c_r y_{t-r} c_q y_{s-q} \\ &= \sum_{r,q} c_r c_q \sigma[(t-r) - (s-q)] \\ &= \sum_{r,q} c_r c_q \sigma[(t-s) - (r-q)]. \end{aligned} \tag{2}$$

The random variable z_t exists as a limit in the mean [when the series in (1) is infinite] if and only if the right-hand side of (2) converges for $s = t$ (Corollary 7.6.1); in that case the right-hand side is indeed $\mathscr{E} z_t z_s$ (Theorem 7.6.4). Equation (2) shows that $\{z_t\}$ is stationary in the wide sense.

The spectral distribution function of $\{z_t\}$ is found from (2), which can be written

$$\begin{aligned} \mathscr{E} z_t z_s &= \sum_{r,q} c_r c_q \int_{-\pi}^{\pi} \cos \lambda[(t-s) - (r-q)]\, dF(\lambda) \\ &= \int_{-\pi}^{\pi} \mathscr{R} \sum_{r,q} c_r c_q e^{i\lambda[(t-s)+(q-r)]}\, dF(\lambda) \\ &= \int_{-\pi}^{\pi} \mathscr{R} e^{i\lambda(t-s)} \sum_{r,q} c_q e^{i\lambda q} c_r e^{-i\lambda r}\, dF(\lambda) \\ &= \int_{-\pi}^{\pi} \cos \lambda(t-s) \left| \sum_r c_r e^{i\lambda r} \right|^2 dF(\lambda). \end{aligned} \tag{3}$$

The spectral distribution function of $\{z_t\}$, therefore, is

$$H(\lambda) = \int_{-\pi}^{\lambda} \left| \sum_r c_r e^{i\nu r} \right|^2 dF(\nu), \tag{4}$$

and

$$\mathscr{E} z_t z_s = \int_{-\pi}^{\pi} \cos \lambda(t-s)\, dH(\lambda). \tag{5}$$

If $\{y_t\}$ has a spectral density $F'(\lambda) = f(\lambda)$, then $\{z_t\}$ has a spectral density

$$H'(\lambda) = h(\lambda) = \left| \sum_r c_r e^{i\lambda r} \right|^2 f(\lambda). \tag{6}$$

The function $\sum_r c_r e^{i\lambda r}$ is sometimes called the *frequency response function* or *transfer function*, and $|\sum_r c_r e^{i\lambda r}|^2$ is sometimes called the *power transfer function.*

THEOREM 7.5.1. *If $\{y_t\}$ has spectral distribution function $F(\lambda)$, then $\{z_t\}$ defined by* (1) *has spectral distribution function* (4). *If $\{y_t\}$ has spectral density $f(\lambda)$, then $\{z_t\}$ has spectral density* (6).

7.5.2. Moving Average Processes

Suppose $\{y_t\}$ is a moving average of uncorrelated random variables

$$y_t = \sum_r \gamma_r v_{t-r}, \tag{7}$$

where $\mathscr{E} v_t = 0$, $\mathscr{E} v_t^2 = 1$, and $\mathscr{E} v_t v_s = 0$, $t \neq s$. A necessary and sufficient condition for the series in (7) to converge in the mean is

$$\sum_r \gamma_r^2 < \infty \tag{8}$$

(Corollary 7.6.1). The process $\{v_t\}$ has spectral density $1/(2\pi)$, and the process $\{y_t\}$ has spectral density

$$f(\lambda) = \frac{1}{2\pi} \left| \sum_r \gamma_r e^{i\lambda r} \right|^2. \tag{9}$$

The process $\{y_t\}$ is called a *moving average process.* The covariance function of $\{y_t\}$ is

$$\begin{aligned} \mathscr{E} y_t y_{t+h} &= \mathscr{E} \sum_{r,s} \gamma_r \gamma_s v_{t-r} v_{t+h-s} \\ &= \sum_r \gamma_r \gamma_{r+h} \end{aligned} \tag{10}$$

because $\mathscr{E} v_{t-r} v_{t+h-s} = 1$ if $t - r = t + h - s$ and is 0 otherwise.

Conversely, if a stationary process $\{y_t\}$ has a spectral density $f(\lambda)$, then it can be represented by (7), for the nonnegative square root of $\pi g(\lambda) = 2\pi f(\lambda)$ has a Fourier representation

$$\begin{aligned} \sqrt{\pi g(\lambda)} &= \sum_{r=-\infty}^{\infty} \gamma_r^* e^{i\lambda r} \\ &= a_0 + \sum_{r=1}^{\infty} (a_r \cos \lambda r + b_r \sin \lambda r), \end{aligned} \tag{11}$$

where $a_0 = \gamma_0^*$, $a_r = \gamma_r^* + \gamma_{-r}^*$ and $b_r = i(\gamma_r^* - \gamma_{-r}^*)$, $r = 1, 2, \ldots$. Since $g(\lambda) = 2f(\lambda)$ is even, $b_r = 0$ and $\gamma_r^* = \gamma_{-r}^*$, and $\gamma_r^* = \frac{1}{2}a_r(r \neq 0)$ is real. [Note that if y_t is defined by (7) when $\gamma_r = \gamma_{-r}$ is not true, $\sum_r \gamma_r e^{i\lambda r}$ is not real.] There exists a sequence of uncorrelated random variables $\{V_t\}$ such that y_t can be expressed as $\sum_{r=-\infty}^{\infty} \gamma_r^* V_{t-r}$ with $\mathscr{E} V_t = 0$ and $\mathscr{E} V_t^2 = 1$.

We can sketch the definition of $\{V_t\}$ by using the spectral representation of $\{y_t\}$. Suppose $g(\lambda) > 0$, $0 \leq \lambda \leq \pi$. Then define

$$V_t = \int_0^\pi \frac{\cos \lambda t}{\sqrt{\pi g(\lambda)}}\, dC(\lambda) + \int_0^\pi \frac{\sin \lambda t}{\sqrt{\pi g(\lambda)}}\, dS(\lambda). \tag{12}$$

Then

(13)

$$\sum_{r=-\infty}^{\infty} \gamma_r^* V_{t-r} = \int_0^\pi \frac{\sum_{r=-\infty}^{\infty} \gamma_r^* \cos \lambda(t-r)}{\sqrt{\pi g(\lambda)}}\, dC(\lambda) + \int_0^\pi \frac{\sum_{r=-\infty}^{\infty} \gamma_r^* \sin \lambda(t-r)}{\sqrt{\pi g(\lambda)}}\, dS(\lambda)$$

$$= \int_0^\pi \frac{\sum_{r=-\infty}^{\infty} \gamma_r^*(\cos \lambda t \cos \lambda r + \sin \lambda t \sin \lambda r)}{\sqrt{\pi g(\lambda)}}\, dC(\lambda)$$

$$+ \int_0^\pi \frac{\sum_{r=-\infty}^{\infty} \gamma_r^*(\sin \lambda t \cos \lambda r - \cos \lambda t \sin \lambda r)}{\sqrt{\pi g(\lambda)}}\, dS(\lambda)$$

$$= \int_0^\pi \cos \lambda t\, dC(\lambda) + \int_0^\pi \sin \lambda t\, dS(\lambda)$$

$$= y_t$$

since

$$\sum_{r=-\infty}^{\infty} \gamma_r^* \cos \lambda r = \sqrt{\pi g(\lambda)}, \qquad \sum_{r=-\infty}^{\infty} \gamma_r^* \sin \lambda r = 0 \tag{14}$$

by $\gamma_r^* = \gamma_{-r}^*$. Thus (12) gives the desired sequence. For more details see Doob (1953), Chapter X, Section 8.

If $\{y_t\}$ has a spectral density $f(\lambda)$, if $f(\lambda) > 0$ almost everywhere in $[0, \pi]$, and if

$$\int_{-\pi}^{\pi} \log f(\lambda)\, d\lambda > -\infty, \tag{15}$$

then there exist (real) constants $\{\gamma_r\}$ and random variables $\{v_t\}$ such that

$$y_t = \sum_{r=0}^{\infty} \gamma_r v_{t-r}, \qquad t = \dots, -1, 0, 1, \dots . \tag{16}$$

The point is that the sum (16) is singly infinite. (See Section 7.6.3.)

Now consider a finite moving average

$$y_t = \sum_{r=0}^{q} \alpha_r v_{t-r}, \tag{17}$$

where $\alpha_0 = 1$ and $\mathscr{E}v_t^2 = \sigma^2$. Then

$$\begin{aligned} f(\lambda) &= \frac{\sigma^2}{2\pi}\left|\sum_{r=0}^{q} \alpha_r e^{i\lambda r}\right|^2 \\ &= \frac{\sigma^2}{2\pi}\left|\sum_{r=0}^{q} \alpha_r e^{-i\lambda r}\right|^2 \\ &= \frac{\sigma^2}{2\pi}\left|\sum_{r=0}^{q} \alpha_r e^{i\lambda(q-r)}\right|^2 \\ &= \frac{\sigma^2}{2\pi}\prod_{j=1}^{q} |e^{i\lambda} - z_j|^2, \end{aligned} \tag{18}$$

where $z_1, \dots, z_q$ are the q roots of

$$\sum_{r=0}^{q} \alpha_r z^{q-r} = 0. \tag{19}$$

If $\alpha_q \neq 0$, then the q roots are all different from 0. For finite q we shall use $|\sum_{r=0}^{q} \alpha_r e^{i\lambda(q-r)}|^2$ with $\alpha_0 = 1$ as the standard form of $f(\lambda)$. As was shown in Section 5.7.1, any process with a finite number of covariances different from 0 has the same covariance sequence as a suitably chosen finite-order moving average process. Then the spectral density can be written in the form of (18).

If $q = 1$,

$$y_t = v_t + \alpha_1 v_{t-1}, \tag{20}$$

$$\begin{aligned} f(\lambda) &= \frac{\sigma^2}{2\pi}|e^{i\lambda} + \alpha_1|^2 \\ &= \frac{\sigma^2}{2\pi}(e^{i\lambda} + \alpha_1)(e^{-i\lambda} + \alpha_1) \\ &= \frac{\sigma^2}{2\pi}(1 + \alpha_1^2 + 2\alpha_1 \cos \lambda). \end{aligned} \tag{21}$$

Since $\cos\lambda$ varies from -1 to 1 in $[-\pi, 0]$ and decreases to -1 in $[0, \pi]$, $f(\lambda)$ increases from $\sigma^2(1-\alpha_1)^2/(2\pi)$ to $\sigma^2(1+\alpha_1)^2/(2\pi)$ in $[-\pi, 0]$ and decreases to $\sigma^2(1-\alpha_1)^2/(2\pi)$ in $[0, \pi]$ if $\alpha_1 > 0$; if $\alpha_1 < 0$, $f(\lambda)$ decreases in $[-\pi, 0]$ and increases in $[0, \pi]$. If $\alpha_1 > 0$, low frequencies are emphasized; if $\alpha_1 < 0$, high frequencies are emphasized. We can also write the spectral density as

$$f(\lambda) = \frac{\sigma^2}{2\pi}|e^{i\lambda} + \alpha_1|^2 = \frac{\sigma^2}{2\pi}|1 + \alpha_1 e^{-i\lambda}|^2 = \frac{\sigma^2\alpha_1^2}{2\pi}\left|\frac{1}{\alpha_1} + e^{-i\lambda}\right|^2 = \frac{\sigma^2\alpha_1^2}{2\pi}\left|e^{i\lambda} + \frac{1}{\alpha_1}\right|^2 \tag{22}$$

because $|e^{i\lambda}| = 1$. The last form corresponds to $v_t^* + (1/\alpha_1)v_{t-1}^*$, where v_t^* has variance $\sigma^2\alpha_1^2$; this process has the same covariance function as (20). If $\alpha_1 \neq \pm 1$, the latter moving average is different from the former. [The spectral density can also be written $\sigma^2 |\alpha_1 e^{i\lambda} + 1|^2/(2\pi)$ corresponding to $\alpha_1\tilde{v}_t + \tilde{v}_{t-1}$, where $\tilde{v}_t$ has variance σ^2.]

If $q = 2$,

$$y_t = v_t + \alpha_1 v_{t-1} + \alpha_2 v_{t-2}, \tag{23}$$

$$\begin{aligned} f(\lambda) &= \frac{\sigma^2}{2\pi}|e^{i2\lambda} + \alpha_1 e^{i\lambda} + \alpha_2|^2 \\ &= \frac{\sigma^2}{2\pi}(e^{i2\lambda} + \alpha_1 e^{i\lambda} + \alpha_2)(e^{-i2\lambda} + \alpha_1 e^{-i\lambda} + \alpha_2) \\ &= \frac{\sigma^2}{2\pi}[1 + \alpha_1^2 + \alpha_2^2 + 2\alpha_1(1+\alpha_2)\cos\lambda + 2\alpha_2\cos 2\lambda] \\ &= \frac{\sigma^2}{2\pi}[\alpha_1^2 + (1-\alpha_2)^2 + 2\alpha_1(1+\alpha_2)\cos\lambda + 4\alpha_2\cos^2\lambda] \\ &= \frac{\sigma^2}{2\pi}\left\{4\alpha_2\left[\cos\lambda + \frac{\alpha_1(1+\alpha_2)}{4\alpha_2}\right]^2 + \frac{(1-\alpha_2)^2(4\alpha_2 - \alpha_1^2)}{4\alpha_2}\right\}. \end{aligned} \tag{24}$$

If $\alpha_1(1+\alpha_2) > 4|\alpha_2|$, $f(0) = \sigma^2(1+\alpha_1+\alpha_2)^2/(2\pi)$ is the maximum of $f(\lambda)$ and $f(\pm\pi) = \sigma^2(1-\alpha_1+\alpha_2)^2/(2\pi)$ are minima; if $\alpha_1(1+\alpha_2) < -4|\alpha_2|$, $f(0)$ is the minimum and $f(\pm\pi)$ are the maxima. If $|\alpha_1(1+\alpha_2)| < 4|\alpha_2|$,

$\cos \lambda = -\alpha_1(1 + \alpha_2)/(4\alpha_2)$ for a value of λ in $[0, \pi]$, say λ_0, and at $\lambda = -\lambda_0$; if $\alpha_2 > 0$ (implying that $\alpha_1^2 < 4\alpha_2$ and the roots of the associated polynomial are complex), $f(\pm\lambda_0)$ are minima and $f(0)$ and $f(\pm\pi)$ are relative maxima; if $\alpha_2 < 0$ (implying $4\alpha_2 < \alpha_1^2$ and the roots are real), $f(\pm\lambda_0)$ are maxima and $f(0)$ and $f(\pm\pi)$ are relative minima. If z_1 and z_2 are the roots of

$$z^2 + \alpha_1 z + \alpha_2 = 0, \tag{25}$$

then

$$f(\lambda) = \frac{\sigma^2}{2\pi} |e^{i\lambda} - z_1|^2 \, |e^{i\lambda} - z_2|^2. \tag{26}$$

Since $|e^{i\lambda}| = 1$, the term $|e^{i\lambda} - z_1|$ can be replaced by $|1 - z_1 e^{-i\lambda}| = |1 - \bar{z}_1 e^{i\lambda}| = |z_1| \cdot |e^{i\lambda} - 1/\bar{z}_1|$, where $\bar{z}_1$ is conjugate to z_1, and $|e^{i\lambda} - z_2|$ can be replaced by $|z_2| \cdot |e^{i\lambda} - 1/\bar{z}_2|$. Thus $f(\lambda)$ can alternatively be written

$$\begin{aligned}
f(\lambda) &= \frac{\sigma^2 |z_1|^2}{2\pi} \left| e^{i\lambda} - \frac{1}{\bar{z}_1} \right|^2 |e^{i\lambda} - z_2|^2 \\
&= \frac{\sigma^2 |z_1|^2}{2\pi} \left| e^{i2\lambda} - \left(\frac{1}{\bar{z}_1} + z_2\right) e^{i\lambda} + \frac{z_2}{\bar{z}_1} \right|^2, \\
f(\lambda) &= \frac{\sigma^2 |z_2|^2}{2\pi} |e^{i\lambda} - z_1|^2 \left| e^{i\lambda} - \frac{1}{\bar{z}_2} \right|^2 \\
&= \frac{\sigma^2 |z_2|^2}{2\pi} \left| e^{i2\lambda} - \left(z_1 + \frac{1}{\bar{z}_2}\right) e^{i\lambda} + \frac{z_1}{\bar{z}_2} \right|^2, \\
f(\lambda) &= \frac{\sigma^2 |z_1|^2 |z_2|^2}{2\pi} \left| e^{i\lambda} - \frac{1}{\bar{z}_1} \right|^2 \left| e^{i\lambda} - \frac{1}{\bar{z}_2} \right|^2 \\
&= \frac{\sigma^2 \alpha_2^2}{2\pi} \left| e^{i2\lambda} + \frac{\alpha_1}{\alpha_2} e^{i\lambda} + \frac{1}{\alpha_2} \right|^2.
\end{aligned} \tag{27}$$

If z_1 and z_2 are real the alternative forms correspond to alternative moving averages. The four moving averages are different if $z_1 \neq z_2$ and $z_i \neq \pm 1$; there are three different moving averages if $z_1 = z_2 \neq \pm 1$; there are two different moving averages if $z_1 = \pm 1$ and $z_2 \neq \pm 1$ or if $z_1 \neq \pm 1$ and $z_2 = \pm 1$; and there is only one moving average if $z_1 = \pm 1$ and $z_2 = \pm 1$. The spectral density is the product of two densities of the type for $q = 1$.

If z_1 and z_2 are conjugate complex, say $\gamma e^{i\theta}$ and $\gamma e^{-i\theta}$ $(0 < \theta < \pi)$, then $z_2 + 1/\bar{z}_1$ and $z_1 + 1/\bar{z}_2$ are not real if $\gamma \neq 1$, and the first two forms of $f(\lambda)$ in (27) do not correspond to moving averages with real coefficients; the two moving averages with real coefficients are different. All forms of $f(\lambda)$ and the moving averages are the same if $\gamma = 1$ (that is, $\alpha_2 = 1$). When the roots are conjugate complex

$$(28)\quad f(\lambda) = \frac{\sigma^2}{2\pi} |e^{i\lambda} - \gamma e^{i\theta}|^2 \, |e^{i\lambda} - \gamma e^{-i\theta}|^2$$

$$= \frac{\sigma^2}{2\pi} |e^{i2\lambda} - 2\gamma e^{i\lambda} \cos\theta + \gamma^2|^2$$

$$= \frac{\sigma^2}{2\pi} [(1 - \gamma^2)^2 + 4\gamma^2(\cos\theta - \cos\lambda)^2 - 4\gamma(1 - \gamma)^2 \cos\theta \cos\lambda]$$

$$= \frac{\sigma^2}{2\pi}\left[4\gamma^2\left(\cos\lambda - \frac{1 + \gamma^2}{2\gamma}\cos\theta\right)^2 + (1 - \gamma^2)^2 \sin^2\theta\right].$$

If γ is near 1, then $f(\lambda)$ tends to be small for $\lambda = \pm\theta$. In fact, the minimum of (28) occurs for $\cos\lambda = (1 + \gamma^2)\cos\theta/(2\gamma)$ if the latter is less than 1 in absolute value.

For general q the spectral density is the product of terms such as (21) and (28). Let $z_j = \gamma_j e^{i\theta_j}$. (If $0 < \theta_j < \pi$, then $\theta_j = -\theta_k$ and $\gamma_j = \gamma_k$ for some k.) Then

$$(29)\qquad \sum_{r=0}^{q} \alpha_r \gamma_j^{q-r} e^{i\theta_j(q-r)} = 0.$$

If γ_j is close to 1 (that is, if z_j lies close to the unit circle in the complex plane), then

$$(30)\qquad f(\theta_j) = \frac{\sigma^2}{2\pi} \left| \sum_{r=0}^{q} \alpha_r e^{i\theta_j(q-r)} \right|^2$$

will be close to 0. Thus frequencies near θ_j will have low intensity.

In the general case the factor $|e^{i\lambda} - z_j|$ in $f(\lambda)$ as expressed in (18) can be replaced by

$$(31)\qquad |z_j| \cdot |1 - e^{i\lambda}/z_j| = |z_j| \cdot |e^{i\lambda} - 1/\bar{z}_j|,$$

where $\bar{z}_j$ is conjugate to z_j. If all the roots are real, distinct, and different from ± 1, there are 2^q different ways of expressing $f(\lambda)$ corresponding to different moving averages. The number of different moving averages in general will depend on the number of roots equal to 1 in absolute value, the multiplicities of distinct roots and the number of complex conjugate roots. We shall not enumerate the possibilities for $q > 2$.

It will be convenient to write a finite moving average in the form so that no root of (19) is greater than 1 in absolute value. (Note that a root of absolute value 1 is permitted for the moving average.)

The moving average (17) can be written

$$(32) \qquad \begin{aligned} y_t &= \left(\sum_{r=0}^{q} \alpha_r \mathscr{L}^r\right) v_t \\ &= \left(\sum_{r=0}^{q} \alpha_r \mathscr{P}^{q-r}\right) \mathscr{L}^q v_t \\ &= \left(\sum_{r=0}^{q} \alpha_r \mathscr{P}^{q-r}\right) v_{t-q}, \end{aligned}$$

where $\mathscr{L}$ and $\mathscr{P}$ are operators such that $\mathscr{L} v_t = v_{t-1}$ and $\mathscr{P} v_t = v_{t+1}$. If the roots of (19) are less than 1 in absolute value, we can invert (32) to obtain

$$(33) \qquad \begin{aligned} v_t &= \left(\sum_{r=0}^{q} \alpha_r \mathscr{L}^r\right)^{-1} y_t \\ &= \prod_{j=1}^{q} (1 - z_j \mathscr{L})^{-1} y_t \\ &= \prod_{j=1}^{q} \sum_{r=0}^{\infty} (z_j \mathscr{L})^r y_t. \end{aligned}$$

If

$$(34) \qquad \left(\sum_{r=0}^{q} \alpha_r z^r\right)^{-1} = \sum_{r=0}^{\infty} \gamma_r z^r,$$

then (33) is

$$(35) \qquad v_t = \sum_{r=0}^{\infty} \gamma_r y_{t-r},$$

or for $\gamma_0 = 1$

$$(36) \qquad y_t = v_t - \sum_{r=1}^{\infty} \gamma_r y_{t-r}.$$

Thus

$$(37) \qquad \mathscr{E}(y_t \mid y_{t-1}, y_{t-2}, \ldots) = -\sum_{r=1}^{\infty} \gamma_r y_{t-r};$$

this is the predictor of y_t on the basis of $y_{t-1}, y_{t-2}, \ldots$ that is best in the sense of minimizing the mean square error.

7.5.3. Autoregressive Processes

In Chapter 5 we studied stationary processes $\{y_t\}$ which satisfy equations

$$\sum_{r=0}^{p} \beta_r y_{t-r} = u_t, \tag{38}$$

where $\{u_t\}$ is a process of uncorrelated random variables with variance σ^2. (To discuss second-order moment properties we do not need to assume the u_t's are independently and identically distributed.) For convenience we assume $\mathscr{E}u_t = 0 = \mathscr{E}y_t$, and $\beta_0 = 1$. If the p roots $x_1, \ldots, x_p$ of the associated polynomial equation

$$\sum_{r=0}^{p} \beta_r x^{p-r} = 0 \tag{39}$$

are all less than 1 in absolute value, (38) can be inverted to give

$$y_t = \sum_{r=0}^{\infty} \delta_r u_{t-r}, \tag{40}$$

and this is a moving average process. If $f(\lambda)$ is the spectral density of $\{y_t\}$, then

$$\left| \sum_{r=0}^{p} \beta_r e^{i\lambda r} \right|^2 f(\lambda) = \frac{\sigma^2}{2\pi}, \tag{41}$$

since the right-hand side of (41) is the spectral density of $\{u_t\}$. Thus

$$f(\lambda) = \frac{\sigma^2}{2\pi \left| \sum_{r=0}^{p} \beta_r e^{i\lambda r} \right|^2} = \frac{\sigma^2}{2\pi \left| \sum_{r=0}^{p} \beta_r e^{i\lambda(p-r)} \right|^2}. \tag{42}$$

As noted in Section 5.2.1, stationarity of $\{y_t\}$ and $\sigma^2 > 0$ implies that no root of (39) can have absolute value 1; this is equivalent to saying that the denominator of $f(\lambda)$ has no zero, and hence that $f(\lambda)$ is integrable.

If $p = 1$,

$$y_t + \beta_1 y_{t-1} = u_t, \tag{43}$$

$$f(\lambda) = \frac{\sigma^2}{2\pi \, |e^{i\lambda} + \beta_1|^2} = \frac{\sigma^2}{2\pi(1 + \beta_1^2 + 2\beta_1 \cos \lambda)}. \tag{44}$$

If $\beta_1 > 0$, this spectral density has its minimum at $\lambda = 0$, namely $\sigma^2/[2\pi(1+\beta_1)^2]$, and its maximum at $\lambda = \pm\pi$, namely $\sigma^2/[2\pi(1-\beta_1)^2]$; it decreases in $[-\pi, 0]$ and increases in $[0, \pi]$; the high frequencies are emphasized. If $\beta_1 < 0$, the maximum is at $\lambda = 0$, and the minimum at $\lambda = \pm\pi$; low frequencies are emphasized.

If $p = 2$,

$$y_t + \beta_1 y_{t-1} + \beta_2 y_{t-2} = u_t, \tag{45}$$

$$\begin{aligned} f(\lambda) &= \frac{\sigma^2}{2\pi\,|e^{i2\lambda} + \beta_1 e^{i\lambda} + \beta_2|^2} \\ &= \frac{\sigma^2}{2\pi[1 + \beta_1^2 + \beta_2^2 + 2\beta_1(1+\beta_2)\cos\lambda + 2\beta_2\cos 2\lambda]} \\ &= \frac{\sigma^2}{2\pi[\beta_1^2 + (1-\beta_2)^2 + 2\beta_1(1+\beta_2)\cos\lambda + 4\beta_2\cos^2\lambda]} \\ &= \frac{\sigma^2}{2\pi\left\{4\beta_2\left[\cos\lambda + \dfrac{\beta_1(1+\beta_2)}{4\beta_2}\right]^2 + \dfrac{(1-\beta_2)^2(4\beta_2 - \beta_1^2)}{4\beta_2}\right\}}. \end{aligned} \tag{46}$$

If $\beta_1(1+\beta_2) > 4\,|\beta_2|$, $f(0) = \sigma^2/[2\pi(1+\beta_1+\beta_2)^2]$ is the minimum of $f(\lambda)$ and $f(\pm\pi) = \sigma^2/[2\pi(1-\beta_1+\beta_2)^2]$ are maxima; if $\beta_1(1+\beta_2) < -4\,|\beta_2|$, $f(0)$ is the maximum and $f(\pm\pi)$ are minima. If $|\beta_1(1+\beta_2)| < 4\,|\beta_2|$, $\cos\lambda = -\beta_1(1+\beta_2)/(4\beta_2)$ for a value of λ in $[0, \pi]$, say λ_0, and at $\lambda = -\lambda_0$; if $\beta_2 > 0$ (implying that $\beta_1^2 < 4\beta_2$ and the roots are complex), $f(\pm\lambda_0)$ are maxima and $f(\pm\pi)$ and $f(0)$ are relative minima; if $\beta_2 < 0$ (implying that $4\beta_2 < \beta_1^2$ and the roots are real), $f(\pm\lambda_0)$ are minima, and $f(\pm\pi)$ and $f(0)$ are relative maxima. If the roots are conjugate complex, say $\rho e^{\pm i\psi}$, then

$$\begin{aligned} f(\lambda) &= \frac{\sigma^2}{2\pi\left[4\rho^2\left(\cos\lambda - \dfrac{1+\rho^2}{2\rho}\cos\psi\right)^2 + (1-\rho^2)^2\sin^2\psi\right]} \\ &= \frac{\sigma^2}{2\pi[(1-\rho^2)^2 + 4\rho^2(\cos\psi - \cos\lambda)^2 - 4\rho(1-\rho)^2\cos\psi\cos\lambda]}. \end{aligned} \tag{47}$$

If ρ is near 1, $f(\lambda)$ tends to be large near $\lambda = \pm\psi$.

In general the spectral density is the product of terms like (44) and (47). If the root $x_j = \rho_j e^{i\psi_j}$ and ρ_j is near 1, then $f(\lambda)$ tends to be large near $\lambda = \pm\psi_j$.

The spectral density (42) can be written

$$f(\lambda) = \frac{\sigma^2}{2\pi\left|\sum_{r=0}^{p} \beta_r e^{i\lambda(p-r)}\right|^2} \tag{48}$$

$$= \frac{\sigma^2}{2\pi \prod_{j=1}^{p} |e^{i\lambda} - x_j|^2}.$$

Each term $|e^{i\lambda} - x_j|$ in (48) can be replaced by $|x_j| \cdot |1 - e^{i\lambda}/x_j| = |x_j| \cdot |e^{i\lambda} - 1/\bar{x}_j|$. There are alternative ways of expressing $f(\lambda)$ just as for the moving average process; here, however, no root has absolute value 1. If the roots of (39) are all less than 1 in absolute value, $|1/\bar{x}_j| > 1$; then only (48) is an expression corresponding to an associated polynomial equation with all roots less than 1 in absolute value.

7.5.4. Autoregressive Processes with Moving Average Residuals

The moving average and stochastic difference equation models can be combined by letting the disturbance in the stochastic difference equation be a finite moving average. The model is

$$\sum_{r=0}^{p} \beta_r y_{t-r} = \sum_{s=0}^{q} \alpha_s v_{t-s}, \tag{49}$$

where $\mathscr{E} v_t = 0$ (for convenience), $\mathscr{E} v_t^2 = \sigma^2$, $\mathscr{E} v_t v_s = 0$, $t \neq s$, $\beta_0 = \alpha_0 = 1$. If $f(\lambda)$ is the spectral density of $\{y_t\}$, then

$$\left|\sum_{r=0}^{p} \beta_r e^{i\lambda r}\right|^2 f(\lambda) = \sigma^2 \left|\sum_{s=0}^{q} \alpha_s e^{i\lambda s}\right|^2 \Big/ (2\pi), \tag{50}$$

since the spectral density of $\{v_t\}$ is $\sigma^2/(2\pi)$. The spectral density of $\{y_t\}$ is

$$f(\lambda) = \frac{\sigma^2 \left|\sum_{s=0}^{q} \alpha_s e^{i\lambda(q-s)}\right|^2}{2\pi \left|\sum_{r=0}^{p} \beta_r e^{i\lambda(p-r)}\right|^2}. \tag{51}$$

This can be written as a rational function of $e^{i\lambda}$. The numerator and denominator can be written in various ways. We shall require the roots of the polynomial associated with the denominator to be less than 1 in absolute value and the roots of the polynomial associated with the numerator to be not greater than 1 in absolute value. (For example, 1 could be a root for the numerator but not for the denominator if $\sigma^2 > 0$; see Theorem 5.2.2.) We shall also assume that these two polynomials have no root in common. (If there were common roots, the corresponding factors in numerator and denominator could be deleted.)

An arbitrary continuous spectral density can be approximated by a rational spectral density. In fact, an arbitrary continuous spectral density can be approximated by the spectral density of a moving average. This follows from the Weierstrass (Trigonometric) Approximation Theorem, which states that if $k(\lambda)$ is a continuous function in $[-\pi, \pi]$ with $k(\pi) = k(-\pi)$, then for every ε (> 0) there exists a real-valued trigonometric polynomial

$$h_m(\lambda) = \sum_{j=-m}^{m} c_j e^{i\lambda j} = a_0 + \sum_{j=1}^{m} (a_j \cos \lambda j + b_j \sin \lambda j), \tag{52}$$

where $a_0, a_1, \ldots, a_m, b_1, \ldots, b_m$ are real, such that

$$|k(\lambda) - h_m(\lambda)| < \varepsilon, \qquad -\pi \leq \lambda \leq \pi. \tag{53}$$

[See Lukacs (1960), Appendix C, for example.] If $k(\lambda) = k(-\lambda)$, then $h_m(\lambda)$ can be taken to be symmetric (for if the approximating polynomial, say $h_m^*(\lambda)$, were not symmetric, one could take $h_m(\lambda) = \frac{1}{2}[h_m^*(\lambda) + h_m^*(-\lambda)]$.) Then $b_1 = \cdots = b_m = 0$ and $c_j = c_{-j}$ is real, $j = 0, 1, \ldots, m$.

THEOREM 7.5.2. *If a spectral density $f(\lambda)$ is continuous, for any arbitrary $\varepsilon > 0$ there is a spectral density of the form*

$$h_m(\lambda) = \sum_{j=-m}^{m} c_j e^{i\lambda j} = \sum_{j=-m}^{m} c_j \cos \lambda j, \tag{54}$$

where $c_j = c_{-j}$ is real, $h_m(\lambda) > \frac{1}{2}\varepsilon$, and $|f(\lambda) - h_m(\lambda)| < \varepsilon$, $-\pi \leq \lambda \leq \pi$.

PROOF. Let $k(\lambda) = k(-\lambda) = f(\lambda) + 3\varepsilon/4$ and let $h_m(\lambda) = h_m(-\lambda)$ satisfy the Weierstrass Approximation Theorem with ε replaced by $\varepsilon/4$ in (53). Then

$$|f(\lambda) - h_m(\lambda)| \leq |f(\lambda) - k(\lambda)| + |k(\lambda) - h_m(\lambda)| < \tfrac{3}{4}\varepsilon + \tfrac{1}{4}\varepsilon = \varepsilon, \qquad -\pi \leq \lambda \leq \pi. \tag{55}$$

COROLLARY 7.5.1. *If a spectral density $f(\lambda)$ is continuous, then for any arbitrary $\varepsilon > 0$, there is a finite moving average process* (17) *with positive spectral density* (18), *say $h_q(\lambda)$, such that $|f(\lambda) - h_q(\lambda)| < \varepsilon$, $-\pi \leq \lambda \leq \pi$.*

PROOF. It was shown in Section 5.7.1 that an arbitrary covariance sequence which has only a finite number of nonzero covariances is the covariance sequence of a finite moving average process. By Theorem 7.5.2 an arbitrary continuous spectral density may be approximated by a positive spectral density (54).

COROLLARY 7.5.2. *If a spectral density $f(\lambda)$ is continuous, then for any arbitrary $\varepsilon > 0$, there is an autoregressive process $\{y_t\}$ satisfying* (38) *with spectral density* (42), *say $g_p(\lambda)$, such that $|f(\lambda) - g_p(\lambda)| < \varepsilon$, $-\pi \leq \lambda \leq \pi$.*

PROOF. Let $f_c(\lambda) = f(\lambda)$ if $f(\lambda) \geq c$ and $f_c(\lambda) = c$ if $f(\lambda) \leq c$ $(-\pi \leq \lambda \leq \pi)$, where $0 < c \leq \frac{1}{2}\varepsilon$. Then $|f(\lambda) - f_c(\lambda)| \leq \frac{1}{2}\varepsilon$. Since $f_c(\lambda)$ is continuous and positive in $[-\pi, \pi]$, $1/f_c(\lambda) = g(\lambda)$, say, is continuous, positive, and bounded and is the spectral density of a stationary process. By Corollary 7.5.1 there is a positive spectral density of a moving average process, say $k_q(\lambda)$, such that $|g(\lambda) - k_q(\lambda)| < \varepsilon'$, $-\pi \leq \lambda \leq \pi$. Then $1/k_q(\lambda) = g_p(\lambda)$ is the spectral density of an autoregressive process with $p = q$ and

$$
(56) \qquad |f_c(\lambda) - g_p(\lambda)| = \left| \frac{1}{g(\lambda)} - \frac{1}{k_q(\lambda)} \right| = \frac{|g(\lambda) - k_q(\lambda)|}{g(\lambda) k_q(\lambda)} < \tfrac{1}{2}\varepsilon,
$$

if

$$
(57) \qquad \varepsilon' \leq \frac{\frac{1}{2}\varepsilon \min g(\lambda)}{\max f(\lambda) + \frac{1}{2}\varepsilon}.
$$

This proves the corollary.

Thus a continuous spectral density can be approximated arbitrarily closely by the spectral density of a moving average process and it can be approximated by the spectral density of an autoregressive process. The corresponding covariance sequence can also be approximated arbitrarily closely. (See Problem 28.)

7.5.5. Effects of Linear Operations

Linear operations, such as differencing and smoothing, were considered in Chapter 3. The effects of these operations on the covariance structure can be studied by means of spectral densities.

COROLLARY 7.5.3. *If* $P(z) = \sum_{r=0}^{q} c_r z^r$, *then the spectral density* $h(\lambda)$ *of*

$$z_t = P(\mathscr{L})y_t \tag{58}$$

is

$$h(\lambda) = |P(e^{i\lambda})|^2 f(\lambda), \tag{59}$$

where $f(\lambda)$ *is the spectral density of* $\{y_t\}$ *and* $\mathscr{L}$ *is the lag operator.*

Consider the process obtained by applying the difference operator $\Delta = \mathscr{P} - 1$ to $\{y_t\}$ r times, that is,

$$z_t = \Delta^r y_t = (1 - \mathscr{L})^r y_{t+r} = (1 - \mathscr{L})^r \mathscr{P}^r y_t. \tag{60}$$

Then if $h(\lambda)$ is the spectral density of $\{z_t\}$

$$\begin{aligned} \frac{h(\lambda)}{f(\lambda)} &= |1 - e^{i\lambda}|^{2r} = |e^{i\frac{1}{2}\lambda} - e^{-i\frac{1}{2}\lambda}|^{2r} \\ &= 2^{2r} \sin^{2r} \tfrac{1}{2}\lambda \\ &= |e^{i\lambda} - 2 + e^{-i\lambda}|^r \\ &= 2^r(1 - \cos \lambda)^r. \end{aligned} \tag{61}$$

The effect of differencing is to emphasize high frequencies.

Addition of adjacent terms is the operation $\mathscr{P} + 1$; consider repeating the operation r times,

$$x_t = (1 + \mathscr{L})^r y_{t+r} = (1 + \mathscr{L})^r \mathscr{P}^r y_t. \tag{62}$$

Then if $k(\lambda)$ is the spectral density of $\{x_t\}$,

$$\begin{aligned} \frac{k(\lambda)}{f(\lambda)} &= |1 + e^{i\lambda}|^{2r} = |e^{i\frac{1}{2}\lambda} + e^{-i\frac{1}{2}\lambda}|^{2r} \\ &= 2^{2r} \cos^{2r} \tfrac{1}{2}\lambda \\ &= |e^{i\lambda} + 2 + e^{-i\lambda}|^r \\ &= 2^r(1 + \cos \lambda)^r. \end{aligned} \tag{63}$$

The effect of summing is to emphasize low frequencies.

Suppose both differencing and simple smoothing are done. If

$$z_t = (1 - \mathscr{L})^r (1 + \mathscr{L})^s y_t \tag{64}$$

then

$$\frac{h(\lambda)}{f(\lambda)} = 2^{r+s}(1 - \cos \lambda)^r (1 + \cos \lambda)^s. \tag{65}$$

Call this $\phi(\lambda)$. Then

(66)
$$\frac{d}{d\lambda}\phi(\lambda) = 2^{r+s}(1-\cos\lambda)^{r-1}(1+\cos\lambda)^{s-1}\sin\lambda\,[r(1+\cos\lambda)-s(1-\cos\lambda)],$$
$$r \geq 1, s \geq 1,$$

(67)
$$\frac{d^2}{d\lambda^2}\phi(\lambda) = 2^{r+s}(1-\cos\lambda)^{r-2}(1+\cos\lambda)^{s-2}\sin^2\lambda\{[r(1+\cos\lambda)-s(1-\cos\lambda)]^2$$
$$-\cos\lambda[r(1+\cos\lambda)-s(1-\cos\lambda)]-(r+s)\sin^2\lambda\},$$
$$r \geq 2, \quad s \geq 2.$$

The first derivative has a zero at

$$\cos\lambda = \frac{s-r}{s+r} = \frac{1-r/s}{1+r/s}, \tag{68}$$

and at this value of λ, say λ_0, the second derivative is

$$-2^{r+s}(1-\cos\lambda_0)^{r-2}(1+\cos\lambda_0)^{s-2}(r+s)\sin^4\lambda_0 = -\frac{4^{r+s}r^r s^s}{(r+s)^{r+s-1}}. \tag{69}$$

Thus $h(\lambda)/f(\lambda)$ has a maximum at λ_0. The other zeros of the first derivative are zeros of $h(\lambda)/f(\lambda)$. By choosing r and s properly we can make λ_0 arbitrarily close to any desired frequency, and if r and s are large the maximum of (65) will be sharp. Then $h(\lambda)$ will have a maximum near λ_0 and the corresponding process will be primarily a cosine wave with this frequency.

As one further example, consider the moving average

$$z_t = \frac{1}{2m+1}\sum_{s=-m}^{m} y_{t+s}. \tag{70}$$

Then

$$\begin{aligned} \frac{h(\lambda)}{f(\lambda)} &= \frac{1}{(2m+1)^2}\left|\sum_{s=-m}^{m} e^{i\lambda s}\right|^2 \\ &= \frac{1}{(2m+1)^2}\left|e^{-i\lambda m}\sum_{s=0}^{2m} e^{i\lambda s}\right|^2 \\ &= \frac{1}{(2m+1)^2}\left|\frac{1-e^{i\lambda(2m+1)}}{1-e^{i\lambda}}\right|^2 \\ &= \frac{1}{(2m+1)^2}\left|\frac{e^{i\frac{1}{2}\lambda(2m+1)}-e^{-i\frac{1}{2}\lambda(2m+1)}}{e^{i\frac{1}{2}\lambda}-e^{-i\frac{1}{2}\lambda}}\right|^2 \\ &= \left[\frac{\sin\lambda(m+\frac{1}{2})}{(2m+1)\sin\frac{1}{2}\lambda}\right]^2. \end{aligned} \tag{71}$$

The effect of smoothing is to increase the spectral density near $\lambda = 0$ (low frequencies) and to decrease it near $\lambda = k2\pi/(2m + 1)$, $k = 1, 2, \ldots$.

7.6. HILBERT SPACE AND PREDICTION THEORY

7.6.1. Hilbert Space

The developments of the last two sections can be clarified by putting them into the appropriate geometrical context, which is a Hilbert space. A finite-dimensional Euclidean space is a Hilbert space; the general theory of Hilbert space extends many of the ideas and results of Euclidean geometry to possibly infinite-dimensional spaces.

We use the doubly-infinite sequence of random variables $\{y_t\}$ to generate the relevant Hilbert space. The sequence of random variables constitutes a stochastic process stationary to the second order. For convenience we let $\mathscr{E}y_t = 0$ because we are concerned only with second-order moments. Let $\mathscr{E}y_t y_s = \sigma(t - s)$. The random variables y_t, $t = \ldots, -1, 0, 1, \ldots$, are elements or points of the Hilbert space. Finite linear combinations, $\sum_{t \in S} c_t y_t$, where S is a finite set of (positive and negative) integers, are also points of the space. The Hilbert space is completed by adding all random variables that are limits in quadratic mean of sequences of the random variables already in the space.

If x and y are two elements of this Hilbert space, then $\mathscr{E}xy$ is called the *inner product*, $\sqrt{\mathscr{E}x^2}$ is called the *norm* of x, and $\sqrt{\mathscr{E}(x - y)^2}$ is the *distance* between x and y. If $\mathscr{E}(x - y)^2 = 0$, we consider x and y to be the same element. A sequence of elements $\{x^{(n)}, n = 1, 2, \ldots\}$ *converges* if and only if it satisfies the *Cauchy criterion*, namely, given an $\varepsilon > 0$, there is an $N(\varepsilon)$ such that for $m > N(\varepsilon)$ and $n > N(\varepsilon)$

$$\sqrt{\mathscr{E}(x^{(n)} - x^{(m)})^2} < \varepsilon. \tag{1}$$

The Hilbert space spanned by $\{y_t\}$, which we are considering, is *complete* in the sense that the limits of Cauchy sequences are points in the space. In the Hilbert space we shall write $x^{(n)} \to x$ to mean $\mathscr{E}(x^{(n)} - x)^2 \to 0$ as $n \to \infty$. The space is also said to be *closed*.

THEOREM 7.6.1. *If $\mathscr{E}(x^{(n)} - x^{(m)})^2 \to 0$ as $n, m \to \infty$, there exists a random variable x such that $\mathscr{E}(x^{(n)} - x)^2 \to 0$ as $n \to \infty$.*

PROOF. The Tchebycheff inequality implies that $x^{(n)} - x^{(m)}$ converges to 0 stochastically as $n, m \to \infty$. Then there exists a subsequence n_k and a random variable x such that $x^{(n_k)}$ converges to x with probability 1 [Loève (1963), Section 6.3]. For every fixed m, $x^{(n_k)} - x^{(m)}$ converges to $x - x^{(m)}$ with probability 1. Then the Fatou-Lebesgue Theorem [Loève (1963), Section 7.2] and the fact that $\mathscr{E}(x^{(n_k)} - x^{(m)})^2 \to 0$ as $n_k, m \to \infty$ imply

$$\mathscr{E}(x - x^{(m)})^2 \leq \varliminf_{k\to\infty} \mathscr{E}(x^{(n_k)} - x^{(m)})^2 \to 0 \tag{2}$$

as $m \to \infty$. Q.E.D.

A useful criterion for convergence is the following theorem.

THEOREM 7.6.2. *If $\mathscr{E}x^{(n)}x^{(m)} \to \sigma^2 < \infty$ as $n, m \to \infty$, then $\mathscr{E}(x^{(n)} - x^{(m)})^2 \to 0$.*

PROOF.

$$\mathscr{E}(x^{(n)} - x^{(m)})^2 = \mathscr{E}x^{(n)^2} + \mathscr{E}x^{(m)^2} - 2\mathscr{E}x^{(n)}x^{(m)} \to 0. \tag{3}$$

COROLLARY 7.6.1. If $\mathscr{E}x^{(n)}x^{(m)} \to \sigma^2 < \infty$ *as* $n, m \to \infty$, *there exists a random variable x such that $\mathscr{E}(x^{(n)} - x)^2 \to 0$ as $n \to \infty$.*

In general a Hilbert space is a complete inner-product linear vector space. A *linear vector space* consists of elements or points or vectors $x, y, z, \ldots$ with 'addition' of two elements $x + y$ defined to yield a third element and 'multiplication' by a real† number α to yield another element; these operations satisfy the following conditions:

$$\begin{aligned} &x + y = y + x, && x + (y + z) = (x + y) + z, \\ &\alpha(x + y) = \alpha x + \alpha y, && (\alpha + \beta)x = \alpha x + \beta x, \\ &\alpha(\beta x) = (\alpha\beta)x, && 1x = x, \end{aligned} \tag{4}$$

for all elements x, y, and z in the vector space and all real numbers α and β. Furthermore, there exists a unique element designated 0 and called the zero (not to be confused with the number zero) such that $x + 0 = x$ and $0x = 0$ for every element x in the vector space.

An *inner product* is a real-valued function of two elements, designated by (x, y), which satisfies the following conditions:

$$\begin{aligned} &(x, y) = (y, x), && (x + y, z) = (x, z) + (y, z), \\ &(\alpha x, y) = \alpha(x, y), && (x, x) \geq 0, \\ &(x, x) > 0 \quad \text{if} \quad x \neq 0. \end{aligned} \tag{5}$$

In Euclidean space $\boldsymbol{x}'\boldsymbol{y}$ is an inner product; for the finite linear combinations of a set of random variables and their limits $\mathscr{E}xy$ is an inner product.

THEOREM 7.6.3. *(Cauchy-Schwarz Inequality)*

$$(x, y) \leq \sqrt{(x, x)}\sqrt{(y, y)}. \tag{6}$$

† In a complex Hilbert space the number is complex.

PROOF. The theorem follows from

$$0 \leq (\alpha x + \beta y, \alpha x + \beta y) = \alpha^2(x, x) + 2\alpha\beta(x, y) + \beta^2(y, y) \tag{7}$$

by setting $\alpha = \sqrt{(y, y)}$ and $\beta = -\sqrt{(x, x)}$.

From an inner product a *norm* can be defined as $\sqrt{(x, x)}$, often denoted as $\|x\|$. It satisfies the conditions

$$\begin{aligned} &\|x\| \geq 0, \qquad &&\|x\| > 0 \quad \text{if} \quad x \neq 0, \\ &\|\alpha x\| = |\alpha| \, \|x\|, \qquad &&\|x + y\| \leq \|x\| + \|y\|. \end{aligned} \tag{8}$$

The last (triangle) inequality follows from the Cauchy-Schwarz inequality by

$$\begin{aligned} \|x + y\|^2 &= \|x\|^2 + \|y\|^2 + 2(x, y) \\ &\leq \|x\|^2 + \|y\|^2 + 2\,\|x\| \cdot \|y\| = (\|x\| + \|y\|)^2. \end{aligned} \tag{9}$$

Then $\|x - y\|$ is defined as the *distance* between x and y and is called the *metric* of the space.

A sequence $\{x^{(n)}\}$ is said to be a *Cauchy sequence* if

$$\lim_{\substack{n\to\infty \\ m\to\infty}} \|x^{(n)} - x^{(m)}\| = 0. \tag{10}$$

An element x is the limit of a sequence $\{x^{(n)}\}$ if

$$\lim_{n\to\infty} \|x^{(n)} - x\| = 0. \tag{11}$$

A normed linear vector space is *complete* if there exists a limit in the space to every Cauchy sequence.

For any Hilbert space the inner product of two elements is a continuous function of these two elements.

THEOREM 7.6.4. *If* $x^{(n)} \to x$ *and* $y^{(n)} \to y$ *as* $n \to \infty$, *then*

$$(x^{(n)}, y^{(n)}) \to (x, y). \tag{12}$$

PROOF. We have

$$\begin{aligned} |(x^{(n)}, y^{(n)}) - (x, y)| &= |(x^{(n)} - x, y^{(n)} - y) \\ &\qquad + (x, y^{(n)} - y) + (x^{(n)} - x, y)| \\ &\leq |(x^{(n)} - x, y^{(n)} - y)| \\ &\qquad + |(x, y^{(n)} - y)| + |(x^{(n)} - x, y)| \\ &\leq \|x^{(n)} - x\| \cdot \|y^{(n)} - y\| \\ &\qquad + \|x\| \cdot \|y^{(n)} - y\| + \|x^{(n)} - x\| \cdot \|y\| \end{aligned} \tag{13}$$

by the Cauchy-Schwarz inequality, and the right-hand side of (13) converges to 0. Q.E.D.

COROLLARY 7.6.2. *If* $x^{(n)} \to x$, *then* $\|x^{(n)}\| \to \|x\|$.

When we take (x, y) to be $\mathscr{E}xy$ in the Hilbert space generated by $\{y_t\}$ Theorem 7.6.4 justifies the limits of variances and covariances in Sections 7.4 and 7.5.

7.6.2. Projections and Linear Prediction

In terms of the stochastic process $\{y_t\}$ with mean $\mathscr{E}y_t = 0$ and covariance function $\mathscr{E}y_t y_s = \sigma(t - s)$, we may be interested in predicting y_t from knowledge of $y_{t-1}, y_{t-2}, \ldots, y_{t-p}$ (a part of the past) or from $y_{t-1}, y_{t-2}, \ldots$ (the entire past). If $\hat{y}_t$ is the prediction, we shall take $\mathscr{E}(y_t - \hat{y}_t)^2$ as the criterion to be made small. This mean square error of prediction is minimized if $\hat{y}_t$ is taken as the conditional expectation of y_t given $y_{t-1}, y_{t-2}, \ldots, y_{t-p}$ or $y_{t-1}, y_{t-2}, \ldots$ as the case may be. This conditional expectation will, of course, depend on the distribution of $y_t, y_{t-1}, \ldots, y_{t-p}$ or the probability measure of $y_t, y_{t-1}, \ldots$. If the distribution is normal or the process is Gaussian, the conditional expectation will be a linear function of $y_{t-1}, y_{t-2}, \ldots, y_{t-p}$ or $y_{t-1}, y_{t-2}, \ldots$, but in other instances it may not be a linear function.

We shall consider prediction of y_t by a linear function, $\hat{y}_t = \sum_{s=1}^{p} c_s y_{t-s}$ in the case of prediction from $y_{t-1}, \ldots, y_{t-p}$ or by a limit $\hat{y}_t$ of such linear functions. Then $\mathscr{E}(y_t - \hat{y}_t)^2$ depends only on $\{\sigma(h)\}$. It will be convenient to put this problem into the geometry of Hilbert space.

A *linear manifold* is a nonempty subset of the Hilbert space such that for every real α and β $\alpha x + \beta y$ is in the subset if x and y are. The linear manifold is *closed* if it contains the limit of every Cauchy sequence in it. A finite or infinite set of points in the Hilbert space may generate a closed linear manifold, which then consists of all finite linear combinations of these points and their limits; the original set of points is said to span the closed linear manifold. A closed linear manifold is a Hilbert space, and will be called a *subspace*.

Since $\mathscr{E}(y_t - \hat{y}_t)^2$ is $\|y_t - \hat{y}_t\|^2$, finding the best linear prediction of y_t is the problem of finding the point in the linear manifold spanned by the y_s's ($s < t$) which is closest (in the distance defined by the norm) to the point y_t. For a T-dimensional Euclidean space with norm $\sqrt{\mathbf{x}'\mathbf{x}}$, this problem was solved in Section 2.2. The vector in the linear manifold spanned by the p columns of $\mathbf{z}$ closest to $\mathbf{y}$ was a uniquely defined vector $\mathbf{z}$ such that $\mathbf{y} - \mathbf{z}$ was orthogonal to the p columns of $\mathbf{z}$ (and hence to every vector in the linear manifold). We shall now show that this solution holds in a general Hilbert space.

In terms of the inner product (x, y) two elements x and y are *orthogonal* if $(x, y) = 0$. We may write $x \perp y$.

THEOREM 7.6.5. (*Projection Theorem*) *If $\mathscr{M}$ is a subspace of a Hilbert space $\mathscr{H}$ and y is an element in $\mathscr{H}$, there exists a unique element x in $\mathscr{M}$ such that $\|y - x\| = \min_{z \in \mathscr{M}} \|y - z\|$; if y is not in $\mathscr{M}$, $y - x$ is orthogonal to every $z \in \mathscr{M}$.*

PROOF. Let $\min_{z \in \mathscr{M}} \|y - z\| = d$ and let $\{x^{(n)}\}$ be a sequence such that $\|y - x^{(n)}\| \to d$. (Such a sequence exists by the definition of the minimum.) The existence of the element x will follow from the demonstration that $\{x^{(n)}\}$ is a Cauchy sequence. We have

$$\|(y - x^{(n)}) - (y - x^{(m)})\|^2 + \|(y - x^{(n)}) + (y - x^{(m)})\|^2 = 2\,\|y - x^{(n)}\|^2 + 2\,\|y - x^{(m)}\|^2. \tag{14}$$

The left-hand side is

$$\|x^{(m)} - x^{(n)}\|^2 + 4\,\|y - \tfrac{1}{2}(x^{(n)} + x^{(m)})\|^2; \tag{15}$$

since $\frac{1}{2}(x^{(n)} + x^{(m)}) \in \mathscr{M}$, the second term in (15) is at least $4d^2$. Thus (14) implies

$$\|x^{(m)} - x^{(n)}\|^2 \leq 2\,\|y - x^{(n)}\|^2 + 2\,\|y - x^{(m)}\|^2 - 4d^2, \tag{16}$$

and the right-hand side of (16) converges to 0. Thus $\{x^{(n)}\}$ is a Cauchy sequence and a limit x exists; by Corollary 7.6.2 $\|y - x\| = d$.

For an arbitrary element $z \in \mathscr{M}$

$$\begin{aligned} d^2 \leq \|y - x - \alpha z\|^2 &= \|y - x\|^2 - 2(y - x, \alpha z) + \|\alpha z\|^2 \\ &= d^2 - 2\alpha(y - x, z) + \alpha^2\,\|z\|^2. \end{aligned} \tag{17}$$

If $y \notin \mathscr{M}$ and hence $y - x \neq 0$, the inequality can hold for all α only if $(y - x, z) = 0$; that is, if $y - x$ is orthogonal to z.

If there were another element x' such that $\|y - x'\|^2 = d^2$, then $(y - x', z) = 0$ for every $z \in \mathscr{M}$; for $z = x - x'$ we would have $0 = (y - x', x - x') - (y - x, x - x') = (x - x', x - x') = \|x - x'\|^2$, which implies $x = x'$. Q.E.D.

The point x is called the *projection* of y on the subspace $\mathscr{M}$ and $y - x$ shall be termed the *residual* of y from $\mathscr{M}$.

If $\mathscr{M}$ is a subspace of a Hilbert space $\mathscr{H}$, then the set of all elements orthogonal to $\mathscr{M}$ is also a subspace, say $\mathscr{M}^\perp$. (It is closed by continuity of the inner product and completeness of $\mathscr{H}$.) Every element y in $\mathscr{H}$ can be written uniquely as $x + z$, where $x \in \mathscr{M}$ and $z \in \mathscr{M}^\perp$. We write this as $\mathscr{H} = \mathscr{M} \oplus \mathscr{M}^\perp$.

In terms of Hilbert space linear prediction is simply a projection. The Hilbert space $\mathscr{H}$ is the closed linear manifold spanned by $y_t, y_{t-1}, \ldots,$ and the inner product is $\mathscr{E}xy$. Let $\mathscr{M}_{t-1,p}$ be the subspace spanned by $y_{t-1}, \ldots, y_{t-p}$. The best prediction of y_t by a linear combination $c_1 y_{t-1} + \cdots + c_p y_{t-p}$ is the projection of y_t on $\mathscr{M}_{t-1,p}$, say $\hat{y}_{t,p}$; it minimizes

$$\mathscr{E}\left(y_t - \sum_{j=1}^{p} c_j y_{t-j}\right)^2 = \sigma(0) - 2\sum_{j=1}^{p} c_j \sigma(j) + \sum_{i,j=1}^{p} c_i c_j \sigma(j-i). \tag{18}$$

The problem is analogous to that treated in Chapter 2. The solution is given by the normal equations

$$\begin{bmatrix} \sigma(0) & \sigma(1) & \cdots & \sigma(p-1) \\ \sigma(1) & \sigma(0) & \cdots & \sigma(p-2) \\ \cdot & \cdot & & \cdot \\ \cdot & \cdot & & \cdot \\ \cdot & \cdot & & \cdot \\ \sigma(p-1) & \sigma(p-2) & \cdots & \sigma(0) \end{bmatrix} \begin{bmatrix} c_1 \\ c_2 \\ \cdot \\ \cdot \\ \cdot \\ c_p \end{bmatrix} = \begin{bmatrix} \sigma(1) \\ \sigma(2) \\ \cdot \\ \cdot \\ \cdot \\ \sigma(p) \end{bmatrix}. \tag{19}$$

Let $\mathscr{M}_{t-1}$ be the closed linear manifold spanned by $y_{t-1}, y_{t-2}, \ldots$. The best linear prediction of y_t by $y_{t-1}, y_{t-2}, \ldots$ is the projection of y_t on $\mathscr{M}_{t-1}$, say $\hat{y}_t$.

THEOREM 7.6.6.

$$\lim_{p\to\infty} \mathscr{E}(\hat{y}_t - \hat{y}_{t,p})^2 = 0. \tag{20}$$

PROOF. Let $\mathscr{E}(y_t - \hat{y}_{t,p})^2 = d_p$, and let $\mathscr{E}(y_t - \hat{y}_t)^2 = d$. Since $\mathscr{M}_{t-1,p} \subset \mathscr{M}_{t-1,p+1}$ and $\mathscr{M}_{t-1} = \bigcup_{p=1}^{\infty} \mathscr{M}_{t-1,p}$, there is a sequence $\{x^{(p)}\}$ with $x^{(p)} \in \mathscr{M}_{t-1,p}$ such that $\mathscr{E}(x^{(p)} - \hat{y}_t)^2 \to 0$ as $p \to \infty$. Since $\mathscr{E}(x^{(p)} - y_t)^2 \geq d_p \geq d$ and

$$\mathscr{E}(x^{(p)} - y_t)^2 \leq \mathscr{E}(x^{(p)} - \hat{y}_t)^2 + \mathscr{E}(\hat{y}_t - y_t)^2, \tag{21}$$

which tends to d, we have

$$\lim_{p\to\infty} \mathscr{E}(x^{(p)} - y_t)^2 = \lim_{p\to\infty} d_p = d. \tag{22}$$

We can also write

$$\mathscr{E}(x^{(p)} - y_t)^2 = \mathscr{E}(x^{(p)} - \hat{y}_{t,p})^2 + \mathscr{E}(\hat{y}_{t,p} - y_t)^2 \tag{23}$$

because $y_t - \hat{y}_{t,p}$ is orthogonal to every vector in $\mathscr{M}_{t,p}$. This implies $\mathscr{E}(x^{(p)} - \hat{y}_{t,p})^2 \to 0$ inasmuch as the other two terms in (23) approach d. The theorem follows from

$$\mathscr{E}(\hat{y}_t - \hat{y}_{t,p})^2 \leq \mathscr{E}(\hat{y}_t - x^{(p)})^2 + \mathscr{E}(x^{(p)} - \hat{y}_{t,p})^2. \tag{24}$$

This theorem shows that the prediction based on the finite past approximates the prediction based on the infinite past. In practice the statistician would like a large sample to estimate the covariance structure as well as for prediction on the basis of this estimated covariance structure.

The variance of the residual may be any nonnegative number not exceeding $\sigma(0)$. If $\mathscr{E}(y_t - \hat{y}_t)^2 = \sigma^2 = 0$, the process is called *deterministic*. This means that y_t can be predicted perfectly by an element of $\mathscr{M}_{t-1}$. In most cases this element is an infinite linear combination $\sum_{s=1}^{\infty} c_s y_{t-s}$. If $\mathscr{E}(y_t - \hat{y}_t)^2 = \sigma^2 > 0$, the process is called *regular*.

Since $y_t - \hat{y}_t$ is orthogonal to every point in $\mathscr{M}_{t-1}$, we can write

$$y_t = \hat{y}_t + u_t, \tag{25}$$

where $u_t = y_t - \hat{y}_t$ is uncorrelated with $y_{t-1}, y_{t-2}, \ldots$. The random variable u_t may be called the *innovation* or disturbance.

7.6.3. The Wold Decomposition

The so-called Wold decomposition [Wold (1954)] clarifies the structure of a stationary process.

THEOREM 7.6.7. *If $\{y_t\}$ is a regular stationary stochastic process with $\mathscr{E}y_t = 0$, it can be written as*

$$y_t = \sum_{s=0}^{\infty} \gamma_s u_{t-s} + v_t, \tag{26}$$

where $\sum_{s=0}^{\infty} \gamma_s^2 < \infty$, $\gamma_0 = 1$, $\mathscr{E}u_s = \mathscr{E}v_s = 0$, $\mathscr{E}u_s^2 = \sigma^2$, $\mathscr{E}u_s u_t = 0$, $s \neq t$, and $\mathscr{E}u_s v_t = 0$, $u_s \in \mathscr{M}_s$ and $v_t \in \mathscr{M}_{-\infty} = \bigcap_{s=0}^{\infty} \mathscr{M}_{t-s}$. The sequences $\{\gamma_s\}$ and $\{u_s\}$ are unique.

PROOF. Let $u_s = y_s - \hat{y}_s$, $s = t, t-1, \ldots$, and $\sigma^2 = \mathscr{E}u_s^2$. Since u_s is orthogonal to $\mathscr{M}_{s-1}$, $\mathscr{E}u_s u_r = 0$, $r < s$. Let $\gamma_s = \mathscr{E}y_t u_{t-s}/\sigma^2$, $s = 1, 2, \ldots$. Then

$$0 \leq \mathscr{E}\left(y_t - \sum_{s=0}^{m} \gamma_s u_{t-s}\right)^2 = \mathscr{E}y_t^2 - \sigma^2 \sum_{s=0}^{m} \gamma_s^2, \tag{27}$$

and hence $\sum_{s=0}^{\infty} \gamma_s^2 < \infty$. Therefore $\sum_{s=0}^{\infty} \gamma_s u_{t-s}$ converges in the mean and is in the subspace spanned by $u_t, u_{t-1}, \ldots$. Define v_t by

$$v_t = y_t - \sum_{s=0}^{\infty} \gamma_s u_{t-s}. \tag{28}$$

Then $\mathscr{E}v_t u_r = \mathscr{E}y_t u_r - \sigma^2 \gamma_{t-r} = 0$ for $r \leq t$; $\mathscr{E}v_t u_r = 0$ for $r > t$ because u_r is orthogonal to all points in $\mathscr{M}_t$ and $v_t \in \mathscr{M}_t$. Since v_t is orthogonal to u_t, it is in

$\mathscr{M}_{t-1}$. By induction it is in $\mathscr{M}_s$ for $s \leq t$ and hence in $\mathscr{M}_{-\infty} = \bigcap_{s=0}^{\infty} \mathscr{M}_{t-s}$. The resolution is unique because $u_s \in \mathscr{M}_s$ and u_s is orthogonal to $\mathscr{M}_{s-1}$ and $\gamma_s = \mathscr{E} y_t u_{t-s}/\sigma^2$. Q.E.D.

THEOREM 7.6.8. *The process* $\{v_t\}$ *defined in Theorem* 7.6.7 *is deterministic, and the process* $\{w_t\}$, *where* $w_t = \sum_{s=0}^{\infty} \gamma_s u_{t-s}$, *is regular.*

PROOF. To prove $\{v_t\}$ is deterministic we shall show that the closed linear manifold spanned by $v_t, v_{t-1}, \ldots$, say $\mathscr{M}_{tv}$, is $\mathscr{M}_{-\infty}$; that is, $\mathscr{M}_{tv} = \mathscr{M}_{t-1,v} = \cdots = \mathscr{M}_{-\infty}$; then v_t is a (possibly infinite) linear combination of $v_{t-1}, v_{t-2}, \ldots$.

LEMMA 7.6.1.

$$\mathscr{M}_{-\infty} = \mathscr{M}_u^{\perp}. \tag{29}$$

PROOF. Here $\mathscr{M}_u^{\perp}$ is the subspace orthogonal to $\mathscr{M}_u$, the subspace spanned by $\{u_t\}$. If $x \in \mathscr{M}_{-\infty} = \bigcap_{t=-\infty}^{\infty} \mathscr{M}_t$, then $x \in \mathscr{M}_t$ and hence is orthogonal to u_{t+1} for every t; $x \in \mathscr{M}_u^{\perp}$. Conversely, suppose $x \in \mathscr{M}_u^{\perp}$. Since $\mathscr{H} = \bigcup_{t=-\infty}^{\infty} \mathscr{M}_t$, $x \in \mathscr{M}_t$ for some t. Because $x \perp u_t$, $x \in \mathscr{M}_{t-1}$, and by induction $x \in \mathscr{M}_s$ for $s \leq t$. Moreover, $x \in \mathscr{M}_s$ for $s > t$ because $\mathscr{M}_t \subseteq \mathscr{M}_s$. This proves the lemma.

LEMMA 7.6.2.

$$\mathscr{M}_{tu} = \mathscr{M}_{tw}. \tag{30}$$

PROOF. Since $w_t = \sum_{s=0}^{\infty} \gamma_s u_{t-s}$, the subspace $\mathscr{M}_{tw}$ spanned by $w_t, w_{t-1}, \ldots$ is contained in the subspace $\mathscr{M}_{tu}$ spanned by $u_t, u_{t-1}, \ldots$. Conversely, $u_t \in \mathscr{M}_t = \mathscr{M}_{tw} \oplus \mathscr{M}_{tv}$ and $u_t \perp \mathscr{M}_{tv}$ and therefore $u_t \in \mathscr{M}_{tw}$. This proves the lemma.

We now complete the proof of Theorem 7.6.8. Since $v_s \in \mathscr{M}_{-\infty}$ for every s, $\mathscr{M}_{tv} \subseteq \mathscr{M}_{-\infty}$. Since $\mathscr{M}_{-\infty} = \bigcap_{t=-\infty}^{\infty} \mathscr{M}_t$, $x \in \mathscr{M}_{-\infty}$ implies $x \in \mathscr{M}_t$; since $x \perp \mathscr{M}_{tu} = \mathscr{M}_{tw}$, $x \in \mathscr{M}_{tv}$. This shows $\mathscr{M}_{tv} = \mathscr{M}_{-\infty}$ and $\{v_t\}$ is deterministic. Since $w_t = u_t + \sum_{s=1}^{\infty} \gamma_s u_{t-s}$ and $u_t \perp \sum_{s=1}^{\infty} \gamma_s u_{t-s} \in \mathscr{M}_{t-1,u}$, $\mathscr{E}(w_t - \hat{w}_t)^2 = \sigma^2 > 0$. Thus $\{w_t\}$ is regular. Q.E.D.

The process $\{v_t\}$ is often called purely deterministic. The process $\{w_t\}$ is called *purely indeterministic* because it does not have a deterministic component. Since $y_t = u_t + \sum_{s=1}^{\infty} \gamma_s u_{t-s} + v_t$ and $u_t \perp (\sum_{s=1}^{\infty} \gamma_s u_{t-s} + v_t) \in \mathscr{M}_{t-1}$, the projection on $\mathscr{M}_{t-1}$ is $\hat{y}_t = \sum_{s=1}^{\infty} \gamma_s u_{t-s} + v_t$; this is the best linear prediction of y_t and $\mathscr{E}(y_t - \hat{y}_t)^2 = \mathscr{E} u_t^2 = \sigma^2$. By Lemma 7.6.2 $\hat{y}_t \in \mathscr{M}_{t-1,w} \oplus \mathscr{M}_{t-1,v} = \mathscr{M}_{t-1}$ and hence $\hat{y}_t$ is implicitly a linear function of $y_{t-1}, y_{t-2}, \ldots$, or a limit of such.

The spectral density of $\{w_t\}$ is $\sigma^2 |\sum_{s=0}^{\infty} \gamma_s e^{i\lambda s}|^2/(2\pi)$. The process $\{v_t\}$ has no absolutely continuous component; its spectral distribution function consists of the jump and singular components. We shall not prove the last assertion. [See Doob (1953) Chapter XII, Section 4.]

Kolmogorov (1941b) has shown that the variance of u_t (the error of linear prediction) is

$$\sigma^2 = 2\pi \exp\left[\frac{1}{2\pi}\int_{-\pi}^{\pi} \log F'(\lambda)\, d\lambda\right]. \tag{31}$$

In fact, $\{y_t\}$ is regular if and only if $F'(\lambda) > 0$ almost everywhere $(-\pi \le \lambda \le \pi)$ and

$$\int_{-\pi}^{\pi} \log F'(\lambda)\, d\lambda > -\infty. \tag{32}$$

The argument involves expanding $\frac{1}{2}\log F'(\lambda) = \log\sqrt{F'(\lambda)}$ in a Fourier series, say $\sum_{n=-\infty}^{\infty} a_n e^{i\lambda n}$ with $a_{-n} = a_n$ real (because $\log F'(\lambda)$ is real and symmetric); then

$$\Gamma(z) = \sum_{s=0}^{\infty} \gamma_s z^s = \exp\left[a_0 + 2\sum_{n=1}^{\infty} a_n z^n\right] \tag{33}$$

and $F'(\lambda) = |\Gamma(e^{-i\lambda})|^2$.

If the process $\{y_t\}$ has a spectral density $f(\lambda) = F'(\lambda)$ and if (32) holds, we have

$$y_t = \sum_{s=0}^{\infty} \gamma_s u_{t-s} \tag{34}$$

with $\gamma_0 = 1$ and $\sum_{s=0}^{\infty} \gamma_s^2 < \infty$. Then the best linear prediction of y_t on the basis of $u_{t-1}, u_{t-2}, \ldots$ is $\sum_{s=1}^{\infty} \gamma_s u_{t-s}$. We now ask the question of whether we can write

$$u_t = \sum_{s=0}^{\infty} \beta_s y_{t-s} \tag{35}$$

with $\beta_0 = 1$. If so, the best linear prediction of y_t is $-\sum_{s=1}^{\infty} \beta_s y_{t-s}$, which is explicit in $y_{t-1}, y_{t-2}, \ldots$. If the representation (35) exists, we can write

$$u_t = \sum_{s=0}^{\infty} \beta_s \sum_{r=0}^{\infty} \gamma_r u_{t-r-s} = \sum_{q=0}^{\infty}\sum_{r=0}^{q} \beta_{q-r}\gamma_r u_{t-q}. \tag{36}$$

Here $\beta_0\gamma_0 = 1$ and

$$\sum_{r=0}^{q} \beta_{q-r}\gamma_r = 0, \qquad q = 1, 2, \ldots. \tag{37}$$

More precisely, suppose $\Gamma(z) = \sum_{s=0}^{\infty} \gamma_s z^s$ is analytic and nonzero in $|z| \leq 1$. Then

$$\frac{1}{\Gamma(z)} = \mathrm{B}(z) = \sum_{r=0}^{\infty} \beta_r z^r \tag{38}$$

is analytic and nonzero in $|z| \leq 1$. Hence

$$\mathscr{E}\left(\sum_{s=0}^{\infty} \beta_s y_{t-s}\right)^2 = \int_{-\pi}^{\pi} |\mathrm{B}(e^{-i\lambda})|^2 \, dF(\lambda) \tag{39}$$

exists and is well-defined because $|\mathrm{B}(e^{-i\lambda})|$ is bounded. Hence, (35) exists as a limit in mean square. If

$$\begin{aligned} u_t &= \int_0^{\pi} \cos \lambda t \, dC_u(\lambda) + \int_0^{\pi} \sin \lambda t \, dS_u(\lambda) \\ &= \mathscr{R} \int_0^{\pi} e^{i\lambda t} \, d[C_u(\lambda) - iS_u(\lambda)], \end{aligned} \tag{40}$$

where $\mathscr{E}C_u^2(\lambda) = \mathscr{E}S_u^2(\lambda) = \sigma^2\lambda/\pi$, let

$$y_t = \mathscr{R} \int_0^{\pi} e^{i\lambda t} \mathrm{B}^{-1}(e^{-i\lambda}) \, d[C_u(\lambda) - iS_u(\lambda)]. \tag{41}$$

Then

$$\begin{aligned} \sum_{s=0}^{\infty} \beta_s y_{t-s} &= \mathscr{R} \int_0^{\pi} \sum_{s=0}^{\infty} \beta_s e^{i\lambda(t-s)} \mathrm{B}^{-1}(e^{-i\lambda}) \, d[C_u(\lambda) - iS_u(\lambda)] \\ &= \mathscr{R} \int_0^{\pi} e^{i\lambda t} \, d[C_u(\lambda) - iS_u(\lambda)]. \end{aligned} \tag{42}$$

It remains to verify that the spectral density of (41) is

$$\frac{\sigma^2}{2\pi} |\mathrm{B}^{-1}(e^{-i\lambda})|^2 = f(\lambda). \tag{43}$$

We can put the problem of prediction in most cases as finding $b_1, b_2, \ldots$ to minimize

$$\mathscr{E}\left(y_t - \sum_{s=1}^{\infty} b_s y_{t-s}\right)^2. \tag{44}$$

The normal equations are

$$\sum_{s=1}^{\infty} \sigma(r-s) b_s = \sigma(r), \qquad r = 1, 2, \ldots. \tag{45}$$

These are equivalent to

$$(46) \qquad \int_{-\pi}^{\pi} e^{i\lambda r} f(\lambda)\left(\sum_{s=1}^{\infty} b_s e^{-i\lambda s} - 1\right) d\lambda = 0, \qquad r = 1, 2, \ldots .$$

Thus we want to find a function $B(e^{-i\lambda})$, where

$$(47) \qquad B(z) = \sum_{s=1}^{\infty} b_s z^s,$$

such that the positive Fourier coefficients of $f(\lambda)[B(e^{-i\lambda}) - 1]$ are 0.

More definitive results were given by Akutowicz (1957), which we report without proof.

THEOREM 7.6.9. *If $f(\lambda)$ is bounded,* (35) *holds with* $\sum_{s=0}^{\infty} |\beta_s|^2 < \infty$ *if and only if* $\Gamma(z) = \sum_{s=0}^{\infty} \gamma_s z^s$, $|z| < 1$, *is such that*

$$(48) \qquad \int_0^{2\pi} \frac{1}{|\Gamma(\rho e^{i\lambda})|^2} \, d\lambda$$

is bounded as $\rho \to 1$.

The function $1/\Gamma(z)$, regular for $|z| < 1$, is said to belong to Hardy class H^2 if (48) is bounded as $\rho \to 1$. [See Zygmund (1959), Volume I, Chapter VII, Section 7; see also Grenander and Rosenblatt (1957), p. 288.]

THEOREM 7.6.10. *Under the conditions of Theorem* 7.6.9 *the optimal linear predictor is*

$$(49) \qquad -\sum_{r=0}^{\infty} \beta_r y_{t-r}.$$

Prediction theory was developed independently by Kolmogorov (1941a), (1941b) and Wiener (1949) (originally in a 1942 publication of restricted distribution). Parzen (1961a) has exploited the Hilbert space approach.

7.7. SOME LIMIT THEOREMS

The distributions of many statistics of interest in time series analysis are too complicated to derive in useful forms. In many cases, however, limiting distributions can be obtained; these may be used as approximations to the exact distributions when the number of observations T is large. In this section we study some limit theorems of general applicability.

A method of obtaining the limiting distribution of a sequence of random variables is to find the limiting distribution of an approximating sequence (as in Section 5.5).

THEOREM 7.7.1. *Let*

$$S_T = Z_{kT} + X_{kT}, \qquad T = 1, 2, \ldots, k = 1, 2, \ldots . \tag{1}$$

Suppose that for every $\delta > 0$ *and* $\varepsilon > 0$ *there exists a* k_0 *such that for* $k > k_0$

$$\Pr\{|X_{kT}| > \delta\} < \varepsilon \tag{2}$$

for all T *and suppose*

$$\Pr\{Z_{kT} \leq z\} = F_{kT}(z) \to F_k(z), \tag{3}$$

as $T \to \infty$, *and*

$$\lim_{k\to\infty} F_k(z) = F(z) \tag{4}$$

at every continuity point of $F(z)$. *Then*

$$\lim_{T\to\infty} \Pr\{S_T \leq z\} = F(z) \tag{5}$$

at every continuity point of $F(z)$.

PROOF. The condition (2) is that as $k \to \infty$ X_{kT} converges stochastically to 0 uniformly in T. Since $F(z)$ is continuous at z, for an arbitrary $\varepsilon > 0$ there is a $\delta > 0$ such that

$$|F(z \pm \delta) - F(z)| < \tfrac{1}{2}\varepsilon, \tag{6}$$

and such that $z - \delta$ and $z + \delta$ are continuity points of $F(z)$; then there is a k_1 such that for $k > k_1$

$$|F_k(z \pm \delta) - F(z)| < \varepsilon. \tag{7}$$

Choose the δ to satisfy (6) and choose k so $k > k_0$ and $k > k_1$; hold δ and k fixed. Then for $T > T_k$ (which depends on ε)

$$|F_{kT}(z) - F_k(z)| < \varepsilon, \tag{8}$$

$$|F_{kT}(z + \delta) - F_k(z + \delta)| < \varepsilon, \tag{9}$$

$$|F_{kT}(z - \delta) - F_k(z - \delta)| < \varepsilon. \tag{10}$$

Now

$$\begin{aligned} \Pr\{S_T \leq z\} &= \Pr\{Z_{kT} + X_{kT} \leq z\} \\ &\leq \Pr\{Z_{kT} \leq z + \delta, |X_{kT}| \leq \delta\} + \Pr\{|X_{kT}| > \delta\} \\ &\leq F_{kT}(z + \delta) + \varepsilon. \end{aligned} \tag{11}$$

Thus

$$\Pr\{S_T \le z\} \le F(z) + 3\varepsilon. \tag{12}$$

Similarly,

$$\Pr\{S_T \le z\} \ge F(z) - 3\varepsilon. \tag{13}$$

This proves the theorem.

COROLLARY 7.7.1. *Suppose for a sequence* $\{M_k\}$

$$\mathscr{E} X_{kT}^2 \le M_k, \tag{14}$$

$$\lim_{k\to\infty} M_k = 0, \tag{15}$$

and suppose (3) *and* (4) *hold. Then* (5) *holds.*

PROOF. The Tchebycheff inequality implies (2). Then Theorem 7.7.1 implies the corollary.

This result was given by Anderson (1959) and in a somewhat different form by Diananda (1953).

Because we use central limit theorems frequently, we give several here for reference. Theorem 7.7.4 follows from Theorem 7.7.2.

THEOREM 7.7.2. (*Lindeberg Central Limit Theorem*) *Let* $w_1^T, \ldots, w_T^T$ *be a set of independent random variables,* $T = 1, 2, \ldots$. *Let the distribution of* w_t^T *be* $F_t^T(w)$ *and* $\mathscr{E} w_t^T = 0$, $t = 1, \ldots, T$, *with*

$$\sum_{t=1}^{T} \operatorname{Var} w_t^T = 1. \tag{16}$$

A sufficient condition that the distribution of $\sum_{t=1}^T w_t^T$ *converge to* $N(0, 1)$ *is that*

$$\sum_{t=1}^{T} \int_{|w|>\delta} w^2\, dF_t^T(w) \to 0 \tag{17}$$

as $T \to \infty$ *for every* $\delta > 0$.

The condition (17) is also necessary for the distribution of $\sum_{t=1}^T w_t^T$ to converge to $N(0, 1)$ if $\max_{t=1,\ldots,T} \mathscr{E}(w_t^T)^2 \to 0$. [See Loève (1963), Section 21.2.]

THEOREM 7.7.3. (*Liapounov's Central Limit Theorem*) *Let* $w_1^T, \ldots, w_T^T$ *be a set of independent random variables,* $T = 1, 2, \ldots$, *such that* $\mathscr{E} w_t^T = 0$, $t = 1, \ldots, T$, *and* (16) *holds. A sufficient condition that the distribution of* $\sum_{t=1}^T w_t^T$

converge to $N(0, 1)$ *is that*

$$\sum_{t=1}^{T} \mathscr{E} |w_t^T|^{2+\delta} \to 0 \tag{18}$$

for some $\delta > 0$.

THEOREM 7.7.4. (*Identically distributed random variables*) *If* $u_1, u_2, \ldots$ *are independently and identically distributed with* $\mathscr{E}u_t = 0$ *and* $\mathscr{E}u_t^2 = \sigma^2$, *then the distribution of* $\sum_{t=1}^T u_t/(\sigma\sqrt{T})$ *converges to* $N(0, 1)$ *as* $T \to \infty$.

From Theorem 7.7.1 and Corollary 7.7.1 we can derive central limit theorems for certain stationary stochastic processes. One case is that of finitely dependent variables, in which case the variables are independent if the difference in indices is sufficiently great. A finite moving average, $y_t = \sum_{s=0}^q \alpha_s v_{t-s}$, where the v_t's are independently and identically distributed, has this property.

THEOREM 7.7.5. *Let* $y_1, y_2, \ldots$ *be a stationary stochastic process such that for every integer* n *and integers* $t_1, \ldots, t_n$ $(0 < t_1 < \cdots < t_n)$ $y_{t_1}, \ldots, y_{t_n}$ *is distributed independently of* $y_1, \ldots, y_{t_1-m-1}$ *and* $y_{t_n+m+1}, \ldots$. *If* $\mathscr{E}y_t = 0$ *and* $\mathscr{E}y_t^2 < \infty$, *then* $\sum_{t=1}^T y_t/\sqrt{T}$ *has a limiting normal distribution with mean* 0 *and variance*

$$\mathscr{E}y_1^2 + 2\mathscr{E}y_1y_2 + \cdots + 2\mathscr{E}y_1y_{m+1}. \tag{19}$$

PROOF. Let

$$Y_T = \frac{1}{\sqrt{T}} \sum_{t=1}^{T} y_t. \tag{20}$$

Then

$$\mathscr{E}Y_T = 0, \tag{21}$$

$$\begin{aligned} \mathscr{E}Y_T^2 &= \frac{1}{T} \sum_{t,s=1}^{T} \mathscr{E}y_t y_s \\ &= \frac{1}{T} [T\mathscr{E}y_1^2 + 2(T-1)\mathscr{E}y_1y_2 \\ &\quad + 2(T-2)\mathscr{E}y_1y_3 + \cdots + 2(T-m)\mathscr{E}y_1y_{m+1}], \end{aligned} \tag{22}$$

and

$$\lim_{T\to\infty} \mathscr{E}Y_T^2 = \mathscr{E}y_1^2 + 2\mathscr{E}y_1y_2 + \cdots + 2\mathscr{E}y_1y_{m+1} = W, \tag{23}$$

say. For given integral $k > 2m$, let $T = kN + r$, where $0 \leq r < k$; that is, $N = [T/k]$. Let

$$(24)\quad U_{kT} = \frac{1}{\sqrt{T}}[(y_1 + \cdots + y_{k-m}) + (y_{k+1} + \cdots + y_{2k-m}) + \cdots + (y_{k(N-1)+1} + \cdots + y_{kN-m})],$$

$$(25)\quad X_{kT} = \frac{1}{\sqrt{T}}[(y_{k-m+1} + \cdots + y_k) + (y_{2k-m+1} + \cdots + y_{2k}) + \cdots + (y_{kN-m+1} + \cdots + y_{kN})],$$

$$(26)\quad R_{kT} = \frac{1}{\sqrt{T}}(y_{kN+1} + \cdots + y_T),$$

for $k \leq T$. Let $U_{kT} = Y_T$, $X_{kT} = R_{kT} = 0$ for $k > T$.

U_{kT} is $1/\sqrt{T}$ times the sum of N independently and identically distributed terms. For the jth term

$$(27)\quad \mathscr{E}(y_{k(j-1)+1} + \cdots + y_{kj-m}) = 0,$$

$$\begin{aligned}(28)\quad \mathscr{E}(y_{k(j-1)+1} + \cdots + y_{kj-m})^2 &= \mathscr{E}(y_1 + \cdots + y_{k-m})^2 \\ &= (k - m)\mathscr{E}y_1^2 + 2(k - m - 1)\mathscr{E}y_1y_2 + \cdots + 2(k - 2m)\mathscr{E}y_1y_{m+1} \\ &= kV_k,\end{aligned}$$

say. By Theorem 7.7.4, the sum of these terms divided by $\sqrt{N}$ (that is, $U_{kT}\sqrt{T/N}$) has a limiting normal distribution with mean 0 and variance kV_k as $T \to \infty$; that is, $N \to \infty$. Thus $U_{kT}\sqrt{T/(Nk)}$ has a limiting normal distribution with mean 0 and variance V_k.

The mean square of R_{kT} is bounded since

$$(29)\quad \mathscr{E}R_{kT}^2 \leq \frac{r^2}{T}\mathscr{E}y_1^2 \leq \frac{k^2}{T}\mathscr{E}y_1^2.$$

We have used the fact that $|\mathscr{E}y_ty_s| \leq \mathscr{E}y_1^2$. As $T \to \infty$, this goes to 0 and R_{kT} converges stochastically to 0. Thus $U_{kT} + R_{kT}$ has a limiting normal distribution with mean 0 and variance V_k.

Now for $k \leq T$

$$(30)\quad \mathscr{E}X_{kT}^2 = \frac{N}{T}\mathscr{E}(y_1 + \cdots + y_m)^2 \leq \frac{1}{k}\mathscr{E}(y_1 + \cdots + y_m)^2,$$

and $\mathscr{E}X_{kT}^2 = 0$ for $k > T$. Thus for k sufficiently large this is arbitrarily small uniformly in T. Note that $V_k \to W$ as $k \to \infty$. Application of Corollary 7.7.1 completes the proof.

THEOREM 7.7.6. *Let $\mathbf{y}_1, \mathbf{y}_2, \ldots$ be a sequence of vector random variables satisfying the conditions of Theorem 7.7.5 where $\mathscr{E}y_t^2 < \infty$ is replaced by $\mathscr{E}\mathbf{y}_t'\mathbf{y}_t < \infty$. Then $(1/\sqrt{T})\sum_{t=1}^T \mathbf{y}_t$ has a limiting normal distribution with mean $\mathbf{0}$ and covariance matrix*

$$\mathscr{E}\mathbf{y}_1\mathbf{y}_1' + \mathscr{E}\mathbf{y}_1\mathbf{y}_2' + \mathscr{E}\mathbf{y}_2\mathbf{y}_1' + \cdots + \mathscr{E}\mathbf{y}_1\mathbf{y}_{m+1}' + \mathscr{E}\mathbf{y}_{m+1}\mathbf{y}_1'. \tag{31}$$

PROOF. The theorem follows from Theorem 7.7.5 and the following theorem.

THEOREM 7.7.7. *If a sequence of random matrices $\boldsymbol{W}_T$, $T = 1, 2, \ldots$, is such that every linear combination of elements* $\operatorname{tr}\boldsymbol{\Psi}\boldsymbol{W}_T$ *has a limiting normal distribution with mean 0 and variance* $\operatorname{tr}\boldsymbol{\Sigma\Psi A\Psi}'$, *then $\boldsymbol{W}_T$ has a limiting multivariate normal distribution with mean $\mathbf{0}$ and variances and covariances*

$$\mathscr{E}w_{gi}w_{hj} = a_{gh}\sigma_{ij}. \tag{32}$$

PROOF. The condition of the theorem implies that

$$\lim_{T\to\infty} \mathscr{E}e^{it\operatorname{tr}\boldsymbol{\Psi}\boldsymbol{W}_T} = e^{-\frac{1}{2}t^2\operatorname{tr}\boldsymbol{\Sigma\Psi A\Psi}'} \tag{33}$$

for every real t and hence (for $t = 1$) that

$$\lim_{T\to\infty} \mathscr{E}\exp\left[i\sum_{j,g}\psi_{jg}w_{gjT}\right] = \exp\left[-\tfrac{1}{2}\sum_{k,j,g,h}\sigma_{kj}\psi_{jg}a_{gh}\psi_{kh}\right], \tag{34}$$

which is the characteristic function of the multivariate normal distribution. Since this characteristic function is continuous at $\boldsymbol{\Psi} = \mathbf{0}$, the result follows from the so-called continuity theorem. [See Cramér (1946), p. 102, for example.]

These theorems can be extended to include a linear process. [See also Marsaglia (1954) and Parzen (1957a).]

THEOREM 7.7.8. *If*

$$y_t = \sum_{s=-\infty}^{\infty} \gamma_s v_{t-s}, \tag{35}$$

where $\{v_t\}$ is a sequence of independently and identically distributed random variables with $\mathscr{E}v_t = 0$ and $\mathscr{E}v_t^2 = \sigma^2$, and $\sum_{s=-\infty}^{\infty}|\gamma_s| < \infty$, then $\sum_{t=1}^T y_t/\sqrt{T}$ has a limiting normal distribution with mean 0 and variance

$$\sum_{r=-\infty}^{\infty}\sigma(r), \tag{36}$$

where

$$\sigma(r) = \sigma^2\sum_{s=-\infty}^{\infty}\gamma_s\gamma_{s+r}. \tag{37}$$

PROOF. Let

$$y_{t,k} = \sum_{s=-k}^{k} \gamma_s v_{t-s}, \tag{38}$$

$$u_{t,k} = \sum_{s=k+1}^{\infty} \gamma_s v_{t-s}, \tag{39}$$

$$u^*_{t,k} = \sum_{s=-\infty}^{-(k+1)} \gamma_s v_{t-s}. \tag{40}$$

Then $y_t = y_{t,k} + u_{t,k} + u^*_{t,k}$. Let

$$Z_{kT} = \frac{1}{\sqrt{T}} \sum_{t=1}^{T} y_{t,k}, \tag{41}$$

$$X_{kT} = \frac{1}{\sqrt{T}} \sum_{t=1}^{T} u_{t,k}, \tag{42}$$

$$X^*_{kT} = \frac{1}{\sqrt{T}} \sum_{t=1}^{T} u^*_{t,k}. \tag{43}$$

Then $\sum_{t=1}^{T} y_t/\sqrt{T} = Z_{kT} + X_{kT} + X^*_{kT}$. We have

$$\begin{aligned} \mathscr{E} X^2_{kT} &= \frac{1}{T} \mathscr{E}\left(\sum_{t=1}^{T} u_{t,k}\right)^2 \\ &= \frac{1}{T} \sum_{t,s=1}^{T} \mathscr{E} u_{t,k} u_{s,k} \\ &= \frac{1}{T} \sum_{t,s=1}^{T} \sum_{p,r=k+1}^{\infty} \gamma_p \gamma_r \mathscr{E} v_{t-p} v_{s-r} \\ &= \frac{\sigma^2}{T} \sum_{t,s=1}^{T} \sum_{p=\max(0,t-s)+k+1}^{\infty} \gamma_p \gamma_{p+s-t} \\ &\le \frac{\sigma^2}{T} \sum_{t,s=1}^{T} \sum_{p=\max(0,t-s)+k+1}^{\infty} |\gamma_p| \cdot |\gamma_{p+s-t}| \\ &\le \sigma^2 \sum_{q=k+1}^{\infty} \sum_{p=k+1}^{\infty} |\gamma_p| \cdot |\gamma_q| \\ &= \sigma^2 \left(\sum_{p=k+1}^{\infty} |\gamma_p|\right)^2. \end{aligned} \tag{44}$$

Since $\sum_{p=1}^{\infty}|\gamma_p| < \infty$, (44) converges to 0 as $k \to \infty$. Similarly

$$\mathscr{E}X_{kT}^{*2} \le \sigma^2\left(\sum_{p=-\infty}^{-(k+1)}|\gamma_p|\right)^2, \tag{45}$$

which converges to 0 as $k \to \infty$. Then

$$\begin{aligned}\mathscr{E}(X_{kT}+X_{kT}^*)^2 &\le 2[\mathscr{E}X_{kT}^2+\mathscr{E}X_{kT}^{*2}]\\ &\le 2\sigma^2\left[\left(\sum_{p=k+1}^{\infty}|\gamma_p|\right)^2+\left(\sum_{p=-\infty}^{-(k+1)}|\gamma_p|\right)^2\right],\end{aligned} \tag{46}$$

which converges to 0 as $k \to \infty$. Theorem 7.7.8 follows from Corollary 7.7.1 and Theorem 7.7.5.

Similar theorems (for $y_t = \sum_{s=0}^{\infty}\gamma_s v_{t-s}$) were proved by Moran (1947) in a somewhat different way and by Diananda (1953).

THEOREM 7.7.9. *Let $y_1, y_2, \ldots$ be a sequence of random variables such that for every n and $t_1, \ldots, t_n$ $(0 < t_1 < \cdots < t_n)$ $y_{t_1}, \ldots, y_{t_n}$ is distributed independently of $y_1, \ldots, y_{t_1-m-1}$ and $y_{t_n+m+1}, \ldots$. Let $\mathscr{E}y_t = 0$ and $\mathscr{E}|y_t|^{2+\delta} < M$, $t = 1, 2, \ldots$, for some $\delta > 0$ and some M. Let $a_1, a_2, \ldots$ be a sequence of constants such that $|a_t| < L$, $t = 1, 2, \ldots$, for some L and such that the following limit exists:*

$$\lim_{T\to\infty}\frac{1}{T}\sum_{t,s=1}^{T}a_t a_s \mathscr{E}y_t y_s. \tag{47}$$

Then $\sum_{t=1}^{T}a_t y_t/\sqrt{T}$ has a limiting normal distribution with mean 0 and variance given by (47).

PROOF. Let $y_t^* = a_t y_t$, $t = 1, 2, \ldots$. The proof proceeds according to the proof of Theorem 7.7.5 with y_t replaced by y_t^*. The random variables $y_{k(j-1)+1}^* + \cdots + y_{kj-m}^*$, $j = 1, \ldots, N$, are independently, but not identically, distributed; however, their $(2+\delta)$th absolute moments are uniformly bounded by $(k-m)^{2+\delta}L^{2+\delta}M$ (by Minkowski's inequality). The mean square of the difference between $\sum_{t=1}^{T}y_t^*/\sqrt{T}$ and U_{kT}^* [which is (24) with y_t replaced by y_t^*] has a bound [approximately $m^2L^2(M+1)/k$] that goes to 0 uniformly as $k \to \infty$; $\lim_{k\to\infty}\lim_{T\to\infty}\operatorname{Var}U_{kT}^*$ is (47). The Liapounov Central Limit Theorem (Theorem 7.7.3) completes the proof.

COROLLARY 7.7.2. *Let $y_1, y_2, \ldots$ be a sequence of random variables such that for every n and $t_1, \ldots, t_n$ $(0 < t_1 < \cdots < t_n)$ $y_{t_1}, \ldots, y_{t_n}$ is distributed independently of $y_1, \ldots, y_{t_1-m-1}$ and $y_{t_n+m+1}, \ldots$. Suppose $\mathscr{E}y_t = 0$, $\mathscr{E}y_t y_s = \sigma(t-s)$, and $\mathscr{E}|y_t|^{2+\delta} < M$, $t = 1, 2, \ldots$, for some $\delta > 0$ and some M. Then*

$\sum_{t=1}^{T} y_t/\sqrt{T}$ *has a limiting normal distribution with mean* 0 *and variance* $\sigma(0) + 2\sum_{h=1}^{m} \sigma(h)$.

The proof of the theorem would be slightly easier if $\delta = 2$. Note that these results apply to regression estimates.

REFERENCES

A thorough reference book on stochastic processes is Doob (1953); Loève (1963) also gives a general treatment. The reader is further referred to Bartlett (1966), Yaglom (1962), Parzen (1962), and Rosenblatt (1962).

Section 7.2. Kolmogorov (1933).

Section 7.3. Hobson (1907), Loève (1963).

Section 7.4. Doob (1953).

Section 7.5. Doob (1953), Lukacs (1960).

Section 7.6. Akutowicz (1957), Doob (1953), Grenander and Rosenblatt (1957), Kolmogorov (1941a), (1941b), Loève (1963), Parzen (1961a), Wiener (1949), Wold (1954), Zygmund (1959).

Section 7.7. T. W. Anderson (1959), Cramér (1946), Diananda (1953), Loève (1963), Marsaglia (1954), Moran (1947), Parzen (1957a).

PROBLEMS

1. (Sec. 7.2.1.) Prove that a process stationary in the strict sense and having a finite variance is stationary in the wide sense.

2. (Sec. 7.2.1.) Prove that if $\{y_t\}$ and $\{z_t\}$ are two independent stationary processes, then $\{y_t + z_t\}$ is a stationary process.

3. (Sec. 7.2.2.) Let $y_t = (-1)^t y$, where y is a random variable.

(a) What are necessary and sufficient conditions on the distribution of y so $\{y_t\}$ is a strictly stationary process?

(b) What are necessary and sufficient conditions on the distribution of y so $\{y_t\}$ is a process stationary in the wide sense?

4. (Sec. 7.2.2.) Show that

$$y_t = \alpha \cos \lambda t + \beta \sin \lambda t + u_t,$$

where the u_t's are independent and identically distributed, does not constitute a stationary process if $\lambda \neq 0$.

5. (Sec. 7.2.2.) Give an example of A_j and B_j, $j = 1, \ldots, q$, such that (13) is not stationary in the strict sense.

6. (Sec. 7.2.2.) Show that

$$y_t = \rho \cos(\lambda t + v),$$

where v is uniformly distributed on $(0, 2\pi)$, is a stationary stochastic process.

7. (Sec. 7.2.2.) Show that $\{y_t\}$ defined by (43) is strictly stationary.

8. (Sec. 7.2.2.) Let

$$y_t = \sum_{j=1}^{q} D_j \cos(\lambda_j t - v_j),$$

where $D_1, \ldots, D_q, v_1, \ldots, v_q$ are independently distributed with $\mathscr{E} D_j^2 = 2\sigma_j^2$ and v_j having a uniform distribution on $(0, 2\pi)$. Show $\{y_t\}$ has covariance function (44).

9. (Sec. 7.3.) Let $\{y_t\}$ and $\{z_t\}$ be independent stationary processes with spectral densities $f_y(\lambda)$ and $f_z(\lambda)$. Show that $\{x_t\} = \{y_t + z_t\}$ has a spectral density $f_y(\lambda) + f_z(\lambda)$. (See Problem 2.)

10. (Sec. 7.3.) Let $y_t = (-1)^t y$, where y is a random variable such that $\{y_t\}$ is stationary in the wide sense. (See Problem 3.) Find $\sigma(h)$ and $F(\lambda)$.

11. (Sec. 7.3.) Graph $\cos 2\pi t$ and $\cos(4 \cdot 2\pi t)$ and show that they agree at $t = 0, \pm 1/5, \pm 2/5, \ldots$.

12. (Sec. 7.3.) In the case of a complex-valued process let $\sigma(k) = \alpha(k) + i\beta(k)$. Show $\alpha(-k) = \alpha(k)$ and $\beta(-k) = -\beta(k)$. Show

$$f(\lambda) = \frac{1}{2\pi}\left\{\sigma(0) + 2\sum_{k=1}^{\infty} [\alpha(k)\cos\lambda k + \beta(k)\sin\lambda k]\right\}.$$

13. (Sec. 7.4.1.) Let

(i) $$0 = \lambda_0 < \lambda_1 < \cdots < \lambda_n = \pi.$$

If $h(\lambda)$ is continuous in $[0, \pi]$, it is uniformly continuous and

(ii) $$\max h(\lambda) - \min h(\lambda) < \varepsilon, \qquad \lambda_{j-1} \le \lambda \le \lambda_j,$$

if $\lambda_j - \lambda_{j-1} < \delta$ (suitably chosen). Show that for this δ

(iii) $$\sum_{j=1}^{n} \max_{\lambda_{j-1}\le\lambda\le\lambda_j} h(\lambda)[H(\lambda_j) - H(\lambda_{j-1})] - \sum_{j=1}^{n} \min_{\lambda_{j-1}\le\lambda\le\lambda_j} h(\lambda)[H(\lambda_j) - H(\lambda_{j-1})] < \varepsilon[H(\pi) - H(0)].$$

Use these facts to show the existence of the Riemann-Stieltjes integral

(iv) $$\int_0^{\pi} h(\lambda)\, dH(\lambda).$$

14. (Sec. 7.4.1.) Let

$$0 = \lambda_0 < \lambda_1 < \cdots < \lambda_n = \pi,$$

$$\lambda_0 \leq \lambda_1' \leq \lambda_1, \ldots, \lambda_{n-1} \leq \lambda_n' \leq \lambda_n,$$

$$\lambda_0 \leq \lambda_1'' \leq \lambda_1, \ldots, \lambda_{n-1} \leq \lambda_n'' \leq \lambda_n.$$

Let $C(\lambda)$ be a stochastic process of uncorrelated increments, $0 \leq \lambda \leq \pi$, with covariance function $H(\lambda)$. Show that

$$\mathscr{E}\left\{\sum_{j=1}^{n} h(\lambda_j')[C(\lambda_j) - C(\lambda_{j-1})] - \sum_{j=1}^{n} h(\lambda_j'')[C(\lambda_j) - C(\lambda_{j-1})]\right\}^2$$

$$= \sum_{j=1}^{n} [h(\lambda_j') - h(\lambda_j'')]^2[H(\lambda_j) - H(\lambda_{j-1})].$$

15. (Sec. 7.4.1.) Let

$$0 = \lambda_0 < \lambda_1 < \cdots < \lambda_{2n} = \pi,$$

$$S_n = \sum_{j=1}^{n} h(\lambda_{2j})[C(\lambda_{2j}) - C(\lambda_{2(j-1)})],$$

$$S_{2n} = \sum_{i=1}^{2n} h(\lambda_i)[C(\lambda_i) - C(\lambda_{i-1})].$$

Show that

$$\mathscr{E}(S_{2n} - S_n)^2 \leq \max_{j=1,\ldots,n} [h(\lambda_{2j-1}) - h(\lambda_{2j})]^2[H(\pi) - H(0)].$$

16. (Sec. 7.4.1.) Prove (21) in detail.

17. (Sec. 7.4.1.) Show that the definition of the Riemann-Stieltjes stochastic integral can be used if $h(\lambda)$ has a finite number of discontinuities and if $h(\lambda)$ and $H(\lambda)$ have no common point of discontinuity.

18. (Sec. 7.4.1.) Let $C(\lambda)$ be a stochastic process of uncorrelated increments, $0 \leq \lambda \leq \pi$, with covariance function $H(\lambda)$, and let $k(\lambda)$ be the step function

$$k(\lambda) = c_j, \qquad a_{j-1} < \lambda \leq a_j, j = 1, \ldots, n,$$

where $0 = a_0 < a_1 < \cdots < a_n = \pi$ with $k(0) = c_1$. Assume $H(\lambda)$ is continuous at $a_0, a_1, \ldots, a_n$. Show that the Riemann-Stieltjes stochastic integral (as defined in Problem 17) is

$$\int_0^\pi k(\lambda)\, dC(\lambda) = \sum_{j=1}^{n} c_j[C(a_j) - C(a_{j-1})].$$

19. (Sec. 7.4.1.) Let $h(\lambda)$ satisfy

$$\int_0^\pi h^2(\lambda)\, dH(\lambda) < \infty.$$

The Lebesgue-Stieltjes stochastic integral is defined as a limit in the mean of

$$\int_0^\pi h_n(\lambda)\, dC(\lambda),$$

where $\{h_n(\lambda)\}$ is a sequence of step functions such that

$$\lim_{n\to\infty} \int_0^\pi |h_n(\lambda) - h(\lambda)|^2\, dH(\lambda) = 0.$$

Show that when the Riemann-Stieltjes stochastic integral is defined it is the same as the Lebesgue-Stieltjes stochastic integral (with probability 1).

20. (Sec. 7.5.1.) Show that

$$\sum_{r=-\infty}^{\infty} |c_r| < \infty$$

is a sufficient condition for the convergence of the right-hand side of (2).

21. (Sec. 7.5.2.) Let the roots of $z^2 + c_1 z + c_2 = 0$ be $\rho e^{i\theta}$ and $\rho e^{-i\theta}$, and let

$$y_t = v_t + c_1 v_{t-1} + c_2 v_{t-2},$$

where $\mathscr{E} v_t = 0$, $\mathscr{E} v_t^2 = \sigma^2$ and $\mathscr{E} v_t v_s = 0$, $t \neq s$. Show

$$\sigma(0) = (1 + 4\rho^2 \cos^2\theta + \rho^4)\sigma^2,$$
$$\sigma(1) = -2\rho(1 + \rho^2)\sigma^2 \cos\theta,$$
$$\sigma(2) = \rho^2\sigma^2.$$

22. (Sec. 7.5.3.) Show that if y_t satisfies the stochastic difference equation

$$\prod_{j=1}^{p} (1 - x_j \mathscr{L}) y_t = u_t$$

(where $|x_j| < 1$), then

$$\prod_{j=1}^{r} \left(1 - \frac{1}{\bar{x}_j}\mathscr{L}\right) \prod_{j=r+1}^{p} (1 - x_j \mathscr{L}) y_t = u_t^*$$

is a sequence of uncorrelated random variables with constant variance; if an x_j, $j = 1, \ldots, r$, is complex, its conjugate root $\bar{x}_j = x_i$, $i \neq j$, is included in the first r roots.

23. (Sec. 7.5.3.) Show that

$$f(\lambda) = \sum_{i=1}^{p} \frac{c_i(1 - x_i^2)}{2\pi(1 - 2x_i \cos\lambda + x_i^2)},$$

where

$$\sigma(h) = \sum_{i=1}^{p} c_i x_i^h$$

and $|x_i| < 1$, $i = 1, \ldots, p$.

24. (Sec. 7.5.3.) Let

$$y_t - \tfrac{3}{4}y_{t-1} + \tfrac{1}{8}y_{t-2} = u_t.$$

(a) Find x_1 and x_2, the roots of the associated polynomial equation.

(b) Write the covariance function $\sigma(h)$.

(c) Write the spectral density $f(\lambda)$.

(d) Express the spectral density in three other ways (corresponding to replacing x_1 and/or x_2 by $1/\bar{x}_1$ and/or $1/\bar{x}_2$, respectively).

(e) Write the stochastic difference equation and associated polynomial equation corresponding to each form of $f(\lambda)$.

25. (Sec. 7.5.3.) Let

$$y_t - 1.4y_{t-1} + 0.98y_{t-2} = u_t.$$

Find (a) the roots x_1 and x_2 of the associated polynomial equation; (b) $\sigma(h)$; and (c) $f(\lambda)$. (d) Express $f(\lambda)$ in three other ways (replacing x_1 and/or x_2 by $1/\bar{x}_1$ and/or $1/\bar{x}_2$, respectively), and (e) write the stochastic difference equation and associated polynomial equation corresponding to each form of $f(\lambda)$ admitting such equations with real coefficients.

26. (Sec. 7.5.3.) Write $\sigma(h)$ and $f(\lambda)$ for $y_t + \tfrac{1}{2}y_{t-1} = u_t$.

27. (Sec. 7.5.3.) Write $\sigma(h)$ and $f(\lambda)$ for $y_t - \tfrac{1}{2}y_{t-1} = u_t$.

28. (Sec. 7.5.4.) Let $f(\lambda)$ be a continuous spectral density. Show that given any ε (> 0), there is a finite moving average process $\sum_{j=0}^{m} \alpha_j v_{t-j}$ with spectral density $f_m(\lambda)$ such that if $\{\sigma(h)\}$ and $\{\sigma_m(h)\}$ are the corresponding covariance sequences,

$$\text{(i)} \qquad |\sigma(h) - \sigma_m(h)| < 2\pi\varepsilon.$$

Show that there is an autoregressive process such that (i) holds for $f_m(\lambda)$ and $\{\sigma_m(h)\}$ corresponding to the autoregressive process.

29. (Sec. 7.5.4.) Let

$$z_t = y_t + w_t,$$

where

$$y_t + \beta y_{t-1} = u_t$$

and the u_t's and the w_t's are independently distributed with means 0 and variances $\mathscr{E}u_t^2 = \sigma^2$ and $\mathscr{E}w_t^2 = \tau^2$. Show that the spectral density of $\{z_t\}$ is

$$\frac{\sigma^2 + (2\beta\cos\lambda + 1 + \beta^2)\tau^2}{2\pi(2\beta\cos\lambda + 1 + \beta^2)}.$$

Find a process $\{z_t^*\}$ satisfying

$$z_t^* + \beta z_{t-1}^* = v_t + \alpha v_{t-1},$$

where $\mathscr{E}v_t = 0$, $\mathscr{E}v_t^2 = K$, and $\mathscr{E}v_t v_s = 0$, $t \neq s$, which has the same spectral density.

30. (Sec. 7.5.5.) The smoothing formula based on fitting a second-degree polynomial to five points is

$$z_t = -\tfrac{3}{35}y_{t-2} + \tfrac{12}{35}y_{t-1} + \tfrac{17}{35}y_t + \tfrac{12}{35}y_{t+1} - \tfrac{3}{35}y_{t+2}.$$

When $\mathscr{E}y_t = 0$, $\mathscr{E}y_t^2 = \sigma^2$, $\mathscr{E}y_t y_s = 0$, $t \neq s$, find the spectral density and covariance sequence of $\{z_t\}$.

31. (Sec. 7.5.5.) The coefficients for the smoothing formula based on fitting a polynomial of degree $2k$ (or $2k + 1$) to $2k + 3$ points are defined in (19), (20), and (21) of Section 3.3. Show that the spectral density of the residuals from smoothing an uncorrelated sequence with variance σ^2 is

$$\frac{\sigma^2}{2\pi} C^2 (2 \sin \tfrac{1}{2}\lambda)^{4k+4},$$

where C is defined by (21) in Section 3.3.

32. (Sec. 7.5.5.) Give explicitly the spectral density and covariance sequence of $\{\Delta^2 y_t\}$, when $\mathscr{E}y_t = 0$, $\mathscr{E}y_t^2 = \sigma^2$, $\mathscr{E}y_t y_s = 0$, $t \neq s$, and graph them.

33. (Sec. 7.5.5.) Show that the spectral density of the residuals from the smoothing by $\sum_{s=-m}^{m} y_{t+s}/(2m + 1)$ for $\{y_t\}$ an uncorrelated sequence with variance σ^2 is

$$\frac{\sigma^2}{2\pi}\left(1 - \frac{\sin \lambda(m + \frac{1}{2})}{(2m + 1)\sin \frac{1}{2}\lambda}\right)^2.$$

34. (Sec. 7.6.1.) Prove that in a linear vector space to every point x there corresponds a unique point $-x$ such that $x + (-x) = 0$.

35. (Sec. 7.6.1.) Prove that $(x, x) = 0$ if and only if $x = 0$.

36. (Sec. 7.6.1.) Show that a finite-dimensional Euclidean space is a linear vector space. Show that $\boldsymbol{x}'\boldsymbol{y}$ is an inner product.

37. (Sec. 7.6.1.) Show that the set of all finite linear combinations of y_t, $t = \ldots, -1, 0, 1, \ldots,$ constitutes a linear vector space. Show that $\mathscr{E}xy$ is an inner product.

38. (Sec. 7.6.1.) Show that (29) and (32) of Section 7.2 satisfy the condition of Corollary 7.6.1.

39. (Sec. 7.6.2.) Verify that $\mathscr{M}^{\perp}$ and $\mathscr{M} \oplus \mathscr{M}^{\perp}$ are subspaces if $\mathscr{M}$ is a subspace of a Hilbert space.

40. (Sec. 7.6.3.) Verify that the spectral density of (41) is (43).

CHAPTER 8

The Sample Mean, Covariances, and Spectral Density

8.1. INTRODUCTION

A stationary Gaussian process is described completely by its mean $\mathscr{E}y_t = \mu$ and its covariance sequence $\mathscr{E}(y_t - \mu)(y_{t+h} - \mu) = \sigma(h)$, $h = 0, 1, \ldots,$ or equivalently by its mean and spectral function $F(\lambda)$, or equivalently by its mean and spectral density when the latter exists. If the process is not Gaussian, these quantities are also of importance, although they may not determine the process uniquely. In this chapter we consider sample analogues of these quantities. The mean of a sample, $y_1, \ldots, y_T$, say $\bar{y}$, is an unbiased estimate of the population mean μ. If the population mean is known, a sample covariance is defined by a sum of products $\sum_{t=1}^{T-h} (y_t - \mu)(y_{t+h} - \mu)$ divided by the number of observations T or the number of terms in the sum $T - h$. If μ is unknown, a sample covariance can be defined in terms of deviations from the sample mean $y_t - \bar{y}$, but other definitions are also considered.

Trigonometric coefficients are defined by $A(\lambda) = (2/T)\sum_{t=1}^{T} (y_t - \mu) \cos \lambda t$ and $B(\lambda) = (2/T)\sum_{t=1}^{T} (y_t - \mu) \sin \lambda t$ when μ is known; μ can be replaced by $\bar{y}$ when μ is unknown. The sample spectral density $I(\lambda)$ is a multiple of the sample squared amplitude $R^2(\lambda) = A^2(\lambda) + B^2(\lambda)$, specifically $I(\lambda) = [T/(8\pi)]R^2(\lambda)$.

We find the first- and second-order moments of these sample quantities; these moments of most of the sample quantities are complicated functions of first-, second-, and fourth-order moments of the process. In Section 8.3 we find limits of appropriately normalized moments. The sample mean, covariances, correlations, and trigonometric coefficients are asymptotically normally distributed under suitable conditions (Section 8.4).

The sample spectral density $I(\lambda)$ is distributed in the limit as χ^2 with 2 degrees of freedom multiplied by $\frac{1}{2}f(\lambda)$. Thus $I(\lambda)$ is not a consistent estimate of $f(\lambda)$. In Chapter 9 estimation of $f(\lambda)$ will be considered in greater generality.

8.2. DEFINITIONS OF THE SAMPLE MEAN, COVARIANCES, AND SPECTRAL DENSITY, AND THEIR MOMENTS

8.2.1. Definitions and Relations

Let $y_1, \ldots, y_T$ be T consecutive observations on a stochastic process $\{y_t\}$, $t = \ldots, -1, 0, 1, \ldots$, that is stationary in the wide sense. The mean and covariance sequence are

$$\mathscr{E}y_t = \mu, \qquad t = \ldots, -1, 0, 1, \ldots, \tag{1}$$

$$\begin{aligned} \operatorname{Cov}(y_t, y_{t+h}) &= \mathscr{E}(y_t - \mu)(y_{t+h} - \mu) \\ &= \sigma(h), \qquad t = \ldots, -1, 0, 1, \ldots, \\ & \qquad h = 0, \pm 1, \pm 2, \ldots, \end{aligned} \tag{2}$$

with $\sigma(h) = \sigma(-h)$. We assume the spectral function is absolutely continuous with spectral density $f(\lambda)$, $-\pi \leq \lambda \leq \pi$. Then

$$\begin{aligned} \sigma(h) &= \int_{-\pi}^{\pi} \cos \lambda h\, f(\lambda)\, d\lambda \\ &= \int_{-\pi}^{\pi} e^{i\lambda h} f(\lambda)\, d\lambda, \qquad h = 0, \pm 1, \pm 2, \ldots. \end{aligned} \tag{3}$$

If

$$\sum_{h=-\infty}^{\infty} |\sigma(h)| = \sigma(0) + 2 \sum_{h=1}^{\infty} |\sigma(h)| < \infty, \tag{4}$$

the spectral density is continuous and

$$f(\lambda) = \frac{1}{2\pi} \sum_{h=-\infty}^{\infty} \sigma(h) \cos \lambda h = \frac{1}{2\pi} \sum_{h=-\infty}^{\infty} \sigma(h) e^{i\lambda h}. \tag{5}$$

An unbiased estimate of μ is

$$\bar{y} = \frac{1}{T} \sum_{t=1}^{T} y_t. \tag{6}$$

An unbiased estimate of $\sigma(h)$ is

$$(7) \qquad C_h = C_{-h} = \frac{1}{T-h}\sum_{t=1}^{T-h}(y_t - \mu)(y_{t+h} - \mu), \qquad h = 0, 1, \ldots, T-1,$$

if μ is known. If μ is unknown, an analogue may be used as an estimate, namely,

$$(8) \qquad C_h^* = C_{-h}^* = \frac{1}{T-h}\sum_{t=1}^{T-h}(y_t - \bar{y})(y_{t+h} - \bar{y}), \qquad h = 0, 1, \ldots, T-1.$$

Other analogues are possible, such as

$$(9) \qquad \begin{aligned} \tilde{C}_h &= \tilde{C}_{-h} \\ &= \frac{1}{T-h}\sum_{t=1}^{T-h}(y_t - \bar{y}_h)(y_{t+h} - \bar{y}_{h+}) \\ &= \frac{1}{T-h}\left[\sum_{t=1}^{T-h} y_t y_{t+h} - (T-h)\bar{y}_h\bar{y}_{h+}\right], \qquad h = 0, 1, \ldots, T-2, \end{aligned}$$

where

$$(10) \qquad \bar{y}_h = \frac{1}{T-h}\sum_{t=1}^{T-h} y_t, \qquad h = 0, 1, \ldots, T-2,$$

$$(11) \qquad \bar{y}_{h+} = \frac{1}{T-h}\sum_{t=1}^{T-h} y_{t+h}, \qquad h = 0, 1, \ldots, T-2.$$

It is also convenient to consider

$$(12) \qquad \begin{aligned} c_h &= c_{-h} \\ &= \frac{1}{T}\sum_{t=1}^{T-h}(y_t - \mu)(y_{t+h} - \mu) \\ &= \frac{T-h}{T} C_h, \qquad h = 0, 1, \ldots, T-1, \end{aligned}$$

and

$$(13) \qquad \begin{aligned} c_h^* &= c_{-h}^* \\ &= \frac{1}{T}\sum_{t=1}^{T-h}(y_t - \bar{y})(y_{t+h} - \bar{y}) \\ &= \frac{T-h}{T} C_h^*, \qquad h = 0, 1, \ldots, T-1. \end{aligned}$$

Corresponding to definitions in Chapter 4 let

$$A(\lambda) = \frac{2}{T}\sum_{t=1}^{T}(y_t - \mu)\cos\lambda t \tag{14}$$

$$= \frac{2}{T}\sum_{t=1}^{T} y_t \cos\lambda t - \mu\frac{2}{T}\sum_{t=1}^{T}\cos\lambda t, \qquad -\pi \le \lambda \le \pi,$$

$$B(\lambda) = \frac{2}{T}\sum_{t=1}^{T}(y_t - \mu)\sin\lambda t \tag{15}$$

$$= \frac{2}{T}\sum_{t=1}^{T} y_t \sin\lambda t - \mu\frac{2}{T}\sum_{t=1}^{T}\sin\lambda t, \qquad -\pi \le \lambda \le \pi,$$

$$R^2(\lambda) = A^2(\lambda) + B^2(\lambda), \qquad -\pi \le \lambda \le \pi. \tag{16}$$

The last, $R^2(\lambda)$, is the sample spectrogram. Note that $A(\lambda) = A(-\lambda)$, $B(\lambda) = -B(-\lambda)$, and $R^2(\lambda) = R^2(-\lambda)$. We may also define

$$A^*(\lambda) = \frac{2}{T}\sum_{t=1}^{T}(y_t - \bar{y})\cos\lambda t \tag{17}$$

$$= \frac{2}{T}\sum_{t=1}^{T} y_t \cos\lambda t - \bar{y}\frac{2}{T}\sum_{t=1}^{T}\cos\lambda t, \qquad -\pi \le \lambda \le \pi,$$

$$B^*(\lambda) = \frac{2}{T}\sum_{t=1}^{T}(y_t - \bar{y})\sin\lambda t \tag{18}$$

$$= \frac{2}{T}\sum_{t=1}^{T} y_t \sin\lambda t - \bar{y}\frac{2}{T}\sum_{t=1}^{T}\sin\lambda t, \qquad -\pi \le \lambda \le \pi,$$

$$R^{*2}(\lambda) = A^{*2}(\lambda) + B^{*2}(\lambda), \qquad -\pi \le \lambda \le \pi. \tag{19}$$

Because of the orthogonality properties of $\cos 2\pi jt/T$ and $\sin 2\pi jt/T$, $t = 1, \ldots, T$,

$$A\left(\frac{2\pi j}{T}\right) = A^*\left(\frac{2\pi j}{T}\right),$$

$$B\left(\frac{2\pi j}{T}\right) = B^*\left(\frac{2\pi j}{T}\right), \tag{20}$$

$$R^2\left(\frac{2\pi j}{T}\right) = R^{*2}\left(\frac{2\pi j}{T}\right), \qquad j = 1, \ldots, [\tfrac{1}{2}T],$$

and $A^*(0) = B(0) = B^*(0) = 0$, $A(0) = 2(\bar{y} - \mu)$.

The sample spectral density is

$$I(\lambda) = \frac{T}{8\pi} R^2(\lambda) = \frac{1}{2\pi T}\left|\sum_{t=1}^{T} (y_t - \mu)e^{i\lambda t}\right|^2, \qquad -\pi \le \lambda \le \pi, \tag{21}$$

$[I(\lambda) = I(-\lambda),\ -\pi \le \lambda \le \pi]$ when μ is known. When μ is not known, we define

$$I^*(\lambda) = \frac{T}{8\pi} R^{*2}(\lambda) = \frac{1}{2\pi T}\left|\sum_{t=1}^{T} (y_t - \bar{y})e^{i\lambda t}\right|^2, \qquad -\pi \le \lambda \le \pi, \tag{22}$$

$[I^*(\lambda) = I^*(-\lambda),\ -\pi \le \lambda \le \pi]$ although other definitions may be considered [corresponding to (9) for the covariance sequence]. $I(\lambda)$ and $I^*(\lambda)$ are to be considered as estimates of $f(\lambda)$.

THEOREM 8.2.1.

$$I(\lambda) = \frac{1}{2\pi} \sum_{r=-(T-1)}^{T-1} c_r \cos \lambda r, \qquad -\pi \le \lambda \le \pi, \tag{23}$$

$$I^*(\lambda) = \frac{1}{2\pi} \sum_{r=-(T-1)}^{T-1} c_r^* \cos \lambda r, \qquad -\pi \le \lambda \le \pi. \tag{24}$$

PROOF. We have

$$\begin{aligned} I(\lambda) &= \frac{T}{8\pi}\left(\frac{2}{T}\right)^2 \left[\sum_{t,s=1}^{T} (y_t - \mu)(y_s - \mu) \cos \lambda t \cos \lambda s \right. \\ &\qquad \left. + \sum_{t,s=1}^{T} (y_t - \mu)(y_s - \mu) \sin \lambda t \sin \lambda s \right] \\ &= \frac{1}{2\pi T} \sum_{t,s=1}^{T} (y_t - \mu)(y_s - \mu) \cos \lambda(t - s) \\ &= \frac{1}{2\pi T} \sum_{r=-(T-1)}^{T-1} \sum_{s \in S_r} (y_{s+r} - \mu)(y_s - \mu) \cos \lambda r, \end{aligned} \tag{25}$$

where $S_r = \{1, 2, \ldots, T - r\}$, $r \ge 0$, and $S_r = \{1 - r, 2 - r, \ldots, T\}$, $r \le 0$. Thus (23) follows. Replacement of μ by $\bar{y}$ gives (24).

THEOREM 8.2.2.

$$\begin{aligned} \int_{-\pi}^{\pi} \cos \lambda h\, I(\lambda)\, d\lambda &= c_h, \qquad & h &= 0, \pm 1, \ldots, \pm(T-1), \\ &= 0, & h &= \pm T, \pm(T+1), \ldots; \end{aligned} \tag{26}$$

$$\begin{aligned} \int_{-\pi}^{\pi} \cos \lambda h\, I^*(\lambda)\, d\lambda &= c_h^*, \qquad & h &= 0, \pm 1, \ldots, \pm(T-1), \\ &= 0, & h &= \pm T, \pm(T+1), \ldots. \end{aligned} \tag{27}$$

PROOF. These equations follow from the orthogonality properties of $\cos \lambda h$ and Theorem 8.2.1.

The covariances and spectral density are related in the sample in the same way as in the population; (23) and (24) are analogues of (5), and (26) and (27) correspond to (3).

8.2.2. Moments of the Sample Mean and Covariances

The mean and variance of an arbitrary linear combination

$$L = \sum_{t=1}^{T} k_t y_t \tag{28}$$

are

$$\mathscr{E} L = \mu \sum_{t=1}^{T} k_t, \tag{29}$$

$$\begin{aligned} \operatorname{Var} L &= \mathscr{E}\left[\sum_{t=1}^{T} k_t (y_t - \mu)\right]^2 \\ &= \sum_{t,s=1}^{T} \sigma(t-s) k_t k_s \\ &= \sum_{r=-(T-1)}^{T-1} \sigma(r) \sum_{s \in S_r} k_{s+r} k_s \\ &= \int_{-\pi}^{\pi} \left| \sum_{t=1}^{T} k_t e^{i\lambda t} \right|^2 f(\lambda)\, d\lambda, \end{aligned} \tag{30}$$

where S_r is defined in the proof of Theorem 8.2.1. In order for L to be an unbiased estimate of μ

$$\sum_{t=1}^{T} k_t = 1. \tag{31}$$

For a given covariance matrix $\boldsymbol{\Sigma} = [\sigma(t-s)]$, the variance of L satisfying (31) is minimized by $\boldsymbol{k} = (k_1, \ldots, k_T)'$ defined by $\boldsymbol{k} = [1/(\boldsymbol{\varepsilon}'\boldsymbol{\Sigma}^{-1}\boldsymbol{\varepsilon})]\boldsymbol{\Sigma}^{-1}\boldsymbol{\varepsilon}$, where $\boldsymbol{\varepsilon} = (1, \ldots, 1)'$. This will define the mean if $\boldsymbol{k} = (1/T)\boldsymbol{\varepsilon}$, that is, if $\boldsymbol{\varepsilon}$ is a characteristic vector of $\boldsymbol{\Sigma}$, that is, if $\sum_{s=1}^{T} \sigma(t-s)$ does not depend on t. (See Problems 24 and 25 of Chapter 6.) In general, this will not be the case.

THEOREM 8.2.3. *The variance of the mean $\bar{y}$ is*

$$\begin{aligned}\text{Var}\,\bar{y} &= \frac{1}{T^2}\sum_{t,s=1}^{T}\sigma(t-s) \\ &= \frac{1}{T}\sum_{r=-(T-1)}^{T-1}\left(1-\frac{|r|}{T}\right)\sigma(r) \\ &= \int_{-\pi}^{\pi}\left|\frac{1}{T}\sum_{t=1}^{T}e^{i\lambda t}\right|^2 f(\lambda)\,d\lambda \\ &= \int_{-\pi}^{\pi}\left(\frac{\sin\frac{1}{2}\lambda T}{T\sin\frac{1}{2}\lambda}\right)^2 f(\lambda)\,d\lambda.\end{aligned} \tag{32}$$

We have used $\sum_{t=1}^{T} e^{i\lambda t} = e^{i\lambda}(1-e^{i\lambda T})/(1-e^{i\lambda}) = e^{i\frac{1}{2}\lambda(T+1)}\sin\frac{1}{2}\lambda T/\sin\frac{1}{2}\lambda$.

An estimate of a covariance or of the spectral density at a given point will usually be a quadratic form

$$Q = \sum_{s,t=1}^{T} a_{st}(y_s-\mu)(y_t-\mu). \tag{33}$$

If μ is unknown, the matrix (a_{st}) will be chosen so Q does not depend on μ (that is, $\sum_{t=1}^{T} a_{st} = 0,\ s = 1, \ldots, T$).

The variance of a quadratic form will involve the fourth-order moments. We shall assume that the fourth-order moments exist; that is, $\mathscr{E}(y_t-\mu)^4 < \infty$. We also assume that the fourth-order moments correspond to stationarity, that is, that

$$\begin{aligned}\mathscr{E}(y_t-\mu)(y_{t+h}-\mu)(y_{t+r}-\mu)(y_{t+s}-\mu)& \\ = \mathscr{E}(y_0-\mu)(y_h-\mu)(y_r-\mu)(y_s-\mu)&, \\ t = \ldots, -1, 0, 1, \ldots&.\end{aligned} \tag{34}$$

Let (34) be $\nu(h, r, s)$. If $\{y_t\}$ is Gaussian, this fourth-order moment is

$$\sigma(h)\sigma(r-s) + \sigma(r)\sigma(h-s) + \sigma(s)\sigma(h-r). \tag{35}$$

In general, let

$$\begin{aligned}\kappa(h, r, s) &= \nu(h, r, s) - [\sigma(h)\sigma(r-s) + \sigma(r)\sigma(h-s) + \sigma(s)\sigma(h-r)] \\ &= \mathscr{E}(y_t-\mu)(y_{t+h}-\mu)(y_{t+r}-\mu)(y_{t+s}-\mu) \\ &\qquad - [\sigma(h)\sigma(r-s) + \sigma(r)\sigma(h-s) + \sigma(s)\sigma(h-r)], \\ &\qquad\qquad t = \ldots, -1, 0, 1, \ldots,\end{aligned} \tag{36}$$

which is the fourth-order cumulant. [See Problem 2 for symmetries in $\nu(h, r, s)$ and $\kappa(h, r, s)$.]

THEOREM 8.2.4. *The mean and variance of the quadratic form Q defined in (33) are*

$$\mathscr{E}Q = \sum_{s,t=1}^{T} a_{st}\sigma(t-s) \tag{37}$$

$$= \sum_{r=-(T-1)}^{T-1} \sigma(r) \sum_{s\in S_r} a_{s,s+r}$$

$$= \int_{-\pi}^{\pi} \sum_{s,t=1}^{T} a_{st} e^{i\lambda(t-s)} f(\lambda)\, d\lambda$$

$$= \int_{-\pi}^{\pi} A(\lambda, \lambda) f(\lambda)\, d\lambda,$$

$$\operatorname{Var} Q = \sum_{s,t,s',t'=1}^{T} a_{st}a_{s't'}[2\sigma(s-s')\sigma(t-t') + \kappa(t-s, s'-s, t'-s)] \tag{38}$$

$$= 2 \sum_{r,p=-(T-1)}^{T-1} \sigma(r)\sigma(p) \sum_{s\in S_r} \sum_{t\in S_p} a_{st}a_{s+r,t+p}$$

$$+ \sum_{r,p=-(T-1)}^{T-1} \sum_{s\in S_r} \sum_{t\in S_p} a_{st}a_{s+r,t+p}\kappa(t-s, r, t-s+p)$$

$$= 2\int_{-\pi}^{\pi}\int_{-\pi}^{\pi} \left| \sum_{s,t=1}^{T} a_{st} e^{i(\lambda s - \nu t)} \right|^2 f(\lambda) f(\nu)\, d\lambda\, d\nu$$

$$+ \sum_{s,t,s',t'=1}^{T} a_{st}a_{s't'}\kappa(t-s, s'-s, t'-s)$$

$$= 2\int_{-\pi}^{\pi}\int_{-\pi}^{\pi} |A(\lambda, \nu)|^2 f(\lambda) f(\nu)\, d\lambda\, d\nu$$

$$+ \sum_{s,t,s',t'=1}^{T} a_{st}a_{s't'}\kappa(t-s, s'-s, t'-s),$$

where

$$A(\lambda, \nu) = \sum_{s,t=1}^{T} a_{st} e^{i(\lambda s - \nu t)} \tag{39}$$

and $S_r = \{1, 2, \ldots, T - r\}$, $r \geq 0$, *and* $S_r = \{1 - r, 2 - r, \ldots, T\}$, $r \leq 0$.

PROOF. The alternative forms of $\mathscr{E}Q$ follow directly. The expected value of Q^2 is

(40)

$$\begin{aligned}\mathscr{E}Q^2 &= \sum_{s,t,s',t'=1}^{T} a_{st}a_{s't'}\mathscr{E}(y_s - \mu)(y_t - \mu)(y_{s'} - \mu)(y_{t'} - \mu) \\ &= \sum_{s,t,s',t'=1}^{T} a_{st}a_{s't'}[\sigma(s - t)\sigma(s' - t') \\ &\quad + \sigma(s - s')\sigma(t - t') + \sigma(s - t')\sigma(t - s') + \kappa(t - s, s' - s, t' - s)].\end{aligned}$$

We observe that

$$\sum_{s,t,s',t'=1}^{T} a_{st}a_{s't'}\sigma(s - s')\sigma(t - t') = \sum_{s,t,s',t'=1}^{T} a_{st}a_{s't'}\sigma(s - t')\sigma(t - s') \tag{41}$$

because $a_{s't'} = a_{t's'}$. The first two forms of Var Q follow directly. The spectral representation of $\sigma(h)$ yields

$$\begin{aligned}\text{Var } Q = 2\int_{-\pi}^{\pi}\int_{-\pi}^{\pi} &\sum_{s,t,s',t'=1}^{T} a_{st}a_{s't'}e^{i[\lambda(s-s')-\nu(t-t')]}f(\lambda)f(\nu)\,d\lambda\,d\nu \\ &+ \sum_{s,t,s',t'=1}^{T} a_{st}a_{s't'}\kappa(t - s, s' - s, t' - s),\end{aligned} \tag{42}$$

since $f(\nu) = f(-\nu)$. This yields the last two forms of Var Q.

Let

$$P = \sum_{s,t=1}^{T} b_{st}(y_s - \mu)(y_t - \mu). \tag{43}$$

THEOREM 8.2.5.

$$(44)\quad \operatorname{Cov}(Q, P) = \sum_{s,t,s,t'=1}^{T} a_{st} b_{s't'}[2\sigma(s-s')\sigma(t-t') + \kappa(t-s, s'-s, t'-s)]$$

$$= 2\sum_{r,p=-(T-1)}^{T-1} \sigma(r)\sigma(p) \sum_{s\in S_r}\sum_{t\in S_p} a_{st} b_{s+r,t+p}$$

$$+ \sum_{r,p=-(T-1)}^{T-1} \sum_{s\in S_r}\sum_{t\in S_p} a_{st} b_{s+r,t+p} \kappa(t-s, r, t-s+p)$$

$$= 2\int_{-\pi}^{\pi}\int_{-\pi}^{\pi} \sum_{s,t=1}^{T} a_{st} e^{i(\lambda s - \nu t)} \sum_{s',t'=1}^{T} b_{s't'} e^{-i(\lambda s' - \nu t')} f(\lambda) f(\nu)\, d\lambda\, d\nu$$

$$+ \sum_{s,t,s',t'=1}^{T} a_{st} b_{s't'} \kappa(t-s, s'-s, t'-s)$$

$$= 2\int_{-\pi}^{\pi}\int_{-\pi}^{\pi} A(\lambda, \nu)\, \bar{B}(\lambda, \nu) f(\lambda) f(\nu)\, d\lambda\, d\nu$$

$$+ \sum_{s,t,s',t'=1}^{T} a_{st} b_{s't'} \kappa(t-s, s'-s, t'-s),$$

where

$$(45)\quad \bar{B}(\lambda, \nu) = \sum_{s',t'=1}^{T} b_{s't'} e^{-i(\lambda s' - \nu t')}.$$

PROOF.

$$(46)\quad \mathscr{E}QP = \sum_{s,t,s',t'=1}^{T} a_{st} b_{s't'} \mathscr{E}(y_s - \mu)(y_t - \mu)(y_{s'} - \mu)(y_{t'} - \mu)$$

$$= \sum_{s,t,s',t'=1}^{T} a_{st} b_{s't'}[\sigma(s-t)\sigma(s'-t') + \sigma(s-s')\sigma(t-t')$$

$$+ \sigma(s-t')\sigma(t-s') + \kappa(t-s, s'-s, t'-s)];$$

from this follow the first two forms of Cov (Q, P). We also find

$$(47)\quad \operatorname{Cov}(Q, P) = 2\int_{-\pi}^{\pi}\int_{-\pi}^{\pi} \sum_{s,t,s',t'=1}^{T} a_{st} b_{s't'}\, e^{i[\lambda(s-s') - \nu(t-t')]} f(\lambda) f(\nu)\, d\lambda\, d\nu$$

$$+ \sum_{s,t,s',t'=1}^{T} a_{st} b_{s't'} \kappa(t-s, s'-s, t'-s);$$

this gives the last two forms of Cov (Q, P). Note that Theorems 8.2.4 and 8.2.5 are generalizations of Lemmas 3.4.3, 3.4.4, and 3.4.5.

If the process is Gaussian, the fourth-order cumulants $\kappa(h, r, s) = 0$. (They may be 0 even if the process is not Gaussian.)

We now turn to the first- and second-order moments of the estimates of $\sigma(h)$. Rather than use Theorems 8.2.4 and 8.2.5, we shall evaluate the means, variances, and covariances directly (because each covariance can be written as a single sum instead of a double sum).

The estimate C_h of $\sigma(h)$ when μ is known is unbiased, for

$$\mathscr{E}C_h = \frac{1}{T-h}\sum_{t=1}^{T-h}\mathscr{E}(y_t-\mu)(y_{t+h}-\mu) = \sigma(h), \qquad h = 0, 1, \ldots, T-1, \tag{48}$$

and hence

$$\mathscr{E}c_h = \frac{T-h}{T}\sigma(h) = \left(1-\frac{h}{T}\right)\sigma(h), \qquad h = 0, 1, \ldots, T-1. \tag{49}$$

The bias of c_h is small for h small and is relatively large for h large (when the relative sampling variability is great).

The expected value of C_h^* $(h \geq 0)$ is

(50)

$$\begin{aligned}\mathscr{E}C_h^* &= \frac{1}{T-h}\mathscr{E}\left\{\sum_{t=1}^{T-h}(y_t-\mu)(y_{t+h}-\mu) - \frac{1}{T}\sum_{t=1}^{T-h}\sum_{s=1}^{T}(y_t-\mu)(y_s-\mu)\right.\\ &\qquad \left. -\frac{1}{T}\sum_{t=1}^{T-h}\sum_{s=1}^{T}(y_{t+h}-\mu)(y_s-\mu) + \frac{T-h}{T^2}\sum_{t,s=1}^{T}(y_t-\mu)(y_s-\mu)\right\}\\ &= \sigma(h) - \frac{1}{T(T-h)}\sum_{t=1}^{T-h}\sum_{s=1}^{T}[\sigma(t-s)+\sigma(t+h-s)] + \frac{1}{T^2}\sum_{t,s=1}^{T}\sigma(t-s).\end{aligned}$$

Straightforward, but laborious, calculation yields

$$\mathscr{E}C_0^* = \sigma(0) - \frac{1}{T}\left\{\sigma(0) + 2\sum_{r=1}^{T-1}\left(1-\frac{r}{T}\right)\sigma(r)\right\}, \tag{51}$$

$$\begin{aligned}\mathscr{E}C_h^* &= \sigma(h) - \frac{1}{T}\left\{\sigma(0) + 2\sum_{r=1}^{h}\left[1-\frac{rh}{T(T-h)}\right]\sigma(r)\right.\\ &\qquad \left.+ 2\sum_{r=h+1}^{T-h-1}\left[1-\frac{rh}{T(T-h)}-\frac{r-h}{T-h}\right]\sigma(r) + 2\sum_{r=T-h}^{T-1}\frac{(T-r)h}{T(T-h)}\sigma(r)\right\},\\ &\qquad\qquad 1 \leq h < T-h-1,\end{aligned} \tag{52}$$

$$(53)\quad \mathscr{E}C_h^* = \sigma(h) - \frac{1}{T}\left\{\sigma(0) + 2\sum_{r=1}^{h}\left[1 - \frac{rh}{T(T-h)}\right]\sigma(r) + 2\sum_{r=h+1}^{T-1}\frac{(T-r)h}{T(T-h)}\sigma(r)\right\},$$
$$1 \leq h = T - h - 1,$$

$$(54)\quad \mathscr{E}C_h^* = \sigma(h) - \frac{1}{T}\left\{\sigma(0) + 2\sum_{r=1}^{T-h-1}\left[1 - \frac{rh}{T(T-h)}\right]\sigma(r) + 2\sum_{r=T-h}^{h}\frac{r}{T}\sigma(r) + 2\sum_{r=h+1}^{T-1}\frac{(T-r)h}{T(T-h)}\sigma(r)\right\},$$
$$T - h - 1 < h < T - 1,$$

$$(55)\quad \mathscr{E}C_{T-1}^* = \sigma(T-1) - \frac{1}{T}\left\{\sigma(0) + 2\sum_{r=1}^{T-1}\frac{r}{T}\sigma(r)\right\}.$$

The estimates are biased, but the bias is of order $1/T$, as shall be shown in Section 8.3.

The expected value of C_h^* can also be expressed in terms of the spectral density. From (50) we have

$$(56)\quad \mathscr{E}C_h^* = \int_{-\pi}^{\pi}\left\{\cos\lambda h - \frac{1}{T(T-h)}\sum_{t=1}^{T-h}\sum_{s=1}^{T}[e^{i\lambda(t-s)} + e^{i\lambda(t-s+h)}] + \frac{1}{T^2}\sum_{s,t=1}^{T}e^{i\lambda(t-s)}\right\}f(\lambda)\,d\lambda$$
$$= \int_{-\pi}^{\pi}\left\{\cos\lambda h - 2\frac{\sin\frac{1}{2}\lambda T\sin\frac{1}{2}\lambda(T-h)}{T\sin\frac{1}{2}\lambda\,(T-h)\sin\frac{1}{2}\lambda}\cos\tfrac{1}{2}\lambda h + \left(\frac{\sin\frac{1}{2}\lambda T}{T\sin\frac{1}{2}\lambda}\right)^2\right\}f(\lambda)\,d(\lambda).$$

The expected value of $\tilde{C}_h$ is

$$(57)\quad \mathscr{E}\tilde{C}_h = \frac{1}{T-h}\mathscr{E}\sum_{t=1}^{T-h}[(y_t-\mu) - (\bar{y}_h - \mu)][(y_{t+h}-\mu) - (\bar{y}_{h+} - \mu)]$$
$$= \frac{1}{T-h}\mathscr{E}\left[\sum_{t=1}^{T-h}(y_t-\mu)(y_{t+h}-\mu) - (T-h)(\bar{y}_h-\mu)(\bar{y}_{h+}-\mu)\right]$$
$$= \sigma(h) - \frac{1}{(T-h)^2}\mathscr{E}\sum_{t=1}^{T-h}\sum_{s=h+1}^{T}(y_t-\mu)(y_s-\mu)$$
$$= \sigma(h) - \frac{1}{(T-h)^2}\sum_{t=1}^{T-h}\sum_{s=h+1}^{T}\sigma(t-s).$$

Since $\tilde{C}_0 = C_0^*$, $\mathscr{E}\tilde{C}_0$ is given by (51). Evaluation of (57) yields

$$\mathscr{E}\tilde{C}_h = \sigma(h) - \frac{1}{T-h}\left\{\left(1 - \frac{h}{T-h}\right)\sigma(0) + 2\left(1 - \frac{h}{T-h}\right)\sum_{r=1}^{h}\sigma(r) + \sum_{r=h+1}^{T-2h-1} 2\left(1 - \frac{r}{T-h}\right)\sigma(r) + \sum_{r=T-2h}^{T-1}\left(1 - \frac{r-h}{T-h}\right)\sigma(r)\right\}, \quad 1 \le h < T - 2h - 1. \tag{58}$$

Other cases are considered in Problem 10.

We can also express these expected values in terms of the spectral density as

$$\begin{aligned}\mathscr{E}\tilde{C}_h &= \int_{-\pi}^{\pi}\left[\cos \lambda h - \frac{1}{(T-h)^2}\sum_{t=1}^{T-h}\sum_{s=h+1}^{T} e^{i\lambda(t-s)}\right] f(\lambda)\, d\lambda \\ &= \int_{-\pi}^{\pi} \cos \lambda h \left\{1 - \left[\frac{\sin \frac{1}{2}\lambda(T-h)}{(T-h)\sin \frac{1}{2}\lambda}\right]^2\right\} f(\lambda)\, d\lambda.\end{aligned} \tag{59}$$

We now turn to the evaluation of the variances and covariances of these estimates of $\sigma(0), \sigma(1), \ldots, \sigma(T-1)$.

THEOREM 8.2.6.

(60)

$$\begin{aligned}&(T-h)(T-g)\operatorname{Cov}(C_h, C_g) \\ &= \int_{-\pi}^{\pi}\int_{-\pi}^{\pi} \frac{\sin \frac{1}{2}(\lambda-\nu)(T-h)}{\sin \frac{1}{2}(\lambda-\nu)} \frac{\sin \frac{1}{2}(\lambda-\nu)(T-g)}{\sin \frac{1}{2}(\lambda-\nu)} [e^{i\frac{1}{2}(\lambda+\nu)(g-h)} + e^{-i\frac{1}{2}(\lambda+\nu)(g+h)}] \\ &\quad \times f(\lambda)f(\nu)\, d\lambda\, d\nu + \sum_{t=1}^{T-h}\sum_{s=1}^{T-g} \kappa(h, s-t, s-t+g) \\ &= 2\int_{-\pi}^{\pi}\int_{-\pi}^{\pi} \frac{\sin \frac{1}{2}(\lambda-\nu)(T-h)\sin \frac{1}{2}(\lambda-\nu)(T-g)}{\sin^2 \frac{1}{2}(\lambda-\nu)} \\ &\quad \times \cos \tfrac{1}{2}(\lambda+\nu)h \cos \tfrac{1}{2}(\lambda+\nu)g\, f(\lambda)f(\nu)\, d\lambda\, d\nu \\ &\quad + \sum_{t=1}^{T-h}\sum_{s=1}^{T-g} \kappa(h, s-t, s-t+g) \\ &= \sum_{r=\max(g-h,0)+1}^{T-h-1} (T-h-r)[\sigma(r)\sigma(r+h-g) \\ &\quad + \sigma(r-g)\sigma(r+h) + \kappa(h, -r, g-r)] \\ &\quad + \sum_{r=\min(g-h,0)}^{\max(g-h,0)} [T - \max(g,h)][\sigma(r)\sigma(r+h-g) \\ &\quad + \sigma(r-g)\sigma(r+h) + \kappa(h, -r, g-r)] \\ &\quad + \sum_{r=-(T-g-1)}^{\min(g-h,0)-1} (T-g-|r|)[\sigma(r)\sigma(r+h-g) \\ &\quad + \sigma(r-g)\sigma(r+h) + \kappa(h, -r, g-r)], \qquad 0 \le h < T, \quad 0 \le g < T.\end{aligned}$$

PROOF. We have

$$(61)\quad (T-h)(T-g)\mathscr{E}C_hC_g$$

$$= \mathscr{E}\sum_{t=1}^{T-h}\sum_{s=1}^{T-g}(y_t-\mu)(y_{t+h}-\mu)(y_s-\mu)(y_{s+g}-\mu)$$

$$= \sum_{t=1}^{T-h}\sum_{s=1}^{T-g}\nu(h, s-t, s-t+g)$$

$$= \sum_{t=1}^{T-h}\sum_{s=1}^{T-g}[\sigma(h)\sigma(g) + \sigma(t-s)\sigma(t+h-s-g)$$

$$+ \sigma(t-s-g)\sigma(t+h-s) + \kappa(h, s-t, s-t+g)],$$

$$0 \le h < T, \quad 0 \le g < T,$$

and

$$(62)\quad (T-h)(T-g)\operatorname{Cov}(C_h, C_g)$$

$$= \sum_{t=1}^{T-h}\sum_{s=1}^{T-g}[\sigma(t-s)\sigma(t+h-s-g)$$

$$+ \sigma(t-s-g)\sigma(t+h-s) + \kappa(h, s-t, s-t+g)]$$

$$= \int_{-\pi}^{\pi}\int_{-\pi}^{\pi}\sum_{t=1}^{T-h}\sum_{s=1}^{T-g}[e^{i\lambda(t-s)-i\nu(t-s+h-g)}$$

$$+ e^{i\lambda(t-s-g)-i\nu(t-s+h)}]f(\lambda)f(\nu)\,d\lambda\,d\nu$$

$$+ \sum_{t=1}^{T-h}\sum_{s=1}^{T-g}\kappa(h, s-t, s-t+g)$$

$$= \int_{-\pi}^{\pi}\int_{-\pi}^{\pi}\frac{\sin\frac{1}{2}(\lambda-\nu)(T-h)\sin\frac{1}{2}(\lambda-\nu)(T-g)}{\sin\frac{1}{2}(\lambda-\nu)\sin\frac{1}{2}(\lambda-\nu)}$$

$$\times [e^{i\frac{1}{2}(\lambda+\nu)(g-h)} + e^{-i\frac{1}{2}(\lambda+\nu)(g+h)}]f(\lambda)f(\nu)\,d\lambda\,d\nu$$

$$+ \sum_{t=1}^{T-h}\sum_{s=1}^{T-g}\kappa(h, s-t, s-t+g),$$

and the other forms of (60) follow.

In particular

(63)

$$(T-h)\operatorname{Var} C_h$$

$$= \sum_{r=-(T-h-1)}^{T-h-1} \left(1 - \frac{|r|}{T-h}\right)[\sigma^2(r) + \sigma(r+h)\sigma(r-h) + \kappa(h, -r, h-r)]$$

$$= \frac{1}{T-h}\int_{-\pi}^{\pi}\int_{-\pi}^{\pi}\left[\frac{\sin\frac{1}{2}(\lambda-\nu)(T-h)}{\sin\frac{1}{2}(\lambda-\nu)}\right]^2[1+e^{-i(\nu+\lambda)h}]f(\lambda)f(\nu)\,d\lambda\,d\nu$$

$$+ \sum_{t,s=1}^{T-h}\kappa(h, s-t, s-t+h)$$

$$= 2\int_{-\pi}^{\pi}\int_{-\pi}^{\pi}\frac{\sin^2\frac{1}{2}(\lambda-\nu)(T-h)}{(T-h)\sin^2\frac{1}{2}(\lambda-\nu)}\cos^2\tfrac{1}{2}(\lambda+\nu)h\,f(\lambda)f(\nu)\,d\lambda\,d\nu$$

$$+ \sum_{t,s=1}^{T-h}\kappa(h, s-t, s-t+h), \qquad h = 0, 1, \ldots, T-1.$$

If the process is Gaussian, the fourth-order cumulants vanish in the above expressions.

Evaluation of the covariances of the sample covariances when deviations from the sample mean are used is more difficult. We consider

(64)
$$\tilde{C}_h = \frac{1}{T-h}\sum_{t=1}^{T-h}(y_t - \bar{y}_h)(y_{t+h} - \bar{y}_{h+})$$

$$= \frac{1}{T-h}\left[\sum_{t=1}^{T-h} y_t y_{t+h} - \frac{1}{T-h}\sum_{t,s=1}^{T-h} y_t y_{s+h}\right].$$

Then

(65)

$$(T-h)(T-g)\operatorname{Cov}(\tilde{C}_h, \tilde{C}_g)$$

$$= \sum_{t=1}^{T-h}\sum_{t'=1}^{T-g}[\sigma(t-t')\sigma(t+h-t'-g)$$

$$+ \sigma(t-t'-g)\sigma(t+h-t') + \kappa(h, t'-t, t'-t+g)]$$

$$- \frac{1}{T-g}\sum_{t=1}^{T-h}\sum_{t',s'=1}^{T-g}[\sigma(t-t')\sigma(t+h-s'-g)$$

$$+ \sigma(t-s'-g)\sigma(t+h-t') + \kappa(h, t'-t, s'-t+g)]$$

$$-\frac{1}{T-h}\sum_{t,s=1}^{T-h}\sum_{t'=1}^{T-g}[\sigma(t-t')\sigma(s+h-t'-g)$$

$$+\sigma(t-t'-g)\sigma(s+h-t')+\kappa(s+h-t,t'-t,t'+g-t)]$$

$$+\frac{1}{(T-h)(T-g)}\sum_{t,s=1}^{T-h}\sum_{t',s'=1}^{T-g}[\sigma(t-t')\sigma(s+h-s'-g)$$

$$+\sigma(t-s'-g)\sigma(s+h-t')+\kappa(s+h-t,t'-t,s'+g-t)]$$

$$=\int_{-\pi}^{\pi}\int_{-\pi}^{\pi}\left[\frac{\sin\frac{1}{2}(\lambda-\nu)(T-h)}{\sin\frac{1}{2}(\lambda-\nu)}-\frac{\sin\frac{1}{2}\lambda(T-h)\sin\frac{1}{2}\nu(T-h)}{(T-h)\sin\frac{1}{2}\lambda\sin\frac{1}{2}\nu}\right]$$

$$\times\left[\frac{\sin\frac{1}{2}(\lambda-\nu)(T-g)}{\sin\frac{1}{2}(\lambda-\nu)}-\frac{\sin\frac{1}{2}\lambda(T-g)\sin\frac{1}{2}\nu(T-g)}{(T-g)\sin\frac{1}{2}\lambda\sin\frac{1}{2}\nu}\right]$$

$$\times[e^{i\frac{1}{2}(\lambda+\nu)(g-h)}+e^{-i\frac{1}{2}(\lambda+\nu)(g+h)}]f(\lambda)f(\nu)\,d\lambda\,d\nu$$

$$+\sum_{t=1}^{T-h}\sum_{t'=1}^{T-g}\Bigg[\kappa(h,t'-t,t'-t+g)-\frac{1}{T-g}\sum_{s'=1}^{T-g}\kappa(h,t'-t,s'-t+g)$$

$$-\frac{1}{T-h}\sum_{s=1}^{T-h}\kappa(s+h-t,t'-t,t'+g-t)$$

$$+\frac{1}{(T-h)(T-g)}\sum_{s=1}^{T-h}\sum_{s'=1}^{T-g}\kappa(s+h-t,t'-t,s'+g-t)\Bigg],$$

$$0\le h<T,\quad 0\le g<T.$$

(See Problem 11.) Covariances of C_h^* can be found similarly.

8.2.3. Moments of the Sample Spectral Density

Now let us consider the first- and second-order moments of the sample spectral density $I(\lambda)$ defined by (21) when the mean μ is known. The first-order moment is

$$(66)\quad \mathscr{E}I(\lambda)=\frac{1}{2\pi}\sum_{r=-(T-1)}^{T-1}\left(1-\frac{|r|}{T}\right)\sigma(r)\cos\lambda r$$

$$=\frac{1}{2\pi}\int_{-\pi}^{\pi}\sum_{r=-(T-1)}^{T-1}\left(1-\frac{|r|}{T}\right)\cos\nu r\cos\lambda r\,f(\nu)\,d\nu,\qquad -\pi\le\lambda\le\pi,$$

since $\mathscr{E}c_r = (T - |r|)\sigma(r)/T$. This can also be written

$$\mathscr{E}I(\lambda) = \frac{1}{2\pi T}\int_{-\pi}^{\pi}\sum_{t,s=1}^{T}\cos\nu(t-s)\,\cos\lambda(t-s)\,f(\nu)\,d\nu \tag{67}$$

$$= \frac{1}{2\pi T}\int_{-\pi}^{\pi}\left|\sum_{t=1}^{T}e^{i(\nu-\lambda)t}\right|^2 f(\nu)\,d\nu$$

$$= \frac{1}{2\pi T}\int_{-\pi}^{\pi}\left[\frac{\sin\frac{1}{2}(\nu-\lambda)T}{\sin\frac{1}{2}(\nu-\lambda)}\right]^2 f(\nu)\,d\nu.$$

THEOREM 8.2.7.

$$\mathscr{E}I(\lambda) = \int_{-\pi}^{\pi}k_T(\nu-\lambda)f(\nu)\,d\nu, \tag{68}$$

where

$$k_T(\nu) = \frac{\sin^2\frac{1}{2}\nu T}{2\pi T\sin^2\frac{1}{2}\nu}. \tag{69}$$

The kernel $k_T(\nu)$ is symmetric, has its maximum at $\nu = 0$, and is periodic with period 2π; $k_T(0) = T/(2\pi)$. The kernel is called Fejér's kernel [Lanczos (1956), Chapter IV, Section 2, for example]. The expected value of $I(\lambda)$ is not $f(\lambda)$, but a weighted integral of $f(\lambda)$ [or $g(\lambda)$]. Since $f(\lambda)$ is an infinite series with $\sigma(r)$, $r = 0, 1, \ldots$, as coefficients, in general $I(\lambda)$ cannot be an unbiased estimate since $I(\lambda)$ involves only T of the c_r's. We shall show in the next section that $\lim_{T\to\infty}\mathscr{E}I(\lambda) = f(\lambda)$.

We note also that

$$\mathscr{E}I^*(\lambda) = \frac{1}{2\pi}\sum_{h=-(T-1)}^{T-1}\left(1-\frac{|h|}{T}\right)\mathscr{E}C_h^*\cos\lambda h \tag{70}$$

$$= \frac{1}{2\pi}\sum_{h=-(T-1)}^{T-1}\left(1-\frac{|h|}{T}\right)\sigma(h)\cos\lambda h$$

$$- \frac{1}{2\pi T}\sum_{h=-(T-1)}^{T-1}\left(1-\frac{|h|}{T}\right)\cos\lambda h\sum_{r=-(T-1)}^{T-1}C(h,r)\sigma(r),$$

$$-\pi \le \lambda \le \pi,$$

where the coefficients $C(h, r)$ are given in (51) to (55).

To evaluate this expected value as an integral of the spectral density we write

$$(71)\quad \mathscr{E}I^*(\lambda) = \frac{1}{2\pi T}\mathscr{E}\sum_{s,t=1}^{T}(y_t y_s - y_t\bar{y} - \bar{y}y_s + \bar{y}^2)e^{i\lambda(t-s)}$$

$$= \frac{1}{2\pi T}\Bigg[\sum_{s,t=1}^{T}\sigma(t-s)e^{i\lambda(t-s)} - \frac{1}{T}\sum_{t,r=1}^{T}\sigma(t-r)e^{i\lambda t}\frac{\sin\frac{1}{2}\lambda T}{\sin\frac{1}{2}\lambda}e^{-i\frac{1}{2}\lambda(T+1)}$$

$$- \frac{1}{T}\sum_{r,s=1}^{T}\sigma(r-s)e^{-i\lambda s}\frac{\sin\frac{1}{2}\lambda T}{\sin\frac{1}{2}\lambda}e^{i\frac{1}{2}\lambda(T+1)} + \operatorname{Var}\bar{y}\left(\frac{\sin\frac{1}{2}\lambda T}{\sin\frac{1}{2}\lambda}\right)^2\Bigg]$$

$$= \frac{1}{2\pi T}\int_{-\pi}^{\pi}\Bigg[\sum_{s,t=1}^{T}e^{i(\lambda-\nu)(t-s)} - \frac{1}{T}\frac{\sin\frac{1}{2}\lambda T}{\sin\frac{1}{2}\lambda}\Bigg\{\sum_{t,r=1}^{T}e^{i(\lambda-\nu)t+i\nu r-i\frac{1}{2}\lambda(T+1)}$$

$$+\sum_{r,s=1}^{T}e^{i(\nu-\lambda)s-i\nu r+i\frac{1}{2}\lambda(T+1)}\Bigg\} + \left(\frac{\sin\frac{1}{2}\lambda T}{\sin\frac{1}{2}\lambda}\right)^2\left(\frac{\sin\frac{1}{2}\nu T}{T\sin\frac{1}{2}\nu}\right)^2\Bigg]f(\nu)\,d\nu$$

$$= \frac{1}{2\pi T}\int_{-\pi}^{\pi}\left\{\frac{\sin\frac{1}{2}(\nu-\lambda)T}{\sin\frac{1}{2}(\nu-\lambda)} - \frac{\sin\frac{1}{2}\lambda T}{\sin\frac{1}{2}\lambda}\,\frac{\sin\frac{1}{2}\nu T}{T\sin\frac{1}{2}\nu}\right\}^2 f(\nu)\,d\nu, \qquad \lambda \neq 0.$$

We now turn to the covariances of $I(\lambda)$.

THEOREM 8.2.8.

$$(72)\quad \operatorname{Cov}[I(\lambda), I(\lambda')]$$

$$= \left[\int_{-\pi}^{\pi}k_T(\nu+\lambda, \nu-\lambda')f(\nu)\,d\nu\right]^2 + \left[\int_{-\pi}^{\pi}k_T(\nu+\lambda, \nu+\lambda')f(\nu)\,d\nu\right]^2$$

$$+ \frac{1}{4\pi^2T^2}\sum_{t,s,t',s'=1}^{T}\cos\lambda(t-s)\cos\lambda'(t'-s')\,\kappa(s-t, t'-t, s'-t),$$

where

$$(73)\qquad k_T(\lambda, \nu) = \frac{\sin\frac{1}{2}\lambda T\sin\frac{1}{2}\nu T}{2\pi T\sin\frac{1}{2}\lambda\sin\frac{1}{2}\nu}.$$

PROOF. Since (for $\mu = 0$)

$$(74)\qquad 2\pi TI(\lambda) = \sum_{t,s=1}^{T}y_t y_s(\cos\lambda t\cos\lambda s + \sin\lambda t\sin\lambda s)$$

$$= \sum_{t,s=1}^{T}y_t y_s\cos\lambda(t-s)$$

$$= \sum_{t,s=1}^{T}y_t y_s e^{i\lambda(t-s)}$$

and

$$\sigma(p-r) = \int_{-\pi}^{\pi} \cos \nu(p-r)\, f(\nu)\, d\nu = \int_{-\pi}^{\pi} e^{i\nu(p-r)} f(\nu) d\nu, \tag{75}$$

we have

(76)

$$\begin{aligned}
&(2\pi T)^2 \operatorname{Cov}[I(\lambda), I(\lambda')] \\
&= \sum_{t,s,t',s'=1}^{T} e^{i\lambda(t-s)} e^{i\lambda'(t'-s')} \operatorname{Cov}(y_t y_s, y_{t'} y_{s'}) \\
&= \sum_{t,s,t',s'=1}^{T} e^{i\lambda(t-s)+i\lambda'(t'-s')} [\sigma(t-t')\sigma(s-s') \\
&\quad + \sigma(t-s')\sigma(s-t') + \kappa(s-t, t'-t, s'-t)] \\
&= \int_{-\pi}^{\pi}\int_{-\pi}^{\pi} \sum_{t,s,t',s'=1}^{T} e^{i\lambda(t-s)+i\lambda'(t'-s')} [e^{i\nu(t-t')} e^{i\mu(s-s')} \\
&\quad + e^{i\nu(t-s')} e^{i\mu(s-t')}] f(\nu) f(\mu)\, d\nu\, d\mu \\
&\quad + \sum_{t,s,t',s'=1}^{T} \cos \lambda(t-s) \cos \lambda'(t'-s')\, \kappa(s-t, t'-t, s'-t) \\
&= \int_{-\pi}^{\pi}\int_{-\pi}^{\pi} \left[\frac{\sin \frac{1}{2}(\lambda+\nu)T}{\sin \frac{1}{2}(\lambda+\nu)} \frac{\sin \frac{1}{2}(\mu-\lambda)T}{\sin \frac{1}{2}(\mu-\lambda)} \frac{\sin \frac{1}{2}(\lambda'-\nu)T}{\sin \frac{1}{2}(\lambda'-\nu)} \frac{\sin \frac{1}{2}(\lambda'+\mu)T}{\sin \frac{1}{2}(\lambda'+\mu)} \right. \\
&\quad \left. + \frac{\sin \frac{1}{2}(\lambda+\nu)T}{\sin \frac{1}{2}(\lambda+\nu)} \frac{\sin \frac{1}{2}(\mu-\lambda)T}{\sin \frac{1}{2}(\mu-\lambda)} \frac{\sin \frac{1}{2}(\lambda'-\mu)T}{\sin \frac{1}{2}(\lambda'-\mu)} \frac{\sin \frac{1}{2}(\lambda'+\nu)T}{\sin \frac{1}{2}(\lambda'+\nu)} \right] f(\nu) f(\mu)\, d\nu\, d\mu \\
&\quad + \sum_{t,s,t',s'=1}^{T} \cos \lambda(t-s) \cos \lambda'(t'-s')\, \kappa(s-t, t'-t, s'-t) \\
&= \left\{ \int_{-\pi}^{\pi} \frac{\sin \frac{1}{2}(\nu+\lambda)T}{\sin \frac{1}{2}(\nu+\lambda)} \frac{\sin \frac{1}{2}(\nu-\lambda')T}{\sin \frac{1}{2}(\nu-\lambda')} f(\nu)\, d\nu \right\}^2 \\
&\quad + \left\{ \int_{-\pi}^{\pi} \frac{\sin \frac{1}{2}(\nu+\lambda)T}{\sin \frac{1}{2}(\nu+\lambda)} \frac{\sin \frac{1}{2}(\nu+\lambda')T}{\sin \frac{1}{2}(\nu+\lambda')} f(\nu)\, d\nu \right\}^2 \\
&\quad + \sum_{t,s,t',s'=1}^{T} \cos \lambda(t-s) \cos \lambda'(t'-s')\, \kappa(s-t, t'-t, s'-t),
\end{aligned}$$

from which (72) follows.

An alternative form for the covariance is

$$(77)\quad \operatorname{Cov}[I(\lambda), I(\lambda')] = \frac{1}{4\pi^2} \sum_{r,p=-(T-1)}^{T-1} \operatorname{Cov}(c_r, c_p) \cos \lambda r \cos \lambda' p$$

$$= \frac{1}{4\pi^2 T^2} \sum_{r,p=-(T-1)}^{T-1} \sum_{t=1}^{T-|r|} \sum_{s=1}^{T-|p|} [\sigma(t-s)\sigma(t-s+|r|-|p|)$$

$$+ \sigma(t-s-|p|)\sigma(t-s+|r|)$$

$$+ \kappa(|r|, s-t, s-t+|p|)] \cos \lambda r \cos \lambda' p.$$

Since $k_T(\lambda, \lambda) = k_T(\lambda)$, we can write (72) for $\lambda = \lambda'$ as

(78)

$$\operatorname{Var}[I(\lambda)] = \left[\int_{-\pi}^{\pi} k_T(\nu+\lambda, \nu-\lambda) f(\nu)\, d\nu\right]^2 + \left[\int_{-\pi}^{\pi} k_T(\nu-\lambda) f(\nu)\, d\nu\right]^2$$

$$+ \frac{1}{4\pi^2 T^2} \sum_{t,s,t',s'=1}^{T} \cos \lambda(t-s) \cos \lambda(t'-s')\, \kappa(s-t, t'-t, s'-t)$$

$$= \left[\int_{-\pi}^{\pi} k_T(\nu+\lambda, \nu-\lambda) f(\nu)\, d\nu\right]^2 + [\mathscr{E} I(\lambda)]^2$$

$$+ \frac{1}{4\pi^2 T^2} \sum_{t,s,t',s'=1}^{T} \cos \lambda(t-s) \cos \lambda(t'-s')\, \kappa(s-t, t'-t, s'-t).$$

It will be useful to know the covariances of the cosine and sine transforms $A(\lambda)$ and $B(\lambda)$.

THEOREM 8.2.9.

$$(79)\quad \operatorname{Cov}[A(\lambda), A(\lambda')] = \frac{4\pi}{T} \int_{-\pi}^{\pi} [k_T(\nu-\lambda', \nu+\lambda) \cos \tfrac{1}{2}(\lambda'+\lambda)(T+1)$$

$$+ k_T(\nu-\lambda', \nu-\lambda) \cos \tfrac{1}{2}(\lambda'-\lambda)(T+1)] f(\nu)\, d\nu,$$

$$(80)\quad \operatorname{Cov}[A(\lambda), B(\lambda')] = \frac{4\pi}{T} \int_{-\pi}^{\pi} [k_T(\nu-\lambda', \nu+\lambda) \sin \tfrac{1}{2}(\lambda'+\lambda)(T+1)$$

$$+ k_T(\nu-\lambda', \nu-\lambda) \sin \tfrac{1}{2}(\lambda'-\lambda)(T+1)] f(\nu)\, d\nu,$$

$$(81)\quad \operatorname{Cov}[B(\lambda), B(\lambda')] = \frac{4\pi}{T} \int_{-\pi}^{\pi} [k_T(\nu+\lambda, \nu+\lambda') \cos \tfrac{1}{2}(\lambda-\lambda')(T+1)$$

$$- k_T(\nu+\lambda, \nu-\lambda') \cos \tfrac{1}{2}(\lambda+\lambda')(T+1)] f(\nu)\, d\nu.$$

PROOF. We have

(82)

$$
\begin{aligned}
&\mathscr{E}A(\lambda)A(\nu) + i\mathscr{E}A(\lambda)B(\nu) = \mathscr{E}A(\lambda)[A(\nu) + iB(\nu)] \\
&= \frac{4}{T^2}\sum_{t,s=1}^{T} \sigma(t-s)\cos\lambda t\, e^{i\nu s} \\
&= \frac{4}{T^2}\int_{-\pi}^{\pi}\sum_{t,s=1}^{T} e^{i\mu(t-s)+i\nu s}\cos\lambda t f(\mu)\,d\mu \\
&= \frac{4}{T^2}\int_{-\pi}^{\pi}\sum_{t=1}^{T} e^{i\mu t}\cos\lambda t\,\frac{\sin\frac{1}{2}(\nu-\mu)T}{\sin\frac{1}{2}(\nu-\mu)}\, e^{i\frac{1}{2}(\nu-\mu)(T+1)} f(\mu)\,d\mu \\
&= \frac{4}{T^2}\int_{-\pi}^{\pi}\frac{\sin\frac{1}{2}(\nu-\mu)T}{\sin\frac{1}{2}(\nu-\mu)}\, e^{i\frac{1}{2}(\nu-\mu)(T+1)}\sum_{t=1}^{T} e^{i\mu t}\tfrac{1}{2}[e^{i\lambda t}+e^{-i\lambda t}] f(\mu)\,d\mu \\
&= \frac{2}{T^2}\int_{-\pi}^{\pi}\frac{\sin\frac{1}{2}(\nu-\mu)T}{\sin\frac{1}{2}(\nu-\mu)}\, e^{i\frac{1}{2}\nu(T+1)}\left[\frac{\sin\frac{1}{2}(\lambda+\mu)T}{\sin\frac{1}{2}(\lambda+\mu)}\, e^{i\frac{1}{2}\lambda(T+1)}\right. \\
&\qquad\qquad\left. + \frac{\sin\frac{1}{2}(\lambda-\mu)T}{\sin\frac{1}{2}(\lambda-\mu)}\, e^{-i\frac{1}{2}\lambda(T+1)}\right] f(\mu)\,d\mu \\
&= \frac{2}{T^2}\int_{-\pi}^{\pi}\frac{\sin\frac{1}{2}(\nu-\mu)T}{\sin\frac{1}{2}(\nu-\mu)}\left[\frac{\sin\frac{1}{2}(\lambda+\mu)T}{\sin\frac{1}{2}(\lambda+\mu)}\, e^{i\frac{1}{2}(\nu+\lambda)(T+1)}\right. \\
&\qquad\qquad\left. + \frac{\sin\frac{1}{2}(\lambda-\mu)T}{\sin\frac{1}{2}(\lambda-\mu)}\, e^{i\frac{1}{2}(\nu-\lambda)(T+1)}\right] f(\mu)\,d\mu
\end{aligned}
$$

and

(83)

$$
\begin{aligned}
&\mathscr{E}A(\lambda)B(\nu) + i\mathscr{E}B(\lambda)B(\nu) = \mathscr{E}[A(\lambda) + iB(\lambda)]B(\nu) \\
&= \frac{4}{T^2}\sum_{t,s=1}^{T} \sigma(t-s)e^{i\lambda t}\sin\nu s \\
&= \frac{4}{T^2}\int_{-\pi}^{\pi}\sum_{t,s=1}^{T} e^{i\mu(t-s)+i\lambda t}\sin\nu s f(\mu)\,d\mu \\
&= \frac{4}{T^2}\int_{-\pi}^{\pi}\sum_{s=1}^{T} e^{-i\mu s}\sin\nu s\,\frac{\sin\frac{1}{2}(\lambda+\mu)T}{\sin\frac{1}{2}(\lambda+\mu)}\, e^{i\frac{1}{2}(\lambda+\mu)(T+1)} f(\mu)\,d\mu \\
&= \frac{4}{T^2}\int_{-\pi}^{\pi}\frac{\sin\frac{1}{2}(\lambda+\mu)T}{\sin\frac{1}{2}(\lambda+\mu)}\, e^{i\frac{1}{2}(\lambda+\mu)(T+1)}\sum_{s=1}^{T} e^{-i\mu s}\frac{1}{2i}[e^{i\nu s} - e^{-i\nu s}] f(\mu)\,d\mu
\end{aligned}
$$

$$= \frac{2}{iT^2}\int_{-\pi}^{\pi}\frac{\sin\frac{1}{2}(\lambda+\mu)T}{\sin\frac{1}{2}(\lambda+\mu)}e^{i\frac{1}{2}\lambda(T+1)}\left[\frac{\sin\frac{1}{2}(\nu-\mu)T}{\sin\frac{1}{2}(\nu-\mu)}e^{i\frac{1}{2}\nu(T+1)} - \frac{\sin\frac{1}{2}(\nu+\mu)T}{\sin\frac{1}{2}(\nu+\mu)}e^{-i\frac{1}{2}\nu(T+1)}\right]f(\mu)\,d\mu$$

$$= \frac{2}{iT^2}\int_{-\pi}^{\pi}\frac{\sin\frac{1}{2}(\lambda+\mu)T}{\sin\frac{1}{2}(\lambda+\mu)}\left[\frac{\sin\frac{1}{2}(\nu-\mu)T}{\sin\frac{1}{2}(\nu-\mu)}e^{i\frac{1}{2}(\lambda+\nu)(T+1)} - \frac{\sin\frac{1}{2}(\nu+\mu)T}{\sin\frac{1}{2}(\nu+\mu)}e^{i\frac{1}{2}(\lambda-\nu)(T+1)}\right]f(\mu)\,d\mu.$$

The theorem follows from (82) and (83) by consideration of real and imaginary parts.

8.3. ASYMPTOTIC MEANS AND COVARIANCES OF THE SAMPLE MEAN, COVARIANCES, AND SPECTRAL DENSITY

8.3.1. The Sample Mean

The sample mean $\bar{y} = \sum_{t=1}^{T} y_t/T$ is an unbiased estimate of $\mu = \mathscr{E}y_t$ for a process stationary in the wide sense; that is, $\mathscr{E}\bar{y} = \mu$. The variance of $\bar{y}$ was given in Section 8.2.2 as a linear combination of $\sigma(0), \ldots, \sigma(T-1)$ and as a weighted integral of $f(\lambda)$; we have

$$T \operatorname{Var} \bar{y} = \sum_{r=-(T-1)}^{T-1}\left(1 - \frac{|r|}{T}\right)\sigma(r) = \int_{-\pi}^{\pi}\frac{\sin^2 \frac{1}{2}\lambda T}{T \sin^2 \frac{1}{2}\lambda} f(\lambda)\,d\lambda. \tag{1}$$

THEOREM 8.3.1. *If*

$$\sum_{r=-\infty}^{\infty} \sigma(r) < \infty, \tag{2}$$

then

$$\lim_{T\to\infty} T \operatorname{Var} \bar{y} = \sum_{r=-\infty}^{\infty} \sigma(r). \tag{3}$$

If $f(\lambda)$ is continuous at $\lambda = 0$,

$$\lim_{T\to\infty} T \operatorname{Var} \bar{y} = 2\pi f(0). \tag{4}$$

The proof using the sum in (1) is a special case of the fact that the Cesàro summation of a convergent sum is the sum. This is stated equivalently in the following lemma:

LEMMA 8.3.1. *If* $\sum_{r=1}^{\infty} a_r$ *is convergent, then*

$$\lim_{T\to\infty} \sum_{r=1}^{T-1}\left(1-\frac{r}{T}\right)a_r = \sum_{r=1}^{\infty} a_r. \tag{5}$$

PROOF OF LEMMA. The convergence of $\sum_{r=1}^{\infty} a_r$ implies that for any positive ε there exists a number T_ε such that for every $T \geq T_\varepsilon$

$$\left|\sum_{r=T}^{\infty} a_r\right| < \tfrac{1}{2}\varepsilon; \tag{6}$$

the convergence also implies that there exists a number M such that

$$\left|\sum_{r=T}^{\infty} a_r\right| < M, \qquad T = 1, 2, \ldots. \tag{7}$$

Then for $T > T_\varepsilon$

$$\begin{aligned}
\left|\sum_{r=1}^{\infty} a_r - \sum_{r=1}^{T-1}\left(1-\frac{r}{T}\right)a_r\right| &= \left|\frac{T\sum_{r=T}^{\infty} a_r + \sum_{r=1}^{T-1} r a_r}{T}\right| \\
&= \frac{\left|\sum_{r=1}^{T-1}\sum_{s=r+1}^{\infty} a_s + \sum_{s=1}^{\infty} a_s\right|}{T} \\
&\leq \frac{\sum_{r=1}^{T_\varepsilon-1}\left|\sum_{s=r+1}^{\infty} a_s\right| + \sum_{r=T_\varepsilon}^{T-1}\left|\sum_{s=r+1}^{\infty} a_s\right| + \left|\sum_{r=1}^{\infty} a_r\right|}{T} \\
&< \frac{T_\varepsilon M}{T} + \tfrac{1}{2}\varepsilon.
\end{aligned} \tag{8}$$

For $T > T_\varepsilon$ and $T > T_\varepsilon' = 2T_\varepsilon M/\varepsilon$ (8) is less than ε. This proves the lemma.

Note that the absolute convergence of (2) is unnecessarily strong for (3).

The second form of the limiting variance follows from the first if the Fourier series for $f(\lambda)$ converges to $f(0)$ at $\lambda = 0$. However, we shall prove the result directly from the second form of (1) because the techniques will be useful. The weighted integral involves Fejér's kernel

$$k_T(\lambda) = \frac{\sin^2 \frac{1}{2}\lambda T}{2\pi T \sin^2 \frac{1}{2}\lambda}. \tag{9}$$

The integral of $k_T(\lambda)$ over $(-a, a)$ is

$$\begin{aligned} \int_{-a}^{a} k_T(\lambda)\, d\lambda &= \int_{-a}^{a} \frac{\sin^2 \frac{1}{2}\lambda T}{2\pi T \sin^2 \frac{1}{2}\lambda}\, d\lambda \\ &= \int_{-a}^{a} \frac{1}{2\pi T} \sum_{s,t=1}^{T} e^{i\lambda(s-t)}\, d\lambda \\ &= \int_{-a}^{a} \frac{1}{2\pi} \sum_{r=-(T-1)}^{T-1} \left(1 - \frac{|r|}{T}\right) \cos \lambda r\, d\lambda \\ &= \frac{1}{\pi}\left[a + 2\sum_{r=1}^{T-1} \frac{T-r}{Tr} \sin ar\right]. \end{aligned} \tag{10}$$

In particular

$$\int_{-\pi}^{\pi} k_T(\lambda)\, d\lambda = 1. \tag{11}$$

Since $|\sin x| \le 1$,

$$k_T(\lambda) \le \frac{1}{2\pi T \sin^2 \frac{1}{2}\lambda}, \tag{12}$$

and hence

$$\lim_{T\to\infty} k_T(\lambda) = 0, \qquad \lambda \ne 0. \tag{13}$$

In fact, since $\sin \frac{1}{2}\lambda$ is increasing in $[0, \pi)$, the convergence is uniform in any interval $[a, \pi]$ for $a > 0$; that is,

$$k_T(\lambda) \le \frac{1}{2\pi T \sin^2 \frac{1}{2}a}, \qquad 0 < a \le |\lambda| \le \pi. \tag{14}$$

Thus

$$\int_{a}^{\pi} k_T(\lambda)\, d\lambda \le \int_{a}^{\pi} \frac{d\lambda}{2\pi T \sin^2 \frac{1}{2}a} = \frac{\pi - a}{2\pi T \sin^2 \frac{1}{2}a} \to 0 \tag{15}$$

as $T \to \infty$. This proves the following lemma:

LEMMA 8.3.2.

$$\lim_{T\to\infty} \int_{-a}^{a} k_T(\lambda)\, d\lambda = 1, \qquad 0 < a \le \pi, \tag{16}$$

$$\lim_{T\to\infty} \int_{-\pi}^{-a} k_T(\lambda)\, d\lambda = \lim_{T\to\infty} \int_{a}^{\pi} k_T(\lambda)\, d\lambda = 0, \qquad 0 < a \le \pi. \tag{17}$$

The relevant part of Theorem 8.3.1 results from the following lemma (which is stated generally in order that it may be used in other contexts):

LEMMA 8.3.3. *If* $\int_{-\pi}^{\pi} |h(\lambda)|\, d\lambda < \infty$, *if* $h(\lambda)$ *is continuous at* $\lambda = \nu$, *if for some* T_0

$$\int_{-\pi}^{\pi} l_T(\lambda)\, d\lambda = 1, \qquad T = T_0, \ldots, \tag{18}$$

if there is a number K and integer T_1 *such that*

$$\int_{-\pi}^{\pi} |l_T(\lambda)|\, d\lambda \leq K, \qquad T = T_1, \ldots, \tag{19}$$

and if for some T_2

$$|l_T(\lambda)| \leq m_T n(a), \qquad |\lambda - \nu| \geq a, \; T = T_2, \ldots, \tag{20}$$

for every $a > 0$, *and* $n(a)$ *is a monotonically nonincreasing function of* a *for* $0 < a < \pi$, *and* $m_T \to 0$ *as* $T \to \infty$, *then*

$$\lim_{T \to \infty} \int_{-\pi}^{\pi} h(\lambda) l_T(\lambda)\, d\lambda = h(\nu). \tag{21}$$

PROOF. Suppose $-\pi < \nu < \pi$. Given a positive ε, there is a δ such that $|h(\lambda) - h(\nu)| < \varepsilon$ for $|\lambda - \nu| < \delta$. From (18) and (20) we obtain

$$\lim_{T \to \infty} \int_{\nu-\delta}^{\nu+\delta} l_T(\lambda)\, d\lambda = 1 \tag{22}$$

by an analogue of the proof of Lemma 8.3.2. Then

$$\begin{aligned} \left| \int_{\nu-\delta}^{\nu+\delta} h(\lambda) l_T(\lambda)\, d\lambda - h(\nu) \int_{\nu-\delta}^{\nu+\delta} l_T(\lambda)\, d\lambda \right| &= \left| \int_{\nu-\delta}^{\nu+\delta} [h(\lambda) - h(\nu)] l_T(\lambda)\, d\lambda \right| \\ &\leq \int_{\nu-\delta}^{\nu+\delta} |h(\lambda) - h(\nu)|\, |l_T(\lambda)|\, d\lambda \leq K\varepsilon, \qquad T = T_1, \ldots . \end{aligned} \tag{23}$$

We have for $T \geq T_2$

$$\left| \int_{-\pi}^{\nu-\delta} h(\lambda) l_T(\lambda)\, d\lambda + \int_{\nu+\delta}^{\pi} h(\lambda) l_T(\lambda)\, d\lambda \right| \leq m_T \left[\int_{-\pi}^{\nu-\delta} |h(\lambda)|\, d\lambda + \int_{\nu+\delta}^{\pi} |h(\lambda)|\, d\lambda \right] n(\delta), \tag{24}$$

which approaches 0 as $T \to \infty$. The proofs for $\nu = -\pi$ and $\nu = \pi$ proceed similarly. Note that (18) implies (19) for $K = 1$ if $l_T(\lambda) \geq 0$.

The proof of (4) is the special case of Lemma 8.3.3 when $h(\lambda) = 2\pi f(\lambda)$ and $l_T(\lambda) = k_T(\lambda)$.

The alternative conditions of Theorem 8.3.1 [$\sum_{r=-\infty}^{\infty} \sigma(r) < \infty$ and continuity of $f(\lambda)$ at $\lambda = 0$] are implied by

$$\sum_{r=-\infty}^{\infty} |\sigma(r)| < \infty. \tag{25}$$

8.3.2. The Sample Covariances

If the mean μ is known, C_h, defined by (7) of Section 8.2, provides an unbiased estimate of $\sigma(h)$, $h = 0, 1, \ldots, T - 1$. The estimate $c_h = (T - |h|)C_h/T$ has the bias $- |h|\, \sigma(h)/T$, and for any h the limit of this is 0 as $T \to \infty$.

It was shown in Section 8.2.2 that when μ is unknown, expectations of the estimates C_h^* and $\tilde{C}_h$ are linear combinations of $\sigma(0), \ldots, \sigma(T - 1)$; hence, in general they are biased. We shall show that the bias is of order $1/T$. In fact, the bias is approximately $1/T$ times

$$-\frac{1}{T}\sum_{t,s=1}^{T} \sigma(t - s) = -\sum_{r=-(T-1)}^{T-1} \left(1 - \frac{|r|}{T}\right)\sigma(r), \tag{26}$$

the limit of which is $- \sum_{r=-\infty}^{\infty} \sigma(r)$ if this last series converges. The difference between $\sigma(h)$ and $\mathscr{E}C_h^*$ as given in the first two lines of (50) in Section 8.2 consists of three terms. The last term is $-1/T$ times (26). The second and third terms are $-\sum_{t=1}^{T-h} \sum_{s=1}^{T} [\sigma(t - s) + \sigma(t + h - s)]/[T(T - h)]$. Since

$$\begin{aligned}
&\left|\sum_{t=1}^{T-h}\sum_{s=1}^{T} \sigma(t - s) - \sum_{t,s=1}^{T} \sigma(t - s)\right| \\
&\quad= \left|\sum_{t=T-h+1}^{T}\sum_{s=1}^{T} \sigma(t - s)\right| \\
&\quad= \left|\sum_{r=-(h-1)}^{-1} (h - |r|)\sigma(r) + h\sum_{r=0}^{T-h}\sigma(r) + \sum_{r=T-h+1}^{T} (T - r)\sigma(r)\right| \\
&\quad\to \left|h\sum_{r=0}^{\infty} \sigma(r) + \sum_{r=1}^{h-1} (h - r)\sigma(r)\right|
\end{aligned} \tag{27}$$

as $T \to \infty$ if $\sum_{r=0}^{\infty} \sigma(r) < \infty$, T times the bias converges to $-\sum_{r=-\infty}^{\infty} \sigma(r)$. Other estimates, such as $\tilde{C}_h$, can be handled similarly.

THEOREM 8.3.2. *If* $\sum_{r=-\infty}^{\infty} \sigma(r) < \infty$,

$$\lim_{T\to\infty} T[\mathscr{E}C_h^* - \sigma(h)] = \lim_{T\to\infty} T[\mathscr{E}\tilde{C}_h - \sigma(h)] = -\sum_{r=-\infty}^{\infty} \sigma(r). \tag{28}$$

If $f(\lambda)$ is continuous at $\lambda = 0$,

$$\lim_{T\to\infty} T[\mathscr{E}C_h^* - \sigma(h)] = \lim_{T\to\infty} T[\mathscr{E}\tilde{C}_h - \sigma(h)] = -2\pi f(0). \tag{29}$$

We now turn to the covariances of the estimates. Most of these results have been given by Bartlett (1935), (1946), and Parzen (1957b). We have from Theorem 8.2.6

$$\begin{aligned}(30)\quad (T-h)\operatorname{Cov}(C_h, C_g) &= \sum_{r=0}^{g-h}[\sigma(r)\sigma(r+h-g) + \sigma(r-g)\sigma(r+h) + \kappa(h,-r,g-r)] \\ &\quad + \sum_{r=g-h+1}^{T-h-1}\left(1 - \frac{r-(g-h)}{T-g}\right) \\ &\quad \times [\sigma(r)\sigma(r+h-g) + \sigma(r-g)\sigma(r+h) + \kappa(h,-r,g-r)] \\ &\quad + \sum_{r=-(T-g-1)}^{-1}\left(1 - \frac{|r|}{T-g}\right) \\ &\quad \times [\sigma(r)\sigma(r+h-g) + \sigma(r-g)\sigma(r+h) + \kappa(h,-r,g-r)], \\ &\qquad 0 \le h \le g < T.\end{aligned}$$

The limit of (30) is

$$\begin{aligned}(31)\quad \lim_{T\to\infty}(T-h)\operatorname{Cov}(C_h, C_g) \\ = \sum_{r=-\infty}^{\infty}[\sigma(r)\sigma(r+h-g) + \sigma(r-g)\sigma(r+h) + \kappa(h,-r,g-r)],\end{aligned}$$

if the series converge. Since the right-hand side of (31) is symmetric in h and g [because of the symmetry in $\kappa(r, s, t)$ as indicated in Problem 2], the restriction $h \le g$ may be dropped.

The integral form of the covariance when $\kappa(r, s, t) = 0$ is

$$\begin{aligned}(32)\quad (T-h)\operatorname{Cov}(C_h, C_g) &= 2\int_{-\pi}^{\pi}\int_{-\pi}^{\pi}\frac{\sin\frac{1}{2}(\lambda-\nu)(T-h)\sin\frac{1}{2}(\lambda-\nu)(T-g)}{(T-g)\sin^2\frac{1}{2}(\lambda-\nu)} \\ &\quad \times \cos\tfrac{1}{2}(\lambda+\nu)g\cos\tfrac{1}{2}(\lambda+\nu)h\, f(\lambda)f(\nu)\, d\lambda\, d\nu \\ &= 2\int_{-\pi}^{\pi}\int_{-\pi-\nu}^{\pi-\nu}\frac{\sin\frac{1}{2}\mu(T-h)\sin\frac{1}{2}\mu(T-g)}{(T-g)\sin^2\frac{1}{2}\mu} \\ &\quad \times \cos(\nu+\tfrac{1}{2}\mu)g\cos(\nu+\tfrac{1}{2}\mu)h\, f(\nu+\mu)f(\nu)\, d\mu\, d\nu \\ &= 2\int_{-\pi}^{\pi}\int_{-\pi}^{\pi}\frac{\sin\frac{1}{2}\mu(T-h)\sin\frac{1}{2}\mu(T-g)}{(T-g)\sin^2\frac{1}{2}\mu} \\ &\quad \times \cos(\nu+\tfrac{1}{2}\mu)g\cos(\nu+\tfrac{1}{2}\mu)h\, f(\nu+\mu)f(\nu)\, d\mu\, d\nu, \\ &\qquad 0 \le h < T,\ 0 \le g < T,\end{aligned}$$

because $f(\lambda \pm 2\pi) = f(\lambda)$, $\sin \frac{1}{2}(\lambda \pm 2\pi)k = (-1)^k \sin \frac{1}{2}\lambda k$, and $\cos \frac{1}{2}(\lambda \pm 2\pi)k = (-1)^k \cos \frac{1}{2}\lambda k$. [In the second expression of (32) the integration from $-\pi - \nu$ to $-\pi$ is written as an integration from $\pi - \nu$ to π by replacing μ by $\mu - 2\pi$.] If we assume $f(\lambda)$ is continuous, $-\pi \leq \lambda \leq \pi$, it is uniformly continuous. Similarly, $\cos \frac{1}{2}\lambda g$ and $\cos \frac{1}{2}\lambda h$ are uniformly continuous. Then

$$(33) \qquad 2\int_{-\pi}^{\pi} \cos(\nu + \tfrac{1}{2}\mu)g \cos(\nu + \tfrac{1}{2}\mu)h\, f(\nu + \mu) f(\nu)\, d\nu = m(\mu),$$

say, is a uniformly continuous function of μ. The integral (32) can be written

$$(34) \qquad \int_{-\pi}^{\pi} \frac{\sin \frac{1}{2}\mu(T-h) \sin \frac{1}{2}\mu(T-g)}{(T-g)\sin^2 \frac{1}{2}\mu} m(\mu)\, d\mu$$
$$= \int_{-\pi}^{\pi} \frac{\sin^2 \frac{1}{2}\mu(T-g)}{(T-g)\sin^2 \frac{1}{2}\mu} \cos \tfrac{1}{2}\mu(g-h)\, m(\mu)\, d\mu$$
$$+ \int_{-\pi}^{\pi} \frac{\sin \frac{1}{2}\mu(T-g) \sin \frac{1}{2}\mu(g-h)}{(T-g)\sin^2 \frac{1}{2}\mu} \cos \tfrac{1}{2}\mu(T-g)\, m(\mu)\, d\mu.$$

The first integral on the right-hand side of (34) converges to $2\pi m(0)$. The absolute value of the second integral is less than $\max_{-\pi \leq \mu \leq \pi} |m(\mu)|$ times

$$(35) \qquad \int_{-\delta^*}^{\delta^*} \frac{\sin \frac{1}{2}\mu(T-g) \sin \frac{1}{2}\mu(g-h)}{(T-g)\sin^2 \frac{1}{2}\mu} d\mu + 2\int_{\delta^*}^{\pi} \frac{1}{(T-g)\sin^2 \frac{1}{2}\delta^*} d\mu$$
$$\leq 2\,|g-h|\,\delta^* + 2\frac{\pi - \delta^*}{(T-g)\sin^2 \frac{1}{2}\delta^*}$$

for arbitrary δ^* $(0 < \delta^* \leq \pi)$ since

$$(36) \qquad \left|\frac{\sin \frac{1}{2}\mu k}{\sin \frac{1}{2}\mu}\right| = \left|\frac{e^{i\frac{1}{2}\mu k} - e^{-i\frac{1}{2}\mu k}}{e^{i\frac{1}{2}\mu} - e^{-i\frac{1}{2}\mu}}\right| = \left|\sum_{j=0}^{k-1} e^{i\mu j}\right| \leq k;$$

the right-hand side of (35) is made arbitrarily small by making δ^* sufficiently small and T sufficiently large. Thus the limit of the second term on the right-hand side of (34) is 0.

THEOREM 8.3.3. *If $f(\lambda)$ is continuous, $-\pi \leq \lambda \leq \pi$, and if the condition $|\sum_{r=-\infty}^{\infty} \kappa(h, -r, g-r)| < \infty$ holds, then*

$$(37) \qquad \lim_{T\to\infty} (T-h)\operatorname{Cov}(C_h, C_g)$$
$$= 4\pi \int_{-\pi}^{\pi} \cos \nu g \cos \nu h\, f^2(\nu)\, d\nu + \sum_{r=-\infty}^{\infty} \kappa(h, -r, g-r);$$

if $\sum_{r=-\infty}^{\infty} \sigma^2(r) < \infty$ *and* $|\sum_{r=-\infty}^{\infty} \kappa(h, -r, g - r)| < \infty$, *then*

$$\lim_{T\to\infty} (T - h) \operatorname{Cov}(C_h, C_g) \tag{38}$$

$$= \sum_{r=-\infty}^{\infty} [\sigma(r)\sigma(r + h - g) + \sigma(r - g)\sigma(r + h) + \kappa(h, -r, g - r)].$$

If the process is Gaussian, $\kappa(r, s, t) = 0$.

If $\{y_t\}$ is a moving average process, that is,

$$y_t - \mu = \sum_{s=-\infty}^{\infty} \gamma_s v_{t-s}, \qquad t = \ldots, -1, 0, 1, \ldots, \tag{39}$$

where $\mathscr{E} v_t = 0$, $\mathscr{E} v_t^2 = \sigma^2$, and $\mathscr{E} v_t v_s = 0$, $t \neq s$, and $\sum_{s=-\infty}^{\infty} \gamma_s^2 < \infty$, then, if $\mathscr{E} v_t^4 < \infty$, the fourth-order central moments of $\{y_t\}$ are

$$\nu(h, g, k) = \sum_{n,s,r,p=-\infty}^{\infty} \gamma_n \gamma_s \gamma_r \gamma_p \mathscr{E} v_{t-n} v_{t+h-s} v_{t+g-r} v_{t+k-p}. \tag{40}$$

If the v_t's are independent (or their moments to fourth-order agree with independence), each expected value in the above sum is 0 unless one of the following three pairs holds:

$$\begin{aligned} t - n &= t + h - s, & t + g - r &= t + k - p, \\ t - n &= t + g - r, & t + h - s &= t + k - p, \\ t - n &= t + k - p, & t + h - s &= t + g - r; \end{aligned} \tag{41}$$

that is, one of the pairs

$$\begin{aligned} n &= s - h, & r - g &= p - k, \\ n &= r - g, & s - h &= p - k, \\ n &= p - k, & s - h &= r - g. \end{aligned} \tag{42}$$

Then $\nu(h, g, k)$ is

$$\begin{aligned} \nu(h, g, k) = {} & \sum_{s=-\infty}^{\infty} \gamma_s \gamma_{s-h} \gamma_{s-h+g} \gamma_{s-h+k} \mathscr{E} v_s^4 + \sum_{\substack{s,r=-\infty \\ s-h \neq r-g}}^{\infty} \gamma_s \gamma_{s-h} \gamma_r \gamma_{r-g+k} \sigma^4 \\ & + \sum_{\substack{s,r=-\infty \\ s-h \neq r-g}}^{\infty} \gamma_s \gamma_{s-h+k} \gamma_r \gamma_{r-g} \sigma^4 + \sum_{\substack{s,p=-\infty \\ s-h \neq p-k}}^{\infty} \gamma_s \gamma_{s-h+g} \gamma_p \gamma_{p-k} \sigma^4. \end{aligned} \tag{43}$$

If $\{v_t\}$ is stationary, $\mathscr{E}v_t^4$ does not depend on t. We have

(44) $$\kappa(h, g, k) = \nu(h, g, k) - [\sigma(h)\sigma(g-k) + \sigma(g)\sigma(h-k) + \sigma(k)\sigma(h-g)]$$

$$= \sum_{s=-\infty}^{\infty} \gamma_s \gamma_{s-h} \gamma_{s-h+g} \gamma_{s-h+k} (\mathscr{E}v_s^4 - 3\sigma^4)$$

$$= \sum_{s=-\infty}^{\infty} \gamma_s \gamma_{s-h} \gamma_{s-h+g} \gamma_{s-h+k}\, \kappa_4$$

$$= \kappa_4 \sum_{t=-\infty}^{\infty} \gamma_t \gamma_{t+h} \gamma_{t+g} \gamma_{t+k},$$

where $\kappa_4 = \mathscr{E}v_s^4 - 3\sigma^4$ is the fourth-order cumulant of v_s and

(45) $$\sigma(h) = \sigma^2 \sum_{t=-\infty}^{\infty} \gamma_t \gamma_{t+h}.$$

Then

(46) $$\sum_{r=-\infty}^{\infty} \kappa(h, -r, g-r) = \sum_{r=-\infty}^{\infty} \kappa(h, r, r-g) = \kappa_4 \sum_{r,s=-\infty}^{\infty} \gamma_s \gamma_{s-h} \gamma_{s-h+r} \gamma_{s-h+r-g}$$

$$= \kappa_4 \sum_{s,t=-\infty}^{\infty} \gamma_s \gamma_{s-h} \gamma_t \gamma_{t-g}$$

$$= \kappa_4 \frac{\sigma(h)\sigma(g)}{\sigma^4}.$$

COROLLARY 8.3.1. *If* $\{y_t\}$ *is generated by* (39), *where* $\mathscr{E}v_t = 0$, $\mathscr{E}v_t^2 = \sigma^2$, $\mathscr{E}v_t v_s = 0$, $t \neq s$, $\mathscr{E}v_t v_s v_r v_q = 0$, $t \neq s$, $t \neq r$, $t \neq q$, $\mathscr{E}v_t^4 = 3\sigma^4 + \kappa_4 < \infty$, $\mathscr{E}v_t^2 v_s^2 = \sigma^4$, $t \neq s$, *and* $\sum_{t=-\infty}^{\infty} |\gamma_t| < \infty$, *then*

(47) $$\lim_{T\to\infty} (T-h) \operatorname{Cov}(C_h, C_g)$$

$$= \sum_{r=-\infty}^{\infty} [\sigma(r+h)\sigma(r+g) + \sigma(r-g)\sigma(r+h)] + \frac{\kappa_4}{\sigma^4}\sigma(h)\sigma(g)$$

$$= 4\pi \int_{-\pi}^{\pi} \cos \nu g \cos \nu h\, f^2(\nu)\, d\nu + \frac{\kappa_4}{\sigma^4}\sigma(h)\sigma(g)$$

$$= 4\pi \int_{-\pi}^{\pi} \cos \nu g \cos \nu h\, f^2(\nu)\, d\nu + \frac{\kappa_4}{\sigma^4} \int_{-\pi}^{\pi} \cos \nu g\, f(\nu)\, d\nu \int_{-\pi}^{\pi} \cos \nu h\, f(\nu)\, d\nu.$$

The condition $\sum_{t=-\infty}^{\infty} |\gamma_t| < \infty$ implies $\sum_{t=-\infty}^{\infty} \gamma_t^2 < \infty$, $\sum_{h=-\infty}^{\infty} \sigma^2(h) < \infty$,

$$\sum_{r=-\infty}^{\infty} \sigma(r) = \sigma^2 \sum_{r=-\infty}^{\infty} \sum_{s=-\infty}^{\infty} \gamma_s \gamma_{s+r} = \sigma^2\left(\sum_{s=-\infty}^{\infty} \gamma_s\right)^2 < \infty, \tag{48}$$

and

$$f(\lambda) = \frac{\sigma^2}{2\pi}\left|\sum_{s=-\infty}^{\infty} \gamma_s e^{i\lambda s}\right|^2 \tag{49}$$

is continuous. Thus Corollary 8.3.1 follows from Theorem 8.3.3. Also Theorem 8.3.2 holds.

For any process for which $\sum_{r=-\infty}^{\infty} \sigma^2(r) < \infty$ and $\left|\sum_{r=-\infty}^{\infty} \kappa(h, -r, h - r)\right| < \infty$, the limiting variance of $\sqrt{T-h}\, C_h$ is

$$\begin{aligned}\lim_{T\to\infty} (T-h) \operatorname{Var} C_h &= \sum_{r=-\infty}^{\infty} [\sigma^2(r) + \sigma(r)\sigma(r+2h)] + \sum_{r=-\infty}^{\infty} \kappa(h, -r, h-r) \\ &= 2\pi\int_{-\pi}^{\pi} [1 + e^{i2\nu h}] f^2(\nu)\, d\nu + \sum_{r=-\infty}^{\infty} \kappa(h, -r, h-r) \\ &= 4\pi\int_{-\pi}^{\pi} \cos^2 \nu h\, f^2(\nu)\, d\nu + \sum_{r=-\infty}^{\infty} \kappa(h, -r, h-r).\end{aligned} \tag{50}$$

The term $\sum_{r=-\infty}^{\infty} \sigma^2(r)$ appears in each limiting variance regardless of h; the second term

$$\sum_{r=-\infty}^{\infty} \sigma(r-h)\sigma(r+h) = \sigma(-h)\sigma(h) + 2\sum_{r=1}^{\infty} \sigma(r-h)\sigma(r+h) \tag{51}$$

in absolute value is less than or equal to

$$\sigma^2(h) + 2\sum_{r=1}^{\infty} |\sigma(r-h)\sigma(r+h)| \le \sigma^2(h) + 2\sigma(0)\sum_{s=h+1}^{\infty} |\sigma(s)|, \tag{52}$$

which converges to 0 as $h \to \infty$ if $\sum_{r=-\infty}^{\infty} |\sigma(r)| < \infty$. If $\kappa(r, s, t) = 0$, the limiting variance tends to $\sum_{r=-\infty}^{\infty} \sigma^2(r)$ as h tends to ∞.

Now we turn our attention to estimates of the covariances when the mean is unknown, such as

$$\tilde{C}_h = \frac{1}{T-h}\sum_{t=1}^{T-h} y_t y_{t+h} - \bar{y}_h \bar{y}_{h+} \tag{53}$$
$$= C_h - \frac{1}{(T-h)^2}\sum_{s,t=1}^{T-h}(y_s-\mu)(y_{t+h}-\mu).$$

Since $\tilde{C}_h$ is invariant with respect to change of location ($y_t - \mu \to y_t$), we shall assume $\mu = 0$. From the expressions of (62) and (65) of Section 8.2 we obtain

$$(T-g)[(T-h)\operatorname{Cov}(C_g, C_h) - (T-h)\operatorname{Cov}(\tilde{C}_g, \tilde{C}_h)] \tag{54}$$
$$= \frac{1}{T-g}\sum_{t=1}^{T-h}\sum_{t',s'=1}^{T-g}[\sigma(t-t')\sigma(t+h-s'-g)$$
$$+ \sigma(t-s'-g)\sigma(t+h-t') + \kappa(h, t'-t, s'-t+g)]$$
$$+ \frac{1}{T-h}\sum_{t,s=1}^{T-h}\sum_{t'=1}^{T-g}[\sigma(t-t')\sigma(s+h-t'-g)$$
$$+ \sigma(t-t'-g)\sigma(s+h-t') + \kappa(s+h-t, t'-t, t'+g-t)]$$
$$- \frac{1}{(T-h)(T-g)}\sum_{t,s=1}^{T-h}\sum_{t',s'=1}^{T-g}[\sigma(t-t')\sigma(s+h-s'-g)$$
$$+ \sigma(t-s'-g)\sigma(s+h-t') + \kappa(s+h-t, t'-t, s'+g-t)]$$
$$= \frac{1}{T-g}\sum_{t=1}^{T-h}\sum_{q,r=1-t}^{T-g-t}[\sigma(r)\sigma(q+g-h) + \sigma(q+g)\sigma(r-h)$$
$$+ \kappa(h, r, q+g)]$$
$$+ \frac{1}{T-h}\sum_{t'=1}^{T-g}\sum_{q,r=1-t'}^{T-h-t'}[\sigma(r)\sigma(q+h-g) + \sigma(r-g)\sigma(q+h)$$
$$+ \kappa(q-r+h, -r, g-r)]$$
$$- \frac{1}{(T-h)(T-g)}\sum_{t,s=1}^{T-h}\left[\sum_{r=1-t}^{T-g-t}\sigma(r)\sum_{q=1-s}^{T-g-s}\sigma(q+g-h)\right.$$
$$\left. + \sum_{r=1-}^{T-g-t}\sigma(r+g)\sum_{q=1-s}^{T-g-s}\sigma(q-h) + \sum_{q,r=1-t}^{T-g-t}\kappa(s+h-t, r, q+g)\right].$$

If $\sum_{r=-\infty}^{\infty} \sigma(r) < \infty$, then for any $\varepsilon > 0$ there exists an N such that for all $k > N$ $2\left|\sum_{r=k}^{\infty} \sigma(r)\right| < \varepsilon$. Then

$$\left|\sum_{r=1-t}^{T-g-t} \sigma(r) - \sum_{r=-\infty}^{\infty} \sigma(r)\right| < \varepsilon \tag{55}$$

if $1 - t < -N$ and $T - g - t > N$; that is, if $N + 1 < t < T - g - N$. Thus if T is sufficiently large (with g and h fixed), the sum $\sum_{r=1-t}^{T-g-t} \sigma(r)$ will satisfy (55) for an arbitrarily large proportion of t in the range 1 to $T - h$. Similarly the sum $\sum_{q=1-t}^{T-g-t} \sigma(q + g - h)$ will be within ε of $\sum_{s=-\infty}^{\infty} \sigma(s)$ for an arbitrarily large proportion of t if T is large enough. Therefore, the first triple sum on the right-hand side of (54) is arbitrarily close to $[\sum_{s=-\infty}^{\infty} \sigma(s)]^2$. Similarly each sum involving products of covariances has this limit. Corresponding analysis of the fourth-order cumulant terms shows that if

$$\left|\sum_{r,q=-\infty}^{\infty} \kappa(h, r, q)\right| < \infty \tag{56}$$

the first sum involving fourth-order cumulants has a limit. In particular, if $\{y_t\}$ is a linear process generated by (39) with $\sum_{s=-\infty}^{\infty} |\gamma_s| < \infty$, the sum involving $\kappa(h, r, q + g)$ has the limit

$$\lim_{T\to\infty} \frac{\kappa_4}{T - g} \sum_{t=1}^{T-h} \sum_{q,r=1-t}^{T-g-t} \sum_{s=-\infty}^{\infty} \gamma_s \gamma_{s+h} \gamma_{s+r} \gamma_{s+q+g} = \kappa_4 \sum_{s=-\infty}^{\infty} \gamma_s \gamma_{s+h} \left(\sum_{r=-\infty}^{\infty} \gamma_r\right)^2. \tag{57}$$

The sum involving $\kappa(q - r + h, -r, g - r)$ has the limit

$$\lim_{T\to\infty} \frac{\kappa_4}{T - h} \sum_{t'=1}^{T-g} \sum_{q,r=1-t'}^{T-h-t'} \sum_{s=-\infty}^{\infty} \gamma_s \gamma_{s+q-r+h} \gamma_{s-r} \gamma_{s-r+g} = \kappa_4 \sum_{s=-\infty}^{\infty} \gamma_s \gamma_{s+g} \left(\sum_{r=-\infty}^{\infty} \gamma_r\right)^2. \tag{58}$$

The sum involving $\kappa(s + h - t, r, q + g)$ has the limit

$$\lim_{T\to\infty} \frac{\kappa_4}{(T - h)(T - g)} \sum_{t,s=1}^{T-h} \sum_{q,r=1-t}^{T-g-t} \sum_{p=-\infty}^{\infty} \gamma_p \gamma_{p+s-t+h} \gamma_{p+r} \gamma_{p+q+g} = 0. \tag{59}$$

Corollary 8.3.2. *If the moments of $\{y_t\}$ to fourth-order agree with stationarity and if $\sum_{r=-\infty}^{\infty} \sigma(r) < \infty$, then*

$$\lim_{T\to\infty} (T - g)[(T - h)\operatorname{Cov}(C_g, C_h) - (T - h)\operatorname{Cov}(\bar{C}_g, \bar{C}_h)] = 2\left[\sum_{r=-\infty}^{\infty} \sigma(r)\right]^2 + \sum_{q,r=-\infty}^{\infty} [\kappa(h, r, q) + \kappa(g, r, q)] \tag{60}$$

if $\left|\sum_{s,r,q=-\infty}^{\infty} \kappa(s, r, q)\right| < \infty$. If $\{y_t\}$ is a linear process with $\sum_{t=-\infty}^{\infty} |\gamma_t| < \infty$ and $|\kappa_4| < \infty$, the last two sums in (60) are given by (57) and (58), respectively. The first term on the right-hand side of (60) is $8\pi^2 f^2(0)$.

Thus the difference between $(T-h)\operatorname{Cov}(C_g, C_h)$ and $(T-h)\operatorname{Cov}(\tilde{C}_g, \tilde{C}_h)$ is of order $1/(T-g)$ if $\sum_{r=-\infty}^{\infty} \sigma(r) < \infty$, if $\left|\sum_{s,r,q=-\infty}^{\infty} \kappa(s,r,q)\right| < \infty$, and if the moments of $\{y_t\}$ to fourth-order agree with stationarity. For large samples we can use $\operatorname{Cov}(C_g, C_h)$ as an approximation to $\operatorname{Cov}(\tilde{C}_g, \tilde{C}_h)$. $\operatorname{Cov}(C_g^*, C_h^*)$ can be treated similarly.

8.3.3. The Sample Spectral Density

We now consider the asymptotic means and variances of $I(\lambda)$ and $I^*(\lambda)$. We note that $\sigma(0) = \int_{-\pi}^{\pi} f(\lambda)\, d\lambda < \infty$.

THEOREM 8.3.4. *If* $\sum_{r=-\infty}^{\infty} \sigma(r) \cos \lambda r$ *converges,*

$$\lim_{T\to\infty} \mathscr{E} I(\lambda) = \frac{1}{2\pi} \sum_{r=-\infty}^{\infty} \sigma(r) \cos \lambda r; \tag{61}$$

if $f(\nu)$ *is continuous at* $\nu = \lambda$,

$$\lim_{T\to\infty} \mathscr{E} I(\lambda) = f(\lambda). \tag{62}$$

PROOF. The theorem follows from (66) of Section 8.2 and Lemma 8.3.1 and from (67) of Section 8.2 and Lemma 8.3.3.

THEOREM 8.3.5. *If* $\sum_{r=-\infty}^{\infty} |r\sigma(r)| < \infty$,

$$\lim_{T\to\infty} T[\mathscr{E} I(\lambda) - f(\lambda)] = -\frac{1}{\pi} \sum_{r=1}^{\infty} r\sigma(r) \cos \lambda r. \tag{63}$$

PROOF. From (66) of Section 8.2 we have

$$T[\mathscr{E} I(\lambda) - f(\lambda)] = -\frac{1}{\pi}\left[\sum_{r=1}^{T-1} r\sigma(r) \cos \lambda r + T \sum_{r=T}^{\infty} \sigma(r) \cos \lambda r\right]. \tag{64}$$

Since

$$\left| T \sum_{r=T}^{\infty} \sigma(r) \cos \lambda r \right| \le \sum_{r=T}^{\infty} T\, |\sigma(r)| \le \sum_{r=T}^{\infty} |r| \cdot |\sigma(r)|, \tag{65}$$

the theorem follows.

To study the asymptotic behavior of $\mathscr{E} I^*(\lambda)$ we prove the following lemma:

LEMMA 8.3.4. *If* $f(\nu)$ *is bounded in* $(\lambda - \delta, \lambda + \delta)$ *and* $(\lambda' - \delta, \lambda' + \delta)$ *for some* $\delta > 0$ *and* $\lambda \ne \lambda'$, *then*

$$\lim_{T\to\infty} \int_{-\pi}^{\pi} k_T(\nu - \lambda, \nu - \lambda') f(\nu)\, d\nu = 0, \qquad \lambda \ne \lambda', \quad |\lambda - \lambda'| \ne 2\pi. \tag{66}$$

PROOF. We have

$$|k_T(\nu - \lambda, \nu - \lambda')| \le \frac{1}{2\pi T \sin^2 \frac{1}{2}\delta^*}, \qquad \begin{array}{l} \delta^* \le |\nu - \lambda| \le 2\pi - \delta^*, \\ \delta^* \le |\nu - \lambda'| \le 2\pi - \delta^*, \end{array} \tag{67}$$

for every $\delta^* > 0$. If $\delta^* \le \frac{1}{2}|\lambda - \lambda'|$, then $|\nu - \lambda| < \delta^*$ implies $|\nu - \lambda'| > \frac{1}{2}|\lambda - \lambda'|$ and

$$\begin{aligned} |k_T(\nu - \lambda, \nu - \lambda')| &\le \left| \frac{\sin \frac{1}{2}(\nu - \lambda)T}{T \sin \frac{1}{2}(\nu - \lambda)} \right| \frac{1}{2\pi \sin(|\lambda - \lambda'|/4)} \\ &\le \frac{1}{2\pi \sin(|\lambda - \lambda'|/4)}. \end{aligned} \tag{68}$$

Since δ^* can be arbitrarily small, the lemma results.

THEOREM 8.3.6. *If $f(\nu)$ is continuous at $\nu = 0$ and at $\nu = \lambda$,*

$$\lim_{T \to \infty} T[\mathscr{E}I^*(\lambda) - \mathscr{E}I(\lambda)] = 0, \qquad \lambda \ne 0. \tag{69}$$

PROOF. From (71) of Section 8.2 we obtain

$$\begin{aligned} T[\mathscr{E}I^*(\lambda) - \mathscr{E}I(\lambda)] \\ = \int_{-\pi}^{\pi} \left[-2 \frac{\sin \frac{1}{2}\lambda T}{\sin \frac{1}{2}\lambda} k_T(\nu - \lambda, \nu) + \frac{1}{T} \frac{\sin^2 \frac{1}{2}\lambda T}{\sin^2 \frac{1}{2}\lambda} k_T(\nu) \right] f(\nu)\, d\nu, \\ \lambda \ne 0. \end{aligned} \tag{70}$$

Lemma 8.3.4 implies the first term on the right has a limit of 0. The second term is $(1/T)$ $[\sin^2 \frac{1}{2}\lambda T / \sin^2 \frac{1}{2}\lambda]$ times an integral that converges to $f(0)$. Q.E.D.

Note that $I^*(0) = 0$ while $\mathscr{E}I(0) \to f(0)$.

We now apply Lemma 8.3.4 to

$$\begin{aligned} \operatorname{Cov}[I(\lambda), I(\lambda')] \\ = \left[\int_{-\pi}^{\pi} k_T(\nu + \lambda, \nu - \lambda') f(\nu)\, d\nu \right]^2 + \left[\int_{-\pi}^{\pi} k_T(\nu + \lambda, \nu + \lambda') f(\nu)\, d\nu \right]^2 \\ + \frac{1}{4\pi^2 T^2} \sum_{t,s,t',s'=1}^{T} \cos \lambda(t - s) \cos \lambda'(t' - s') \kappa(s - t, t' - t, s' - t) \end{aligned} \tag{71}$$

to obtain the following theorem:

THEOREM 8.3.7.

$$\lim_{T \to \infty} \operatorname{Var} I(0) = 2f^2(0) + \lim_{T \to \infty} \frac{1}{4\pi^2 T^2} \sum_{t,s,t',s'=1}^{T} \kappa(s - t, t' - t, s' - t) \tag{72}$$

if $f(\nu)$ is continuous at $\nu = 0$ and the limit on the right exists;

$$(73)\quad \lim_{T\to\infty} \operatorname{Var} I(\pi) = \lim_{T\to\infty} \operatorname{Var} I(-\pi)$$
$$= 2f^2(\pi) + \lim_{T\to\infty} \frac{1}{4\pi^2T^2} \sum_{t,s,t',s'=1}^{T} (-1)^{t+s+t'+s'}\kappa(s-t, t'-t, s'-t)$$

if $f(\nu)$ is continuous at $\nu = \pi$ and the limit on the right exists;

$$(74)\quad \lim_{T\to\infty} \operatorname{Var} I(\lambda)$$
$$= f^2(\lambda) + \lim_{T\to\infty} \frac{1}{4\pi^2T^2} \sum_{t,s,t',s'=1}^{T} \cos\lambda(t-s)\cos\lambda(t'-s')\kappa(s-t, t'-t, s'-t),$$
$$\lambda \neq 0, \pm\pi,$$

if $f(\nu)$ is continuous at $\nu = \lambda$ and the limit on the right exists; and

$$(75)\quad \lim_{T\to\infty} \operatorname{Cov}[I(\lambda), I(\lambda')]$$
$$= \lim_{T\to\infty} \frac{1}{4\pi^2T^2} \sum_{t,s,t',s'=1}^{T} \cos\lambda(t-s)\cos\lambda'(t'-s')\kappa(s-t, t'-t, s'-t),$$
$$\lambda \neq \pm\lambda',$$

if $f(\nu)$ is bounded in $(\lambda-\delta, \lambda+\delta)$ and $(\lambda'-\delta, \lambda'+\delta)$ for some $\delta > 0$ and the limit on the right of (75) *exists.*

LEMMA 8.3.5. *If*

$$(76)\qquad \lim_{T\to\infty} \frac{1}{T} \sum_{r,u,w=-(T-1)}^{T-1} |\kappa(r, u, w)| = 0,$$

then

$$(77)\quad \lim_{T\to\infty} \frac{1}{4\pi^2T^2} \sum_{t,s,t',s'=1}^{T} \cos\lambda(t-s)\cos\lambda'(t'-s')\kappa(s-t, t'-t, s'-t) = 0.$$

PROOF. We have

$$(78)\quad \frac{1}{T^2}\left| \sum_{t,s,t',s'=1}^{T} \cos\lambda(t-s)\cos\lambda'(t'-s')\kappa(s-t, t'-t, s'-t)\right|$$
$$\leq \frac{1}{T^2} \sum_{t,s,t',s'=1}^{T} |\kappa(s-t, t'-t, s'-t)|$$
$$\leq \frac{1}{T^2} \sum_{t=1}^{T} \sum_{r,u,w=-(T-1)}^{T-1} |\kappa(r, u, w)|$$
$$= \frac{1}{T} \sum_{r,u,w=-(T-1)}^{T-1} |\kappa(r, u, w)|$$

by the substitution $s = t + r$, $t' = t + u$, $s' = t + w$. Q.E.D.

COROLLARY 8.3.3. *If* (76) *holds*

$$\lim_{T\to\infty} \operatorname{Var} I(0) = 2f^2(0), \qquad \lim_{T\to\infty} \operatorname{Var} I(\pm\pi) = 2f^2(\pi), \tag{79}$$

if $f(\nu)$ *is continuous at* $\nu = 0$ *and* $\nu = \pi$;

$$\lim_{T\to\infty} \operatorname{Var} I(\lambda) = f^2(\lambda), \qquad \lambda \neq 0, \pm\pi, \tag{80}$$

if $f(\nu)$ *is continuous at* $\nu = \lambda$; *and*

$$\lim_{T\to\infty} \operatorname{Cov}\,[I(\lambda), I(\lambda')] = 0, \qquad \lambda \neq \pm\lambda', \tag{81}$$

if $f(\nu)$ *is bounded in* $(\lambda - \delta, \lambda + \delta)$ *and* $(\lambda' - \delta, \lambda' + \delta)$ *for some* $\delta > 0$.

It is of considerable importance to note that the variance of $I(\lambda)$ does not go to 0 as $T \to \infty$. In fact, as we shall see later, $I(\lambda)$ is not a consistent estimate of $f(\lambda)$ [since $2I(\lambda)/f(\lambda)$ has a limiting χ^2-distribution]. Moreover, $I(\lambda)$ and $I(\lambda')$ for $\lambda \neq \pm\lambda'$ are asymptotically uncorrelated. This suggests averaging $I(\lambda)$ over values in an interval in order to improve the estimate of $f(\lambda)$.

The fourth-order cumulant term in the covariance of $I(\lambda)$ and $I(\lambda')$ is 0 if the process is Gaussian. If $\{y_t - \mu\}$ is a linear process, the cumulant term is $1/T$ times

$$\begin{aligned}
(82)\quad &\frac{1}{4\pi^2 T} \sum_{t,s,t',s'=1}^{T} \cos \lambda(t-s) \cos \lambda'(t'-s')\kappa(s-t, t'-t, s'-t) \\
&= \frac{\kappa_4}{4\pi^2 T} \sum_{t,s,t',s'=1}^{T} \cos \lambda(t-s) \cos \lambda'(t'-s') \sum_{p=-\infty}^{\infty} \gamma_{p+t}\gamma_{p+s}\gamma_{p+t'}\gamma_{p+s'} \\
&= \frac{\kappa_4}{4\pi^2 T} \sum_{r,r'=-(T-1)}^{T-1} \cos \lambda r \cos \lambda' r' \sum_{s\in S_r} \sum_{s'\in S_{r'}} \sum_{p=-\infty}^{\infty} \gamma_{p+s+r}\gamma_{p+s}\gamma_{p+s'+r'}\gamma_{p+s'},
\end{aligned}$$

where $S_r = \{1, 2, \ldots, T-r\}$ for $r \geq 0$ and $S_r = \{1-r, 2-r, \ldots, T\}$ for $r \leq 0$. The absolute value of (82) is not greater than

$$\begin{aligned}
(83)\quad &\frac{|\kappa_4|}{4\pi^2 T} \sum_{t,s,t',s'=1}^{T} \sum_{p=-\infty}^{\infty} |\gamma_{p+t}| \cdot |\gamma_{p+s}| \cdot |\gamma_{p+t'}| \cdot |\gamma_{p+s'}| \\
&\leq \frac{|\kappa_4|}{4\pi^2 T} \sum_{t=1}^{T} \sum_{s,t',s',p=-\infty}^{\infty} |\gamma_{p+t}| \cdot |\gamma_{p+s}| \cdot |\gamma_{p+t'}| \cdot |\gamma_{p+s'}| \\
&= \frac{|\kappa_4|}{4\pi^2} \left(\sum_{p=-\infty}^{\infty} |\gamma_p| \right)^4,
\end{aligned}$$

which is finite if $\sum_{p=-\infty}^{\infty} |\gamma_p| < \infty$ and $|\kappa_4| < \infty$. Thus the absolute value of (82) is bounded by an absolutely convergent sum.

COROLLARY 8.3.4. *If* $\{y_t\}$ *is generated by* $y_t = \mu + \sum_{s=-\infty}^{\infty} \gamma_s v_{t-s}$, *where* $\mathscr{E} v_t = 0$, $\mathscr{E} v_t^2 = \sigma^2$, $\mathscr{E} v_t v_s = 0$, $t \neq s$, $\mathscr{E} v_t v_s v_r v_q = 0$, $t \neq s$, $t \neq r$, $t \neq q$, $\mathscr{E} v_t^4 = 3\sigma^4 + \kappa_4 < \infty$, $\mathscr{E} v_t^2 v_s^2 = \sigma^4$, $t \neq s$, *and* $\sum_{t=-\infty}^{\infty} |\gamma_t| < \infty$, *then if* $f(\nu)$ *is continuous at* $\nu = 0$ *and* $\nu = \pi$,

$$\lim_{T\to\infty} \operatorname{Var} I(0) = 2f^2(0), \qquad \lim_{T\to\infty} \operatorname{Var} I(\pm\pi) = 2f^2(\pi); \tag{84}$$

if $f(\nu)$ *is continuous at* $\nu = \lambda$,

$$\lim_{T\to\infty} \operatorname{Var} I(\lambda) = f^2(\lambda), \qquad \lambda \neq 0, \pm\pi; \tag{85}$$

and if $f(\nu)$ *is bounded in* $(\lambda - \delta, \lambda + \delta)$ *and* $(\lambda' - \delta, \lambda' + \delta)$ *for some* $\delta > 0$,

$$\lim_{T\to\infty} \operatorname{Cov}\,[I(\lambda), I(\lambda')] = 0, \qquad \lambda \neq \pm\lambda'. \tag{86}$$

It may be of interest to evaluate the limit of (82). The triple summation for $r, r' \geq 0$ on the right-hand side of (82) divided by T is

$$\begin{aligned}
\frac{1}{T}\sum_{s=1}^{T-r}\sum_{s'=1}^{T-r'}\sum_{p=-\infty}^{\infty} \gamma_{p+s+r}\gamma_{p+s}\gamma_{p+s'+r'}\gamma_{p+s'}
&= \frac{1}{T}\sum_{s=1}^{T-r}\sum_{s'=1}^{T-r'}\sum_{q=-\infty}^{\infty} \gamma_{q+r'}\gamma_q\gamma_{q+s-s'+r}\gamma_{q+s-s'} \\
&= \sum_{q=-\infty}^{\infty} \gamma_{q+r'}\gamma_q \frac{1}{T}\sum_{s=1}^{T-r}\sum_{s'=1}^{T-r'} \gamma_{q+s-s'+r}\gamma_{q+s-s'}.
\end{aligned} \tag{87}$$

For given q

$$\begin{aligned}
\lim_{T\to\infty} \frac{1}{T}\sum_{s=1}^{T-r}\sum_{s'=1}^{T-r'} \gamma_{q+s-s'+r}\gamma_{q+s-s'} &= \lim_{T\to\infty} \sum_{n=-(T-r'-1)}^{T-r-1} \left(1 - \frac{|n|}{T}\right)\gamma_{q+n+r}\gamma_{q+n} \\
&= \sum_{n=-\infty}^{\infty} \gamma_{q+n+r}\gamma_{q+n} \\
&= \sum_{t=-\infty}^{\infty} \gamma_{t+r}\gamma_t \\
&= \frac{\sigma(r)}{\sigma^2}.
\end{aligned} \tag{88}$$

Thus the limit of (87) is

$$\frac{\sigma(r)}{\sigma^2}\sum_{q=-\infty}^{\infty}\gamma_{q+r'}\gamma_q = \frac{\sigma(r)\sigma(r')}{\sigma^4}. \tag{89}$$

The derivation for $r, r' \geq 0$ was a convenience; the result holds for every pair of integers r, r'. Finally, the limit of (82) is

$$\begin{aligned}\frac{\kappa_4}{4\pi^2\sigma^4}\lim_{T\to\infty}\sum_{r,r'=-(T-1)}^{T-1}\cos\lambda r\cos\lambda' r'\ \sigma(r)\sigma(r') &= \frac{\kappa_4}{4\pi^2\sigma^4}\sum_{r,r'=-\infty}^{\infty}\cos\lambda r\cos\lambda' r'\ \sigma(r)\sigma(r') \\ &= \frac{\kappa_4}{\sigma^4}f(\lambda)f(\lambda').\end{aligned} \tag{90}$$

These steps are justified by the convergence of $\sum_{q=-\infty}^{\infty}|\gamma_q|$.

We now consider the limiting covariances of

$$I^*(\lambda) = \frac{1}{2\pi}\sum_{r=-(T-1)}^{T-1}\left(1-\frac{|r|}{T}C_r^*\right)\cos\lambda r. \tag{91}$$

Then

$$\begin{aligned}T\{\mathrm{Cov}\,[I(\lambda), I(\lambda')] - \mathrm{Cov}\,[I^*(\lambda), I^*(\lambda')]\} &= \frac{1}{4\pi^2}\sum_{r,r'=-(T-1)}^{T-1}\left(1-\frac{|r|}{T}\right) \\ &\times\left(1-\frac{|r'|}{T}\right)\cos\lambda r\cos\lambda' r'\ T[\mathrm{Cov}\,(C_r, C_{r'}) - \mathrm{Cov}\,(C_r^*, C_{r'}^*)].\end{aligned} \tag{92}$$

If $\kappa(g, h, r) = 0$ and if we replace C_h^* by $\tilde{C}_h$, then (92) is approximated by

$$\begin{aligned}\frac{1}{4\pi^2 T}\sum_{r,r'=-(T-1)}^{T-1}\cos\lambda r\cos\lambda' r'\ 2\left[\sum_{q=-\infty}^{\infty}\sigma(q)\right]^2 &= \frac{1}{2\pi^2}\left[\sum_{q=-\infty}^{\infty}\sigma(q)\right]^2\frac{1}{T}\sum_{r,r'=-(T-1)}^{T-1}e^{i\lambda r+i\lambda' r'} \\ &= \frac{1}{2\pi^2}\left[\sum_{q=-\infty}^{\infty}\sigma(q)\right]^2\frac{1}{T}\frac{\sin\frac{1}{2}\lambda(2T-1)}{\sin\frac{1}{2}\lambda}\frac{\sin\frac{1}{2}\lambda'(2T-1)}{\sin\frac{1}{2}\lambda'} \\ &\to 0, \qquad \lambda, \lambda' \neq 0.\end{aligned} \tag{93}$$

This is a heuristic argument.

Now let us find the limiting variances and covariances of $\sqrt{T}\,A(\lambda)$ and $\sqrt{T}\,B(\lambda)$, from Section 8.2.3.

THEOREM 8.3.8. *If $f(\nu)$ is continuous at $\nu = 0$,*

$$\lim_{T\to\infty} T\mathscr{E}A^2(0) = 8\pi f(0), \tag{94}$$

$$\lim_{T\to\infty} T\mathscr{E}B^2(0) = 0; \tag{95}$$

if $f(\nu)$ is continuous at $\nu = \pi$,

$$\lim_{T\to\infty} T\mathscr{E}A^2(\pm\pi) = 8\pi f(\pi), \tag{96}$$

$$\lim_{T\to\infty} T\mathscr{E}B^2(\pm\pi) = 0; \tag{97}$$

if $f(\nu)$ is continuous at $\nu = \lambda$,

$$\lim_{T\to\infty} T\mathscr{E}A^2(\lambda) = 4\pi f(\lambda), \qquad \lambda \neq 0, \quad \pm\pi, \tag{98}$$

$$\lim_{T\to\infty} T\mathscr{E}B^2(\lambda) = 4\pi f(\lambda), \qquad \lambda \neq 0, \quad \pm\pi; \tag{99}$$

if $f(\nu)$ is continuous at $\nu = \lambda$ and $\nu = \lambda'$,

$$\lim_{T\to\infty} T\mathscr{E}A(\lambda)A(\lambda') = 0, \qquad \lambda \neq \pm\lambda', \tag{100}$$

$$\lim_{T\to\infty} T\mathscr{E}B(\lambda)B(\lambda') = 0, \qquad \lambda \neq \pm\lambda', \tag{101}$$

$$\lim_{T\to\infty} T\mathscr{E}A(\lambda)B(\lambda') = 0. \tag{102}$$

If $\sqrt{T}\,A(\lambda)$ and $\sqrt{T}\,B(\lambda)$ have a limiting normal distribution, then $\sqrt{T}\,A(\lambda)/\sqrt{4\pi f(\lambda)}$ and $\sqrt{T}\,B(\lambda)/\sqrt{4\pi f(\lambda)}$, $\lambda \neq 0, \pm\pi$, have a limiting normal distribution with means 0, variances 1, and correlation 0, and

$$T\,\frac{A^2(\lambda) + B^2(\lambda)}{4\pi f(\lambda)} = T\,\frac{R^2(\lambda)}{4\pi f(\lambda)} = \frac{2I(\lambda)}{f(\lambda)} \tag{103}$$

has a limiting χ^2-distribution with 2 degrees of freedom.

8.4. ASYMPTOTIC DISTRIBUTIONS OF THE SAMPLE MEAN, COVARIANCES, AND SPECTRAL DENSITY

8.4.1. The Sample Mean

A central limit theorem was proved in Section 7.7 (Theorem 7.7.8).

THEOREM 8.4.1. *If* $y_t = \mu + \sum_{s=-\infty}^{\infty} \gamma_s v_{t-s}$, *where* $\{v_t\}$ *consists of independently and identically distributed random variables with* $\mathscr{E} v_t = 0$ *and* $\mathscr{E} v_t^2 = \sigma^2$ *and* $\sum_{t=-\infty}^{\infty} |\gamma_t| < \infty$, *then* $\sqrt{T}\,(\bar{y} - \mu)$ *has a limiting normal distribution with mean* 0 *and variance*

$$\sigma^2\left(\sum_{s=-\infty}^{\infty} \gamma_s\right)^2 = \sum_{r=-\infty}^{\infty} \sigma(r). \tag{1}$$

There are alternative conditions for the asymptotic normality of the sample mean; one will be given in Section 8.4.3. The result is useful because it permits the use of normal theory for inferences about μ when T is large.

8.4.2. The Sample Covariances

For large samples statistical inference for the covariances may be based on the asymptotic normality of the sample covariances.

THEOREM 8.4.2. *If* $y_t = \mu + \sum_{s=-\infty}^{\infty} \gamma_s v_{t-s}$, *where* $\{v_t\}$ *consists of independently and identically distributed random variables with* $\mathscr{E} v_t = 0$, $\mathscr{E} v_t^2 = \sigma^2$, *and* $\mathscr{E} v_t^4 = 3\sigma^4 + \kappa_4 < \infty$, *and* $\sum_{t=-\infty}^{\infty} |\gamma_t| < \infty$, *then* $\sqrt{T}\,[C_0 - \sigma(0)], \ldots, \sqrt{T}\,[C_n - \sigma(n)]$ *have a limiting normal distribution with means* 0 *and covariances*

$$\begin{aligned} \lim_{T\to\infty} T \operatorname{Cov}(C_h, C_g) &= \sum_{r=-\infty}^{\infty} [\sigma(r+h)\sigma(r+g) + \sigma(r-g)\sigma(r+h)] + \frac{\kappa_4}{\sigma^4}\sigma(h)\sigma(g) \\ &= 4\pi \int_{-\pi}^{\pi} \cos \nu g \cos \nu h\, f^2(\nu)\, d\nu + \frac{\kappa_4}{\sigma^4}\sigma(h)\sigma(g). \end{aligned} \tag{2}$$

PROOF. Suppose $\mu = 0$. Let

$$y_{t,k} = \sum_{s=-k}^{k} \gamma_s v_{t-s}, \tag{3}$$

$$\begin{aligned} C_{h,k} &= \frac{1}{T-h} \sum_{t=1}^{T-h} y_{t,k} y_{t+h,k} \\ &= \frac{1}{T-h} \sum_{t=1}^{T-h} \sum_{s,s'=-k}^{k} \gamma_s \gamma_{s'} v_{t-s} v_{t+h-s'}, \qquad h = 0, 1, \ldots, T-1, \end{aligned} \tag{4}$$

$$\begin{aligned}(5)\qquad \sigma(h, k) &= \mathscr{E} y_{t,k} y_{t+h,k} \\ &= \sum_{s,s'=-k}^{k} \gamma_s \gamma_{s'} \mathscr{E} v_{t-s} v_{t+h-s'} \\ &= \sigma^2 \sum_{s=-k}^{k-h} \gamma_s \gamma_{s+h}, \qquad h = 0, 1, \ldots, 2k, \\ &= 0, \qquad h = 2k+1, \ldots .\end{aligned}$$

Then $\mathscr{E} C_{h,k} = \sigma(h, k)$ and

$$(6)\quad \lim_{T\to\infty} T \operatorname{Cov}(C_{h,k}, C_{g,k})$$

$$= \sum_{r=-\infty}^{\infty} [\sigma(r+h, k)\sigma(r+g, k) + \sigma(r-g, k)\sigma(r+h, k)] + \frac{\kappa_4}{\sigma^4}\sigma(h, k)\sigma(g, k).$$

(Note that the sum on the right has only a finite number of nonzero terms.) Since $y_{t,k} y_{t+h,k}$ is independent of $y_{s,k} y_{s+h,k}$ if t and s differ by more than $2k + h$, $\{y_{t,k} y_{t+h,k}\}$ is a finitely-dependent stationary process with mean $\sigma(h, k)$ and variance

$$\begin{aligned}(7)\quad \operatorname{Var} y_{t,k} y_{t+h,k} &= \sigma^2(0, k) + \sigma^2(h, k) + \kappa_4 \sum_{s=-k}^{k-h} \gamma_s^2 \gamma_{s+h}^2, \qquad h = 0, 1, \ldots, 2k, \\ &= \sigma^2(0, k), \qquad h = 2k+1, \ldots .\end{aligned}$$

By Theorem 7.7.6 $\sqrt{T}[C_{0,k} - \sigma(0, k)], \ldots, \sqrt{T}[C_{n,k} - \sigma(n, k)]$ have a limiting normal distribution with means 0 and covariances given by (6).

Now consider

$$\begin{aligned}(8)\qquad \sqrt{T}(C_h - C_{h,k}) &= \frac{\sqrt{T}}{T-h} \sum_{t=1}^{T-h} (y_t y_{t+h} - y_{t,k} y_{t+h,k}) \\ &= \frac{\sqrt{T}}{T-h} \sum_{t=1}^{T-h} (u_{t,k} y_{t+h,k} + y_{t,k} u_{t+h,k} + u_{t,k} u_{t+h,k}),\end{aligned}$$

where

$$(9)\qquad u_{t,k} = y_t - y_{t,k} = \sum_{|s|>k} \gamma_s v_{t-s}.$$

Let

$$S_1 = \frac{\sqrt{T}}{T-h}\sum_{t=1}^{T-h} u_{t,k}y_{t+h,k}, \tag{10}$$

$$S_2 = \frac{\sqrt{T}}{T-h}\sum_{t=1}^{T-h} y_{t,k}u_{t+h,k}, \tag{11}$$

$$S_3 = \frac{\sqrt{T}}{T-h}\sum_{t=1}^{T-h} u_{t,k}u_{t+h,k}. \tag{12}$$

Let $\gamma_s' = \gamma_s$ for $|s| \leq k$ and $\gamma_s' = 0$ for $|s| > k$, and let $\gamma_s^* = 0$ for $|s| \leq k$ and $\gamma_s^* = \gamma_s$ for $|s| > k$. Then (for $0 \leq h \leq k$)

$$\mathscr{E}S_1 = \sqrt{T}\,\sigma^2 \sum_{s=-\infty}^{\infty} \gamma_s^*\gamma_{s+h}', \tag{13}$$

$$\mathscr{E}S_2 = \sqrt{T}\,\sigma^2 \sum_{s=-\infty}^{\infty} \gamma_s'\gamma_{s+h}^*, \tag{14}$$

$$\mathscr{E}S_3 = \sqrt{T}\,\sigma^2 \sum_{s=-\infty}^{\infty} \gamma_s^*\gamma_{s+h}^*. \tag{15}$$

Note

$$\mathscr{E}S_1 + \mathscr{E}S_2 + \mathscr{E}S_3 = \sqrt{T}\,\mathscr{E}C_h - \sqrt{T}\,\mathscr{E}C_{h,k}. \tag{16}$$

The variance of S_3 is

$$\begin{aligned}
\text{Var}\, S_3 &= \frac{T}{(T-h)^2}\sum_{t,t'=1}^{T-h} \text{Cov}\,(u_{t,k}u_{t+h,k},\, u_{t',k}u_{t'+h,k}) \\
&= \frac{T}{(T-h)^2}\sum_{t,t'=1}^{T-h}\ \sum_{s,r,s',r'=-\infty}^{\infty} \gamma_s^*\gamma_r^*\gamma_{s'}^*\gamma_{r'}^*\,\text{Cov}\,(v_{t-s}v_{t+h-r},\, v_{t'-s'}v_{t'+h-r'}) \\
&= \frac{T}{(T-h)^2}\sum_{t,t'=1}^{T-h}\Bigg[\sigma^4 \sum_{s,r=-\infty}^{\infty} \gamma_s^*\gamma_{s-t+t'}^*\gamma_r^*\gamma_{r-t+t'}^* \\
&\quad + \sigma^4 \sum_{s,r=-\infty}^{\infty} \gamma_s^*\gamma_{s-t+t'+h}^*\gamma_r^*\gamma_{r-t+t'-h}^* + \kappa_4 \sum_{s=-\infty}^{\infty} \gamma_s^*\gamma_{s-t+t'}^*\gamma_{s+h}^*\gamma_{s-t+t'+h}^*\Bigg].
\end{aligned} \tag{17}$$

Thus for $T \geq 2h$

$$(18)\quad \operatorname{Var} S_3 \leq \frac{T}{T-h}\left[\sigma^4 \sum_{p,s,r=-\infty}^{\infty} |\gamma_s^*|\cdot|\gamma_{s+p}^*|\cdot|\gamma_r^*|\cdot|\gamma_{r+p}^*| + \sigma^4 \sum_{p,s,r=-\infty}^{\infty} |\gamma_s^*|\cdot|\gamma_{s+p+h}^*|\cdot|\gamma_r^*|\cdot|\gamma_{r+p-h}^*| + \kappa_4 \sum_{p,s=-\infty}^{\infty} |\gamma_s^*|\cdot|\gamma_{s+p}^*|\cdot|\gamma_{s+h}^*|\cdot|\gamma_{s+p+h}^*|\right]$$

$$\leq 2\left[2\sigma^4 \sum_{q,p,s,r=-\infty}^{\infty} |\gamma_s^*|\cdot|\gamma_p^*|\cdot|\gamma_r^*|\cdot|\gamma_q^*| + \kappa_4 \sum_{q,s=-\infty}^{\infty} |\gamma_s^*|\cdot|\gamma_{s+h}^*|\cdot|\gamma_q^*|\cdot|\gamma_{q+h}^*|\right]$$

$$= 2\left[2\sigma^4\left(\sum_{|s|>k} |\gamma_s|\right)^4 + \kappa_4\left(\sum_{\substack{|s|>k \\ |s+h|>k}} |\gamma_s \gamma_{s+h}|\right)^2\right],$$

which does not depend on T and goes to 0 as $k \to \infty$. Similarly for $T \geq 2h$

(19)

$$\operatorname{Var} S_1 \leq 2\left[2\sigma^4 \sum_{q,p,s,r=-\infty}^{\infty} |\gamma_q^*|\cdot|\gamma_p'|\cdot|\gamma_s^*|\cdot|\gamma_r'| + \kappa_4 \sum_{q,s=-\infty}^{\infty} |\gamma_s^*|\cdot|\gamma_{s+h}'|\cdot|\gamma_q^*|\cdot|\gamma_{q+h}'|\right]$$

$$\leq 2\left[2\sigma^4\left(\sum_{|p|\leq k} |\gamma_p|\right)^2\left(\sum_{|s|>k} |\gamma_s|\right)^2 + \kappa_4\left(\sum_{|s|>k} |\gamma_s \gamma_{s+h}|\right)^2\right],$$

and $\operatorname{Var} S_2$ has the same bound. Then

$$(20)\qquad \operatorname{Var}(S_1 + S_2 + S_3) \leq 3(\operatorname{Var} S_1 + \operatorname{Var} S_2 + \operatorname{Var} S_3).$$

By Corollary 7.7.1 the limiting distribution of the $\sqrt{T}\,[C_h - \sigma(h)]$ is the limit with respect to k of the limiting distribution of the $\sqrt{T}\,[C_{h,k} - \sigma(h, k)]$. This proves the theorem.

COROLLARY 8.4.1. *Under the conditions of Theorem* 8.4.2 *the limiting distributions of* $\sqrt{T}\,[C_0^* - \sigma(0)], \ldots, \sqrt{T}\,[C_n^* - \sigma(n)]$, *of* $\sqrt{T}\,[\tilde{C}_0 - \sigma(0)], \ldots, \sqrt{T}\,[\tilde{C}_n - \sigma(n)]$, *of* $\sqrt{T}\,[c_0 - \sigma(0)], \ldots, \sqrt{T}\,[c_n - \sigma(n)]$, *of* $\sqrt{T}\,[c_0^* - \sigma(0)], \ldots, \sqrt{T}\,[c_n^* - \sigma(n)]$, *and of* $\sqrt{T}\,[\tilde{c}_0 - \sigma(0)], \ldots, \sqrt{T}\,[\tilde{c}_n - \sigma(n)]$ *are normal with means* 0 *and covariances given by* (2).

8.4.3. The Trigonometric Coefficients

The trigonometric coefficients

$$A(\lambda) = \frac{2}{T}\sum_{t=1}^{T}(y_t - \mu)\cos\lambda t, \qquad -\pi \leq \lambda \leq \pi, \tag{21}$$

$$B(\lambda) = \frac{2}{T}\sum_{t=1}^{T}(y_t - \mu)\sin\lambda t, \qquad -\pi \leq \lambda \leq \pi, \tag{22}$$

are linear functions of $y_1 - \mu, \ldots, y_T - \mu$ with means 0 and variances and covariances given in Section 8.2.3. If $\{y_t\}$ is a Gaussian process, any set of these trigonometric coefficients is normally distributed. (If the number of these coefficients is greater than T, the distribution is singular.) We shall show in this section that under specified general conditions $\sqrt{T}\,A(\lambda)$ and $\sqrt{T}\,B(\lambda)$ are asymptotically normally distributed with means 0, covariances 0, and variances given in Section 8.3.3.

THEOREM 8.4.3. *If $y_t = \mu + \sum_{s=-\infty}^{\infty}\gamma_s v_{t-s}$, where $\{v_t\}$ consists of independently distributed random variables and v_t has mean 0, variance σ^2, and distribution function $F_t(v)$, and $\sum_{t=-\infty}^{\infty}|\gamma_t| < \infty$, and if*

$$\sup_{t=1,2,\ldots}\int_{|v|>c} v^2\,dF_t(v) \to 0 \tag{23}$$

as $c \to \infty$, then $\sqrt{T}\,A(\lambda_1), \sqrt{T}\,B(\lambda_1), \ldots, \sqrt{T}\,A(\lambda_n), \sqrt{T}\,B(\lambda_n)$, $0 < \lambda_j < \pi$, $\lambda_i \neq \lambda_j$, $i \neq j$, $i, j = 1, \ldots, n$, have a limiting normal distribution with means 0, covariances 0, and variances $4\pi f(\lambda_1), 4\pi f(\lambda_1), \ldots, 4\pi f(\lambda_n), 4\pi f(\lambda_n)$, respectively.

PROOF. Suppose $\mu = 0$. Let

$$A_k(\lambda) = \frac{2}{T}\sum_{t=1}^{T} y_{t,k}\cos\lambda t. \tag{24}$$

The limiting distribution of $\sqrt{T}\,A(\lambda)$ as $T \to \infty$ is the same as the limit as $k \to \infty$ of the limiting distribution of $\sqrt{T}\,A_k(\lambda)$ as $T \to \infty$, for

$$\begin{aligned} \mathscr{E}T[A(\lambda) - A_k(\lambda)]^2 &= \frac{4}{T}\mathscr{E}\left(\sum_{t=1}^{T} u_{t,k}\cos\lambda t\right)^2 \\ &= \frac{4}{T}\mathscr{E}\sum_{t,t'=1}^{T}\cos\lambda t\cos\lambda t' \sum_{|s|,|s'|>k}\gamma_s\gamma_{s'}v_{t-s}v_{t'-s'} \\ &= \frac{4}{T}\sum_{|s|,|s'|>k}\gamma_s\gamma_{s'}\sigma^2\sum_t \cos\lambda t\cos\lambda(t-s+s'), \end{aligned} \tag{25}$$

where the sum on t is for values of t and $t - s + s'$ between 1 and T. The interchange of summation on s and s' and expectation is justified by the fact that

$$\mathscr{E}\left|\sum_{s,s'} \gamma_s\gamma_{s'}v_s v_{s'}\right| \leq \mathscr{E}\sum_{s,s'} |\gamma_s| \cdot |\gamma_{s'}| \cdot |v_s v_{s'}|, \tag{26}$$

all terms in the right-hand sum are nonnegative, and $\sum_{s=-\infty}^{\infty} |\gamma_s| < \infty$. Then

$$\begin{aligned}\mathscr{E}T[A(\lambda) - A_k(\lambda)]^2 &\leq 4\sigma^2 \max_{t=1,\ldots,T} \cos^2 \lambda t \sum_{|s|,|s'|>k} |\gamma_s| \cdot |\gamma_{s'}| \\ &\leq 4\sigma^2\left(\sum_{|s|>k} |\gamma_s|\right)^2,\end{aligned} \tag{27}$$

which goes to 0 as $k \to \infty$; we apply Corollary 7.7.1.

We write

$$\begin{aligned}A_k(\lambda) &= \frac{2}{T}\sum_{t=1}^{T}\sum_{s=-k}^{k} \gamma_s v_{t-s} \cos \lambda t \\ &= \frac{2}{T}\left\{\sum_{r=k+1}^{T-k}\left[\sum_{s=-k}^{k} \gamma_s \cos \lambda(r+s)\right]v_r + \sum_{r=1-k}^{k}\left[\sum_{s=-r+1}^{k} \gamma_s \cos \lambda(r+s)\right]v_r \right. \\ &\quad \left. + \sum_{r=T-k+1}^{T+k}\left[\sum_{s=-k}^{T-r} \gamma_s \cos \lambda(r+s)\right]v_r\right\}.\end{aligned} \tag{28}$$

The sum of squares of the coefficients of $v_{1-k}, \ldots, v_k, v_{T-k+1}, \ldots, v_{T+k}$ times T converges to 0 as T increases; thus the limiting distribution of $\sqrt{T}\,A_k(\lambda)$ is the limiting distribution of $2\sigma\sqrt{\sum_{r=k+1}^{T-k} [\sum_{s=-k}^{k} \gamma_s \cos \lambda(r+s)]^2/T}$ times $\sum_{r=k+1}^{T-k} w_r^T$, where

$$w_r^T = \frac{\displaystyle\sum_{s=-k}^{k} \gamma_s \cos \lambda(r+s)}{\sigma\sqrt{\displaystyle\sum_{r=k+1}^{T-k}\left[\sum_{s=-k}^{k} \gamma_s \cos \lambda(r+s)\right]^2}}\, v_r, \tag{29}$$

$$r = k+1, \ldots, T-k; \; T = 2k+1, \ldots.$$

Then the Lindeberg central limit theorem (Theorem 7.7.2) implies that $\sum_{r=k+1}^{T-k} w_r^T$ has a limiting normal distribution (according to the proof of Theorem 2.6.1) and the multiplicative factor converges to a constant. Then we apply Corollary 7.7.1 to show that $\sqrt{T}\,A(\lambda)$ has a limiting normal distribution. The theorem follows by treating an arbitrary linear combination

$$\begin{aligned}&\sqrt{T}\,[a_1A(\lambda_1) + b_1B(\lambda_1) + \cdots + a_nA(\lambda_n) + b_nB(\lambda_n)] \\ &= \frac{2}{\sqrt{T}}\sum_{t=1}^{T} y_t[a_1 \cos \lambda_1 t + b_1 \sin \lambda_1 t + \cdots + a_n \cos \lambda_n t + b_n \sin \lambda_n t],\end{aligned} \tag{30}$$

since the characteristic function of the limiting distribution of the $2n$ variables is determined by the characteristic functions of all linear combinations (Theorem 7.7.7).

COROLLARY 8.4.2. *If* $y_t = \mu + \sum_{s=-\infty}^{\infty} \gamma_s v_{t-s}$, *where* $\{v_t\}$ *consists of independently distributed random variables and* v_t *has mean* 0, *variance* σ^2, *and distribution function* $F_t(v)$, *and* $\sum_{t=-\infty}^{\infty} |\gamma_t| < \infty$, *and if* (23) *holds as* $c \to \infty$, *then* $\sqrt{T}\, A^*(\lambda_1)$, $\sqrt{T}\, B^*(\lambda_1), \ldots, \sqrt{T}\, A^*(\lambda_n)$, $\sqrt{T}\, B^*(\lambda_n)$, $0 < \lambda_j < \pi$, $\lambda_i \neq \lambda_j$, $i \neq j$, $i, j = 1, \ldots, n$, *have a limiting normal distribution with means* 0, *covariances* 0, *and variances* $4\pi f(\lambda_1), 4\pi f(\lambda_1), \ldots, 4\pi f(\lambda_n), 4\pi f(\lambda_n)$, *respectively.*

PROOF. The corollary follows from the theorem because

$$\sqrt{T}\, A^*(\lambda) = \sqrt{T}\, A(\lambda) - \sqrt{T}\,(\bar{y} - \mu)\left\{\frac{2}{T}\sum_{t=1}^{T} \cos \lambda t\right\}, \tag{31}$$

$$\sqrt{T}\, B^*(\lambda) = \sqrt{T}\, B(\lambda) - \sqrt{T}\,(\bar{y} - \mu)\left\{\frac{2}{T}\sum_{t=1}^{T} \sin \lambda t\right\}, \tag{32}$$

and the terms in braces approach 0 as $T \to \infty$.

The theorem and corollary could be proved with (23) replaced by the assumption that $\mathscr{E}\,|v_t|^{2+\delta} < M$ for some $\delta > 0$. [See Olshen (1967) for weaker conditions.]

Sometimes we are interested in $\bar{A}(\lambda) = A(\lambda)/\sqrt{c_0}$, $\bar{B}(\lambda) = B(\lambda)/\sqrt{c_0}$, and $\bar{I}(\lambda) = I(\lambda)/c_0$. Since $c_0 \to \sigma(0)$ in probability and $\mathscr{E}A(\lambda) = \mathscr{E}B(\lambda) = 0$, the limiting distribution of the set $\sqrt{T}\,\bar{A}(\lambda_j), \sqrt{T}\,\bar{B}(\lambda_j), j = 1, \ldots, n$, is the same as that of $\sqrt{T}\,A(\lambda_j), \sqrt{T}\,B(\lambda_j), j = 1, \ldots, n$, except that for each j $f(\lambda_j)$ is replaced by $\bar{f}(\lambda_j)$. Similarly, the limiting distribution of $\bar{I}(\lambda_1), \ldots, \bar{I}(\lambda_n)$ is that of $I(\lambda_1), \ldots, I(\lambda_n)$ with $f(\lambda_j)$ replaced by $\bar{f}(\lambda_j)$ for each j.

8.4.4. The Sample Spectral Density

Since

$$I(\lambda) = \frac{T}{8\pi}\,[A^2(\lambda) + B^2(\lambda)], \tag{33}$$

we can obtain the limiting distribution of the sample spectral density $I(\lambda)$ from the limiting distribution of the trigonometric coefficients $A(\lambda)$ and $B(\lambda)$.

THEOREM 8.4.4. *If* $y_t = \mu + \sum_{s=-\infty}^{\infty} \gamma_s v_{t-s}$, *where* $\{v_t\}$ *consists of independently distributed random variables and* v_t *has mean* 0, *variance* σ^2, *and distribution function* $F_t(v)$, *and* $\sum_{t=-\infty}^{\infty} |\gamma_t| < \infty$, *and if* (23) *holds as* $c \to \infty$, *then* $2I(\lambda_1)/f(\lambda_1), \ldots, 2I(\lambda_n)/f(\lambda_n)$, $0 < \lambda_j < \pi$, $\lambda_i \neq \lambda_j$, $i \neq j$, $i, j = 1, \ldots, n$, *have a limiting distribution in which these n variables are independent, each having a* χ^2*-distribution with* 2 *degrees of freedom.*

The theorem follows from the general result that if the random vector $\boldsymbol{X}_T$ is a vector-valued function $\boldsymbol{h}(\boldsymbol{Z}_T)$ of a random vector $\boldsymbol{Z}_T$, if $\boldsymbol{h}(z)$ is continuous, and if the limiting distribution of $\boldsymbol{Z}_T$ is the distribution of a vector $\boldsymbol{Z}$, then the limiting distribution of $\boldsymbol{X}_T$ is the distribution of $\boldsymbol{h}(\boldsymbol{Z})$. [See Mann and Wald (1943a) for example.]

Theorem 8.4.4 holds also for $I(\lambda_1), \ldots, I(\lambda_n)$ replaced by $I^*(\lambda_1), \ldots, I^*(\lambda_n)$.

For a large sample we can say that $I(\lambda)$ is distributed approximately as $f(\lambda)\chi_2^2/2$, where χ_2^2 has a χ^2-distribution with 2 degrees of freedom. This shows, of course, that $I(\lambda)$ is not a consistent estimate of $f(\lambda)$. However, the fact that the $I(\lambda)$'s for different values of λ are asymptotically independent suggests that averaging $I(\lambda)$ over values of λ in an interval will give an estimate of the corresponding average of $f(\lambda)$ with a small variance. If $f(\lambda)$ varies little over the range of λ, the estimate will be reasonable. This idea will be developed in detail in the next chapter.

A somewhat different approach has been used by Bartlett (1966), Section 9.2.2.

THEOREM 8.4.5. *If $y_t = \mu + \sum_{s=-\infty}^{\infty} \gamma_s v_{t-s}$, where $\mathscr{E} v_t = 0$, $\mathscr{E} v_t^2 = \sigma^2$, and $\mathscr{E} v_t v_s = 0$, $t \neq s$, and $\sum_{t=-\infty}^{\infty} |\gamma_t| \sqrt{|t|} < \infty$, then the absolute value of the difference between $I(\lambda)$ and $2\pi f(\lambda) I_v(\lambda)/\sigma^2$, where*

$$I_v(\lambda) = \frac{1}{2\pi T}\left[\left(\sum_{t=1}^{T} v_t \cos \lambda t\right)^2 + \left(\sum_{t=1}^{T} v_t \sin \lambda t\right)^2\right], \tag{34}$$

has an expected value that is less than a constant times $1/\sqrt{T}$. If the v_t's are independent and $\mathscr{E} v_t^4 = 3\sigma^4 + \kappa_4 < \infty$, then the mean square of the difference is less than a constant times $1/T$.

PROOF. Let

(35)

$$\begin{aligned}
\sqrt{T}\, A_y(\lambda) + i\sqrt{T}\, B_y(\lambda) &= \frac{2}{\sqrt{T}} \sum_{t=1}^{T} (y_t - \mu) e^{i\lambda t} \\
&= \frac{2}{\sqrt{T}} \sum_{t=1}^{T} \sum_{s=-\infty}^{\infty} \gamma_s v_{t-s} e^{i\lambda t} \\
&= \frac{2}{\sqrt{T}} \sum_{s=-\infty}^{\infty} \gamma_s \sum_{r=1-s}^{T-s} v_r e^{i\lambda(r+s)} \\
&= \frac{2}{\sqrt{T}} \sum_{s=-\infty}^{\infty} \gamma_s e^{i\lambda s}\left[\sum_{r=1}^{T} v_r e^{i\lambda r} - \sum_{r=1}^{-s} v_r e^{i\lambda r} + \sum_{r=T+1}^{T-s} v_r e^{i\lambda r} \right. \\
&\qquad \left. - \sum_{r=T-s+1}^{T} v_r e^{i\lambda r} + \sum_{r=1-s}^{0} v_r e^{i\lambda r}\right],
\end{aligned}$$

where either of the last pairs of sums is 0 if the upper index is smaller than the lower (that is, $\sum_{r=T-s+1}^{T} v_r e^{i\lambda r} \equiv \sum_{r=1-s}^{0} v_r e^{i\lambda r} \equiv 0$ if $s \leq 0$ and $\sum_{r=1}^{-s} v_r e^{i\lambda r} \equiv \sum_{r=T+1}^{T-s} v_r e^{i\lambda r} \equiv 0$ if $s \geq 0$). Then

$$\sqrt{T}\left[A_y(\lambda) - A_v(\lambda)\sum_{s=-\infty}^{\infty}\gamma_s e^{i\lambda s}\right] + i\sqrt{T}\left[B_y(\lambda) - B_v(\lambda)\sum_{s=-\infty}^{\infty}\gamma_s e^{i\lambda s}\right] \tag{36}$$

$$= \frac{2}{\sqrt{T}}\sum_{s=1}^{\infty}\gamma_s e^{i\lambda s}\left[\sum_{r=1-s}^{0} v_r e^{i\lambda r} - \sum_{r=T-s+1}^{T} v_r e^{i\lambda r}\right]$$

$$+ \frac{2}{\sqrt{T}}\sum_{s=-\infty}^{-1}\gamma_s e^{i\lambda s}\left[\sum_{r=T+1}^{T-s} v_r e^{i\lambda r} - \sum_{r=1}^{-s} v_r e^{i\lambda r}\right]$$

$$= \frac{2}{\sqrt{T}}\sum_{s=-\infty}^{\infty}\gamma_s e^{i\lambda s} g_s(\lambda),$$

say. Then the expected value of the absolute value squared of (36) is

$$\frac{4}{T}\mathscr{E}\sum_{s,t=-\infty}^{\infty}\gamma_s\gamma_t e^{i\lambda(s-t)} g_s(\lambda)\bar{g}_t(\lambda) \leq \frac{4}{T}\sum_{s,t=-\infty}^{\infty}|\gamma_s|\cdot|\gamma_t|\,\mathscr{E}\,|g_s(\lambda)\bar{g}_t(\lambda)| \tag{37}$$

$$\leq \frac{4}{T}\sum_{s,t=-\infty}^{\infty}|\gamma_s|\cdot|\gamma_t|\sqrt{\mathscr{E}\,|g_s(\lambda)|^2}\sqrt{\mathscr{E}\,|g_t(\lambda)|^2}$$

$$= \frac{4}{T}\left(\sum_{s=-\infty}^{\infty}|\gamma_s|\sqrt{\mathscr{E}\,|g_s(\lambda)|^2}\right)^2$$

$$\leq \frac{4}{T}\left(\sigma\sum_{s=-\infty}^{\infty}|\gamma_s|\sqrt{2\,|s|}\right)^2$$

$$= \frac{8\sigma^2}{T}\left(\sum_{s=-\infty}^{\infty}|\gamma_s|\sqrt{|s|}\right)^2$$

since

$$\mathscr{E}\,|g_s(\lambda)|^2 = \mathscr{E}\left|\sum_{r=1-s}^{0} v_r e^{i\lambda r} - \sum_{r=T-s+1}^{T} v_r e^{i\lambda r}\right|^2 \tag{38}$$

$$= 2s\sigma^2, \qquad 0 < s < T,$$

$$= 2T\sigma^2, \qquad T \leq s,$$

and

$$\mathscr{E}\,|g_s(\lambda)|^2 = \mathscr{E}\left|\sum_{r=T+1}^{T-s} v_r e^{i\lambda r} - \sum_{r=1}^{-s} v_r e^{i\lambda r}\right|^2 \tag{39}$$

$$= 2\,|s|\,\sigma^2, \qquad -T < s < 0,$$

$$= 2T\sigma^2, \qquad s \leq -T.$$

Similarly the expected value of the fourth power of the absolute value of (36) is (when the v_t's are independent and $\mathscr{E}v_t^4 < \infty$)

$$(40)\quad \frac{16}{T^2}\mathscr{E} \sum_{s,s',t,t'=-\infty}^{\infty} \gamma_s\gamma_{s'}\gamma_t\gamma_{t'}e^{i\lambda(s+s'-t-t')}g_s(\lambda)g_{s'}(\lambda)\bar{g}_t(\lambda)\bar{g}_{t'}(\lambda)$$

$$\leq \frac{16}{T^2}\sum_{s,s',t,t'=-\infty}^{\infty} |\gamma_s|\cdot|\gamma_{s'}|\cdot|\gamma_t|\cdot|\gamma_{t'}|$$

$$\times [\mathscr{E}\,|g_s(\lambda)|^4]^{\frac{1}{4}}[\mathscr{E}\,|g_{s'}(\lambda)|^4]^{\frac{1}{4}}[\mathscr{E}\,|g_t(\lambda)|^4]^{\frac{1}{4}}[\mathscr{E}\,|g_{t'}(\lambda)|^4]^{\frac{1}{4}}$$

$$= \frac{16}{T^2}\left(\sum_{s=-\infty}^{\infty}|\gamma_s|\,[\mathscr{E}\,|g_s(\lambda)|^4]^{\frac{1}{4}}\right)^4$$

$$\leq \frac{48}{T^2}\left[(\mathscr{E}v_t^4)^{\frac{1}{4}}\sum_{s=-\infty}^{\infty}|\gamma_s|\sqrt{2\,|s|}\right]^4.$$

Let the right-hand side of (36) be U. Then

$$(41)\qquad \sqrt{T}\,[A_y(\lambda) + iB_y(\lambda)] = \sqrt{T}\,[A_v(\lambda) + iB_v(\lambda)]\sum_{s=-\infty}^{\infty}\gamma_s e^{i\lambda s} + U,$$

and

$$(42)\quad T[A_y^2(\lambda) + B_y^2(\lambda)] = T[A_v^2(\lambda) + B_v^2(\lambda)]\left|\sum_{s=-\infty}^{\infty}\gamma_s e^{i\lambda s}\right|^2$$

$$+ \sqrt{T}\left[A_v(\lambda)\left(\bar{U}\sum_{s=-\infty}^{\infty}\gamma_s e^{i\lambda s} + U\sum_{s=-\infty}^{\infty}\gamma_s e^{-i\lambda s}\right)\right.$$

$$\left. + iB_v(\lambda)\left(\bar{U}\sum_{s=-\infty}^{\infty}\gamma_s e^{i\lambda s} - U\sum_{s=-\infty}^{\infty}\gamma_s e^{-i\lambda s}\right)\right] + |U|^2.$$

The expected value of $|U|^2$ was bounded in (37). The expected value of the absolute value of the second term on the right-hand side of (42) is less than the square root of the expected value of $T[A_v^2(\lambda) + B_v^2(\lambda)]$, which is less than $4\sigma^2$, times the square root of $4\mathscr{E}\,|U|^2\,|\sum_{s=-\infty}^{\infty}\gamma_s e^{i\lambda s}|^2$, which is less than $8\pi f(\lambda)/\sigma^2$ times the right-hand side of (37). This proves the first part of the theorem.

From (42) we obtain

$$(43)\quad (8\pi)^2[I(\lambda) - 2\pi f(\lambda)I_v(\lambda)/\sigma^2]^2$$
$$= \left\{2\sqrt{T}\left[A_v(\lambda)\mathscr{R}\left(\bar{U}\sum_{s=-\infty}^{\infty}\gamma_s e^{i\lambda s}\right) - B_v(\lambda)\mathscr{I}\left(\bar{U}\sum_{s=-\infty}^{\infty}\gamma_s e^{i\lambda s}\right)\right] + |U|^2\right\}^2$$
$$\leq 4T[|A_v(\lambda)| + |B_v(\lambda)|]^2\, |U|^2 \left|\sum_{s=-\infty}^{\infty}\gamma_s e^{i\lambda s}\right|^2$$
$$+ 4\sqrt{T}\,[|A_v(\lambda)| + |B_v(\lambda)|]\, |U|^3 \left|\sum_{s=-\infty}^{\infty}\gamma_s e^{i\lambda s}\right| + |U|^4.$$

The second part of the theorem follows from the above, Problem 33, (40), and the fact that $\mathscr{E}\,[\sqrt{T}\,A_v(\lambda)]^4$ and $\mathscr{E}\,[\sqrt{T}\,B_v(\lambda)]^4$ are bounded.

A. M. Walker (1965) has shown that the probability limit of the difference between $I(\lambda)$ and $2\pi f(\lambda)\,I_v(\lambda)/\sigma^2$ is 0, where $y_t = \sum_{s=0}^{\infty}\gamma_s v_{t-s}$ and $\sum_{s=0}^{\infty}|\gamma_s| < \infty$. Olshen (1967) has given more results along this line.

8.4.5. The Sample Correlations

If the sample covariances have a limiting normal distribution, then the sample correlations do also. However, the existence of the limiting normal distribution of the correlations can be proved under weaker conditions. We shall do that in this section. The result was given in Theorem 5.7.1 for the finite moving average process.

If $r_h = c_h/c_0$ and $\rho_h = \sigma(h)/\sigma(0)$, we can write

$$(44)\qquad \sqrt{T}\,(r_h - \rho_h) = \frac{\sqrt{T}\,(c_h - \rho_h c_0)}{c_0}.$$

The denominator behaves like $\sigma(0)$ and the numerator has limiting covariance

$$(45)\quad \lim_{T\to\infty} T\,\mathrm{Cov}\,[(c_h - \rho_h c_0), (c_g - \rho_g c_0)]$$
$$= \lim_{T\to\infty} T\,\mathrm{Cov}\,(c_h, c_g) - \rho_h \lim_{T\to\infty} T\,\mathrm{Cov}\,(c_0, c_g)$$
$$- \rho_g \lim_{T\to\infty} T\,\mathrm{Cov}\,(c_0, c_h) + \rho_g\rho_h \lim_{T\to\infty} T\,\mathrm{Var}\,c_0$$
$$= \sum_{r=-\infty}^{\infty}[\sigma(r+g)\sigma(r+h) + \sigma(r-g)\sigma(r+h) - 2\rho_h\sigma(r)\sigma(r+g)$$
$$- 2\rho_g\sigma(r)\sigma(r+h) + 2\rho_g\rho_h\sigma^2(r) + \kappa(h, -r, g-r) - \rho_h\kappa(0, -r, g-r)$$
$$- \rho_g\kappa(h, -r, -r) + \rho_g\rho_h\kappa(0, -r, -r)]$$

$$= \sum_{r=-\infty}^{\infty} [\mathscr{E} y_t y_{t+h} y_{t-r} y_{t+g-r} - \rho_h \mathscr{E} y_t^2 y_{t-r} y_{t+g-r} - \rho_g \mathscr{E} y_t y_{t+h} y_{t-r}^2 + \rho_g \rho_h \mathscr{E} y_t^2 y_{t-r}^2]$$

$$= \sum_{r=-\infty}^{\infty} \mathscr{E} y_t y_{t-r} (y_{t+h} - \rho_h y_t)(y_{t+g-r} - \rho_g y_{t-r})$$

for $\mathscr{E} y_t = 0$ if the series converge. (See Section 8.3.2.) If $y_t = \mu + \sum_{s=-\infty}^{\infty} \gamma_s v_{t-s}$, $\mathscr{E} v_t = 0$, $\mathscr{E} v_t^2 = \sigma^2$ and $\mathscr{E} v_t^4 = 3\sigma^4 + \kappa_4 < \infty$, the v_t's are independent, and $\sum_{s=-\infty}^{\infty} \gamma_s^2 < \infty$, then

$$\begin{aligned}(46)\quad \sum_{r=-\infty}^{\infty} [\kappa(h, -r, g-r) &- \rho_h \kappa(0, -r, g-r) \\ &- \rho_g \kappa(h, -r, -r) + \rho_g \rho_h \kappa(0, -r, -r)] \\ &= \frac{\kappa_4}{\sigma^4} [\sigma(h)\sigma(g) - \rho_h \sigma(0)\sigma(g) - \rho_g \sigma(0)\sigma(h) + \rho_g \rho_h \sigma^2(0)] \\ &= 0.\end{aligned}$$

Then the limiting covariance (45) does not depend on κ_4. When $\sqrt{T}\,[c_h - \sigma(h)]$, $\sqrt{T}\,[c_g - \sigma(g)]$, and $\sqrt{T}\,[c_0 - \sigma(0)]$ have a limiting normal distribution, $\sqrt{T}\,(r_h - \rho_h)$ and $\sqrt{T}\,(r_g - \rho_g)$ have a limiting normal distribution with means 0 and covariances

$$\begin{aligned}(47)\quad w_{gh} &= \sum_{r=-\infty}^{\infty} (\rho_{r+g}\rho_{r+h} + \rho_{r-g}\rho_{r+h} - 2\rho_h \rho_r \rho_{r+g} - 2\rho_g \rho_r \rho_{r+h} + 2\rho_g \rho_h \rho_r^2) \\ &= \frac{4\pi}{\sigma^2(0)} \int_{-\pi}^{\pi} (\cos \nu h - \rho_h)(\cos \nu g - \rho_g) f^2(\nu)\, d\nu\end{aligned}$$

from the integral form of the limiting covariances. Since this limiting distribution does not depend on κ_4, it may be expected that the result may be obtained by assuming only second-order moments.

THEOREM 8.4.6. *Let* $y_t = \mu + \sum_{s=-\infty}^{\infty} \gamma_s v_{t-s}$, *where* $\sum_{h=-\infty}^{\infty} |\gamma_h| < \infty$, $\sum_{h=-\infty}^{\infty} |h|\, \gamma_h^2 < \infty$ *and* $\{v_t\}$ *consists of independently and identically distributed random variables with* $\mathscr{E} v_t = 0$ *and* $\mathscr{E} v_t^2 = \sigma^2 < \infty$. *Let* $r_h = c_h/c_0$, $h = 1, \ldots, m$. *Then the joint distribution of* $\sqrt{T}(r_1 - \rho_1), \ldots, \sqrt{T}(r_m - \rho_m)$, *where* $\rho_h = \sigma(h)/\sigma(0) = \sum_{s=-\infty}^{\infty} \gamma_s \gamma_{s+h} / \sum_{s=-\infty}^{\infty} \gamma_s^2$, *tends to* $N(\mathbf{0}, \boldsymbol{W})$ *when* $T \to \infty$, *where* $\boldsymbol{W} = (w_{gh})$ *is given by* (47).

COROLLARY 8.4.3. *Under the conditions of Theorem* 8.4.6 *the joint distribution of* $\sqrt{T}(r_h^* - \rho_h)$, $h = 1, \ldots, m$, *where* $r_h^* = c_h^*/c_0^*$, *tends to* $N(\mathbf{0}, \boldsymbol{W})$ *when* $T \to \infty$, *where* $\boldsymbol{W} = (w_{gh})$ *is given by* (47).

PROOF. We shall first prove that $\sqrt{T}(r_l - \rho_l)$ has $N(0, w_{ll})$ as its limiting distribution. This will follow from the fact that the limiting distribution of

$$z_T^{(l)} = \frac{1}{\sqrt{T}}\left\{\sum_{t=1}^{T-l} y_t y_{t+l} - \rho_l \sum_{t=1}^{T} y_t^2\right\}, \tag{48}$$

where we assume $\mu = 0$, is $N(0, \sigma^4 s_l)$ with

$$s_l = \left(\sum_{h=-\infty}^{\infty} \gamma_h^2\right)^2 w_{ll}. \tag{49}$$

Let

$$z_{T,k}^{(l)} = \frac{1}{\sqrt{T}}\left\{\sum_{t=1}^{T-l} y_{t,k} y_{t+l,k} - \rho_{l,k} \sum_{t=1}^{T} y_{t,k}^2\right\}, \tag{50}$$

where $y_{t,k}$ is defined in (3) and

$$\rho_{l,k} = \sum_{h=-k}^{k-l} \gamma_h \gamma_{h+l} \Big/ \sum_{h=-k}^{k} \gamma_h^2. \tag{51}$$

Substituting for $y_{t,k}$ in (50) we have

$$z_{T,k}^{(l)} = \frac{1}{\sqrt{T}}\left[\sum_{t=1}^{T-l} \sum_{h,g=-k}^{k} \gamma_h \gamma_g v_{t-h} v_{t+l-g} - \rho_{l,k} \sum_{t=1}^{T} \sum_{h,g=-k}^{k} \gamma_h \gamma_g v_{t-h} v_{t-g}\right]. \tag{52}$$

Let $z_{T,k}^{(l)*}$ denote the expression obtained from $z_{T,k}^{(l)}$ by omitting all terms containing v_s^2 $(1 - k \le s \le T + k)$, namely,

$$z_{T,k}^{(l)*} = \frac{1}{\sqrt{T}}\left[\sum_{t=1}^{T-l} \sum_{h=-k}^{k} \sum_{\substack{g=-k-l \\ h \neq g}}^{k-l} \gamma_h \gamma_{g+l} v_{t-h} v_{t-g} - \rho_{l,k} \sum_{t=1}^{T} \sum_{\substack{h,g=-k \\ h \neq g}}^{k} \gamma_h \gamma_g v_{t-h} v_{t-g}\right]. \tag{53}$$

LEMMA 8.4.1. *Under the conditions of Theorem* 8.4.6 *the limiting distribution of* $z_{T,k}^{(l)*}$ *as* $T \to \infty$ *is* $N(0, \sigma^4 s_{l,k})$, *where*

$$s_{l,k} = \sum_{r=1}^{2k+l} \delta_{r,k}^{(l)2}, \tag{54}$$

$$\delta_{r,k}^{(l)} = \sum_{h=-k}^{k} [\gamma_h' \gamma_{h+l+r}' + \gamma_h' \gamma_{h+l-r}' - \rho_{l,k}(\gamma_h' \gamma_{h+r}' + \gamma_h' \gamma_{h-r}')], \tag{55}$$

and $\gamma_h' = \gamma_h$ *when* $|h| \le k$ *and* $\gamma_h' = 0$ *when* $|h| > k$.

PROOF. This was proved in Section 5.7.3.

Let $z_T^{(l)*}$ denote the expression obtained by substituting $\sum_{s=-\infty}^{\infty} \gamma_s v_{t-s}$ for y_t in (48) and omitting all terms containing v_r^2 $(-\infty < r < \infty)$, namely

$$(56)\qquad z_T^{(l)*} = \frac{1}{\sqrt{T}}\left[\sum_{t=1}^{T-l} \sum_{\substack{h,g=-\infty \\ h\neq g}}^{\infty} \gamma_h \gamma_{g+l} v_{t-h} v_{t-g} - \rho_l \sum_{t=1}^{T} \sum_{\substack{h,g=-\infty \\ h\neq g}}^{\infty} \gamma_h \gamma_g v_{t-h} v_{t-g}\right].$$

LEMMA 8.4.2. *Under the conditions of Theorem 8.4.6 the limiting distribution of $z_T^{(l)*}$ as $T \to \infty$ is $N(0, \sigma^4 s_l)$, where s_l is given by (49).*

PROOF. The limit of $N(0, \sigma^4 s_{l,k})$ (which is the limiting distribution of $z_{T,k}^{(l)*}$ as $T \to \infty$) is $N(0, \sigma^4 s_l)$ as $k \to \infty$ since

$$(57)\qquad \lim_{k\to\infty} s_{l,k} = \frac{1}{2}\sum_{r=-\infty}^{\infty}\left\{\sum_{h=-\infty}^{\infty}[\gamma_h\gamma_{h+l+r} + \gamma_h\gamma_{h+l-r} - \rho_l(\gamma_h\gamma_{h+r} + \gamma_h\gamma_{h-r})]\right\}^2,$$

which is equivalent to (49). Lemma 8.4.2 then follows from Corollary 7.7.1 when we show that

$$(58)\qquad \mathscr{E}\,|R_{T,k}^{(l)*}|^2 \le M_k,$$

$$(59)\qquad \lim_{k\to\infty} M_k = 0,$$

where

$$(60)\qquad R_{T,k}^{(l)*} = z_T^{(l)*} - z_{T,k}^{(l)*}.$$

Let

$$(61)\qquad R_{T,k}^{(l)} = z_T^{(l)} - z_{T,k}^{(l)}.$$

Then

$$\begin{aligned}(62)\qquad \sqrt{T}\, R_{T,k}^{(l)} &= \sum_{t=1}^{T-l}[(y_{t,k} + u_{t,k})(y_{t+l,k} + u_{t+l,k}) - y_{t,k}y_{t+l,k}] \\ &\quad + \sum_{t=1}^{T}[-\rho_l(y_{t,k} + u_{t,k})^2 + \rho_{l,k}y_{t,k}^2] \\ &= \sum_{=1}^{T-l}[u_{t,k}y_{t+l,k} + y_{t,k}u_{t+l,k} + u_{t,k}u_{t+l,k}] \\ &\quad + \sum_{t=1}^{T}[(\rho_{l,k} - \rho_l)y_{t,k}^2 - 2\rho_l y_{t,k}u_{t,k} - \rho_l u_{t,k}^2] \\ &= \sum_{r=1}^{6} T_r,\end{aligned}$$

say, where

$$T_1 = \sum_{t=1}^{T-l} u_{t,k} y_{t+l,k}, \quad \ldots, \quad T_6 = -\rho_l \sum_{t=1}^{T} u_{t,k}^2. \tag{63}$$

Hence from (53), (56), and (60) we have

$$\sqrt{T}\, R_{T,k}^{(l)*} = \sum_{r=1}^{6} T_r^*, \tag{64}$$

where T_r^* is obtained from T_r by expressing the latter as a linear combination of a finite or countably infinite number of terms of the form $\sum_{t=1}^{T-l} v_{t-h} v_{t-g}$ or $\sum_{t=1}^{T} v_{t-h} v_{t-g}$, and omitting the contributions from all terms for which $h = g$.

Now we have

$$\frac{1}{T} \mathscr{E} T_r^{*2} \leq M_{r,k}, \qquad \lim_{k\to\infty} M_{r,k} = 0, \qquad r = 1, \ldots, 6. \tag{65}$$

For example,

$$\mathscr{E} T_3^{*2} = \sum_{\substack{|h|,|g+l|>k \\ h\neq g}} \gamma_h \gamma_{g+l} \sum_{\substack{|h'|,|g'+l|>k \\ h'\neq g'}} \gamma_{h'} \gamma_{g'+l} \sum_{t,t'=1}^{T-l} \mathscr{E} v_{t-h} v_{t-g} v_{t'-h'} v_{t'-g'}. \tag{66}$$

If

$$\mathscr{E} v_{t-h} v_{t-g} v_{t'-h'} v_{t'-g'} \neq 0, \tag{67}$$

we must have either $t - h = t' - h'$, $t - g = t' - g'$, and therefore $t' - t = h' - h = g' - g$, or $t - h = t' - g'$, $t - g = t' - h'$, and therefore $t' - t = g' - h = h' - g$, the value of t' for a nonzero contribution to the final summation in (66), if any, being thus uniquely determined by t. It follows that

$$\begin{aligned} \mathscr{E} T_3^{*2} &\leq (T - l)\sigma^4 \sum_{|h|,|h'|,|g+l|,|g'+l|>k} |\gamma_h| \cdot |\gamma_{h'}| \cdot |\gamma_{g+l}| \cdot |\gamma_{g'+l}| \\ &= (T - l)\sigma^4 \left\{ \sum_{|h|>k} |\gamma_h| \right\}^4, \end{aligned} \tag{68}$$

giving

$$\frac{1}{T} \mathscr{E} T_3^{*2} \leq \sigma^4 \left\{ \sum_{|h|>k} |\gamma_h| \right\}^4, \tag{69}$$

which shows that (65) holds for $r = 3$. A similar type of argument applies for $r = 1, 2, 5, 6$, and for $r = 4$ we use $\lim_{k\to\infty} \rho_{l,k} = \rho_l$. We then obtain (58) from (65) by using the inequality

$$\mathscr{E}\left(\sum_{r=1}^{6} T_r^* \right)^2 \leq 6 \sum_{r=1}^{6} \mathscr{E} T_r^{*2}. \tag{70}$$

This proves Lemma 8.4.2.

LEMMA 8.4.3. *Under the conditions of Theorem* 8.4.6 *the limiting distribution of* $z_T^{(l)}$ *is* $N(0, \sigma^4 s_l)$ *as* $T \to \infty$.

PROOF. We have

(71)
$$\begin{aligned} z_T^{(l)} - z_T^{(l)*} &= \frac{1}{\sqrt{T}}\left[\sum_{t=1}^{T-l}\sum_{h=-\infty}^{\infty}\gamma_h\gamma_{h+l}v_{t-h}^2 - \rho_l\sum_{t=1}^{T}\sum_{h=-\infty}^{\infty}\gamma_h^2 v_{t-h}^2\right] \\ &= \frac{1}{\sqrt{T}}\left[\sum_{h=-\infty}^{\infty}\gamma_h\gamma_{h+l}\sum_{u=1-h}^{T-l-h}v_u^2 - \rho_l\sum_{h=-\infty}^{\infty}\gamma_h^2\sum_{u=1-h}^{T-h}v_u^2\right] \\ &= \frac{1}{\sqrt{T}}\left[\sum_{h=-\infty}^{\infty}\gamma_h\gamma_{h+l}\left(\sum_{t=1}^{T}v_t^2 + U_{T,h}^{(l)}\right) - \rho_l\sum_{h=-\infty}^{\infty}\gamma_h^2\left(\sum_{t=1}^{T}v_t^2 + U_{T,h}^{(0)}\right)\right] \\ &= \frac{1}{\sqrt{T}}\left[\sum_{h=-\infty}^{\infty}\gamma_h\gamma_{h+l}U_{T,h}^{(l)} - \rho_l\sum_{h=-\infty}^{\infty}\gamma_h^2 U_{T,h}^{(0)}\right], \end{aligned}$$

where

(72)
$$U_{T,h}^{(l)} = \sum_{t=1-h}^{T-l-h} v_t^2 - \sum_{t=1}^{T} v_t^2.$$

Then

(73)
$$\mathscr{E}\,|U_{T,h}^{(l)}| \le (2\,|h| + l)\sigma^2$$

and

(74)
$$\mathscr{E}\sqrt{T}\,|z_T^{(l)} - z_T^{(l)*}| \le \sigma^2\left[\sum_{h=-\infty}^{\infty}|\gamma_h|\cdot|\gamma_{h+l}|\,(2\,|h| + l) + |\rho_l|\sum_{h=-\infty}^{\infty}\gamma_h^2\, 2\,|h|\right] < \infty$$

since

(75)
$$\begin{aligned} \left[\sum_{h=-\infty}^{\infty}|h|\cdot|\gamma_h|\cdot|\gamma_{h+l}|\right]^2 &\le \sum_{h=-\infty}^{\infty}|h|\,\gamma_h^2\sum_{h=-\infty}^{\infty}|h|\,\gamma_{h+l}^2 \\ &= \sum_{h=-\infty}^{\infty}|h|\,\gamma_h^2\sum_{g=-\infty}^{\infty}|g-l|\,\gamma_g^2 < \infty. \end{aligned}$$

Thus

(76)
$$\operatorname*{plim}_{T\to\infty}\,[z_T^{(l)} - z_T^{(l)*}] = 0.$$

This proves Lemma 8.4.3.

LEMMA 8.4.4. *Under the conditions of Theorem* 8.4.6 *the limiting distribution of* $\sqrt{T}(r_l - \rho_l)$ *as* $T \to \infty$ *is* $N(0, w_{ll})$.

PROOF. This follows from Lemma 8.4.3, the fact that

$$\sqrt{T}\,(r_l - \rho_l) = \sqrt{T}\,\frac{c_l - \rho_l c_0}{c_0} \tag{77}$$

and

$$\operatorname*{plim}_{T\to\infty} c_0 = \sigma(0) = \sigma^2 \sum_{h=-\infty}^{\infty} \gamma_h^2. \tag{78}$$

The last follows by writing

$$\frac{1}{T}\sum_{t=1}^{T} y_t^2 = \frac{1}{T}\sum_{t=1}^{T} y_{t,k}^2 + \frac{2}{T}\sum_{t=1}^{T} y_{t,k}u_{t,k} + \frac{1}{T}\sum_{t=1}^{T} u_{t,k}^2, \tag{79}$$

and using

$$\frac{1}{T}\mathscr{E}\sum_{t=1}^{T} u_{t,k}^2 = \sigma^2 \sum_{|h|>k} \gamma_h^2 \to 0 \tag{80}$$

as $k \to \infty$ and

$$\frac{1}{T}\mathscr{E}\left|\sum_{t=1}^{T} y_{t,k}u_{t,k}\right| \le \mathscr{E}\,|y_{t,k}u_{t,k}| \le [\mathscr{E} y_{t,k}^2 \mathscr{E} u_{t,k}^2]^{\frac{1}{2}} = \sigma^2\left\{\sum_{|h|\le k}\gamma_h^2 \sum_{|h|>k}\gamma_h^2\right\}^{\frac{1}{2}} \to 0 \tag{81}$$

as $k \to \infty$. It was shown in Section 5.7.3 that

$$\operatorname*{plim}_{T\to\infty} \frac{1}{T}\sum_{t=1}^{T} y_{t,k}^2 = \sigma^2 \sum_{h=-k}^{k} \gamma_h^2 \to \sigma(0) \tag{82}$$

as $k \to \infty$. This proves Lemma 8.4.4.

To complete the proof of Theorem 8.4.6 we observe that a similar argument can be carried through to show that the limiting distribution of $\sum_{l=1}^{m} k_l\sqrt{T}$ $(r_l - \rho_l)$, $\boldsymbol{k} = (k_1, \ldots, k_m)'$ being an arbitrary set of constants, is $N(0, \boldsymbol{k}'\boldsymbol{W}\boldsymbol{k})$; it will then follow by Theorem 7.7.7 that the joint limiting distribution of $\sqrt{T}(r_l - \rho_l)$, $l = 1, \ldots, m$ is $N(\boldsymbol{0}, \boldsymbol{W})$.

The corollary follows from Theorem 8.4.6 and Theorem 8.4.1 since $\sqrt{T}(c_l - c_l^*)$ and hence $\sqrt{T}(r_l - r_l^*)$ converge stochastically to 0 as $T \to \infty$.

Theorem 8.4.6, which assumes the existence of only the second-order moment of v_t, was proved by T. W. Anderson and A. M. Walker (1964). In previous work the existence of higher moments of v_t had always been assumed. For example, Mann and Wald (1943b) suppose that $\mathscr{E}\,|v_t|^r < \infty$ for all $r > 0$ when dealing with a stochastic difference equation. Diananda (1953) supposes that $\mathscr{E}v_t^4 < \infty$, this being a weakening of the condition $\mathscr{E}v_t^6 < \infty$ previously used by Hoeffding and Robbins (1948); and A. M. Walker (1954), in extending Diananda's method to obtain the result for any linear process, retains his condition $\mathscr{E}v_t^4 < \infty$ and also requires that $\sum_{h=-\infty}^{\infty} |h\gamma_h| < \infty$. (Diananda and A. M. Walker considered a generalization of the linear process in which $\{v_t\}$ is merely a finitely-dependent stationary process.) The fact that the contribution of $\mathscr{E}v_t^4$ to (45) automatically vanishes, as was first noted by Bartlett (1946), p. 29, makes it reasonable to expect that the finiteness of $\mathscr{E}v_t^4$ can be dispensed with. Indeed, T. W. Anderson (1959), Section 4, proved a result equivalent to the asymptotic normality of r_1 for the stochastic difference equation of order 1. The argument here is essentially a generalization of that method.

8.5 EXAMPLES

An idea of the behavior of the sample covariance sequence or correlation sequence from a simple stationary random process can be obtained by considering the artificial series generated by Wold (1965) by use of random numbers as described in Section A.2 of the Appendix. The sample correlation sequences and normalized sample spectral densities are given for the cases of a second-order autoregressive process and are graphed in Section A.2. It will be observed how irregular they are. The population spectral densities (and the spectral densities defined by the estimated second-order autoregressive process) are plotted in Figures A.2.7, A.2.8, and A.2.9.

The sample spectral density (times a constant) of the Beveridge trend-free wheat price series (described in Section 4.5) is tabulated in Table A.1.3 of the Appendix and plotted in Figure A.1.3 of the Appendix. These also show considerable variation.

Another example of a sample spectral density is that of the sunspot data (described in Section 5.9) as given in Table A.3.2 and graphed in Figure A.3.1. It is not quite as variable as the others, and has a fairly well-defined peak at a frequency of about 0.09 (a period of about 11 years).

8.6 DISCUSSION

The information obtained from an observed time series, which is considered to be generated by a stochastic process stationary in the wide sense, can be summarized by the sample mean and alternatively by the sample covariance sequence, the sample variance and the sample correlation sequence, the sample spectral density, or the sample variance and normalized sample spectral density.

The sample covariances and spectral densities are subject to considerable sampling variability. The covariances or correlations for small lags give pertinent information about dependence in the process, but those with greater lags are less informative, the lesser dependence being small relative to the sampling variability.

As will be seen in the next chapter, the sample spectral density can be smoothed to obtain an estimate of the population spectral density with relatively small variability. The sample spectral distribution function, $\hat{F}_T(\nu) = \int_{-\pi}^{\nu} I(\lambda)\, d\lambda$, is a consistent estimate of $F(\nu)$ at a continuity point and $\sqrt{T}[\hat{F}_T(\nu) - F(\nu)]$ has a limiting normal distribution. (See Problem 26 of Chapter 9.) The covariance function of the limiting normal distribution function (for nonnegative values of ν) is similar to that for empirical cumulative distribution functions $(0 \le \nu \le \pi)$ except for the fact that the spectral distribution function is not necessarily 1 at $\nu = \pi$; adaptations of tests of goodness of fit can be used. [See Chapter 6 of Grenander and Rosenblatt (1957).]

REFERENCES

Section 8.2. Lanczos (1956).

Section 8.3. Bartlett (1935), (1946), Parzen (1957b).

Section 8.4. T. W. Anderson (1959), T. W. Anderson and A. M. Walker (1964), Bartlett (1946), (1966), Diananda (1953), Hoeffding and Robbins (1948), Mann and Wald (1943a), (1943b), Olshen (1967), A. M. Walker (1954), (1965).

Section 8.5. Wold (1965).

Section 8.6. Grenander and Rosenblatt (1957).

PROBLEMS

1. (Sec. 8.2.2.) Show that if $\sum_{s=1}^{T} \sigma(t-s)$ does not depend on t, $t = 1, \ldots, T$, then $\sigma(h) = \sigma(T-h)$, $h = 1, \ldots, T-1$. [*Hint:*

$$\sum_{s=1}^{T} \sigma(t-s) = \sigma(0) + 2 \sum_{h=1}^{\min(t-1,T-t)} \sigma(h) + \sum_{h=\min(t,T-t+1)}^{\max(t-1,T-t)} \sigma(h), \quad t = 2, \ldots, T-1,$$

$$\sum_{s=1}^{T} \sigma(1-s) = \sum_{h=0}^{T-1} \sigma(h) = \sum_{s=1}^{T} \sigma(T-s).\Bigg]$$

2. (Sec. 8.2.2.) Find the symmetries in the fourth-order cumulant. (a) Show

$$\kappa(h, r, s) = \kappa(h, s, r) = \kappa(r, h, s) = \kappa(r, s, h)$$
$$= \kappa(s, h, r) = \kappa(s, r, h).$$

(b) Show

$$\kappa(h, r, s) = \kappa(-h, r - h, s - h)$$
$$= \kappa(-r, h - r, s - r)$$
$$= \kappa(-s, h - s, r - s).$$

(c) Find 15 other forms of $\kappa(h, r, s)$.

3. (Sec. 8.2.2.) Find $\mathscr{E}Q_1^*$ defined circularly when $\{y_t\}$ is a stochastic process stationary in the wide sense.

4. (Sec. 8.2.2.) Find $\mathscr{E}Q_1^*$ defined according to successive differences when $\{y_t\}$ is a process stationary in the wide sense.

5. (Sec. 8.2.2.) Let

$$Q = a_{11}y_1^2 + a_{22}y_2^2 + 2a_{12}y_1y_2,$$
$$\tilde{Q} = \tfrac{1}{2}(a_{11} + a_{22})(y_1^2 + y_2^2) + 2a_{12}y_1y_2,$$

where $\mathscr{E}y_1 = \mathscr{E}y_2 = 0$, $\mathscr{E}y_1^2 = \mathscr{E}y_2^2 = \sigma(0)$, $\mathscr{E}y_1y_2 = \sigma(1)$, and y_1 and y_2 have a bivariate normal distribution. Show that

$$\mathscr{E}Q - \mathscr{E}\tilde{Q} = 0,$$
$$\text{Var } Q - \text{Var } \tilde{Q} = (a_{11} - a_{22})^2[\sigma^2(0) - \sigma^2(1)].$$

6. (Sec. 8.2.2.) Let

$$Q = 3y_1^2 + y_2^2 + 2y_3^2,$$
$$\tilde{Q} = 2y_1^2 + 2y_2^2 + 2y_3^2,$$

where $\mathscr{E}y_t = 0$, $\mathscr{E}y_ty_{t+h} = \sigma(h)$, $t = 1, 2, 3$, $h = 0, 1, 2$, and y_1, y_2, y_3 have a joint normal distribution. Show that

$$\mathscr{E}Q - \mathscr{E}\tilde{Q} = 0,$$
$$\text{Var } Q - \text{Var } \tilde{Q} = 4[\sigma^2(0) - 3\sigma^2(1) + 2\sigma^2(2)].$$

Show that the difference in variances can be positive, zero, or negative. [*Hint:* Use the first-order stationary Gaussian process to show the last result.]

7. (Sec. 8.2.2.) Derive Lemmas 3.4.3, 3.4.4, and 3.4.5 as special cases of Theorems 8.2.4 and 8.2.5.

8. (Sec. 8.2.2.) Verify (51) to (55).

9. (Sec. 8.2.2.) Verify (58).

10. (Sec. 8.2.2.) Verify

$$\mathscr{E}\tilde{C}_h = \sigma(h) - \frac{1}{T-h}\left[\left(1 - \frac{h}{T-h}\right)\sigma(0) + 2\left(1 - \frac{h}{T-h}\right)\sum_{r=1}^{h}\sigma(r) + \sum_{r=h+1}^{T-1}\left(1 - \frac{r-h}{T-h}\right)\sigma(r)\right], \qquad 1 \le h = T - 2h - 1,$$

$$\mathscr{E}\tilde{C}_h = \sigma(h) - \frac{1}{T-h}\left[\left(1 - \frac{h}{T-h}\right)\sigma(0) + \left(2 - \frac{2h}{T-h}\right)\sum_{r=1}^{T-2h-1}\sigma(r) + \sum_{r=T-2h}^{h}\left(1 - \frac{h-r}{T-h}\right)\sigma(r) + \sum_{r=h+1}^{T-1}\left(1 - \frac{r-h}{T-h}\right)\sigma(r)\right],$$
$$1 \le T - 2h - 1 < h < T - h - 1,$$

$$\mathscr{E}\tilde{C}_h = \sigma(h) - \frac{1}{T-h}\left[\sum_{r=2h-T+1}^{h}\left(1 - \frac{h-r}{T-h}\right)\sigma(r) + \sum_{r=h+1}^{T-1}\left(1 - \frac{r-h}{T-h}\right)\sigma(r)\right],$$
$$T - h - 1 \le h < T - 1.$$

11. (Sec. 8.2.2.) Deduce the second part of (65) from the first part.

12. (Sec. 8.2.2.) Show that if $y_t = \mu + \sum_{s=-\infty}^{\infty} \gamma_s v_{t-s}$, $t = \ldots, -1, 0, 1, \ldots,$ with $\sum_{r=-\infty}^{\infty} |\gamma_r| < \infty$ and v_r's independent, $\mathscr{E}v_r = 0$, and $\mathscr{E}v_r^3 = \kappa_3$

(i) $$\operatorname{Cov}(\bar{y}, C_0) = \frac{\kappa_3}{T}\sum_{h=-(T-1)}^{T-1}\left(1 - \frac{|h|}{T}\right)\sum_{r=-\infty}^{\infty}\gamma_r^2\gamma_{r+h}.$$

Show

(ii) $$\lim_{T\to\infty}\operatorname{Cov}(\bar{y}, C_0) = 0.$$

13. (Sec. 8.2.3.) Show

$$k_T'(\nu) = \frac{\sin\frac{1}{2}\nu T[T\cos\frac{1}{2}\nu T\sin\frac{1}{2}\nu - \sin\frac{1}{2}\nu T\cos\frac{1}{2}\nu]}{2\pi T\sin^3\frac{1}{2}\nu}.$$

14. (Sec. 8.2.3.) Show

$$k_T(\nu \pm 2\pi n) = k_T(\nu), \qquad n = 1, 2, \ldots.$$

15. (Sec. 8.2.3.) Show

$$\frac{d}{d\lambda}\frac{\sin\frac{1}{2}\lambda T}{\sin\frac{1}{2}\lambda} = \frac{T\cos\frac{1}{2}\lambda T\sin\frac{1}{2}\lambda - \sin\frac{1}{2}\lambda T\cos\frac{1}{2}\lambda}{2\sin^2\frac{1}{2}\lambda}.$$

16. (Sec. 8.2.3.) Show

$$\frac{\sin\frac{1}{2}(\lambda \pm 2\pi k)T}{\sin\frac{1}{2}(\lambda \pm 2\pi k)} = (-1)^{k(T-1)}\frac{\sin\frac{1}{2}\lambda T}{\sin\frac{1}{2}\lambda}, \qquad k = 1, 2, \ldots.$$

17. (Sec. 8.2.3.) If $\{y_t\}$ consists of independent random variables with $\mathscr{E}y_t = 0$, $\mathscr{E}y_t^2 = \sigma^2$, and $\mathscr{E}y_t^4 = 3\sigma^4 + \kappa_4$, show

$$\operatorname{Var} I(\lambda) = \frac{\sigma^4}{(2\pi)^2} + \frac{\kappa_4}{(2\pi)^2 T} + \frac{\sigma^4}{(2\pi)^2 T^2} \frac{\sin^2 \lambda T}{\sin^2 \lambda}, \qquad \lambda \neq 0, \pm\pi$$

$$\operatorname{Var} I(0) = \operatorname{Var} I(\pm\pi) = \frac{2\sigma^4}{(2\pi)^2} + \frac{\kappa_4}{(2\pi)^2 T},$$

$$\operatorname{Cov}[I(\lambda), I(\lambda')] = \frac{\kappa_4}{(2\pi)^2 T} + \frac{\sigma^4}{(2\pi)^2 T^2}\left[\frac{\sin^2 \frac{1}{2}(\lambda + \lambda')T}{\sin^2 \frac{1}{2}(\lambda + \lambda')} + \frac{\sin^2 \frac{1}{2}(\lambda - \lambda')T}{\sin^2 \frac{1}{2}(\lambda - \lambda')}\right], \quad \lambda \neq \pm\lambda'.$$

18. (Sec. 8.3.1.) Verify the conditions of Theorem 8.3.1 for the first-order autoregressive process when $|\beta_1| < 1$.

19. (Sec. 8.3.1.) Verify the conditions of Theorem 8.3.1 for a wide sense stationary process generated by a stochastic difference equation.

20. (Sec. 8.3.1.) Let $a_1, a_2, \ldots$ be a sequence of numbers; let

$$s_n = a_1 + a_2 + \cdots + a_n,$$

$$\bar{s}_n = \frac{s_1 + s_2 + \cdots + s_n}{n}.$$

Show

$$\bar{s}_n = \sum_{r=0}^{n-1}\left(1 - \frac{r}{n}\right) a_{r+1}.$$

[*Hint:* Use induction.]

21. (Sec. 8.3.1.) Prove that if $h(\lambda)$ is continuous at $\lambda = \nu$, if $|h(\lambda)|$ is bounded, if

$$\int_{-\pi}^{\pi} l_T(\lambda)\, d\lambda = 1, \qquad T = T_0, \ldots,$$

if there is a number K such that

$$\int_{-\pi}^{\pi} |l_T(\lambda)|\, d\lambda \leq K, \qquad T = T_1, \ldots,$$

and if

$$|l_T(\lambda)| \leq m_T n(\lambda), \qquad \lambda \neq \nu,\ T = T_2, \ldots,$$

where

$$\int_{-\pi}^{\nu - a} n(\lambda)\, d\lambda + \int_{\nu + a}^{\pi} n(\lambda)\, d\lambda < \infty$$

for every $a > 0$ and $m_T \to 0$ as $T \to \infty$, then

$$\lim_{T\to\infty} \int_{-\pi}^{\pi} h(\lambda) l_T(\lambda)\, d\lambda = h(\nu).$$

22. (Sec. 8.3.2.) Prove (28) for $\tilde{C}_h$.

23. (Sec. 8.3.2.) Prove (29).

24. (Sec. 8.3.2.) Show for $g - h$ (≥ 0) even

$$\int_{-\pi}^{\pi} \frac{\sin \frac{1}{2}\mu(T-h) \sin \frac{1}{2}\mu(T-g)}{(T-g)\sin^2 \frac{1}{2}\mu}\, d\mu = 2\pi.$$

25. (Sec. 8.3.2.) Prove $\sum_{t=-\infty}^{\infty} |\gamma_t| < \infty$ implies $\sum_{t=-\infty}^{\infty} \gamma_t^2 < \infty$. [*Hint:* The first condition implies $\lim_{t\to\pm\infty} |\gamma_t| = 0$.]

26. (Sec. 8.3.2.) Prove $\sum_{h=-\infty}^{\infty} \sigma^2(h) < \infty$ if $y_t = \sum_{s=-\infty}^{\infty} \gamma_s v_{t-s}$, $\sum_{s=-\infty}^{\infty} |\gamma_s| < \infty$, $\mathscr{E}v_t = 0$, $\mathscr{E}v_t^2 = \sigma^2$, $\mathscr{E}v_t v_s = 0$, $t \neq s$.

27. (Sec. 8.3.2.) Prove $\sum_{t=-\infty}^{\infty} |\gamma_t| < \infty$ implies $f(\lambda) = \sigma^2 |\sum_{s=-\infty}^{\infty} \gamma_s e^{i\lambda s}|^2/(2\pi)$ is continuous.

28. (Sec. 8.3.2.) Suppose $y_t = \sum_{s=1}^{\infty} \gamma_s v_{t-s}$, $t = \ldots, -1, 0, 1, \ldots,$ where $\gamma_t = (-1)^t/t$, $\mathscr{E}v_t = 0$, $\mathscr{E}v_t^2 = \sigma^2$, $\mathscr{E}v_t v_s = 0$, $t \neq s$.

(a) Show

$$\sigma(h) = (-1)^h \frac{\sigma^2}{h} \sum_{t=1}^{h} \frac{1}{t}, \qquad h = 1, 2, \ldots.$$

(b) Show $\sum_{h=-\infty}^{\infty} \sigma^2(h) < \infty$.

(c) Show $\sum_{h=-\infty}^{\infty} \sigma(h) < \infty$.

29. (Sec. 8.3.2.) Show $\sum_{h=-\infty}^{\infty} \sigma(h) < \infty$ does not imply $\sum_{h=-\infty}^{\infty} \sigma^2(h) < \infty$.

30. (Sec. 8.3.2.) Show that $|\sum_{s,r,q=-\infty}^{\infty} \kappa(s, r, q)| < \infty$ implies (56) for any h. Show that the condition implies the limits of the three fourth-order cumulant sums in (54).

31. (Sec. 8.3.2.) Find

$$\lim_{T\to\infty} (T-g)[(T-h)\operatorname{Cov}(C_g, C_h) - (T-h)\operatorname{Cov}(\tilde{C}_g, \tilde{C}_h)]$$

by use of the integral forms of the covariances when $\kappa(r, s, t) = 0$ and if $f(\lambda)$ is continuous for $-\pi \leq \lambda \leq \pi$.

32. (Sec. 8.4.3.) Show that $\sum_{t=1}^{T} a_t y_t/\sqrt{T}$ has a limiting normal distribution if $|a_t| \leq 1$ and $\{y_t\}$ consists of independently and identically distributed random variables with $\mathscr{E}y_t = 0$ and $\mathscr{E}y_t^2 = \sigma^2 < \infty$. [*Hint:* Show that $\{a_t y_t\}$ satisfies the Lindeberg condition. If $\sum_{t=1}^{\infty} a_t^2 < \infty$, the variance of the limiting normal distribution is 0.]

33. (Sec. 8.4.4.) Show

$$(\mathscr{E}\,|X_1X_2X_3X_4|)^4 \leq \mathscr{E}X_1^4\mathscr{E}X_2^4\mathscr{E}X_3^4\mathscr{E}X_4^4.$$

[*Hint:* Use $(\mathscr{E}\,|Y_1Y_2|)^2 \leq \mathscr{E}Y_1^2\mathscr{E}Y_2^2$ where $Y_1 = X_1X_2$ and $Y_2 = X_3X_4$.]

34. (Sec. 8.4.5.) Suppose $y_t = \sum_{s=-\infty}^{\infty} \gamma_s v_{t-s}$, where $\{v_t\}$ are independently and identically distributed with $\mathscr{E}v_t = 0$, $\mathscr{E}v_t^2 = \sigma^2$, $\mathscr{E}v_t^4 = 3\sigma^4 + \kappa_4 < \infty$. Suppose $\operatorname{plim}_{T\to\infty} \sum_{t,s=1}^{T} a_{ts}^{(T)} y_t y_s / \sum_{t,s=1}^{T} b_{ts}^{(T)} y_t y_s = \alpha/\beta$. Under what conditions is the variance of the asymptotic distribution of the ratio independent of κ_4?

CHAPTER 9

Estimation of the Spectral Density

9.1. INTRODUCTION

If a stationary stochastic process has a spectral distribution function which is absolutely continuous, the spectral density function determines the variance and covariances of the process. It is desired to estimate the spectral density using T observations, $y_1, \ldots, y_T$, from the process. In Chapter 8 the sample spectral density, $I(\lambda) = [1/(2\pi)] \sum_{r=-(T-1)}^{T-1} \cos \lambda r \, c_r$, was studied. It was shown that $\mathscr{E}I(\lambda) \to f(\lambda)$, the population spectral density, as $T \to \infty$; but the variance of $I(\lambda)$ does not approach 0, and $I(\lambda)$ is not a satisfactory estimate of $f(\lambda)$.

In this chapter we consider estimating $f(\lambda)$ at $\lambda = \nu$ by $\hat{f}(\nu) = [1/(2\pi)] \sum_{r=-(T-1)}^{T-1} w_r \cos \nu r \, C_r = [1/(2\pi)] \sum_{r=-(T-1)}^{T-1} w_r^* \cos \nu r \, c_r$, where the $w_r = (T - |r|)w_r^*/T$ are suitably chosen numbers which may depend on T. The estimate may be equivalently written as $\int_{-\pi}^{\pi} w^*(\lambda \mid \nu) I(\lambda) \, d\lambda$, a weighted average of $I(\lambda)$; the weighting function $w^*(\lambda \mid \nu) = [1/(2\pi)] \sum_{r=-(T-1)}^{T-1} w_r^* \cos \lambda r \cos \nu r$ may depend on T. Such estimates are introduced in Section 9.2, and examples are given. The expected value of such an estimate is $\int_{-\pi}^{\pi} w(\lambda \mid \nu) f(\lambda) \, d\lambda$, where $w(\lambda \mid \nu) = [1/(2\pi)] \sum_{r=-(T-1)}^{T-1} w_r \cos \lambda r \cos \nu r$ may depend on T. A sequence of estimates may be a consistent estimator of $f(\nu)$. In Section 9.3.2 a method is given for defining such a sequence, and the asymptotic bias is evaluated. In Section 9.3.3 the asymptotic variances and covariances of such estimates are obtained. These estimates are asymptotically normally distributed (Section 9.4) under appropriate conditions.

Throughout this chapter it is assumed that $\sigma(0) = \int_{-\pi}^{\pi} f(\lambda) \, d\lambda < \infty$. Since $C_r = C_{-r}$, $\cos \lambda r = \cos(-\lambda r)$, etc., we shall always take $w_r = w_{-r}$ and $w_r^* = w_{-r}^*$ (with no loss of generality, but considerable gain in convenience).

If $\mathscr{E}y_t = \mu$ is unknown, C_r^*, c_r^*, $\tilde{C}_r$, and $\tilde{c}_r$ can be substituted into the above definitions to obtain $I^*(\lambda)$ and $\tilde{I}(\lambda)$. These have the same asymptotic properties.

If $\hat{f}(\nu) \geq 0$, $-\pi \leq \nu \leq \pi$, it is the spectral density of a finite moving average process, since $w_r C_r = w_{-r} C_{-r}$ is real, $r = 0, 1, \ldots, T-1$. If $w_r = 0$, $r = K+1, \ldots, T-1$, then the corresponding moving average process is the average of $K+1$ uncorrelated variables. If $w^*(\lambda \mid 0) \geq 0$, $-\pi \leq \lambda \leq \pi$, then $\hat{f}(\nu) \geq 0$, $-\pi \leq \nu \leq \pi$.

9.2. ESTIMATES BASED ON SAMPLE COVARIANCES

9.2.1. Quadratic Forms as Estimates

Since a population covariance $\sigma(h)$ is the expected value of a quadratic form in $y_t - \mu$ and the population spectral density $f(\lambda)$ at a point ν is a linear combination of covariances, it seems natural to estimate $f(\nu)$ by a quadratic form in the observations. For the present we shall assume that μ is known and is 0. Let the quadratic form be

$$\sum_{s,t=1}^{T} w_{st} y_s y_t. \tag{1}$$

[If μ is known but is not 0, we replace y_t in (1) by $y_t - \mu$.] The expected value of this quadratic form is

$$\begin{aligned} \mathscr{E} \sum_{s,t=1}^{T} w_{st} y_s y_t &= \sum_{s,t=1}^{T} w_{st} \sigma(s-t) \\ &= \frac{1}{2\pi} \sum_{r=-(T-1)}^{T-1} w_r \sigma(r), \end{aligned} \tag{2}$$

where

$$\begin{aligned} w_r &= 2\pi \sum_{t=1}^{T-r} w_{t,t+r} \\ &= w_{-r}, \qquad\qquad r = 0, 1, \ldots, T-1; \end{aligned} \tag{3}$$

the expected value of the quadratic form is a linear function of $\sigma(0), \sigma(1), \ldots, \sigma(T-1)$, the population covariances estimable from $y_1, \ldots, y_T$. The expected value can also be written

$$\begin{aligned} \mathscr{E} \sum_{s,t=1}^{T} w_{st} y_s y_t &= \frac{1}{2\pi} \sum_{r=-(T-1)}^{T-1} w_r \int_{-\pi}^{\pi} \cos \lambda r \, f(\lambda) \, d\lambda \\ &= \frac{1}{2\pi} \int_{-\pi}^{\pi} \sum_{r=-(T-1)}^{T-1} w_r \cos \lambda r \, f(\lambda) \, d\lambda \\ &= \int_{-\pi}^{\pi} w(\lambda) f(\lambda) \, d\lambda, \end{aligned} \tag{4}$$

where

$$w(\lambda) = \frac{1}{2\pi}\sum_{r=-(T-1)}^{T-1} w_r \cos \lambda r = \sum_{s,t=1}^{T} w_{st} \cos \lambda(s-t); \tag{5}$$

the expected value of the quadratic form is a weighted average of the spectral density.

It is clear that as far as the expected value goes the estimate (1) could be replaced by a linear combination of sample covariances,

$$W = \frac{1}{2\pi}\sum_{r=-(T-1)}^{T-1} w_r C_r, \tag{6}$$

where

$$\begin{aligned} C_r &= \frac{1}{T-r}\sum_{t=1}^{T-r} y_t y_{t+r} \\ &= C_{-r}, \qquad\qquad r = 0, 1, \ldots, T-1. \end{aligned} \tag{7}$$

The variance of a quadratic function (1) satisfying (3) for given $w_0, w_1, \ldots, w_{T-1}$ may be greater than, equal to, or less than the variance of (6). See Problems 5 and 6 of Chapter 8 for examples of these different cases. It seems reasonable, however, that in many cases the variance of an estimate (6) will be less than the variances of other estimates (1) satisfying (3), but what these cases might be is an open question. In any event the estimates (6) are relatively easy to compute since only $C_0, C_1, \ldots, C_{T-1}$ are needed.

For large samples it is possible to argue that estimates (6) are as good as any others. If $\sum_{s,t=1}^{T} w_{st}^T y_s y_t$ is a sequence of estimates, $T = 1, 2, \ldots$, then $[1/(2\pi)]\sum_{r=-(T-1)}^{T-1} w_r^T C_r$, where $w_r^T = w_{-r}^T = 2\pi \sum_{t=1}^{T-r} w_{t,t+r}^T$, is a sequence of estimates with the same sequence of expected values for any stochastic process stationary in the wide sense. Grenander and Rosenblatt (1957), Section 4.2, have shown that if

$$\lim_{T\to\infty} \mathscr{E}\sum_{s,t=1}^{T} w_{st}^T y_s y_t = f(\nu) \tag{8}$$

for some ν and every spectral density $f(\lambda)$ in some class (such as the class of all linear processes), then the second sequence has a limiting variance no greater than the limiting variance of the first sequence for every linear process $y_t = \sum_{s=-\infty}^{\infty} \gamma_s v_{t-s}$, where the first four moments of $\{v_t\}$ are finite and are the moments of a process of stationary independent random variables with

$$\lim_{s\to\infty} s^{2+\delta}\gamma_s = 0, \qquad \lim_{s\to-\infty} |s|^{2+\delta}\gamma_s = 0, \tag{9}$$

for some $\delta > 0$, and for which $f(\lambda) > 0$ for all λ, $-\pi \leq \lambda \leq \pi$. (We shall not prove this result.) At least asymptotically, estimates based on sample covariances have minimum variance. From now on we shall consider only estimates which are linear combinations of sample covariances.

These estimates can also be expressed in terms of the sample spectral density since (Theorem 8.2.2)

$$(10)\qquad \int_{-\pi}^{\pi} \cos \lambda r\, I(\lambda)\, d\lambda = c_r = \frac{T-|r|}{T} C_r, \qquad r = 0, \pm 1, \ldots, \pm(T-1),$$
$$= 0, \qquad r = \pm T, \pm(T+1), \ldots .$$

We have

$$(11)\qquad W = \frac{1}{2\pi} \sum_{r=-(T-1)}^{T-1} w_r C_r = \frac{1}{2\pi} \sum_{r=-(T-1)}^{T-1} w_r \frac{T}{T-|r|} c_r$$
$$= \frac{1}{2\pi} \sum_{r=-(T-1)}^{T-1} w_r^* c_r,$$

where

$$(12)\qquad w_r^* = \frac{T}{T-|r|} w_r, \qquad r = 0, \pm 1, \ldots, \pm(T-1).$$

Then

$$(13)\qquad W = \frac{1}{2\pi} \sum_{r=-(T-1)}^{T-1} w_r^* \int_{-\pi}^{\pi} \cos \lambda r\, I(\lambda)\, d\lambda$$
$$= \frac{1}{2\pi} \int_{-\pi}^{\pi} \sum_{r=-(T-1)}^{T-1} w_r^* \cos \lambda r\, I(\lambda)\, d\lambda$$
$$= \int_{-\pi}^{\pi} w^*(\lambda) I(\lambda)\, d\lambda,$$

where

$$(14)\qquad w^*(\lambda) = \frac{1}{2\pi} \sum_{r=-(T-1)}^{T-1} w_r^* \cos \lambda r = \frac{1}{2\pi} \sum_{r=-(T-1)}^{T-1} w_r \frac{T}{T-|r|} \cos \lambda r.$$

Note that $w^*(\lambda) \equiv w(\lambda)$ only when $w_r = w_r^* = 0$, $r \neq 0$. The weighting function $w^*(\lambda)$ is pertinent in describing the estimate as a weighted average of the sample spectral density, and the weighting function $w(\lambda)$ is useful to describe the expected value of the estimate as a weighted average of the population spectral density.

The variance of an estimate such as (11) can be found from the results of Section 8.2. Let $G = [1/(2\pi)] \sum_{r=-(T-1)}^{T-1} g_r C_r$ be another estimate. Then

$$(15)\quad \operatorname{Cov}(G, W) = \frac{1}{(2\pi)^2} \operatorname{Cov}\left(g_0 C_0 + 2\sum_{r=1}^{T-1} g_r C_r,\ w_0 C_0 + 2\sum_{s=1}^{T-1} w_s C_s\right)$$

$$= \frac{1}{(2\pi)^2}\Bigg[g_0 w_0 \operatorname{Var} C_0 + 2w_0 \sum_{r=1}^{T-1} g_r \operatorname{Cov}(C_r, C_0)$$

$$+ 2g_0 \sum_{s=1}^{T-1} w_s \operatorname{Cov}(C_0, C_s) + 4\sum_{r,s=1}^{T-1} g_r w_s \operatorname{Cov}(C_r, C_s)\Bigg].$$

If the process is Gaussian,

$$(16)\quad \operatorname{Cov}(G, W)$$

$$= \frac{1}{(2\pi)^2}\int_{-\pi}^{\pi}\int_{-\pi}^{\pi}\Bigg\{2g_0 w_0 \frac{\sin^2 \frac{1}{2}(\nu - \nu')T}{T^2 \sin^2 \frac{1}{2}(\nu - \nu')}$$

$$+ 2w_0 \sum_{r=1}^{T-1} g_r \frac{\sin \frac{1}{2}(\nu - \nu')(T - r) \sin \frac{1}{2}(\nu - \nu')T}{(T - r)T \sin^2 \frac{1}{2}(\nu - \nu')} 2e^{-i\frac{1}{2}(\nu+\nu')r}$$

$$+ 2g_0 \sum_{s=1}^{T-1} w_s \frac{\sin \frac{1}{2}(\nu - \nu')T \sin \frac{1}{2}(\nu - \nu')(T - s)}{T(T - s) \sin^2 \frac{1}{2}(\nu - \nu')} [e^{i\frac{1}{2}(\nu+\nu')s} + e^{-i\frac{1}{2}(\nu+\nu')s}]$$

$$+ 4\sum_{r,s=1}^{T-1} g_r w_s \frac{\sin \frac{1}{2}(\nu - \nu')(T - r) \sin \frac{1}{2}(\nu - \nu')(T - s)}{(T - r)(T - s) \sin^2 \frac{1}{2}(\nu - \nu')}$$

$$\times [e^{i\frac{1}{2}(\nu+\nu')(s-r)} + e^{-i\frac{1}{2}(\nu+\nu')(s+r)}]\Bigg\} f(\nu) f(\nu')\, d\nu\, d\nu'$$

$$= \int_{-\pi}^{\pi}\int_{-\pi}^{\pi} g^*(\lambda) w^*(\lambda') \left\{\left[\int_{-\pi}^{\pi} k_T(\nu + \lambda, \nu - \lambda') f(\nu)\, d\nu\right]^2\right.$$

$$\left. + \left[\int_{-\pi}^{\pi} k_T(\nu + \lambda, \nu + \lambda') f(\nu)\, d\nu\right]^2\right\} d\lambda\, d\lambda'$$

$$= 2\int_{-\pi}^{\pi}\int_{-\pi}^{\pi}\int_{-\pi}^{\pi}\int_{-\pi}^{\pi} g^*(\lambda) w^*(\lambda') k_T(\nu + \lambda, \nu - \lambda')$$

$$\times k_T(\nu' + \lambda, \nu' - \lambda') f(\nu) f(\nu')\, d\lambda\, d\lambda'\, d\nu\, d\nu'$$

$$= 2\int_{-\pi}^{\pi}\int_{-\pi}^{\pi} G_T^*(\nu, \nu') W_T^*(\nu, \nu') f(\nu) f(\nu')\, d\nu\, d\nu',$$

where

$$G_T^*(\nu, \nu') = \int_{-\pi}^{\pi} g^*(\lambda) k_T(\nu + \lambda, \nu' + \lambda)\, d\lambda, \tag{17}$$

$$W_T^*(\nu, \nu') = \int_{-\pi}^{\pi} w^*(\lambda') k_T(\nu + \lambda', \nu' + \lambda')\, d\lambda', \tag{18}$$

and $k_T(\lambda, \nu)$ is given by (73) of Section 8.2. [The first form of Cov (G, W) in (16) is obtained from Theorem 8.2.6, and the second form is obtained from Theorem 8.2.8 and (13) of this section.] The non-Gaussian part of Cov (G, W) is

$$\begin{aligned}
(19)\quad & \frac{1}{(2\pi)^2}\left\{\frac{g_0 w_0}{T^2}\sum_{t,p=1}^{T} \kappa(0, p - t, p - t)\right. \\
& + 2\frac{w_0}{T}\sum_{r=1}^{T-1} \frac{g_r}{T - r}\sum_{t=1}^{T-r}\sum_{p=1}^{T} \kappa(r, p - t, p - t) \\
& + 2\frac{g_0}{T}\sum_{s=1}^{T-1}\frac{w_s}{T - s}\sum_{t=1}^{T}\sum_{p=1}^{T-s} \kappa(0, p - t, p - t + s) \\
& \left. + 4\sum_{r,s=1}^{T-1} \frac{g_r w_s}{(T - r)(T - s)}\sum_{t=1}^{T-r}\sum_{p=1}^{T-s} \kappa(r, p - t, p - t + s)\right\} \\
& = \frac{1}{4\pi^2 T^2}\int_{-\pi}^{\pi}\int_{-\pi}^{\pi} g^*(\lambda) w^*(\lambda') \sum_{t,s,t',s'=1}^{T} \cos \lambda(t - s) \cos \lambda'(t' - s') \\
& \quad \times \kappa(s - t, t' - t, s' - t)\, d\lambda\, d\lambda'.
\end{aligned}$$

9.2.2. Features of Estimating the Spectral Density

If we use the expected value of a statistic as a measure of location, then we would like the expected value of a linear combination of sample covariances W, which is also a weighted average of the sample spectral density, to be close to the value of the population density $f(\lambda)$ at $\lambda = \nu$ when that is what we are estimating. The properties of the expected value can be considered in terms of the coefficients $\{w_r\}$ or in terms of the weighting function $w(\lambda)$. If we are estimating $f(\nu)$, we want $w(\lambda)$ to peak up at $\lambda = \nu$. At the same time we want the variance to be small. In a sense these two desiderata are competitive. We shall consider the asymptotic bias and variance of a sequence of such estimates in Section 9.3.

An estimate is a linear combination of sample covariances. The number of these in the particular estimate $I(\nu)$ is T. As T increases, the sampling variability in each sample covariance decreases, but this effect is balanced by the increase in the number of sample covariances. We shall study estimates which control the increase in the number of sample covariances and the effect of those with sampling variability rather large relative to expected values.

The estimates can also be considered as weighted averages of the sample spectral density. Since the values of the sample spectral density at different points are asymptotically uncorrelated, it may be expected that a weighted average of the values at different points will have a small asymptotic variance. To control the asymptotic variance the sequence of weighting functions $w^*(\lambda)$ should not peak up too rapidly at the point at which the spectral density is being estimated. This is the case when the number of cosine functions in the weighting function increases slowly with T.

If

$$\hat{f}(0) = \frac{1}{2\pi} \sum_{r=-(T-1)}^{T-1} w_r C_r \tag{20}$$

($w_r = w_{-r}$) is an estimate of $f(0)$, then a corresponding estimate of $f(\nu)$ is

$$\hat{f}(\nu) = \frac{1}{2\pi} \sum_{r=-(T-1)}^{T-1} w_r \cos \nu r \, C_r. \tag{21}$$

The weighting functions of $\hat{f}(0)$ are

$$w(\lambda \mid 0) = \frac{1}{2\pi} \sum_{r=-(T-1)}^{T-1} w_r \cos \lambda r, \tag{22}$$

$$w^*(\lambda \mid 0) = \frac{1}{2\pi} \sum_{r=-(T-1)}^{T-1} w_r^* \cos \lambda r, \tag{23}$$

and the weighting functions of $\hat{f}(\nu)$ are

$$\begin{aligned} w(\lambda \mid \nu) &= \frac{1}{2\pi} \sum_{r=-(T-1)}^{T-1} w_r \cos \lambda r \cos \nu r \\ &= \tfrac{1}{2}[w(\lambda - \nu \mid 0) + w(\lambda + \nu \mid 0)], \end{aligned} \tag{24}$$

$$\begin{aligned} w^*(\lambda \mid \nu) &= \frac{1}{2\pi} \sum_{r=-(T-1)}^{T-1} w_r^* \cos \lambda r \cos \nu r \\ &= \tfrac{1}{2}[w^*(\lambda - \nu \mid 0) + w^*(\lambda + \nu \mid 0)]. \end{aligned} \tag{25}$$

Thus, specification of an estimate of $f(0)$ implies an estimate of $f(\nu)$ for every ν. Note that $\hat{f}(\nu) = \hat{f}(-\nu)$, $w(\lambda \mid \nu) = w(\lambda \mid -\nu)$, and $w^*(\lambda \mid \nu) = w^*(\lambda \mid -\nu)$.

Since $I(\lambda)$, $f(\lambda)$, and $w(\lambda \mid \nu)$ and $w^*(\lambda \mid \nu)$ for all ν are even and periodic with period 2π,

$$\int_{-\pi}^{\pi} w(\lambda \mid \nu) f(\lambda)\, d\lambda = \int_{-\pi}^{\pi} w(\lambda - \nu \mid 0) f(\lambda)\, d\lambda, \tag{26}$$

$$\int_{-\pi}^{\pi} w^*(\lambda \mid \nu) I(\lambda)\, d\lambda = \int_{-\pi}^{\pi} w^*(\lambda - \nu \mid 0) I(\lambda)\, d\lambda. \tag{27}$$

We shall usually require $\int_{-\pi}^{\pi} w(\lambda \mid \nu)\, d\lambda = 1$ and $\int_{-\pi}^{\pi} w^*(\lambda \mid \nu)\, d\lambda = 1$ for every ν. This implies that $w_0 = w_0^* = 1$.

The functions $w(\lambda \mid \nu)$ and $w^*(\lambda \mid \nu)$ are called *windows* because in a sense they determine the parts of $f(\lambda)$ and $I(\lambda)$ that are "seen" by the expected value of the estimate and by the estimate, respectively. The typical shape of the weighting function is one large lobe centered at ν, flanked by smaller lobes.

9.2.3. Examples of Estimates of the Spectral Density

The possible estimates are linear combinations of the sample covariances $C_0, C_1, \ldots, C_{T-1}$, and the possible windows are trigonometric polynomials of degree at most $T - 1$. It will be helpful to the reader to plot the sequences $\{w_r\}$ and $\{w_r^*\}$ as well as the functions $w(\lambda \mid 0)$ and $w^*(\lambda \mid 0)$ in each case. Many of the kernels involve $k_T(\lambda) = \sin^2 \frac{1}{2}\lambda T/[2\pi T \sin^2 \frac{1}{2}\lambda]$ and $h_{2T-1}(\lambda) = \sin \frac{1}{2}\lambda(2T-1)/[2\pi \sin \frac{1}{2}\lambda]$.

A. *The Sample Spectral Density*. The sample spectral density $I(\nu)$ is of the above form with $w_r = 1 - |r|/T$ for $r = 0, \pm 1, \ldots, \pm(T-1)$ and $w_r^* = 1$ for $r = 0, \pm 1, \ldots, \pm(T-1)$. The windows are

$$\begin{aligned} w(\lambda \mid 0) &= \frac{1}{2\pi} \sum_{r=-(T-1)}^{T-1} \left(1 - \frac{|r|}{T}\right) \cos \lambda r \\ &= k_T(\lambda) = \frac{\sin^2 \frac{1}{2}\lambda T}{2\pi T \sin^2 \frac{1}{2}\lambda}, \end{aligned} \tag{28}$$

$$\begin{aligned} w^*(\lambda \mid 0) &= \frac{1}{2\pi} \sum_{r=-(T-1)}^{T-1} \cos \lambda r \\ &= h_{2T-1}(\lambda) = \frac{\sin \frac{1}{2}\lambda(2T-1)}{2\pi \sin \frac{1}{2}\lambda}. \end{aligned} \tag{29}$$

[Note that $w^*(\lambda \mid 0)$ is not the Dirac delta function because $I(\nu) = \hat{f}(\nu)$ is a linear combination of only T cosine functions.] The window $w(\lambda \mid 0) = k_T(\lambda)$ is nonnegative and has a maximum of $T/(2\pi)$ at $\lambda = 0$ and minima of 0 at $\lambda = \pm k 2\pi/T$, $k = 1, \ldots, \frac{1}{2}(T-1)$ or $\frac{1}{2}T$, with relative maxima between the

0's; the relative maxima occur at the roots ($\lambda \neq 0$) of $\tan \frac{1}{2}\lambda T = T \tan \frac{1}{2}\lambda$, which are approximately $\pm(2k+1)\pi/T$ for $k = 1, \ldots, \frac{1}{2}(T-2)$ or $\frac{1}{2}(T-1)$. (If T is odd, a relative maximum occurs at $\lambda = \pm\pi$.) The window $w^*(\lambda \mid 0) = \sin \frac{1}{2}\lambda(2T-1)/[2\pi \sin \frac{1}{2}\lambda]$ has a maximum of $(2T-1)/(2\pi)$ at $\lambda = 0$ and has 0's at $\lambda = \pm k2\pi/(2T-1)$ for $k = 1, \ldots, T-1$, with relative minima and maxima alternating between 0's; the relative minima and maxima occur at the roots ($\lambda \neq 0$) of $\tan \frac{1}{2}\lambda(2T-1) = (2T-1) \tan \frac{1}{2}\lambda$. (If T is odd, a relative maximum occurs at $\lambda = \pm\pi$; if T is even, a relative minimum occurs at $\lambda = \pm\pi$.) These windows are graphed in Figures 9.1 and 9.2.

The windows graphed in Figures 9.1 and 9.2 have features that are representative of weighting functions. There is one large lobe centered at 0, and the side lobes decrease in height as the centers of the lobes depart from 0. A kernel that is nonnegative, such as $k_T(\lambda)$, insures that the weighted integral of a nonnegative function [$f(\lambda)$ or $I(\lambda)$] is nonnegative. A kernel that takes on negative as well as positive values, such as that graphed in Figure 9.2, permits a

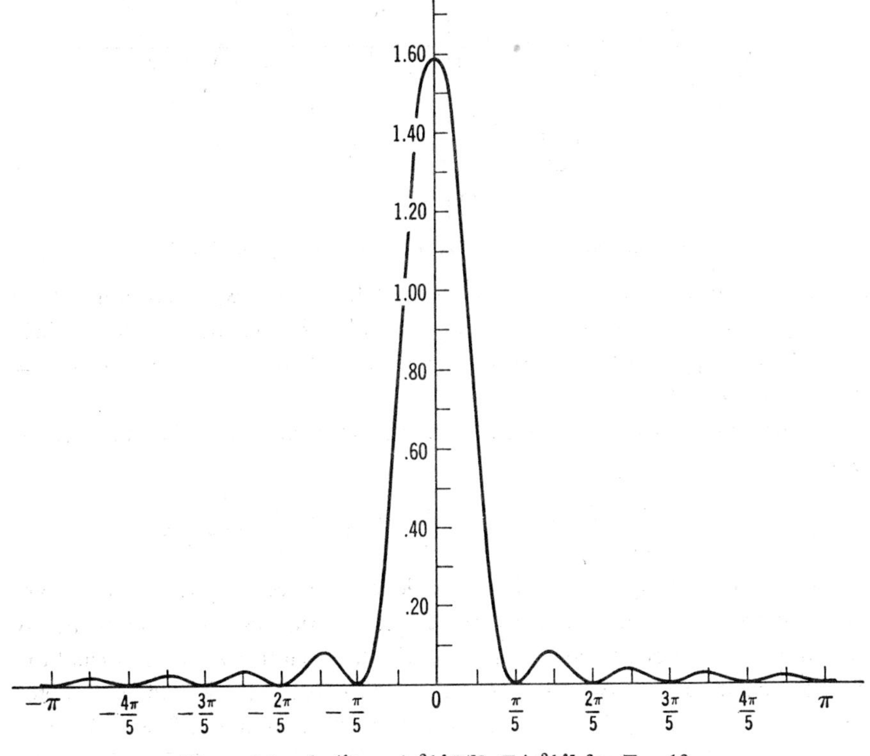

Figure 9.1. $k_T(\lambda) = \sin^2 \frac{1}{2}\lambda T/[2\pi T \sin^2 \frac{1}{2}\lambda]$ for $T = 10$.

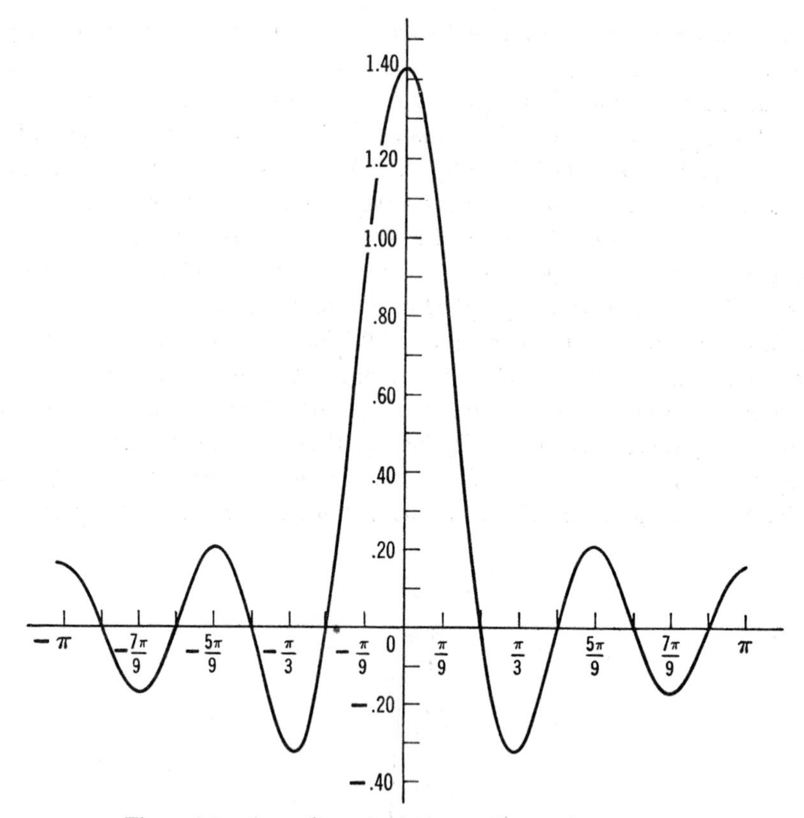

Figure 9.2. $h_{2T-1}(\lambda) = \sin\frac{1}{2}\lambda(2T-1)/[2\pi\sin\frac{1}{2}\lambda]$ for $T = 5$.

weighted integral to be negative; such an estimate or expected value of an estimate is disadvantageous because $f(\lambda)$ and $I(\lambda)$ are nonnegative. Since $w_0 = w_0^* = 1$, $\int_{-\pi}^{\pi} w(\lambda \mid 0)\, d\lambda = \int_{-\pi}^{\pi} k_T(\lambda)\, d\lambda = 1$ and $\int_{-\pi}^{\pi} w^*(\lambda \mid 0)\, d\lambda = \int_{-\pi}^{\pi} h_{2T-1}(\lambda)\, d\lambda = 1$.

B. *Truncated.* The series for the sample spectral density is truncated to obtain

$$\hat{f}(\nu) = \frac{1}{2\pi} \sum_{r=-K}^{K} \cos \nu r\, c_r = \frac{1}{2\pi} \sum_{r=-K}^{K} \left(1 - \frac{|r|}{T}\right) \cos \nu r\, C_r \tag{30}$$

for some integer $K \le T - 1$. The sample covariances $C_{K+1}, \ldots, C_{T-1}$ are omitted in this estimate. These are the sample covariances based on a relatively small number of products and hence contributing relatively little information. The coefficients are $w_r = 1 - |r|/T$ for $r = 0, \pm 1, \ldots, \pm K$ and $w_r = 0$ for $r = \pm(K+1), \ldots, \pm(T-1)$; the other coefficients are $w^* = 1$ for $r = 0, \pm 1, \ldots, \pm K$ and $w_r^* = 0$ for $r = \pm(K+1), \ldots, \pm(T-1)$. The windows are

$$w(\lambda \mid 0) = \frac{K+1}{T} k_{K+1}(\lambda) + \left(1 - \frac{K+1}{T}\right) h_{2K+1}(\lambda), \tag{31}$$

$$w^*(\lambda \mid 0) = h_{2K+1}(\lambda). \tag{32}$$

The maximum of $w(\lambda \mid 0)$ is

$$\frac{2K+1}{2\pi} - \frac{K(K+1)}{2\pi T} \tag{33}$$

at $\lambda = 0$ [which is $(\frac{3}{4}T + \frac{1}{2})/(2\pi)$ for $K = \frac{1}{2}T$, for example]. The window takes on both positive and negative values. The window $w^*(\lambda \mid 0)$ has a maximum of $(2K+1)/(2\pi)$ at $\lambda = 0$ and has 0's at $\lambda = \pm k2\pi/(2K+1)$ for $k = 1, \ldots, K$, with relative minima and maxima between the 0's. The smaller K is, the less is the peak of $w(\lambda \mid 0)$ and $w^*(\lambda \mid 0)$ and the farther apart are the 0's.

C. *Bartlett*.† The sampling error due to the higher-order sample covariances in the truncated estimate can be reduced by use of a damping factor to obtain

$$\begin{aligned} \hat{f}(\nu) &= \frac{1}{2\pi} \sum_{r=-K}^{K} \left(1 - \frac{|r|}{K}\right) \cos \nu r \, C_r \\ &= \frac{1}{2\pi} \sum_{r=-K}^{K} \frac{1 - \dfrac{|r|}{K}}{1 - \dfrac{|r|}{T}} \cos \nu r \, c_r. \end{aligned} \tag{34}$$

The coefficients are $w_r = 1 - |r|/K$ for $r = 0, \pm 1, \ldots, \pm K$ and $w_r = 0$ for $r = \pm(K+1), \ldots, \pm(T-1)$, as compared to $1 - |r|/T$ and 0, respectively, for the simple truncation; the other coefficients are $w_r^* = (1 - |r|/K)/(1 - |r|/T)$ for $r = 0, \pm 1, \ldots, \pm K$ and $w_r^* = 0$ for $r = \pm(K+1), \ldots, \pm(T-1)$, as compared to 1 and 0, respectively, for the simple truncation. The unbiased estimates $\{C_r\}$ are damped linearly to 0 at $r = \pm K$. The windows are

$$w(\lambda \mid 0) = k_K(\lambda) = \frac{\sin^2 \frac{1}{2}\lambda K}{2\pi K \sin^2 \frac{1}{2}\lambda}, \tag{35}$$

$$\begin{aligned} w^*(\lambda \mid 0) &= \frac{1}{2\pi} \frac{T}{K} \sum_{r=-K}^{K} \frac{K - |r|}{T - |r|} e^{i\lambda r} \\ &= \frac{1}{2\pi} \frac{T}{K} \sum_{r=-K}^{K} \left(1 + \frac{K - T}{T - |r|}\right) e^{i\lambda r} \\ &= \frac{T}{K} \left\{ h_{2K+1}(\lambda) - \frac{T-K}{2\pi} \sum_{r=-K}^{K} \frac{\cos \lambda r}{T - |r|} \right\}, \end{aligned} \tag{36}$$

† Some authors do not distinguish between Example C, given by Bartlett (1950), and Example D, a modification permitting easier mathematical treatment.

which cannot be written in a simple form. (See Problem 7.) Some properties of $w(\lambda \mid 0)$ are described in terms of the equivalent $w^*(\lambda \mid 0)$ in Example D.

Bartlett (1950) observed that the sampling variability of an estimate of the spectral density could be reduced by breaking the observed time series $y_1, \ldots, y_T$ into m segments, each of length K ($T = mK$), calculating the sample spectral density for each segment, and averaging these to obtain

$$\text{(37)} \quad \frac{1}{2\pi}\frac{1}{m}\sum_{g=1}^{m}\left\{\frac{1}{K}\sum_{t=1}^{K} y_{(g-1)K+t}^2 + 2\sum_{r=1}^{K-1}\cos\nu r \frac{1}{K}\sum_{t=1}^{K-r} y_{(g-1)K+t}\, y_{(g-1)K+t+r}\right\}$$

$$= \frac{1}{2\pi}\frac{1}{T}\sum_{t=1}^{T} y_t^2 + \frac{2}{2\pi}\sum_{r=1}^{K-1}\left(1 - \frac{r}{K}\right)\cos\nu r \sum_{g=1}^{m}\sum_{t=1}^{K-r} y_{(g-1)K+t}\, y_{(g-1)K+t+r}/[m(K-r)].$$

Each double sum in (37) is an unbiased estimate of the corresponding $\sigma(r)$, but some available products $y_t y_{t+r}$ are not used; when each such unbiased estimate is replaced by the corresponding C_r, (34) results.

D. *Modified Bartlett.* A further damping is effected by applying the factor $1 - |r|/T$ to the preceding estimate to obtain

$$\text{(38)} \quad \hat{f}(\nu) = \frac{1}{2\pi}\sum_{r=-K}^{K}\left(1 - \frac{|r|}{K}\right)\left(1 - \frac{|r|}{T}\right)\cos\nu r\, C_r$$

$$= \frac{1}{2\pi}\sum_{r=-K}^{K}\left(1 - \frac{|r|}{K}\right)\cos\nu r\, c_r.$$

The coefficients are $w_r = (1 - |r|/K)(1 - |r|/T)$ for $r = 0, \pm 1, \ldots, \pm K$ and $w_r = 0$ for $r = \pm(K+1), \ldots, \pm(T-1)$, compared to $1 - |r|/K$ and 0, respectively, for Bartlett's estimate; the other coefficients are $w_r^* = 1 - |r|/K$ for $r = 0, \pm 1, \ldots, \pm K$ and $w_r^* = 0$ for $r = \pm(K+1), \ldots, \pm(T-1)$, compared to $(1 - |r|/K)/(1 - |r|/T)$ and 0, respectively, for Bartlett's estimate. The windows are

$$\text{(39)} \quad w(\lambda \mid 0) = \frac{1}{2\pi}\sum_{r=-K}^{K}\left[1 - |r|\left(\frac{1}{K} + \frac{1}{T}\right) + \frac{r^2}{TK}\right]\cos\lambda r$$

$$= \frac{\cos^2 \frac{1}{2}\lambda K}{2\pi T \sin^2 \frac{1}{2}\lambda} + \frac{\sin^2 \frac{1}{2}\lambda K}{2\pi K \sin^2 \frac{1}{2}\lambda} - \frac{\sin\lambda K \cos\frac{1}{2}\lambda}{4\pi K T \sin^3 \frac{1}{2}\lambda},$$

$$\text{(40)} \quad w^*(\lambda \mid 0) = k_K(\lambda) = \frac{\sin^2 \frac{1}{2}\lambda K}{2\pi K \sin^2 \frac{1}{2}\lambda}.$$

The window $w^*(\lambda \mid 0)$ is nonnegative, which fact assures that the estimate $\hat{f}(\nu)$ is nonnegative. The maximum of $w^*(\lambda \mid 0)$ is $K/(2\pi)$ at $\lambda = 0$, and the minima of 0 occur at $\lambda = \pm k2\pi/K$ for $k = 1, \ldots, \frac{1}{2}(K-1)$ or $\frac{1}{2}K$. Grenander and Rosenblatt (1957), p. 146, and Hannan (1960), p. 60, have considered this modification of Bartlett's estimate in order to include it in a general asymptotic theory.

E. *Daniell.* Instead of trying to estimate $f(\lambda)$ at a point $\lambda = \nu$ one might try to estimate the average of $f(\lambda)$ over an interval. [See discussion by Daniell of Bartlett (1946).] If the interval is $[\nu - b, \nu + b]$, $0 < b < \pi$, the integral $\int_{\nu-b}^{\nu+b} f(\lambda)\, d\lambda/(2b)$ is a weighted integral $\int_{-\pi}^{\pi} g(\lambda - \nu) f(\lambda)\, d\lambda$, where

$$
\begin{aligned}
(41) \qquad g(\lambda) &= \frac{1}{2b}, & -b \le \lambda \le b, \\
&= 0, & b < |\lambda|.
\end{aligned}
$$

(If ν is not in the interval $[-\pi + b, \pi - b]$, we use the facts that $f(\lambda)$ and $I(\lambda)$ are periodic with period 2π.) For the purpose of estimation one might use

$$
\begin{aligned}
(42) \qquad \int_{-\pi}^{\pi} g(\lambda - \nu) I(\lambda)\, d\lambda &= \frac{1}{2b} \int_{\nu-b}^{\nu+b} I(\lambda)\, d\lambda \\
&= \frac{1}{4\pi b}\left[c_0 \int_{\nu-b}^{\nu+b} d\lambda + 2 \sum_{r=1}^{T-1} c_r \int_{\nu-b}^{\nu+b} \cos \lambda r\, d\lambda\right] \\
&= \frac{1}{2\pi}\left[c_0 + 2 \sum_{r=1}^{T-1} \frac{\sin(\nu + b)r - \sin(\nu - b)r}{2br} c_r\right] \\
&= \frac{1}{2\pi}\left[c_0 + 2 \sum_{r=1}^{T-1} \frac{2 \cos \nu r \sin br}{2br} c_r\right] \\
&= \frac{1}{2\pi} \sum_{r=-(T-1)}^{T-1} \frac{\sin br}{br} \cos \nu r\, c_r \\
&= \frac{1}{2\pi} \sum_{r=-(T-1)}^{T-1} \frac{\sin br}{br}\left(1 - \frac{|r|}{T}\right) \cos \nu r\, C_r,
\end{aligned}
$$

where $\sin br/(br) = 1$ for $r = 0$. The coefficients are the Fourier coefficients of $g(\lambda)$, namely, $w_0^* = \int_{-\pi}^{\pi} g(\lambda)\, d\lambda = 1$ and

$$
(43) \qquad w_r^* = \int_{-\pi}^{\pi} g(\lambda) \cos \lambda r\, d\lambda = \frac{1}{2b} \int_{-b}^{b} \cos \lambda r\, d\lambda = \frac{\sin br}{br}, \qquad r \ne 0.
$$

[Note that $g(\lambda)$ is not identical to $w^*(\lambda \mid 0)$.]

The resulting window $w^*(\lambda \mid 0)$ is the linear combination of 1, $\cos \lambda, \ldots,$ $\cos \lambda(T-1)$ that best approximates $g(\lambda)$ in the sense of minimizing the average squared error

$$\int_{-\pi}^{\pi} |g(\lambda) - w^*(\lambda \mid 0)|^2 \, d\lambda. \tag{44}$$

The approximation of $g(\lambda)$ by $w^*(\lambda \mid 0)$ may not be good at all values of λ. In fact, it will not be very good at the discontinuities of $g(\lambda)$, namely, at $\lambda = \pm b$. The oscillation of the approximating harmonic series near the discontinuities is known as the *Gibbs phenomenon* [Lanczos (1956), Chapter 4, and Hamming (1962), Chapter 22]; it can be reduced by multiplying the w_r^*'s by weights.

The series (42) may be truncated to $K = [\pi/b]$; then $w_r^* = 0$ for $r = \pm(K+1), \ldots, \pm(T-1)$. This estimate is sometimes referred to as "rectangular." The window $w(\lambda \mid 0)$ differs from $g(\lambda)$ because of the weights $1 - |r|/T$ as well as the fact that only a finite number of coefficients are involved.

F. *General Blackman-Tukey Estimates.* Blackman and Tukey (1959) have suggested two estimates (called hanning and hamming by them) which are special cases except for the terms $r = \pm K$

$$\begin{aligned}\hat{f}(\nu) &= \frac{1}{2\pi} \sum_{r=-K}^{K} \left(1 - 2a + 2a \cos \frac{\pi r}{K}\right)\left(1 - \frac{|r|}{T}\right) \cos \nu r \, C_r \\ &= \frac{1}{2\pi} \sum_{r=-K}^{K} \left(1 - 2a + 2a \cos \frac{\pi r}{K}\right) \cos \nu r \, c_r\end{aligned} \tag{45}$$

for $0 \leq a \leq \frac{1}{4}$. The coefficients $w_r^* = 1 - 2a + 2a \cos(\pi r/K)$, $r = 0, \pm 1, \ldots,$ $\pm K$, decrease from 1 at $r = 0$ to $1 - 4a$ at $r = \pm K$ and $w_r^* = 0$ for $r = \pm(K+1), \ldots, \pm(T-1)$. The estimate can also be written

$$\begin{aligned}\hat{f}(\nu) &= \frac{1-2a}{2\pi} \sum_{r=-K}^{K} \cos \nu r \, c_r + \frac{2a}{2\pi} \sum_{r=-K}^{K} \cos \frac{\pi r}{K} \cos \nu r \, c_r \\ &= \frac{1-2a}{2\pi} \sum_{r=-K}^{K} \cos \nu r \, c_r + \frac{a}{2\pi} \sum_{r=-K}^{K} \cos \left(\nu - \frac{\pi}{K}\right) r \, c_r + \frac{a}{2\pi} \sum_{r=-K}^{K} \cos \left(\nu + \frac{\pi}{K}\right) r \, c_r \\ &= a\hat{f}^K\left(\nu - \frac{\pi}{K}\right) + (1-2a)\hat{f}^K(\nu) + a\hat{f}^K\left(\nu + \frac{\pi}{K}\right),\end{aligned} \tag{46}$$

where $\hat{f}^K(\nu)$ is the truncated sample spectral density (Example B), $\hat{f}^K(\pi + \lambda) = \hat{f}^K(\pi - \lambda)$ and $\hat{f}^K(-\pi - \lambda) = \hat{f}^K(-\pi + \lambda)$. Thus $\hat{f}(\nu)$ is calculated by computing $\hat{f}^K(\nu)$ and then taking the moving average (46). The spectral windows can be calculated from the spectral windows for the truncated sample spectral density. They are

(47)
$$\begin{aligned}
w(\lambda \mid 0) &= \frac{a}{2\pi} \sum_{r=-K}^{K} \left(1 - \frac{|r|}{T}\right) \cos\left(\lambda - \frac{\pi}{K}\right) r + \frac{1-2a}{2\pi} \sum_{r=-K}^{K} \left(1 - \frac{|r|}{T}\right) \cos \lambda r \\
&\quad + \frac{a}{2\pi} \sum_{r=-K}^{K} \left(1 - \frac{|r|}{T}\right) \cos\left(\lambda + \frac{\pi}{K}\right) r \\
&= a\left\{\frac{K+1}{T} k_{K+1}\left(\lambda - \frac{\pi}{K}\right) + \left(1 - \frac{K+1}{T}\right) h_{2K+1}\left(\lambda - \frac{\pi}{K}\right)\right\} \\
&\quad + (1-2a)\left\{\frac{K+1}{T} k_{K+1}(\lambda) + \left(1 - \frac{K+1}{T}\right) h_{2K+1}(\lambda)\right\} \\
&\quad + a\left\{\frac{K+1}{T} k_{K+1}\left(\lambda + \frac{\pi}{K}\right) + \left(1 - \frac{K+1}{T}\right) h_{2K+1}\left(\lambda + \frac{\pi}{K}\right)\right\},
\end{aligned}$$

(48) $w^*(\lambda \mid 0)$
$$\begin{aligned}
&= \frac{a}{2\pi} \sum_{r=-K}^{K} \cos\left(\lambda - \frac{\pi}{K}\right) r + \frac{1-2a}{2\pi} \sum_{r=-K}^{K} \cos \lambda r + \frac{a}{2\pi} \sum_{r=-K}^{K} \cos\left(\lambda + \frac{\pi}{K}\right) r \\
&= a h_{2K+1}\left(\lambda - \frac{\pi}{K}\right) + (1-2a) h_{2K+1}(\lambda) + a h_{2K+1}\left(\lambda + \frac{\pi}{K}\right).
\end{aligned}$$

The maximum of $w^*(\lambda \mid 0)$ is $[(2K+1)(1-2a) - 2a]/(2\pi)$ at $\lambda = 0$.

G. *Hanning*. A case of the Blackman-Tukey estimate of special interest is $a = 1/4$. Then

(49)
$$\begin{aligned}
\hat{f}(\nu) &= \frac{1}{2\pi} \sum_{r=-K}^{K} \left(\tfrac{1}{2} + \tfrac{1}{2} \cos \frac{\pi}{K} r\right) \cos \nu r \, c_r \\
&= \tfrac{1}{4}\hat{f}^K\left(\nu - \frac{\pi}{K}\right) + \tfrac{1}{2}\hat{f}^K(\nu) + \tfrac{1}{4}\hat{f}^K\left(\nu + \frac{\pi}{K}\right) \\
&= \frac{1}{2}\left[\tfrac{1}{2}\hat{f}^K\left(\nu - \frac{\pi}{K}\right) + \tfrac{1}{2}\hat{f}^K(\nu)\right] + \frac{1}{2}\left[\tfrac{1}{2}\hat{f}^K(\nu) + \tfrac{1}{2}\hat{f}^K\left(\nu + \frac{\pi}{K}\right)\right].
\end{aligned}$$

This estimate can be computed as a simple average of a simple average of the truncated sample spectral density. The weights $w_r^* = \frac{1}{2}(1 + \cos \pi r/K) = \cos^2 \pi r/(2K)$, $r = 0, \pm 1, \ldots, \pm K$, decrease from 1 at $r = 0$ to 0 at $r = \pm K$. The 0's of $w^*(\lambda \mid 0)$ are at $\lambda = \pm k\pi/K$ for $k = 2, 3, \ldots, K$; the maximum of $w^*(\lambda \mid 0)$ is $K/(2\pi)$ at $\lambda = 0$. The $w^*(\lambda \mid 0)$ of the modified Bartlett estimate (with the same value of K) has the same maximum, but the 0's are at $\pm k\pi/K$ for $k = 2, 4, \ldots, K - 1$ or K. The hanning window has all of the 0's of the modified Bartlett window and has additional 0's near the relative maxima of the modified Bartlett window (except the maximum at the origin); this fact suggests that the hanning side lobes are smaller than those of the modified Bartlett window.

H. *Hamming*. Blackman and Tukey (1959) suggest $a = 0.23$ to reduce the size of the first lobes, resulting in

$$\hat{f}(\nu) = \frac{1}{2\pi} \sum_{r=-K}^{K} \left(0.54 + 0.46 \cos \frac{\pi}{K} r\right) \cos \nu r \, c_r \tag{50}$$

$$= 0.23 \hat{f}^K\left(\nu - \frac{\pi}{K}\right) + 0.54 \hat{f}^K(\nu) + 0.23 \hat{f}^K\left(\nu + \frac{\pi}{K}\right).$$

I. *Parzen*. The modification of Bartlett's estimate is the case of $w_r^* = 1 - (|r|/K)^q$ when $q = 1$. Parzen (1957b) has suggested other powers. In the case of $q = 2$

$$\hat{f}(\nu) = \frac{1}{2\pi} \sum_{r=-K}^{K} \left(1 - \frac{r^2}{K^2}\right)\left(1 - \frac{|r|}{T}\right) \cos \nu r \, C_r \tag{51}$$

$$= \frac{1}{2\pi} \sum_{r=-K}^{K} \left(1 - \frac{r^2}{K^2}\right) \cos \nu r \, c_r.$$

The window operating on the sample spectral density is

$$w^*(\lambda \mid 0) = \frac{1}{2\pi} \sum_{r=-K}^{K} \left(1 - \frac{r^2}{K^2}\right) \cos \lambda r \tag{52}$$

$$= \frac{1}{2\pi}\left\{\frac{\sin \lambda K \cos \frac{1}{2}\lambda}{2K^2 \sin^3 \frac{1}{2}\lambda} - \frac{\cos \lambda K}{K \sin^2 \frac{1}{2}\lambda}\right\}.$$

Its maximum at $\lambda = 0$ is $[1/(2\pi)][\frac{4}{3}K - 1/(3K)]$.

J. *Parzen.* Another suggestion by Parzen (1961b) is for K even

(53)

$$\begin{aligned}
\hat{f}(\nu) &= \frac{1}{2\pi}\sum_{r=-\frac{1}{2}K}^{\frac{1}{2}K}\left(1-6\frac{r^2}{K^2}+6\frac{|r|^3}{K^3}\right)\left(1-\frac{|r|}{T}\right)\cos\nu r\, C_r \\
&\quad+\frac{2}{2\pi}\sum_{r=-K}^{-\frac{1}{2}K-1}\left(1-\frac{|r|}{K}\right)^3\left(1-\frac{|r|}{T}\right)\cos\nu r\, C_r \\
&\quad+\frac{2}{2\pi}\sum_{r=\frac{1}{2}K+1}^{K}\left(1-\frac{r}{K}\right)^3\left(1-\frac{r}{T}\right)\cos\nu r\, C_r \\
&=\frac{1}{2\pi}\sum_{r=-\frac{1}{2}K}^{\frac{1}{2}K}\left(1-6\frac{r^2}{K^2}+6\frac{|r|^3}{K^3}\right)\cos\nu r\, c_r+\frac{2}{2\pi}\sum_{r=-K}^{-\frac{1}{2}K-1}\left(1-\frac{|r|}{K}\right)^3\cos\nu r\, c_r \\
&\quad+\frac{2}{2\pi}\sum_{r=\frac{1}{2}K+1}^{K}\left(1-\frac{r}{K}\right)^3\cos\nu r\, c_r \\
&=\frac{2}{2\pi}\sum_{r=-K}^{K}\left(1-\frac{|r|}{K}\right)^3\cos\nu r\, c_r-\frac{1}{2\pi}\sum_{r=-\frac{1}{2}K}^{\frac{1}{2}K}\left(1-2\frac{|r|}{K}\right)^3\cos\nu r\, c_r.
\end{aligned}$$

The function which is $1-6y^2+6|y|^3$ for $|y|\le\frac{1}{2}$, $2(1-|y|)^3$ for $\frac{1}{2}\le|y|\le 1$, and 0 for $1\le|y|$ is proportional to the density function of the mean of four observations on the uniform distribution from -1 to 1, just as the function which is $1-|y|$ for $|y|\le 1$ and 0 for $1\le|y|$ (related to the modified Bartlett estimate) is proportional to the density function of the mean of two such observations. (See Problem 9.) The characteristic function of this uniform distribution is $(\sin t)/t$. The window operating on the sample spectral density is then related to $[(\sin t)/t]^4$ for the Parzen estimate, as the window for the modified Bartlett estimate is related to $[(\sin t)/t]^2$.

Letting $K=2M$ and using the result of Problem 12, we find the Parzen window from

(54)
$$\begin{aligned}
2\pi w^*(\lambda\mid 0) &= 2\sum_{r=-2M}^{2M}\left(1-\frac{|r|}{2M}\right)^3\cos\lambda r-\sum_{r=-M}^{M}\left(1-\frac{|r|}{M}\right)^3\cos\lambda r \\
&=1+\frac{1}{2M^3}\sum_{r=1}^{2M}(2M-r)^3\cos\lambda r-\frac{2}{M^3}\sum_{r=1}^{M}(M-r)^3\cos\lambda r \\
&=1+\frac{1}{2M^3}\left\{-\frac{8M^3}{2}+\frac{2\sin^2\lambda M+12M^2}{4\sin^2\frac{1}{2}\lambda}-\frac{3\sin^2\lambda M}{4\sin^4\frac{1}{2}\lambda}\right\} \\
&\quad-\frac{2}{M^3}\left\{-\frac{M^3}{2}+\frac{2\sin^2\frac{1}{2}\lambda M+3M^2}{4\sin^2\frac{1}{2}\lambda}-\frac{3\sin^2\frac{1}{2}\lambda M}{4\sin^4\frac{1}{2}\lambda}\right\} \\
&=\frac{1}{M^3}\left\{\frac{\sin^2\lambda M-4\sin^2\frac{1}{2}\lambda M}{4\sin^2\frac{1}{2}\lambda}+\frac{12\sin^2\frac{1}{2}\lambda M-3\sin^2\lambda M}{8\sin^4\frac{1}{2}\lambda}\right\}.
\end{aligned}$$

If we substitute

$$\sin^2 \lambda M = (2 \sin \tfrac{1}{2}\lambda M \cos \tfrac{1}{2}\lambda M)^2 = 4 \sin^2 \tfrac{1}{2}\lambda M(1 - \sin^2 \tfrac{1}{2}\lambda M), \tag{55}$$

we obtain

$$\begin{aligned} w^*(\lambda \mid 0) &= \frac{1}{2\pi M^3}\left\{\frac{3}{2}\left(\frac{\sin \frac{1}{2}\lambda M}{\sin \frac{1}{2}\lambda}\right)^4 - \frac{\sin^4 \frac{1}{2}\lambda M}{\sin^2 \frac{1}{2}\lambda}\right\} \\ &= \frac{2\pi}{M}\left\{\frac{3}{2} k_M^2(\lambda) - \frac{\sin^2 \frac{1}{2}\lambda M}{M^2} h_M^2(\lambda)\right\} \\ &= \frac{8}{2\pi K^3}\left\{\frac{3}{2}\frac{\sin^4 (\lambda K/4)}{\sin^4 \frac{1}{2}\lambda} - \frac{\sin^4 (\lambda K/4)}{\sin^2 \frac{1}{2}\lambda}\right\} \\ &= \frac{2\pi}{K}\left\{3k_{\frac{1}{2}K}^2(\lambda) - 8\frac{\sin^2 (\lambda K/4)}{K^2} h_{\frac{1}{2}K}^2(\lambda)\right\}. \end{aligned} \tag{56}$$

For large K and small λ, the first term in (56) dominates. The window $w^*(\lambda \mid 0)$ is nonnegative; hence the estimate $\hat{f}(\nu)$ is nonnegative.

K. *Averaging for Discrete Values of* λ. In some cases $I(\lambda)$ may be calculated for $\lambda = 2\pi j/T$, $j = 0, 1, \ldots, [\frac{1}{2}T]$; in particular, the calculations are made for these values if T is large and the Fast Fourier Transform (Section 4.3.5) is used. An estimate closely related to the Daniell estimate (Example E) but based on $I(2\pi j/T)$, $j = 0, 1, \ldots, [\frac{1}{2}T]$, is

$$\begin{aligned} \hat{f}\left(\frac{2\pi k}{T}\right) &= \frac{1}{2n+1}\sum_{j=k-n}^{k+n} I\left(\frac{2\pi j}{T}\right) \\ &= \frac{1}{2\pi}\sum_{r=-(T-1)}^{T-1} c_r \frac{1}{2n+1}\sum_{j=k-n}^{k+n} e^{i2\pi jr/T} \\ &= \frac{1}{2\pi}\sum_{r=-(T-1)}^{T-1} c_r e^{i2\pi kr/T} \frac{1}{2n+1}\sum_{h=-n}^{n} e^{i2\pi hr/T} \\ &= \frac{1}{2\pi}\sum_{r=-(T-1)}^{T-1} \frac{2\pi}{2n+1} h_{2n+1}\left(\frac{2\pi r}{T}\right)\cos\frac{2\pi kr}{T}\, c_r. \end{aligned} \tag{57}$$

For $\nu = 2\pi k/T$, $k = 0, 1, \ldots, [\frac{1}{2}T]$, the estimate uses

$$\begin{aligned} w_r^* &= \frac{2\pi}{2n+1} h_{2n+1}\left(\frac{2\pi r}{T}\right) \\ &= \frac{1}{2n+1}\frac{\sin \pi(2n+1)r/T}{\sin \pi r/T}, \qquad r = 0, \pm 1, \ldots, \pm(T-1), \end{aligned} \tag{58}$$

which are the Daniell weights (Example E) with $br = \pi(2n + 1)r/T$ replaced by $(2n + 1)\sin \pi r/T$ in the denominator.

9.3. ASYMPTOTIC MEANS AND COVARIANCES OF SPECTRAL ESTIMATES

9.3.1. Estimating a Given Average of the Spectral Density

The expected values of estimates of densities were treated in Section 9.2. The variances and covariances of the estimates can be found from (16) and (19) of Section 9.2 by appropriate substitutions. However, the second-order moments are difficult to interpret. Instead of studying such moments we shall consider the limiting values (as $T \to \infty$) of appropriately normalized estimates.

Consider a sequence of estimates

$$\hat{f}_T(\nu) = \frac{1}{2\pi} \sum_{r=-(T-1)}^{T-1} w_r^* \cos \nu r \, c_{rT}, \tag{1}$$

where $\{w_r^*\}$ is a fixed sequence ($w_r^* = w_{-r}^*$) such that

$$\sum_{r=-\infty}^{\infty} |w_r^*| < \infty. \tag{2}$$

We write c_{rT} for c_r to indicate that T observations are used. Let

$$w^*(\lambda \mid \nu) = \frac{1}{2\pi} \sum_{r=-\infty}^{\infty} w_r^* \cos \lambda r \cos \nu r. \tag{3}$$

Then $w^*(\lambda \mid \nu) = w(\lambda \mid \nu)$, where

$$w(\lambda \mid \nu) = \frac{1}{2\pi} \lim_{T \to \infty} \sum_{r=-(T-1)}^{T-1} w_r^* \left(1 - \frac{|r|}{T}\right) \cos \lambda r \cos \nu r \tag{4}$$

(by Lemma 8.3.1). Then

$$\begin{aligned} \lim_{T \to \infty} \mathscr{E} \hat{f}_T(\nu) &= \frac{1}{2\pi} \lim_{T \to \infty} \sum_{r=-(T-1)}^{T-1} w_r^* \left(1 - \frac{|r|}{T}\right) \cos \nu r \, \sigma(r) \\ &= \frac{1}{2\pi} \sum_{r=-\infty}^{\infty} w_r^* \cos \nu r \, \sigma(r) \\ &= \int_{-\pi}^{\pi} w(\lambda \mid \nu) f(\lambda) \, d\lambda. \end{aligned} \tag{5}$$

We see that such a sequence of estimates is an asymptotically unbiased estimate of a weighted average of $f(\lambda)$, namely, the right-hand side of (5). Since (2) implies $w(\lambda \mid \nu)$ is continuous, the weighted average of $f(\lambda)$ is not identically $f(\nu)$.

The variance of $\hat{f}_T(\nu)$ is of order $1/T$. Let

$$\hat{f}_{TK}(\nu) = \frac{1}{2\pi} \sum_{r=-K}^{K} w_r^* \cos \nu r \, c_{rT}. \tag{6}$$

Application of Theorem 8.3.3 yields

$$\begin{aligned}
(7)\quad & \lim_{T\to\infty} T \operatorname{Cov}[\hat{f}_{TK}(\nu), \hat{f}_{TK}(\nu')] \\
&= \frac{1}{(2\pi)^2} \sum_{g,h=-K}^{K} w_g^* w_h^* \cos \nu g \cos \nu' h \\
&\quad \times \left[4\pi \int_{-\pi}^{\pi} \cos \lambda g \cos \lambda h \, f^2(\lambda)\, d\lambda + \sum_{r=-\infty}^{\infty} \kappa(g, -r, h-r)\right] \\
&= 4\pi \int_{-\pi}^{\pi} w_K^*(\lambda \mid \nu) w_K^*(\lambda \mid \nu') f^2(\lambda)\, d\lambda \\
&\quad + \frac{1}{(2\pi)^2} \sum_{g,h=-K}^{K} w_g^* w_h^* \cos \nu g \cos \nu' h \sum_{r=-\infty}^{\infty} \kappa(g, -r, h-r),
\end{aligned}$$

where

$$w_K^*(\lambda \mid \nu) = \frac{1}{2\pi} \sum_{g=-K}^{K} w_g^* \cos \nu g \cos \lambda g. \tag{8}$$

If $\int_{-\pi}^{\pi} f^2(\lambda)\, d\lambda < \infty$, then (2) implies that the limit on K of the first part of the right-hand side of (7) is $4\pi \int_{-\pi}^{\pi} w^*(\lambda \mid \nu) w^*(\lambda \mid \nu') f^2(\lambda) d\lambda$. If the quantity $\left|\sum_{r=-\infty}^{\infty} \kappa(g, -r, h-r)\right|$ is bounded (uniformly in g and h), the limit on K of the second part exists. By an analogue of Corollary 7.7.1 (stated as Problem 15) the limit on K of the left-hand side of (7) is $\lim_{T\to\infty} T \operatorname{Cov}[\hat{f}_T(\nu), \hat{f}_T(\nu')]$.

THEOREM 9.3.1. *If (2) holds, then the limit of $\mathscr{E}\hat{f}_T(\nu)$, defined by (1), is $\int_{-\pi}^{\pi} w(\lambda \mid \nu) f(\lambda)\, d\lambda$, where $w(\lambda \mid \nu) = w^*(\lambda \mid \nu)$ is defined by (3). If (2) holds, if $f(\lambda)$ is continuous, $-\pi \le \lambda \le \pi$, and if $\left|\sum_{r=-\infty}^{\infty} \kappa(g, -r, h-r)\right|$ is uniformly bounded, then*

$$\begin{aligned}
(9)\quad \lim_{T\to\infty} T \operatorname{Cov}[\hat{f}_T(\nu), \hat{f}_T(\nu')] &= 4\pi \int_{-\pi}^{\pi} w(\lambda \mid \nu) w(\lambda \mid \nu') f^2(\lambda)\, d\lambda \\
&\quad + \frac{1}{(2\pi)^2} \sum_{g,h=-\infty}^{\infty} w_g^* w_h^* \cos \nu g \cos \nu' h \sum_{r=-\infty}^{\infty} \kappa(g, -r, h-r).
\end{aligned}$$

If

$$y_t = \mu + \sum_{s=-\infty}^{\infty} \gamma_s v_{t-s} \tag{10}$$

is a linear process, then

$$\sum_{r=-\infty}^{\infty} \kappa(g, -r, h-r) = \kappa_4 \frac{\sigma(g)\sigma(h)}{\sigma^4}, \tag{11}$$

and the second term on the right-hand side of (9) can be simplified.

COROLLARY 9.3.1. *If* $\{y_t\}$ *is generated by* (10), *where* $\mathscr{E}v_t = 0$, $\mathscr{E}v_t^2 = \sigma^2$, $\mathscr{E}v_tv_s = 0$, $t \neq s$, $\mathscr{E}v_tv_sv_rv_q = 0$, $t \neq s$, $t \neq r$, $t \neq q$, $\mathscr{E}v_t^4 = 3\sigma^4 + \kappa_4 < \infty$, $\mathscr{E}v_t^2v_s^2 = \sigma^4$, $t \neq s$, *and* $\sum_{t=-\infty}^{\infty} |\gamma_t| < \infty$, *and if* (2) *holds, then*

$$\lim_{T\to\infty} T \operatorname{Cov}[\hat{f}_T(\nu), \hat{f}_T(\nu')] \tag{12}$$

$$= 4\pi \int_{-\pi}^{\pi} w(\lambda\,|\,\nu)w(\lambda\,|\,\nu')f^2(\lambda)\,d\lambda + \frac{\kappa_4}{\sigma^4}\left[\int_{-\pi}^{\pi} w(\lambda\,|\,\nu)f(\lambda)\,d\lambda\right]\left[\int_{-\pi}^{\pi} w(\lambda\,|\,\nu')f(\lambda)\,d\lambda\right]$$

$$= 4\pi \int_{-\pi}^{\pi} w(\lambda\,|\,\nu)w(\lambda\,|\,\nu')f^2(\lambda)\,d\lambda + \frac{\kappa_4}{\sigma^4} \lim_{T\to\infty} \mathscr{E}\hat{f}_T(\nu) \lim_{T\to\infty} \mathscr{E}\hat{f}_T(\nu').$$

It was shown in Section 8.4.2 (Theorem 8.4.2) that a finite set of sample covariances is asymptotically normally distributed for a linear process composed of independent and identically distributed variables. Hence, $\sqrt{T}\,[\hat{f}_{TK}(\nu) - \mathscr{E}\hat{f}_{TK}(\nu)]$ has a limiting normal distribution as $T \to \infty$ for fixed K. Application of Corollary 7.7.1 shows that $\hat{f}_T(\nu)$ has an asymptotic normal distribution.

THEOREM 9.3.2. *If* y_t *is generated by* (10), *where* $\{v_t\}$ *consists of independently and identically distributed random variables with* $\mathscr{E}v_t = 0$, $\mathscr{E}v_t^2 = \sigma^2$, $\mathscr{E}v_t^4 = 3\sigma^4 + \kappa_4 < \infty$, *and* $\sum_{t=-\infty}^{\infty} |\gamma_t| < \infty$, *and if* (2) *holds, then* $\sqrt{T}\,[\hat{f}_T(\nu_1) - \int_{-\pi}^{\pi} w(\lambda\,|\,\nu_1)f(\lambda)\,d\lambda], \ldots, \sqrt{T}\,[\hat{f}_T(\nu_n) - \int_{-\pi}^{\pi} w(\lambda\,|\,\nu_n)f(\lambda)\,d\lambda]$ *have a limiting normal distribution with means* 0 *and covariances given by* (12).

Let $\hat{f}_T^*(\nu)$ be defined by (1) with c_{rT} replaced by c_{rT}^*, where c_{rT}^* denotes the sample rth covariance using deviations about the sample mean (and dividing by T). Then $\mathscr{E}\hat{f}_T^*(\nu)$ has the limit (5), and Theorem 9.3.1, Corollary 9.3.1, and Theorem 9.3.2 apply to $\hat{f}_T^*(\nu)$. Similarly c_{rT} can be replaced by C_{rT}, C_{rT}^*, $\tilde{C}_{rT}$, or $\tilde{c}_{rT}$ and the same limiting results hold.

9.3.2. Asymptotic Bias

If one wishes to estimate consistently $f(\lambda)$ at a value $\lambda = \nu$ rather than an average of $f(\lambda)$, one must choose a sequence of estimates for which the corresponding sequence of expected values converges to $f(\nu)$ for a wide class of densities $f(\lambda)$. We let the coefficients $w_r = (T - |r|)w_r^*/T$ depend on T. We can write an estimate as

$$\hat{f}_T(\nu) = \frac{1}{2\pi} \sum_{r=-(T-1)}^{T-1} w_{rT} \cos \nu r \, C_{rT} \tag{13}$$

$$= \frac{1}{2\pi} \sum_{r=-(T-1)}^{T-1} w_{rT}^* \cos \nu r \, c_{rT}$$

$$= \int_{-\pi}^{\pi} w_T^*(\lambda \mid \nu) I_T(\lambda) \, d\lambda,$$

where $w_{rT} = (T - |r|)w_{rT}^*/T$ and

$$w_T^*(\lambda \mid \nu) = \frac{1}{2\pi} \sum_{r=-(T-1)}^{T-1} w_{rT}^* \cos \lambda r \cos \nu r. \tag{14}$$

We write $I_T(\lambda)$ for $I(\lambda)$ to indicate that T observations are used. For each T, $\{w_{rT}^*\}$ is a sequence of constants, $r = 0, 1, \ldots$ ($w_{rT}^* = 0$ for $T \le r$ and $w_{-rT}^* = w_{rT}^*$). Then $w_T^*(\lambda \mid \nu)$ is a sequence of functions. The expected value of $\hat{f}_T(\nu)$ is

$$\mathscr{E}\hat{f}_T(\nu) = \frac{1}{2\pi} \sum_{r=-(T-1)}^{T-1} w_{rT} \cos \nu r \, \sigma(r) \tag{15}$$

$$= \int_{-\pi}^{\pi} w_T(\lambda \mid \nu) f(\lambda) \, d\lambda,$$

where

$$w_T(\lambda \mid \nu) = \frac{1}{2\pi} \sum_{r=-(T-1)}^{T-1} w_{rT} \cos \lambda r \cos \nu r. \tag{16}$$

We want

$$\lim_{T \to \infty} \mathscr{E}\hat{f}_T(\nu) = f(\nu) \tag{17}$$

for all spectral densities in some class of interest.

Most of the estimates described in Section 9.2.3 are defined by coefficients of the form

$$(18)\qquad \begin{aligned} w_{rT}^* &= k(r/K_T), & r &= 0, \pm 1, \ldots, \pm K_T, \\ &= 0, & r &= \pm(K_T + 1), \ldots, \pm(T-1), \end{aligned}$$

where we have written K_T for K to indicate the dependence on T. The function $k(x)$ satisfies

$$(19)\qquad k(x) = k(-x),$$

$$(20)\qquad k(0) = 1,$$

so that $w_{0T}^* = w_{0T} = 1$ and hence $\int_{-\pi}^{\pi} w_T^*(\lambda \mid \nu)\, d\lambda = \int_{-\pi}^{\pi} w_T(\lambda \mid \nu)\, d\lambda = 1$. The function $k(x)$ is to be continuous at $x = 0$. The asymptotic bias of the estimate depends on the smoothness of $k(x)$ at $x = 0$. Parzen (1957b) has developed a theory of asymptotic bias, variance, and mean square error for the model. We follow his treatment.

Suppose for some numbers $q > 0$ and $k > 0$

$$(21)\qquad \lim_{x \to 0} \frac{1 - k(x)}{|x|^q} = k.$$

(q is the largest exponent for which $k < \infty$.) If q is an even integer, k is the negative of the qth derivative of $k(x)$ at $x = 0$ divided by $q!$; if q is an odd integer, k is the negative of the qth right-hand derivative of $k(x)$ at $x = 0$ divided by $q!$. That $k > 0$ implies that in an interval about 0 $k(x)$ decreases as x moves away from 0. For small $|x|$, $k(x)$ is approximately $1 - k\,|x|^q$. If $k(x) = \int_{-\infty}^{\infty} K(v) e^{ivx}\, dv$ and $q = 2$, then k is $\frac{1}{2} \int_{-\infty}^{\infty} v^2 K(v)\, dv$. [Under suitable conditions $K(v) = \lim w_T^*(v/K_T \mid 0)/K_T$ as $K_T \to \infty$; see Problem 17.] In a sense k indicates the spread of the family of windows.

The asymptotic bias of an estimate of the spectral density also depends on the smoothness of the spectral density. Suppose

$$(22)\qquad \sum_{r=-\infty}^{\infty} |r|^p\, |\sigma(r)| < \infty$$

for some $p > 0$. Let

$$(23)\qquad f^{(s)}(\lambda) = \frac{(-i)^s}{2\pi} \sum_{r=-\infty}^{\infty} r^s e^{-i\lambda r} \sigma(r)$$

for integral $s \leq p$. Then $f^{(s)}(\lambda)$ is the sth derivative of $f(\lambda)$. The condition (22) implies that all derivatives of order no greater than p are bounded and continuous. Let $\{K_T\}$ be a sequence of integers depending on T such that $K_T \to \infty$

and $K_T^m/T \to 0$ as $T \to \infty$, where $m = \min(p, q)$. Then (when $w_{rT}^* = 0$ for $|r| > K_T$)

$$
\begin{aligned}
(24) \qquad K_T^m[\mathscr{E}\hat{f}_T(\nu) - f(\nu)] &= \frac{K_T^m}{2\pi} \sum_{r=-K_T}^{K_T} \left[k\left(\frac{r}{K_T}\right)\left(1 - \frac{|r|}{T}\right) - 1 \right] \cos \nu r \, \sigma(r) \\
&\quad - \frac{2K_T^m}{2\pi} \sum_{r=K_T+1}^{\infty} \cos \nu r \, \sigma(r) \\
&= \frac{K_T^m}{2\pi} \sum_{r=-K_T}^{K_T} \left[k\left(\frac{r}{K_T}\right) - 1 \right] \cos \nu r \, \sigma(r) \\
&\quad - \frac{K_T^m}{\pi} \sum_{r=1}^{K_T} k\left(\frac{r}{K_T}\right) \frac{r}{T} \cos \nu r \, \sigma(r) \\
&\quad - \frac{K_T^m}{\pi} \sum_{r=K_T+1}^{\infty} \cos \nu r \, \sigma(r).
\end{aligned}
$$

The first term on the extreme right-hand side of (24) can be written

$$
(25) \qquad \frac{1}{2\pi} \sum_{r=-K_T^*}^{K_T^*} \frac{k\left(\frac{r}{K_T}\right) - 1}{\left|\frac{r}{K_T}\right|^m} |r|^m \cos \nu r \, \sigma(r) + \frac{1}{\pi} \sum_{r=K_T^*+1}^{K_T} \frac{k\left(\frac{r}{K_T}\right) - 1}{\left(\frac{r}{K_T}\right)^m} r^m \cos \nu r \, \sigma(r)
$$

for any integer $K_T^* < K_T$. Because of (21), for any $\varepsilon > 0$ we can choose δ so

$$
(26) \qquad \left| \frac{k\left(\frac{r}{K_T}\right) - 1}{\left|\frac{r}{K_T}\right|^q} + k \right| < \varepsilon, \qquad \left|\frac{r}{K_T}\right| < \delta;
$$

thus set $K_T^* = [\delta K_T]$. Then, if $m = q \le p$, the first term in (25) is within $\varepsilon' = \varepsilon \sum_{r=-\infty}^{\infty} |r|^q \, |\sigma(r)|/(2\pi)$ of

$$
(27) \qquad -\frac{k}{2\pi} \sum_{r=-K^*}^{K_T^*} |r|^q \cos \nu r \, \sigma(r),
$$

which converges to $-kf^{[q]}(\nu)$, where

$$
(28) \qquad f^{[q]}(\nu) = \frac{1}{2\pi} \sum_{r=-\infty}^{\infty} |r|^q \cos \nu r \, \sigma(r), \qquad 0 \le q < \infty.
$$

If $m = p < q$, the first term in (25) is within ε' (with $\delta \leq 1$) of

$$-\frac{k}{2\pi} \sum_{r=-K_T^*}^{K_T^*} \left| \frac{r}{K_T} \right|^{q-p} |r|^p \cos \nu r \, \sigma(r), \tag{29}$$

which in absolute value is no greater than

$$\delta^{q-p} \frac{k}{2\pi} \sum_{r=-\infty}^{\infty} |r|^p \, |\sigma(r)| \tag{30}$$

and hence arbitrarily small. If $|k(x)| \leq M$, then

$$\frac{|1 - k(x)|}{|x|^m} \leq \frac{M+1}{\delta^m}, \qquad |x| \geq \delta. \tag{31}$$

The second term in (25) is in absolute value no greater than

$$\frac{M+1}{\pi \delta^m} \sum_{r=[\delta K_T]+1}^{\infty} r^m \, |\sigma(r)|, \tag{32}$$

which converges to 0. The second term on the extreme right-hand side of (24) is in absolute value no greater than

$$M \frac{K_T^m}{\pi T} \sum_{r=1}^{K_T} r \, |\sigma(r)|, \tag{33}$$

which converges to 0 if $p \geq 1$ and $K_T^m/T \to 0$. If $p \leq 1$, then (33) is not greater than

$$\frac{M}{\pi} \frac{K_T^{m+1-p}}{T} \sum_{r=1}^{K_T} r^p \, |\sigma(r)|, \tag{34}$$

which converges to 0 if $K_T^{m+1-p}/T \to 0$. The third term on the extreme right-hand side of (24) is in absolute value no greater than

$$\frac{K_T^m}{\pi} \sum_{r=K_T+1}^{\infty} |\sigma(r)| \leq \frac{1}{\pi} \sum_{r=K_T+1}^{\infty} |r|^p \, |\sigma(r)|, \tag{35}$$

which converges to 0 by (22).

THEOREM 9.3.3. *Let*

$$\hat{f}_T(\nu) = \frac{1}{2\pi} \sum_{r=-K_T}^{K_T} k\left(\frac{r}{K_T}\right) \cos \nu r \, c_{rT}, \tag{36}$$

where $k(x) = k(-x)$, $k(0) = 1$, $|k(x)| \leq M$ *for some* M *and all* $|x| \leq 1$, *and for some* $q > 0$ *and* $k > 0$ $\lim_{x \to 0} [1 - k(x)]/|x|^q = k$. *Let* $\{K_T\}$ *be a sequence of integers such that* $K_T \to \infty$ *as* $T \to \infty$. *If*

$$\sum_{r=-\infty}^{\infty} |r|^p \, |\sigma(r)| < \infty \tag{37}$$

for $p \geq q$, *then*

$$\lim_{T\to\infty} K_T^q \, [\mathscr{E}\hat{f}_T(\nu) - f(\nu)] = -\frac{k}{2\pi} \sum_{r=-\infty}^{\infty} |r|^q \cos \nu r \, \sigma(r) \tag{38}$$

if $p \geq 1$ *and* $K_T^q/T \to 0$ *or if* $p \leq 1$ *and* $K_T^{q+1-p}/T \to 0$. *If* (37) *holds for* $p < q$, *then*

$$\lim_{T\to\infty} K_T^p[\mathscr{E}\hat{f}_T(\nu) - f(\nu)] = 0 \tag{39}$$

if $p \geq 1$ *and* $K_T^p/T \to 0$ *or if* $p \leq 1$ *and* $K_T/T \to 0$.

The conditions of the theorem are summarized in Table 9.3.1.

TABLE 9.3.1

CONDITIONS FOR THEOREM 9.3.3

	$p \leq 1$	$p \geq 1$
$q \leq p$	$\frac{K_T^{q+1-p}}{T} \to 0$	$\frac{K_T^q}{T} \to 0$
$q > p$	$\frac{K_T}{T} \to 0$	$\frac{K_T^p}{T} \to 0$

Theorem 9.3.3 indicates that if the degree of smoothness in the weighting function, namely q, is more than the degree of smoothness in the density, namely p, the bias is $o(K_T^{-p})$. If q is no greater than p, the bias is approximately $-kf^{[q]}(\nu)/K_T^q$; that is, the bias decreases proportionally to $1/K_T^q$. In either case (for $p > 0$ and $q > 0$) $\hat{f}_T(\nu)$ is asymptotically unbiased; the theorem evaluates the order of the bias. Given a class of processes characterized by the number p, the bias is controlled (in the sense of Theorem 9.3.3) by choosing a kernel $k(x)$ with $q > p$; if $q \leq p$, the bias is controlled by choosing a kernel $k(x)$ with small k. When q is even, the right-hand side of (38) is $(-1)^{\frac{1}{2}q+1}kf^{(q)}(\nu)$.

The characteristics of the relevant examples given in Section 9.2.3 are given in Table 9.3.2.

If $\mathscr{E}y_t = \mu$ is not known, $f(\lambda)$ at $\lambda = \nu$ may be estimated by $\hat{f}_T^*(\nu)$, which corresponds to $\hat{f}_T(\nu)$ with c_{rT} replaced by c_{rT}^*, the covariances based on

TABLE 9.3.2

CHARACTERISTICS OF EXAMPLES

		$k(x)$ for $\lvert x\rvert \le 1$	q	k
B.	Truncated	1	∞	
D.	Modified Bartlett	$1 - \lvert x\rvert$	1	1
E.	Truncated Daniell ($b = \pi/K_T$)	$\sin \pi x/(\pi x)$	2	$\pi^2/6$
F.	Blackman-Tukey	$1 - 2a + 2a \cos \pi x$	2	$\pi^2 a$
G.	Hanning	$\frac{1}{2} + \frac{1}{2}\cos \pi x$	2	$\pi^2/4$
H.	Hamming	$0.54 + 0.46 \cos \pi x$	2	$0.23\pi^2$
I.	Parzen	$1 - x^2$	2	1
J.	Parzen	$1 - 6x^2 + 6\lvert x\rvert^3,\ \lvert x\rvert \le \frac{1}{2}$, $2(1-\lvert x\rvert)^3,\ \frac{1}{2} \le \lvert x\rvert \le 1$	2	6

deviations from the sample mean; or c_{rT} may be replaced by $\tilde{c}_{rT}$. It is shown at the end of Section 9.4 that if $K_T^{m+1}/T \to 0$ as $T \to \infty$, then the bias of $\hat{f}_T^*(\nu)$ is equivalent to the bias of $\hat{f}_T(\nu)$.

9.3.3. Asymptotic Variances and Covariances

We now turn our attention to the asymptotic variances and covariances of estimates (13), where $w_{rT}^* = k(r/K_T)$, $r = 0, \pm 1, \ldots, \pm K_T$, and $w_{rT}^* = 0$, $r = \pm(K_T + 1), \ldots, \pm(T-1)$, $k(x) = k(-x)$, and $K_T \to \infty$ and $K_T/T \to 0$ as $T \to \infty$. We assume $k(x)$ is continuous in $-1 \le x \le 1$ and $\sum_{r=-\infty}^{\infty} |\sigma(r)| < \infty$.

We have from Section 8.2.2

$$
\begin{aligned}
(40)\quad & T \operatorname{Cov}(c_{gT}, c_{hT}) \\
&= \frac{1}{T}\Big[\mathscr{E} \sum_{t,s} (y_t - \mu)(y_{t+h} - \mu)(y_s - \mu)(y_{s+g} - \mu) \\
&\qquad - (T - |g|)(T - |h|)\sigma(g)\sigma(h)\Big] \\
&= \frac{1}{T}\sum_{t,s} [\sigma(t-s)\sigma(t+h-s-g) + \sigma(t-s-g)\sigma(t-s+h) \\
&\qquad + \kappa(h, s-t, s-t+g)] \\
&= \sum_{r=-(T-1)}^{T-1} \phi_T(r; g, h)[\sigma(r)\sigma(r+h-g) + \sigma(r-g)\sigma(r+h) \\
&\qquad + \kappa(h, -r, g-r)], \qquad g, h = 0, \pm 1, \ldots, \pm(T-1),
\end{aligned}
$$

where the sum on s and t is over the ranges max $(1, 1-g) \le s \le \min(T-g, T)$ and max $(1, 1-h) \le t \le \min(T-h, T)$ and $T\phi_T(r; g, h)$ is the number of pairs (s, t) satisfying the above inequalities with $t - s = r$, $r = -(T-1), \ldots, T-1$. The exact values of $T\phi_T(r; g, h)$ are stated in Problem 19. [For some triplets (r, g, h), $\phi_T(r; g, h) = 0$.] Note that

$$0 \le \phi_T(r; g, h) \le 1 \tag{41}$$

and

$$\lim_{T\to\infty} \phi_T(r; g, h) = 1 \tag{42}$$

for every r, g, and h. In fact,

$$\phi_T(r; g, h) \ge 1 - \frac{|r| + |g| + |h|}{T}. \tag{43}$$

Then

$$\begin{aligned} \frac{T}{K_T}\operatorname{Cov}[\hat{f}_T(\lambda), \hat{f}_T(\nu)] &= \frac{T}{(2\pi)^2 K_T}\sum_{g,h=-K_T}^{K_T} k\left(\frac{g}{K_T}\right)k\left(\frac{h}{K_T}\right)\cos\lambda g\cos\nu h\operatorname{Cov}(c_{gT}, c_{hT}) \\ &= \frac{1}{(2\pi)^2 K_T}\sum_{g,h=-K_T}^{K_T} k\left(\frac{g}{K_T}\right)k\left(\frac{h}{K_T}\right)e^{i(\lambda g+\nu h)} \\ &\quad\times \sum_{r=-(T-1)}^{T-1}\phi_T(r; g, h)[\sigma(r)\sigma(r+h-g) \\ &\quad + \sigma(r-g)\sigma(r+h) + \kappa(h, -r, g-r)]. \end{aligned} \tag{44}$$

Consider

$$\begin{aligned} &\frac{1}{(2\pi)^2 K_T}\sum_{g,h=-K_T}^{K_T}\sum_{r=-(T-1)}^{T-1}\phi_T(r; g, h)k\left(\frac{g}{K_T}\right)k\left(\frac{h}{K_T}\right)e^{i(\lambda g+\nu h)}\sigma(r-g)\sigma(r+h) \\ &= \frac{1}{(2\pi)^2 K_T}\sum_{u,v=-(K_T+T-1)}^{K_T+T-1}\ \sum_{r=\max[u-K_T, v-K_T, -(T-1)]}^{\min[u+K_T, v+K_T, T-1]}\phi_T(r; r-u, v-r) \\ &\qquad\times k\left(\frac{r-u}{K_T}\right)k\left(\frac{v-r}{K_T}\right)e^{i(\nu v-\lambda u)+i(\lambda-\nu)r}\sigma(u)\sigma(v); \end{aligned} \tag{45}$$

the sum on r is 0 if the stated lower limit is greater than the stated upper limit.

The difference between (45) and

$$(46)\qquad \frac{1}{(2\pi)^2 K_T}\sum_{u,v=-m}^{m}\sum_{r=\max[u-K_T,v-K_T,-(T-1)]}^{\min[u+K_T,v+K_T,T-1]} \phi_T(r;\, r-u,\, v-r) \times k\left(\frac{r-u}{K_T}\right)k\left(\frac{v-r}{K_T}\right)e^{i(\nu v-\lambda u)+i(\lambda-\nu)r}\sigma(u)\sigma(v)$$

is less in absolute value than

$$(47)\qquad \left(2+\frac{1}{K_T}\right)\frac{4}{(2\pi)^2}\sum_{u=-\infty}^{\infty}\sum_{v=m+1}^{\infty}\sup_{-1\le x\le 1}k^2(x)\,|\sigma(u)|\,|\sigma(v)|,$$

which is arbitrarily small if m $(\le K_T)$ is sufficiently large and if $|k(x)|$ is bounded, since $\sum_{r=-\infty}^{\infty}|\sigma(r)|<\infty$. If $k(x)$ is continuous, $-1\le x\le 1$, then for $|u|\le m$, $|v|\le m$, and K_T sufficiently large

$$(48)\qquad \left|k\left(\frac{r-u}{K_T}\right)k\left(\frac{v-r}{K_T}\right)-k^2\left(\frac{r}{K_T}\right)\right|<\varepsilon$$

for r such that $-K_T\le r-u\le K_T$, $-K_T\le v-r\le K_T$, and $-K_T\le r\le K_T$. For $|u|\le m$, $|v|\le m$, $|r|\le m+K_T$

$$(49)\qquad \phi_T(r;\, r-u,\, v-r)\ge 1-\frac{5m+3K_T}{T}.$$

The difference between (46) and

$$(50)\qquad \frac{1}{(2\pi)^2}\sum_{u,v=-m}^{m}\sum_{r=-K_T}^{K_T}\frac{1}{K_T}k^2\left(\frac{r}{K_T}\right)e^{i(\nu v-\lambda u)+i(\lambda-\nu)r}\sigma(u)\sigma(v)$$
$$=\left[\sum_{r=-K_T}^{K_T}\frac{1}{K_T}k^2\left(\frac{r}{K_T}\right)e^{i(\lambda-\nu)r}\right]\left[\frac{1}{2\pi}\sum_{u=-m}^{m}e^{-i\lambda u}\sigma(u)\right]\left[\frac{1}{2\pi}\sum_{v=-m}^{m}e^{i\nu v}\sigma(v)\right]$$

is arbitrarily small if T is sufficiently large. If $\lambda-\nu=0$ or $\pm 2\pi$, the limit of the sum on r as $T\to\infty$ is

$$(51)\qquad \lim_{K_T\to\infty}\sum_{r=-K_T}^{K_T}\frac{1}{K_T}k^2\left(\frac{r}{K_T}\right)=\int_{-1}^{1}k^2(x)\,dx;$$

for m sufficiently large, the limit of (50) is arbitrarily close to

$$(52)\qquad f^2(\nu)\int_{-1}^{1}k^2(x)\,dx.$$

If $\lambda - \nu \neq 0, \pm 2\pi$, then $0 < |\lambda - \nu| < 2\pi$, and by Problem 23 (an analogue of the Riemann-Lebesgue Lemma)

$$\lim_{K_T \to \infty} \sum_{r=-K_T}^{K_T} \frac{1}{K_T} k^2\left(\frac{r}{K_T}\right) e^{i(\lambda-\nu)r} = 0; \tag{53}$$

hence the limit of (50) is 0.

Next from (44) we consider

(54)

$$\frac{1}{(2\pi)^2 K_T} \sum_{g,h=-K_T}^{K_T} \sum_{r=-(T-1)}^{T-1} \phi_T(r; g, h) k\left(\frac{g}{K_T}\right) k\left(\frac{h}{K_T}\right) e^{i(\lambda g + \nu h)} \sigma(r)\sigma(r + h - g)$$

$$= \frac{1}{(2\pi)^2 K_T} \sum_{u=-(T-1)}^{T-1} \sum_{v=u-2K_T}^{u+2K_T} \sum_{s=\max(u,v)-K_T}^{\min(u,v)+K_T} \phi_T(u; u - s, v - s)$$

$$\times k\left(\frac{u-s}{K_T}\right) k\left(\frac{v-s}{K_T}\right) e^{i(\lambda u + \nu v) - i(\lambda+\nu)s} \sigma(u)\sigma(v).$$

Then (54) is approximated by

$$\frac{1}{(2\pi)^2 K_T} \sum_{u,v=-m}^{m} \sum_{s=\max(u,v)-K_T}^{\min(u,v)+K_T} \phi_T(u; u - s, v - s) \tag{55}$$

$$\times k\left(\frac{u-s}{K_T}\right) k\left(\frac{v-s}{K_T}\right) e^{i(\lambda u + \nu v) - i(\lambda+\nu)s} \sigma(u)\sigma(v),$$

which is in turn approximated by

$$\frac{1}{(2\pi)^2} \sum_{u,v=-m}^{m} \sum_{s=-K_T}^{K_T} \frac{1}{K_T} k^2\left(\frac{s}{K_T}\right) e^{i(\lambda u + \nu v) - i(\lambda+\nu)s} \sigma(u)\sigma(v) \tag{56}$$

since $k[(u - s)/K_T]k[(v - s)/K_T]$ can be approximated by $k^2(-s/K_T) = k^2(s/K_T)$ for large K_T. Then if $\lambda + \nu = 0$ or $\pm 2\pi$, the limit of (56) is

$$f^2(\nu) \int_{-1}^{1} k^2(x)\, dx; \tag{57}$$

if $\nu + \lambda \neq 0, \pm 2\pi$, the limit of (56) is 0.

Finally, from (44) we consider

(58)

$$\frac{1}{(2\pi)^2 K_T} \sum_{g,h=-K_T}^{K_T} \sum_{r=-(T-1)}^{T-1} \phi_T(r; g, h) k\left(\frac{g}{K_T}\right) k\left(\frac{h}{K_T}\right) e^{i(\lambda g + \nu h)} \kappa(h, -r, g - r).$$

It is in absolute value less than

$$\frac{1}{(2\pi)^2 K_T} \sup_{-1 \leq x \leq 1} k^2(x) \sum_{r,s,t=-\infty}^{\infty} |\kappa(r, s, t)|. \tag{59}$$

If the triple sum is finite, the limit of (59) is 0.

THEOREM 9.3.4. *Let $\hat{f}_T(\nu)$ be defined by* (36), *where $k(x) = k(-x)$ and $k(x)$ is continuous in $-1 \leq x \leq 1$. Suppose*

$$\sum_{r=-\infty}^{\infty} |\sigma(r)| < \infty, \tag{60}$$

$$\sum_{r,s,t=-\infty}^{\infty} |\kappa(r, s, t)| < \infty. \tag{61}$$

Let $\{K_T\}$ be a sequence of integers such that $K_T \to \infty$ and $K_T/T \to 0$ as $T \to \infty$. Then

$$\lim_{T\to\infty} \frac{T}{K_T} \operatorname{Var} \hat{f}_T(0) = 2f^2(0) \int_{-1}^{1} k^2(x)\, dx, \tag{62}$$

$$\lim_{T\to\infty} \frac{T}{K_T} \operatorname{Var} \hat{f}_T(\pm\pi) = 2f^2(\pi) \int_{-1}^{1} k^2(x)\, dx, \tag{63}$$

$$\lim_{T\to\infty} \frac{T}{K_T} \operatorname{Var} \hat{f}_T(\nu) = f^2(\nu) \int_{-1}^{1} k^2(x)\, dx, \qquad \nu \neq 0, \pm\pi, \tag{64}$$

$$\lim_{T\to\infty} \frac{T}{K_T} \operatorname{Cov}\,[\hat{f}_T(\lambda), \hat{f}_T(\nu)] = 0, \qquad \nu \neq \pm\lambda. \tag{65}$$

Theorem 9.3.4 indicates that the variance of $\hat{f}_T(\nu)$ is of the order K_T/T. Since $K_T \to \infty$, the ratio K_T/T will be greater than K/T for arbitrary K and for all sufficiently large T. Thus the variance of a consistent estimate of the form (36) will be larger asymptotically than an estimate of the type considered in Section 9.3.1.

For a given sequence of integers $\{K_T\}$ the various estimates are to be compared on the basis of $\int_{-1}^{1} k^2(x)\, dx$. Table 9.3.3 gives the values of the integral for several kernels given as examples. For the kernels with $q = 2$, the bias characteristic k and the integral tend to vary inversely.

Note from (44) of Section 8.3 that (61) is satisfied if $y_t = \mu + \sum_{s=-\infty}^{\infty} \gamma_s v_{t-s}$, where $\sum_{s=-\infty}^{\infty} |\gamma_s| < \infty$ and the first four moments of $\{v_t\}$ correspond to stationarity and independence.

TABLE 9.3.3

INTEGRALS OF KERNELS

		$k(x)$ for $\lvert x\rvert \le 1$	$\int_{-1}^{1} k^2(x)\,dx$
B.	Truncated	1	2
D.	Modified Bartlett	$1 - \lvert x\rvert$	$2/3$
E.	Truncated Daniell ($b = \pi/K_T$)	$\sin \pi x/(\pi x)$	0.90282336
F.	Blackman-Tukey	$1 - 2a + 2a \cos \pi x$	$2(1 - 4a + 6a^2)$
G.	Hanning	$\frac{1}{2} + \frac{1}{2}\cos \pi x$	$3/4 = 0.75$
H.	Hamming	$0.54 + 0.46 \cos \pi x$	0.7948
I.	Parzen	$1 - x^2$	$16/15 = 1.0667$
J.	Parzen	$1 - 6x^2 + 6\lvert x\rvert^3,\ \lvert x\rvert \le \frac{1}{2}$, $2(1 - \lvert x\rvert)^3,\ \frac{1}{2} \le \lvert x\rvert \le 1$	$151/280 = 0.5393$

If μ is not known, c_{rT} may be replaced by c^*_{rT} or $\tilde{c}_{rT}$ in the definition of $\hat{f}_T(\nu)$. It is shown at the end of Section 9.4 that Theorem 9.3.4 applies to these estimates, too.

9.3.4. Asymptotic Mean Square Error

From Theorems 9.3.3 and 9.3.4 we see that if T is large under appropriate conditions the mean square error of an estimate of the spectral density at a point ν,

$$\mathscr{E}[\hat{f}_T(\nu) - f(\nu)]^2 = \operatorname{Var} \hat{f}_T(\nu) + [\mathscr{E}\hat{f}_T(\nu) - f(\nu)]^2, \tag{66}$$

is approximately ($\nu \neq 0, \pm\pi$)

$$\frac{K_T}{T} f^2(\nu) \int_{-1}^{1} k^2(x)\,dx + \left(\frac{1}{K_T^q}\right)^2 k^2\{f^{[q]}(\nu)\}^2, \tag{67}$$

where

$$f^{[q]}(\nu) = \frac{1}{2\pi} \sum_{r=-\infty}^{\infty} |r|^q \cos \nu r\, \sigma(r). \tag{68}$$

It will be noted that K_T appears in the numerator of the approximate variance of $\hat{f}_T(\nu)$ and in the denominator of the bias. Thus the larger K_T is (relative to T)

the larger the variance is and the smaller the bias is. In this way one can say that variance and bias work against each other. For the two terms in (67) to be of the same order K_T^{2q+1} must be of order T. (If the two terms are of different orders, one term will dominate and it will be of greater order than if the two terms were of the same order.) We may take K_T as $[\gamma T^{1/(2q+1)}]$. Then

$$\lim_{T\to\infty} T^{2q/(2q+1)}\mathscr{E}[\hat{f}_T(\nu) - f(\nu)]^2 = \gamma f^2(\nu)\int_{-1}^{1} k^2(x)\,dx + \frac{1}{\gamma^{2q}}\{f^{[q]}(\nu)\}^2 k^2. \tag{69}$$

The appropriate conditions are that $k(x) = k(-x)$ is continuous in $-1 \le x \le 1$, that (21) holds for $q > 0$ and $k > 0$, that (37) holds for $p = q$, that $\sum_{r=-\infty}^{\infty} |\sigma(r)| < \infty$, and that $\sum_{r,s,t=-\infty}^{\infty} |\kappa(r, s, t)| < \infty$. If q is an even integer, $(-1)^{\frac{1}{2}q} f^{[q]}(\nu) = f^{(q)}(\nu)$, the qth derivative of $f(\lambda)$ at $\lambda = \nu$ [see (23)]; the condition that (37) holds for $p = q$ implies that the qth derivative exists and is continuous at all ν.

The order of the mean square error is $T^{-2q/(2q+1)}$. The larger q is the smaller is this order. Thus the Blackman-Tukey and Parzen estimates are to be preferred to the modified Bartlett estimate.

The two terms on the right-hand side of (69) depend on the constant γ, the two characteristics of the kernel, $\int_{-1}^{1} k^2(x)\,dx$ and k^2, and on $f(\nu)$ and $f^{[q]}(\nu)$. In terms of this asymptotic theory we can compare estimates with kernels with a given characteristic exponent q. We can say that an estimate with kernel $k(x)$ is asymptotically at least as good as an estimate with kernel $k^*(x)$ and corresponding characteristic k^* if

$$k^2 \le k^{*2}, \qquad \int_{-1}^{1} k^2(x)\,dx \le \int_{-1}^{1} k^{*2}(x)\,dx. \tag{70}$$

The estimate based on $k(x)$ is asymptotically better than that based on $k^*(x)$ if at least one of the inequalities in (70) is strict. The estimate based on $k^*(x)$ is asymptotically admissible if there is no $k(x)$ for which (70) holds with at least one strict inequality. Of the five examples with $q = 2$, there is no estimate asymptotically better than another. See Table 9.3.4.

The theory of this section is asymptotic. It can be expected that very large values of T are needed for the asymptotic theory to be a good approximation.

For small or moderate values of T the windows $w_T(\lambda \mid \nu)$ and $w_T^*(\lambda \mid \nu)$ will give weight to values of λ near ν because of the main lobes and to values away from ν because of the side lobes. The range of values to which considerable weight is given is called the bandwidth. Several definitions have been given, but we shall not pursue this matter. See Jenkins (1961) and Parzen (1961b).

TABLE 9.3.4

CHARACTERISTICS OF KERNELS IN MEAN SQUARE ERRORS OF ESTIMATES

		k	k^2	$\int_{-1}^{1} k^2(x)\,dx$
E.	Truncated Daniell ($b = \pi/K_T$)	$\pi^2/6$	$\pi^4/36 = 2.7058$	0.90282336
F.	Blackman-Tukey	$\pi^2 a$	$\pi^4 a^2$	$2(1 - 4a + 6a^2)$
G.	Hanning	$\pi^2/4$	$\pi^4/16 = 6.0881$	$3/4 = 0.75$
H.	Hamming	$0.23\pi^2$	$0.0529\pi^4 = 5.1529$	0.7948
I.	Parzen	1	1	$16/15 = 1.0667$
J.	Parzen	6	36	$151/280 = 0.5393$

9.4. ASYMPTOTIC NORMALITY OF SPECTRAL ESTIMATES

The spectral estimates treated in Section 9.3 are asymptotically normally distributed under suitable conditions. We shall now show that $\sqrt{T/K_T}\,[\hat{f}_T(\nu) - \mathscr{E}\hat{f}_T(\nu)]$ has a limiting normal distribution. We assume

$$y_t = \sum_{s=-\infty}^{\infty} \gamma_s v_{t-s}, \tag{1}$$

where $\sum_{s=-\infty}^{\infty} |\gamma_s| < \infty$ and $\{v_t\}$ is a sequence of independently and identically distributed random variables with $\mathscr{E}v_t = 0$, $\mathscr{E}v_t^2 = \sigma^2$, and $\mathscr{E}v_t^4 = 3\sigma^4 + \kappa_4 < \infty$. We shall find the limiting distribution of the difference between

$$U_T = \sqrt{\frac{T}{K_T}} \sum_{g=1}^{K_T} k\left(\frac{g}{K_T}\right) \cos \nu g \, \frac{1}{T} \sum_{t=1}^{T-g} y_t y_{t+g} \tag{2}$$

and $\mathscr{E}U_T$. The difference between $\sqrt{T/K_T}\,[\hat{f}_T(\nu) - \mathscr{E}\hat{f}_T(\nu)]$ and $(U_T - \mathscr{E}U_T)/\pi$ is

$$\frac{k(0)}{2\pi} \frac{1}{\sqrt{TK_T}} \sum_{t=1}^{T} [y_t^2 - \sigma(0)], \tag{3}$$

the variance of which goes to 0 as $T \to \infty$. Let

$$y_{t,n} = \sum_{s=-n}^{n} \gamma_s v_{t-s}, \tag{4}$$

$$u_{t,n} = \sum_{|s|>n} \gamma_s v_{t-s}, \tag{5}$$

$$U_{Tn} = \sqrt{\frac{T}{K_T}} \sum_{g=1}^{K_T} k\left(\frac{g}{K_T}\right) \cos \nu g \, \frac{1}{T} \sum_{t=1}^{T-g} y_{t,n} y_{t+g,n}. \tag{6}$$

Then

$$U_T - U_{Tn} = \frac{1}{\sqrt{TK_T}} \sum_{g=1}^{K_T} k\left(\frac{g}{K_T}\right) \cos \nu g \sum_{t=1}^{T-g} (y_t y_{t+g} - y_{t,n} y_{t+g,n}) \tag{7}$$

$$= S_1 + S_2 + S_3,$$

where

$$S_1 = \frac{1}{\sqrt{TK_T}} \sum_{g=1}^{K_T} k\left(\frac{g}{K_T}\right) \cos \nu g \sum_{t=1}^{T-g} u_{t,n} y_{t+g,n}, \tag{8}$$

$$S_2 = \frac{1}{\sqrt{TK_T}} \sum_{g=1}^{K_T} k\left(\frac{g}{K_T}\right) \cos \nu g \sum_{t=1}^{T-g} y_{t,n} u_{t+g,n}, \tag{9}$$

$$S_3 = \frac{1}{\sqrt{TK_T}} \sum_{g=1}^{K_T} k\left(\frac{g}{K_T}\right) \cos \nu g \sum_{t=1}^{T-g} u_{t,n} u_{t+g,n}. \tag{10}$$

Let $\gamma_s' = \gamma_s$ for $|s| \leq n$ and $\gamma_s' = 0$ for $|s| > n$, and let $\gamma_s^* = 0$ for $|s| \leq n$ and $\gamma_s^* = \gamma_s$ for $|s| > n$. The variance of S_3 is

$$\begin{aligned}
&\frac{1}{TK_T} \sum_{g,h=1}^{K_T} k\left(\frac{g}{K_T}\right) k\left(\frac{h}{K_T}\right) \cos \nu g \cos \nu h \sum_{s=1}^{T-g} \sum_{t=1}^{T-h} \operatorname{Cov}(u_{s,n} u_{s+g,n}, u_{t,n} u_{t+h,n}) \\
&= \frac{1}{TK_T} \sum_{g,h=1}^{K_T} k\left(\frac{g}{K_T}\right) k\left(\frac{h}{K_T}\right) \cos \nu g \cos \nu h \\
&\quad \times \sum_{s=1}^{T-g} \sum_{t=1}^{T-h} \sum_{p,q,r,m=-\infty}^{\infty} \gamma_p^* \gamma_q^* \gamma_r^* \gamma_m^* \operatorname{Cov}(v_{s-p} v_{s+g-q}, v_{t-r} v_{t+h-m}) \\
&= \frac{1}{TK_T} \sum_{g,h=1}^{K_T} k\left(\frac{g}{K_T}\right) k\left(\frac{h}{K_T}\right) \cos \nu g \cos \nu h \\
&\quad \times \sum_{s=1}^{T-g} \sum_{t=1}^{T-h} \Bigg[\sigma^4 \sum_{p,q=-\infty}^{\infty} \gamma_p^* \gamma_{p-s+t}^* \gamma_q^* \gamma_{q-s+t+h-g}^* \\
&\quad + \sigma^4 \sum_{p,q=-\infty}^{\infty} \gamma_p^* \gamma_{p-s+t+h}^* \gamma_q^* \gamma_{q-s+t-g}^* \\
&\quad + \kappa_4 \sum_{p=-\infty}^{\infty} \gamma_p^* \gamma_{p+g}^* \gamma_{p-s+t}^* \gamma_{p-s+t+h}^* \Bigg].
\end{aligned} \tag{11}$$

Then

$$(12)\qquad \operatorname{Var} S_3 \le \frac{\sup_{-1\le x\le 1} k^2(x)}{K_T} \sum_{g,h=1}^{K_T} \Bigg[\sigma^4 \sum_{p,q,r=-\infty}^{\infty} |\gamma_p^*| \cdot |\gamma_{p+r}^*| \cdot |\gamma_q^*| \cdot |\gamma_{q+r+h-g}^*|$$

$$+ \sigma^4 \sum_{p,q,r=-\infty}^{\infty} |\gamma_p^*| \cdot |\gamma_{p+r+h}^*| \cdot |\gamma_q^*| \cdot |\gamma_{q+r-g}^*|$$

$$+ \kappa_4 \sum_{p,r=-\infty}^{\infty} |\gamma_p^*| \cdot |\gamma_{p+g}^*| \cdot |\gamma_{p+r}^*| \cdot |\gamma_{p+r+h}^*| \Bigg]$$

$$\le \sup_{-1\le x\le 1} k^2(x)(2\sigma^4 + \kappa_4) \sum_{p,q,r,m=-\infty}^{\infty} |\gamma_p^*| \cdot |\gamma_q^*| \cdot |\gamma_r^*| \cdot |\gamma_m^*|$$

$$= \sup_{-1\le x\le 1} k^2(x)(2\sigma^4 + \kappa_4) \left(\sum_{|s|>n} |\gamma_s| \right)^4,$$

which does not depend on T and goes to 0 as $n \to \infty$. Similarly

$$(13)\qquad \operatorname{Var} S_1 = \frac{1}{TK_T} \sum_{g,h=1}^{K_T} k\left(\frac{g}{K_T}\right) k\left(\frac{h}{K_T}\right) \cos \nu g \cos \nu h$$

$$\times \sum_{s=1}^{T-g} \sum_{t=1}^{T-h} \sum_{p,q,r,m=-\infty}^{\infty} \gamma_p^* \gamma_q' \gamma_r^* \gamma_m' \operatorname{Cov}(v_{s-p} v_{s+g-q}, v_{t-r} v_{t+h-m})$$

$$\le \sup_{-1\le x\le 1} k^2(x)(2\sigma^4 + \kappa_4) \left(\sum_{|s|\le n} |\gamma_s| \right)^2 \left(\sum_{|s|>n} |\gamma_s| \right)^2$$

and $\operatorname{Var} S_2$ has the same bound. Then

$$(14)\qquad \mathscr{E}[(U_T - \mathscr{E}U_T) - \mathscr{E}(U_{Tn} - \mathscr{E}U_{Tn})]^2 = \operatorname{Var}(S_1 + S_2 + S_3)$$

$$\le 3(\operatorname{Var} S_1 + \operatorname{Var} S_2 + \operatorname{Var} S_3).$$

By Corollary 7.7.1 the limiting distribution of $U_T - \mathscr{E}U_T$ is the limit on n of the limiting distribution of $U_{Tn} - \mathscr{E}U_{Tn}$.

U_{Tn} is the real part of

$$(15)\quad \frac{1}{\sqrt{TK_T}} \sum_{g=1}^{K_T} k\left(\frac{g}{K_T}\right) e^{ivg} \sum_{t=1}^{T-g} y_{t,n} y_{t+g,n}$$

$$= \frac{1}{\sqrt{TK_T}} \sum_{g=1}^{K_T} k\left(\frac{g}{K_T}\right) e^{ivg} \sum_{t=1}^{T-g} \sum_{r,s=-n}^{n} \gamma_s \gamma_r v_{t-s} v_{t+g-r}$$

$$= \frac{1}{\sqrt{TK_T}} \sum_{g=1}^{K_T} k\left(\frac{g}{K_T}\right) e^{ivg} \sum_{r,s=-n}^{n} \sum_{q=1-s}^{T-g-s} \gamma_s \gamma_r v_q v_{q+g+s-r}$$

$$= \frac{1}{\sqrt{TK_T}} \sum_{r,s=-n}^{n} \sum_{h=s-r+1}^{K_T+s-r} \sum_{q=1-s}^{T-h-r} k\left(\frac{h+r-s}{K_T}\right) e^{ivh} e^{ivr} \gamma_r e^{-ivs} \gamma_s v_q v_{q+h}$$

$$= \frac{1}{\sqrt{TK_T}} \sum_{r,s=-n}^{n} \gamma_r e^{ivr} \gamma_s e^{-ivs} \sum_{h=s-r+1}^{K_T+s-r} k\left(\frac{h+r-s}{K_T}\right) e^{ivh} \sum_{q=1-s}^{T-h-r} v_q v_{q+h}.$$

The difference between the real parts of (15) and

$$(16)\quad \frac{1}{\sqrt{TK_T}} \sum_{r,s=-n}^{n} \gamma_r e^{ivr} \gamma_s e^{-ivs} \sum_{h=1}^{K_T-2n} k\left(\frac{h+r-s}{K_T}\right) e^{ivh} \sum_{q=1}^{T-h} v_q v_{q+h}$$

has a mean square error that goes to 0 as $T \to \infty$. For given r and s the difference between the summands in (15) and (16) consists of the terms in the sums on h and q that are included in one expression and not in the other. The number of such terms is less than $AK_T n + BTn + Cn^2$ for suitable A, B, and C (see Problems 27 and 28); the terms are uncorrelated (see Problem 29); and the expected value of the square of the real part of each such term is at most

$$(17)\quad \left(\max_{-n \le r \le n} |\gamma_r|\right)^4 \sup_{-1 \le x \le 1} k^2(x)\sigma^4/(TK_T).$$

Hence the expected value of the square of the difference for each r and s goes to 0 as $T \to \infty$.

If $k(x)$ is continuous in $-1 \le x \le 1$, then $k[(h+r-s)/K_T]$ is arbitrarily close to $k(h/K_T)$ for K_T sufficiently large and $|r| \le n$, $|s| \le n$, $|h| \le K_T$, and $|h+r-s| \le K_T$. Thus the difference between the real parts of (16) and

$$(18)\quad \frac{1}{\sqrt{TK_T}} f_{(n)}(\nu) \sum_{h=1}^{K_T-2n} k\left(\frac{h}{K_T}\right) e^{ivh} \sum_{q=1}^{T-h} v_q v_{q+h}$$

$$= \frac{f_{(n)}(\nu)}{\sqrt{TK_T}} \sum_{q=1}^{T-1} \sum_{h=1}^{\min(K_T-2n,\,T-q)} k\left(\frac{h}{K_T}\right) e^{ivh} v_q v_{q+h},$$

where $f_{(n)}(\nu) = |\sum_{r=-n}^{n} \gamma_r e^{i\nu r}|^2$, has a mean square error that is arbitrarily small (see Problem 30). The difference between the real part of (18) and

$$\frac{f_{(n)}(\nu)}{\sqrt{T}} \sum_{q=1}^{T} \frac{1}{\sqrt{K_T}} \sum_{h=1}^{K_T} k\left(\frac{h}{K_T}\right) \cos \nu h \, v_q v_{q+h} = f_{(n)}(\nu) \frac{1}{\sqrt{T}} \sum_{q=1}^{T} W_{qT}, \tag{19}$$

where

$$W_{qT} = \frac{1}{\sqrt{K_T}} \sum_{h=1}^{K_T} k\left(\frac{h}{K_T}\right) \cos \nu n \, v_q v_{q+h}, \qquad q = 1, \ldots, T, \tag{20}$$

has a mean square error that goes to 0 as T increases. The process $\{W_{qT}\}$ is stationary and finitely dependent, with

$$\mathscr{E} W_{qT} = 0, \tag{21}$$

$$\begin{aligned} \mathscr{E} W_{qT}^2 &= \frac{1}{K_T} \sum_{g,h=1}^{K_T} k\left(\frac{g}{K_T}\right) k\left(\frac{h}{K_T}\right) \cos \nu g \cos \nu h \, \mathscr{E} v_q^2 v_{q+g} v_{q+h} \\ &= \frac{\sigma^4}{K_T} \sum_{h=1}^{K_T} k^2\left(\frac{h}{K_T}\right) \cos^2 \nu h \\ &= \tfrac{1}{2}\sigma^4 \sum_{h=1}^{K_T} k^2\left(\frac{h}{K_T}\right) (1 + \cos 2\nu h)/K_T, \end{aligned} \tag{22}$$

$$\begin{aligned} \mathscr{E} W_{qT} W_{q+r,T} &= \frac{1}{K_T} \sum_{g,h=1}^{K_T} k\left(\frac{g}{K_T}\right) k\left(\frac{h}{K_T}\right) \cos \nu g \cos \nu h \, \mathscr{E} v_q v_{q+g} v_{q+r} v_{q+h+r} \\ &= 0, \qquad r \neq 0. \end{aligned} \tag{23}$$

Hence the variance of (19) is $f_{(n)}^2(\nu)$ times (22), the limit of which is

$$\lim_{K_T \to \infty} \tfrac{1}{2}\sigma^4 \sum_{h=1}^{K_T} k^2\left(\frac{h}{K_T}\right) \frac{1}{K_T} = \tfrac{1}{2}\sigma^4 \int_0^1 k^2(x)\, dx \tag{24}$$

if $\nu \neq 0, \pm\pi$ and is twice (24) if $\nu = 0$ or $\pm\pi$. (See Problem 23.)

Let $\{N_T\}$ be a sequence of integers such that $K_T/N_T \to 0$ and $N_T/T \to 0$ as $T \to \infty$. (Let $N_T = [\sqrt{TK_T}]$, for example.) Let M_T be the largest integer in T/N_T. Let

$$Z_{jT} = \frac{1}{\sqrt{N_T}} [W_{(j-1)N_T+1,T} + \cdots + W_{jN_T - K_T, T}], \qquad j = 1, \ldots, M_T. \tag{25}$$

Then $Z_{1T}, \ldots, Z_{M_T T}$ are independently and identically distributed with $\mathscr{E} Z_{jT} = 0$ and $\mathscr{E} Z_{jT}^2$ given by $1 - K_T/N_T$ times (22). Moreover,

$$\mathscr{E} Z_{1T}^4 = \frac{1}{N_T^2} \sum_{t,s,r,q=1}^{N_T - K_T} \mathscr{E} W_{tT} W_{sT} W_{rT} W_{qT} \tag{26}$$

is uniformly bounded. Although W_{tT} is quadratic in the v_r's, any fourth-order moment of the W_{tT}'s is bounded because only moments of the v_r's to order four are involved since in each W_{tT} the products of v_r's involve different indices r and the v_r's are independent. Since $\mathscr{E} W_{tT} W_{sT} W_{rT} W_{qT} = 0$ if the smallest index is different from the other three indices, we need consider only terms $\mathscr{E} W_{tT}^4$, $\mathscr{E} W_{tT}^3 W_{sT}$, $s > t$, and $\mathscr{E} W_{tT}^2 W_{sT} W_{qT}$, $s > t, q > t$. There are $N_T - K_T$ terms $\mathscr{E} W_{tT}^4$, at most $2(N_T - K_T)(N_T - K_T - 1)$ terms $\mathscr{E} W_{tT}^3 W_{sT}$, $s > t$, and at most $3(N_T - K_T)(N_T - K_T - 1)$ terms $\mathscr{E} W_{tT}^2 W_{sT}^2$, $s > t$. Now consider $\mathscr{E} W_{tT}^2 W_{sT} W_{qT}$, $s > t, q > t, s \neq q$. Such a term is 0 if $|s - q| > K_T$; thus the number of nonzero terms is at most $6(N_T - K_T)(N_T - K_T - 1)K_T$. However, $|\mathscr{E} W_{tT}^2 W_{sT} W_{qT}| \leq 4\sigma^8 \sup_{-1 \leq x \leq 1} k^4(x)/K_T$. Each type of term is uniformly bounded; see Problem 31. Therefore $\mathscr{E} Z_{jT}^4$ is bounded (uniformly in T). Then

$$\frac{1}{\sqrt{M_T}} \sum_{j=1}^{M_T} Z_{jT} \tag{27}$$

has a limiting normal distribution with mean 0 and variance (24) if $\nu \neq 0, \pm\pi$ [or twice (24) for $\nu = 0, \pm\pi$] by the Liapounov Central Limit Theorem. (In Theorem 7.7.3 we take w_t^T as $Z_{tT}/[M_T \mathscr{E} Z_{tT}^2]^{\frac{1}{2}}$ and $\delta = 2$.) Since

$$\frac{1}{\sqrt{T}} \sum_{t=1}^{T} W_{tT} - \frac{1}{\sqrt{M_T}} \sum_{j=1}^{M_T} Z_{jT} \tag{28}$$

has a mean square error that approaches 0 as $T \to \infty$, the limiting distribution of $\sum_{t=1}^{T} W_{tT}/\sqrt{T}$ is that of (27).

THEOREM 9.4.1. *Let*

$$\hat{f}_T(\nu) = \frac{1}{2\pi} \sum_{r=-K_T}^{K_T} k\left(\frac{r}{K_T}\right) \cos \nu r \, c_{rT}, \tag{29}$$

where $k(x) = k(-x)$, $k(x)$ is continuous in $-1 \leq x \leq 1$, and $\{K_T\}$ is a sequence of integers such that $K_T \to \infty$ and $K_T/T \to 0$ as $T \to \infty$. Let

$$y_t = \sum_{s=-\infty}^{\infty} \gamma_s v_{t-s}, \tag{30}$$

where $\sum_{s=-\infty}^{\infty} |\gamma_s| < \infty$ and $\{v_t\}$ is a process of independently and identically distributed variables with $\mathscr{E} v_t = 0$, $\mathscr{E} v_t^2 = \sigma^2$, and $\mathscr{E} v_t^4 < \infty$. Then $\sqrt{T/K_T}$ $[\hat{f}_T(\nu) - \mathscr{E}\hat{f}_T(\nu)]$ has a limiting normal distribution with variance given by Theorem 9.3.4.

The estimates at a fixed number of values of ν, say $\hat{f}_T(\nu_1), \ldots, \hat{f}_T(\nu_n)$, similarly have an asymptotic joint normal distribution.

The conditions of Theorem 9.4.1 can be modified to eliminate the condition $\mathscr{E}v_t^4 < \infty$. Cheong and Hannan (1968) have shown that if $\sum_{t=1}^T (v_t^2 - \sigma^2)/\sqrt{TK_T}$ converges to 0 stochastically and $\sum_{s=-\infty}^{\infty} |s|\, \gamma_s^2 < \infty$, then asymptotic normality results. Rosenblatt (1959) has demonstrated normality under other conditions.

Now we consider

$$(31)\qquad \sqrt{\frac{T}{K_T}}\,[\hat{f}_T(\nu) - f(\nu)] = \sqrt{\frac{T}{K_T}}\,[\hat{f}_T(\nu) - \mathscr{E}\hat{f}_T(\nu)] + \sqrt{\frac{T}{K_T}}\,[\mathscr{E}\hat{f}_T(\nu) - f(\nu)].$$

Let p and q be defined by Theorem 9.3.3. If $p < q$ and $K_T/T \to 0$ if $p \leq 1$ and $K_T^p/T \to 0$ if $p \geq 1$, then the second term on the right-hand side of (31) converges to 0 if $\lim_{T\to\infty} T/K_T^{2p+1}$ is finite. If $q \leq p$ and $K_T^{q+1-p}/T \to 0$ if $p \leq 1$ and $K_T^q/T \to 0$ if $p \geq 1$, then the second term converges to 0 if $T/K_T^{2q+1} \to 0$. In either case the limiting distribution of the left-hand side is the limiting distribution of the first term on the right-hand side.

COROLLARY 9.4.1. *If the conditions of Theorem* 9.4.1 *hold, if the conditions of Theorem* 9.3.3 *hold, and if* $\lim_{T\to\infty} T/K_T^{2p+1}$ *is finite if* $p < q$ *or* $T/K_T^{2q+1} \to 0$ *as* $T \to \infty$ *if* $q \leq p$, *then* $\sqrt{T/K_T}\,[\hat{f}_T(\nu) - f(\nu)]$ *has a limiting normal distribution with mean* 0 *and variance given by Theorem* 9.3.4.

Note that for the bias to become negligible compared to the random part of (31), K_T must increase (in comparison to T) faster than it increases to make the variance and squared bias of the same order. The estimates at a fixed number of values of ν, say $\hat{f}_T(\nu_1), \ldots, \hat{f}_T(\nu_n)$, similarly have an asymptotic joint normal distribution.

Corollary 9.4.1 can be restated to the effect that for $f(\nu) > 0$

$$(32)\qquad \sqrt{\frac{T}{K_T}}\,\frac{\hat{f}_T(\nu) - f(\nu)}{\tau f(\nu)} = \sqrt{\frac{T}{K_T}}\,\frac{1}{\tau}\left[\frac{\hat{f}_T(\nu)}{f(\nu)} - 1\right]$$

has a limiting normal distribution with mean 0 and variance 1, where $\tau^2 = \int_{-1}^{1} k^2(x)\,dx$ $(\nu \neq 0, \pm\pi)$. Let $t(\varepsilon)$ be the number such that the probability is $1 - \varepsilon$ of a standard normal variate falling into the interval $[-t(\varepsilon), t(\varepsilon)]$. The event that (32) falls in this interval can be rewritten

$$(33)\qquad \frac{\hat{f}_T(\nu)}{1 + \tau t(\varepsilon)\sqrt{\dfrac{K_T}{T}}} \leq f(\nu) \leq \frac{\hat{f}_T(\nu)}{1 - \tau t(\varepsilon)\sqrt{\dfrac{K_T}{T}}}, \qquad \nu \neq 0, \pm\pi.$$

This constitutes a confidence interval for $f(\nu)$ with confidence approximately $1 - \varepsilon$ for large T (under the conditions of Corollary 9.4.1).

We can use the theorem that if $Y_T = g(X_T)$, where $\operatorname{plim}_{T\to\infty} X_T = \mu$, $\beta_T(X_T - \mu)$ has a limiting normal distribution with mean 0 and variance σ^2 ($\beta_T \to \infty$) as $T \to \infty$, and $g(x)$ has the derivative $g'(\mu)$ at $x = \mu$, then $\beta_T[Y_T - g(\mu)]$ has a limiting normal distribution with mean 0 and variance $\sigma^2[g'(\mu)]^2$. If $X_T = \hat{f}_T(\nu)$ and $g(x) = \log x$, then $\mu = f(\nu)$ and $g(\mu) = \log f(\nu)$ and $g'(\mu) = 1/f(\nu)$ if $f(\nu) > 0$.

THEOREM 9.4.2. *Under the conditions of Corollary* 9.4.1 *if* $f(\nu) > 0$

$$\sqrt{\frac{T}{K_T}}\,[\log \hat{f}_T(\nu) - \log f(\nu)] = \sqrt{\frac{T}{K_T}}\,\log\frac{\hat{f}_T(\nu)}{f(\nu)} \tag{34}$$

has a limiting normal distribution with mean 0 *and variance* $\tau^2 = \int_{-1}^{1} k^2(x)\,dx$ *if* $\nu \neq 0, \pm\pi$ *and variance* $2\tau^2$ *if* $\nu = 0, \pm\pi$.

This leads to confidence intervals

$$\log \hat{f}_T(\nu) - \tau t(\varepsilon)\sqrt{\frac{K_T}{T}} \leq \log f(\nu) \leq \log \hat{f}_T(\nu) + \tau t(\varepsilon)\sqrt{\frac{K_T}{T}}, \tag{35}$$

or

$$\hat{f}_T(\nu)e^{-\tau t(\varepsilon)\sqrt{K_T/T}} \leq f(\nu) \leq \hat{f}_T(\nu)e^{\tau t(\varepsilon)\sqrt{K_T/T}}. \tag{36}$$

Theorem 9.4.2 suggests that an informative graphical representation of the estimation of the spectral density is to plot $\log \hat{f}_T(\nu)$ against ν because (except at $\nu = 0$ and $\pm\pi$) the asymptotic standard deviation is constant and known, namely $\tau\sqrt{K_T/T}$.

It has been suggested by Blackman and Tukey (1959), p. 22, that the distribution of $\hat{f}_T(\nu)$, which is a quadratic form in the observations, be approximated by the distribution of $\beta f(\nu)\chi^2_\alpha$, where χ^2_α denotes a χ^2-variable with α degrees of freedom, and α and β are chosen so the first two moments of $\beta f(\nu)\chi^2_\alpha$ are approximately those of $\hat{f}_T(\nu)$. Thus, for $\nu \neq 0, \pm\pi$ we want

$$\beta\alpha = 1, \tag{37}$$

$$2\beta^2\alpha = \frac{K_T}{T}\int_{-1}^{1} k^2(x)\,dx. \tag{38}$$

Then

$$\alpha = \frac{1}{\beta} = \frac{2T}{K_T\displaystyle\int_{-1}^{1} k^2(x)\,dx}, \tag{39}$$

and for $f(\nu) > 0$

$$\frac{2T\hat{f}_T(\nu)}{K_T f(\nu)\int_{-1}^{1} k^2(x)\,dx} \tag{40}$$

is taken to have a χ^2-distribution with degrees of freedom equal to α given by (39).

If the mean μ is not known, estimates of $f(\lambda)$ may be based on c_{rT}^*'s or $\dot{\bar{c}}_{rT}$'s. Let

$$\hat{f}_T^*(\nu) = \frac{1}{2\pi}\sum_{r=-K_T}^{K_T} k\left(\frac{r}{K_T}\right)\cos \nu r\, c_{rT}^*. \tag{41}$$

Let $m = \min(p,q)$. Then

$$\begin{aligned}(42)\quad K_T^m[\mathscr{E}\hat{f}_T^*(\nu) - f(\nu)] &\\ = K_T^m[\mathscr{E}\hat{f}_T(\nu) - f(\nu)] &+ \frac{K_T^m}{2\pi}\sum_{r=-K_T}^{K_T} k\left(\frac{r}{K_T}\right)\cos \nu r\, \mathscr{E}(c_{rT}^* - c_{rT})\\ = K_T^m[\mathscr{E}\hat{f}_T(\nu) - f(\nu)] &+ \frac{1}{2\pi}\frac{K_T^m}{T}\sum_{r=-K_T}^{K_T} k\left(\frac{r}{K_T}\right)\cos \nu r\, T\mathscr{E}(c_{rT}^* - c_{rT}).\end{aligned}$$

The absolute value of the second term on the extreme right-hand side of (42) is not greater than

$$\frac{1}{2\pi}\frac{K_T^m}{T}\sup_{-1\le x\le 1}|k(x)|\sum_{r=-K_T}^{K_T}|T\mathscr{E}(c_{rT}^* - c_{rT})|. \tag{43}$$

From Theorem 8.3.2 we see that if $\sum_{r=-\infty}^{\infty}|\sigma(r)| < \infty$, then $|T(\mathscr{E}c_{rT}^* - \mathscr{E}c_{rT})| \le 3\sum_{s=-\infty}^{\infty}|\sigma(s)|$ for all T and r. Hence, (43) is not greater than

$$\frac{3}{2\pi}\sup_{-1\le x\le 1}|k(x)|\sum_{r=-\infty}^{\infty}|\sigma(r)|\frac{K_T^m}{T}(2K_T + 1), \tag{44}$$

the limit of which is 0 if $K_T^{m+1}/T \to 0$ as $T \to \infty$.

THEOREM 9.4.3. *Let $\hat{f}_T^*(\nu)$ be defined by* (41), *where $k(x) = k(-x)$, $k(0) = 1$, $|k(x)| \le M$ for some M and all $|x| \le 1$, and for some $q > 0$ and $k > 0$, $\lim_{x\to 0}[1 - k(x)]/|x|^q = k$. Let $\{K_T\}$ be a sequence of integers such that $K_T \to \infty$ as $T \to \infty$. If*

$$\sum_{r=-\infty}^{\infty}|r|^p|\sigma(r)| < \infty \tag{45}$$

for $p \ge q$ and $p \ge 1$ and if $K_T^{q+1}/T \to 0$ as $T \to \infty$, then

$$\lim_{T\to\infty} K_T^q[\mathscr{E}\hat{f}_T^*(\nu) - f(\nu)] = -\frac{k}{2\pi}\sum_{r=-\infty}^{\infty}|r|^q\cos \nu r\,\sigma(r). \tag{46}$$

If (45) *holds for* $1 \le p < q$ *and if* $K_T^{p+1}/T \to 0$ *as* $T \to \infty$, *then*

$$\lim_{T\to\infty} K_T^p[\mathscr{E}\hat{f}_T^*(\nu) - f(\nu)] = 0. \tag{47}$$

The covariances of $\hat{f}_T^*(\nu)$ are given by

$$\frac{T}{K_T}\operatorname{Cov}[\hat{f}_T^*(\lambda), \hat{f}_T^*(\nu)] = \frac{1}{(2\pi)^2 K_T}\sum_{g,h=-K_T}^{K_T} k\left(\frac{g}{K_T}\right)k\left(\frac{h}{K_T}\right)\cos\lambda g\cos\nu h\, T\operatorname{Cov}(c_{gT}^*, c_{hT}^*). \tag{48}$$

Hence for $2K_T < T$

$$\left|\frac{T}{K_T}\operatorname{Cov}[\hat{f}_T^*(\lambda), \hat{f}_T^*(\nu)] - \frac{T}{K_T}\operatorname{Cov}[\hat{f}_T(\lambda), \hat{f}_T(\nu)]\right| \le \frac{1}{(2\pi)^2 K_T}\sup_{-1\le x\le 1} k^2(x)\sum_{g,h=-K_T}^{K_T}|T\operatorname{Cov}(c_{gT}^*, c_{hT}^*) - T\operatorname{Cov}(c_{gT}, c_{hT})|. \tag{49}$$

We shall now take $\mu = 0$ for convenience. Since

$$\begin{aligned} c_{gT}^* &= c_{gT} - \frac{1}{T^2}\sum_{t=1}^{T} y_t\left(\sum_{s=g+1}^{T} y_s + \sum_{s=1}^{T-g} y_s - \frac{T-g}{T}\sum_{s=1}^{T} y_s\right) \\ &= c_{gT} - \frac{1}{T^2}\sum_{t=1}^{T}\sum_{s=g+1}^{T-g} y_t y_s - \frac{g}{T^3}\sum_{t,s=1}^{T} y_t y_s, \qquad 0 \le 2g \le T-1, \end{aligned} \tag{50}$$

we consider for $0 \le h \le T-1$, $0 \le 2g \le T-1$

$$\begin{aligned} T\operatorname{Cov}&\left(c_{hT}, \frac{1}{T^2}\sum_{t=1}^{T}\sum_{s=g+1}^{T-g} y_t y_s + \frac{g}{T^3}\sum_{t,s=1}^{T} y_t y_s\right) \\ &= \frac{1}{T^2}\sum_{r=1}^{T-h}\sum_{t=1}^{T}\left\{\sum_{s=g+1}^{T-g}\mathscr{E}[y_r y_{r+h} - \sigma(h)][y_t y_s - \sigma(t-s)]\right. \\ &\qquad \left. + \frac{g}{T}\sum_{s=1}^{T}\mathscr{E}[y_r y_{r+h} - \sigma(h)][y_t y_s - \sigma(t-s)]\right\} \\ &= \frac{1}{T^2}\sum_{r=1}^{T-h}\sum_{t=1}^{T}\left\{\sum_{s=g+1}^{T-g}[\sigma(r-t)\sigma(r+h-s) + \sigma(r-s)\sigma(r+h-t)\right. \\ &\qquad + \kappa(h, t-r, s-r)] + \frac{g}{T}\sum_{s=1}^{T}[\sigma(r-t)\sigma(r+h-s) \\ &\qquad \left. + \sigma(r-s)\sigma(r+h-t) + \kappa(h, t-r, s-r)]\right\}, \end{aligned} \tag{51}$$

which is in absolute value less than

$$\frac{2}{T}\left\{2\left[\sum_{t=-\infty}^{\infty}|\sigma(t)|\right]^2+\sum_{r,s,t=-\infty}^{\infty}|\kappa(r,s,t)|\right\}. \tag{52}$$

In Theorem 9.3.4 the sums within the braces are assumed finite. Consider

(53)

$$T\operatorname{Cov}\left[\frac{1}{T^2}\sum_{t=1}^{T}\left(\sum_{s=g+1}^{T-g}y_ty_s+\frac{g}{T}\sum_{s=1}^{T}y_ty_s\right),\frac{1}{T^2}\sum_{r=1}^{T}\left(\sum_{q=h+1}^{T-h}y_ry_q+\frac{h}{T}\sum_{q=1}^{T}y_ry_q\right)\right]$$

$$\begin{aligned}
&=\frac{1}{T^3}\sum_{t,r=1}^{T}\left\{\sum_{s=g+1}^{T-g}\sum_{q=h+1}^{T-h}\mathscr{E}[y_ty_s-\sigma(t-s)][y_ry_q-\sigma(r-q)]\right.\\
&\quad+\frac{h}{T}\sum_{s=g+1}^{T-g}\sum_{q=1}^{T}\mathscr{E}[y_ty_s-\sigma(t-s)][y_ry_q-\sigma(r-q)]\\
&\quad+\frac{g}{T}\sum_{s=1}^{T}\sum_{q=h+1}^{T-h}\mathscr{E}[y_ty_s-\sigma(t-s)][y_ry_q-\sigma(r-q)]\\
&\quad\left.+\frac{gh}{T^2}\sum_{s,q=1}^{T}\mathscr{E}[y_ty_s-\sigma(t-s)][y_ry_q-\sigma(r-q)]\right\}\\
&=\frac{1}{T^3}\sum_{t,r=1}^{T}\left\{\sum_{s=g+1}^{T-g}\sum_{q=h+1}^{T-h}[\sigma(t-r)\sigma(s-q)\right.\\
&\quad+\sigma(t-q)\sigma(s-r)+\kappa(s-t,r-t,q-t)]\\
&\quad+\frac{h}{T}\sum_{s=g+1}^{T-g}\sum_{q=1}^{T}[\sigma(t-r)\sigma(s-q)\\
&\quad+\sigma(t-q)\sigma(s-r)+\kappa(s-t,r-t,q-t)]\\
&\quad+\frac{g}{T}\sum_{s=1}^{T}\sum_{q=h+1}^{T-h}[\sigma(t-r)\sigma(s-q)\\
&\quad+\sigma(t-q)\sigma(s-r)+\kappa(s-t,r-t,q-t)]\\
&\quad+\frac{gh}{T^2}\sum_{s,q=1}^{T}[\sigma(t-r)\sigma(s-q)\\
&\quad\left.+\sigma(t-q)\sigma(s-r)+\kappa(s-t,r-t,q-t)]\right\},
\end{aligned}$$

which is in absolute value less than twice (52). Therefore for $2K_T < T$ the right-hand side of (49) is less than

$$8\frac{(2K_T+1)^2}{(2\pi)^2K_TT}\sup_{-1\le x\le 1}k^2(x)\left\{2\left[\sum_{t=-\infty}^{\infty}|\sigma(t)|\right]^2+\sum_{r,s,t=-\infty}^{\infty}|\kappa(r,s,t)|\right\}. \tag{54}$$

THEOREM 9.4.4. *Under the assumptions of Theorem* 9.3.4, *the conclusions of Theorem* 9.3.4 *hold for* $\hat{f}_T^*(\nu)$ *defined by* (41).

We can write

$$\begin{aligned}\sqrt{\frac{T}{K_T}}[\hat{f}_T(\nu)-\hat{f}_T^*(\nu)] \\ = \sqrt{\frac{T}{K_T}}\frac{1}{2\pi}\sum_{r=-K_T}^{K_T}k\left(\frac{r}{K_T}\right)\cos\nu r\frac{1}{T^2}\sum_{t=1}^{T}y_t\left\{\sum_{s=|r|+1}^{T-|r|}y_s+\frac{|r|}{T}\sum_{s=1}^{T}y_s\right\}\end{aligned} \tag{55}$$

for $T > 2K_T$. The expected value of the square of (55) is

$$\begin{aligned}&\frac{1}{(2\pi)^2}\frac{T}{K_T}\sum_{r,r'=-K_T}^{K_T}k\left(\frac{r}{K_T}\right)k\left(\frac{r'}{K_T}\right)\cos\nu r\cos\nu r' \\ &\times\mathscr{E}\frac{1}{T^4}\sum_{t,t'=1}^{T}\left\{\sum_{s=|r|+1}^{T-|r|}\sum_{s'=|r'|+1}^{T-|r'|}y_ty_sy_{t'}y_{s'}+\frac{|r'|}{T}\sum_{s=|r|+1}^{T-|r|}\sum_{s'=1}^{T}y_ty_sy_{t'}y_{s'}\right. \\ &\left.+\frac{|r|}{T}\sum_{s=1}^{T}\sum_{s'=|r'|+1}^{T-r'}y_ty_sy_{t'}y_{s'}+\frac{|rr'|}{T^2}\sum_{s,s'=1}^{T}y_ty_sy_{t'}y_{s'}\right\} \\ &\le 4\frac{(2K_T+1)^2}{(2\pi)^2K_TT}\sup_{-1\le x\le 1}k^2(x)\left\{3\left[\sum_{t=-\infty}^{\infty}|\sigma(t)|\right]^2+\sum_{r,s,t=-\infty}^{\infty}|\kappa(r,s,t)|\right\}.\end{aligned} \tag{56}$$

THEOREM 9.4.5. *Under the assumptions of Theorem* 9.4.1 $\sqrt{T/K_T}\,[\hat{f}_T^*(\nu_1)-\mathscr{E}\hat{f}_T^*(\nu_1)],\ldots,\sqrt{T/K_T}\,[\hat{f}_T^*(\nu_n)-\mathscr{E}\hat{f}_T^*(\nu_n)]$ *have a limiting normal distribution with means* 0 *and variances and covariances given by Theorem* 9.3.4.

As was seen in Section 7.5.1, the spectral density of the process $\{z_t\}$ defined by

$$z_t=\sum_{r=0}^{q}c_ry_{t-r},\qquad t=\ldots,-1,0,1,\ldots, \tag{57}$$

is

$$h(\lambda)=\left|\sum_{r=0}^{q}c_re^{i\lambda r}\right|^2f(\lambda). \tag{58}$$

For some purposes it may be useful to apply the moving average (57) to the observed series, $y_1, \ldots, y_T$, obtaining a new series $z_{q+1}, \ldots, z_T$, to estimate $h(\nu)$ on the basis of this new series, and then estimate $f(\nu)$ by $\hat{h}(\nu)/|\sum_{r=0}^{q} c_r e^{i\nu r}|^2$. The construction of the moving average is called *prewhitening*. The purpose is to obtain a process with a relatively smooth spectral density $h(\lambda)$, and this can be done if one has some a priori knowledge of the peaks and troughs of $f(\lambda)$.

To estimate the spectral distribution function $G(\nu)$ [$= F(\nu) - F(-\nu)$ at points of continuity of $F(\lambda)$] we can use

$$\hat{G}(\nu) = 2\int_0^{\nu} I(\lambda)\, d\lambda = \frac{1}{\pi} \sum_{r=-(T-1)}^{T-1} c_r \frac{\sin \nu r}{r}, \tag{59}$$

where $(\sin \nu r)/r = \nu$ for $r = 0$. See Problem 26.

We can estimate the normalized spectral density by

$$\hat{\bar{f}}_T(\nu) = \frac{\hat{f}_T(\nu)}{c_0} = \frac{1}{2\pi} \sum_{h=-K_T}^{K_T} k\left(\frac{h}{K_T}\right) \cos \nu h \; r_h \tag{60}$$

or by $\hat{\bar{f}}{}^*_T(\nu) = \hat{f}^*_T(\nu)/c^*_0$. Then

$$\sqrt{\frac{T}{K_T}}\,[\hat{\bar{f}}_T(\nu) - \bar{f}(\nu)] = \sqrt{\frac{T}{K_T}}\left\{\frac{f(\nu)/\sigma(0) + [\hat{f}_T(\nu) - f(\nu)]/\sigma(0)}{1 + [c_0 - \sigma(0)]/\sigma(0)} - \frac{f(\nu)}{\sigma(0)}\right\} \tag{61}$$

has the limiting distribution of

$$\sqrt{\frac{T}{K_T}}\left\{\frac{\hat{f}_T(\nu) - f(\nu)}{\sigma(0)} - \frac{f(\nu)}{\sigma(0)}\,\frac{c_0 - \sigma(0)}{\sigma(0)}\right\}. \tag{62}$$

If $\sqrt{T/K_T}\,[c_0 - \sigma(0)] = \sqrt{T}\,[c_0 - \sigma(0)]/\sqrt{K_T}$ converges to 0 in probability, (62) will have the limiting distribution of $\sqrt{T/K_T}\,[\hat{f}_T(\nu) - f(\nu)]/\sigma(0)$.

9.5. EXAMPLES

In Section 8.5 it was pointed out that the sample spectral densities of the data given in the appendices are very irregular; they are not good estimates of the corresponding population spectral densities. In each of those cases an estimate of $f(\lambda)$ [or $\bar{f}(\lambda)$] according to Section 9.2.3 has been obtained. For the artificially generated series Wold estimated $2\pi\bar{f}(\lambda)$ by the Bartlett estimate (C of

Section 9.2.3) with $K = 20$. The results are tabulated in Table A.2.4 and graphed in Figures A.2.4, A.2.5, and A.2.6. The population normalized spectral densities (multiplied by 2π) are given in Figures A.2.7, A.2.8, and A.2.9 (as well as the normalized densities corresponding to the estimated second-order autoregressive process). Note that the larger $\sqrt{\beta_2} = \gamma$ is, the greater is the peak of $f(\lambda)$; that is, the more pronounced is the oscillatory nature of the process.

It will be observed that the estimated spectral density in the case $\gamma = 0.25$ is very close to the density of the process itself. In each of the other two cases the maximum of the estimated density occurs at the frequency of the maximum of the density of the process and in the case of $\gamma = 0.7$ the values of the maxima of the estimated and actual densities agree. The estimated density for $\gamma = 0.9$ has many relative maxima that the process density does not have. These features reflect the fact that the nearer the process is to an uncorrelated sequence, which has a flat spectral density, the better it can be estimated.

Plotting the spectral densities on a logarithmic scale is done because then the asymptotic standard deviation is independent of the value of the density. However, it has the effect that it exaggerates the visual effects of variations where $\hat{f}_T(\lambda)$ is small and the variations are small. The Bartlett window (with $q = 1$) does not yield as smooth estimates as other windows (with $q = 2$); this is discussed further in Section 9.6.

The normalized spectral density (multiplied by 2π) for the Beveridge trend-free wheat price index was estimated by means of the Blackman-Tukey window with $a = 0.25$ (hanning, G of Section 9.2.3) for $K = 10$, 20, and 30. The results are tabulated in Table A.1.4 and graphed in Figure A.1.4. It will be observed that the smaller K is the smoother is the estimating function. For $K = 10$ there is one relative maximum; for $K = 20$ there are two marked relative maxima and one trivial one; and for $K = 30$ there are six relative maxima. The maximum for $K = 20$ and 30 occurs at a frequency of 0.065 (period of 15.4 years) and the secondary maximum occurs at a frequency of about 0.185 (period of 5.4 years). For $K = 30$ the other relative maxima are at frequencies of 0.285, 0.365, 0.430, and 0.490 (periods of 3.5, 2.74, 2.33, and 2.04 years, respectively).

Estimates of the spectral density for the annual sunspot series (1749 to 1924) are graphed in Figure A.3.2; Parzen's window (J of Section 9.2.3) was used with $K = 20$, 40, and 60. It will be observed that there is a marked peak at a frequency slightly less than 0.09 (period about 11 years) and a secondary peak at about twice that frequency. These observations indicate a cycle with a period of about 11 years; the secondary frequency indicates that the cycle is not a pure trigonometric function.

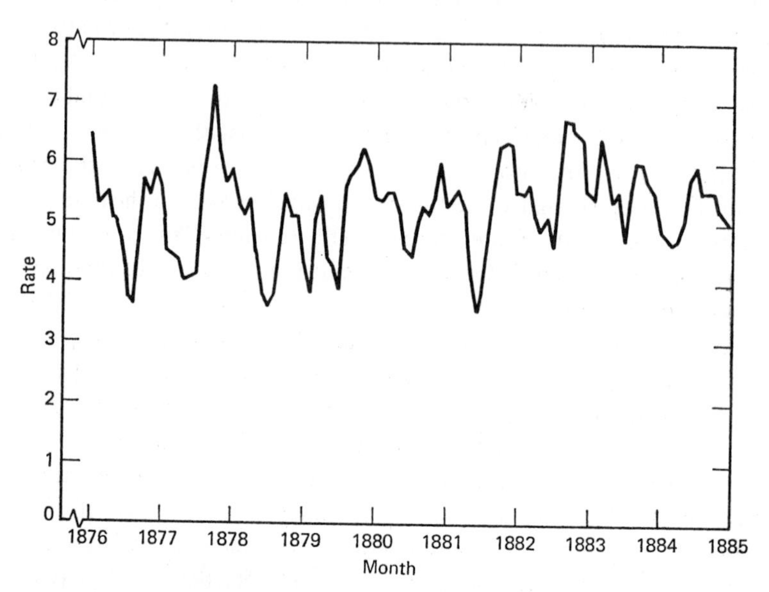

Figure 9.3. New York Commercial Paper Rate (%), 1876–1885.

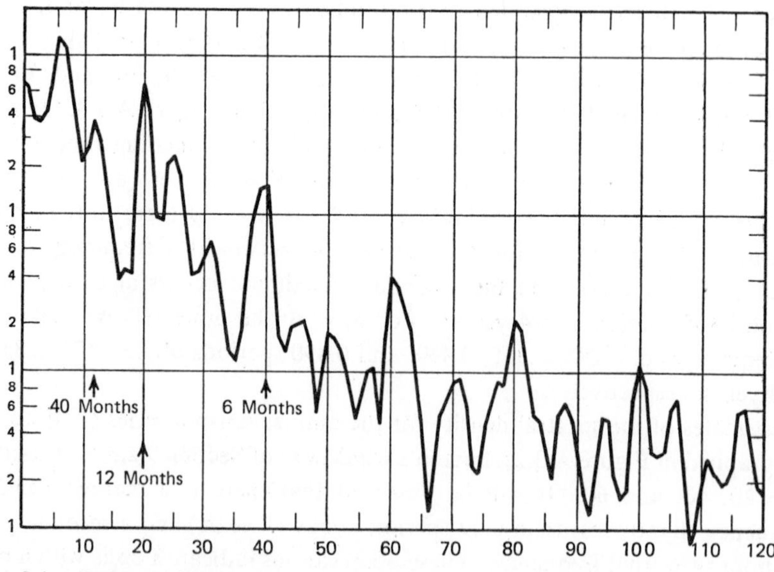

Figure 9.4. Estimated Spectral Density of New York Commercial Paper Rate, 1876–1914.

As one further example, we consider the New York Commercial Paper Rate studied by Granger (1964) based on monthly data from Macaulay (1938) for the period 1876-1914. A segment of the observed times series is plotted in Figure 9.3, and the estimated spectral density is given in Figure 9.4 (120 on the horizontal scale corresponding to $\lambda = \pi$). The Blackman-Tukey estimate with $a = 0.25$ (hanning, G of Section 9.2.3) and $K = 120$ was used. It will be observed that there are peaks at frequencies $j/12$, $j = 1, 2, 3, 4, 5$, corresponding to the seasonal variation (periods that divide 12). In addition there is a peak at a frequency corresponding to a period of 40 months, which is interpreted as the business cycle. The scale in Figure 9.4 is logarithmic.

9.6. DISCUSSION

The spectral density (when it exists) of a weakly stationary stochastic process specifies the covariance properties of the process as the Fourier transform of the covariance sequence; the spectral density is interpreted in terms of the variances (or power) of sinusoidal components with random amplitudes (as well as random phases). Use of an observed time series to estimate the spectral density can be considered as a nonparametric method; a finite set of observations is used to estimate a function over $[0, \pi]$ which cannot be characterized by a finite set of parameters. The procedure is used when the investigator does not wish to specify a parametric form with a specified number of parameters for the spectral density or covariance sequence, such as was done in Chapter 5. The spectral analysis is more flexible than parametric inference, but also less precise. In some sense it can be thought of as exploratory or "data-analytic."

In order that the estimation of the spectrum be informative the observed series must be reasonably long. How long it must be depends on features of the spectral density of the process. The spectral density of an uncorrelated sequence can be estimated by a relatively short series, while a density with many peaks will require a rather large number of observations; in fact, a rather large value of T may be required for the asymptotic theory to be a good approximation.

The choice of K (or K_T) depends on the nature of the spectral density being estimated and the sample size. The smoother the density the smaller K can be (with resulting smaller variance). However, since the density is unknown, the "optimal" K is unknown. Several values of K may be tried. As is seen from the examples, too small a K may lead to peaks being overlooked, and too large a K may yield an estimate with too much irregularity.

It has been observed that if $\hat{f}_T(\lambda) \geq 0$, $-\pi \leq \lambda \leq \pi$, then one can write

$$\hat{f}_T(\lambda) = \frac{\sigma^2}{2\pi}\left|\sum_{j=0}^{K_T} \hat{\alpha}_j e^{i\lambda j}\right|^2, \tag{1}$$

where $\hat{\alpha}_0 = 1$ and $\hat{\alpha}_1, \ldots, \hat{\alpha}_{K_T}$ are real. It then follows from Section 7.5 that $\hat{f}_T(\lambda)$ can be written as a polynomial in $\cos \lambda$ of degree K_T. Hence, there are at most $K_T - 1$ relative maxima and minima of $\hat{f}_T(\lambda)$ in $(0, \pi)$. If $\hat{f}_T(\lambda)$ is not necessarily nonnegative, this is not the case; for example, the Bartlett estimate of the spectral density may be negative because the window is not nonnegative.

Because the flat density of uncorrelated variables is easiest to estimate, a transformation is often used to bring about a density that is flatter. More precisely, a linear transformation of the time series corresponds to multiplying the spectral density by the transfer function. If something is known about the density to be estimated, a transformation can be performed so that the resulting spectral density is flatter; this density is estimated and the result divided by the transfer function to obtain an estimate of the original density. Such a procedure was discussed at the end of Section 9.4.

It should be noted that usually it is computationally more efficient to use the Fast Fourier Transform (as described in Section 4.3.5) for calculations rather than the covariances. When the shape of the spectral density is of interest the normalized spectral density $\bar{f}(\lambda) = f(\lambda)/\sigma(0)$ may be estimated; the same asymptotic theory applies except that $f(\lambda)$ is replaced by $\bar{f}(\lambda)$.

Most of the population, sample, and estimated spectral densities have been plotted on logarithmic scales because the asymptotic sampling variability of the latter two is constant. However, there are also reasons for plotting these densities on arithmetic scales. The height of the density at each frequency corresponds to the variance of the amplitudes near that frequency. Over a range of frequencies for which the density is small, variations of the density are unimportant, but the logarithmic scale exaggerates such variations; thus the values of the smaller relative maxima on the logarithmic scale should be discounted from their visual impressions. Since an estimate $\hat{f}_T(\nu)$ is an arithmetic mean of $I_T(\lambda)$, it is easier to relate an estimated spectral density to the sample density on an arithmetic scale.

In some of the examples one or more relative maxima may appear at multiples of the frequency of the absolute maximum. Since it takes several terms in a Fourier series to approximate a cyclical function or sequence that is not trigonometric, these secondary peaks in the graph may only indicate the non-sinusoidal character of the main cyclical component.

As was noted in the introduction to this chapter, if the spectral density estimate considered here is nonnegative, it is the spectral density of a finite moving average process. An alternative method of estimating the spectral density is to estimate the coefficients of an autoregressive process that approximates the sampled process and use the corresponding spectral density as the estimate. The asymptotic behavior of such estimates is somewhat similar to that of the moving-average type estimates, but the theory has not been worked out to the point that the two methods are exactly comparable. The example given here on the Beveridge trend-free wheat price index suggests that fewer lags may be necessary, but that if too few lags are used the estimate may be misleading. In this case the fitted second-order process has a spectral density whose maximum does not agree with any maxima of the densities of the fitted higher-order processes. However, advantages of this method are that the resulting coefficients yield a predicting function and there is a rationale for determining the number of lags to use (Section 5.6).

REFERENCES

Section 9.2. Bartlett (1946), (1950), Blackman and Tukey (1959), Grenander and Rosenblatt (1957), Hamming (1962), Hannan (1960), Lanczos (1956), Parzen (1957b), (1961b).

Section 9.3. Jenkins (1961), Parzen (1957b), (1961b).

Section 9.4. Blackman and Tukey (1959), Cheong and Hannan (1968), Rosenblatt (1959).

Section 9.5. Granger (1964), Macaulay (1938).

PROBLEMS

1. (Sec. 9.2.1.) Derive the second form of Cov (G, W) in (16) directly from the first form, and derive the second form of (19) directly from the first form. [*Hint:* Start from the second form, express $\sin \frac{1}{2}xT/\sin \frac{1}{2}x$ in (16) for various x, and integrate with respect to λ and λ'.]

2. (Sec. 9.2.3.) Show

(i) $$\sum_{r=1}^{K} \cos \lambda r = \frac{\cos \frac{1}{2}\lambda(K+1) \sin \frac{1}{2}\lambda K}{\sin \frac{1}{2}\lambda},$$

(ii) $$\sum_{r=1}^{K} \sin \lambda r = \frac{\sin \frac{1}{2}\lambda(K+1) \sin \frac{1}{2}\lambda K}{\sin \frac{1}{2}\lambda},$$

(iii) $$\sum_{r=-K}^{K} \cos \lambda r = \frac{\sin \frac{1}{2}\lambda(2K+1)}{\sin \frac{1}{2}\lambda} = 2\pi h_{2K+1}(\lambda),$$

(iv) $$\sum_{r=-K}^{K} |r| \cos \lambda r = K \frac{\sin \frac{1}{2}\lambda(2K+1)}{\sin \frac{1}{2}\lambda} - \frac{\sin^2 \frac{1}{2}\lambda K}{\sin^2 \frac{1}{2}\lambda}$$

$$= K \frac{\sin \frac{1}{2}\lambda(2K+1)}{\sin \frac{1}{2}\lambda} + \frac{\cos \lambda K - 1}{2 \sin^2 \frac{1}{2}\lambda},$$

(v) $$\sum_{r=-K}^{K} r^2 \cos \lambda r = K^2 \frac{\sin \frac{1}{2}\lambda(2K+1)}{\sin \frac{1}{2}\lambda} + (K - \tfrac{1}{2}) \frac{\cos \lambda K}{\sin^2 \frac{1}{2}\lambda} - \frac{\sin \frac{1}{2}\lambda(2K-1)}{2 \sin^3 \frac{1}{2}\lambda}.$$

[*Hint:* $\sum \cos \lambda r + i \sum \sin \lambda r = \sum e^{i\lambda r}$; $\sum r \cos \lambda r = (d/d\lambda) \sum \sin \lambda r$; $\sum r^2 \cos \lambda r = -(d^2/d\lambda^2) \sum \cos \lambda r$.]

3. (Sec. 9.2.3.) For

$$\hat{f}(\nu) = \frac{1}{2\pi} \sum_{r=-K}^{K} \cos \nu r C_r$$

find $\{w_r\}$, $\{w_r^*\}$, and $w(\lambda \mid 0)$.

4. (Sec. 9.2.3.) Verify

$$\int_{-\pi}^{\pi} w^*(\lambda \mid \nu) I(\lambda)\, d\lambda = I(\nu)$$

for $w^*(\lambda \mid \nu)$ of Example A.

5. (Sec. 9.2.3.) Show that $k_T(\lambda)$ has minima at $\lambda = \pm j2\pi/T$, $j = 1, \ldots, [\frac{1}{2}T]$, maximum at $\lambda = 0$ and relative maxima at values of λ for which $\tan \frac{1}{2}\lambda T = T \tan \frac{1}{2}\lambda$.

6. (Sec. 9.2.3.) Show that $h_{2T-1}(\lambda)$ has 0's at $\lambda = \pm k2\pi/(2T-1)$ for $k = 1, \ldots, T-1$, maximum at $\lambda = 0$, and relative minima and maxima at the roots ($\lambda \neq 0$) of $\tan \frac{1}{2}\lambda(2T-1) = (2T-1) \tan \frac{1}{2}\lambda$.

7. (Sec. 9.2.3.) Show

$$\sum_{r=-K}^{K} \frac{e^{i\lambda r}}{T - |r|} = \frac{1}{T} + \sum_{r=1}^{K} \frac{e^{i\lambda r} + e^{-i\lambda r}}{T - r}$$

$$= \frac{1}{T} + e^{i\lambda T} \sum_{s=T-K}^{T-1} \frac{e^{-i\lambda s}}{s} + e^{-i\lambda T} \sum_{s=T-K}^{T-1} \frac{e^{i\lambda s}}{s}$$

$$= \frac{1}{T} + 2 \cos \lambda T \sum_{s=T-K}^{T-1} \frac{1}{s} - ie^{i\lambda T} \int_0^{\lambda} \frac{e^{-iu(T-K)} - e^{-iuT}}{1 - e^{-iu}}\, du$$

$$+ ie^{-i\lambda T} \int_0^{\lambda} \frac{e^{iu(T-K)} - e^{iuT}}{1 - e^{iu}}\, du.$$

8. (Sec. 9.2.3.) Verify (39).

9. (Sec. 9.2.3.) Let X_1, X_2, X_3, X_4 be independently distributed, each with density

$$\begin{aligned} f(x) &= \tfrac{1}{2}, & -1 \le x \le 1, \\ &= 0, & |x| > 1. \end{aligned}$$

Show that the densities of $X_1 + X_2$, $X_1 + X_2 + X_3$, and $X_1 + X_2 + X_3 + X_4$ are, respectively,

$$\begin{aligned} f_2(x) &= \tfrac{1}{4}(2 - |x|), & -2 \le x \le 2, \\ &= 0, & |x| \ge 2, \\ f_3(x) &= \frac{3 - x^2}{8}, & -1 \le x \le 1, \\ &= \frac{(3 - |x|)^2}{16}, & 1 \le |x| \le 3, \\ &= 0, & |x| \ge 3, \\ f_4(x) &= \frac{32 - 12x^2 + 3|x|^3}{96}, & -2 \le x \le 2, \\ &= \frac{(4 - |x|)^3}{96}, & 2 \le |x| \le 4, \\ &= 0, & |x| \ge 4. \end{aligned}$$

Verify that the characteristic function of $f(x)$ is $(\sin t)/t$ and, hence, that the characteristic function of $f_j(x)$ is $[(\sin t)/t]^j$.

10. (Sec. 9.2.3.) Let X_1, X_2, X_3, X_4 be independently distributed over the integers, each with probability function

$$\begin{aligned} \Pr\{X = r\} &= \frac{1}{2K + 1}, & r = 0, \pm 1, \ldots, \pm K, \\ &= 0, & |r| > K. \end{aligned}$$

Show that the probability functions of $X_1 + X_2$, $X_1 + X_2 + X_3$, and $X_1 + X_2 + X_3 + X_4$ are, respectively,

$$
\begin{aligned}
p_2(r) &= \frac{2K + 1 - |r|}{(2K + 1)^2}, & r &= 0, \pm 1, \dots, \pm 2K, \\
&= 0, & |r| &> 2K, \\
p_3(r) &= \frac{(2K + 1)^2 - K(K + 1) - r^2}{(2K + 1)^3}, & r &= 0, \pm 1, \dots, \pm K, \\
&= \frac{[3(K + \frac{1}{2}) - |r|]^2 - \frac{1}{4}}{2(2K + 1)^3}, & r &= \pm K, \dots, \pm 3K, \\
&= 0, & |r| &> 3K, \\
p_4(r) &= \frac{4(2K + 1)^3 + 2(2K + 1) - 3|r| - 6(2K + 1)r^2 + 3|r|^3}{6(2K + 1)^4}, \\
& & r &= 0, \pm 1, \dots, \pm 2K, \\
&= \frac{[2(2K + 1) - |r|]^3 - [2(2K + 1) - |r|]}{6(2K + 1)^4}, & r &= \pm 2K, \dots, \pm 4K, \\
&= 0, & |r| &> 4K
\end{aligned}
$$

Verify that the characteristic function of the initial probability function is $\sin \frac{1}{2}t(2K + 1)/[(2K + 1) \sin \frac{1}{2}t]$ and hence that the characteristic function of $p_j(r)$ is $\{\sin \frac{1}{2}t(2K + 1)/[(2K + 1) \sin \frac{1}{2}t]\}^j$.

11. (Sec. 9.2.3.) Verify the "summation by parts" formula

$$\sum_{r=1}^{L} a_r(b_{r+1} - b_r) = a_L b_{L+1} - a_0 b_1 - \sum_{r=1}^{L} b_r(a_r - a_{r-1}).$$

12. (Sec. 9.2.3.) Verify

$$\sum_{r=1}^{L} (L - r)^3 \cos \lambda r = -\tfrac{1}{2}L^3 + \frac{2 \sin^2 \frac{1}{2}\lambda L + 3L^2}{4 \sin^2 \frac{1}{2}\lambda} - \frac{3 \sin^2 \frac{1}{2}\lambda L}{4 \sin^4 \frac{1}{2}\lambda}.$$

[*Hint:* Use summation by parts according to Problem 11 successively with (1) $a_r = (L - r)^3$, $b_r = \sin \frac{1}{2}\lambda(2r - 1)/(2 \sin \frac{1}{2}\lambda)$, (2) $a_r = 3r^2 - 3(2L + 1)r + 3L^2 + 3L + 1$, $b_r = -\cos \lambda(r - 1)/(4 \sin^2 \frac{1}{2}\lambda)$, (3) $a_r = 6(L - r + 1)$, $b_r = -\sin \frac{1}{2}\lambda(2r - 3)/(8 \sin^3 \frac{1}{2}\lambda)$, and (4) $a_r = 6$, $b_r = \cos \lambda(r - 2)/(16 \sin^4 \frac{1}{2}\lambda)$.]

13. (Sec. 9.2.3.) Use summation by parts according to Problem 11 to answer Problem 2.

14. (Sec. 9.2.3.) Graph $\{w_r\}$ and $\{w_r^*\}$ for $r = 0, 1, \dots, T - 1$ for Examples A, B, C, D, E with $K = [\pi/b]$, G, H, I, and J. Let $T = 20$ and $K = 6$; use the same scales for all examples.

15. (Sec. 9.3.1.) Let

$$S_T = Z_{kT} + X_{kT}, \quad T = 1, 2, \ldots, \quad k = 1, 2, \ldots,$$

be a set of random variables such that

$$\mathscr{E}X_{kT}^2 \le M_k,$$

$$\lim_{k\to\infty} M_k = 0,$$

$$\mathscr{E}Z_{kT}^2 = L_{kT} \to L_k \quad \text{as} \quad T \to \infty,$$

$$\lim_{k\to\infty} L_k = L.$$

Show that

$$\lim_{T\to\infty} \mathscr{E}S_T^2 = L.$$

16. (Sec. 9.3.1.) Use the second forms of (16) and (19) of Section 9.2 to prove (9) of Theorem 9.3.1.

17. (Sec. 9.3.2.) Show that

$$\lim_{K_T\to\infty} \frac{1}{K_T} w_T^*\left(\frac{v}{K_T}\bigg| 0\right) = \frac{1}{2\pi}\int_{-1}^{1} k(x)e^{ivx}\,dx$$

if $k(x)$ is Riemann-integrable in $-1 \le x \le 1$.

18. (Sec. 9.3.2.) Verify Table 9.3.2.

19. (Sec. 9.3.3.) Show that

$$\begin{aligned}
T\phi_T(r;\, g, h) &= T - \tfrac{1}{2}[|g| + |h| + (g - h)] + r,\\
&\qquad r = -\{T - \tfrac{1}{2}[|g| + |h| + (g - h)]\}, \ldots, \tfrac{1}{2}(g - h) - \tfrac{1}{2}\big||g| - |h|\big|,\\
&= T - \max(|g|, |h|),\\
&\qquad r = \tfrac{1}{2}(g - h) - \tfrac{1}{2}\big||g| - |h|\big|, \ldots, \tfrac{1}{2}(g - h) + \tfrac{1}{2}\big||g| - |h|\big|,\\
&= T - \tfrac{1}{2}[|g| + |h| - (g - h)] - r,\\
&\qquad r = \tfrac{1}{2}(g - h) + \tfrac{1}{2}\big||g| - |h|\big|, \ldots, T - \tfrac{1}{2}[|g| + |h| - (g - h)],
\end{aligned}$$

and that $T\phi_T(r;\, g, h)$ is 0 for all other values of r.

20 (Sec. 9.3.3.) Prove (43).

21. (Sec. 9.3.3. For $-1 \le a < b \le 1$ let

$$\begin{aligned}
g(x) &= C > 0, & a < x \le b,\\
&= 0, & \text{elsewhere.}
\end{aligned}$$

Show that

$$\lim_{K\to\infty} \sum_{r=-K}^{K} \frac{1}{K} g\left(\frac{r}{K}\right) e^{idr} = 0$$

for $d \ne k2\pi$, $k = 0, \pm 1, \ldots$. [*Hint:* By summation of e^{idr} show

$$\left|\sum_{r=-K}^{K} \frac{1}{K} g\left(\frac{r}{K}\right) e^{idr}\right| \le \frac{C}{K}\frac{1}{|\sin \frac{1}{2}d|}.\Big]$$

22. (Sec. 9.3.3.) Let $h(x)$ be a step function defined for $-1 \le x \le 1$; that is,

$$h(x) = C_j, \qquad a_{j-1} < x \le a_j, \quad j = 1, \ldots, n,$$

where $-1 = a_0 < a_1 < \cdots < a_n = 1$, and $h(-1) = C_1$. Show that

$$\lim_{K \to \infty} \sum_{r=-K}^{K} \frac{1}{K} h\left(\frac{r}{K}\right) e^{idr} = 0$$

for $d \ne k2\pi$, $k = 0, \pm 1, \ldots$. [*Hint:* Apply Problem 21.]

23. (Sec. 9.3.3.) Let $f(x)$ be continuous in $-1 \le x \le 1$. Show that

$$\lim_{K \to \infty} \sum_{r=-K}^{K} \frac{1}{K} f\left(\frac{r}{K}\right) e^{idr} = 0$$

for $d \ne k2\pi$, $k = 0, \pm 1, \ldots$. [*Hint:* Approximate $f(x)$ by a step function and apply Problem 22.]

24. (Sec. 9.3.3.) Verify Table 9.3.3 except for the truncated Daniell kernel. (See Problem 25.)

25. (Sec. 9.3.3.) Show

$$\int_{-1}^{1} \left(\frac{\sin \pi z}{\pi z}\right)^2 dz = \frac{2}{\pi} \int_0^{2\pi} \frac{\sin y}{y}\, dy$$

$$= 4 \sum_{\nu=0}^{\infty} (-1)^{\nu} \frac{(2\pi)^{2\nu}}{(2\nu + 1)(2\nu + 1)\,!}.$$

[*Hint:* Transform $\sin^2 \pi z$ to $\frac{1}{2}(1 - \cos y)$ and integrate the second term by parts.]

26. (Sec. 9.3.3.) Let

$$\hat{G}(\nu) = 2 \int_0^{\nu} I(\lambda)\, d\lambda = \frac{1}{\pi}\left\{\nu c_0 + 2 \sum_{r=1}^{T-1} \frac{\sin \nu r}{r} c_r\right\}.$$

Show that under appropriate conditions

$$\lim_{T \to \infty} T\mathscr{E}[\hat{G}(\nu) - G(\nu)][\hat{G}(\nu') - G(\nu')] = 8\pi \int_0^{\min(\nu, \nu')} f^2(\lambda)\, d\lambda.$$

[*Hint:* The proof can be patterned after that of Problem 16.]

27. (Sec. 9.4.) Show that if $|r| \le n$ and $|s| \le n$ the number of terms in the difference

$$\sum_{h=s-r+1}^{K_T+s-r} \sum_{q=1-s}^{T-h-r} b_{h,q,r,s} - \sum_{h=1}^{K_T-2n} \sum_{q=1}^{T-h} b_{h,q,r,s}$$

is less than $8Tn + 4K_Tn + 32n^2$. [*Hint:* Show that for a value of h between $1 + 2n$ and $K_T - 2n$ the number of terms in the difference is at most $4n$ and that for a value of q between $1 + n$ and $T - n$ the number of terms in the difference is at most $8n$.]

28. (Sec. 9.4.) Show that a sharper bound for Problem 27 is $8Tn - 4n$.

29. (Sec. 9.4.) Show that the variables $v_q v_{q+h}$, $q = 1, \ldots, T$, $h = 1, \ldots, K_T$, are uncorrelated if $\mathscr{E} v_q v_r v_s v_t = 0$ unless q, r, s, t are equal in pairs. Assume $\mathscr{E} v_q = 0$ and $\mathscr{E} v_q v_r = 0$ for $q \neq r$.

30. (Sec. 9.4.) Show that

$$\mathscr{E} \frac{1}{TK_T} \sum_{g,h=1}^{K_T - 2n} a_g a_h e^{i\lambda(g-h)} \sum_{q=1}^{T-h} \sum_{s=1}^{T-g} v_q v_{q+h} v_s v_{s+g} = \frac{1}{TK_T} \sum_{h=1}^{K_T-2n} a_h^2 (T-h)\sigma^4,$$

where $\mathscr{E} v_t = 0$, $\mathscr{E} v_t^2 = \sigma^2$, and $\mathscr{E} v_q v_r v_s v_t = 0$ unless q, r, s, t are equal in pairs.

31. (Sec. 9.4.) Show

$$\begin{aligned}
\mathscr{E} W_{tT}^4 &= \frac{\kappa_4(\kappa_4 + 3\sigma^4)}{K_T^2} \sum_{h=1}^{K_T} k^4\left(\frac{h}{K_T}\right) \cos^4 \nu h + 3\sigma^4(\kappa_4 + 3\sigma^4) \left[\frac{1}{K_T} \sum_{h=1}^{K_T} k^2\left(\frac{h}{K_T}\right) \cos^2 \nu h\right]^2 \\
&\leq (\kappa_4 + 3\sigma^4) \sup k^4(x) \left\{\frac{|\kappa_4|}{K_T} + 3\sigma^4\right\},
\end{aligned}$$

$$\begin{aligned}
\mathscr{E} W_{tT}^2 W_{rT}^2 &= \sigma^8 \left[\frac{1}{K_T} \sum_{h=1}^{K_T} k^2\left(\frac{h}{K_T}\right) \cos^2 \nu h\right]^2 \\
&\leq \sigma^8 \sup k^4(x), \qquad r > t + K_T,
\end{aligned}$$

$$\begin{aligned}
\mathscr{E} W_{tT}^2 W_{t+r,T}^2 &= \frac{\sigma^8}{K_T^2} \left[\sum_{h=1}^{K_T} k^2\left(\frac{h}{K_T}\right) \cos^2 \nu h\right]^2 \\
&\quad + \frac{\sigma^4(\kappa_4 + 2\sigma^4)}{K_T^2} k^2\left(\frac{r}{K_T}\right) \cos^2 \nu r \sum_{j=1}^{K_T} k^2\left(\frac{j}{K_T}\right) \cos^2 \nu j \\
&\quad + \frac{\sigma^4 \kappa_4}{K_T^2} \sum_{h=r+1}^{K_T} k^2\left(\frac{h}{K_T}\right) k^2\left(\frac{h-r}{K_T}\right) \cos^2 \nu h \cos^2 \nu(h-r) \\
&\quad + \frac{2\sigma^8}{K_T^2} \left[\sum_{h=r+1}^{K_T} k\left(\frac{h}{K_T}\right) k\left(\frac{h-r}{K_T}\right) \cos \nu h \cos \nu(h-r)\right]^2 \\
&\quad + \frac{2\sigma^2 \nu_3^2}{K_T^2} k\left(\frac{r}{K_T}\right) \cos \nu r \sum_{h=r+1}^{K_T} k\left(\frac{h}{K_T}\right) k^2\left(\frac{h-r}{K_T}\right) \cos \nu h \cos^2 \nu(h-r) \\
&\leq \sup k^4(x) \left\{3\sigma^8 + \frac{1}{K_T}[\sigma^4 |\kappa_4 + 2\sigma^4| + \sigma^4 |\kappa_4| + 2\sigma^2 \nu_3^2]\right\}, \qquad 1 \leq r \leq K_T,
\end{aligned}$$

$$|\mathscr{E} W_{tT}^2 W_{rT} W_{sT}| \leq \frac{4}{K_T} \sigma^8 \sup k^4(x), \qquad r > t,\ s > t,\ r \neq s,$$

$$|\mathscr{E} W_{tT}^3 W_{rT}| \leq \frac{6}{K_T} \nu_3^2 \sigma^2 \sup k^4(x), \qquad r > t,$$

$$\mathscr{E} W_{tT} W_{sT} W_{rT} W_{qT} = 0, \qquad \text{otherwise.}$$

32. (Sec. 9.4.) If $q \leq p$ and $\sqrt{T/K_T}/K_T^q \to \beta$ as $T \to \infty$, find the limiting distribution of $\sqrt{T/K_T}[\hat{f}_T(\nu) - f(\nu)]$ under appropriate conditions and state those conditions. Of what use, if any, is this result?

33. (Sec. 9.4.) Compare the confidence intervals (33) and (36).

34. (Sec. 9.4.) Find a confidence interval for $f(\nu)$ of confidence $1 - \varepsilon$ from (40). Relate this interval to (33).

35. (Sec. 9.4.) Prove that the convergences (28) in Theorem 8.3.2 are uniform.

CHAPTER 10

Linear Trends with Stationary Random Terms

10.1. INTRODUCTION

In Chapters 2, 3, and 4 we considered problems of statistical inference in models based on trend functions linear in the coefficients with error terms independent and with the same variance; in Chapter 5 the assumption of independent error terms was relaxed to permit the random element to satisfy a stochastic difference equation. Now we shall treat the estimation of the coefficients when the error covariance function is arbitrary (Sections 10.2.1 and 10.2.2) and on an asymptotic basis when the random term is a stationary stochastic process (Sections 10.2.3 and 10.2.4). Since least squares estimates are available regardless of the knowledge of the random part, it is of particular interest to study the relationship between the least squares estimates and the best linear unbiased estimates.

It was pointed out by T. W. Anderson (1948) that if the independent variables constituted characteristic vectors of the error covariance matrix, the tests such as considered in Section 6.3 were similar and optimal, and the maximum likelihood estimates of the regression coefficients under normality were shown to be the least squares estimates (Section 2.4). Watson (1952), (1955) and Watson and Hannan (1956) studied the efficiency of least squares estimates; the inequality given by Watson shows least squares estimates are efficient only if the independent variables constitute characteristic vectors. Magness and McGuire (1962) rediscovered the conditions, proving sufficiency and necessity. More intensive studies of the conditions and surveys of the literature have been made by Watson (1967) and Zyskind (1967).

Grenander (1954), Rosenblatt (1956), and Grenander and Rosenblatt (1957) in Chapter 7 treated the corresponding problems in the asymptotic case. In

Section 10.2.3 their results are developed in a form that is similar to the case of finite T. In Section 10.2.4 it is shown that under general circumstances the coefficients may be treated as normally distributed if T is sufficiently large.

In Chapters 8 and 9 we studied the estimation of covariances and the spectral density when the mean of the process was known or when it was estimated as a constant. In Section 10.3 we consider the estimation of covariances and the spectral density when the mean of the process is a linear regression function and the coefficients are estimated by least squares. The treatment is based on Hannan (1958).

Tests of independence were developed in Chapter 6 in some generality, but the distributions given explicitly and the tables of significance points were based on the mean being known or estimated by a constant. In Section 10.4 we carry this study on to the case of using residuals from estimated regression.

10.2. EFFICIENT ESTIMATION OF TREND FUNCTIONS

10.2.1. Efficient Estimation by Least Squares

We consider a trend function that is linear in the unknown coefficients. As was seen in Section 2.4, the best linear unbiased estimates of the coefficients (the Markov estimates) involve the covariance matrix (except for a scale factor), which frequently is unknown. The least squares estimates are always available and yield unbiased estimates; however, they may not have minimum variances (among unbiased linear estimates). In this section we shall compare least squares estimates with Markov estimates.

Let $\boldsymbol{y} = (y_1, \ldots, y_T)'$ have mean vector and covariance matrix

$$\mathscr{E}\boldsymbol{y} = \boldsymbol{Z\beta}, \tag{1}$$

$$\mathscr{E}(\boldsymbol{y} - \boldsymbol{Z\beta})(\boldsymbol{y} - \boldsymbol{Z\beta})' = \boldsymbol{\Sigma}, \tag{2}$$

respectively, where $\boldsymbol{Z} = (z_{it})'$ is a $T \times p$ known matrix of rank p ($\leq T$), $\boldsymbol{\beta}$ is a p-component vector of coefficients to be estimated from an observation on $\boldsymbol{y}$, and $\boldsymbol{\Sigma}$ is a positive definite matrix. The Markov estimate (when $\boldsymbol{\Sigma}$ is known) and the least squares estimate of $\boldsymbol{\beta}$ are

$$\boldsymbol{b} = (\boldsymbol{Z}'\boldsymbol{\Sigma}^{-1}\boldsymbol{Z})^{-1}\boldsymbol{Z}'\boldsymbol{\Sigma}^{-1}\boldsymbol{y}, \tag{3}$$

$$\boldsymbol{b}^* = (\boldsymbol{Z}'\boldsymbol{Z})^{-1}\boldsymbol{Z}'\boldsymbol{y}, \tag{4}$$

with covariance matrices

$$\mathscr{E}(\boldsymbol{b} - \boldsymbol{\beta})(\boldsymbol{b} - \boldsymbol{\beta})' = (\boldsymbol{Z}'\boldsymbol{\Sigma}^{-1}\boldsymbol{Z})^{-1}, \tag{5}$$

$$\mathscr{E}(\boldsymbol{b}^* - \boldsymbol{\beta})(\boldsymbol{b}^* - \boldsymbol{\beta})' = (\boldsymbol{Z}'\boldsymbol{Z})^{-1}\boldsymbol{Z}'\boldsymbol{\Sigma}\boldsymbol{Z}(\boldsymbol{Z}'\boldsymbol{Z})^{-1}, \tag{6}$$

respectively. The Markov estimate and the least squares estimate of any linear combination $\boldsymbol{\gamma}'\boldsymbol{\beta}$ are $\boldsymbol{\gamma}'\boldsymbol{b}$ and $\boldsymbol{\gamma}'\boldsymbol{b}^*$ with variances

$$\mathscr{E}(\boldsymbol{\gamma}'\boldsymbol{b} - \boldsymbol{\gamma}'\boldsymbol{\beta})^2 = \boldsymbol{\gamma}'(\boldsymbol{Z}'\boldsymbol{\Sigma}^{-1}\boldsymbol{Z})^{-1}\boldsymbol{\gamma}, \tag{7}$$

$$\mathscr{E}(\boldsymbol{\gamma}'\boldsymbol{b}^* - \boldsymbol{\gamma}'\boldsymbol{\beta})^2 = \boldsymbol{\gamma}'(\boldsymbol{Z}'\boldsymbol{Z})^{-1}\boldsymbol{Z}'\boldsymbol{\Sigma}\boldsymbol{Z}(\boldsymbol{Z}'\boldsymbol{Z})^{-1}\boldsymbol{\gamma}, \tag{8}$$

respectively. The estimates are unbiased and $\text{Var}\,(\boldsymbol{\gamma}'\boldsymbol{b}) \le \text{Var}\,(\boldsymbol{\gamma}'\boldsymbol{b}^*)$; that is,

$$\boldsymbol{\gamma}'(\boldsymbol{Z}'\boldsymbol{\Sigma}^{-1}\boldsymbol{Z})^{-1}\boldsymbol{\gamma} \le \boldsymbol{\gamma}'(\boldsymbol{Z}'\boldsymbol{Z})^{-1}\boldsymbol{Z}'\boldsymbol{\Sigma}\boldsymbol{Z}(\boldsymbol{Z}'\boldsymbol{Z})^{-1}\boldsymbol{\gamma} \tag{9}$$

for every $\boldsymbol{\gamma}$.

In Section 2.4 it was shown that the Markov and least squares estimates are identical if $\boldsymbol{Z} = \boldsymbol{V}^*\boldsymbol{C}$, where the p columns of $\boldsymbol{V}^*\,(T \times p)$ are characteristic vectors of $\boldsymbol{\Sigma}$ and $\boldsymbol{C}$ is nonsingular. We shall show that the Markov and least squares estimates are identical only if $\boldsymbol{Z}$ can be expressed as $\boldsymbol{V}^*\boldsymbol{C}$.

THEOREM 10.2.1. *The least squares estimate* (4) *is the same as the Markov estimate* (3) *for all* $\boldsymbol{y}$ *if and only if* $\boldsymbol{Z} = \boldsymbol{V}^*\boldsymbol{C}$, *where the* p *columns of* $\boldsymbol{V}^*$ *are* p *linearly independent characteristic vectors of* $\boldsymbol{\Sigma}$ *and* $\boldsymbol{C}$ *is a nonsingular matrix.*

PROOF. The estimates are the same if and only if $\boldsymbol{b}^* = \boldsymbol{b}$ is an identity in $\boldsymbol{y}$; that is, if and only if

$$\boldsymbol{Z}(\boldsymbol{Z}'\boldsymbol{Z})^{-1} = \boldsymbol{\Sigma}^{-1}\boldsymbol{Z}(\boldsymbol{Z}'\boldsymbol{\Sigma}^{-1}\boldsymbol{Z})^{-1}. \tag{10}$$

Multiplication of (10) on the left by $\boldsymbol{\Sigma}$ and on the right by $\boldsymbol{Z}'\boldsymbol{Z}$ yields

$$\boldsymbol{\Sigma}\boldsymbol{Z} = \boldsymbol{Z}(\boldsymbol{Z}'\boldsymbol{\Sigma}^{-1}\boldsymbol{Z})^{-1}\boldsymbol{Z}'\boldsymbol{Z}. \tag{11}$$

There is a nonsingular $p \times p$ matrix $\boldsymbol{P}$ and a (nonsingular) diagonal matrix $\boldsymbol{D}$ such that

$$\boldsymbol{P}'\boldsymbol{Z}'\boldsymbol{Z}\boldsymbol{P} = \boldsymbol{I}, \tag{12}$$

$$\boldsymbol{P}'\boldsymbol{Z}'\boldsymbol{\Sigma}^{-1}\boldsymbol{Z}\boldsymbol{P} = \boldsymbol{D}^{-1}. \tag{13}$$

(See Problem 30 of Chapter 6.) Then multiplication of (11) on the right by $\boldsymbol{P}$ gives

$$\boldsymbol{\Sigma}(\boldsymbol{Z}\boldsymbol{P}) = (\boldsymbol{Z}\boldsymbol{P})\boldsymbol{D}. \tag{14}$$

Thus the columns of $\boldsymbol{Z}\boldsymbol{P}$ are characteristic vectors of $\boldsymbol{\Sigma}$ and the diagonal elements of $\boldsymbol{D}$ are the corresponding characteristic roots. The theorem follows with $\boldsymbol{Z}\boldsymbol{P} = \boldsymbol{V}^*$ and $\boldsymbol{C} = \boldsymbol{P}^{-1}$.

Since the minimum variance unbiased linear estimates are unique, any unbiased linear estimate has minimum variance only if it is the Markov estimate.

The least squares estimates have minimum variance only if they satisfy the condition of Theorem 10.2.1. The condition for the least squares estimates to be identical to the Markov estimates for all components of $\boldsymbol{\beta}$ (that is, all vectors $\boldsymbol{\gamma}$) is equivalently stated that p linearly independent linear combinations of the columns of $\boldsymbol{Z}$ are characteristic vectors of $\boldsymbol{\Sigma}$. Another way of stating the criterion is that

$$\boldsymbol{\Sigma Z} = \boldsymbol{ZG}, \tag{15}$$

where $\boldsymbol{G}$ is nonsingular, for (15) is the same as (14) with $\boldsymbol{G} = \boldsymbol{PDP}^{-1}$.

We now consider stating the criterion in terms of a given set of characteristic vectors of $\boldsymbol{\Sigma}$ and shall derive the result from the covariance matrices of the estimates. Let the characteristic roots of $\boldsymbol{\Sigma}$ be $\lambda_1 \geq \lambda_2 \geq \cdots \geq \lambda_T$. Let $\boldsymbol{V}$ be a matrix whose columns are characteristic vectors of $\boldsymbol{\Sigma}$ normalized so $\boldsymbol{V}'\boldsymbol{V} = \boldsymbol{I}$; that is, $\boldsymbol{\Sigma V} = \boldsymbol{V\Lambda}$, where $\boldsymbol{\Lambda}$ is the diagonal matrix of characteristic roots. If the roots are different, the characteristic vectors are uniquely determined. Any $T \times p$ matrix $\boldsymbol{Z}$ can be written as $\boldsymbol{Z} = \boldsymbol{VA}$ for a suitable $T \times p$ matrix $\boldsymbol{A}$. Since $\boldsymbol{V}'\boldsymbol{\Sigma V} = \boldsymbol{\Lambda}$ and $\boldsymbol{V}'\boldsymbol{\Sigma}^{-1}\boldsymbol{V} = \boldsymbol{\Lambda}^{-1}$,

$$\boldsymbol{Z}'\boldsymbol{Z} = \boldsymbol{A}'\boldsymbol{V}'\boldsymbol{V}\boldsymbol{A} = \boldsymbol{A}'\boldsymbol{A} = \sum_{r=1}^{T} \boldsymbol{a}_r\boldsymbol{a}_r', \tag{16}$$

$$\boldsymbol{Z}'\boldsymbol{\Sigma Z} = \boldsymbol{A}'\boldsymbol{V}'\boldsymbol{\Sigma V}\boldsymbol{A} = \boldsymbol{A}'\boldsymbol{\Lambda A} = \sum_{r=1}^{T} \lambda_r \boldsymbol{a}_r\boldsymbol{a}_r', \tag{17}$$

$$\boldsymbol{Z}'\boldsymbol{\Sigma}^{-1}\boldsymbol{Z} = \boldsymbol{A}'\boldsymbol{V}'\boldsymbol{\Sigma}^{-1}\boldsymbol{V}\boldsymbol{A} = \boldsymbol{A}'\boldsymbol{\Lambda}^{-1}\boldsymbol{A} = \sum_{r=1}^{T} \frac{1}{\lambda_r} \boldsymbol{a}_r\boldsymbol{a}_r', \tag{18}$$

where $\boldsymbol{A} = (\boldsymbol{a}_1, \ldots, \boldsymbol{a}_T)'$.

If there are multiple roots, the characteristic vectors are not uniquely determined. Suppose the distinct roots are $\nu_1 > \nu_2 > \cdots > \nu_H$ and

$$\boldsymbol{\Lambda} = \begin{bmatrix} \nu_1 \boldsymbol{I} & \boldsymbol{0} & \cdots & \boldsymbol{0} \\ \boldsymbol{0} & \nu_2 \boldsymbol{I} & \cdots & \boldsymbol{0} \\ \cdot & \cdot & & \cdot \\ \cdot & \cdot & & \cdot \\ \cdot & \cdot & & \cdot \\ \boldsymbol{0} & \boldsymbol{0} & \cdots & \nu_H \boldsymbol{I} \end{bmatrix}, \tag{19}$$

where the identity matrices are of order $m_1, m_2, \ldots, m_H$, respectively. Partition $\boldsymbol{V} = (\boldsymbol{V}^{(1)}\boldsymbol{V}^{(2)} \cdots \boldsymbol{V}^{(H)})$ and $\boldsymbol{A}' = (\boldsymbol{A}^{(1)\prime}\boldsymbol{A}^{(2)\prime} \cdots \boldsymbol{A}^{(H)\prime})$ similarly. Then

$Z = \sum_{h=1}^{H} V^{(h)} A^{(h)}$,

$$Z'Z = \sum_{h=1}^{H} A^{(h)\prime} A^{(h)}, \tag{20}$$

$$Z'\Sigma Z = \sum_{h=1}^{H} \nu_h A^{(h)\prime} A^{(h)}, \tag{21}$$

$$Z'\Sigma^{-1} Z = \sum_{h=1}^{H} \frac{1}{\nu_h} A^{(h)\prime} A^{(h)}. \tag{22}$$

The indeterminacy in $V^{(h)}$ is that it may be replaced by $V^{(h)} Q^{(h)}$, where $Q^{(h)}$ is an orthogonal matrix of order m_h. Then $A^{(h)}$ is replaced by $Q^{(h)\prime} A^{(h)}$, but (20), (21), and (22) are unchanged. (For a given $V^{(h)}$, $A^{(h)} = V^{(h)\prime} Z$.) We shall now state the condition for the least squares estimates to be the same as the Markov estimates in terms of the ranks of the $A^{(h)}$'s. The result in this form will be used in the asymptotic theory (Section 10.2.3).

THEOREM 10.2.2. *The covariance matrix of the least squares estimate* (6) *is identical to the covariance matrix of the Markov estimate* (5) *if and only if the sum of the ranks of the* $A^{(h)}$*'s in* $Z = \sum_{h=1}^{H} V^{(h)} A^{(h)}$ *is* p.

PROOF. There is a nonsingular matrix P such that $P'Z'ZP = I$ and $P'Z'\Sigma ZP = D$, where D is diagonal and $d_{11} \geq \cdots \geq d_{pp} > 0$. Then

$$I = P'Z'ZP = \sum_{h=1}^{H} C^{(h)}, \tag{23}$$

$$D = P'Z'\Sigma ZP = \sum_{h=1}^{H} \nu_h C^{(h)}, \tag{24}$$

$$P'Z'\Sigma^{-1} ZP = \sum_{h=1}^{H} \frac{1}{\nu_h} C^{(h)}, \tag{25}$$

where $C^{(h)} = P' A^{(h)\prime} A^{(h)} P$ is a positive semidefinite matrix, $h = 1, \ldots, H$. The two covariance matrices are equal if and only if (25) is D^{-1}. The diagonal elements of (23), (24), and (25) are

$$1 = \sum_{h=1}^{H} c_{ii}^{(h)}, \qquad i = 1, \ldots, p, \tag{26}$$

$$d_{ii} = \sum_{h=1}^{H} \nu_h c_{ii}^{(h)}, \qquad i = 1, \ldots, p, \tag{27}$$

$$\sum_{h=1}^{H} \frac{1}{\nu_h} c_{ii}^{(h)}, \qquad i = 1, \ldots, p. \tag{28}$$

Since $C^{(h)}$ is positive semidefinite, $c_{ii}^{(h)} \geq 0$; (26) implies that for each i, $c_{ii}^{(1)}, \ldots, c_{ii}^{(H)}$ are a set of probabilities. The right-hand side of (27) is $\mathscr{E}X_i$, where $\Pr\{X_i = \nu_h\} = c_{ii}^{(h)}$, $h = 1, \ldots, H$, and (28) is $\mathscr{E}X_i^{-1}$, $i = 1, \ldots, p$. If (25) is D^{-1}, then (28) is $1/d_{ii}$. Since $\mathscr{E}X_i^{-1} = 1/\mathscr{E}X_i$ for $X_i > 0$ only if X_i is a positive constant with probability 1, $c_{ii}^{(h)} = 1$ for one value of h and $c_{ii}^{(h)} = 0$ for other values of h. (See Lemma 10.2.1 below.) In the H positive semidefinite matrices $C^{(1)}, \ldots, C^{(H)}$ there are p diagonal elements equal to 1 and all other diagonal elements are equal to 0. Because the ν_h's and d_{ii}'s are ordered similarly, the $c_{ii}^{(h)}$'s that are 1 for a given h are for consecutive values of i. Since $C^{(h)}$ is positive semidefinite, if $c_{ii}^{(h)} = 0$, then $c_{ij}^{(h)} = c_{ji}^{(h)} = 0$, $j = 1, \ldots, p$. Thus $C^{(h)}$ has at most one diagonal block that is not $\mathbf{0}$. Then (23) implies the other off-diagonal elements are 0, and the rank of $C^{(h)}$ is the number of diagonal elements equal to 1. Hence, the sum of the ranks of the positive semidefinite $C^{(h)}$'s is p. The rank of $C^{(h)}$ is the rank of $A^{(h)}$, $h = 1, \ldots, H$.

Now let us assume that the sum of the ranks of the $C^{(h)}$'s is p. From (23) and (24) we find $\sum_{g \neq h} (\nu_g - \nu_h) C^{(g)} = D - \nu_h I$. The rank of the left-hand side is not greater than the sum of the ranks of the $H - 1$ $C^{(g)}$'s ($g \neq h$), which is p minus the rank of $C^{(h)}$. Thus the number of d_{ii}'s that are equal to ν_h is at least the rank of $C^{(h)}$; since this holds for every h, the number is exactly the rank of $C^{(h)}$. Let the nonnull $C^{(h)}$'s be $L_1, \ldots, L_G$ of ranks $r_1, \ldots, r_G$, respectively, and let the corresponding ν_h's be $u_1, \ldots, u_G$. Then (27) is $d_{ii} = \sum_{g=1}^{G} u_g l_{ii}^{(g)}$. For $i = 1, \ldots, r_1$ $d_{ii} = u_1$, which is larger than the other u_g's. Thus for these values of i $l_{ii}^{(1)} = 1$ and $l_{ii}^{(g)} = 0$, $g \neq 1$. The previous argument shows that the upper left-hand corner of L_1 is I and the upper left-hand corner of L_g is $\mathbf{0}$, $g \neq 1$. Then the rest of L_1 is composed of 0's because the rank of L_1 is the order of I. A similar argument applied to the next set of equations shows that L_2 has I in the next diagonal block and 0's elsewhere. Continuation of the argument shows that each L_g has a diagonal block of I of order equal to the rank of L_g. Q.E.D.

LEMMA 10.2.1. *If $X > 0$, $\mathscr{E}X\mathscr{E}X^{-1} \geq 1$, and $\mathscr{E}X\mathscr{E}X^{-1} = 1$ if and only if X is a positive constant with probability* 1.

PROOF.† We have

$$0 \leq \mathscr{E}\left(\frac{1}{\sqrt{X}} - \frac{\sqrt{X}}{\mathscr{E}X}\right)^2 \tag{29}$$
$$= \mathscr{E}\frac{1}{X} - 2\frac{1}{\mathscr{E}X} + \mathscr{E}\frac{X}{(\mathscr{E}X)^2}$$
$$= \mathscr{E}\frac{1}{X} - \frac{1}{\mathscr{E}X}.$$

Q.E.D.

† A considerable discussion, including alternative proofs, has been carried on in *The American Statistician*, beginning February, 1966.

The condition that the sum of the ranks of $A^{(1)}, \ldots, A^{(H)}$ is p is equivalent to the condition that the columns of

$$Z = \sum_{h=1}^{H} V^{(h)}A^{(h)} = \sum_{h=1}^{H} (V^{(h)}Q^{(h)})(Q^{(h)\prime}A^{(h)}) \tag{30}$$

form a set of p characteristic vectors of $\mathbf{\Sigma}$ multiplied by a nonsingular matrix. For $Q^{(h)}$ can be taken so that the number of rows of $Q^{(h)\prime}A^{(h)}$ that do not consist entirely of 0's is equal to the rank of $A^{(h)}$.

Now suppose Z does not satisfy the condition of Theorem 10.2.1 (or Theorem 10.2.2); that is, there are fewer than p linearly independent linear combinations of the columns of Z that are characteristic vectors of $\mathbf{\Sigma}$. Let p_1 be the maximum number of such linear combinations. Let the linear combinations be ZG_1, where G_1 is $p \times p_1$. Let G_2 be a $p \times (p - p_1)$ matrix of rank $p - p_1$ such that $(ZG_2)'(ZG_1) = G_2'Z'ZG_1 = \mathbf{0}$, and let $G = (G_1\ G_2)$. Then ZG consists of p_1 columns that are characteristic vectors of $\mathbf{\Sigma}$ and $p - p_1$ columns orthogonal to the first p_1 columns; no linear combination of the last $p - p_1$ columns is a characteristic vector of $\mathbf{\Sigma}$. We have $ZG_1 = V_1^*$ and $\mathbf{\Sigma} V_1^* = V_1^*\mathbf{\Lambda}_1^*$, where $\mathbf{\Lambda}_1^*$ is diagonal. From the fact that $G_2'Z'ZG_1 = \mathbf{0}$ we find that

$$\begin{aligned}(ZG_2)'\mathbf{\Sigma}(ZG_1) &= G_2'Z'\mathbf{\Sigma}V_1^* = G_2'Z'V_1^*\mathbf{\Lambda}_1^* \\ &= G_2'Z'ZG_1\mathbf{\Lambda}_1^* = \mathbf{0},\end{aligned} \tag{31}$$

$$\begin{aligned}(ZG_2)'\mathbf{\Sigma}^{-1}(ZG_1) &= G_2'Z'\mathbf{\Sigma}^{-1}V_1^* = G_2'Z'V_1^*(\mathbf{\Lambda}_1^*)^{-1} \\ &= G_2'Z'ZG_1(\mathbf{\Lambda}_1^*)^{-1} = \mathbf{0}.\end{aligned} \tag{32}$$

Let $ZG = Z(G_1\ G_2) = (X_1\ X_2) = X$, and

$$G^{-1}\boldsymbol{\beta} = \boldsymbol{\gamma} = \begin{pmatrix}\boldsymbol{\gamma}^{(1)} \\ \boldsymbol{\gamma}^{(2)}\end{pmatrix}. \tag{33}$$

Then $Z\boldsymbol{\beta} = X\boldsymbol{\gamma} = X_1\boldsymbol{\gamma}^{(1)} + X_2\boldsymbol{\gamma}^{(2)}$, $Zb = Xc$, and $Zb^* = Xc^*$, where $c = G^{-1}b$ is the Markov estimate of $\boldsymbol{\gamma}$,

$$\begin{aligned}c &= \left[\begin{pmatrix}X_1' \\ X_2'\end{pmatrix}\mathbf{\Sigma}^{-1}(X_1\ X_2)\right]^{-1}\begin{pmatrix}X_1' \\ X_2'\end{pmatrix}\mathbf{\Sigma}^{-1}y \\ &= \begin{pmatrix}V_1^{*\prime}y \\ (X_2'\mathbf{\Sigma}^{-1}X_2)^{-1}X_2'\mathbf{\Sigma}^{-1}y\end{pmatrix},\end{aligned} \tag{34}$$

and $c^* = G^{-1}b^*$ is the least squares estimate of $\boldsymbol{\gamma}$,

$$c^* = \left[\begin{pmatrix}X_1' \\ X_2'\end{pmatrix}(X_1\ X_2)\right]^{-1}\begin{pmatrix}X_1' \\ X_2'\end{pmatrix}y = \begin{pmatrix}V_1^{*\prime}y \\ (X_2'X_2)^{-1}X_2'y\end{pmatrix}. \tag{35}$$

The least squares estimate of $\gamma^{(1)}$ is the same as the Markov estimate of $\gamma^{(1)}$, which is orthogonal to the least squares and Markov estimates of $\gamma^{(2)}$. See Watson (1967) and Zyskind (1967) [following T. W. Anderson (1948)] for further study of these problems.

10.2.2. Measuring the Efficiency of Linear Estimates

We now turn to measuring the loss in accuracy of estimation due to using least squares estimates when they are different from the Markov estimates. The covariance matrix of the least squares estimate $\boldsymbol{b}^*$ is at least as large as the covariance matrix of the Markov estimate $\boldsymbol{b}$ in the sense that the difference of the matrices is positive semidefinite; that is

$$\gamma'[(Z'Z)^{-1}Z'\Sigma Z(Z'Z)^{-1} - (Z'\Sigma^{-1}Z)^{-1}]\gamma \geq 0 \tag{36}$$

for every vector γ. This is a consequence of (9). The larger the covariance matrix of $\boldsymbol{b}^*$ is compared to that of $\boldsymbol{b}$ the less efficient is the estimation by least squares. We shall take as our measure of efficiency the ratio of the determinants of the covariance matrices

$$\begin{aligned} \text{Eff}\,(\boldsymbol{b}^*) &= \frac{|\mathscr{E}(\boldsymbol{b} - \boldsymbol{\beta})(\boldsymbol{b} - \boldsymbol{\beta})'|}{|\mathscr{E}(\boldsymbol{b}^* - \boldsymbol{\beta})(\boldsymbol{b}^* - \boldsymbol{\beta})'|} \\ &= \frac{|(Z'\Sigma^{-1}Z)^{-1}|}{|(Z'Z)^{-1}Z'\Sigma Z(Z'Z)^{-1}|} \\ &= \frac{|Z'Z|^2}{|Z'\Sigma Z| \cdot |Z'\Sigma^{-1}Z|}. \end{aligned} \tag{37}$$

A geometrical interpretation of the covariance matrices is in terms of the concentration ellipsoids of $\boldsymbol{b}$ and $\boldsymbol{b}^*$, respectively,

$$\begin{aligned} (\boldsymbol{x} - \boldsymbol{\beta})'[\mathscr{E}(\boldsymbol{b} - \boldsymbol{\beta})(\boldsymbol{b} - \boldsymbol{\beta})']^{-1}(\boldsymbol{x} - \boldsymbol{\beta}) \\ = (\boldsymbol{x} - \boldsymbol{\beta})'Z'\Sigma^{-1}Z(\boldsymbol{x} - \boldsymbol{\beta}) = p + 2, \end{aligned} \tag{38}$$

$$\begin{aligned} (\boldsymbol{x} - \boldsymbol{\beta})'[\mathscr{E}(\boldsymbol{b}^* - \boldsymbol{\beta})(\boldsymbol{b}^* - \boldsymbol{\beta})']^{-1}(\boldsymbol{x} - \boldsymbol{\beta}) \\ = (\boldsymbol{x} - \boldsymbol{\beta})'Z'Z(Z'\Sigma Z)^{-1}Z'Z(\boldsymbol{x} - \boldsymbol{\beta}) = p + 2. \end{aligned} \tag{39}$$

Uniform distributions over the interiors of (38) and (39) have the same mean vectors and covariance matrices as $\boldsymbol{b}$ and $\boldsymbol{b}^*$, respectively. The inequality (36) has the geometrical meaning that the ellipsoid (38) is entirely within the ellipsoid

(39). The numerator of the first and second ratios in (37) is proportional to the squared volume of the ellipsoid (38), and the denominator is proportional to the squared volume of the ellipsoid (39). [See T. W. Anderson (1958).]

THEOREM 10.2.3. Eff $(\boldsymbol{b}^*) \leq 1$, *and* Eff $(\boldsymbol{b}^*) = 1$ *if and only if the least squares estimate is identical to the Markov estimate.*

To facilitate the study of the efficiency of least squares estimates we desire some algebraic results.

LEMMA 10.2.2. *If* $\boldsymbol{Z} = \boldsymbol{WQ}$, *where* $\boldsymbol{Q}$ *is nonsingular, the efficiency of the least squares estimates of the coefficients of regression on* $\boldsymbol{Z}$ *is the same as the efficiency of the least squares estimates of the coefficients of regression on* $\boldsymbol{W}$.

PROOF. The efficiency of the least squares estimates of the coefficients of regression on $\boldsymbol{Z}$ is (37). Substitution of $\boldsymbol{Z} = \boldsymbol{WQ}$ yields

$$(40) \qquad \frac{|\boldsymbol{Q}'\boldsymbol{W}'\boldsymbol{WQ}|^2}{|\boldsymbol{Q}'\boldsymbol{W}'\boldsymbol{\Sigma}\boldsymbol{WQ}| \cdot |\boldsymbol{Q}'\boldsymbol{W}'\boldsymbol{\Sigma}^{-1}\boldsymbol{WQ}|} = \frac{|\boldsymbol{W}'\boldsymbol{W}|^2}{|\boldsymbol{W}'\boldsymbol{\Sigma}\boldsymbol{W}| \cdot |\boldsymbol{W}'\boldsymbol{\Sigma}^{-1}\boldsymbol{W}|},$$

which is the efficiency of the least squares estimates of the coefficients of regression on $\boldsymbol{W}$. Q.E.D.

LEMMA 10.2.3. *If* $\boldsymbol{Z} = \boldsymbol{KW}$ *and* $\boldsymbol{K}'\boldsymbol{\Sigma}\boldsymbol{K} = \boldsymbol{\Psi}$, *where* $\boldsymbol{K}$ *is orthogonal, the efficiency of the least squares estimates of the coefficients of regression on* $\boldsymbol{Z}$ *with covariance matrix* $\boldsymbol{\Sigma}$ *is the same as the efficiency of the least squares estimates of the coefficients of regression on* $\boldsymbol{W}$ *with covariance matrix* $\boldsymbol{\Psi}$.

PROOF. The efficiency of the least squares estimates of the coefficients of regression on $\boldsymbol{Z}$ with covariance matrix $\boldsymbol{\Sigma}$ is (37). Substitution of $\boldsymbol{Z} = \boldsymbol{KW}$ and $\boldsymbol{K}'\boldsymbol{\Sigma}\boldsymbol{K} = \boldsymbol{\Psi}$ (that is, $\boldsymbol{\Sigma} = \boldsymbol{K\Psi K}'$) yields

$$(41) \qquad \frac{|\boldsymbol{W}'\boldsymbol{K}'\boldsymbol{KW}|^2}{|\boldsymbol{W}'\boldsymbol{K}'\boldsymbol{K\Psi K}'\boldsymbol{KW}| \cdot |\boldsymbol{W}'\boldsymbol{K}'\boldsymbol{K\Psi}^{-1}\boldsymbol{K}'\boldsymbol{KW}|} = \frac{|\boldsymbol{W}'\boldsymbol{W}|^2}{|\boldsymbol{W}'\boldsymbol{\Psi}\boldsymbol{W}| \cdot |\boldsymbol{W}'\boldsymbol{\Psi}^{-1}\boldsymbol{W}|},$$

which is the efficiency of the least squares estimates of the coefficients of regression on $\boldsymbol{W}$ with covariance matrix $\boldsymbol{\Psi}$. Q.E.D.

Suppose there are p_1 $(0 < p_1 < p)$ linearly independent linear combinations of the columns of $\boldsymbol{Z}$ that are characteristic vectors of $\boldsymbol{\Sigma}$, say $\boldsymbol{ZG}_1 = \boldsymbol{V}_1^*$; here p_1 is the maximum number of such linearly independent linear combinations. Let $\boldsymbol{G}_2$ $[p \times (p - p_1)]$ be chosen so $\boldsymbol{ZG}_2$ is orthogonal to $\boldsymbol{ZG}_1$. Then (31) and (32) hold. Let $(\boldsymbol{ZG}_1\ \boldsymbol{ZG}_2) = \boldsymbol{ZG} = \boldsymbol{W} = (\boldsymbol{W}_1\ \boldsymbol{W}_2)$. Then the efficiency of the least squares estimate is

$$
(42) \qquad \frac{\begin{vmatrix} W_1'W_1 & 0 \\ 0 & W_2'W_2 \end{vmatrix}^2}{\begin{vmatrix} W_1'\Sigma W_1 & 0 \\ 0 & W_2'\Sigma W_2 \end{vmatrix} \begin{vmatrix} W_1'\Sigma^{-1}W_1 & 0 \\ 0 & W_2'\Sigma^{-1}W_2 \end{vmatrix}}
$$

$$
= \frac{|W_1'W_1|^2}{|W_1'\Sigma W_1| \cdot |W_1'\Sigma^{-1}W_1|} \cdot \frac{|W_2'W_2|^2}{|W_2'\Sigma W_2| \cdot |W_2'\Sigma^{-1}W_2|}
$$

$$
= \frac{|W_2'W_2|^2}{|W_2'\Sigma W_2| \cdot |W_2'\Sigma^{-1}W_2|}
$$

because $W_1 = V_1^*$, where V_1^* consists of characteristic vectors of Σ [that is, $W_1'\Sigma W_1 = \Lambda_1^*$, $W_1'\Sigma^{-1}W_1 = (\Lambda_1^*)^{-1}$]. The efficiency of the least squares estimates is the efficiency of the least squares estimates of the coefficients of regression on the part of Z that is orthogonal to the part of Z whose coefficients are efficiently estimated. The geometric interpretation is that there is a p_1-dimensional hyperplane through β whose intersections with the two ellipsoids (38) and (39) coincide; the ratio of volumes depends on the lengths of the principal axes orthogonal to these p_1 dimensions.

LEMMA 10.2.4. *The efficiency of the least squares estimates of the coefficients of the regression on Z is*

$$
(43) \qquad \frac{1}{|W'\Lambda W| \cdot |W'\Lambda^{-1}W|},
$$

where $W = V'ZP$ and P is a nonsingular matrix such that

$$
(44) \qquad W'W = P'Z'ZP = I.
$$

PROOF. Lemmas 10.2.2 and 10.2.3.

We may be interested in lower bounds for the efficiency of least squares estimates when some knowledge about Z or Σ is available. Suppose Σ is known, but that fact was not used in the estimation; how inefficient can the least squares estimates be? The problem is to find the minimum of (43) over $T \times p$ matrices W such that $W'W = I$. [The minimization of (43) can be carried out over a compact set of W, and hence the minimum exists and is attained.]

If $p = 1$, (43) is

$$\text{(45)} \qquad \operatorname{Eff}(b^*) = \frac{1}{w'\Lambda w \cdot w'\Lambda^{-1}w} = \frac{1}{\sum_{t=1}^{T} \lambda_t w_t^2 \sum_{t=1}^{T} w_t^2/\lambda_t}$$

$$= \frac{1}{\sum_{j=1}^{H} \nu_j P_j \cdot \sum_{j=1}^{H} P_j/\nu_j},$$

where the different values of the λ_t's are $\nu_1 > \nu_2 > \cdots > \nu_H$, and $P_j = \sum_t w_t^2$, where the sum is over the values of t for which $\lambda_t = \nu_j$. The denominator is of the form $\mathscr{E}X \cdot \mathscr{E}X^{-1}$, where X is a random variable with $\Pr\{X = \nu_j\} = P_j$, $j = 1, \ldots, H$. We can apply the so-called Kantorovich (1948) inequality.

LEMMA 10.2.5. (*Kantorovich Inequality*) *If X is a random variable such that $0 < m \le X \le M$, then*

$$\text{(46)} \qquad \mathscr{E}X \cdot \mathscr{E}X^{-1} \le \frac{(m + M)^2}{4mM} = \frac{1}{4}\left(\frac{m}{M} + \frac{M}{m} + 2\right).$$

PROOF. For $0 < m \le x \le M$,

$$\text{(47)} \qquad 0 \le (M - x)(x - m) = (M + m - x)x - Mm,$$

which implies

$$\text{(48)} \qquad \frac{1}{x} \le \frac{M + m - x}{Mm}.$$

The expectation of $1/X$ multiplied by $\mathscr{E}X$, therefore, satisfies

$$\text{(49)} \qquad \mathscr{E}X \cdot \mathscr{E}X^{-1} \le \frac{1}{Mm}(M + m - \mathscr{E}X)\mathscr{E}X.$$

Then (49) and

$$\text{(50)} \qquad 0 \le [\mathscr{E}X - \tfrac{1}{2}(M + m)]^2 = (\mathscr{E}X)^2 - (M + m)\mathscr{E}X + \frac{(M + m)^2}{4}$$

imply (46). Q.E.D.

THEOREM 10.2.4. *If $p = 1$ and λ_1 and λ_T are the maximum and minimum characteristic roots of $\mathbf{\Sigma}$*

$$\text{(51)} \qquad \operatorname{Eff}(b^*) \ge \frac{4\lambda_1\lambda_T}{(\lambda_1 + \lambda_T)^2}.$$

The upper bound in Lemma 10.2.5 is attained for the random variable for which $\Pr\{X = m\} = \Pr\{X = M\} = \frac{1}{2}$. The lower bound in (51) is attained for $P_1 = P_H$ and $P_j = 0, j = 2, \ldots, H - 1$, that is, for z being the average of

TABLE 10.1

MINIMUM EFFICIENCY OF THE SCALAR LEAST SQUARES ESTIMATE

λ_T/λ_1	0.1	0.2	0.3	0.4	0.5	0.6	0.7	0.8	0.9	1.0
min Eff (b^*)	0.331	0.556	0.710	0.816	0.889	0.938	0.969	0.988	0.997	1.000

two characteristic vectors corresponding to the largest and smallest characteristic roots.

The lower bound to Eff (b^*) can be written $4(\lambda_T/\lambda_1)/[1 + (\lambda_T/\lambda_1)]^2$ and is seen to be a function of λ_T/λ_1, the ratio of the smallest to largest characteristic root of $\mathbf{\Sigma}$. Table 10.1 indicates values of the minimum of Eff (b^*) for various values of λ_T/λ_1. It will be seen that the possible loss in efficiency due to using the least squares estimate is not very great if the ratio λ_T/λ_1 is $\frac{1}{2}$ or more.

For $p \geq 2$, there is no available bound that is satisfactory in the sense that it is valid and is attained. One valid lower bound is

$$\text{Eff}(\boldsymbol{b}^*) \geq \left[\frac{4\lambda_1\lambda_T}{(\lambda_1 + \lambda_T)^2}\right]^p, \tag{52}$$

but it will not be attained in general. Watson (1967) has obtained the lower bound

$$\text{Eff}(\boldsymbol{b}^*) \geq \frac{4\lambda_1\lambda_2 \cdots \lambda_p\lambda_{T-p+1} \cdots \lambda_T}{(\lambda_1\lambda_2 \cdots \lambda_p + \lambda_{T-p+1} \cdots \lambda_T)^2} \tag{53}$$

by considering compounds of the matrices involved; (53) will not be attained in general. Watson gives an example where the bound (52) is less than the bound (53) and an example where the reverse is true. [See also Watson (1955).]

THEOREM 10.2.5. *A necessary and sufficient condition that the least squares estimates be efficient for all* $\boldsymbol{Z}$ *is that* $\mathbf{\Sigma}$ *be a scalar multiple of* $\boldsymbol{I}$.

PROOF. The only matrices for which all vectors are characteristic vectors are multiples of $\boldsymbol{I}$.

As an example, let $\boldsymbol{A}_0 = \boldsymbol{I}$, $\boldsymbol{A}_j = \frac{1}{2}(\boldsymbol{B}^j + \boldsymbol{B}'^j)$, where

$$\boldsymbol{B} = \begin{bmatrix} 0 & 1 & 0 & \cdots & 0 & 0 \\ 0 & 0 & 1 & \cdots & 0 & 0 \\ 0 & 0 & 0 & \cdots & 0 & 0 \\ \cdot & \cdot & \cdot & & \cdot & \cdot \\ \cdot & \cdot & \cdot & & \cdot & \cdot \\ \cdot & \cdot & \cdot & & \cdot & \cdot \\ 0 & 0 & 0 & \cdots & 0 & 1 \\ 1 & 0 & 0 & \cdots & 0 & 0 \end{bmatrix}, \tag{54}$$

for $j = 1, \ldots, [\frac{1}{2}T]$. If $T = 2K$,

$$B^K = B'^K = \begin{pmatrix} 0 & I \\ I & 0 \end{pmatrix}, \tag{55}$$

where the submatrices are of order K. Let

$$\mathbf{\Sigma} = \sum_{j=0}^{[\frac{1}{2}T]} \gamma_j A_j \tag{56}$$

as in Chapter 6. The characteristic roots are $\sum_j \gamma_j$ corresponding to the characteristic vector $\sqrt{1/T}\,\boldsymbol{\varepsilon} = \sqrt{1/T}\,(1, \ldots, 1)'$, $\sum_j (-1)^j \gamma_j$ corresponding to the characteristic vector $\sqrt{1/T}\,(-1, 1, \ldots, 1)'$ if T is even, and $\sum_j \gamma_j \cos 2\pi jk/T$ corresponding to $\sqrt{2/T}\,(\cos 2\pi k/T, \cos 4\pi k/T, \ldots, 1)'$ and $\sqrt{2/T}\,(\sin 2\pi k/T, \sin 4\pi k/T, \ldots, 0)'$, $k = 1, 2, \ldots, [\frac{1}{2}(T-1)]$. For each k ($k = 1, 2, \ldots, [\frac{1}{2}(T-1)]$) the root is double and the two characteristic vectors listed may be replaced by any two linear combinations of them, where the linear combinations constitute an orthogonal transformation of order two. Least squares estimation is efficient if and only if the p columns of $\boldsymbol{Z}$ are linearly independent linear combinations of p of these characteristic vectors.

10.2.3. The Asymptotic Efficiency of Least Squares Estimates

Let $\{u_t\}$ be a stochastic process with $\mathscr{E}u_t = 0$ and $\mathscr{E}u_t u_s = \sigma(t-s)$. Let

$$y_t = \sum_{j=1}^{p} \beta_j z_{jt} + u_t, \qquad t = 1, 2, \ldots . \tag{57}$$

The sequence $y_1, \ldots, y_T$ for any T may constitute a vector $\boldsymbol{y}$ which satisfies (1) and (2) with $\boldsymbol{\beta} = (\beta_1, \ldots, \beta_p)'$, $\boldsymbol{Z} = (z_{it})'$, and $\mathbf{\Sigma} = [\sigma(t-s)]$. The Markov and least squares estimates of $\boldsymbol{\beta}$ and their covariance matrices are given by (3), (4), (5), and (6), respectively. We ask under what conditions the two covariance matrices are asymptotically equivalent in the sense that

(58)

$$\lim_{T\to\infty} \text{normalized } \{\mathscr{E}(\boldsymbol{b}^* - \boldsymbol{\beta})(\boldsymbol{b}^* - \boldsymbol{\beta})'\} = \lim_{T\to\infty} \text{normalized } \{\mathscr{E}(\boldsymbol{b} - \boldsymbol{\beta})(\boldsymbol{b} - \boldsymbol{\beta})'\},$$

where the normalization of the two covariance matrices is the same and is such that the limits are nontrivial. We raise the question for one given covariance sequence $\{\sigma(h)\}$ and also for the class of all covariance functions with continuous, positive spectral densities. The questions were first considered by Grenander (1954), Rosenblatt (1956), and Grenander and Rosenblatt (1957).

In order to answer such questions we need to make some assumptions about the sequences of independent variables. Let

$$
\begin{aligned}
(59) \qquad a_{ij}^T(h) &= \sum_{t=1}^{T-h} z_{i,t+h} z_{jt}, \qquad & h = 0, 1, \ldots, \\
&= \sum_{t=1-h}^{T} z_{i,t+h} z_{jt}, \qquad & h = 0, -1, \ldots .
\end{aligned}
$$

Then $a_{ij}^T(h) = a_{ji}^T(-h)$. We make the following assumptions:

ASSUMPTION 10.2.1. *$a_{ii}^T(0) \to \infty$ as $T \to \infty$, $i = 1, \ldots, p$.*

ASSUMPTION 10.2.2.

$$
(60) \qquad \lim_{T\to\infty} \frac{z_{i,T+1}^2}{a_{ii}^T(0)} = 0, \qquad i = 1, \ldots, p.
$$

ASSUMPTION 10.2.3. *The limit of*

$$
(61) \qquad \frac{a_{ij}^T(h)}{\sqrt{a_{ii}^T(0) a_{jj}^T(0)}} = r_{ij}^T(h)
$$

as $T \to \infty$ exists for every i, j, and h, $i, j = 1, \ldots, p$ and $h = 0, \pm 1, \ldots$.

Let

$$
(62) \qquad \lim_{T\to\infty} r_{ij}^T(h) = \rho_{ij}(h), \qquad i, j = 1, \ldots, p, \quad h = 0, \pm 1, \ldots,
$$

and let $\mathbf{R}(h) = [\rho_{ij}(h)]$.

ASSUMPTION 10.2.4. *$\mathbf{R}(0)$ is nonsingular.*

The point of the first assumption is to ensure that the effect of each z_{it} sequence grows unboundedly, and the point of the second assumption is to prevent the last z_{it}^2 from being an appreciable part of the sum of squares for large T. The third assumption is that the correlations between independent variables for all sufficiently large T are approximately fixed values. The fourth assumption is for convenience.

The sequence $\{\mathbf{R}(h)\}$ is a positive definite Hermitian sequence of matrices; that is

$$
(63) \qquad \sum_{h,k} \mathbf{c}'\mathbf{R}(h-k)\bar{\mathbf{c}} x_h x_k \geq 0,
$$

where the sum is over any finite set of indices, $\{x_h\}$ is any sequence of real numbers, and $\mathbf{c}$ is any p-component complex vector. It follows that there is a

Hermitian matrix function $M(\lambda)$ with positive semidefinite increments such that

$$\mathbf{R}(h) = \int_{-\pi}^{\pi} e^{i\lambda h}\, dM(\lambda). \tag{64}$$

(See Section 7.3.)

Define

$$D_T = \begin{bmatrix} \sqrt{a_{11}^T(0)} & 0 & \cdots & 0 \\ 0 & \sqrt{a_{22}^T(0)} & \cdots & 0 \\ \cdot & \cdot & & \cdot \\ \cdot & \cdot & & \cdot \\ \cdot & \cdot & & \cdot \\ 0 & 0 & \cdots & \sqrt{a_{pp}^T(0)} \end{bmatrix}. \tag{65}$$

Then

$$\mathbf{R}(h) = \lim_{T\to\infty} D_T^{-1} A_T(h) D_T^{-1}, \tag{66}$$

where $A_T(h) = [a_{ij}^T(h)]$.

To find the limits of

$$D_T^{-1} Z'\Sigma Z D_T^{-1} = D_T^{-1} \sum_{h=-(T-1)}^{T-1} \sigma(h) A_T(h) D_T^{-1} \tag{67}$$

and

$$D_T^{-1} Z'\Sigma^{-1} Z D_T^{-1} \tag{68}$$

we make the following assumption on the covariance sequence:

ASSUMPTION 10.2.5.

$$\sigma(h) = \int_{-\pi}^{\pi} e^{i\lambda h} f(\lambda)\, d\lambda, \tag{69}$$

where $f(\lambda)$ *is continuous in* $[-\pi, \pi]$.

We shall find these limits by approximating the spectral density $f(\lambda)$ by trigonometric polynomials $f_L(\lambda) = f_L(-\lambda)$ and $f_U(\lambda) = f_U(-\lambda)$, where

$$f_L(\lambda) \le f(\lambda) \le f_U(\lambda), \qquad -\pi \le \lambda \le \pi. \tag{70}$$

Since $0 \le f(\lambda) \le f_U(\lambda)$, $f_U(\lambda)$ is a spectral density and $\sigma_U(h) = \int_{-\pi}^{\pi} e^{i\lambda h} f_U(\lambda)\, d\lambda$ is a covariance sequence. If $0 \le f_L(\lambda)$, $f_L(\lambda)$ is a spectral density and $\sigma_L(h) = \int_{-\pi}^{\pi} e^{i\lambda h} f_L(\lambda)\, d\lambda$ constitutes a covariance sequence, but for some arguments it is not necessary to have $f_L(\lambda)$ nonnegative.

LEMMA 10.2.6. *The inequalities* (70) *imply*

$$x'\Sigma_L x \le x'\Sigma x \le x'\Sigma_U x \tag{71}$$

for every vector x, *and* (70) *and* $f_L(\lambda) > 0$ *imply*

$$x'\Sigma_U^{-1}x \leq x'\Sigma^{-1}x \leq x'\Sigma_L^{-1}x \tag{72}$$

for every vector x, *with* $\Sigma_L = [\sigma_L(t-s)]$ *and* $\Sigma_U = [\sigma_U(t-s)]$.

PROOF. We have

$$\begin{aligned} x'\Sigma x &= \sum_{t,s=1}^{T} \sigma(t-s)x_t x_s \\ &= \int_{-\pi}^{\pi} \sum_{t,s=1}^{T} e^{i\lambda(t-s)} x_t x_s f(\lambda)\, d\lambda \\ &= \int_{-\pi}^{\pi} \left| \sum_{t=1}^{T} x_t e^{i\lambda t} \right|^2 f(\lambda)\, d\lambda, \end{aligned} \tag{73}$$

with similar expressions for $x'\Sigma_L x$ and $x'\Sigma_U x$. [If $f_L(\lambda)$ is negative for some λ, $\sigma_L(u)$ is not a covariance sequence, but (71) and (74) hold, nevertheless.] Then (70) implies

$$\begin{aligned} \int_{-\pi}^{\pi} \left| \sum_{t=1}^{T} x_t e^{i\lambda t} \right|^2 f_L(\lambda)\, d\lambda &\leq \int_{-\pi}^{\pi} \left| \sum_{t=1}^{T} x_t e^{i\lambda t} \right|^2 f(\lambda)\, d\lambda \\ &\leq \int_{-\pi}^{\pi} \left| \sum_{t=1}^{T} x_t e^{i\lambda t} \right|^2 f_U(\lambda)\, d\lambda, \end{aligned} \tag{74}$$

which is equivalent to (71). The inequality $x'\Sigma x \leq x'\Sigma_U x$ for all x is equivalent to the roots of $|\Sigma - \theta\Sigma_U| = 0$ being less than or equal to 1, and they are positive. These are the roots of $|\Sigma_U^{-1} - \theta\Sigma^{-1}| = 0$, and thus the first inequality in (72) holds for all x. The second inequality follows similarly. Q.E.D.

THEOREM 10.2.6. *Under Assumptions* 10.2.1, 10.2.2, 10.2.3, *and* 10.2.5

$$\lim_{T\to\infty} D_T^{-1}Z'\Sigma Z D_T^{-1} = 2\pi \int_{-\pi}^{\pi} f(\lambda)\, dM(\lambda). \tag{75}$$

PROOF. The idea is that the left-hand side of (75) is $\sum_{h=-\infty}^{\infty} \sigma(h)\mathbf{R}(h)$, which can be written $\sum_{h=-\infty}^{\infty} \int_{-\pi}^{\pi} \sigma(h)e^{i\lambda h}\, dM(\lambda)$ and which is the right-hand side of (75). The difficulties of the proof are due to the fact that the number of terms in the sum (67) increases indefinitely, and that we assume $f(\lambda)$ is continuous instead of assuming $\sum_{h=-\infty}^{\infty}|\sigma(h)| < \infty$. [The assumption of continuity of $f(\lambda)$ is made because it is also used in the proof of Theorem 10.2.7.]

By Theorem 7.5.2 for an arbitrary ε (> 0) there are two trigonometric polynomials, $f_L(\lambda) = \sum_{k=-K}^{K} a_k e^{i\lambda k}$ and $f_U(\lambda) = \sum_{k=-K}^{K} b_k e^{i\lambda k}$ with $a_k = a_{-k}$ and $b_k = b_{-k}$, $k = 1, \ldots, K$, such that (70) holds and

$$f_U(\lambda) - f_L(\lambda) \le \varepsilon, \qquad -\pi \le \lambda \le \pi. \tag{76}$$

Then

$$\begin{aligned} \boldsymbol{x}'\boldsymbol{\Sigma}_U \boldsymbol{x} &= \int_{-\pi}^{\pi} \left| \sum_{t=1}^{T} x_t e^{i\lambda t} \right|^2 f_U(\lambda)\, d\lambda \\ &= \int_{-\pi}^{\pi} \sum_{k=-K}^{K} \sum_{t,s=1}^{T} b_k x_t x_s e^{i\lambda(k+t-s)}\, d\lambda \\ &= 2\pi \sum_{k=-K}^{K} b_k \sum_{t \in S_k} x_{t+k} x_t, \end{aligned} \tag{77}$$

where $S_k = \{1, \ldots, T-k\}$, $k \ge 0$, and $S_k = \{1-k, \ldots, T\}$, $k \le 0$. For any $\boldsymbol{\gamma}$ of p components let $x_t = (z_{1t}, \ldots, z_{pt})\boldsymbol{D}_T^{-1}\boldsymbol{\gamma}$. Then

$$\boldsymbol{\gamma}'\boldsymbol{D}_T^{-1}\boldsymbol{Z}'\boldsymbol{\Sigma}\boldsymbol{Z}\boldsymbol{D}_T^{-1}\boldsymbol{\gamma} \le 2\pi \sum_{k=-K}^{K} b_k \boldsymbol{\gamma}'\boldsymbol{D}_T^{-1}\boldsymbol{A}_T(k)\boldsymbol{D}_T^{-1}\boldsymbol{\gamma}. \tag{78}$$

Therefore

$$\begin{aligned} \overline{\lim_{T\to\infty}}\, \boldsymbol{\gamma}'\boldsymbol{D}_T^{-1}\boldsymbol{Z}'\boldsymbol{\Sigma}\boldsymbol{Z}\boldsymbol{D}_T^{-1}\boldsymbol{\gamma} &\le 2\pi \sum_{k=-K}^{K} b_k \boldsymbol{\gamma}'\mathbf{R}(k)\boldsymbol{\gamma} \\ &= 2\pi\boldsymbol{\gamma}' \sum_{k=-K}^{K} b_k \int_{-\pi}^{\pi} e^{i\lambda k}\, d\boldsymbol{M}(\lambda)\boldsymbol{\gamma} \\ &= 2\pi\boldsymbol{\gamma}' \int_{-\pi}^{\pi} f_U(\lambda)\, d\boldsymbol{M}(\lambda)\boldsymbol{\gamma}. \end{aligned} \tag{79}$$

Similarly

$$2\pi\boldsymbol{\gamma}' \int_{-\pi}^{\pi} f_L(\lambda)\, d\boldsymbol{M}(\lambda)\boldsymbol{\gamma} \le \underline{\lim_{T\to\infty}}\, \boldsymbol{\gamma}'\boldsymbol{D}_T^{-1}\boldsymbol{Z}'\boldsymbol{\Sigma}\boldsymbol{Z}\boldsymbol{D}_T^{-1}\boldsymbol{\gamma}. \tag{80}$$

Since ε is arbitrary,

$$\lim_{T\to\infty} \boldsymbol{\gamma}'\boldsymbol{D}_T^{-1}\boldsymbol{Z}'\boldsymbol{\Sigma}\boldsymbol{Z}\boldsymbol{D}_T^{-1}\boldsymbol{\gamma} = 2\pi\boldsymbol{\gamma}' \int_{-\pi}^{\pi} f(\lambda)\, d\boldsymbol{M}(\lambda)\boldsymbol{\gamma}. \tag{81}$$

Since this holds for every vector $\boldsymbol{\gamma}$, the theorem follows.

ASSUMPTION 10.2.6. *If* $f(\lambda)$ *is defined by* (69), $f(\lambda) > 0$, $-\pi \le \lambda \le \pi$.

THEOREM 10.2.7. *Under Assumptions* 10.2.1, 10.2.2, 10.2.3, 10.2.5, *and* 10.2.6,

$$\lim_{T\to\infty} \boldsymbol{D}_T^{-1}\boldsymbol{Z}'\boldsymbol{\Sigma}^{-1}\boldsymbol{Z}\boldsymbol{D}_T^{-1} = \frac{1}{2\pi} \int_{-\pi}^{\pi} \frac{1}{f(\lambda)}\, d\boldsymbol{M}(\lambda). \tag{82}$$

PROOF. By Corollary 7.5.2 there are spectral densities of autoregressive processes, $f_L(\lambda) = [2\pi\,|\sum_{k=0}^{K} a_k e^{i\lambda k}|^2]^{-1}$ and $f_U(\lambda) = [2\pi\,|\sum_{k=0}^{K} b_k e^{i\lambda k}|^2]^{-1}$ such that (70) holds and for an arbitrary $\varepsilon > 0$

$$\frac{1}{f_L(\lambda)} - \frac{1}{f_U(\lambda)} \leq \varepsilon, \qquad -\pi \leq \lambda \leq \pi. \tag{83}$$

The function $f_U(\lambda)$ is the spectral density of the process $\{w_t\}$ satisfying the stochastic difference equation

$$\sum_{k=0}^{K} b_k w_{t-k} = v_t, \qquad t = \ldots, -1, 0, 1, \ldots, \tag{84}$$

and $\{v_t\}$ is a sequence of uncorrelated random variables with means 0 and variances 1. The elements σ_{TU}^{ts} of $\mathbf{\Sigma}_U^{-1}$ are

$$\begin{aligned} \sigma_{TU}^{ts} &= \sum_{k=0}^{\min(t,s)-1} b_k b_{k+|t-s|} \\ &= \sigma_{TU}^{T+1-t,\,T+1-s}, \qquad s, t = 1, \ldots, K, \end{aligned} \tag{85}$$

$$\begin{aligned} \sigma_{TU}^{ts} &= \sum_{k=0}^{K-|t-s|} b_k b_{k+|t-s|}, \qquad 0 \leq |t-s| \leq K, \\ &= 0, \qquad K < |t-s|, \end{aligned} \tag{86}$$

for $K < \max(t, s)$ and $\min(t, s) \leq T - K$ with $T > 2K$. (See Section 6.2.) Let

$$\boldsymbol{B} = \begin{pmatrix} b_{11} & 0 & 0 & \cdots & 0 & 0 & 0 & \cdots & 0 & 0 \\ b_{21} & b_{22} & 0 & \cdots & 0 & 0 & 0 & \cdots & 0 & 0 \\ b_{31} & b_{32} & b_{33} & \cdots & 0 & 0 & 0 & \cdots & 0 & 0 \\ \vdots & \vdots & \vdots & & \vdots & \vdots & \vdots & & \vdots & \vdots \\ b_{K1} & b_{K2} & b_{K3} & \cdots & b_{KK} & 0 & 0 & \cdots & 0 & 0 \\ b_K & b_{K-1} & b_{K-2} & \cdots & b_1 & b_0 & 0 & \cdots & 0 & 0 \\ 0 & b_K & b_{K-1} & \cdots & b_2 & b_1 & b_0 & \cdots & 0 & 0 \\ \vdots & \vdots & \vdots & & \vdots & \vdots & \vdots & & \vdots & \vdots \\ 0 & 0 & 0 & \cdots & 0 & 0 & 0 & \cdots & b_0 & 0 \\ 0 & 0 & 0 & \cdots & 0 & 0 & 0 & \cdots & b_1 & b_0 \end{pmatrix}, \tag{87}$$

where $b_{11}, b_{21}, b_{22}, \ldots, b_{KK}$ are chosen so that

$$\boldsymbol{\Sigma}_U^{-1} = \boldsymbol{B}'\boldsymbol{B}. \tag{88}$$

[A triangular $\boldsymbol{B}$ is defined uniquely by (88); the last $T - K$ rows of the right-hand side of (87) are explicitly the last $T - K$ rows of this matrix $\boldsymbol{B}$ determined by (86) and the second set of (85).] Then

$$\begin{aligned} \boldsymbol{x}'\boldsymbol{\Sigma}_U^{-1}\boldsymbol{x} &= \boldsymbol{x}'\boldsymbol{B}'\boldsymbol{B}\boldsymbol{x} \\ &= \sum_{i=1}^{K}\left(\sum_{j=1}^{i} b_{ij}x_j\right)^2 + \sum_{t=K+1}^{T}\left(\sum_{j=0}^{K} b_j x_{t-j}\right)^2 \\ &= \sum_{j,k=0}^{K}\sum_{t=K+1}^{T} b_j b_k x_{t-j}x_{t-k} + \sum_{i=1}^{K}\left(\sum_{j=1}^{i} b_{ij}x_j\right)^2. \end{aligned} \tag{89}$$

For any $\boldsymbol{\gamma}$ of p components let $x_t = (z_{1t}, \ldots, z_{pt})\boldsymbol{D}_T^{-1}\boldsymbol{\gamma}$. Then

$$\begin{aligned} \boldsymbol{\gamma}'\boldsymbol{D}_T^{-1}\boldsymbol{Z}'\boldsymbol{\Sigma}^{-1}\boldsymbol{Z}\boldsymbol{D}_T^{-1}\boldsymbol{\gamma} \geq \boldsymbol{\gamma}'\boldsymbol{D}_T^{-1}\boldsymbol{Z}'\boldsymbol{\Sigma}_U^{-1}\boldsymbol{Z}\boldsymbol{D}_T^{-1}\boldsymbol{\gamma} &= \boldsymbol{x}'\boldsymbol{B}'\boldsymbol{B}\boldsymbol{x} \\ &\geq \sum_{j,k=0}^{K}\sum_{t=K+1}^{T} b_j b_k x_{t-j}x_{t-k} \end{aligned} \tag{90}$$

and

$$\begin{aligned} \varliminf_{T\to\infty} \boldsymbol{\gamma}'\boldsymbol{D}_T^{-1}\boldsymbol{Z}'\boldsymbol{\Sigma}^{-1}\boldsymbol{Z}\boldsymbol{D}_T^{-1}\boldsymbol{\gamma} &\geq \lim_{T\to\infty}\sum_{j,k=0}^{K} b_j b_k \sum_{t=K+1}^{T} \boldsymbol{\gamma}'\boldsymbol{D}_T^{-1}z_{t-j}z_{t-k}'\boldsymbol{D}_T^{-1}\boldsymbol{\gamma} \\ &= \lim_{T\to\infty}\sum_{j,k=0}^{K} b_j b_k \boldsymbol{\gamma}'\boldsymbol{D}_T^{-1}\boldsymbol{A}_T(k-j)\boldsymbol{D}_T^{-1}\boldsymbol{\gamma} \\ &= \sum_{j,k=0}^{K} b_j b_k \boldsymbol{\gamma}'\mathbf{R}(k-j)\boldsymbol{\gamma} \\ &= \boldsymbol{\gamma}'\sum_{j,k=0}^{K} b_j b_k \int_{-\pi}^{\pi} e^{i\lambda(k-j)}\, d\boldsymbol{M}(\lambda)\boldsymbol{\gamma} \\ &= \boldsymbol{\gamma}'\int_{-\pi}^{\pi}\left|\sum_{j=0}^{K} b_j e^{i\lambda j}\right|^2 d\boldsymbol{M}(\lambda)\boldsymbol{\gamma} \\ &= \frac{1}{2\pi}\boldsymbol{\gamma}'\int_{-\pi}^{\pi}\frac{1}{f_U(\lambda)}\, d\boldsymbol{M}(\lambda)\boldsymbol{\gamma}. \end{aligned} \tag{91}$$

Similarly,

$$\varlimsup_{T\to\infty} \boldsymbol{\gamma}'\boldsymbol{D}_T^{-1}\boldsymbol{Z}'\boldsymbol{\Sigma}^{-1}\boldsymbol{Z}\boldsymbol{D}_T^{-1}\boldsymbol{\gamma} \leq \frac{1}{2\pi}\boldsymbol{\gamma}'\int_{-\pi}^{\pi}\frac{1}{f_L(\lambda)}\, d\boldsymbol{M}(\lambda)\boldsymbol{\gamma}. \tag{92}$$

Since ε is arbitrary,

$$\lim_{T\to\infty} \boldsymbol{\gamma}' \boldsymbol{D}_T^{-1} \boldsymbol{Z}' \boldsymbol{\Sigma}^{-1} \boldsymbol{Z} \boldsymbol{D}_T^{-1} \boldsymbol{\gamma} = \frac{1}{2\pi} \boldsymbol{\gamma}' \int_{-\pi}^{\pi} \frac{1}{f(\lambda)} \, d\boldsymbol{M}(\lambda) \boldsymbol{\gamma}. \tag{93}$$

Since this holds for every vector $\boldsymbol{\gamma}$, the theorem follows.

COROLLARY 10.2.1. *Under Assumptions* 10.2.1 *to* 10.2.5

$$\lim_{T\to\infty} \boldsymbol{D}_T \mathscr{E}(\boldsymbol{b}^* - \boldsymbol{\beta})(\boldsymbol{b}^* - \boldsymbol{\beta})' \boldsymbol{D}_T = 2\pi \mathbf{R}^{-1}(0) \int_{-\pi}^{\pi} f(\lambda) \, d\boldsymbol{M}(\lambda) \mathbf{R}^{-1}(0); \tag{94}$$

under Assumptions 10.2.1 *to* 10.2.6

$$\lim_{T\to\infty} \boldsymbol{D}_T \mathscr{E}(\boldsymbol{b} - \boldsymbol{\beta})(\boldsymbol{b} - \boldsymbol{\beta})' \boldsymbol{D}_T = \left[\frac{1}{2\pi} \int_{-\pi}^{\pi} \frac{1}{f(\lambda)} \, d\boldsymbol{M}(\lambda) \right]^{-1}. \tag{95}$$

Note that $\lim_{T\to\infty} \boldsymbol{D}_T^{-1} \boldsymbol{Z}' \boldsymbol{Z} \boldsymbol{D}_T^{-1} = \mathbf{R}(0)$.

Let

$$S(u) = \{\lambda \mid 2\pi f(\lambda) \le u\}, \qquad m \le u \le M, \tag{96}$$

$$\boldsymbol{T}(u) = \int_{S(u)} d\boldsymbol{M}(\lambda), \tag{97}$$

where $m = 2\pi \min f(\lambda)$ and $M = 2\pi \max f(\lambda)$, $-\pi \le \lambda \le \pi$. Since $f(\lambda) = f(-\lambda)$, $S(u)$ is a set symmetric about the origin, and hence $\boldsymbol{T}(u)$ and $\boldsymbol{T}(u_2) - \boldsymbol{T}(u_1)$, $m \le u_1 < u_2 \le M$, are real and positive semidefinite. Then

$$\lim_{T\to\infty} \boldsymbol{D}_T^{-1} \boldsymbol{Z}' \boldsymbol{Z} \boldsymbol{D}_T^{-1} = \int_m^M d\boldsymbol{T}(u), \tag{98}$$

$$\lim_{T\to\infty} \boldsymbol{D}_T^{-1} \boldsymbol{Z}' \boldsymbol{\Sigma} \boldsymbol{Z} \boldsymbol{D}_T^{-1} = \int_m^M u \, d\boldsymbol{T}(u), \tag{99}$$

$$\lim_{T\to\infty} \boldsymbol{D}_T^{-1} \boldsymbol{Z}' \boldsymbol{\Sigma}^{-1} \boldsymbol{Z} \boldsymbol{D}_T^{-1} = \int_m^M \frac{1}{u} \, d\boldsymbol{T}(u). \tag{100}$$

There is a nonsingular matrix $\boldsymbol{P}$ such that $\boldsymbol{P}' \mathbf{R}(0) \boldsymbol{P} = \boldsymbol{I}$, $\boldsymbol{P}' \lim_{T\to\infty} \boldsymbol{D}_T^{-1} \boldsymbol{Z}' \boldsymbol{\Sigma} \boldsymbol{Z} \boldsymbol{D}_T^{-1} \boldsymbol{P} = \boldsymbol{D}$, where $\boldsymbol{D}$ is diagonal and $d_{11} \ge \cdots \ge d_{pp}$. Let $\boldsymbol{L}(u) = \boldsymbol{P}' \boldsymbol{T}(u) \boldsymbol{P}$, $m \le u \le M$. Then $\boldsymbol{L}(u)$ and increments of $\boldsymbol{L}(u)$ are positive semidefinite. We have

$$\boldsymbol{I} = \boldsymbol{P}' \lim_{T\to\infty} \boldsymbol{D}_T^{-1} \boldsymbol{Z}' \boldsymbol{Z} \boldsymbol{D}_T^{-1} \boldsymbol{P} = \int_m^M d\boldsymbol{L}(u), \tag{101}$$

$$\boldsymbol{D} = \boldsymbol{P}' \lim_{T\to\infty} \boldsymbol{D}_T^{-1} \boldsymbol{Z}' \boldsymbol{\Sigma} \boldsymbol{Z} \boldsymbol{D}_T^{-1} \boldsymbol{P} = \int_m^M u \, d\boldsymbol{L}(u), \tag{102}$$

$$\boldsymbol{P}' \lim_{T\to\infty} \boldsymbol{D}_T^{-1} \boldsymbol{Z}' \boldsymbol{\Sigma}^{-1} \boldsymbol{Z} \boldsymbol{D}_T^{-1} \boldsymbol{P} = \int_m^M \frac{1}{u} \, d\boldsymbol{L}(u). \tag{103}$$

The two limiting normalized covariance matrices are equal if and only if (103) is $\boldsymbol{D}^{-1}$. The diagonal elements of (101), (102), and (103) are

$$1 = \int_m^M dl_{ii}(u), \qquad i = 1, \ldots, p, \tag{104}$$

$$d_{ii} = \int_m^M u \, dl_{ii}(u), \qquad i = 1, \ldots, p, \tag{105}$$

$$\int_m^M \frac{1}{u} \, dl_{ii}(u), \qquad i = 1, \ldots, p. \tag{106}$$

Since increments in $l_{ii}(u)$ are nonnegative, $l_{ii}(u)$ is a probability distribution function and (105) and (106) are the expected values of a random variable and the reciprocal of that random variable with this distribution function. If (103) is $\boldsymbol{D}^{-1}$, then (106) is $1/d_{ii}$ and Lemma 10.2.1 implies that for each i $l_{ii}(u)$ has one point of increase and the jump is 1 at that point.

Conversely, suppose each $l_{ii}(u)$ has one point of increase; by (104) the jump is 1 at that point. Let the different values of u at which the jumps in the diagonal elements of $\boldsymbol{L}(u)$ occur be $u_1 > \cdots > u_G$. Since the increase in $\boldsymbol{L}(u)$ is positive semidefinite, the contributions to the nondiagonal elements also come only from jumps at these values of u. Let $\boldsymbol{L}_j$ be the increase of $\boldsymbol{L}(u)$ at $u = u_j$, $j = 1, \ldots, G$. Then (101) is

$$\sum_{j=1}^{G} \boldsymbol{L}_j = \boldsymbol{I}. \tag{107}$$

The $\boldsymbol{L}_j$'s correspond to the nonzero $\boldsymbol{C}^{(h)}$'s in the proof of Theorem 10.2.2. The jth diagonal block of $\boldsymbol{L}_j$ is an identity matrix of order equal to the number of diagonal elements of $\boldsymbol{L}(u)$ that increase at u_j and the other blocks are zero matrices. The rank of $\boldsymbol{L}_j$ is the order of the identity submatrix. Then (102) and (103) become

$$\sum_{j=1}^{G} u_j \boldsymbol{L}_j = \boldsymbol{D}, \tag{108}$$

$$\sum_{j=1}^{G} \frac{1}{u_j} \boldsymbol{L}_j = \boldsymbol{D}^{-1}. \tag{109}$$

THEOREM 10.2.8. *A necessary and sufficient condition that the least squares estimates be asymptotically efficient under Assumptions* 10.2.1 *to* 10.2.6 *is that each diagonal element of* $\boldsymbol{L}(u)$ *has one point of increase.*

It should be noted that the number of diagonal elements of $\boldsymbol{L}_j$ equal to 1 is the rank of $\boldsymbol{L}_j$.

COROLLARY 10.2.2. *A necessary and sufficient condition that the least squares estimates be asymptotically efficient is that* $\mathbf{L}(u)$ *has* G, $G \leq p$, *points of increase and the sum of the ranks of the increases is* p.

The sets $S_j = \{\lambda \mid 2\pi f(\lambda) = u_j, -\pi \leq \lambda \leq \pi\}, j = 1, \ldots, G$, are the sets of λ over which $\boldsymbol{M}(\lambda)$ increases. Then

$$\boldsymbol{P}' \int_{S_j} d\boldsymbol{M}(\lambda)\boldsymbol{P} = \boldsymbol{L}_j, \qquad j = 1, \ldots, G. \tag{110}$$

THEOREM 10.2.9. *A necessary and sufficient condition that the least squares estimates be asymptotically efficient under Assumptions* 10.2.1 *to* 10.2.6 *is that* $f(\lambda)$ *takes on* G, $G \leq p$, *values on the set of* λ *over which* $\boldsymbol{M}(\lambda)$ *increases and that the sum of the ranks of* $\int d\boldsymbol{M}(\lambda)$ *over the* G *sets of* λ *for which* $f(\lambda)$ *takes on the* G *values is* p.

The diagonal elements of $\boldsymbol{L}_j$ that are equal to 1 must be contiguous and the diagonal submatrix of such elements constitutes an identity submatrix. Then (108) implies

$$\boldsymbol{D} = \begin{bmatrix} u_1\boldsymbol{I} & \boldsymbol{0} & \cdots & \boldsymbol{0} \\ \boldsymbol{0} & u_2\boldsymbol{I} & \cdots & \boldsymbol{0} \\ \cdot & \cdot & & \cdot \\ \cdot & \cdot & & \cdot \\ \cdot & \cdot & & \cdot \\ \boldsymbol{0} & \boldsymbol{0} & \cdots & u_G\boldsymbol{I} \end{bmatrix}, \tag{111}$$

where the identity matrix multiplied by u_j is of order equal to the rank of $\boldsymbol{L}_j$. Because the identity submatrices in $\boldsymbol{L}_1, \ldots, \boldsymbol{L}_G$ are nonoverlapping,

$$\boldsymbol{L}_i\boldsymbol{L}_j = \boldsymbol{0}, \qquad i \neq j. \tag{112}$$

The sets S_j, $j = 1, \ldots, G$, which are the sets of λ over which $\boldsymbol{M}(\lambda)$ increases, are called the *elements of the spectrum* of $\boldsymbol{M}(\lambda)$, and the union of these sets is called the *spectrum*. Since the set S_j is symmetric, the integral (110) involves only the real part of $\boldsymbol{M}(\lambda)$. In fact (110) is

$$2\boldsymbol{P}' \int_{S_j \cap (0,\pi]} d\mathcal{R}\boldsymbol{M}(\lambda)\boldsymbol{P} = \boldsymbol{L}_j, \qquad 0 \notin S_j, \tag{113}$$

$$\boldsymbol{P}'\mathcal{R}[\boldsymbol{M}(0+) - \boldsymbol{M}(0-)]\boldsymbol{P} + 2\boldsymbol{P}' \int_{S_j \cap (0,\pi]} d\mathcal{R}\boldsymbol{M}(\lambda)\boldsymbol{P} = \boldsymbol{L}_j, \qquad 0 \in S_j. \tag{114}$$

It might be noted that $\int dM(\lambda) = \mathbf{R}(0)$, where the integral is over the union of the G sets of λ, and that as a consequence the G matrices $\int_{S_j} dM(\lambda)$ are orthogonal in the metric of $\mathbf{R}(0)$ by (112).

THEOREM 10.2.10. *A necessary and sufficient condition under Assumptions* 10.2.1 *to* 10.2.4 *that the least squares estimates be asymptotically efficient for all stationary processes with continuous, positive spectral densities is that* $M(\lambda)$ *increases at not more than* p *values of* λ, $0 \leq \lambda \leq \pi$, *and the sum of the ranks of the increases in* $M(\lambda)$ *is* p.

PROOF. Theorem 10.2.9 can hold for all such spectral densities $f(\lambda)$ only if $\int dM(\lambda)$ is positive semidefinite over a number of points not greater than p. Q.E.D.

It should be noted that if the real part of $M(\lambda)$ has no increase over a set of λ, then the imaginary part also has no increase over this set. (See Problem 12.) Thus the criterion can be stated in terms of $M(\lambda)$ or in terms of the real part of $M(\lambda)$.

If the spectral density is unknown in a statistical problem, the Markov estimates are unavailable and least squares estimates might be used. The only way to ensure asymptotic efficiency is to use sequences of independent variates satisfying the conditions of Theorem 10.2.10.

Now let us consider some examples. Suppose

$$z_t = \boldsymbol{\alpha}_0 + \sum_{j=1}^{H} (\boldsymbol{\alpha}_j \cos \nu_j t + \boldsymbol{\beta}_j \sin \nu_j t) + \boldsymbol{\alpha}_{H+1}(-1)^t. \tag{115}$$

Then

$$\lim_{T\to\infty} \frac{1}{T} \sum_{t=1}^{T-h} z_{t+h} z_t' = \boldsymbol{\alpha}_0 \boldsymbol{\alpha}_0' + \frac{1}{2} \sum_{j=1}^{H} [\cos \nu_j h\, (\boldsymbol{\alpha}_j \boldsymbol{\alpha}_j' + \boldsymbol{\beta}_j \boldsymbol{\beta}_j') + \sin \nu_j h\, (\boldsymbol{\beta}_j \boldsymbol{\alpha}_j' - \boldsymbol{\alpha}_j \boldsymbol{\beta}_j')] + (-1)^h \boldsymbol{\alpha}_{H+1} \boldsymbol{\alpha}_{H+1}'. \tag{116}$$

Then $\mathbf{R}(h) = \sum_{j=0}^{H+1} \cos \nu_j h\, M_j + \sum_{j=1}^{H} \sin \nu_j h\, M_j^*$, where $M_0 = \boldsymbol{\Gamma}\boldsymbol{\alpha}_0\boldsymbol{\alpha}_0'\boldsymbol{\Gamma}$, $M_j = \frac{1}{2}\boldsymbol{\Gamma}(\boldsymbol{\alpha}_j\boldsymbol{\alpha}_j' + \boldsymbol{\beta}_j\boldsymbol{\beta}_j')\boldsymbol{\Gamma}$, $M_j^* = \frac{1}{2}\boldsymbol{\Gamma}(\boldsymbol{\beta}_j\boldsymbol{\alpha}_j' - \boldsymbol{\alpha}_j\boldsymbol{\beta}_j')\boldsymbol{\Gamma}$, $M_{H+1} = \boldsymbol{\Gamma}\boldsymbol{\alpha}_{H+1}\boldsymbol{\alpha}_{H+1}'\boldsymbol{\Gamma}$, $\boldsymbol{\Gamma}$ is a diagonal matrix, whose diagonal elements are the reciprocals of the square roots of the diagonal elements of $\frac{1}{2}\sum_{j=0}^{H+1} \boldsymbol{\alpha}_j\boldsymbol{\alpha}_j' + \frac{1}{2}\sum_{j=1}^{H} \boldsymbol{\beta}_j\boldsymbol{\beta}_j'$, $\nu_0 = 1$, and $\nu_{H+1} = \pi$.

As another example, consider the independent variables $z_{jt} = t^{j-1}$, $j = 1, \ldots, p$. Then

$$\begin{aligned} a_{jk}^T(h) &= \sum_{t=1}^{T-h} (t+h)^{j-1} t^{k-1} \\ &= \sum_{t=1}^{T-h} [t^{j+k-2} + (j-1)ht^{j+k-3} + \cdots + h^{j-1}t^{k-1}], \end{aligned} \tag{117}$$

$$h = 0, 1, \ldots, T-1,$$

and $a_{jk}^T(0) = \sum_{t=1}^T t^{j+k-2}$ is approximately $T^{j+k-1}/(j+k-1)$. (See Problem 14.) Thus

$$(118) \quad \rho_{jk}(h) = \lim_{T\to\infty} \frac{a_{jk}^T(h)/T^{j+k-1}}{\sqrt{a_{jj}^T(0)/T^{2j-1}}\sqrt{a_{kk}^T(0)/T^{2k-1}}}$$

$$= \frac{\sqrt{2j-1}\sqrt{2k-1}}{j+k-1}, \qquad j, k = 1, \ldots, p, \quad h = 0, \pm 1, \ldots .$$

This does not depend on h, and hence $\boldsymbol{M}(\lambda)$ has only a jump at $\lambda = 0$ of

$$(119) \qquad \boldsymbol{M}_0 = \left(\frac{\sqrt{2j-1}\sqrt{2k-1}}{j+k-1}\right)$$

of rank p.

If z_{jt} is a polynomial of degree $j-1$ with leading coefficient positive, $j = 1, \ldots, p$, the same sequence $\{\mathbf{R}(h)\}$ is obtained because the leading coefficient dominates. Hence, the same $\boldsymbol{M}(\lambda)$ results. If z_{jt} is a polynomial of degree d_j with leading coefficient positive, $d_1 < d_2 < \cdots < d_p$, then

$$(120) \quad \rho_{jk}(h) = \frac{\sqrt{2d_j+1}\sqrt{2d_k+1}}{d_j+d_k+1}, \qquad j, k = 1, \ldots, p, \quad h = 0, \pm 1, \ldots,$$

$\boldsymbol{M}(\lambda)$ has a jump at $\lambda = 0$, and the j, kth element of $\boldsymbol{M}_0$ is (120).

Now let us combine the examples. Let the z_{jt}'s be divided into $H+2$ sets, each set corresponding to a frequency λ_k, $k = 0, 1, \ldots, H+1$, with $0 = \lambda_0 < \lambda_1 < \cdots < \lambda_{H+1} = \pi$ and m_k z_{jt}'s corresponding to λ_k. Let $z_{jt} = P_j(t)$, $j = 1, \ldots, m_0$, where $P_j(t)$ is of degree d_j and having positive leading coefficient. In the set with index k let the z_{jt}'s be $P_{ki}(t) \cos \lambda_k t$ and $Q_{ki}(t) \sin \lambda_k t$, $i = 1, \ldots, \frac{1}{2}m_k$ with m_k even, and $P_{ki}(t)$ and $Q_{ki}(t)$ of degree d_{ki} and having positive leading coefficients, $k = 1, \ldots, H$. In the last set let the z_{jt}'s be $(-1)^t Q_i(t)$, $i = 1, \ldots, m_{H+1}$, where $Q_i(t)$ is of degree $d_{H+1,i}$ and having positive leading coefficient.

Then $\rho_{gj}(h) = 0$ if g and j are in different sets. For j and k in the first set, $\rho_{jk}(h)$ is given by (120). For g and j in the set with index k, $k = 1, \ldots, H$, $\rho_{gj}(h)$ constitute the $m_k \times m_k$ matrix

$$(121) \qquad \begin{pmatrix} \cos \lambda_k h\, \mathbf{R}_k & -\sin \lambda_k h\, \mathbf{R}_k \\ \sin \lambda_k h\, \mathbf{R}_k & \cos \lambda_k h \mathbf{R}_k \end{pmatrix},$$

where the i, lth element of $\mathbf{R}_k$ is $\sqrt{2d_{ki}+1}\sqrt{2d_{kl}+1}/(d_{ki}+d_{kl}+1)$ and the terms with $\cos \lambda_k t$ are in the first subset and the terms with $\sin \lambda_k t$ are in the second subset. The elements $\rho_{gj}(h)$ in the last set are $(-1)^h\sqrt{2d_{H+1,g}+1}$

$\sqrt{2d_{H+1,j}+1}/(d_{H+1,g}+d_{H+1,j}+1)$. The spectral function has only jumps. The jump at $\lambda = 0$ has elements (120) for $j, k = 1, \ldots, m_0$, and 0's elsewhere. The jump at $\lambda = \pm\lambda_k$ has (121) with $\cos \lambda_k h$ replaced by $\frac{1}{2}$ and $\sin \lambda_k h$ replaced by $\mp \frac{1}{2}i$ as the $(k+1)$st diagonal submatrix and 0's elsewhere. The jump at $\lambda = \pm\pi$ has elements $\frac{1}{2}\sqrt{2d_{H+1,g}+1}\sqrt{2d_{H+1,j}+1}/(d_{H+1,g}+d_{H+1,j}+1)$ in the last diagonal submatrix and 0's elsewhere.

Perhaps the simplest case is $p = 1$, $z_t = 1$, $t = 1, 2, \ldots$. Then $a^T(h) = T - h$, $h = 0, 1, \ldots, T$,

$$\rho(h) = \lim_{T\to\infty} \frac{a^T(h)}{T} = 1, \qquad h = 0, \pm 1, \ldots, \tag{122}$$

and each element of $\boldsymbol{M}(\lambda)$ has the jump of 1 at $\lambda = 0$. The least squares estimate of β in $\mathscr{E} y_t = \beta$ is $b^* = \sum_{t=1}^{T} y_t/T = \bar{y}$, and $\bar{y}$ is asymptotically efficient.

Now we shall give another case which has the limiting correlation sequence (122) and corresponding spectral function $\boldsymbol{M}(\lambda)$, but is not equivalent. Let $\mathscr{E} y_t = \beta w_t$, $t = 1, 2, \ldots$. Let w_t take values a and b, where $a \neq \pm b$ and

$$\frac{m}{n} a^2 + \frac{n-m}{n} b^2 = 1, \tag{123}$$

and $0 < m < n$, where m and n are integers. We construct a sequence $\{w_t\}$ such that the limiting relative frequency of a's is m/n and of b's is $(n-m)/n$. Let $w_t = a$ for $t = 1, \ldots, m$, and $w_t = b$ for $t = m+1, \ldots, n$. Next let $w_t = a$ for $t = n+1, \ldots, n+2m$, and $w_t = b$ for $t = n + 2m + 1, \ldots, n + 2n = 3n$. The sequence $\{w_t\}$ consists of finite subsequences of lengths $n, 2n, 3n, \ldots$. In the jth subsequence the first jm elements are a and the last $j(n-m)$ elements are b. As $T \to \infty$ the limiting relative frequency of a's is m/n because the increment in the number of a's in the $(j+1)$st subsequence is $(j+1)m$ compared with the $\frac{1}{2}j(j+1)n$ elements in the first j subsequences. Hence, $\sum_{t=1}^{T} w_t^2/T$ has (123) as a limit. Moreover, (122) holds for this sequence because $w_t w_{t+h} = ab$ for $2h$ products for w_t in the jth subsequence, while $w_t w_{t+h} = a^2$ for $jm - h$ terms and $w_t w_{t+h} = b^2$ for $j(n-m) - h$ terms for j such that $j > h/m$, $j > h/(n-m)$. Thus $\rho(h) = 1$ and each element of $\boldsymbol{M}(\lambda)$ has a jump of 1 at $\lambda = 0$. This shows that the spectral function (or equivalently the sequence of covariances) does not determine the sequence of independent variables or even the limiting frequencies of their values.

Moreover, one cannot replace the sequence occurring in the expected values by another sequence in the estimates. This sequence $\{w_t\}$ is essentially different from $\{z_t\}$, for

$$\lim_{T\to\infty} \frac{1}{T}\sum_{t=1}^{T} z_t w_t = \frac{m}{n} a + \frac{n-m}{n} b = c, \tag{124}$$

which is different from 1. If $\mathscr{E} y_t = \beta w_t$ and $\sum_{t=1}^{T} z_t y_t / T = \bar{y}$ ($z_t \equiv 1$) were used as an estimate of β, the estimate would have a variance asymptotically equivalent to the variance of $\sum_{t=1}^{T} w_t y_t / T$, but the estimate would be asymptotically biased for $\lim_{T\to\infty} \mathscr{E}\bar{y} = \lim_{T\to\infty} \sum_{t=1}^{T} \beta w_t / T = c\beta \neq \beta$.

The situation here is basically different from the finite-dimensional case. Suppose $\boldsymbol{w}'\boldsymbol{w} = 1$, $\boldsymbol{w}'\boldsymbol{\Sigma}\boldsymbol{w} = \lambda^*$, and $\boldsymbol{w}'\boldsymbol{\Sigma}^{-1}\boldsymbol{w} = 1/\lambda^*$, where λ^* is a simple characteristic root of $\boldsymbol{\Sigma}$ corresponding to $\boldsymbol{v}$; that is, $\boldsymbol{\Sigma}\boldsymbol{v} = \lambda^*\boldsymbol{v}$ and $\boldsymbol{\Sigma}^{-1}\boldsymbol{v} = (1/\lambda^*)\boldsymbol{v}$. Since $\boldsymbol{\Sigma} = \sum_{t=1}^{T} \lambda_t \boldsymbol{v}_t \boldsymbol{v}_t'$, $\boldsymbol{w}'\boldsymbol{\Sigma}\boldsymbol{w} = \sum_{t=1}^{T} \lambda_t (\boldsymbol{v}_t'\boldsymbol{w})^2$ and $\boldsymbol{w}'\boldsymbol{\Sigma}^{-1}\boldsymbol{w} = \sum_{t=1}^{T} (1/\lambda_t)(\boldsymbol{v}_t'\boldsymbol{w})^2$; thus $\boldsymbol{v}_t'\boldsymbol{w} = 1$ for $\lambda_t = \lambda^*$ and $\boldsymbol{v}_t'\boldsymbol{w} = 0$ for $\lambda_t \neq \lambda^*$. Therefore, when $\mathscr{E}\boldsymbol{y} = \beta\boldsymbol{w}$, the estimate $\boldsymbol{v}'\boldsymbol{y}$ is an efficient unbiased estimate of β because $\mathscr{E}\boldsymbol{v}'\boldsymbol{y} = \beta\boldsymbol{v}'\boldsymbol{w} = \beta$ and $\operatorname{Var}(\boldsymbol{v}'\boldsymbol{y}) = \boldsymbol{v}'\boldsymbol{\Sigma}\boldsymbol{v} = \lambda^*$. The analogue of $\boldsymbol{v}$ is $\{1\}$, of $\boldsymbol{v}_t$ is $\{\cos \lambda t, \sin \lambda t\}$, the analogue of $\boldsymbol{\Sigma} = \sum_{t=1}^{T} \lambda_t \boldsymbol{v}_t \boldsymbol{v}_t'$ is $\sigma(t-s) = \int_{-\pi}^{\pi} e^{i\lambda(t-s)} f(\lambda)\, d\lambda$. The equalities $(\boldsymbol{v}_t'\boldsymbol{w})^2 = 1$ for $\lambda_t = \lambda^*$ and $(\boldsymbol{v}_t'\boldsymbol{w})^2 = 0$ for $\lambda_t \neq \lambda^*$ correspond to

$$\lim_{T\to\infty} \int_{-\pi}^{\pi} \frac{1}{T} \left| \sum_{t=1}^{T} w_t e^{i\lambda t} \right|^2 f(\lambda)\, d\lambda = 2\pi f(0) \tag{125}$$

for every continuous $f(\lambda)$.

In an asymptotic sense the functions $\cos \nu t$ and $\sin \nu t$ are characteristic vectors of all covariance matrices of stationary processes, for

$$\begin{aligned} \sum_{s=1}^{T} \sigma(t-s)(\cos \nu s + i \sin \nu s) &= \sum_{s=1}^{T} \sigma(t-s) e^{i\nu s} \\ &= e^{i\nu t} \sum_{s=1}^{T} \sigma(s-t) e^{i\nu(s-t)} \\ &= e^{i\nu t} \sum_{h=-(t-1)}^{T-t} \sigma(h) e^{i\nu h} \\ &\to e^{i\nu t} \sum_{h=-(t-1)}^{\infty} \sigma(h) e^{i\nu h} \end{aligned} \tag{126}$$

as $T \to \infty$. For t sufficiently large the right-hand side of the above is arbitrarily close to $2\pi f(\nu) e^{i\nu t} = 2\pi f(\nu)(\cos \nu t + i \sin \nu t)$.

If the sequence $\{\mathbf{R}(h)\}$ is such that $\boldsymbol{M}(\lambda)$ is absolutely continuous, then the least squares estimates cannot be asymptotically efficient if the spectral density

$f(\lambda)$ takes on more than p values over the spectrum of $M(\lambda)$. For instance, if $\sum_{h=-\infty}^{\infty} |\rho_{ii}(h)| < \infty, i = 1, \ldots, p$, then $\mathbf{R}(h) = \int_{-\pi}^{\pi} e^{i\lambda h} \boldsymbol{m}(\lambda)\, d\lambda$, where $m_{jk}(\lambda) = \sum_{s=-\infty}^{\infty} \rho_{jk}(s) e^{i\lambda s}/(2\pi)$. Then $M(\lambda)$ has no jumps. Roughly speaking, this will be the case if $\{z_t\}$ is a realization of a stationary (vector) process.

10.2.4. Asymptotic Normality of Estimates

Under the Assumptions 10.2.1, 10.2.2, 10.2.3, and 10.2.4 it was shown in Section 2.6 that the least squares estimates have an asymptotic normal distribution if the independent error terms satisfy a Lindeberg-type condition. Here we shall prove the corresponding theorem when the random term is $u_t = \sum_{s=-\infty}^{\infty} \gamma_s v_{t-s}$ and the v_t's satisfy the same Lindeberg-type condition. A similar, but weaker, theorem was proved by Hannan (1961).

THEOREM 10.2.11. *Let* $y_t = \boldsymbol{\beta}' z_t + u_t$, $t = 1, 2, \ldots$, *where* $u_t = \sum_{s=-\infty}^{\infty} \gamma_s v_{t-s}$, *the* v_t*'s are independent,* v_t *has mean* 0, *variance* σ^2, *and distribution function* $F_t(v)$, $t = \ldots, -1, 0, 1, \ldots$, *and* $\sum_{s=-\infty}^{\infty} |\gamma_s| < \infty$. *Suppose Assumptions* 10.2.1 *to* 10.2.4 *are satisfied and*

$$\sup_{t=1,2,\ldots} \int_{|v|>c} v^2\, dF_t(v) \to 0 \tag{127}$$

as $c \to \infty$. *Then* $\boldsymbol{D}_T(\boldsymbol{b}^* - \boldsymbol{\beta})$ *has a limiting normal distribution with mean* $\mathbf{0}$ *and covariance matrix* (94).

PROOF. Since

$$\boldsymbol{D}_T(\boldsymbol{b}^* - \boldsymbol{\beta}) = [\boldsymbol{D}_T^{-1} \boldsymbol{A}_T(0) \boldsymbol{D}_T^{-1}]^{-1} \boldsymbol{D}_T^{-1} \sum_{t=1}^{T} z_t u_t, \tag{128}$$

the limiting distribution of $\boldsymbol{D}_T(\boldsymbol{b}^* - \boldsymbol{\beta})$ is the limiting distribution of $\mathbf{R}^{-1}(0)$ times $\boldsymbol{D}_T^{-1} \sum_{t=1}^{T} z_t u_t = \boldsymbol{c}_T$, say. Let

$$\boldsymbol{c}_{T,k} = \boldsymbol{D}_T^{-1} \sum_{t=1}^{T} z_t \sum_{s=-k}^{k} \gamma_s v_{t-s}. \tag{129}$$

Then

$$\begin{aligned} \mathscr{E}(\boldsymbol{c}_T - \boldsymbol{c}_{T,k})'(\boldsymbol{c}_T - \boldsymbol{c}_{T,k}) &= \operatorname{tr} \boldsymbol{D}_T^{-1} \sum_{t,t'=1}^{T} z_t z_{t'}' \sum_{|s|,|s'|>k} \gamma_s \gamma_{s'} \mathscr{E} v_{t-s} v_{t'-s'} \boldsymbol{D}_T^{-1} \\ &= \sigma^2 \operatorname{tr} \boldsymbol{D}_T^{-1} \sum_{|s|,|s'|>k} \gamma_s \gamma_{s'} \sum_t z_t z_{t-s+s'}' \boldsymbol{D}_T^{-1}, \end{aligned} \tag{130}$$

where the sum on t is for values of t and $t - s + s'$ between 1 and T. (The interchange of summation on s and s' and expectation was justified in a similar proof in Section 8.4.3.) Then

$$\mathscr{E}(\boldsymbol{c}_T - \boldsymbol{c}_{T,k})'(\boldsymbol{c}_T - \boldsymbol{c}_{T,k}) = \sigma^2 \sum_{|s|,|s'|>k} \gamma_s\gamma_{s'} \sum_{i=1}^{p} r_{ii}^T(s-s') \tag{131}$$

$$\leq p\sigma^2\left(\sum_{|s|>k} |\gamma_s|\right)^2$$

which goes to 0 as $k \to \infty$.

For an arbitrary vector $\boldsymbol{\alpha}$ consider

$$\boldsymbol{\alpha}'\boldsymbol{c}_{T,k} = \boldsymbol{\alpha}'\boldsymbol{D}_T^{-1} \sum_{t=1}^{T} \boldsymbol{z}_t \sum_{s=-k}^{k} \gamma_s v_{t-s} \tag{132}$$

$$= \sum_{r=k+1}^{T-k} \boldsymbol{\alpha}'\boldsymbol{D}_T^{-1} \sum_{s=-k}^{k} \gamma_s \boldsymbol{z}_{r+s} v_r + \sum_{r=1-k}^{k} \boldsymbol{\alpha}'\boldsymbol{D}_T^{-1} \sum_{s=1-r}^{k} \gamma_s \boldsymbol{z}_{r+s} v_r + \sum_{r=T-k+1}^{T+k} \boldsymbol{\alpha}'\boldsymbol{D}_T^{-1} \sum_{s=-k}^{T-r} \gamma_s \boldsymbol{z}_{r+s} v_r.$$

The sum of squares of the coefficients of $v_{1-k}, \ldots, v_k, v_{T-k+1}, \ldots, v_{T+k}$ converges to 0 as T increases because of Assumption 10.2.2. Thus the limiting distribution of $\boldsymbol{\alpha}'\boldsymbol{c}_{T,k}$ is the limiting distribution of $\sigma[\sum_{s,s'=-k}^{k} \gamma_s\gamma_{s'}\boldsymbol{\alpha}'\mathbf{R}(s-s')\boldsymbol{\alpha}]^{\frac{1}{2}}$ times $\sum_{r=k+1}^{T-k} w_r^T$, where

$$w_r^T = \frac{\boldsymbol{\alpha}'\boldsymbol{D}_T^{-1} \sum_{s=-k}^{k} \gamma_s \boldsymbol{z}_{r+s}}{\sigma \sqrt{\sum_{r=k+1}^{T-k} \left(\boldsymbol{\alpha}'\boldsymbol{D}_T^{-1} \sum_{s=-k}^{k} \gamma_s \boldsymbol{z}_{r+s}\right)^2}} v_r. \tag{133}$$

The sum of squares of the coefficients of $v_{k+1}, \ldots, v_{T-k}$ in (132) is

$$\sum_{s,s'=-k}^{k} \gamma_s\gamma_{s'}\boldsymbol{\alpha}'\boldsymbol{D}_T^{-1} \sum_{r=k+1}^{T-k} \boldsymbol{z}_{r+s}\boldsymbol{z}_{r+s'}'\boldsymbol{D}_T^{-1}\boldsymbol{\alpha} \tag{134}$$

which has the limit $\sum_{s,s'=-k}^{k} \gamma_s\gamma_{s'}\boldsymbol{\alpha}'\mathbf{R}(s-s')\boldsymbol{\alpha}$. The Lindeberg central limit theorem (Theorem 7.7.2) implies that $\sum_{r=k+1}^{T-k} w_r^T$ has a limiting normal distribution with mean 0 and variance 1 (according to the proof of Theorem 2.6.1). Application of Corollary 7.7.1 yields the theorem.

Theorem 10.2.11 implies that for long time series the least squares coefficients can be treated as if they were normally distributed. The limiting covariance matrix is

$$\lim_{T\to\infty} \boldsymbol{D}_T\mathscr{E}(\boldsymbol{b}^* - \boldsymbol{\beta})(\boldsymbol{b}^* - \boldsymbol{\beta})'\boldsymbol{D}_T = \mathbf{R}^{-1}(0) \sum_{h=-\infty}^{\infty} \sigma(h)\mathbf{R}(h)\mathbf{R}^{-1}(0) \tag{135}$$

if $\sum_{h=-\infty}^{\infty} |\sigma(h)| < \infty$. This can be estimated consistently by estimates analogous to estimates of $2\pi f(\lambda) = \sum_{h=-\infty}^{\infty} \sigma(h) \cos \lambda h$, specifically

$$D_T A_T^{-1}(0) \sum_{h=-K_T}^{K_T} k\left(\frac{h}{K_T}\right) c_{h,T}^* A_T(h) A_T^{-1}(0) D_T, \tag{136}$$

where $c_{h,T}^*$ is the estimate of $\sigma(h)$ based on residuals from $b^{*\prime} z_t$, $t = 1, \ldots, T$, and $k(x)$ is a kernel of the type studied in Section 9.3.2. On this basis tests of hypotheses about $\boldsymbol{\beta}$ can be carried out and confidence regions for components of $\boldsymbol{\beta}$ can be constructed. [See Hannan (1963) for treatment of the multivariate case.]

10.3. ESTIMATION OF THE COVARIANCES AND SPECTRAL DENSITY BASED ON RESIDUALS FROM TRENDS

10.3.1. Estimation of Covariance Matrices

We consider again the model $\mathscr{E}y = Z\boldsymbol{\beta}$ and $\mathscr{E}(y - Z\boldsymbol{\beta})(y - Z\boldsymbol{\beta})' = \boldsymbol{\Sigma}$ and now treat questions of estimation of the covariance matrix $\boldsymbol{\Sigma}$. In Section 10.3.2 we study the asymptotic theory when the covariances correspond to a stationary stochastic process. The asymptotic theory for the sample spectral density and for estimates of the spectral density based on residuals from the trend estimated by least squares is treated in Sections 10.3.3 and 10.3.4, respectively. This asymptotic theory generalizes the treatment in Sections 8.3, 8.4, and 9.4 of these statistics based on residuals from sample means.

The least squares estimate of $\boldsymbol{\beta}$ is

$$b^* = (Z'Z)^{-1}Z'y. \tag{1}$$

The residuals from the estimated regression constitute a T-component vector

$$y - Zb^* = [I - Z(Z'Z)^{-1}Z']y, \tag{2}$$

whose expectation is $\mathbf{0}$ (because $\mathscr{E}b^* = \boldsymbol{\beta}$). The components of this vector may be used for estimating $\boldsymbol{\Sigma}$ or functions of $\boldsymbol{\Sigma}$. If there is no knowledge of the structure of $\boldsymbol{\Sigma}$, the matrix $(y - Zb^*)(y - Zb^*)'$ can be used to estimate $\boldsymbol{\Sigma}$. Its expected value is

$$\mathscr{E}(y - Zb^*)(y - Zb^*)' = [I - Z(Z'Z)^{-1}Z']\boldsymbol{\Sigma}[I - Z(Z'Z)^{-1}Z'], \tag{3}$$

which is different from $\boldsymbol{\Sigma}$. This estimate is of rank $T - p$, while $\boldsymbol{\Sigma}$ is of rank T. If $\boldsymbol{\Sigma}$ has no special structure, the number of parameters in $\boldsymbol{\Sigma}$ is $\frac{1}{2}T(T+1)$, as compared to $T - p$ linearly independent terms in $\boldsymbol{y} - \boldsymbol{Z}\boldsymbol{b}^*$. ($\boldsymbol{y} - \boldsymbol{Z}\boldsymbol{b}^*$ is orthogonal to $\boldsymbol{Z}\boldsymbol{b}^*$.)

We can get some insight into (3) in the case where the least squares estimates are efficient. In that case we can write $\boldsymbol{Z} = \boldsymbol{V}^*\boldsymbol{C}$, where $\boldsymbol{C}$ is nonsingular and the p columns of $\boldsymbol{V}^*$ are p characteristic vectors of $\boldsymbol{\Sigma}$ corresponding to some p characteristic roots, say $\lambda_{s_1}, \ldots, \lambda_{s_p}$. (We assume that in the case of multiple roots the characteristic vectors are defined in such a way that they include the columns of $\boldsymbol{V}^*$.) Let $S = \{s_1, \ldots, s_p\}$ be a subset of $1, \ldots, T$. Then

$$\boldsymbol{I} - \boldsymbol{Z}(\boldsymbol{Z}'\boldsymbol{Z})^{-1}\boldsymbol{Z}' = \boldsymbol{I} - \boldsymbol{V}^*\boldsymbol{V}^{*\prime} = \sum_{s \notin S} \boldsymbol{v}_s\boldsymbol{v}_s', \tag{4}$$

$$\begin{aligned}[\boldsymbol{I} - \boldsymbol{Z}(\boldsymbol{Z}'\boldsymbol{Z})^{-1}\boldsymbol{Z}']\boldsymbol{\Sigma}[\boldsymbol{I} - \boldsymbol{Z}(\boldsymbol{Z}'\boldsymbol{Z})^{-1}\boldsymbol{Z}'] &= \sum_{s \notin S} \boldsymbol{v}_s\boldsymbol{v}_s' \sum_{t=1}^{T} \lambda_t \boldsymbol{v}_t\boldsymbol{v}_t' \sum_{r \notin S} \boldsymbol{v}_r\boldsymbol{v}_r' \\ &= \sum_{t \notin S} \lambda_t \boldsymbol{v}_t\boldsymbol{v}_t'.\end{aligned} \tag{5}$$

If we let $\boldsymbol{y} = \boldsymbol{V}\boldsymbol{x}$ ($\boldsymbol{x} = \boldsymbol{V}'\boldsymbol{y}$), we can write

$$\boldsymbol{y} - \boldsymbol{Z}\boldsymbol{b}^* = \sum_{s \notin S} \boldsymbol{v}_s\boldsymbol{v}_s'\boldsymbol{V}\boldsymbol{x} = \sum_{s \notin S} \boldsymbol{v}_s x_s. \tag{6}$$

and the covariance matrix of $\boldsymbol{x}$ is

$$\mathscr{E}(\boldsymbol{x} - \mathscr{E}\boldsymbol{x})(\boldsymbol{x} - \mathscr{E}\boldsymbol{x})' = \boldsymbol{V}'\boldsymbol{\Sigma}\boldsymbol{V} = \boldsymbol{\Lambda}, \tag{7}$$

the diagonal matrix whose elements are the characteristic roots of $\boldsymbol{\Sigma}$. If $\boldsymbol{y}$ has a normal distribution, $\boldsymbol{x}$ does also and the components of $\boldsymbol{x}$ are independently distributed, the variance of the tth component being λ_t, $t = 1, \ldots, T$. The covariance matrix of $\boldsymbol{b}^* = \boldsymbol{C}^{-1}\boldsymbol{V}^{*\prime}\boldsymbol{y} = \boldsymbol{C}^{-1}\boldsymbol{V}^{*\prime}\boldsymbol{V}\boldsymbol{x}$ is

$$\mathscr{E}(\boldsymbol{b}^* - \boldsymbol{\beta})(\boldsymbol{b}^* - \boldsymbol{\beta})' = \boldsymbol{C}^{-1}\boldsymbol{\Lambda}^*(\boldsymbol{C}')^{-1}, \tag{8}$$

where $\boldsymbol{\Lambda}^*$ is the diagonal matrix whose diagonal elements are $\lambda_{s_1}, \ldots, \lambda_{s_p}$. If the characteristic vectors $\boldsymbol{v}_t$, $t \notin S$, are known, $x_t^2 = (\boldsymbol{v}_t'\boldsymbol{y})^2$ can be used to estimate λ_t, $t \notin S$. (If it is known that the least squares estimates are efficient, that is, that $\boldsymbol{Z}$ is of the form $\boldsymbol{V}^*\boldsymbol{C}$, then the p-dimensional linear space of $\boldsymbol{V}^*$ is determined.) These estimates are useful for estimating the covariance matrix of $\boldsymbol{b}^*$ to the extent that λ_t, $t \in S$, are known functions of λ_t, $t \notin S$, and the matrix $\boldsymbol{C}$ is known. Note that $\boldsymbol{b}^*$ and $\boldsymbol{y} - \boldsymbol{Z}\boldsymbol{b}^*$ are uncorrelated, for

$$\begin{aligned}\mathscr{E}(\boldsymbol{y} - \boldsymbol{Z}\boldsymbol{b}^*)(\boldsymbol{b}^* - \boldsymbol{\beta})' &= [\boldsymbol{I} - \boldsymbol{Z}(\boldsymbol{Z}'\boldsymbol{Z})^{-1}\boldsymbol{Z}']\boldsymbol{\Sigma}\boldsymbol{Z}(\boldsymbol{Z}'\boldsymbol{Z})^{-1} \\ &= (\boldsymbol{I} - \boldsymbol{V}^*\boldsymbol{V}^{*\prime})\boldsymbol{V}^*\boldsymbol{\Lambda}^*\boldsymbol{C}(\boldsymbol{Z}'\boldsymbol{Z})^{-1} \\ &= \boldsymbol{0}.\end{aligned} \tag{9}$$

Let us study in detail the case where the characteristic vectors $\boldsymbol{v}_1, \ldots, \boldsymbol{v}_T$ are known. This is the situation, for instance, when

$$(10) \qquad \boldsymbol{\Sigma}^{-1} = \gamma_0 \boldsymbol{A}_0 + \cdots + \gamma_q \boldsymbol{A}_q,$$

$\boldsymbol{v}_1, \ldots, \boldsymbol{v}_T$ are the characteristic vectors of $\boldsymbol{A}_0, \ldots, \boldsymbol{A}_q$, and $\gamma_0, \ldots, \gamma_q$ are unknown parameters. An example is the circular model with $\boldsymbol{A}_0 = \boldsymbol{I}$ and

$$(11) \qquad \boldsymbol{A}_j = \tfrac{1}{2}(\boldsymbol{B}^j + \boldsymbol{B}'^j), \qquad j = 1, \ldots, g,$$

where $\boldsymbol{B}$ is the matrix with 1's on the diagonal above the main diagonal and 1 in the lower left-hand corner with 0's elsewhere. We take $q \leq [\tfrac{1}{2}T]$. (The characteristic roots of $\boldsymbol{\Sigma}^{-1}$ are $\sum_{j=0}^{q} \gamma_j \cos 2\pi jt/T$, $t = 1, \ldots, T$; the components of $\boldsymbol{v}_t$ are $\cos 2\pi ts/T$ or $\sin 2\pi ts/T$.) For convenience let us assume $\boldsymbol{Z} = \boldsymbol{V}^*$ (that is, $\boldsymbol{C} = \boldsymbol{I}$).

Suppose that the p characteristic roots of $\boldsymbol{\Sigma}$ corresponding to the p columns of $\boldsymbol{V}^*$ are functionally independent of the remaining characteristic roots of $\boldsymbol{\Sigma}$. This may be the case in the circular model when $q = [\tfrac{1}{2}T]$. Then $\lambda_T^{-1} = \sum_{j=0}^{q} \gamma_j$, and

$$(12) \qquad \lambda_s^{-1} = \lambda_{T-s}^{-1} = \sum_{j=0}^{q} \gamma_j \cos \frac{2\pi js}{T}, \qquad s = 1, \ldots, [\tfrac{1}{2}(T-1)].$$

If T is even and $q = \tfrac{1}{2}T$, then $\lambda_q^{-1} = \sum_{j=0}^{q} (-1)^j \gamma_j$. The characteristic roots corresponding to the columns of $\boldsymbol{V}^*$ are functionally independent of the remaining roots if the columns of $\boldsymbol{V}^*$ are possibly $\sqrt{1/T}(1, \ldots, 1)'$ and possibly $\sqrt{1/T}\,(-1, 1, \ldots, 1)'$ if T is even, and pairs of linear combinations of $\sin 2\pi ts/T$ and $\cos 2\pi ts/T$ for some values of t. The variances of the estimates of the components of $\boldsymbol{\beta}$ (namely, $\boldsymbol{v}_s'\boldsymbol{y}$ for $s \in S$) are λ_s.

10.3.2. The Asymptotic Bias of Sample Covariances

We now consider $y_t = \boldsymbol{\beta}' \boldsymbol{z}_t + u_t$, where $\{u_t\}$ is a stochastic process that has mean 0 and is stationary in the wide sense with covariance sequence $\{\sigma(h)\}$ and spectral density $f(\lambda)$. A sample covariance of order h based on a sample of length T is

$$(13) \qquad \begin{aligned} C_h^* &= \frac{1}{T-h} \sum_{t=1}^{T-h} (y_t - \boldsymbol{b}^{*\prime} \boldsymbol{z}_t)(y_{t+h} - \boldsymbol{b}^{*\prime} \boldsymbol{z}_{t+h}) \\ &= C_{-h}^*, \qquad\qquad h = 0, 1, \ldots, T-1, \end{aligned}$$

where

$$(14) \qquad \boldsymbol{b}^* = (\boldsymbol{Z}'\boldsymbol{Z})^{-1}\boldsymbol{Z}'\boldsymbol{y},$$

$Z = (z_1, \ldots, z_T)'$, $y = (y_1, \ldots, y_T)'$, and Z is of rank p. We can write

$$(15)\quad C_h^* = \frac{1}{T-h}\sum_{t=1}^{T-h} y_t y_{t+h} - \frac{1}{T-h}\sum_{t=1}^{T-h} b^{*\prime} z_t y_{t+h} - \frac{1}{T-h}\sum_{t=1}^{T-h} b^{*\prime} z_{t+h} y_t + \frac{1}{T-h}\sum_{t=1}^{T-h} b^{*\prime} z_t z_{t+h}' b^*.$$

THEOREM 10.3.1. *Under Assumptions* 10.2.1 *to* 10.2.5

$$(16)\quad \lim_{T\to\infty} T[\mathscr{E}C_h^* - \sigma(h)] = -4\pi \operatorname{tr} \mathbf{R}^{-1}(0)\int_{-\pi}^{\pi} \cos\lambda h\, f(\lambda)\, dM(\lambda) + 2\pi \operatorname{tr} \mathbf{R}^{-1}(0)\int_{-\pi}^{\pi} f(\lambda)\, dM(\lambda)\mathbf{R}^{-1}(0)\int_{-\pi}^{\pi} e^{i\nu h}\, dM'(\nu).$$

PROOF. From (15) we have [since $y_t - b^{*\prime}z_t = u_t - (b^* - \beta)'z_t$ and $b^* - \beta = (Z'Z)^{-1}\sum_{s=1}^T z_s u_s$]

$$(17)\quad T[\mathscr{E}C_h^* - \sigma(h)] = -\frac{T}{T-h}\sum_{t=1}^{T-h}\sum_{s=1}^{T} z_t'(Z'Z)^{-1}z_s \mathscr{E}u_s u_{t+h} - \frac{T}{T-h}\sum_{t=1}^{T-h}\sum_{s=1}^{T} z_{t+h}'(Z'Z)^{-1}z_s \mathscr{E}u_s u_t + \frac{T}{T-h}\sum_{t=1}^{T-h} z_{t+h}'\mathscr{E}(b^*-\beta)(b^*-\beta)'z_t.$$

The third term on the right-hand side of (17) has the limit

$$(18)\quad \lim_{T\to\infty} \operatorname{tr} \mathscr{E}(b^*-\beta)(b^*-\beta)'\sum_{t=1}^{T-h} z_t z_{t+h}' = \lim_{T\to\infty} \operatorname{tr} \mathscr{E}[D_T(b^*-\beta)(b^*-\beta)'D_T]D_T^{-1}\sum_{t=1}^{T-h} z_t z_{t+h}' D_T^{-1} = 2\pi \operatorname{tr} \mathbf{R}^{-1}(0)\int_{-\pi}^{\pi} f(\lambda)\, dM(\lambda)\mathbf{R}^{-1}(0)\int_{-\pi}^{\pi} e^{i\nu h}\, dM'(\nu)$$

by (62), (64), and (94) of Section 10.2. The first term on the right-hand side of (17) is

$$(19)\quad -\frac{T}{T-h}\operatorname{tr}\sum_{t=1}^{T-h}\sum_{s=1}^{T}(Z'Z)^{-1}z_s z_t'\sigma(t+h-s) = -\frac{T}{T-h}\operatorname{tr}\sum_{r=-(T-h-1)}^{T-1}\sum_{s\in S_r}(Z'Z)^{-1}z_s z_{s+r-h}'\sigma(r),$$

where $S_r = \{h - r + 1, \ldots, T\}$ for $r \leq 0$, $S_r = \{h - r + 1, \ldots, T - r\}$ for $0 \leq r \leq h$, and $S_r = \{1, \ldots, T - r\}$ for $h \leq r$. The second term on the right-hand side of (17) is

$$(20) \quad -\frac{T}{T-h} \operatorname{tr} \sum_{t=1}^{T-h} \sum_{s=1}^{T} (\boldsymbol{Z}'\boldsymbol{Z})^{-1} z_s z'_{t+h} \sigma(t-s)$$
$$= -\frac{T}{T-h} \operatorname{tr} \sum_{r=-(T-h-1)}^{T-1} \sum_{s \in S_r} (\boldsymbol{Z}'\boldsymbol{Z})^{-1} z_s z'_{s+r} \sigma(r-h).$$

If $f(\lambda)$ is a trigonometric polynomial [that is, $\sigma(r) = 0$ for $|r| > H$ for some H], then the limit of (19) is

$$(21) \quad -\sum_{r=-\infty}^{\infty} \sigma(r) \lim_{T\to\infty} \operatorname{tr} \boldsymbol{A}_T^{-1}(0)\boldsymbol{A}_T(h-r) = -\sum_{r=-\infty}^{\infty} \sigma(r) \int_{-\pi}^{\pi} \operatorname{tr} \mathbf{R}^{-1}(0) e^{i\lambda(h-r)}\, d\boldsymbol{M}(\lambda)$$
$$= -2\pi \operatorname{tr} \mathbf{R}^{-1}(0) \int_{-\pi}^{\pi} e^{i\lambda h} f(\lambda)\, d\boldsymbol{M}(\lambda).$$

Similarly, the limit of (20) is

$$(22) \quad -\sum_{r=-\infty}^{\infty} \sigma(r-h) \lim_{T\to\infty} \operatorname{tr} \boldsymbol{A}_T^{-1}(0)\boldsymbol{A}_T(-r)$$
$$= -\sum_{r=-\infty}^{\infty} \sigma(r-h) \int_{-\pi}^{\pi} \operatorname{tr} \mathbf{R}^{-1}(0) e^{-i\lambda r}\, d\boldsymbol{M}(\lambda)$$
$$= -2\pi \operatorname{tr} \mathbf{R}^{-1}(0) \int_{-\pi}^{\pi} e^{-i\lambda h} f(\lambda)\, d\boldsymbol{M}(\lambda).$$

If $f(\lambda)$ is continuous, there are trigonometric polynomials $f_L(\lambda)$ and $f_U(\lambda)$ such that $f_L(\lambda) \leq f(\lambda) \leq f_U(\lambda)$ and $f_U(\lambda) - f_L(\lambda) \leq \varepsilon$. The limits above hold for sequences $\sigma_L(h) = \int_{-\pi}^{\pi} e^{i\lambda h} f_L(\lambda)\, d\lambda$ and $\sigma_U(h) = \int_{-\pi}^{\pi} e^{i\lambda h} f_U(\lambda)\, d\lambda$. Let $(\boldsymbol{Z}'\boldsymbol{Z})^{-1} = \sum_{j=1}^{p} \boldsymbol{\alpha}_j \boldsymbol{\alpha}_j'$ and $x_{tj} = a\boldsymbol{\alpha}_j' z_t + b\boldsymbol{\alpha}_j' z_{t-h}$ for arbitrary a and b, $t = 1, \ldots, T$, with $z_0 = \cdots = z_{-h+1} = \mathbf{0}$. Then $\boldsymbol{x}_j' \boldsymbol{\Sigma}_L \boldsymbol{x}_j \leq \boldsymbol{x}_j' \boldsymbol{\Sigma} \boldsymbol{x}_j \leq \boldsymbol{x}_j' \boldsymbol{\Sigma}_U \boldsymbol{x}_j$ as in Section 10.2.3. The corresponding argument is carried out for

$$(23) \quad \sum_{j=1}^{p} \sum_{t,s=1}^{T} x_{tj} \sigma(t-s) x_{sj}$$
$$= \operatorname{tr} (\boldsymbol{Z}'\boldsymbol{Z})^{-1} \sum_{t,s=1}^{T} [a^2 z_t z_s' + ab(z_t z'_{s-h} + z_{t-h} z_s') + b^2 z_{t-h} z'_{s-h}] \sigma(t-s).$$

Since the limit holds for arbitrary a and b, it holds for the a^2 term, the ab term, and the b^2 term. Then for $f(\lambda)$ continuous the limit of the sum of (19) and (20) is the sum of (21) and (22). Q.E.D.

Now suppose that the least squares estimates are asymptotically efficient. Then $M(\lambda)$ has jumps $\frac{1}{2}M_j - i\frac{1}{2}M_j^*$ at $\lambda = \nu_j$ and $\frac{1}{2}M_j + i\frac{1}{2}M_j^*$ at $\lambda = -\nu_j$, $j = 1, \ldots, H$, if there is no jump at $\lambda = 0, \pm\pi$. (The M_j's and M_j^*'s are real.) The first term on the right-hand side of (16) is

$$-4\pi \operatorname{tr} \mathbf{R}^{-1}(0) \sum_{j=1}^{H} \cos \nu_j h\, f(\nu_j) M_j; \tag{24}$$

the second term is

$$\begin{aligned}2\pi \operatorname{tr} \mathbf{R}^{-1}(0) \sum_{j=1}^{H} f(\nu_j) M_j \mathbf{R}^{-1}(0) \sum_{k=1}^{H} (\cos \nu_k h\, M_k + \sin \nu_k h\, M_k^*) \\ = 2\pi \operatorname{tr} \mathbf{R}^{-1}(0) \sum_{j=1}^{H} \cos \nu_j h\, f(\nu_j) M_j\end{aligned} \tag{25}$$

since

$$\operatorname{tr} \mathbf{R}^{-1}(0) M_j \mathbf{R}^{-1}(0) M_i^* = 0 \tag{26}$$

by virtue of $M_j = M_j'$ and $M_i^* = -M_i^{*\prime}$. (If $\nu_j = \pi$, $M_j^* = \mathbf{0}$)

THEOREM 10.3.2. *Under Assumptions* 10.2.1 *to* 10.2.5, *if* $M(\lambda)$ *has jumps* $\frac{1}{2}M_j - i\frac{1}{2}M_j^*$ *at* $\lambda = \nu_j > 0, j = 1, \ldots, H,$

$$\lim_{T\to\infty} T[\mathscr{E}C_h^* - \sigma(h)] = -2\pi \operatorname{tr} \mathbf{R}^{-1}(0) \sum_{j=1}^{H} \cos \nu_j h\, f(\nu_j) M_j; \tag{27}$$

if $M(\lambda)$ *also has the jump of* M_0 *at* $\lambda = 0$,

$$\lim_{T\to\infty} T[\mathscr{E}C_h^* - \sigma(h)] = -2\pi \operatorname{tr} \mathbf{R}^{-1}(0) \sum_{j=0}^{H} \cos \nu_j h\, f(\nu_j) M_j. \tag{28}$$

Theorem 10.3.2 is a generalization of Theorem 8.3.2. The import of Theorems 10.3.1 and 10.3.2 is that the bias in C_h^* is of the order $1/T$. It should be noted that the asymptotic bias can be estimated consistently.

We shall now show that the limiting distribution of $\sqrt{T}\,[C_h^* - \sigma(h)]$ is the same as the limiting distribution of $\sqrt{T}\,[C_h - \sigma(h)]$ as developed in Section 8.4.2 under suitable conditions.

THEOREM 10.3.3. *Under Assumptions* 10.2.1 *to* 10.2.5

$$\operatorname*{plim}_{T\to\infty} \sqrt{T}\,(C_h^* - C_h) = 0, \tag{29}$$

where $C_h = \sum_{t=1}^{T-h} (y_t - \boldsymbol{\beta}' z_t)(y_{t+h} - \boldsymbol{\beta}' z_{t+h})/(T - h)$.

PROOF. We have

$$\text{(30)} \qquad \sqrt{T}\,(C_h^* - C_h) = -\frac{\sqrt{T}}{T-h}\sum_{t=1}^{T-h}(\boldsymbol{b}^* - \boldsymbol{\beta})'\boldsymbol{z}_t u_{t+h} - \frac{\sqrt{T}}{T-h}\sum_{t=1}^{T-h}(\boldsymbol{b}^* - \boldsymbol{\beta})'\boldsymbol{z}_{t+h} u_t + \frac{\sqrt{T}}{T-h}\sum_{t=1}^{T-h}(\boldsymbol{b}^* - \boldsymbol{\beta})'\boldsymbol{z}_t \boldsymbol{z}_{t+h}'(\boldsymbol{b}^* - \boldsymbol{\beta}),$$

where $u_t = y_t - \boldsymbol{\beta}'\boldsymbol{z}_t$. The last term is

$$\text{(31)} \qquad \frac{\sqrt{T}}{T-h}(\boldsymbol{b}^* - \boldsymbol{\beta})'\boldsymbol{A}_T(h)(\boldsymbol{b}^* - \boldsymbol{\beta}) = \frac{\sqrt{T}}{T-h}(\boldsymbol{b}^* - \boldsymbol{\beta})'\boldsymbol{D}_T[\boldsymbol{D}_T^{-1}\boldsymbol{A}_T(h)\boldsymbol{D}_T^{-1}]\boldsymbol{D}_T(\boldsymbol{b}^* - \boldsymbol{\beta}).$$

Since $\lim_{T\to\infty} \boldsymbol{D}_T^{-1}\boldsymbol{A}_T(h)\boldsymbol{D}_T^{-1} = \mathbf{R}(h)$ and $\mathscr{E}\boldsymbol{D}_T(\boldsymbol{b}^* - \boldsymbol{\beta})(\boldsymbol{b}^* - \boldsymbol{\beta})'\boldsymbol{D}_T$ has a limit as $T \to \infty$, (31) converges stochastically to 0. The square of the second term is

$$\text{(32)} \qquad \frac{T}{(T-h)^2}(\boldsymbol{b}^* - \boldsymbol{\beta})'\boldsymbol{D}_T\left(\boldsymbol{D}_T^{-1}\sum_{t,s=1}^{T-h}\boldsymbol{z}_{t+h}u_t u_s \boldsymbol{z}_{s+h}'\boldsymbol{D}_T^{-1}\right)\boldsymbol{D}_T(\boldsymbol{b}^* - \boldsymbol{\beta}).$$

For any vector $\boldsymbol{\alpha}$

$$\text{(33)} \qquad \mathscr{E}\left(\boldsymbol{\alpha}'\boldsymbol{D}_T^{-1}\sum_{t=1}^{T-h}\boldsymbol{z}_{t+h}u_t\right)^2 = \mathscr{E}\boldsymbol{\alpha}'\boldsymbol{D}_T^{-1}\sum_{t,s=1}^{T-h}\boldsymbol{z}_{t+h}u_t u_s \boldsymbol{z}_{s+h}'\boldsymbol{D}_T^{-1}\boldsymbol{\alpha}$$
$$= \boldsymbol{\alpha}'\boldsymbol{D}_T^{-1}\sum_{t,s=1}^{T-h}\boldsymbol{z}_{t+h}\boldsymbol{z}_{s+h}'\sigma(t-s)\boldsymbol{D}_T^{-1}\boldsymbol{\alpha}$$
$$= \boldsymbol{\alpha}'\boldsymbol{D}_T^{-1}\sum_{r=-(T-h-1)}^{T-h-1}\sum_{s\in S_r}\boldsymbol{z}_{s+h+r}\boldsymbol{z}_{s+h}'\sigma(r)\boldsymbol{D}_T^{-1}\boldsymbol{\alpha}$$

has the limit $\boldsymbol{\alpha}'\sum_{r=-\infty}^{\infty}\sigma(r)\mathbf{R}(r)\boldsymbol{\alpha}$ under the conditions of the theorem. Hence the second term converges stochastically to 0. The first term is evaluated similarly. Q.E.D.

THEOREM 10.3.4. *Under Assumptions* 10.2.1 *to* 10.2.5 *and the conditions of Theorem* 8.4.2 $\sqrt{T}\,[C_0^* - \sigma(0)], \ldots, \sqrt{T}\,[C_n^* - \sigma(n)]$ *have a limiting normal distribution with means* 0 *and covariances given in Theorem* 8.4.2.

Under suitable conditions on the fourth-order moments, it can be shown that the limits of the covariances of $\sqrt{T}\,[C_h^* - \sigma(h)]$ are the same as those of $\sqrt{T}\,[C_h - \sigma(h)]$ (without all of the conditions for asymptotic normality).

10.3.3. The Asymptotic Bias of the Sample Spectral Density

We consider the sample spectral density based on residuals from the least squares regression

$$I^*(\lambda) = \frac{1}{2\pi T}\left|\sum_{t=1}^{T}(y_t - \boldsymbol{b}^{*\prime}\boldsymbol{z}_t)e^{i\lambda t}\right|^2. \tag{34}$$

THEOREM 10.3.5. *Under Assumptions* 10.2.1 *to* 10.2.5, *and the conditions that*

$$\lim_{T\to\infty}\frac{1}{\sqrt{T}}\boldsymbol{D}_T^{-1}\sum_{t=1}^{T}\boldsymbol{z}_t e^{i\lambda t} = \boldsymbol{g}(\lambda) \tag{35}$$

exists for every λ *and* $\sum_{h=-\infty}^{\infty}|\sigma(h)| < \infty$,

$$\begin{aligned}\lim_{T\to\infty}\mathscr{E}\, I^*(\lambda) \\ = f(\lambda) - 2f(\lambda)\boldsymbol{g}'(\lambda)\mathbf{R}^{-1}(0)\overline{\boldsymbol{g}(\lambda)} + \boldsymbol{g}'(\lambda)\mathbf{R}^{-1}(0)\int_{-\pi}^{\pi} f(\nu)\, d\boldsymbol{M}(\nu)\mathbf{R}^{-1}(0)\overline{\boldsymbol{g}(\lambda)}.\end{aligned} \tag{36}$$

PROOF.

$$\begin{aligned} I^*(\lambda) &= \frac{1}{2\pi}\sum_{h=-(T-1)}^{T-1}\left(1 - \frac{|h|}{T}\right)C_h^* \cos\lambda h \\ &= \frac{1}{2\pi T}\sum_{s,t=1}^{T}\hat{u}_s\hat{u}_t e^{i\lambda(t-s)} \\ &= \frac{1}{2\pi T}\left|\sum_{t=1}^{T}\hat{u}_t e^{i\lambda t}\right|^2, \end{aligned} \tag{37}$$

where $\hat{\boldsymbol{u}} = (\hat{u}_1, \ldots, \hat{u}_T)'$ is

$$\hat{\boldsymbol{u}} = [\boldsymbol{I} - \boldsymbol{Z}(\boldsymbol{Z}'\boldsymbol{Z})^{-1}\boldsymbol{Z}']\boldsymbol{u}. \tag{38}$$

Let $\boldsymbol{\theta} = (e^{i\lambda}, e^{i\lambda 2}, \ldots, e^{i\lambda T})'$. Then

$$\begin{aligned} 2\pi I^*(\lambda) &= \frac{1}{T}|\hat{\boldsymbol{u}}'\boldsymbol{\theta}|^2 = \frac{1}{T}\boldsymbol{\theta}'\hat{\boldsymbol{u}}\hat{\boldsymbol{u}}'\overline{\boldsymbol{\theta}} \\ &= \frac{1}{T}\boldsymbol{\theta}'[\boldsymbol{I} - \boldsymbol{Z}(\boldsymbol{Z}'\boldsymbol{Z})^{-1}\boldsymbol{Z}']\boldsymbol{u}\boldsymbol{u}'[\boldsymbol{I} - \boldsymbol{Z}(\boldsymbol{Z}'\boldsymbol{Z})^{-1}\boldsymbol{Z}']\overline{\boldsymbol{\theta}} \\ &= \frac{1}{T}\boldsymbol{\theta}'\boldsymbol{u}\boldsymbol{u}'\overline{\boldsymbol{\theta}} - \frac{1}{T}[\boldsymbol{\theta}'\boldsymbol{Z}(\boldsymbol{Z}'\boldsymbol{Z})^{-1}\boldsymbol{Z}'\boldsymbol{u}\boldsymbol{u}'\overline{\boldsymbol{\theta}} + \boldsymbol{\theta}'\boldsymbol{u}\boldsymbol{u}'\boldsymbol{Z}(\boldsymbol{Z}'\boldsymbol{Z})^{-1}\boldsymbol{Z}'\overline{\boldsymbol{\theta}}] \\ &\quad + \frac{1}{T}\boldsymbol{\theta}'\boldsymbol{Z}(\boldsymbol{Z}'\boldsymbol{Z})^{-1}\boldsymbol{Z}'\boldsymbol{u}\boldsymbol{u}'\boldsymbol{Z}(\boldsymbol{Z}'\boldsymbol{Z})^{-1}\boldsymbol{Z}'\overline{\boldsymbol{\theta}}. \end{aligned} \tag{39}$$

The expected value of the first term on the right-hand side of (39) is

$$2\pi\mathscr{E}I(\lambda) = \sum_{r=-(T-1)}^{T-1}\left(1-\frac{|r|}{T}\right)\sigma(r)\cos\lambda r, \tag{40}$$

which converges to $2\pi f(\lambda)$. The expected value of the last term is

$$\begin{aligned}
&\frac{1}{T}\boldsymbol{\theta}'Z(Z'Z)^{-1}Z'\boldsymbol{\Sigma}Z(Z'Z)^{-1}Z'\bar{\boldsymbol{\theta}} \\
&\qquad = \frac{1}{T}\boldsymbol{\theta}'Z\mathscr{E}(\boldsymbol{b}^*-\boldsymbol{\beta})(\boldsymbol{b}^*-\boldsymbol{\beta})'Z'\bar{\boldsymbol{\theta}} \\
&\qquad = \frac{1}{T}\boldsymbol{\theta}'Z\boldsymbol{D}_T^{-1}[\boldsymbol{D}_T\mathscr{E}(\boldsymbol{b}^*-\boldsymbol{\beta})(\boldsymbol{b}^*-\boldsymbol{\beta})'\boldsymbol{D}_T]\boldsymbol{D}_T^{-1}Z'\bar{\boldsymbol{\theta}} \\
&\qquad = \operatorname{tr}\,[\boldsymbol{D}_T\mathscr{E}(\boldsymbol{b}^*-\boldsymbol{\beta})(\boldsymbol{b}^*-\boldsymbol{\beta})'\boldsymbol{D}_T]\,\frac{1}{T}\boldsymbol{D}_T^{-1}Z'\bar{\boldsymbol{\theta}}\boldsymbol{\theta}'Z\boldsymbol{D}_T^{-1}.
\end{aligned} \tag{41}$$

The limit of the matrix in brackets is given in Corollary 10.2.1. The other matrix in (41) is $\overline{\boldsymbol{g}_T(\lambda)}\boldsymbol{g}_T'(\lambda)$, where

$$\boldsymbol{g}_T(\lambda) = \frac{1}{\sqrt{T}}\boldsymbol{D}_T^{-1}\sum_{t=1}^{T} z_t e^{i\lambda t}, \tag{42}$$

which converges to $\boldsymbol{g}(\lambda)$. Thus the limit of the expected value of the last term of the right-hand side of (39) is

$$2\pi\boldsymbol{g}'(\lambda)\mathbf{R}^{-1}(0)\int_{-\pi}^{\pi} f(\nu)\,d\boldsymbol{M}(\nu)\mathbf{R}^{-1}(0)\overline{\boldsymbol{g}(\lambda)}. \tag{43}$$

The expected value of the middle term is

$$\begin{aligned}
&-\frac{1}{T}[\boldsymbol{\theta}'Z(Z'Z)^{-1}Z'\boldsymbol{\Sigma}\bar{\boldsymbol{\theta}} + \boldsymbol{\theta}'\boldsymbol{\Sigma}Z(Z'Z)^{-1}Z'\bar{\boldsymbol{\theta}}] \\
&= -\frac{1}{T}\sum_{t,s,r=1}^{T} z_t'[\boldsymbol{A}_T(0)]^{-1}z_s\sigma(s-r)[e^{i\lambda(t-r)}+e^{i\lambda(r-t)}] \\
&= -\sum_{h=-(T-1)}^{T-1}\sigma(h)\sum_{r\in S_h}\sum_{t=1}^{T} z_t'\boldsymbol{D}_T^{-1}[\boldsymbol{D}_T^{-1}\boldsymbol{A}_T(0)\boldsymbol{D}_T^{-1}]^{-1}\boldsymbol{D}_T^{-1}z_{r+h}\frac{1}{T}[e^{i\lambda(t-r)}+e^{i\lambda(r-t)}],
\end{aligned} \tag{44}$$

where $S_h = \{1,\ldots,T-h\}$, $h \geq 0$, and $S_h = \{1-h,\ldots,T\}$, $h \leq 0$. This converges to

$$-2\sum_{h=-\infty}^{\infty}\sigma(h)\cos\lambda h\,\boldsymbol{g}'(\lambda)\mathbf{R}^{-1}(0)\overline{\boldsymbol{g}(\lambda)} = -4\pi f(\lambda)\boldsymbol{g}'(\lambda)\mathbf{R}^{-1}(0)\overline{\boldsymbol{g}(\lambda)}. \quad \text{Q.E.D.} \tag{45}$$

We observe that

$$\int_{-\pi}^{\pi} g_T(\lambda)\overline{g_T'(\lambda)}\, d\lambda = \frac{1}{T}\int_{-\pi}^{\pi} D_T^{-1}\sum_{t,s=1}^{T} z_t z_s' D_T^{-1} e^{i\lambda(t-s)}\, d\lambda \tag{46}$$

$$= \frac{2\pi}{T} D_T^{-1} A_T(0) D_T^{-1}.$$

The integral of the square of the absolute value of any component of $g_T(\lambda)$ is $2\pi/T$, which goes to 0 as $T \to \infty$. Thus $g(\lambda) = \mathbf{0}$ for almost all λ. If $\{z_t\}$ behaves roughly like a realization from a stationary stochastic process, then $g(\lambda) = \mathbf{0}$ for all λ; each component of $g(\lambda)$ corresponds to $A(\lambda) + iB(\lambda)$, and a diagonal term of $g(\lambda)\overline{g'(\lambda)}$ corresponds to $R^2(\lambda)$. In such a case $I^*(\lambda)$ is asymptotically unbiased.

COROLLARY 10.3.1. *If* $g(\lambda) = \mathbf{0}$, $\lim_{T\to\infty} \mathscr{E} I^*(\lambda) = f(\lambda)$.

Now let us consider a sequence of independent variables $\{z_t\}$ such that the least squares estimates are asymptotically efficient for every continuous, positive spectral density. Then Theorem 10.2.10 implies that

$$\lim_{T\to\infty} T\int_{-\pi}^{\pi} g_T(\lambda)\overline{g_T'(\lambda)} f(\lambda)\, d\lambda = \sum_{j=1}^{G} f(\nu_j) M_j \tag{47}$$

for every continuous, positive spectral density $f(\lambda)$. There is a matrix P (not necessarily uniquely determined) so $P'\mathbf{R}(0)P = I$ and $P'\sum_{j=1}^{G} f(\nu_j) M_j P$ is diagonal for every set of values $f(\nu_j)$, $j = 1, \ldots, G$. Then $P'M_jP = L_j$ is diagonal and the diagonal elements of L_j are 1 and 0 and $\sum_{j=1}^{G} L_j = I$. Let $g_T^*(\lambda) = P' g_T(\lambda)$. (See Problem 32 of Chapter 6.) Then

$$\lim_{T\to\infty} T\int_{-\pi}^{\pi} g_T^*(\lambda)\overline{g_T^{*\prime}(\lambda)} f(\lambda)\, d\lambda = \sum_{j=1}^{G} f(\nu_j) L_j, \tag{48}$$

where

$$L_j = \begin{bmatrix} \mathbf{0} & \cdots & \mathbf{0} & \mathbf{0} & \mathbf{0} & \cdots & \mathbf{0} \\ \vdots & & \vdots & \vdots & \vdots & & \vdots \\ \mathbf{0} & \cdots & \mathbf{0} & \mathbf{0} & \mathbf{0} & \cdots & \mathbf{0} \\ \mathbf{0} & \cdots & \mathbf{0} & I & \mathbf{0} & \cdots & \mathbf{0} \\ \mathbf{0} & \cdots & \mathbf{0} & \mathbf{0} & \mathbf{0} & \cdots & \mathbf{0} \\ \vdots & & \vdots & \vdots & \vdots & & \vdots \\ \mathbf{0} & \cdots & \mathbf{0} & \mathbf{0} & \mathbf{0} & \cdots & \mathbf{0} \end{bmatrix}, \tag{49}$$

the identity being the jth diagonal submatrix. It therefore seems reasonable to assume

$$\lim_{T\to\infty} \mathbf{g}_T^*(\nu_j) = \begin{bmatrix} \mathbf{0} \\ \cdot \\ \cdot \\ \cdot \\ \mathbf{0} \\ \mathbf{g}_{(j)}^* \\ \mathbf{0} \\ \cdot \\ \cdot \\ \cdot \\ \mathbf{0} \end{bmatrix}, \qquad j = 1, \ldots, G, \tag{50}$$

$$\lim_{T\to\infty} \mathbf{g}_T^*(\lambda) = \mathbf{0}, \qquad \lambda \neq \pm\nu_1, \ldots, \pm\nu_G. \tag{51}$$

Under these conditions $\lim_{T\to\infty} \mathscr{E}I^*(\lambda) = f(\lambda)$, $\lambda \neq \pm\nu_1, \ldots, \pm\nu_G$. Then

$$\begin{aligned} \mathbf{g}'(\nu_j)\mathbf{R}^{-1}(0) \sum_{k=1}^{G} f(\nu_k)\mathbf{M}_k\mathbf{R}^{-1}(0)\overline{\mathbf{g}(\nu_j)} &= f(\nu_j)\mathbf{g}_{(j)}^{*\prime}\overline{\mathbf{g}_{(j)}^*} \\ &= f(\nu_j)\mathbf{g}'(\nu_j)\mathbf{R}^{-1}(0)\overline{\mathbf{g}(\nu_j)}. \end{aligned} \tag{52}$$

Thus we have the following theorem.

THEOREM 10.3.6. *If Assumptions* 10.2.1 *to* 10.2.5 *hold, if* (35) *exists for every* λ, *if* $\sum_{h=-\infty}^{\infty} |\sigma(h)| < \infty$, *if for* $G \leq p$

$$\mathbf{R}(h) = \sum_{j=1}^{G} (\cos \nu_j h\, \mathbf{M}_j + \sin \nu_j h\, \mathbf{M}_j^*), \tag{53}$$

if $\mathbf{g}(\lambda) = \mathbf{0}$, $\lambda \neq \pm\nu_1, \ldots, \pm\nu_G$, *and if* $\mathbf{M}_k\mathbf{R}^{-1}(0)\mathbf{g}(\nu_j) = \mathbf{0}$, $k \neq j$, $k, j = 1, \ldots, G$, *then*

$$\lim_{T\to\infty} \mathscr{E}I^*(\lambda) = f(\lambda), \qquad \lambda \neq \pm\nu_1, \ldots, \pm\nu_G, \tag{54}$$

$$\lim_{T\to\infty} \mathscr{E}I^*(\pm\nu_j) = f(\nu_j)[1 - \mathbf{g}'(\nu_j)\mathbf{R}^{-1}(0)\overline{\mathbf{g}(\nu_j)}], \quad j = 1, \ldots, G. \tag{55}$$

The sample spectral density $I^*(\lambda)$ may be asymptotically biased at the values of λ for which $M(\lambda)$ has jumps, that is, corresponding to the frequencies of the periodicities in $\{z_t\}$. If the factor of $f(\nu_j)$ in (55), namely, $1 - \mathbf{g}'(\nu_j)\mathbf{R}^{-1}(0)\overline{\mathbf{g}(\nu_j)}$, is known and if it is different from 0, one can estimate $f(\nu_j)$ by $I^*(\nu_j)$ divided by the factor.

Now let us consider some examples. Suppose $z_{1t} \equiv 1$ (that is, $\nu_1 = 0$),

$$z_{2k-2,t} = \cos \nu_k t, \qquad k = 2, \ldots, G-1, \tag{56}$$

$$z_{2k-1,t} = \sin \nu_k t, \qquad k = 2, \ldots, G-1, \tag{57}$$

and $z_{2G-2,t} = (-1)^t$, $t = 1, 2, \ldots$. Then

$$\mathbf{R}(h) = \begin{bmatrix} 1 & 0 & 0 & 0 & \cdots & 0 \\ 0 & \cos \nu_1 h & -\sin \nu_1 h & 0 & \cdots & 0 \\ 0 & \sin \nu_1 h & \cos \nu_1 h & 0 & \cdots & 0 \\ 0 & 0 & 0 & \cos \nu_2 h & \cdots & 0 \\ \cdot & \cdot & \cdot & \cdot & & \cdot \\ \cdot & \cdot & \cdot & \cdot & & \cdot \\ \cdot & \cdot & \cdot & \cdot & & \cdot \\ 0 & 0 & 0 & 0 & \cdots & (-1)^h \end{bmatrix}. \tag{58}$$

From Problem 13 we can verify that $\mathbf{g}(\lambda) = \mathbf{0}$ for $\lambda \neq \pm\nu_1, \ldots, \pm\nu_G$, and $\mathbf{g}(0)$ has 1 as its first component, $\mathbf{g}(\nu_k)$ has $1/\sqrt{2}$ as its $(2k-2)$nd component and $i/\sqrt{2}$ as its $(2k-1)$st component, $k = 2, \ldots, G-1$, and $\mathbf{g}(\pi)$ has 1 as its last component, and the other components of $\mathbf{g}(\nu_j)$ are 0, $j = 1, \ldots, G$. Then $\lim_{T\to\infty} \mathscr{E} I^*(\nu_j) = 0$, $j = 1, \ldots, G$.

If either $\cos \nu_k t$ or $\sin \nu_k t$ is omitted from z_t, then $\lim_{T\to\infty} \mathscr{E} I^*(\nu_k) = \frac{1}{2} f(\nu_k)$.

As another example, let $z_t = t$. Then

$$g_T(\lambda) = \frac{\sum_{t=1}^{T} te^{i\lambda t}}{\sqrt{T}\sqrt{T(T+1)(2T+1)/6}} \tag{59}$$

$$= -\frac{1 - (T+1)e^{i\lambda T} + Te^{i\lambda(T+1)}}{4\,T \sin^2 \frac{1}{2}\lambda \sqrt{(T+1)(2T+1)/6}}, \qquad \lambda \neq 0,$$

$$g_T(0) = \frac{T+1}{2\sqrt{(T+1)(2T+1)/6}}. \tag{60}$$

Then $g(\lambda) = 0$, $\lambda \neq 0$, and $g(0) = \sqrt{3}/2$. Hence $\lim_{T\to\infty} \mathscr{E} I^*(\lambda) = f(\lambda)$, $\lambda \neq 0$, and $\lim_{T\to\infty} \mathscr{E} I^*(0) = f(0)/4$.

10.3.4. The Asymptotic Bias of the Estimates of the Spectral Density

Let us consider

$$\hat{f}_T^*(\nu) = \frac{1}{2\pi} \sum_{r=-K_T}^{K_T} k\left(\frac{r}{K_T}\right) \cos \nu r \, c_{rT}^*, \tag{61}$$

where

$$c_{rT}^* = \frac{1}{T}\sum_{t=1}^{T-r}(y_t - b^{*\prime}z_t)(y_{t+r} - b^{*\prime}z_{t+r}) \tag{62}$$
$$= c_{-rT}^*, \qquad r = 0, 1, \ldots, T-1,$$

and $k(x)$ is a symmetric function and $\{K_T\}$ is an increasing sequence of integers of the types treated in Chapter 9. Then

$$\hat{f}_T^*(\nu) - f(\nu) = \hat{f}_T(\nu) - f(\nu) + \hat{f}_T^*(\nu) - \hat{f}_T(\nu), \tag{63}$$

where

$$\hat{f}_T(\nu) = \frac{1}{2\pi}\sum_{r=-K_T}^{K_T} k\left(\frac{r}{K_T}\right)\cos \nu r\, c_{rT}, \tag{64}$$

$$c_{rT} = \frac{1}{T}\sum_{t=1}^{T-r}(y_t - \boldsymbol{\beta}'z_t)(y_{t+r} - \boldsymbol{\beta}'z_{t+r}) = \frac{1}{T}\sum_{t=1}^{T-r} u_t u_{t+r} \tag{65}$$
$$= c_{-rT}, \qquad r = 0, 1, \ldots, T-1.$$

The term $\hat{f}_T(\nu) - f(\nu)$ was studied in Chapter 9. Now we consider

$$\hat{f}_T^*(\nu) - \hat{f}_T(\nu) = \frac{1}{2\pi}\sum_{r=-K_T}^{K_T} k\left(\frac{r}{K_T}\right)\cos \nu r\,(c_{rT}^* - c_{rT}), \tag{66}$$

where for $r \geq 0$

$$c_{rT}^* - c_{rT} = -\frac{1}{T}(b^* - \boldsymbol{\beta})'\sum_{t=1}^{T-r} z_t u_{t+r} - \frac{1}{T}(b^* - \boldsymbol{\beta})'\sum_{t=1}^{T-r} z_{t+r}u_t \tag{67}$$
$$+ \frac{1}{T}(b^* - \boldsymbol{\beta})'\sum_{t=1}^{T-r} z_t z_{t+r}'(b^* - \boldsymbol{\beta}).$$

The last term in (67) multiplied by T is in absolute value

$$\left|\sum_{t=1}^{T-r}(b^* - \boldsymbol{\beta})'z_t z_{t+r}'(b^* - \boldsymbol{\beta})\right| \leq \sqrt{\sum_{t=1}^{T-r}[(b^* - \boldsymbol{\beta})'z_t]^2 \sum_{t=1}^{T-r}[(b^* - \boldsymbol{\beta})'z_{t+r}]^2} \tag{68}$$
$$\leq \sum_{t=1}^{T}[(b^* - \boldsymbol{\beta})'z_t]^2$$
$$= (b^* - \boldsymbol{\beta})'A_T(0)(b^* - \boldsymbol{\beta})$$
$$= \sum_{t,s=1}^{T} z_t' A_T^{-1}(0) z_s u_t u_s.$$

We have used the Cauchy-Schwarz inequality. Let $A_T^{-1}(0) = \boldsymbol{P}'\boldsymbol{P}$ and $z_s^* = \boldsymbol{P}z_s$, $s = 1, \ldots, T$. Then the expected value of the right-hand side of (68) is

$$\mathscr{E} \sum_{t,s=1}^{T} z_t' A_T^{-1}(0) z_s u_t u_s = \sum_{t,s=1}^{T} z_t^{*\prime} z_s^* \sigma(t-s) = \sum_{q=-(T-1)}^{T-1} \sum_{t \in S_q} z_t^{*\prime} z_{t+q}^* \sigma(q), \tag{69}$$

where $S_q = \{1, \ldots, T-q\}$ for $q \geq 0$ and $S_q = \{1-q, \ldots, T\}$ for $q \leq 0$. For any $h \geq 0$

$$\left| \sum_{t=1}^{T-h} z_t^{*\prime} z_{t+h}^* \right| \leq \sum_{i=1}^{p} \left| \sum_{t=1}^{T-h} z_{it}^* z_{i,t+h}^* \right| \leq \sum_{i=1}^{p} \sqrt{\sum_{t=1}^{T-h} z_{it}^{*2}} \sqrt{\sum_{t=1}^{T-h} z_{i,t+h}^{*2}} \leq \sum_{i=1}^{p} \sum_{t=1}^{T} z_{it}^{*2} = p. \tag{70}$$

The expected value of the absolute value of the last term in (67) multiplied by T is bounded by $p \sum_{q=-\infty}^{\infty} |\sigma(q)|$, which we assume to be finite.

The first term on the right-hand side of (67) multiplied by T has expected absolute value

$$\begin{aligned} \mathscr{E} \left| (\boldsymbol{b}^* - \boldsymbol{\beta})' \sum_{t=1}^{T-r} z_t u_{t+r} \right| &= \mathscr{E} \left| \sum_{s=1}^{T} \sum_{t=1}^{T-r} z_s' A_T^{-1}(0) z_t u_s u_{t+r} \right| \\ &= \mathscr{E} \left| \sum_{s=1}^{T} \sum_{t=1}^{T-r} z_s^{*\prime} z_t^* u_s u_{t+r} \right| \\ &= \mathscr{E} \left| \left(\sum_{s=1}^{T} u_s z_s^* \right)' \left(\sum_{t=1}^{T-r} u_{t+r} z_t^* \right) \right| \\ &\leq \sqrt{\mathscr{E} \left(\sum_{s=1}^{T} u_s z_s^* \right)' \left(\sum_{s'=1}^{T} u_{s'} z_{s'}^* \right)} \\ &\quad \times \sqrt{\mathscr{E} \left(\sum_{t=1}^{T-r} u_{t+r} z_t^* \right)' \left(\sum_{t'=1}^{T-r} u_{t'+r} z_{t'}^* \right)} \end{aligned} \tag{71}$$

since $\mathscr{E}\,|\boldsymbol{x}'\boldsymbol{y}| \le \mathscr{E}\sum_{i=1}^{p}|x_i|\,|y_i| \le \sqrt{\mathscr{E}\sum_{i=1}^{p}x_i^2}\sqrt{\mathscr{E}\sum_{i=1}^{p}y_i^2}$. The square of (71) is not greater than

$$(72)\quad \mathscr{E}\sum_{s,s'=1}^{T} z_s^{*\prime} z_{s'}^{*} u_s u_{s'}\, \mathscr{E}\sum_{t,t'=1}^{T-r} z_t^{*\prime} z_{t'}^{*} u_{t+r} u_{t'+r}$$

$$= \sum_{s,s'=1}^{T} z_s' A_T^{-1}(0) z_{s'} \sigma(s'-s) \sum_{t,t'=1}^{T-r} z_t' A_T^{-1}(0) z_{t'} \sigma(t'-t)$$

$$\le \left[p \sum_{q=-\infty}^{\infty} |\sigma(q)|\right]^2.$$

Thus the absolute value of each of the first and second terms on the right-hand side of (67) multiplied by T has an expected value bounded by $p\sum_{q=-\infty}^{\infty}|\sigma(q)|$.

THEOREM 10.3.7

$$(73)\qquad |\mathscr{E}(c_{rT}^{*} - c_{rT})| \le \mathscr{E}\,|c_{rT}^{*} - c_{rT}| \le \frac{3p}{T}\sum_{q=-\infty}^{\infty}|\sigma(q)|.$$

From (66) we have

$$(74)\qquad \mathscr{E}|\hat{f}_T^{*}(\nu) - \hat{f}_T(\nu)| \le \frac{1}{2\pi}\sum_{r=-K_T}^{K_T}\left|k\left(\frac{r}{K_T}\right)\right| \cdot \mathscr{E}|c_{rT}^{*} - c_{rT}|.$$

THEOREM 10.3.8

$$(75)\quad |\mathscr{E}\hat{f}_T^{*}(\nu) - \mathscr{E}\hat{f}_T(\nu)| \le \mathscr{E}\,|\hat{f}_T^{*}(\nu) - \hat{f}_T(\nu)|$$

$$\le \frac{3p}{2\pi T}(2K_T + 1)\sup_{-1\le x\le 1}|k(x)|\sum_{q=-\infty}^{\infty}|\sigma(q)|.$$

THEOREM 10.3.9.

$$(76)\qquad \lim_{T\to\infty}\frac{T}{K_T}|\mathscr{E}\hat{f}_T^{*}(\nu) - \mathscr{E}\hat{f}_T(\nu)| \le \frac{3p}{\pi}\sup_{-1\le x\le 1}|k(x)|\sum_{q=-\infty}^{\infty}|\sigma(q)|.$$

THEOREM 10.3.10. *Suppose* $k(x) = k(-x)$, $k(0) = 1$, $|k(x)| \le M$ *for some* M *and all* $|x| \le 1$; *suppose for some* $q > 0$ *and* $k > 0$, $\lim_{x\to 0}\,[1 - k(x)]/|x|^q = k$. *Let* $\{K_T\}$ *be a sequence of integers such that* $K_T \to \infty$ *as* $T \to \infty$. *Suppose*

$$(77)\qquad \sum_{r=-\infty}^{\infty}|r|^p\,|\sigma(r)| < \infty$$

for some $p \geq q$. *If* $K_T^{q+1}/T \to 0$ *as* $T \to \infty$, *then*

$$\lim_{T\to\infty} \frac{\mathscr{E}\hat{f}_T^*(\nu) - \mathscr{E}\hat{f}_T(\nu)}{\mathscr{E}\hat{f}_T(\nu) - f(\nu)} = 0. \tag{78}$$

PROOF. The theorem follows from Theorems 10.3.9 and 9.3.3.

The implication of the theorem is that if K_T grows slowly relative to T, more precisely if $K_T^{q+1}/T \to 0$, as $T \to \infty$, then the contribution to the bias by the regression is smaller than the bias of the estimate when the mean is known. If $q = 2$ and $p \geq 2$, then the condition is $K_T^3/T \to 0$; if further $K_T = [\gamma T^\alpha]$, the condition is $\alpha < 1/3$.

We also wish to show that the asymptotic distribution of $\hat{f}_T^*(\nu) - f(\nu)$ is the same as the asymptotic distribution of $\hat{f}_T(\nu) - f(\nu)$. Theorem 10.3.8 and the general Tchebycheff inequality imply the following theorem.

THEOREM 10.3.11. *Let* $g(T)$ *be a function of* T *such that* $g(T) \to 0$ *as* $T \to \infty$. *If* $\sum_{r=-\infty}^{\infty} |\sigma(r)| < \infty$ *and* $k(x)$ *is bounded,* $-1 \leq x \leq 1$,

$$\operatorname*{plim}_{T\to\infty} g(T) \frac{T}{K_T} [\hat{f}_T^*(\nu) - \hat{f}_T(\nu)] = 0. \tag{79}$$

COROLLARY 10.3.2. *If* $\sum_{r=-\infty}^{\infty} |\sigma(r)| < \infty$ *and* $k(x)$ *is bounded,* $-1 \leq x \leq 1$,

$$\operatorname*{plim}_{T\to\infty} \sqrt{\frac{T}{K_T}} [\hat{f}_T^*(\nu) - \hat{f}_T(\nu)] = 0. \tag{80}$$

From Corollary 9.4.1 and Corollary 10.3.2 we deduce the following theorem.

THEOREM 10.3.12. *Let* $\hat{f}_T^*(\nu)$ *be defined by* (61) *and* (62), *where* $k(x)$ *is continuous in* $-1 \leq x \leq 1$, *and* $K_T/T \to 0$ *as* $T \to \infty$. *Let* $y_t = \boldsymbol{\beta}'\mathbf{z}_t + u_t$, *where*

$$u_t = \sum_{s=-\infty}^{\infty} \gamma_s v_{t-s}, \tag{81}$$

$\sum_{s=-\infty}^{\infty} |\gamma_s| < \infty$, *and* $\{v_t\}$ *is a process of independently and identically distributed variables with* $\mathscr{E}v_t = 0$, $\mathscr{E}v_t^2 = \sigma^2$, *and* $\mathscr{E}v_t^4 < \infty$. *If the conditions of Theorem* 9.4.3 *hold and if* $\lim_{T\to\infty} T/K_T^{2p+1}$ *is finite if* $p < q$ *or* $T/K_T^{2q+1} \to 0$ *if* $q \leq p$, *then* $\sqrt{T/K_T}\,[\hat{f}_T^*(\nu) - f(\nu)]$ *has a limiting normal distribution with mean* 0 *and variance given by Theorem* 9.3.4.

The asymptotic distribution of the estimate of $f(\nu)$ is not affected by using residuals from a regression fitted by least squares. In fact, Theorem 10.3.11 indicates that when $\hat{f}_T^*(\nu) - \hat{f}_T(\nu)$ is multiplied by T/K_T it may be a nontrivial random variable in the limit while $\hat{f}_T(\nu) - f(\nu)$ is multiplied by $\sqrt{T/K_T}$ to obtain a random variable with a nontrivial limiting distribution; thus, $\hat{f}_T^*(\nu) - \hat{f}_T(\nu)$ is usually of a smaller order of magnitude than $\hat{f}_T(\nu) - f(\nu)$.

Many of the results of Section 10.3 were given by Hannan (1958).

10.4. TESTING INDEPENDENCE

10.4.1. The Case When Least Squares Estimates Are Efficient

In Chapter 6 we considered testing the null hypothesis that the covariance matrix of a set of T observable variables was a multiple of the identity matrix against specified alternatives of serial correlation when the expected values were 0 or constant. In the section we study the problem when the expected values constitute a linear regression function. The problems of testing order of dependence and deciding order of dependence, that were treated in Chapter 6, can similarly be treated in these terms when the expected values constitute a linear regression function.

We assume that the density of $\boldsymbol{y} = (y_1, \ldots, y_T)'$ is

$$K \exp\left[-\tfrac{1}{2}(\gamma_0 Q_0 + \cdots + \gamma_q Q_q)\right], \tag{1}$$

where

$$Q_j = (\boldsymbol{y} - \boldsymbol{Z\beta})' \boldsymbol{A}_j (\boldsymbol{y} - \boldsymbol{Z\beta}), \qquad j = 0, 1, \ldots, q, \tag{2}$$

$\boldsymbol{Z}$ is a $T \times p$ matrix of rank ($p < T - q$) and $\boldsymbol{\beta}$ is a vector of p components. Then $\boldsymbol{y}$ is normally distributed with mean vector

$$\mathscr{E}\boldsymbol{y} = \boldsymbol{Z\beta} \tag{3}$$

and covariance matrix

$$\boldsymbol{\Sigma} = (\gamma_0 \boldsymbol{A}_0 + \cdots + \gamma_q \boldsymbol{A}_q)^{-1}. \tag{4}$$

The constant K is

$$K = (2\pi)^{-\frac{1}{2}T} \left|\gamma_0 \boldsymbol{A}_0 + \cdots + \gamma_q \boldsymbol{A}_q\right|^{\frac{1}{2}}. \tag{5}$$

We are particularly interested in testing the null hypothesis $\gamma_1 = 0$ when $q = 1$ and $\boldsymbol{A}_0 = \boldsymbol{I}$. This is the null hypothesis of independence against the alternative that the covariance matrix is of the form $\boldsymbol{\Sigma} = (\gamma_0 \boldsymbol{I} + \gamma_1 \boldsymbol{A}_1)^{-1}$. It is assumed that $\boldsymbol{A}_1$ and $\boldsymbol{Z}$ are known and γ_0, γ_1, and $\boldsymbol{\beta}$ are unknown. Usually the alternative hypothesis (characterized by $\boldsymbol{A}_1$) is of serial correlation.

First let us consider the case that $\boldsymbol{Z} = \boldsymbol{V}^*\boldsymbol{C}$, where $\boldsymbol{C}$ is nonsingular and $\boldsymbol{V}^*$ consists of p characteristic vectors of $\boldsymbol{\Sigma}$, say $\boldsymbol{v}_s$, $s = s_1, \ldots, s_p$ (with $\boldsymbol{V}^{*\prime}\boldsymbol{V}^* = \boldsymbol{I}$).

The regression function is $Z\boldsymbol{\beta} = V^*C\boldsymbol{\beta} = V^*\boldsymbol{\alpha}$, where $\boldsymbol{\alpha} = C\boldsymbol{\beta}$ is a reparametrization.

It was shown in Section 6.6.2 that in this case

$$a_h = \boldsymbol{v}_{s_h}'\boldsymbol{y}, \qquad h = 1, \ldots, p, \tag{6}$$

$$\begin{aligned} Q_j^* &= \left(\boldsymbol{y} - \sum_{h=1}^{p} a_h \boldsymbol{v}_{s_h}\right)' A_j \left(\boldsymbol{y} - \sum_{h=1}^{p} a_h \boldsymbol{v}_{s_h}\right) \\ &= \boldsymbol{y}'A_j\boldsymbol{y} - \sum_{h=1}^{p} \lambda_{js_h} a_h^2, \qquad j = 0, 1, \ldots, q, \end{aligned} \tag{7}$$

where λ_{js_h} is the characteristic root of A_j corresponding to $\boldsymbol{v}_{s_h}$, are a sufficient set of statistics for β_h, $h = 1, \ldots, p$, and γ_j, $j = 0, 1, \ldots, q$. Here a_h is the least squares and Markov estimate of α_h, $h = 1, \ldots, p$. If $\gamma_{i+1} = \cdots = \gamma_q = 0$, the best test of the null hypothesis $\gamma_i = 0$ is based on Q_i^* and the region of acceptance depends on the conditional distribution of Q_i^* given $Q_0^*, \ldots, Q_{i-1}^*$. In particular, the test of $\gamma_1 = 0$ when $\gamma_2 = \cdots = \gamma_q = 0$ is based on

$$r_1^* = \frac{Q_1^*}{Q_0^*}. \tag{8}$$

The distribution of r_1^* does not depend on $\boldsymbol{\beta}$ (Theorem 6.7.1) or γ_0 (Theorem 6.7.2) when $\gamma_1 = 0$.

THEOREM 10.4.1. *If the density of* $\boldsymbol{y}$ *is* (1) *with* $q = 1$ *and* Q_0 *and* Q_1 *defined by* (2) *with* $\gamma_0 A_0 + \gamma_1 A_1$ *positive definite and if* Z *consists of* p *linearly independent linear combinations of* p *characteristic vectors of* A_0 *and* A_1, *then the uniformly most powerful similar test of the null hypothesis* $\gamma_1 = 0$ *against alternatives* $\gamma_1 < 0$ *at significance level* ε *has rejection region* $r_1^* > c_1^*$, *where* $r_1^* = Q_1^*/Q_0^*$,

$$Q_j^* = (\boldsymbol{y} - Z\boldsymbol{b}^*)'A_j(\boldsymbol{y} - Z\boldsymbol{b}^*), \qquad j = 0, 1, \tag{9}$$

$\boldsymbol{b}^* = (Z'Z)^{-1}Z'\boldsymbol{y}$, *and* c_1^* *is defined so* $\Pr\{r_1^* > c_1^*\} = \varepsilon$ *when the null hypothesis is true. The uniformly most powerful similar test against alternatives* $\gamma_1 > 0$ *has rejection region* $r_1^* < c_1^{*\prime}$, *where* $c_1^{*\prime}$ *is defined so* $\Pr\{r_1^* < c_1^{*\prime}\} = \varepsilon$ *for* $\gamma_1 = 0$. *The uniformly most powerful unbiased test against alternatives* $\gamma_1 \neq 0$ *has rejection region* $r_1^* < c_{L1}^*$ *and* $r_1^* > c_{U1}^*$, *where* c_{L1}^* *and* c_{U1}^* *are defined so* $\Pr\{c_{L1}^* \leq r_1^* \leq c_{U1}^*\} = 1 - \varepsilon$ *and*

$$\int_{c_{L1}^*}^{c_{U1}^*} r_1^* f(r_1^*)\, dr_1^* = (1 - \varepsilon) \int_{-\infty}^{\infty} r_1^* f(r_1^*)\, dr_1^*, \tag{10}$$

where $f(r_1^*)$ *is the density of* r_1^* *when* $\gamma_1 = 0$.

We are particularly interested in cases where $\boldsymbol{A}_0 = \boldsymbol{I}$. Let a matrix of characteristic vectors of $\boldsymbol{A}_1$ be $\boldsymbol{V}$ with $\boldsymbol{V}'\boldsymbol{V} = \boldsymbol{I}$. If we let $\boldsymbol{y} = \boldsymbol{V}\boldsymbol{x}$ ($\boldsymbol{x} = \boldsymbol{V}'\boldsymbol{y}$), then the density of $\boldsymbol{x}$ is $N[\sum_{h=1}^{p} \alpha_h \boldsymbol{v}_{s_h}, (\gamma_0 \boldsymbol{I} + \gamma_1 \boldsymbol{\Lambda})^{-1}]$, where $\boldsymbol{\Lambda}$ is the diagonal matrix with diagonal elements $\lambda_1 \geq \cdots \geq \lambda_T$, the characteristic roots of $\boldsymbol{A}_1$. (The characteristic roots of $\boldsymbol{A}_0 = \boldsymbol{I}$ are all 1.) The quadratic forms are

$$Q_0^* = \boldsymbol{x}'\boldsymbol{x} - \sum_{s \in S} x_s^2 = \sum_{s \notin S} x_s^2, \tag{11}$$

$$Q_1^* = \boldsymbol{x}'\boldsymbol{\Lambda}\boldsymbol{x} - \sum_{s \in S} \lambda_s x_s^2 = \sum_{s \notin S} \lambda_s x_s^2, \tag{12}$$

where $S = \{s_1, \ldots, s_p\}$. Then

$$r_1^* = \frac{\sum_{s \notin S} \lambda_s x_s^2}{\sum_{s \notin S} x_s^2}. \tag{13}$$

If $\gamma_1 = 0$, the distribution of r_1^* does not depend on γ_0. The distribution of r_1^* when $\gamma_1 = 0$ is the distribution of (13) when the x_s's are independently normally distributed with means 0 and variances 1.

The general distributions derived in Section 6.7 are appropriate here with $\lambda_1, \ldots, \lambda_T$ or $\lambda_1, \ldots, \lambda_{T-1}$ (when the mean was involved) replaced by λ_s, $s \notin S$. In particular, if the $T - p$ roots occur in pairs, say $\nu_1 > \cdots > \nu_H$, then $\Pr\{r_1^* \geq R\}$ is given by Theorem 6.7.4. If the roots occur in pairs ($\nu_1 > \cdots > \nu_H$) except for the simple root ν_{H+1} ($< \nu_H$), then $\Pr\{r_1^* \geq R\}$ is given by Theorem 6.7.6.

When $\gamma_1 = 0$

$$\mathscr{E} r_1^{*h} = \frac{\mathscr{E} Q_1^{*h}}{\mathscr{E} Q_0^{*h}}. \tag{14}$$

When $\gamma_0 = 1$, the denominator of (14) is

$$\mathscr{E} Q_0^{*h} = 2^h \frac{\Gamma[\frac{1}{2}(T-p) + h]}{\Gamma[\frac{1}{2}(T-p)]}, \qquad -\tfrac{1}{2}(T-p) < h. \tag{15}$$

The numerator of (14) can be found from the cumulants of Q_1^*. The kth cumulant is $k!$ times

$$\begin{aligned} \nu_k^* &= \frac{2^{k-1}}{k} \sum_{t \notin S} \lambda_t^k \\ &= \frac{2^{k-1}}{k} \sum_{t=1}^{T} \lambda_t^k - \frac{2^{k-1}}{k} \sum_{h=1}^{p} \lambda_{s_h}^k. \end{aligned} \tag{16}$$

Then some moments are given by (126) and (127) of Section 6.7 with ν_k replaced by ν_k^*. If $\lambda_1, \ldots, \lambda_T$ and $\lambda_{s_1}, \ldots, \lambda_{s_p}$ are symmetric about 0, then $\nu_{2l-1}^* = 0$, $l = 1, 2, \ldots$, and $\mathscr{E}Q_1^{*2} = 2\nu_2^*$ and $\mathscr{E}Q_1^{*4} = 24\nu_4^* + 12\nu_2^{*2}$. The approximate distributions described in Section 6.8 can be used.

A case of particular interest is the circular case with $T = ph$ and independent variates $\cos 2\pi jt/T$ and $\sin 2\pi jt/T$. If p is even, the independent variates are taken to be

$$(17)\quad 1, \cos\frac{2\pi}{p}t, \sin\frac{2\pi}{p}t, \cos\frac{4\pi}{p}t, \sin\frac{4\pi}{p}t, \ldots, \cos\frac{2\pi(\frac{1}{2}p-1)}{p}t, \sin\frac{2\pi(\frac{1}{2}p-1)}{p}t, (-1)^t, \qquad t = 1, \ldots, T.$$

The corresponding characteristic roots are

$$(18)\quad 1, \cos\frac{2\pi}{p}, \cos\frac{2\pi}{p}, \cos\frac{4\pi}{p}, \cos\frac{4\pi}{p}, \ldots, \cos\frac{2\pi(\frac{1}{2}p-1)}{p}, \cos\frac{2\pi(\frac{1}{2}p-1)}{p}, -1.$$

Then $\nu_{2l-1}^* = 0$, $l = 1, 2, \ldots$, and

$$(19)\quad \nu_{2l}^* = \frac{2^{2l-1}}{2l}\sum_{t=1}^{T}\cos^{2l}\frac{2\pi}{T}t - \frac{2^{2l-1}}{2l}\sum_{j=1}^{p}\cos^{2l}\frac{2\pi}{p}j = \frac{T}{4l}\sum_{j_T}{}'\binom{2l}{j_T} - \frac{p}{4l}\sum_{j_p}{}'\binom{2l}{j_p},$$

where $\sum'$ indicates the sum over the values of j_T, $j_T = 0, 1, \ldots, 2l$, for which $2(j_T - l)$ is divisible by T and the sum over the values of j_p, $j_p = 0, 1, \ldots, 2l$, for which $2(j_p - l)$ is divisible by p. (See Section 6.7.7.) Then

$$(20)\quad \nu_{2l}^* = \frac{T-p}{4l}\binom{2l}{l} + \frac{T}{2l}\sum_{g=1}^{[2l/T]}\binom{2l}{l-\frac{1}{2}gT} - \frac{p}{2l}\sum_{g=1}^{[2l/p]}\binom{2l}{l-\frac{1}{2}gp}$$

for $T = ph$. If $l < \frac{1}{2}p$, then (20) is

$$(21)\quad \nu_{2l}^* = \frac{T-p}{4l}\binom{2l}{l}, \qquad l < \tfrac{1}{2}p, \quad p = 2, 4, \ldots.$$

(Note $T \geq 2p$.) In particular we have

$$(22)\quad \nu_2^* = \tfrac{1}{2}(T-p), \qquad p = 4, 6, \ldots,$$

$$(23)\quad \nu_4^* = \tfrac{3}{4}(T-p), \qquad p = 6, 8, \ldots.$$

We can also calculate $\nu_2^* = \frac{1}{2}T - 2$ for $p = 2$, $\nu_4^* = \frac{3}{4}T - 4$ for $p = 2$ and 4, $T \geq 6$. Then $\mathscr{E}Q_1^{*2} = T - 4$ for $p = 2$ and $T - p$ for $p = 4, 6, \ldots,$ and $\mathscr{E}Q_1^{*4} = 3(T^2 - 2T - 16)$ for $p = 2, 4, T \geq 6$, and $\mathscr{E}Q_1^{*4} = 3[T^2 - 2(p - 3)T + p(p - 6)]$ for $p = 6, 8, \ldots$. Then $\mathscr{E}r_{2l-1}^* = 0$, $\mathscr{E}r_1^{*2} = (T - 4)/[T(T - 2)]$ for $p = 2$,

$$\mathscr{E}r_1^{*2} = \frac{1}{T - p + 2}, \qquad p = 4, 6, \ldots, \tag{24}$$

$\mathscr{E}r_1^{*4} = 3(T^2 - 2T - 16)/[(T - 2)T(T + 2)(T + 4)]$ for $p = 2$ and $T \geq 6$, $\mathscr{E}r_1^{*4} = 3(T^2 - 2T - 16)/[(T - 4)(T - 2)T(T + 2)]$ for $p = 4$, and

$$\mathscr{E}r_1^{*4} = \frac{3}{(T - p + 2)(T - p + 4)}, \qquad p = 6, 8, \ldots. \tag{25}$$

Let us approximate the density of r_1^* by

$$\frac{\Gamma(2m)}{\Gamma^2(m)} \frac{(b^2 - x^2)^{m-1}}{(2b)^{2m-1}}, \qquad -b \leq x \leq b, \tag{26}$$

whose second moment is $b^2/(2m + 1)$. We shall take b as $\cos 2\pi/T$ and $b^2/(2m + 1)$ as $\mathscr{E}r_1^{*2}$. For $p = 4, 6, \ldots$, $m = \frac{1}{2}b^2(T - p + 2) - \frac{1}{2}$. The fourth moment of (26) is $3b^4/[(2m + 1)(2m + 3)]$. The fourth moment as fitted is

$$\frac{3}{(T - p + 2)[(T - p + 2) + 2/b^2]}, \tag{27}$$

to be compared with (25) for $p = 6, 8, \ldots$; the difference is small for b near 1.

R. L. Anderson and T. W. Anderson (1950) tabulated the values of R for which $\Pr\{r_1^* \geq R\} = 0.05$ and 0.01. This table is given here as Table 10.2.† If semiannual data are used, one fits a constant and $(-1)^t$, which has period $P = 2$. If quarterly data are used, one fits a constant, $(-1)^t$ with period $P = 2$, and $\cos \frac{1}{2}\pi t$ and $\sin \frac{1}{2}\pi t$ with period $P = 4$. If bimonthly data are used, one fits a constant, $(-1)^t$ with period $P = 2$, $\cos 2\pi t/3$ and $\sin 2\pi t/3$ with period $P = 3$, and $\cos \pi t/3$ and $\sin \pi t/3$ with period $P = 6$. If monthly data are used, one fits a constant, $(-1)^t$ with period $P = 2$, and $\cos 2\pi jt/12$ and $\sin 2\pi jt/12$ with periods $P = 12/j$, $j = 1, 2, 3, 4, 5$.

In the table it will be observed that the significance points are nearly the same for all cases with the same number of degrees of freedom, that is, number of roots remaining. The distribution of $r_1^*/(\cos 2\pi/T)$ for $P = 2$ is very close to the distribution of $r_1^*/(\cos \pi/T)$ based on the mean square successive difference. (See Table 6.3.)

† In R. L. Anderson and T. W. Anderson (1950) the entries for $T = 6$ and $P = 2$ were incorrect.

TABLE 10.2

ONE-SIDED SIGNIFICANCE POINTS FOR THE CIRCULAR SERIAL CORRELATION COEFFICIENT BASED ON RESIDUALS FROM A FITTED TRIGONOMETRIC TREND†

No. of Roots	$P = 2$			$P = 2, 4$			$P = 2, 3, 6$			$P = 2, 12/5, 3, 4, 6, 12$		
	T	.05	.01	T	.05	.01	T	.05	.01	T	.05	.01
4	6	0.450	0.490	8	0.636	0.693						
6	8	0.484	0.607				12	0.592	0.744			
8	10	0.453	0.601	12	0.515	0.661						
10	12	0.426	0.572									
12	14	0.402	0.544	16	0.439	0.582	18	0.442	0.592	24	0.441	0.592
14	16	0.382	0.519									
16	18	0.364	0.496	20	0.388	0.523						
18	20	0.348	0.476				24	0.369	0.504			
20	22	0.334	0.458	24	0.351	0.478						
22	24	0.321	0.442									
24	26	0.310	0.427	28	0.323	0.441	30	0.323	0.445	36	0.323	0.445
26	28	0.300	0.414									
28	30	0.290	0.402	32	0.300	0.414						
30	32	0.282	0.390				36	0.291	0.403			
32	34	0.274	0.380	36	0.282	0.391						
34	36	0.266	0.370									
36	38	0.260	0.361	40	0.267	0.371	42	0.267	0.371	48	0.267	0.371
38	40	0.254	0.353									
40	42	0.248	0.345	44	0.254	0.354						
42	44	0.242	0.338				48	0.248	0.346			
44	46	0.237	0.331	48	0.243	0.338						
46	48	0.233	0.324									
48	50	0.228	0.318	52	0.233	0.325	54	0.233	0.325	60	0.233	0.325
50	52	0.224	0.313									
52	54	0.220	0.307	56	0.224	0.313						
54	56	0.216	0.302				60	0.220	0.308			
56	58	0.212	0.297	60	0.216	0.302						
58	60	0.209	0.292									

† P denotes the periods of trigonometric terms fitted.

10.4.2. The General Case

Now let us consider the problem when we do not take into account whether the columns of Z are characteristic vectors of A_1. Let the residuals from the least squares regression be

$$\hat{u} = y - Zb^* = y - Z(Z'Z)^{-1}Z'y = [I - Z(Z'Z)^{-1}Z']y. \tag{28}$$

Then the serial correlation coefficient is

$$(29)\qquad r_1^* = \frac{\hat{u}'A_1\hat{u}}{\hat{u}'\hat{u}} = \frac{y'[I - Z(Z'Z)^{-1}Z']A_1[I - Z(Z'Z)^{-1}Z']y}{y'[I - Z(Z'Z)^{-1}Z'][I - Z(Z'Z)^{-1}Z']y}$$

$$= \frac{y'B'A_1By}{y'B'By},$$

where

$$(30)\qquad B = I - Z(Z'Z)^{-1}Z'.$$

Then B is a $T \times T$ matrix with rank $T - p$ since it is orthogonal to Z. B is idempotent since

$$\begin{aligned}(31)\qquad B^2 &= [I - Z(Z'Z)^{-1}Z'][I - Z(Z'Z)^{-1}Z'] \\ &= I - 2Z(Z'Z)^{-1}Z' + Z(Z'Z)^{-1}Z'Z(Z'Z)^{-1}Z' \\ &= I - 2Z(Z'Z)^{-1}Z' + Z(Z'Z)^{-1}Z' \\ &= I - Z(Z'Z)^{-1}Z' \\ &= B.\end{aligned}$$

The characteristic roots of an idempotent matrix are 1 and 0. The matrix B has the characteristic root of 1 with multiplicity $T - p$, its rank, and the characteristic root 0 with multiplicity p. Since $B'B = B^2$, $B'B$ has the same characteristic roots. Thus, there exists an orthogonal matrix, Q, which will diagonalize $B'B$, that is,

$$(32)\qquad Q'B'BQ = \begin{pmatrix} I & 0 \\ 0 & 0 \end{pmatrix},$$

where I in (32) is of order $T - p$. If $BQ = (C \quad G)$, where C is $T \times (T - p)$, then (32) is

$$(33)\qquad Q'B'BQ = \begin{pmatrix} C' \\ G' \end{pmatrix}(C \quad G) = \begin{pmatrix} C'C & C'G \\ G'C & G'G \end{pmatrix}.$$

Since $G'G = 0$, $G = 0$ and $BQ = (C \quad 0)$. Let $y = Qu$ and partition u as $u = (u^{*\prime} \; u^{**\prime})'$, where u^* consists of $T - p$ components. Then

$$\begin{aligned}(34)\qquad r_1^* &= \frac{u'Q'B'A_1BQu}{u'Q'B'BQu} \\ &= \frac{(u^{*\prime} \; u^{**\prime})\begin{pmatrix} C' \\ 0 \end{pmatrix} A_1 (C \; 0)\begin{pmatrix} u^* \\ u^{**} \end{pmatrix}}{(u^{*\prime} \; u^{**\prime})\begin{pmatrix} C' \\ 0 \end{pmatrix} (C \; 0)\begin{pmatrix} u^* \\ u^{**} \end{pmatrix}} \\ &= \frac{u^{*\prime}C'A_1Cu^*}{u^{*\prime}u^*}.\end{aligned}$$

Under the null hypothesis $\gamma_1 = 0$ the covariance matrix of $\boldsymbol{y}$ is $\sigma^2 \boldsymbol{I}$ and the covariance matrix of $\boldsymbol{u}$ is the same. Since $\boldsymbol{y} = \boldsymbol{Q}\boldsymbol{u}$ and $\boldsymbol{B}\boldsymbol{y} = \boldsymbol{B}\boldsymbol{Q}\boldsymbol{u} = (\boldsymbol{C} \quad \boldsymbol{0})\boldsymbol{u} = \boldsymbol{C}\boldsymbol{u}^*$, then $\boldsymbol{C}'\boldsymbol{B}\boldsymbol{y} = \boldsymbol{C}'\boldsymbol{C}\boldsymbol{u}^* = \boldsymbol{u}^*$. Thus $\mathscr{E}\boldsymbol{u}^* = \boldsymbol{C}'\boldsymbol{B}\mathscr{E}\boldsymbol{y} = \boldsymbol{C}'\boldsymbol{B}\boldsymbol{Z}\boldsymbol{\beta} = \boldsymbol{0}$. Under the null hypothesis the distribution of $\boldsymbol{u}^*$ is $N(\boldsymbol{0}, \sigma^2\boldsymbol{I})$. Let the characteristic roots of $\boldsymbol{C}'\boldsymbol{A}_1\boldsymbol{C}$ be $\nu_1 \geq \cdots \geq \nu_{T-p}$. Since $\boldsymbol{Q}'\boldsymbol{B}'\boldsymbol{A}_1\boldsymbol{B}\boldsymbol{Q}$ is a matrix with $\boldsymbol{C}'\boldsymbol{A}_1\boldsymbol{C}$ in the upper left-hand corner and 0's elsewhere, these are also roots of $\boldsymbol{B}'\boldsymbol{A}_1\boldsymbol{B}$. Moreover, they are roots of $\boldsymbol{B}\boldsymbol{B}'\boldsymbol{A}_1 = \boldsymbol{B}^2\boldsymbol{A}_1 = \boldsymbol{B}\boldsymbol{A}_1$.

THEOREM 10.4.2. *When* $\gamma_1 = 0$, *the distribution of* r_1^* *is the distribution of*

$$\frac{\sum_{t=1}^{T-p} \nu_t x_t^2}{\sum_{t=1}^{T-p} x_t^2}, \tag{35}$$

where $\nu_1 \geq \cdots \geq \nu_{T-p}$ *are* $T-p$ *roots of* $[\boldsymbol{I} - \boldsymbol{Z}(\boldsymbol{Z}'\boldsymbol{Z})^{-1}\boldsymbol{Z}']\boldsymbol{A}_1$ *including all nonzero roots and* $x_1, \ldots, x_{T-p}$ *are independently normally distributed with means* 0 *and variances* 1.

The problem then is to find the characteristic roots of $[\boldsymbol{I} - \boldsymbol{Z}(\boldsymbol{Z}'\boldsymbol{Z})^{-1}\boldsymbol{Z}']\boldsymbol{A}_1 = \boldsymbol{A}_1 - \boldsymbol{Z}(\boldsymbol{Z}'\boldsymbol{Z})^{-1}\boldsymbol{Z}'\boldsymbol{A}_1$. The theory of Sections 6.7 and 6.8 applies. The kth cumulant of the numerator of (35) is $k!$ times

$$\begin{aligned} \frac{2^{k-1}}{k}\sum_{t=1}^{T-p}\nu_t^k &= \frac{2^{k-1}}{k}\operatorname{tr}\{[\boldsymbol{I} - \boldsymbol{Z}(\boldsymbol{Z}'\boldsymbol{Z})^{-1}\boldsymbol{Z}']\boldsymbol{A}_1\}^k \\ &= \frac{2^{k-1}}{k}\operatorname{tr}\{\boldsymbol{A}_1 - \boldsymbol{Z}(\boldsymbol{Z}'\boldsymbol{Z})^{-1}\boldsymbol{Z}'\boldsymbol{A}_1\}^k. \end{aligned} \tag{36}$$

In the expansion of the right-hand side of (36) account must be taken of the possibility that $\boldsymbol{A}_1$ and $\boldsymbol{Z}(\boldsymbol{Z}'\boldsymbol{Z})^{-1}\boldsymbol{Z}'$ do not commute. See Problems 21 and 22. Durbin and Watson (1950) neglected this possible lack of commutativity.

The moments of r_1^* can be calculated from the cumulants. Then a Beta-type distribution can be fitted to the first four moments.

In many cases it is laborious to calculate the characteristic roots of $\boldsymbol{B}\boldsymbol{A}_1$. Durbin and Watson (1950), (1951) found inequalities for the roots and hence for the distribution of r_1^*. We shall now show that the $T-p$ roots of $\boldsymbol{B}\boldsymbol{A}_1$ which are not associated with the p vectors for which $\boldsymbol{B}\boldsymbol{x} = \boldsymbol{0}$ are, respectively, not greater than the $T-p$ largest roots of $\boldsymbol{A}_1$ and not less than the $T-p$ smallest roots of $\boldsymbol{A}_1$. We shall obtain these inequalities by help of two lemmas. Let $\operatorname{ch}_i(\boldsymbol{A})$ denote the ith largest characteristic root of $\boldsymbol{A}$; that is, $\operatorname{ch}_1(\boldsymbol{A}) \geq \cdots \geq \operatorname{ch}_T(\boldsymbol{A})$.

LEMMA 10.4.1. *For a symmetric matrix A and arbitrary vectors $\boldsymbol{\alpha}_1, \ldots, \boldsymbol{\alpha}_{i-1}$*

$$\text{ch}_i(A) \leq \max_{\substack{x'\alpha_j=0 \\ j=1,\ldots,i-1}} \frac{x'Ax}{x'x}. \tag{37}$$

PROOF. If A is diagonal $x'Ax = \sum_{t=1}^{T} \lambda_t x_t^2$, where $\lambda_1 \geq \cdots \geq \lambda_T$ are the characteristic roots of A. Then

$$\begin{aligned} \max_{\substack{x'\alpha_j=0 \\ j=1,\ldots,i-1}} \frac{x'Ax}{x'x} &\geq \max_{\substack{x'\alpha_j=0 \\ j=1,\ldots,i-1 \\ x_{i+1}=\cdots=x_T=0}} \frac{\sum_{t=1}^{T} \lambda_t x_t^2}{\sum_{t=1}^{T} x_t^2} \\ &= \max_{\substack{x'\alpha_j=0 \\ j=1,\ldots,i-1 \\ x_{i+1}=\cdots=x_T=0}} \frac{\sum_{t=1}^{i} \lambda_t x_t^2}{\sum_{t=1}^{i} x_t^2} \\ &\geq \lambda_i \frac{\sum_{t=1}^{i} x_t^2}{\sum_{t=1}^{i} x_t^2} = \text{ch}_i(A). \end{aligned} \tag{38}$$

Since an arbitrary symmetric matrix A is diagonalized by an orthogonal matrix, which does not affect the characteristic roots or $x'x$, the lemma holds for every symmetric matrix. [See Courant and Hilbert (1937).]

A similar argument leads to the following lemma.

LEMMA 10.4.2. *For a symmetric matrix A and arbitrary vectors $\boldsymbol{\alpha}_{i+1}, \ldots, \boldsymbol{\alpha}_T$*

$$\text{ch}_i(A) \geq \min_{\substack{x'\alpha_j=0 \\ j=i+1,\ldots,T}} \frac{x'Ax}{x'x}. \tag{39}$$

THEOREM 10.4.3. *If $\lambda_1 \geq \cdots \geq \lambda_T$ are the characteristic roots of the symmetric matrix A_1, $B = I - Z(Z'Z)^{-1}Z'$, where Z is $T \times p$ and of rank $p\ (< T)$, and $\nu_1 \geq \cdots \geq \nu_{T-p}$ are the characteristic roots of $B'A_1B$ with the omission of p roots of 0, then*

$$\lambda_{i+p} \leq \nu_i \leq \lambda_i, \qquad i = 1, \ldots, T-p. \tag{40}$$

PROOF. The p roots of $\boldsymbol{B}'\boldsymbol{A}_1\boldsymbol{B}$ that are omitted are those corresponding to vectors which are annihilated by $\boldsymbol{B}$. Let $\boldsymbol{H} = \boldsymbol{Z}(\boldsymbol{Z}'\boldsymbol{Z})^{-1}\boldsymbol{Z}'$. Then $\boldsymbol{H}$ is symmetric positive semidefinite of rank p. Since $\boldsymbol{B} + \boldsymbol{H} = \boldsymbol{I}$, $\boldsymbol{H}\boldsymbol{x} = \boldsymbol{0}$ implies $\boldsymbol{x} = \boldsymbol{B}\boldsymbol{x}$. Let $\boldsymbol{v}_1, \ldots, \boldsymbol{v}_T$ be a set (linearly independent) of characteristic vectors of $\boldsymbol{A}_1$. Then

$$
\begin{aligned}
(41) \qquad \operatorname{ch}_i(\boldsymbol{A}_1) &= \max_{\substack{\boldsymbol{x}'\boldsymbol{v}_j=0 \\ j=1,\ldots,i-1}} \frac{\boldsymbol{x}'\boldsymbol{A}_1\boldsymbol{x}}{\boldsymbol{x}'\boldsymbol{x}} \\
&\geq \max_{\substack{\boldsymbol{x}'\boldsymbol{v}_j=0 \\ j=1,\ldots,i-1 \\ \boldsymbol{x}'\boldsymbol{H}=0}} \frac{\boldsymbol{x}'\boldsymbol{A}_1\boldsymbol{x}}{\boldsymbol{x}'\boldsymbol{x}} \\
&= \max_{\substack{\boldsymbol{x}'\boldsymbol{v}_j=0 \\ j=1,\ldots,i-1 \\ \boldsymbol{x}'\boldsymbol{H}=0}} \frac{\boldsymbol{x}'\boldsymbol{B}'\boldsymbol{A}_1\boldsymbol{B}\boldsymbol{x}}{\boldsymbol{x}'\boldsymbol{x}} \geq \nu_i.
\end{aligned}
$$

Similarly

$$
\begin{aligned}
(42) \qquad \operatorname{ch}_{i+p}(\boldsymbol{A}_1) &= \min_{\substack{\boldsymbol{x}'\boldsymbol{v}_j=0 \\ j=i+p+1,\ldots,T}} \frac{\boldsymbol{x}'\boldsymbol{A}_1\boldsymbol{x}}{\boldsymbol{x}'\boldsymbol{x}} \\
&\leq \min_{\substack{\boldsymbol{x}'\boldsymbol{v}_j=0 \\ j=i+p+1,\ldots,T \\ \boldsymbol{x}'\boldsymbol{H}=0}} \frac{\boldsymbol{x}'\boldsymbol{A}_1\boldsymbol{x}}{\boldsymbol{x}'\boldsymbol{x}} \\
&= \min_{\substack{\boldsymbol{x}'\boldsymbol{v}_j=0 \\ j=i+p+1,\ldots,T \\ \boldsymbol{x}'\boldsymbol{H}=0}} \frac{\boldsymbol{x}'\boldsymbol{B}'\boldsymbol{A}_1\boldsymbol{B}\boldsymbol{x}}{\boldsymbol{x}'\boldsymbol{x}} \leq \nu_i.
\end{aligned}
$$

COROLLARY 10.4.1. *Under the conditions of Theorem* 10.4.3

$$
(43) \qquad \sum_{i=1}^{T-p} \lambda_{i+p} x_i^2 \leq \sum_{i=1}^{T-p} \nu_i x_i^2 \leq \sum_{i=1}^{T-p} \lambda_i x_i^2.
$$

THEOREM 10.4.4. *Under the conditions of Theorem* 10.4.3

$$
(44) \qquad \Pr\left\{\frac{\sum_{t=1}^{T-p} \lambda_{t+p} w_t^2}{\sum_{t=1}^{T-p} w_t^2} \geq R\right\} \leq \Pr\left\{\frac{\sum_{t=1}^{T-p} \nu_t w_t^2}{\sum_{t=1}^{T-p} w_t^2} \geq R\right\} \leq \Pr\left\{\frac{\sum_{t=1}^{T-p} \lambda_t w_t^2}{\sum_{t=1}^{T-p} w_t^2} \geq R\right\}.
$$

If $w_1, \ldots, w_{T-p}$ are independently normally distributed with means 0 and variances 1, the left-hand side of (44) is 1 minus the cumulative distribution function of the serial correlation when the independent variables correspond to the largest characteristic roots of A_1 and the right-hand side when the independent variables correspond to the smallest roots. We can do better if we know that some of the independent variables are characteristic vectors of A_1.

THEOREM 10.4.5. *Suppose that p^* linearly independent linear combinations of the columns of Z are p^* characteristic vectors of A_1. Let $\lambda_1' \geq \cdots \geq \lambda_{T-p^*}'$ be the roots of A_1 remaining after deleting the p^* roots corresponding to the p^* vectors. If y is distributed according to $N(Z\boldsymbol{\beta}, \sigma^2 I)$,*

$$\Pr\left\{\frac{\sum_{t=1}^{T-p} \lambda_{t+p-p^*}' w_t^2}{\sum_{t=1}^{T-p} w_t^2} \geq R\right\} \leq \Pr\{r_1^* \geq R\} \leq \Pr\left\{\frac{\sum_{t=1}^{T-p} \lambda_t' w_t^2}{\sum_{t=1}^{T-p} w_t^2} \geq R\right\}, \tag{45}$$

where r_1^* is defined by (29) *and $w_1, \ldots, w_{T-p}$ are independently normally distributed with means* 0 *and variances* 1.

For a given matrix A_1 and certain p^* characteristic vectors in Z one can tabulate the values of R for which the right-hand side of (45) is a given value ε, say R_U. If an observed r_1^* exceeds R_U, it is significant at the significance level ε. Let R_L be the value for which the left-hand side is ε. If an observed r_1^* is less than R_L it is not significant at significance level ε. If the observed r_1^* falls between R_L and R_U the result is indeterminate. This is a test of independence against the alternative of positive dependence ($\gamma_1 = 0$ against $\gamma_1 < 0$).

Durbin and Watson (1951) have tabulated $d_L = 2(1 - R_U)$ and $d_U = 2(1 - R_L)$ in the mean square successive difference case for $\varepsilon = .05, .025, .01$ and $p = 2, 3, 4, 5, 6$ when $\boldsymbol{\varepsilon} = (1, \ldots, 1)'$ is one column of Z. If we write the statistics for the right-hand and left-hand sides of (45) for $p = 2$ as $\sum_{t=2}^{T-2} \lambda_t' w_t^2 + \lambda_1' w_1$ and $\sum_{t=2}^{T-2} \lambda_t' w_t^2 + \lambda_{T-1}' w_1^2$ divided by $\sum_{t=1}^{T-2} w_t^2$, ($w_1, \ldots, w_{T-2}$ being convenient symbols for $T-2$ independent standard normal variables), the difference is

$$\frac{w_1^2(\lambda_1' - \lambda_{T-1}')}{\sum_{t=1}^{T-2} w_t^2} = 2 \cos \frac{\pi}{T} \frac{w_1^2}{\sum_{t=1}^{T-2} w_t^2}, \tag{46}$$

which is roughly $2 \cos (\pi/T)/(T - 2)$. Perusal of the tables of Durbin and Watson shows that the difference between R_U and R_L for $p = 2$ is about this amount.

TABLE 10.3

BOUNDS FOR ONE-SIDED SIGNIFICANCE POINTS FOR THE SERIAL CORRELATION COEFFICIENT BASED ON SUCCESSIVE DIFFERENCES OF RESIDUALS FROM A TREND OF A CONSTANT AND ANOTHER INDEPENDENT VARIABLE

	Significance Levels					
	.05		.025		.01	
T	R_L	R_U	R_L	R_U	R_L	R_U
15	0.32	0.46	0.385	0.525	0.465	0.595
16	0.315	0.45	0.38	0.51	0.455	0.58
17	0.31	0.435	0.375	0.495	0.45	0.465
18	0.305	0.42	0.37	0.485	0.44	0.55
19	0.30	0.41	0.36	0.47	0.435	0.535
20	0.295	0.40	0.36	0.46	0.425	0.525
21	0.29	0.39	0.35	0.45	0.42	0.515
22	0.285	0.38	0.345	0.44	0.415	0.50
23	0.28	0.37	0.34	0.43	0.405	0.49
24	0.275	0.365	0.335	0.42	0.40	0.48
25	0.275	0.355	0.33	0.41	0.395	0.475

A rough rule of thumb is that R_U is approximately R from Table 6.3 (at T) plus $\sum_{j=1}^{p-1} \cos(\pi j/T)/(T-p)$ and R_L is R minus this amount. Table 10.3 gives some significance points for $p = 2$.

A case of particular interest is polynomial regression. Suppose A_1 is the matrix associated with the mean square successive difference [(41) of Section 6.5] and $\boldsymbol{\tau} = (1, 2, \ldots, T)'$. Then a linear trend is $\beta_1\boldsymbol{\varepsilon} + \beta_2\boldsymbol{\tau}$. The vector $\boldsymbol{\tau}$ is nearly a characteristic vector of A_1 with root 1; in fact

$$A_1\boldsymbol{\tau} = \boldsymbol{\tau} + \tfrac{1}{2}\begin{bmatrix} 1 \\ 0 \\ \cdot \\ \cdot \\ \cdot \\ 0 \\ -1 \end{bmatrix}. \tag{47}$$

(See Problem 24.) This suggests that the significance points for residuals from the regression $\beta_1\boldsymbol{\varepsilon} + \beta_2\boldsymbol{\tau}$ are close to R_L for $p = 2$. [Another way of looking at the matter is that the vector with tth component $(T+1-2t)/(T-1)$ is close to the characteristic vector with tth component $\cos \pi(t-\frac{1}{2})/T$.] Consider A_1 of Section 6.5.4. Then

$$A_1\varepsilon = \varepsilon - \tfrac{1}{2}\begin{pmatrix} 1 \\ 0 \\ \cdot \\ \cdot \\ \cdot \\ 0 \\ 1 \end{pmatrix}. \tag{48}$$

Thus ε is almost a characteristic vector of A_1 corresponding to a root 1. This suggests that $y'B'A_1By/y'B'By$ is distributed approximately as $\sum_{t=2}^T \lambda_t w_t^2/\sum_{t=2}^T w_t^2$. In fact, the $T-1$ smallest roots of A_1 and the $T-1$ roots of $B'AB$ are, respectively, $-\frac{1}{2}$ and $-\frac{1}{2}$ for $T=2$, $(0, -\frac{1}{2}\sqrt{2} = -0.7071)$ and $(0, -\frac{2}{3} = -0.6667)$ for $T=3$, and $(\cos 2\pi/5 = 0.3090, \cos 3\pi/5 = -0.3090, \cos 4\pi/5 = -0.8090)$ and $[(\sqrt{5}-1)/4 = 0.3090, -0.2500, -(\sqrt{5}+1)/4 = -0.8090]$ for $T=4$.

Durbin (1970) has made another suggestion for testing independence when the least squares estimates are not efficient. Suppose that $Z = (V_1^* \; Z_2)$, where the p_1 columns of V_1^* are p_1 characteristic vectors of A_1, $V_1^{*\prime}V_1^* = I$, $V_1^{*\prime}Z_2 = 0$, and no linear combination of the $p_2 = p - p_1$ columns of Z_2 is a linear combination of p_2 of the other $T - p_1$ characteristic vectors of A_1. Suppose V_2^* of p_2 characteristic vectors is "close" to Z_2, $V_2^{*\prime}V_2^* = I$, and $V_1^{*\prime}V_2^* = 0$. Let

$$\begin{pmatrix} b^{(1)} \\ b^{(2)} \\ c \end{pmatrix} = \begin{pmatrix} I & 0 & 0 \\ 0 & Z_2'Z_2 & Z_2'V_2^* \\ 0 & V_2^{*\prime}Z_2 & I \end{pmatrix}^{-1} \begin{pmatrix} V_1^{*\prime}y \\ Z_2'y \\ V_2^{*\prime}y \end{pmatrix} \tag{49}$$

$$= \begin{bmatrix} V_1^{*\prime}y \\ \begin{pmatrix} C & D \\ D' & E \end{pmatrix}\begin{pmatrix} Z_2'y \\ V_2^{*\prime}y \end{pmatrix} \end{bmatrix},$$

where

$$\begin{pmatrix} C & D \\ D' & E \end{pmatrix} = \begin{pmatrix} Z_2'Z_2 & Z_2'V_2^* \\ V_2^{*\prime}Z_2 & I \end{pmatrix}^{-1}. \tag{50}$$

Let P and Q be $p_2 \times p_2$ matrices satisfying $PP' = C$ and $QQ' = E$. Let

$$v = y - V_1^* b^{(1)} - Z_2 b^{(2)} - V_2^* c, \tag{51}$$

$$w = (Z_2 - V_2^* V_2^{*\prime} Z_2) P Q^{-1} c. \tag{52}$$

Since

$$\mathscr{E} \begin{bmatrix} b^{(1)} \\ b^{(2)} \\ c \end{bmatrix} = \begin{bmatrix} \beta^{(1)} \\ \beta^{(2)} \\ 0 \end{bmatrix}, \tag{53}$$

$\mathscr{E} v = \mathscr{E} w = 0$. Since the regression vector c is uncorrelated with the residuals, $\mathscr{E} vw' = 0$. The covariance matrix of v is (when y has covariance matrix I)

$$\begin{aligned} \mathscr{E} vv' &= I - (V_1^* \; Z_2 \; V_2^*) \begin{bmatrix} I & 0 & 0 \\ 0 & Z_2' Z_2 & Z_2' V_2^* \\ 0 & V_2^{*\prime} Z_2 & I \end{bmatrix}^{-1} \begin{bmatrix} V_1^{*\prime} \\ Z_2' \\ V_2^{*\prime} \end{bmatrix} \\ &= I - V_1^* V_1^{*\prime} - (Z_2 \; V_2^*) \begin{pmatrix} C & D \\ D' & E \end{pmatrix} \begin{pmatrix} Z_2' \\ V_2^{*\prime} \end{pmatrix}. \end{aligned} \tag{54}$$

The covariance matrix of w is

$$\begin{aligned} \mathscr{E} ww' &= (Z_2 - V_2^* V_2^{*\prime} Z_2) P Q^{-1} Q Q' (Q')^{-1} P' (Z_2' - Z_2' V_2^* V_2^{*\prime}) \\ &= (Z_2 - V_2^* V_2^{*\prime} Z_2) C (Z_2' - Z_2' V_2^* V_2^{*\prime}). \end{aligned} \tag{55}$$

Since $Z_2' V_2^* = -C^{-1} D$ and $E - D' C^{-1} D = I$,

$$\mathscr{E}(v + w)(v + w)' = \mathscr{E} vv' + \mathscr{E} ww' = I - V_1^* V_1^{*\prime} - V_2^* V_2^{*\prime}, \tag{56}$$

which is the same as the covariance matrix of the residuals of the regression on $V^* = (V_1^* \; V_2^*)$. The distribution of

$$\frac{(v + w)' A_1 (v + w)}{(v + w)'(v + w)} \tag{57}$$

under the null hypothesis is the same as the distribution of

$$\frac{x'(I - V^* V^{*\prime}) A_1 (I - V^* V^{*\prime}) x}{x'(I - V^* V^{*\prime})(I - V^* V^{*\prime}) x} = \frac{x' A_1 x - x' V^* \Lambda^* V^{*\prime} x}{x' x - x' V^* V^{*\prime} x}, \tag{58}$$

where x has the distribution $N(0, I)$; this is the serial correlation based on residuals from V^* and Λ^* is the diagonal matrix whose diagonal elements are the characteristic roots corresponding to the columns of V^*.

The serial correlation coefficient based on successive differences may be used. The constant term may be included in the regression, $p_1 = 1$ and $V_1^* = \sqrt{1/T}\,\varepsilon = \sqrt{1/T}(1, \ldots, 1)'$. If the other independent variables which constitute Z_2 are slowly changing, they may be close to V_2^* which consists of the p_2 column vectors with components $\cos \pi s(t - \frac{1}{2})/T$, $t = 1, \ldots, T$, $s = 1, \ldots, p_2$. Then the significance point for (58) is R_L.

REFERENCES

Section 10.1. T. W. Anderson (1948), Grenander (1954), Grenander and Rosenblatt (1957), Hannan (1958), Magness and McGuire (1962), Rosenblatt (1956), Watson (1952), (1955), (1967), Watson and Hannan (1956), Zyskind (1967).

Section 10.2. T. W. Anderson (1948), (1958), Grenander (1954), Grenander and Rosenblatt (1957), Hannan (1961), (1963), Kantorovich (1948), Rosenblatt (1956), Watson (1955), (1967), Zyskind (1967).

Section 10.3. Hannan (1958).

Section 10.4. R. L. Anderson and T. W. Anderson (1950), Courant and Hilbert (1937), Durbin (1970), Durbin and Watson (1950), (1951).

PROBLEMS

1. (Sec. 10.2.1.) Show that the covariance matrix of $\boldsymbol{b}$ and $\boldsymbol{b}^*$ is

$$\begin{pmatrix} \mathscr{E}(\boldsymbol{b} - \boldsymbol{\beta})(\boldsymbol{b} - \boldsymbol{\beta})' & \mathscr{E}(\boldsymbol{b} - \boldsymbol{\beta})(\boldsymbol{b}^* - \boldsymbol{\beta})' \\ \mathscr{E}(\boldsymbol{b}^* - \boldsymbol{\beta})(\boldsymbol{b} - \boldsymbol{\beta})' & \mathscr{E}(\boldsymbol{b}^* - \boldsymbol{\beta})(\boldsymbol{b}^* - \boldsymbol{\beta})' \end{pmatrix}$$

$$= \begin{pmatrix} (\boldsymbol{Z}'\boldsymbol{\Sigma}^{-1}\boldsymbol{Z})^{-1} & (\boldsymbol{Z}'\boldsymbol{\Sigma}^{-1}\boldsymbol{Z})^{-1} \\ (\boldsymbol{Z}'\boldsymbol{\Sigma}^{-1}\boldsymbol{Z})^{-1} & (\boldsymbol{Z}'\boldsymbol{Z})^{-1}\boldsymbol{Z}'\boldsymbol{\Sigma}\boldsymbol{Z}(\boldsymbol{Z}'\boldsymbol{Z})^{-1} \end{pmatrix}$$

$$= \begin{pmatrix} (\boldsymbol{Z}'\boldsymbol{\Sigma}^{-1}\boldsymbol{Z})^{-1} & \boldsymbol{0} \\ \boldsymbol{0} & (\boldsymbol{Z}'\boldsymbol{Z})^{-1} \end{pmatrix} \begin{pmatrix} \boldsymbol{Z}'\boldsymbol{\Sigma}^{-1}\boldsymbol{Z} & \boldsymbol{Z}'\boldsymbol{Z} \\ \boldsymbol{Z}'\boldsymbol{Z} & \boldsymbol{Z}'\boldsymbol{\Sigma}\boldsymbol{Z} \end{pmatrix} \begin{pmatrix} (\boldsymbol{Z}'\boldsymbol{\Sigma}^{-1}\boldsymbol{Z})^{-1} & \boldsymbol{0} \\ \boldsymbol{0} & (\boldsymbol{Z}'\boldsymbol{Z})^{-1} \end{pmatrix}.$$

2. (Sec. 10.2.1.) Prove that (15) for $\boldsymbol{G}$ nonsingular is equivalent to $\boldsymbol{Z} = \boldsymbol{V}^*\boldsymbol{C}$, where $\boldsymbol{\Sigma V}^* = \boldsymbol{V}^*\boldsymbol{\Lambda}$, $\boldsymbol{\Lambda}$ is diagonal, and $\boldsymbol{C}$ is nonsingular.

3. (Sec. 10.2.1.) Prove that if the rank of the $T \times p$ matrix $\boldsymbol{A}$ is r, then there exists a $T \times T$ orthogonal matrix $\boldsymbol{Q}$ such that $\boldsymbol{Q}'\boldsymbol{A}$ has only 0's in the last $T - r$ rows.

4. (Sec. 10.2.1.) Give an algorithm for determining the maximum number of linearly independent linear combinations of the columns of $\boldsymbol{Z}$ that are characteristic vectors of $\boldsymbol{\Sigma}$.

5. (Sec. 10.2.2.) Show that $\boldsymbol{b}^*$ is efficient if and only if the rank of the covariance matrix of $(\boldsymbol{b}' \; \boldsymbol{b}^{*\prime})'$ is p.

6. (Sec. 10.2.3.) Show that the conditions $\sum_{t=1}^{T} z_t^2 \to \infty$ and $z_{T+1}^2/\sum_{t=1}^{T} z_t^2 \to 0$ as $T \to \infty$ do not imply the existence of $\lim_{T\to\infty} \sum_{t=1}^{T} z_t z_{t+h}/\sum_{t=1}^{T} z_t^2$. [*Hint:* Construct an example where the limit does not exist by letting $z_t = \pm 1$ in such a sequence that $\sum_{t=1}^{T-1} z_t z_{t+1}/\sum_{t=1}^{T-1} z_t^2$ attains 0 and $\frac{1}{2}$ for arbitrarily large T.]

7. (Sec. 10.2.3.) Show that if $\sum_{h=-\infty}^{\infty} |\rho_{ii}(h)| < \infty$, $i = 1, \ldots, p$, then there exists a positive semidefinite matrix function $\boldsymbol{m}(\lambda)$ such that

$$\mathbf{R}(h) = \int_{-\pi}^{\pi} e^{i\lambda h}\, \boldsymbol{m}(\lambda)\, d\lambda.$$

8. (Sec. 10.2.3.) Show for $\boldsymbol{A}$ positive definite $\boldsymbol{u}'\boldsymbol{A}\boldsymbol{u} \leq \boldsymbol{u}'\boldsymbol{u}$ for all $\boldsymbol{u}$ implies $\boldsymbol{u}'\boldsymbol{u} \leq \boldsymbol{u}'\boldsymbol{A}^{-1}\boldsymbol{u}$ for all $\boldsymbol{u}$ and vice versa.

9. (Sec. 10.2.3.) Show that $\boldsymbol{x}'\boldsymbol{A}\boldsymbol{x} \leq \boldsymbol{x}'\boldsymbol{B}\boldsymbol{x}$ for all $\boldsymbol{x}$ with $\boldsymbol{A}$ symmetric and $\boldsymbol{B}$ positive definite is equivalent to the roots of $|\boldsymbol{A} - \theta \boldsymbol{B}| = 0$ being less than or equal to 1. [*Hint:* Use Problem 30 of Chapter 6.]

10. (Sec. 10.2.3.) Prove

$$2\pi \int_{-\pi}^{\pi} f(\lambda)\, d\boldsymbol{M}(\lambda) = \int_{m}^{M} u\, d\boldsymbol{T}(u).$$

11. (Sec. 10.2.3.) Let $\boldsymbol{L}_j$ be a $p \times p$ positive semidefinite matrix of rank n_j, $j = 1, \ldots, G$. Show that if $\sum_{j=1}^{G} n_j = p$ and

$$\sum_{j=1}^{G} \boldsymbol{L}_j = \boldsymbol{I},$$

then

$$\boldsymbol{L}_i \boldsymbol{L}_j = \boldsymbol{0}, \qquad i \neq j.$$

[*Hint:* There is an orthogonal matrix $\boldsymbol{Q}_1$ such that $\boldsymbol{Q}_1'\boldsymbol{L}_1\boldsymbol{Q}_1 = \boldsymbol{D}_1$ diagonal. Then argue that $\boldsymbol{D}_1$ is of rank n_1, $\boldsymbol{Q}_1' \sum_{j=2}^{G} \boldsymbol{L}_j \boldsymbol{Q}_1 = \boldsymbol{D}_1^*$ is diagonal, $\boldsymbol{D}_1^*$ is of rank $p - n_1$,

$$\boldsymbol{D}_1 = \begin{pmatrix} \boldsymbol{D}_{11} & \boldsymbol{0} \\ \boldsymbol{0} & \boldsymbol{0} \end{pmatrix}, \qquad \boldsymbol{D}_1^* = \begin{pmatrix} \boldsymbol{0} & \boldsymbol{0} \\ \boldsymbol{0} & \boldsymbol{D}_{12} \end{pmatrix},$$

$\boldsymbol{D}_{11} = \boldsymbol{I}$, $\boldsymbol{D}_{12} = \boldsymbol{I}$, and apply induction.]

12. (Sec. 10.2.3.) Show that $\int_S d\mathscr{R}\boldsymbol{M}(\lambda) = \boldsymbol{0}$ implies $\int_S d\mathscr{I}\boldsymbol{M}(\lambda) = \boldsymbol{0}$. [*Hint:* Use the fact that $\int_S d\boldsymbol{M}(\lambda)$ is Hermitian and positive semidefinite.]

13. (Sec. 10.2.3.) Show for $h = 0, 1, \ldots$

$$\lim_{T\to\infty} \frac{1}{T}\sum_{t=1}^{T-h} \cos \nu t \cos \lambda(t+h) = \tfrac{1}{2}\cos \lambda h, \qquad 0 < \nu = \lambda < \pi,$$
$$= 0, \qquad 0 \le \nu \ne \lambda \le \pi,$$
$$\lim_{T\to\infty} \frac{1}{T}\sum_{t=1}^{T-h} \cos \nu t \sin \lambda(t+h) = \tfrac{1}{2}\sin \lambda h, \qquad 0 < \nu = \lambda < \pi,$$
$$= 0, \qquad 0 \le \nu \ne \lambda \le \pi,$$
$$\lim_{T\to\infty} \frac{1}{T}\sum_{t=1}^{T-h} \sin \nu t \sin \lambda(t+h) = \tfrac{1}{2}\cos \lambda h, \qquad 0 < \nu = \lambda < \pi,$$
$$= 0, \qquad 0 \le \nu \ne \lambda \le \pi.$$

14. (Sec. 10.2.3.) Show

$$\lim_{T\to\infty} \frac{1}{T^{n+1}}\sum_{t=1}^{T} t^n = \frac{1}{n+1}, \qquad n = 0, 1, \ldots .$$

[*Hint:*

$$\int_0^T t^n\, dt < \sum_{t=1}^{T} t^n < \int_1^{T+1} t^n\, dt.\text{]}$$

15. (Sec. 10.2.3.) Show for $n = 0, 1, \ldots$

$$\lim_{T\to\infty} \frac{1}{T^{n+1}}\sum_{t=1}^{T} t^n \cos \nu t = 0, \qquad 0 < \nu < 2\pi,$$
$$\lim_{T\to\infty} \frac{1}{T^{n+1}}\sum_{t=1}^{T} t^n \sin \nu t = 0.$$

[*Hint:* The result for $n = 0$ follows from Lemma 4.4.1 and for $n = 1$ from Problem 2 of Chapter 9. Then prove by induction using Problem 11 of Chapter 9 and

$$2 \sin \tfrac{1}{2}\nu \cos \nu t = \sin \nu(t + \tfrac{1}{2}) - \sin \nu(t - \tfrac{1}{2}),$$
$$2 \sin \tfrac{1}{2}\nu \sin \nu t = \cos \nu(t - \tfrac{1}{2}) - \cos \nu(t + \tfrac{1}{2}).\text{]}$$

16. (Sec. 10.2.3.) Show for $n = 0, 1, \ldots$ and for $h = 0, 1, \ldots$

$$\lim_{T\to\infty} \frac{2}{T^{n+1}}\sum_{t=1}^{T-h} t^n \cos \nu t \cos \lambda(t+h) = \frac{1}{n+1}\cos \lambda h, \qquad 0 < \nu = \lambda < \pi,$$
$$= 0, \qquad 0 \le \nu \ne \lambda \le \pi,$$
$$\lim_{T\to\infty} \frac{2}{T^{n+1}}\sum_{t=1}^{T-h} t^n \cos \nu t \sin \lambda(t+h) = \frac{1}{n+1}\sin \lambda h, \qquad 0 < \nu = \lambda < \pi,$$
$$= 0, \qquad 0 \le \nu \ne \lambda \le \pi,$$
$$\lim_{T\to\infty} \frac{2}{T^{n+1}}\sum_{t=1}^{T-h} t^n \sin \nu t \sin \lambda(t+h) = \frac{1}{n+1}\cos \lambda h, \qquad 0 < \nu = \lambda < \pi,$$
$$= 0, \qquad 0 \le \nu \ne \lambda \le \pi.$$

[*Hint:* Use Problem 2 of Chapter 4 and Problems 14 and 15.]

17. (Sec. 10.3.2.) Prove (26).

18. (Sec. 10.4.1.) Show that when the regression function is $\mu_1 + \mu_2(-1)^t$ and r_1^* is defined circularly $r_1^* = 0$, $T = 4$,

$$\begin{aligned}\Pr\{r_1^* \geq R\} &= \tfrac{1}{2} - R, && -\tfrac{1}{2} \leq R \leq \tfrac{1}{2},\ T = 6,\\ &= \tfrac{1}{2}(1 - \sqrt{2}\,R)^2, && 0 \leq R \leq \tfrac{1}{2}\sqrt{2},\ T = 8.\end{aligned}$$

19. (Sec. 10.4.1.) Show that when the regression function is $\mu_1 + \mu_2(-1)^t + \mu_3 \cos \frac{1}{2}\pi t + \mu_4 \sin \frac{1}{2}\pi t$ and r_1^* is defined circularly

$$\Pr\{r_1^* \geq R\} = \tfrac{1}{2}(1 - \sqrt{2}\,R), \qquad -\tfrac{1}{2}\sqrt{2} \leq R \leq \tfrac{1}{2}\sqrt{2},\ T = 8.$$

20. (Sec. 10.4.2.) Show that the characteristic roots of the $T \times T$ symmetric matrix $\boldsymbol{B}$ are 1 of multiplicity m and 0 of multiplicity $T - m$ if $\boldsymbol{B}^2 = \boldsymbol{B}$ is of rank m. [*Hint:* If $\boldsymbol{w}$ is a characteristic vector of $\boldsymbol{B}$ corresponding to the root μ, it is a characteristic vector of $\boldsymbol{B}^2$ corresponding to the root μ^2.]

21. (Sec. 10.4.2.) Show for arbitrary square matrices $\boldsymbol{A}$ and $\boldsymbol{B}$ of the same order

$$\begin{aligned}\operatorname{tr}(\boldsymbol{A} + \boldsymbol{B})^2 &= \operatorname{tr}(\boldsymbol{A}^2 + 2\boldsymbol{AB} + \boldsymbol{B}^2),\\ \operatorname{tr}(\boldsymbol{A} + \boldsymbol{B})^3 &= \operatorname{tr}(\boldsymbol{A}^3 + 3\boldsymbol{A}^2\boldsymbol{B} + 3\boldsymbol{AB}^2 + \boldsymbol{B}^3),\\ \operatorname{tr}(\boldsymbol{A} + \boldsymbol{B})^4 &= \operatorname{tr}(\boldsymbol{A}^4 + 4\boldsymbol{A}^3\boldsymbol{B} + 4\boldsymbol{A}^2\boldsymbol{B}^2 + 2\boldsymbol{ABAB} + 4\boldsymbol{AB}^3 + \boldsymbol{B}^4).\end{aligned}$$

22. (Sec. 10.4.2.) Show that (36) is

$$\begin{aligned}&\operatorname{tr} \boldsymbol{A}_1 - \operatorname{tr}(\boldsymbol{Z}'\boldsymbol{Z})^{-1}\boldsymbol{Z}'\boldsymbol{A}_1\boldsymbol{Z}, && k = 1,\\ &\operatorname{tr} \boldsymbol{A}_1^2 - 2\operatorname{tr}(\boldsymbol{Z}'\boldsymbol{Z})^{-1}\boldsymbol{Z}'\boldsymbol{A}_1^2\boldsymbol{Z} + \operatorname{tr}[(\boldsymbol{Z}'\boldsymbol{Z})^{-1}\boldsymbol{Z}'\boldsymbol{A}_1\boldsymbol{Z}]^2, && k = 2,\\ &\tfrac{4}{3}\{\operatorname{tr} \boldsymbol{A}_1^3 - 3\operatorname{tr}(\boldsymbol{Z}'\boldsymbol{Z})^{-1}\boldsymbol{Z}'\boldsymbol{A}_1^3\boldsymbol{Z} + 3\operatorname{tr}(\boldsymbol{Z}'\boldsymbol{Z})^{-1}\boldsymbol{Z}'\boldsymbol{A}_1^2\boldsymbol{Z}(\boldsymbol{Z}'\boldsymbol{Z})^{-1}\boldsymbol{Z}'\boldsymbol{A}_1\boldsymbol{Z}\\ &\qquad - \operatorname{tr}[(\boldsymbol{Z}'\boldsymbol{Z})^{-1}\boldsymbol{Z}'\boldsymbol{A}_1\boldsymbol{Z}]^3\}, && k = 3.\end{aligned}$$

23. (Sec. 10.4.2.) Show that if $z_t = t - \frac{1}{2}(T + 1)$, $t = 1, \ldots, T$, and $\boldsymbol{A}_1$ is the matrix associated with the mean square successive difference

$$\begin{aligned}\operatorname{tr} \boldsymbol{A}_1 - \operatorname{tr}(\boldsymbol{Z}'\boldsymbol{Z})^{-1}\boldsymbol{Z}'\boldsymbol{A}_1\boldsymbol{Z} &= 1 - \frac{(T-2)(T+3)}{T(T+1)}\\ &= \frac{6}{T(T+1)}.\end{aligned}$$

24. (Sec. 10.4.2.) Let $\boldsymbol{\tau}^{(j)} = (1, 2^j, 3^j, \ldots, T^j)'$ with $\boldsymbol{\tau}^{(0)} = \boldsymbol{\varepsilon} = (1, 1, \ldots, 1)'$, and let $\boldsymbol{A}_1$ be the matrix associated with the mean square successive difference. Show

$$\boldsymbol{A}_1\boldsymbol{\tau}^{(j)} = \boldsymbol{\tau}^{(j)} + \binom{j}{2}\boldsymbol{\tau}^{(j-2)} + \cdots + \binom{j}{2[\frac{1}{2}j]}\boldsymbol{\tau}^{(j-2[\frac{1}{2}j])} + \frac{1}{2}\begin{bmatrix}1\\0\\ \cdot\\ \cdot\\ \cdot\\0\\T^j - (T+1)^j\end{bmatrix}.$$

25. (Sec. 10.4.2.) Graph $(T + 1 - 2t)/(T - 1)$ and $\cos \pi(t-\frac{1}{2})/T$, $t = 1, \ldots, T$.

26. (Sec. 10.4.2.) Let $\boldsymbol{A}_1$ be the matrix of Section 6.5.4. Let $\boldsymbol{B} = \boldsymbol{I} - (1/T)\boldsymbol{\varepsilon\varepsilon}'$. Show that the characteristic roots of $\boldsymbol{B}'\boldsymbol{A}_1\boldsymbol{B}$ are

$$-\tfrac{1}{2}, 0, \qquad T = 2,$$

$$-\tfrac{2}{3}, 0, 0, \qquad T = 3,$$

$$\frac{\sqrt{5}-1}{4}, -\tfrac{1}{4}, -\frac{\sqrt{5}+1}{4}, 0, \qquad T = 4.$$

APPENDIX A

Some Data and Statistics

A.1. THE BEVERIDGE WHEAT PRICE INDEX

The annual index of prices at which wheat was sold in European markets, 1500-1869, constructed by Beveridge (1921), was described briefly in Section 4.5. The data before and after removing the trend (as explained in Section 4.5) are given in Table A.1.1 and the "trend-free" index is graphed in Figure A.1.1. [Granger and Hughes (1969) have studied the effect of removing the trend.]

The first 60 correlations are tabulated in Table A.1.2 and graphed in Figure A.1.2. Here $r_k^* = C_k^*/C_0^*$.

The spectrogram, $R^2(\nu)$, was computed approximately by Beveridge (1921) for a number of frequencies of the form k/N, where N was an integer (usually divisible by 4) ranging from 276 to 356; subsequently, Beveridge (1922) calculated values for frequencies near those frequencies for which there was a relative maximum. His calculations are given in Table A.1.3. (No attempt was made to correct errors of calculation; none seemed large enough to be misleading.) The sample spectral density is proportional to the spectrogram. The logarithm of the spectrogram is plotted in Figure A.1.3 at frequencies approximately 0.002 apart.

The correlations in Table A.1.2 were used to calculate the estimate $2\pi\hat{f}_T^*(\lambda)$ with the Blackman-Tukey window with $a = 0.25$ (hanning). Three values of K were used: $K = 10, 20$, and 30; the values of λ were $\lambda = 2\pi j/200, j = 1, \ldots,$ 100. The estimates are tabulated in Table A.1.4 and graphed on a logarithmic scale in Figure A.1.4. The normalized spectral density was also estimated by the spectral densities of the fitted autoregressive process as given in Table 5.5 for $p = 2, 6, 8$. The estimates (multiplied by 2π) are graphed on a logarithmic scale in Figure A.1.5.

TABLE A.1.1

THE BEVERIDGE TREND-FREE WHEAT PRICE INDEX

	Actual Index Number	Trend-free Index		Actual Index Number	Trend-free Index		Actual Index Number	Trend-free Index
1500	17	106	1534	16	76	1568	34	77
1501	19	118	1535	22	102	1569	36	80
1502	20	124	1536	22	100	1570	43	93
1503	15	94	1537	16	73	1571	55	112
1504	13	82	1538	19	86	1572	64	131
1505	14	88	1539	17	74	1573	79	158
1506	14	87	1540	17	74	1574	59	113
1507	14	88	1541	19	76	1575	47	89
1508	14	88	1542	20	80	1576	48	87
1509	11	68	1543	24	96	1577	49	87
1510	16	98	1544	28	112	1578	45	79
1511	19	115	1545	36	144	1579	53	90
1512	23	135	1546	20	80	1580	55	90
1513	18	104	1547	14	54	1581	55	87
1514	17	96	1548	18	69	1582	54	83
1515	20	110	1549	27	100	1583	56	85
1516	20	107	1550	29	103	1584	52	76
1517	18	97	1551	36	129	1585	76	110
1518	14	75	1552	29	100	1586	113	161
1519	16	86	1553	27	90	1587	68	97
1520	21	111	1554	30	100	1588	59	84
1521	24	125	1555	38	123	1589	74	106
1522	15	78	1556	50	156	1590	78	111
1523	16	86	1557	24	71	1591	69	97
1524	20	102	1558	25	71	1592	78	108
1525	14	71	1559	30	81	1593	73	100
1526	16	81	1560	31	84	1594	88	119
1527	25.5	129	1561	37	97	1595	98	131
1528	25.8	130	1562	41	105	1596	109	143
1529	26	129	1563	36	90	1597	106	138
1530	26	125	1564	32	78	1598	87	112
1531	29	139	1565	47	112	1599	77	99
1532	20	97	1566	42	100	1600	77	97
1533	18	90	1567	37	86	1601	63	80

TABLE A.1.1 (*continued*)

	Actual Index Number	Trend-free Index		Actual Index Number	Trend-free Index		Actual Index Number	Trend-free Index
1602	70	90	1637	114	105	1672	72	84
1603	70	90	1638	103	97	1673	89	106
1604	63	80	1639	98	93	1674	114	134
1605	61	77	1640	103	99	1675	102	122
1606	66	81	1641	101	99	1676	85	102
1607	78	98	1642	110	107	1677	88	107
1608	93	115	1643	109	106	1678	97	115
1609	97	94	1644	98	96	1679	94	113
1610	77	93	1645	84	82	1680	88	104
1611	83	100	1646	90	88	1681	79	92
1612	81	99	1647	120	116	1682	74	84
1613	82	100	1648	124	122	1683	79	86
1614	78	94	1649	136	134	1684	95	101
1615	75	88	1650	120	119	1685	70	74
1616	80	92	1651	135	136	1686	72	75
1617	87	100	1652	100	102	1687	63	66
1618	72	82	1653	70	72	1688	60	62
1619	65	73	1654	60	63	1689	74	76
1620	74	81	1655	72	76	1690	75	79
1621	91	99	1656	70	75	1691	91	97
1622	115	124	1657	71	77	1692	126	134
1623	99	106	1658	94	103	1693	161	169
1624	99	106	1659	95	104	1694	109	111
1625	115	121	1660	110	120	1695	108	109
1626	101	105	1661	154	167	1696	110	111
1627	90	84	1662	116	126	1697	130	128
1628	95	97	1663	99	108	1698	166	163
1629	108	109	1664	82	91	1699	143	137
1630	147	148	1665	76	85	1700	103	99
1631	112	114	1666	64	73	1701	89	85
1632	108	108	1667	63	74	1702	76	72
1633	99	97	1668	68	80	1703	93	88
1634	96	92	1669	64	74	1704	82	77
1635	102	97	1670	67	78	1705	71	66
1636	105	98	1671	71	83	1706	69	64

TABLE A.1.1 (*continued*)

	Actual Index Number	Trend-free Index		Actual Index Number	Trend-free Index		Actual Index Number	Trend-free Index
1707	75	69	1742	94	90	1777	135	94
1708	134	125	1743	85	81	1778	125	87
1709	183	175	1744	89	84	1779	116	79
1710	113	108	1745	109	102	1780	132	87
1711	108	103	1746	110	102	1781	133	88
1712	121	115	1747	109	100	1782	144	94
1713	139	134	1748	120	109	1783	145	94
1714	109	108	1749	116	104	1784	146	92
1715	90	90	1750	101	90	1785	138	85
1716	88	89	1751	113	99	1786	139	84
1717	88	89	1752	109	95	1787	154	93
1718	93	94	1753	105	90	1788	181	108
1719	106	107	1754	94	80	1789	185	108
1720	89	89	1755	102	85	1790	151	86
1721	79	79	1756	141	117	1791	139	78
1722	91	91	1757	135	112	1792	157	87
1723	96	94	1758	118	95	1793	155	85
1724	111	110	1759	115	91	1794	191	103
1725	112	111	1760	111	88	1795	248	130
1726	104	103	1761	127	100	1796	185	95
1727	94	94	1762	124	97	1797	168	84
1728	98	101	1763	113	88	1798	176	87
1729	88	90	1764	122	95	1799	243	120
1730	94	96	1765	130	101	1800	289	139
1731	81	80	1766	137	106	1801	251	117
1732	77	76	1767	148	113	1802	232	105
1733	84	84	1768	142	108	1803	207	94
1734	92	91	1769	143	108	1804	276	125
1735	96	94	1770	176	131	1805	250	114
1736	102	101	1771	184	136	1806	216	98
1737	95	93	1772	164	119	1807	205	93
1738	98	91	1773	146	106	1808	206	94
1739	125	122	1774	147	105	1809	208	94
1740	162	159	1775	124	88	1810	226	104
1741	113	110	1776	119	84	1811	302	140

TABLE A.1.1 (*continued*)

	Actual Index Number	Trend-free Index		Actual Index Number	Trend-free Index		Actual Index Number	Trend-free Index
1812	261	121	1832	151	82	1851	180	86
1813	207	96	1833	144	80	1852	223	105
1814	209	96	1834	138	78	1853	294	138
1815	280	130	1835	145	82	1854	300	141
1816	381	178	1836	156	88	1855	297	138
1817	266	126	1837	184	102	1856	232	107
1818	197	94	1838	216	117	1857	179	82
1819	177	86	1839	204	107	1858	180	81
1820	170	84	1840	186	95	1859	215	97
1821	152	76	1841	197	101	1860	258	116
1822	156	77	1842	183	92	1861	236	107
1823	141	71	1843	175	88	1862	202	92
1824	142	71	1844	183	92	1863	174	79
1825	137	69	1845	230	115	1864	179	81
1826	161	82	1846	278	139	1865	210	94
1827	189	93	1847	179	90	1866	268	119
1828	226	114	1848	161	80	1867	267	118
1829	194	103	1849	150	74	1868	208	93
1830	217	110	1850	159	78	1869	224	102
1831	199	105						

Note: The "Actual Index Number" is the mean of the index numbers for the countries and places included, each of these being reduced to basis: Mean of prices for 1700–1745 = 100.

The next column, headed "Trend-free Index," shows the actual index number for each year as a percentage of the mean of the actual index numbers for 31 years of which that year is the center.

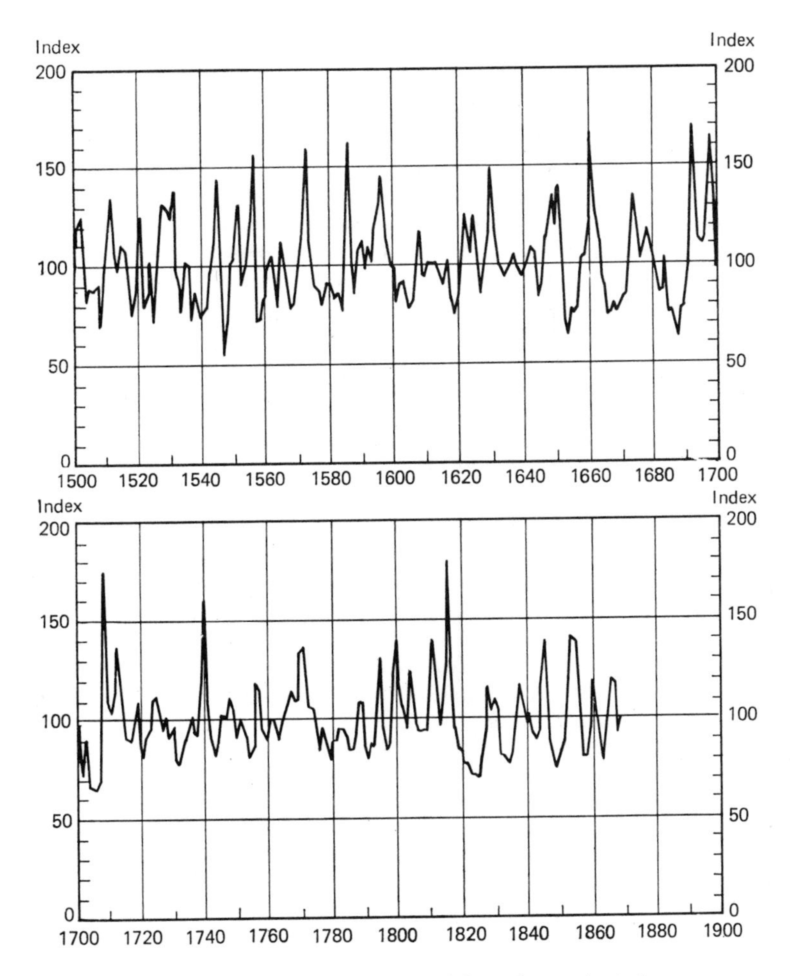

Figure A.1.1. The Beveridge trend-free wheat price index.

TABLE A.1.2

CORRELATION SEQUENCE OF THE BEVERIDGE TREND-FREE WHEAT PRICE INDEX

Order of Correlation k	r_k^*	k	r_k^*	k	r_k^*	k	r_k^*
1	0.562	16	0.158	31	0.060	46	−0.036
2	0.103	17	0.109	32	−0.008	47	−0.013
3	−0.075	18	0.002	33	−0.039	48	0.042
4	−0.092	19	−0.075	34	0.007	49	0.062
5	−0.082	20	−0.062	35	0.056	50	0.065
6	−0.136	21	−0.021	36	0.010	51	0.050
7	−0.211	22	−0.062	37	−0.004	52	0.009
8	−0.261	23	−0.088	38	−0.015	53	−0.027
9	−0.192	24	−0.084	39	−0.047	54	−0.053
10	−0.070	25	−0.076	40	−0.047	55	−0.073
11	−0.003	26	−0.091	41	0.008	56	−0.106
12	−0.015	27	−0.052	42	0.034	57	−0.084
13	−0.012	28	−0.032	43	0.065	58	−0.019
14	0.047	29	−0.012	44	0.099	59	0.003
15	0.101	30	0.059	45	0.009	60	0.010

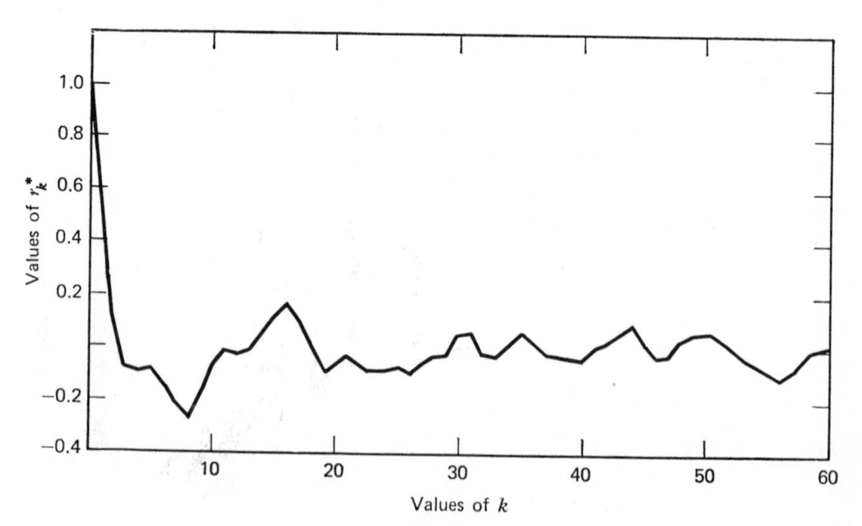

Figure A.1.2. Correlations of the Beveridge trend-free wheat price index.

TABLE A.1.3

SPECTROGRAM OF THE BEVERIDGE TREND-FREE WHEAT PRICE INDEX

Period	k	N	$A\left(\frac{2\pi k}{N}\right)$	$B\left(\frac{2\pi k}{N}\right)$	$\frac{NR^2\left(\frac{2\pi k}{N}\right)}{300}$	Frequency
2.000	150	300	0.11		0.01	0.5000
2.049	164	336	−0.40	−0.09	0.19	0.4881
2.054	148	304	0.48	−0.72	0.77	0.4868
2.061	165	340	0.38	−0.57	0.54	0.4853
2.069	145	300	0.25	0.63	0.46	0.4833
2.074	162	336	−0.61	0.51	0.71	0.4821
2.080	150	312	0.92	−0.50	1.14	0.4808
2.087	138	288	−0.52	−0.11	0.27	0.4792
2.095	147	308	−0.91	0.90	1.69	0.4773
2.105	152	320	0.90	0.07	0.86	0.4750
2.112	136	288	0.90	0.80	1.38	0.4722
2.133	150	320	0.89	0.15	0.84	0.4688
2.154	143	308	0.48	0.23	0.29	0.4643
2.182	132	288	1.32	−0.59	1.99	0.4583
2.200	140	308	−0.13	−0.60	0.39	0.4546
2.222	144	320	−0.32	−0.62	0.52	0.4500
2.261	138	312	0.50	−0.22	0.31	0.4423
2.286	140	320	−0.38	−0.85	0.93	0.4375
2.316	133	308	1.39	−1.05	3.11	0.4318
2.333	132	308	−0.10	−0.25	0.08	0.4286
2.353	136	320	0.90	0.07	0.86	0.4250
2.364	132	312	−0.12	−0.63	0.43	0.4231
2.370	135	320	0.05	−0.28	0.08	0.4219
2.375	128	304	0.29	−0.43	0.27	0.4210
2.381	126	300	−0.19	−1.22	1.53	0.4200
2.385	130	310	−1.00	−0.89	1.86	0.4194
2.391	138	330	−1.30	−0.54	2.18	0.4182
2.395	129	309	−0.72	0.60	0.90	0.4175
2.400	130	312	0.34	0.68	0.60	0.4167
2.412	136	328	−0.08	−0.65	0.47	0.4146
2.417	144	348	0.63	0.57	0.69	0.4138
2.435	138	336	0.44	0.01	0.22	0.4107
2.452	124	304	−1.40	−0.51	2.23	0.4079
2.462	130	320	−0.25	1.49	2.44	0.4062
2.476	126	312	−0.38	0.35	0.27	0.4038

TABLE A.1.3 (*continued*)

Period	k	N	$A\left(\frac{2\pi k}{N}\right)$	$B\left(\frac{2\pi k}{N}\right)$	$\frac{NR^2\left(\frac{2\pi k}{N}\right)}{300}$	Frequency
2.483	116	288	−0.07	0.74	0.53	0.4028
2.500	128	320	−0.24	1.19	1.56	0.4000
2.512	129	324	0.86	0.39	0.97	0.3982
2.516	124	312	0.45	0.24	0.26	0.3974
2.529	119	301	−0.19	−0.31	0.13	0.3954
2.545	132	336	−1.39	−0.81	2.89	0.3929
2.555	126	322	0.38	0.50	0.42	0.3913
2.571	119	306	1.25	0.55	1.91	0.3889
2.588	119	308	0.30	0.43	0.28	0.3864
2.600	120	312	1.02	−0.39	1.25	0.3846
2.615	117	306	−0.75	−0.24	0.63	0.3824
2.625	112	294	−0.45	1.36	2.01	0.3810
2.643	112	296	0.95	−0.62	1.27	0.3784
2.667	117	312	−0.92	1.20	2.38	0.3750
2.687	112	301	1.23	−0.02	1.52	0.3721
2.692	117	315	−0.04	0.23	0.06	0.3714
2.706	119	322	−0.27	1.33	1.97	0.3696
2.714	112	304	0.83	1.17	2.10	0.3684
2.727	110	300	0.86	1.46	2.87	0.3667
2.733	105	287	2.05	1.19	6.16	0.3658
2.735	102	279	2.44	1.23	7.82	0.3656
2.737	114	312	2.23	1.00	6.22	0.3654
2.741	108	296	2.43	0.25	5.86	0.3649
2.750	112	308	0.90	−0.84	1.55	0.3636
2.762	126	348	−0.57	−0.04	0.37	0.3621
2.769	117	324	1.49	0.23	2.28	0.3611
2.778	117	325	1.20	−0.92	2.48	0.3600
2.800	120	336	−1.01	−0.19	1.18	0.3571
2.818	110	310	0.55	1.07	1.49	0.3548
2.833	114	323	0.78	−0.10	0.67	0.3529
2.846	104	296	0.41	0.42	0.34	0.3514
2.857	112	320	0.96	0.21	1.03	0.3500
2.875	112	322	0.35	0.14	0.15	0.3478
2.888	108	312	1.51	0.26	2.43	0.3462
2.895	114	330	−0.69	−1.57	3.21	0.3454
2.909	110	320	0.70	−1.11	1.84	0.3438

TABLE A.1.3 (*continued*)

Period	k	N	$A\left(\frac{2\pi k}{N}\right)$	$B\left(\frac{2\pi k}{N}\right)$	$\frac{NR^2\left(\frac{2\pi k}{N}\right)}{300}$	Frequency
2.933	105	308	−0.04	0.39	0.16	0.3409
2.947	114	336	−0.93	−1.19	2.57	0.3393
2.960	100	296	−0.00	−1.15	1.30	0.3378
3.000	100	300	−0.29	−0.39	0.23	0.3333
3.040	100	304	0.09	0.75	0.58	0.3290
3.077	104	320	0.05	1.18	1.50	0.3250
3.111	108	336	0.91	−0.44	1.15	0.3214
3.143	98	308	2.01	0.23	4.20	0.3182
3.167	96	304	0.46	−1.05	1.33	0.3158
3.200	100	320	0.43	0.95	1.16	0.3125
3.217	92	296	1.25	0.00	1.55	0.3108
3.250	96	312	−1.22	−0.47	1.80	0.3077
3.273	99	324	−0.55	1.18	1.82	0.3056
3.286	98	322	−0.11	0.99	1.07	0.3044
3.304	92	304	0.13	0.75	0.59	0.3026
3.333	96	320	0.90	1.58	3.54	0.3000
3.364	88	296	1.76	0.98	4.00	0.2973
3.375	96	324	0.55	0.92	1.24	0.2963
3.385	91	308	0.35	1.03	1.21	0.2954
3.400	95	323	1.12	2.37	7.41	0.2941
3.407	81	276	2.98	2.81	14.90	0.2935
3.412	102	348	1.27	−3.98	15.53	0.2931
3.417	96	328	3.08	−2.24	15.84	0.2927
3.429	84	288	3.11	−1.40	11.16	0.2917
3.444	90	310	0.09	−0.99	1.03	0.2903
3.455	88	304	0.55	0.29	0.39	0.2895
3.462	91	315	1.57	1.02	4.87	0.2889
3.500	88	308	1.20	−0.94	2.38	0.2857
3.524	84	296	1.41	−1.18	3.31	0.2838
3.538	91	322	0.50	−1.45	2.53	0.2826
3.556	90	320	0.02	−0.43	0.20	0.2812
3.571	91	325	0.80	−0.69	1.21	0.2800
3.600	90	324	−1.03	0.82	1.88	0.2778
3.619	84	304	1.18	1.23	2.94	0.2763
3.636	88	320	1.14	0.13	1.39	0.2750
3.643	84	306	−0.16	0.27	0.10	0.2745

TABLE A.1.3 (*continued*)

Period	k	N	$A\left(\frac{2\pi k}{N}\right)$	$B\left(\frac{2\pi k}{N}\right)$	$\frac{NR^2\left(\frac{2\pi k}{N}\right)}{300}$	Frequency
3.667	84	308	−2.14	−1.07	5.87	0.2727
3.679	84	309	0.34	−1.90	3.83	0.2718
3.692	78	288	1.28	−0.22	1.63	0.2708
3.700	80	296	0.90	−0.59	1.18	0.2703
3.714	84	312	1.15	1.78	4.65	0.2692
3.727	77	287	−0.45	−1.65	2.72	0.2683
3.750	84	315	0.64	−0.06	0.44	0.2667
3.778	81	306	−1.17	−0.68	1.86	0.2647
3.800	80	304	1.60	0.80	3.24	0.2632
3.833	84	322	−1.12	−1.63	4.17	0.2609
3.857	84	324	1.63	0.45	3.08	0.2593
3.888	72	280	−0.15	0.66	0.43	0.2571
3.895	76	296	−0.66	1.00	1.42	0.2568
3.923	78	306	0.64	−1.61	3.06	0.2549
3.962	78	309	−0.67	1.74	3.59	0.2524
4.000	75	300	1.47	−1.13	3.64	0.2500
4.077	78	318	0.57	−0.26	0.41	0.2453
4.111	72	296	1.13	−1.70	4.13	0.2432
4.143	70	290	−0.50	0.23	0.30	0.2414
4.167	78	325	1.21	0.32	1.70	0.2400
4.173	77	322	0.66	−1.46	2.77	0.2391
4.200	70	294	−0.99	−0.41	1.02	0.2381
4.250	76	323	0.50	−2.73	8.32	0.2353
4.286	70	300	−0.65	0.79	1.04	0.2333
4.333	72	312	−1.50	−1.30	4.10	0.2308
4.353	68	296	−2.85	−0.24	8.05	0.2297
4.364	66	288	−2.98	0.75	9.07	0.2292
4.375	72	315	−2.47	0.87	7.19	0.2286
4.385	78	342	−0.50	2.55	7.72	0.2281
4.400	70	308	−1.38	3.27	12.89	0.2273
4.412	68	300	0.08	3.62	13.11	0.2267
4.417	72	318	0.87	3.85	16.48	0.2264
4.429	70	310	1.80	2.41	9.32	0.2258
4.444	72	320	2.15	0.83	5.66	0.2250
4.471	68	304	0.91	0.79	1.48	0.2237
4.500	68	306	1.87	0.72	4.09	0.2222
4.571	70	320	−0.21	0.04	0.22	0.2188

TABLE A.1.3 (*continued*)

Period	k	N	$A\left(\frac{2\pi k}{N}\right)$	$B\left(\frac{2\pi k}{N}\right)$	$\frac{NR^2\left(\frac{2\pi k}{N}\right)}{300}$	Frequency
4.600	70	322	−0.08	1.24	1.65	0.2174
4.667	72	336	0.19	0.93	1.00	0.2143
4.750	64	304	−0.12	2.28	5.28	0.2105
4.800	60	288	2.44	1.08	6.84	0.2083
4.857	63	306	−1.06	−1.30	2.89	0.2059
4.888	64	312	−1.80	2.11	8.00	0.2051
4.933	60	296	1.57	1.58	4.91	0.2027
5.000	60	300	1.85	1.00	4.30	0.2000
5.067	60	304	−0.05	3.98	16.09	0.1974
5.091	66	336	−0.73	5.55	35.05	0.1964
5.100	60	306	5.71	2.98	42.34	0.1961
5.111	63	322	5.70	0.29	34.91	0.1956
5.125	64	328	3.97	2.90	26.38	0.1951
5.143	63	324	2.46	2.46	13.09	0.1944
5.200	60	312	0.02	0.30	0.10	0.1923
5.250	56	294	1.74	1.92	6.56	0.1905
5.333	60	320	0.71	−4.46	21.72	0.1875
5.400	60	324	1.04	3.71	16.06	0.1852
5.415	60	325	4.27	1.90	23.66	0.1846
5.429	56	304	4.72	−0.28	22.61	0.1842
5.455	55	300	1.37	−3.73	15.76	0.1833
5.500	56	308	−1.04	1.49	3.39	0.1818
5.555	54	300	2.40	−0.68	6.23	0.1800
5.600	60	336	0.46	1.21	1.88	0.1786
5.667	54	306	5.31	−1.97	32.72	0.1765
5.692	52	296	2.05	−3.91	19.18	0.1757
5.714	56	320	0.35	−2.13	4.97	0.1750
5.750	56	322	1.39	−0.33	2.18	0.1739
5.800	50	290	3.55	−2.75	19.47	0.1724
5.846	52	304	0.00	−2.29	5.35	0.1710
5.933	60	356	4.37	0.91	23.63	0.1685
6.000	50	300	−3.50	−0.12	12.29	0.1667
6.111	54	330	−0.79	−1.90	4.66	0.1636
6.143	49	301	0.74	−2.96	9.32	0.1628
6.167	48	296	−0.22	−2.94	8.56	0.1622
6.200	50	310	−2.02	−3.38	16.02	0.1613
6.250	52	325	−3.23	−0.11	11.30	0.1600

TABLE A.1.3 (*continued*)

Period	k	N	$A\left(\frac{2\pi k}{N}\right)$	$B\left(\frac{2\pi k}{N}\right)$	$\frac{NR^2\left(\frac{2\pi k}{N}\right)}{300}$	Frequency
6.286	49	308	−1.72	−0.59	3.41	0.1591
6.333	48	304	−1.52	1.29	4.02	0.1579
6.400	50	320	0.80	2.74	8.71	0.1562
6.500	48	312	0.69	−0.73	0.94	0.1538
6.571	49	322	1.49	−0.77	3.02	0.1522
6.667	48	320	0.25	0.21	0.11	0.1500
6.727	44	296	0.08	−0.13	0.02	0.1486
6.750	48	324	−0.20	−1.66	3.01	0.1482
6.800	45	306	0.23	−0.65	0.48	0.1471
6.909	44	304	0.58	2.56	7.00	0.1447
6.933	45	312	1.68	2.01	7.15	0.1442
7.000	44	308	3.10	−2.17	14.74	0.1429
7.143	42	300	1.83	−1.86	6.79	0.1400
7.200	45	324	0.54	−3.93	16.96	0.1389
7.333	42	308	1.52	−2.81	10.46	0.1364
7.400	40	296	−2.33	−2.72	12.65	0.1351
7.417	48	356	1.50	−4.01	21.72	0.1348
7.429	42	312	−3.80	−1.49	17.28	0.1346
7.500	42	315	0.17	1.50	2.40	0.1333
7.600	40	304	−2.33	−1.37	7.43	0.1316
7.667	42	322	−1.46	−2.61	9.57	0.1304
7.750	40	310	1.38	−0.39	2.13	0.1290
7.857	42	330	−0.50	0.28	0.36	0.1273
8.000	39	312	−3.96	1.34	18.63	0.1250
8.091	44	356	4.32	−0.98	23.23	0.1236
8.200	35	287	1.62	−0.64	2.90	0.1220
8.222	36	296	0.19	−0.56	0.34	0.1216
8.333	39	325	0.21	0.91	0.95	0.1200
8.500	38	323	0.17	3.19	10.41	0.1176
8.667	36	312	2.51	−1.01	7.59	0.1154
8.800	35	308	2.97	0.83	9.77	0.1136
9.000	34	306	−1.51	−0.57	2.65	0.1111
9.200	35	322	−0.16	−1.56	2.65	0.1087
9.333	36	336	−0.74	0.64	1.08	0.1071
9.500	32	304	1.08	1.07	2.26	0.1053
9.667	30	290	5.03	0.37	24.55	0.1034
9.750	32	312	4.46	−3.56	33.89	0.1026

TABLE A.1.3 (*continued*)

Period	k	N	$A\left(\frac{2\pi k}{N}\right)$	$B\left(\frac{2\pi k}{N}\right)$	$\frac{NR^2\left(\frac{2\pi k}{N}\right)}{300}$	Frequency
9.818	33	324	1.21	−4.94	27.90	0.1018
10.000	32	320	−1.19	−0.83	2.25	0.1000
10.200	30	306	0.86	−0.22	0.80	0.0980
10.250	32	328	−0.69	1.10	1.84	0.0976
10.400	30	312	1.88	−1.65	6.52	0.0962
10.500	28	294	2.46	−1.82	9.19	0.0952
10.750	28	301	1.47	−3.13	11.98	0.0930
10.800	30	324	1.00	−4.75	25.48	0.0926
11.000	28	308	−3.85	−4.26	33.84	0.0909
11.200	30	336	−2.48	0.55	7.24	0.0893
11.500	28	322	−1.32	−0.66	2.34	0.0870
11.667	24	280	0.46	1.42	2.07	0.0857
12.000	26	312	−2.47	−4.04	23.30	0.0833
12.143	28	340	−0.22	−4.37	21.66	0.0824
12.333	24	296	−2.44	2.74	11.43	0.0811
12.500	26	325	−1.22	2.63	9.13	0.0800
12.667	24	304	2.28	5.19	32.58	0.0789
12.800	25	320	5.70	3.26	46.01	0.0781
12.875	24	309	6.46	0.77	43.58	0.0777
13.000	24	312	4.26	−4.32	38.23	0.0769
13.333	24	320	0.40	0.37	0.32	0.0750
13.500	24	324	2.56	−2.09	11.79	0.0741
13.667	24	328	3.49	−1.34	15.28	0.0732
14.000	22	308	1.15	−1.00	2.38	0.0714
14.500	20	290	−3.78	−0.18	13.82	0.0690
14.667	21	308	−1.50	4.23	20.69	0.0682
15.000	20	300	6.32	−2.66	46.83	0.0667
15.200	20	304	1.19	−8.52	75.04	0.0658
15.250	20	305	−0.28	−8.65	76.17	0.0656
15.286	21	321	−2.35	−7.15	60.62	0.0654
15.333	21	322	−3.89	−6.55	62.29	0.0652
15.500	20	310	−6.92	−2.02	59.11	0.0645
16.000	20	320	−1.46	4.52	24.02	0.0625
16.667	18	300	5.21	−0.39	27.33	0.0600
17.000	18	306	2.56	−6.35	47.84	0.0588
17.333	18	312	−3.04	−6.65	54.55	0.0577
17.500	16	280	−6.18	−4.45	54.12	0.0571

TABLE A.1.3 (*continued*)

Period	k	N	$A\left(\frac{2\pi k}{N}\right)$	$B\left(\frac{2\pi k}{N}\right)$	$\frac{NR^2\left(\frac{2\pi k}{N}\right)}{300}$	Frequency
18.000	17	306	−4.40	1.25	21.29	0.0556
18.500	16	296	−1.46	2.25	7.10	0.0540
19.000	16	304	1.00	−0.23	1.07	0.0526
19.750	16	316	−4.73	−1.59	26.25	0.0506
20.000	16	320	−5.71	1.69	37.88	0.0500
21.000	14	294	0.78	2.61	7.28	0.0476
22.000	14	308	1.87	1.58	6.18	0.0454
23.000	14	322	−2.45	−1.43	8.61	0.0435
24.000	12	288	0.45	5.19	26.10	0.0417
24.667	12	296	4.31	1.99	22.21	0.0405
25.000	13	325	3.86	−0.19	14.94	0.0400
26.000	12	312	1.23	−1.34	3.43	0.0385
27.000	12	324	0.50	−0.33	0.38	0.0370
28.000	11	308	−0.49	0.68	0.72	0.0357
29.000	10	290	1.08	−2.12	5.46	0.0345
30.000	10	300	−1.53	−2.34	7.81	0.0333
31.000	10	310	−1.98	0.13	4.06	0.0322
32.000	10	320	−0.37	0.51	0.42	0.0312
33.000	10	330	0.96	−0.78	1.68	0.0303
34.000	9	306	−3.00	−2.15	13.90	0.0294
35.000	8	280	−4.64	1.79	23.11	0.0286
36.000	8	288	−1.65	4.85	23.29	0.0278
37.000	8	296	2.08	3.92	19.47	0.0270
38.000	8	304	2.99	0.56	9.37	0.0263
40.000	8	320	−1.44	−0.63	2.63	0.0250
41.000	8	328	−1.93	0.93	5.01	0.0244
42.000	7	294	0.93	3.02	9.75	0.0238
44.000	7	308	3.00	−0.14	9.27	0.0227
45.000	7	315	1.69	−1.99	7.14	0.0222
46.000	7	322	0.16	−2.27	5.58	0.0217
48.000	6	288	−0.76	−0.09	0.56	0.0208
50.000	6	300	1.83	2.19	8.14	0.0200
52.000	6	312	4.77	−0.57	24.03	0.0192
53.000	6	318	4.22	−2.60	26.08	0.0189
54.000	6	324	2.84	−4.01	26.09	0.0185
55.000	6	330	3.54	−3.30	25.82	0.0182

TABLE A.1.3 (*continued*)

Period	k	N	$A\left(\frac{2\pi k}{N}\right)$	$B\left(\frac{2\pi k}{N}\right)$	$\frac{NR^2\left(\frac{2\pi k}{N}\right)}{300}$	Frequency
56.000	6	336	3.31	−2.36	18.47	0.0178
58.000	5	290	3.89	1.49	16.82	0.0172
60.000	5	300	−3.08	−0.93	10.32	0.0167
62.000	5	310	−1.62	0.39	2.88	0.0161
64.000	5	320	−0.78	0.13	0.66	0.0156
66.000	5	330	−0.56	−0.56	0.69	0.0152
68.000	5	340	2.90	−1.88	13.58	0.0147
70.000	4	280	−0.69	−0.16	0.47	0.0143
74.000	4	296	−1.20	0.82	2.07	0.0135
76.000	4	304	−0.66	1.17	1.83	0.0132
78.000	4	312	0.58	1.26	2.00	0.0128
80.000	4	320	0.77	0.82	1.34	0.0125
84.000	4	336	0.26	0.69	0.62	0.0119

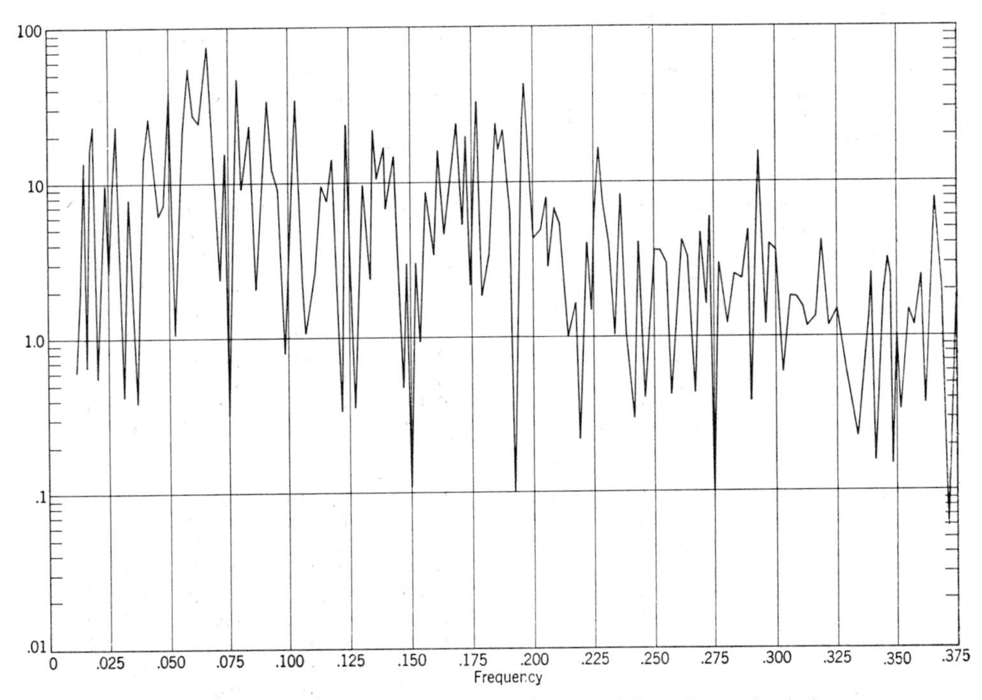

Figure A.1.3. Spectrogram of the Beveridge trend-free wheat price index.

TABLE A.1.4

NORMALIZED BLACKMAN-TUKEY ESTIMATE OF THE SPECTRAL DENSITY

Frequency	K 10	K 20	K 30	Frequency	K 10	K 20	K 30
0.005	0.2751	0.1293	0.0971	0.205	0.1910	0.2156	0.2116
0.010	0.2785	0.1426	0.1152	0.210	0.1836	0.1951	0.1775
0.015	0.2840	0.1648	0.1412	0.215	0.1758	0.1742	0.1495
0.020	0.2914	0.1954	0.1718	0.220	0.1677	0.1546	0.1297
0.025	0.3003	0.2336	0.2059	0.225	0.1593	0.1371	0.1177
0.030	0.3104	0.2781	0.2455	0.230	0.1508	0.1225	0.1115
0.035	0.3211	0.3265	0.2942	0.235	0.1422	0.1110	0.1080
0.040	0.3320	0.3760	0.3544	0.240	0.1337	0.1023	0.1046
0.045	0.3427	0.4232	0.4245	0.245	0.1254	0.0962	0.0997
0.050	0.3527	0.4646	0.4975	0.250	0.1174	0.0919	0.0931
0.055	0.3615	0.4968	0.5625	0.255	0.1097	0.0889	0.0859
0.060	0.3688	0.5175	0.6072	0.260	0.1026	0.0866	0.0800
0.065	0.3743	0.5252	0.6229	0.265	0.0959	0.0848	0.0767
0.070	0.3778	0.5199	0.6066	0.270	0.0898	0.0830	0.0765
0.075	0.3793	0.5027	0.5627	0.275	0.0843	0.0812	0.0788
0.080	0.3785	0.4761	0.5008	0.280	0.0793	0.0792	0.0818
0.085	0.3758	0.4431	0.4330	0.285	0.0748	0.0769	0.0836
0.090	0.3710	0.4071	0.3705	0.290	0.0708	0.0743	0.0828
0.095	0.3646	0.3710	0.3209	0.295	0.0673	0.0712	0.0789
0.100	0.3566	0.3376	0.2873	0.300	0.0641	0.0677	0.0724
0.105	0.3475	0.3085	0.2685	0.305	0.0612	0.0639	0.0647
0.110	0.3375	0.2845	0.2608	0.310	0.0587	0.0597	0.0572
0.115	0.3270	0.2657	0.2593	0.315	0.0563	0.0555	0.0509
0.120	0.3163	0.2515	0.2589	0.320	0.0541	0.0513	0.0464
0.125	0.3056	0.2414	0.2559	0.325	0.0520	0.0476	0.0433
0.130	0.2951	0.2346	0.2483	0.330	0.0500	0.0443	0.0413
0.135	0.2852	0.2305	0.2364	0.335	0.0480	0.0417	0.0398
0.140	0.2759	0.2289	0.2222	0.340	0.0462	0.0398	0.0385
0.145	0.2672	0.2298	0.2097	0.345	0.0444	0.0385	0.0375
0.150	0.2593	0.2330	0.2030	0.350	0.0426	0.0377	0.0369
0.155	0.2520	0.2384	0.2056	0.355	0.0410	0.0371	0.0367
0.160	0.2453	0.2454	0.2187	0.360	0.0394	0.0365	0.0369
0.165	0.2392	0.2533	0.2406	0.365	0.0379	0.0359	0.0372
0.170	0.2333	0.2608	0.2671	0.370	0.0365	0.0350	0.0370
0.175	0.2277	0.2666	0.2919	0.375	0.0352	0.0340	0.0361
0.180	0.2222	0.2692	0.3089	0.380	0.0340	0.0329	0.0344
0.185	0.2165	0.2677	0.3133	0.385	0.0330	0.0317	0.0321
0.190	0.2107	0.2613	0.3032	0.390	0.0320	0.0307	0.0296
0.195	0.2045	0.2500	0.2800	0.395	0.0311	0.0298	0.0275
0.200	0.1980	0.2344	0.2476	0.400	0.0303	0.0293	0.0261

TABLE A.1.4 (*continued*)

Frequency	K 10	K 20	K 30	Frequency	K 10	K 20	K 30
0.405	0.0296	0.0290	0.0258	0.455	0.0232	0.0229	0.0210
0.410	0.0289	0.0290	0.0265	0.460	0.0226	0.0214	0.0186
0.415	0.0283	0.0290	0.0280	0.465	0.0220	0.0201	0.0172
0.420	0.0277	0.0291	0.0299	0.470	0.0214	0.0190	0.0167
0.425	0.0270	0.0290	0.0316	0.475	0.0209	0.0182	0.0169
0.430	0.0264	0.0287	0.0324	0.480	0.0204	0.0177	0.0174
0.435	0.0258	0.0281	0.0320	0.485	0.0201	0.0174	0.0177
0.440	0.0252	0.0271	0.0303	0.490	0.0198	0.0173	0.0178
0.445	0.0245	0.0258	0.0275	0.495	0.0196	0.0172	0.0177
0.450	0.0239	0.0244	0.0242	0.500	0.0196	0.0172	0.0177

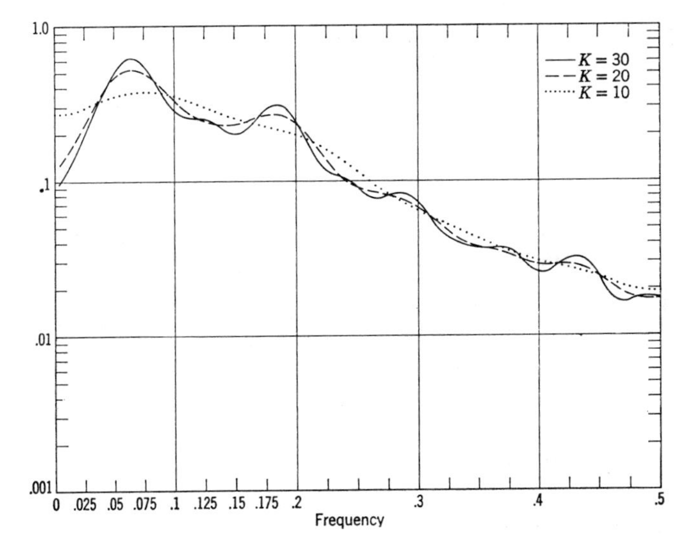

Figure A.1.4. The normalized Blackman-Tukey estimate ($a = 0.25$) of the spectral density of the Beveridge trend-free wheat price index.

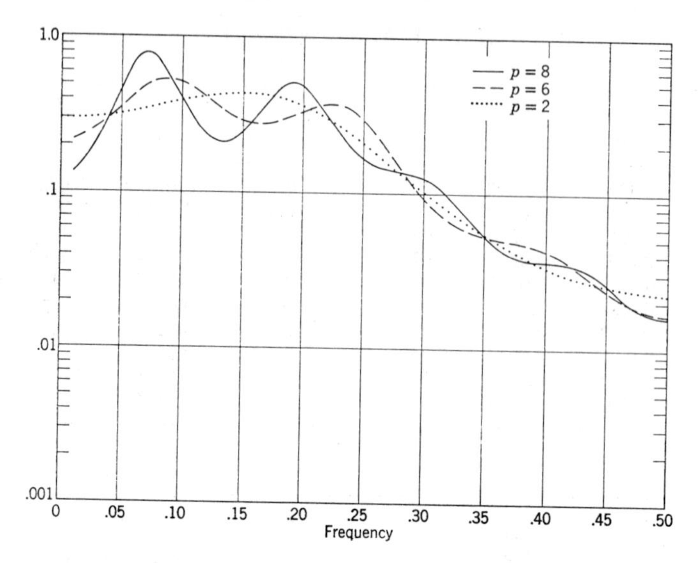

Figure A.1.5. The spectral densities of estimated autoregressive processes for the Beveridge trend-free wheat price index.

A.2. THREE SECOND-ORDER AUTOREGRESSIVE PROCESSES GENERATED BY RANDOM NUMBERS

Many series were generated by Wold (1965) using random numbers according to

$$y_t + \beta_1 y_{t-1} + \beta_2 y_{t-2} = u_t, \tag{1}$$

where $\beta_1 = -\gamma$ and $\beta_2 = \gamma^2$. Then the roots of the associated polynomial equation are $\gamma e^{\pm i2\pi/6}$. The mean of u_t is 0 and the variance is $(1 - \gamma^6)/(1 + \gamma^2)$ in order that the variance of y_t is 1. The u_t's were random normal deviates. Each series was started with $y_{-1} = u_{-1}$ and $y_0 = u_0 + \gamma y_{-1}$; $T = 200$. Wold generated 100 series for each of three values of γ, namely $\gamma = 0.25$, $\gamma = 0.7$, and $\gamma = 0.9$. In Table A.2.1 we give the second series for $\gamma = 0.25$, the second for $\gamma = 0.7$, and the first for $\gamma = 0.9$. They are graphed in Figures A.2.1, A.2.2, and A.2.3. Wold gives some graphs of others.

TABLE A.2.1

THE OBSERVED SERIES

A. $\beta_1 = -0.25$, $\beta_2 = 0.0625$. (Wold's Sample No. 2)

t	y_t	t	y_t	t	y_t	t	y_t
1	1.2268	51	−0.1108	101	0.4048	151	−1.0168
2	0.3973	52	0.8899	102	−1.0996	152	0.7853
3	0.1334	53	0.9237	103	−0.5004	153	−1.4676
4	−0.3104	54	1.2217	104	−1.1657	154	−0.9854
5	−0.4614	55	−0.1803	105	0.3554	155	−1.8534
6	0.8008	56	0.8025	106	−0.5300	156	−1.7285
7	1.7893	57	−1.1287	107	0.2337	157	−1.1020
8	−0.5069	58	−0.0972	108	−1.1382	158	1.2012
9	1.6061	59	2.6301	109	−1.4920	159	1.6259
10	2.4069	60	0.5941	110	−0.7520	160	1.1087
11	−0.2187	61	0.6991	111	0.2857	161	1.8593
12	1.1178	62	0.4784	112	−0.7418	162	1.3407
13	0.3816	63	−3.0912	113	0.4471	163	2.0435
14	−0.4851	64	−0.5962	114	1.3734	164	0.5902
15	−1.3512	65	−1.1888	115	−1.1441	165	−0.9144
16	−0.8527	66	−2.0397	116	−1.5980	166	−1.2349
17	0.1690	67	−0.5262	117	0.3228	167	−1.8843
18	−0.0702	68	−0.7187	118	−0.2360	168	−0.2991
19	−0.8201	69	1.2073	119	−1.2372	169	0.2676
20	−0.4147	70	0.9935	120	−0.5845	170	−0.4726
21	−0.9631	71	−1.2825	121	2.1724	171	1.5744
22	−0.5125	72	−0.6821	122	1.8074	172	0.2603
23	−0.8878	73	−1.3399	123	0.1891	173	−1.5125
24	−1.1251	74	−1.2830	124	0.2040	174	−0.9015
25	0.1423	75	−0.3670	125	−0.2196	175	−1.3728
26	0.0267	76	−1.9386	126	−1.9188	176	1.0273
27	0.2646	77	−0.8190	127	0.8898	177	−0.2447
28	−1.2587	78	−2.0741	128	−1.0344	178	0.2098
29	−0.4948	79	0.3179	129	−0.1844	179	−1.1214
30	0.9793	80	−2.4611	130	−0.6813	180	−0.3899
31	−0.2555	81	−0.4927	131	−0.1136	181	−0.7416
32	−0.8822	82	−0.5582	132	−1.8186	182	−0.1831
33	−0.4405	83	−1.3925	133	0.5493	183	0.7315
34	0.1329	84	−1.0148	134	1.5896	184	0.6945
35	0.3030	85	−0.6616	135	0.1114	185	−0.9982
36	2.6584	86	−1.2155	136	−0.1702	186	0.2781
37	0.0972	87	2.0345	137	0.8204	187	−0.0199
38	−0.3308	88	2.3948	138	−2.2823	188	−0.3416
39	0.6895	89	0.5303	139	0.2210	189	0.2993
40	0.2827	90	−0.0755	140	−0.5385	190	0.9690
41	0.9795	91	−0.3050	141	0.0591	191	−0.9832
42	0.1590	92	0.2302	142	0.0450	192	1.1158
43	−1.7317	93	−0.0099	143	−0.1498	193	−0.7260
44	0.2491	94	−0.4606	144	−2.5719	194	0.7717
45	1.9132	95	0.6743	145	0.9076	195	0.5904
46	0.7171	96	0.7211	146	0.6395	196	1.2021
47	1.1631	97	−0.7078	147	0.9806	197	−0.2385
48	0.7018	98	0.0573	148	0.9813	198	−1.3084
49	0.6421	99	−0.0437	149	−0.0761	199	−0.2723
50	−1.4017	100	0.1556	150	−0.6912	200	−0.6236

$y_{-1} = -0.0729$, $y_0 = 0.5054$

TABLE A.2.1 (*continued*)

B. $\beta_1 = -0.7, \beta_2 = 0.49$. (Wold's Sample No. 2)

t	y_t	t	y_t	t	y_t	t	y_t
1	1.2134	51	−0.7713	101	0.5270	151	−1.4259
2	0.7306	52	0.7760	102	−0.6797	152	0.3695
3	0.0047	53	1.4720	103	−0.8929	153	−0.4131
4	−0.6077	54	1.4803	104	−1.1720	154	−0.9219
5	−0.7256	55	−0.0246	105	0.1055	155	−1.7906
6	0.5013	56	−0.0096	106	0.0993	156	−1.8542
7	1.9442	57	−1.0581	107	0.3260	157	−1.0438
8	0.3980	58	−0.5495	108	−0.7960	158	1.2637
9	0.7894	59	2.1837	109	−1.6632	159	2.3931
10	1.9233	60	1.7427	110	−1.1313	160	1.6726
11	0.3883	61	0.7170	111	0.3249	161	1.3340
12	0.3789	62	−0.0817	112	0.0993	162	0.8640
13	0.1451	63	−2.9210	113	0.4263	163	1.3986
14	−0.4891	64	−1.8409	114	1.2138	164	0.6851
15	−1.3703	65	−0.8355	115	−0.5171	165	−0.9469
16	−1.1521	66	−1.0947	116	−1.9294	166	−1.7676
17	0.1012	67	−0.4288	117	−0.5809	167	−2.0686
18	0.5038	68	−0.3307	118	0.2083	168	−0.5067
19	−0.3252	69	1.0528	119	−0.4882	169	0.8371
20	−0.6444	70	1.4121	120	−0.6738	170	0.3914
21	−1.0141	71	−0.6820	121	1.5455	171	1.2198
22	−0.6302	72	−1.4068	122	2.3861	172	0.5329
23	−0.5947	73	−1.6419	123	0.8122	173	−1.3982
24	−0.8493	74	−1.2459	124	−0.3867	174	−1.6421
25	−0.0111	75	−0.1707	125	−0.8740	175	−1.4496
26	0.3456	76	−1.0378	126	−1.8908	176	0.8325
27	0.4591	77	−0.9262	127	0.1803	177	0.8271
28	−0.8977	78	−1.7189	128	−0.0395	178	0.4369
29	−0.9831	79	−0.1265	129	−0.0130	179	−1.0428
30	0.5643	80	−1.3646	130	−0.5450	180	−1.0205
31	0.4553	81	−0.7803	131	−0.3392	181	−0.7700
32	−0.5585	82	−0.3447	132	−1.4244	182	−0.0564
33	−0.8012	83	−0.8773	133	−0.0401	183	0.9177
34	−0.1381	84	−1.0018	134	1.7319	184	1.0668
35	0.4880	85	−0.6640	135	1.0323	185	−0.5963
36	2.4647	86	−0.8573	136	−0.2043	186	−0.4871
37	1.0511	87	1.5475	137	0.0413	187	−0.1692
38	−0.6219	88	2.9393	138	−1.8527	188	−0.1330
39	−0.3329	89	1.3459	139	−0.6485	189	0.2940
40	0.1428	90	−0.5445	140	−0.1303	190	0.9634
41	1.0182	91	−1.2413	141	0.3912	191	−0.4270
42	0.5887	92	−0.3627	142	0.3350	192	0.3573
43	−1.4436	93	0.2857	143	−0.0820	193	−0.3867
44	−0.7501	94	0.0257	144	−2.2300	194	0.3658
45	1.5649	95	0.5038	145	−0.2982	195	0.7249
46	1.6647	96	0.7555	146	1.0837	196	1.2030
47	1.2739	97	−0.3891	147	1.6008	197	0.0886
48	0.4376	98	−0.4210	148	1.2053	198	−1.4585
49	0.1100	99	−0.1851	149	−0.1471	199	−1.0327
50	−1.3420	100	0.2117	150	−1.1781	200	−0.5138

$y_{-1} = -0.0729, y_0 = 0.4403$

TABLE A.2.1 (*continued*)

C. $\beta_1 = -0.9, \beta_2 = 0.81$. (Wold's Sample No. 1)

t	y_t	t	y_t	t	y_t	t	y_t
1	−0.9606	51	−0.0564	101	−0.8298	151	−0.7219
2	−1.0378	52	0.9502	102	−0.9909	152	−0.9665
3	−0.5484	53	0.2213	103	−0.2148	153	−0.2216
4	−1.0541	54	−0.8695	104	−0.0358	154	0.8067
5	−0.4834	55	−1.4278	105	0.9816	155	1.3913
6	0.6311	56	−1.0505	106	1.5830	156	1.2016
7	1.4114	57	0.4253	107	0.6708	157	−0.5651
8	0.3825	58	0.9019	108	−0.7696	158	−2.4274
9	−1.8239	59	0.7936	109	−1.6945	159	−1.6577
10	−1.4407	60	0.2804	110	0.0077	160	0.9373
11	−0.3132	61	−0.7642	111	1.3638	161	3.0506
12	−0.0874	62	−0.9103	112	1.0386	162	1.8579
13	−0.3918	63	−0.1189	113	−0.3156	163	−1.1132
14	−0.5888	64	1.4175	114	−1.0937	164	−2.9611
15	−0.5612	65	1.7326	115	−1.1822	165	−2.0225
16	0.0961	66	0.3792	116	−0.4685	166	0.5278
17	0.4954	67	−1.2491	117	0.7767	167	1.6486
18	0.1312	68	−0.4060	118	1.5031	168	1.8253
19	−0.5580	69	0.9696	119	1.1802	169	0.4094
20	−0.4685	70	2.1885	120	−0.8624	170	−0.7162
21	−0.1463	71	1.3596	121	−1.4084	171	−0.4403
22	0.5760	72	−0.8464	122	0.6630	172	0.4893
23	−0.0495	73	−1.2709	123	2.0564	173	0.9551
24	−1.0003	74	−0.5210	124	2.0627	174	0.4074
25	−1.8508	75	1.1009	125	0.7909	175	−0.5918
26	−1.0105	76	1.4884	126	−0.9423	176	−1.8504
27	0.7121	77	1.0229	127	−2.0920	177	−1.1401
28	0.4883	78	−0.8339	128	−0.5903	178	0.2450
29	0.2780	79	−2.0272	129	0.8511	179	1.3324
30	−0.2835	80	−1.4314	130	1.4750	180	1.3814
31	−0.5369	81	1.1195	131	0.7712	181	0.8292
32	−0.4541	82	1.5164	132	−1.5542	182	−0.3209
33	−0.2950	83	0.2551	133	−1.9841	183	−0.9560
34	−1.2701	84	−0.9049	134	−0.1077	184	0.4385
35	−1.0686	85	−0.8063	135	0.5974	185	0.8297
36	0.3006	86	0.0818	136	1.4342	186	0.9752
37	1.7958	87	1.8423	137	0.3418	187	−0.5203
38	1.3815	88	1.4011	138	−1.6969	188	−1.4231
39	−1.0157	89	−0.7910	139	−1.7782	189	−0.8991
40	−2.0178	90	−2.1526	140	−0.5346	190	0.3139
41	−0.6857	91	−2.0495	141	0.0004	191	0.7310
42	0.7797	92	−0.5901	142	0.2964	192	0.5303
43	0.8469	93	2.3577	143	0.5667	193	0.0094
44	0.0207	94	1.7757	144	0.4114	194	−0.3464
45	0.0693	95	−0.3490	145	−0.1130	195	−0.7523
46	0.8074	96	−1.9200	146	0.0279	196	−1.2773
47	1.2451	97	−1.3727	147	0.1295	197	−1.1861
48	0.0664	98	0.5940	148	0.2099	198	−0.1403
49	−0.8526	99	1.6419	149	−0.6045	199	1.6277
50	−0.1221	100	0.5217	150	−0.4056	200	1.0874

$y_{-1} = -0.0630,\ y_0 = -0.5200$

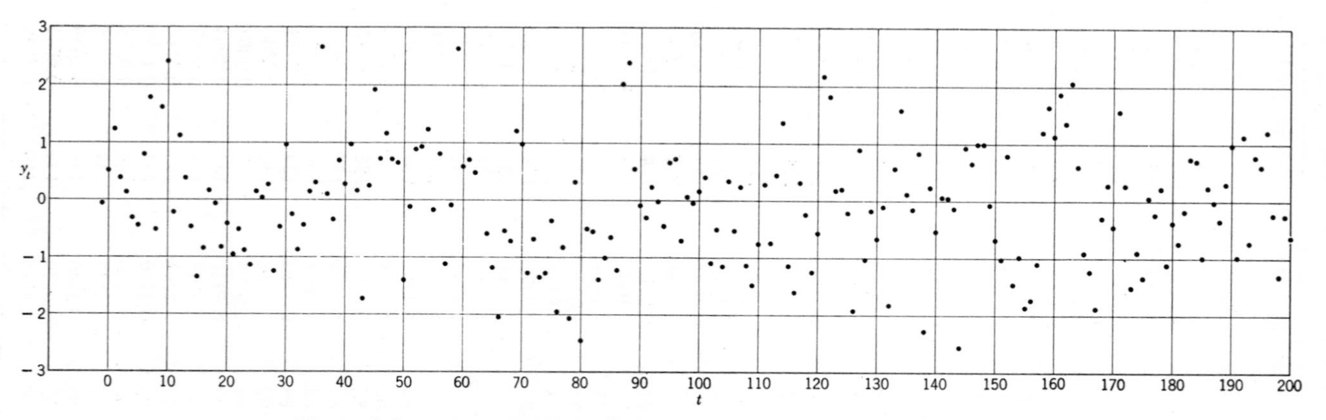

Figure A.2.1. A randomly generated autoregressive process with $\gamma = 0.25$.

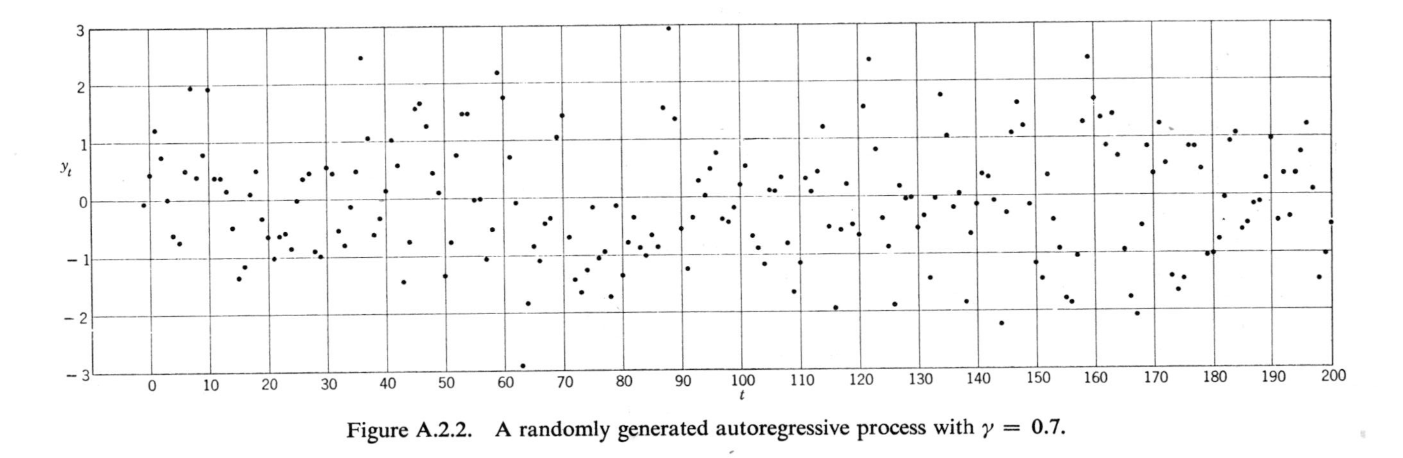

Figure A.2.2. A randomly generated autoregressive process with $\gamma = 0.7$.

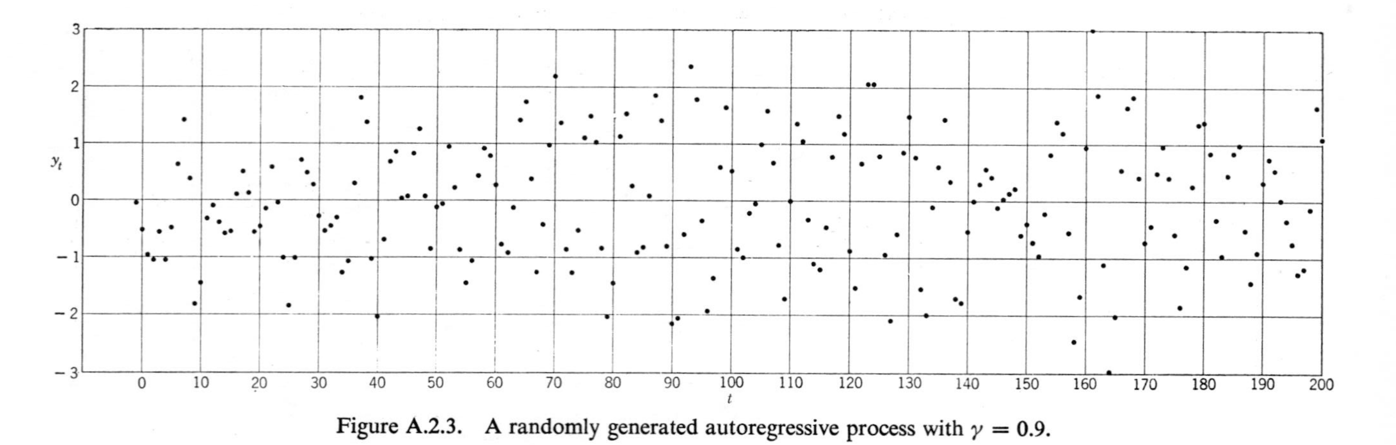

Figure A.2.3. A randomly generated autoregressive process with $\gamma = 0.9$.

The variance and correlation sequence is given for each of these series in Table A.2.2. The sample spectral density divided by the sample variance (and multiplied by 2π) in each case is given in Table A.2.3 and graphed in Figures A.2.4, A.2.5, and A.2.6. The Bartlett estimate (C of Section 9.2.3) of the normalized spectral density with $K = 20$ is tabulated in Table A.2.4 and also

TABLE A.2.2

VARIANCES AND CORRELATIONS

A. $\beta_1 = -0.25$, $\beta_2 = 0.0625$. (Wold's Sample No. 2)

Correlations

h	C_h/C_0	h	C_h/C_0	h	C_h/C_0	h	C_h/C_0
1	0.2473	26	0.0967	51	−0.0142	76	0.0515
2	0.1120	27	−0.1038	52	0.1814	77	−0.0103
3	0.0492	28	0.0269	53	0.0548	78	0.0435
4	−0.0868	29	0.0736	54	−0.0191	79	−0.1610
5	−0.0511	30	−0.0659	55	0.0126	80	−0.0683
6	0.0715	31	−0.0833	56	−0.1752	81	0.0749
7	−0.0224	32	−0.0679	57	−0.0746	82	0.0174
8	0.0167	33	−0.0796	58	0.0725	83	−0.0086
9	−0.0250	34	0.1311	59	0.0299	84	−0.1150
10	0.0291	35	0.1565	60	0.1072	85	−0.1234
11	0.0224	36	0.0443	61	0.0078	86	−0.0354
12	0.0906	37	−0.0264	62	0.0766	87	0.0468
13	0.0936	38	−0.0260	63	0.0151	88	0.1223
14	−0.0458	39	−0.0195	64	−0.1189	89	0.1668
15	−0.0656	40	0.0092	65	0.0401	90	−0.0106
16	−0.0813	41	0.0579	66	−0.0452	91	0.0304
17	−0.0677	42	0.1178	67	−0.1240	92	−0.0031
18	0.0124	43	0.0315	68	−0.1144	93	0.0038
19	−0.0920	44	−0.0223	69	0.0108	94	0.0671
20	−0.0527	45	−0.0206	70	−0.0334	95	−0.0456
21	−0.1819	46	0.0391	71	−0.0599	96	−0.1538
22	−0.1307	47	0.2125	72	−0.0062	97	−0.1372
23	0.0587	48	0.0303	73	0.0478	98	−0.1337
24	0.0278	49	0.1453	74	−0.0907	99	−0.0197
25	0.0023	50	0.0041	75	0.1844	100	0.0565

$C_0 = 1.1248$

plotted in Figures A.2.4, A.2.5, and A.2.6. The population spectral densities and the normalized densities of the fitted second-order autoregressive processes (multiplied by 2π) are graphed in Figures A.2.7, A.2.8, and A.2.9. (In the second case two sample series are treated.) The vertical scale is logarithmic, and the unit on the horizontal scale is $j = 200\lambda/(2\pi)$.

TABLE A.2.2 (*continued*)

B. $\beta_1 = -0.7$, $\beta_2 = 0.49$. (Wold's Sample No. 2)

Correlations

h	C_h/C_0	h	C_h/C_0	h	C_h/C_0	h	C_h/C_0
1	0.5113	26	−0.0110	51	−0.0376	76	0.1672
2	−0.0473	27	−0.0877	52	0.1676	77	0.0497
3	−0.3001	28	0.0399	53	0.1763	78	−0.0920
4	−0.2549	29	0.1242	54	0.0466	79	−0.2366
5	−0.0187	30	−0.0005	55	−0.1064	80	−0.1108
6	0.1613	31	−0.1583	56	−0.2419	81	0.1287
7	0.0997	32	−0.2075	57	−0.1292	82	0.1821
8	−0.0174	33	−0.0826	58	0.0752	83	0.0397
9	−0.1026	34	0.1879	59	0.1397	84	−0.1851
10	−0.0562	35	0.2784	60	0.1258	85	−0.2533
11	0.0579	36	0.1175	61	0.0377	86	−0.1009
12	0.1689	37	−0.0864	62	0.0106	87	0.1201
13	0.1357	38	−0.1646	63	−0.0243	88	0.2470
14	−0.0435	39	−0.1078	64	−0.0454	89	0.1869
15	−0.1557	40	0.0305	65	0.0227	90	−0.0286
16	−0.1329	41	0.1551	66	−0.0369	91	−0.1088
17	−0.0103	42	0.1619	67	−0.1326	92	−0.0552
18	0.0901	43	0.0052	68	−0.1089	93	0.0818
19	0.0174	44	−0.1359	69	0.0122	94	0.1601
20	−0.1108	45	−0.1114	70	0.0313	95	0.0282
21	−0.2329	46	0.0724	71	−0.0160	96	−0.1766
22	−0.1449	47	0.2417	72	−0.0476	97	−0.2563
23	0.0673	48	0.1541	73	−0.0523	98	−0.1860
24	0.1277	49	0.0337	74	−0.0234	99	0.0093
25	0.0629	50	−0.1000	75	0.1654	100	0.1701

$C_0 = 1.0837$

TABLE A.2.2 (*continued*)

C. $\beta_1 = -0.9$, $\beta_2 = 0.81$. (Wold's Sample No. 1)

Correlations

h	C_h/C_0	h	C_h/C_0	h	C_h/C_0	h	C_h/C_0
1	0.5011	26	−0.1094	51	−0.1857	76	−0.0906
2	−0.3858	27	−0.4210	52	−0.2829	77	−0.1983
3	−0.7726	28	−0.3308	53	−0.1160	78	−0.0742
4	−0.3711	29	0.0565	54	0.1330	79	0.1669
5	0.3371	30	0.3689	55	0.2395	80	0.2586
6	0.6311	31	0.3321	56	0.1117	81	0.0823
7	0.2843	32	−0.0101	57	−0.0918	82	−0.2097
8	−0.2853	33	−0.3317	58	−0.1697	83	−0.3125
9	−0.5178	34	−0.3582	59	−0.0819	84	−0.1023
10	−0.2291	35	−0.0762	60	0.0897	85	0.2119
11	0.2813	36	0.2481	61	0.1699	86	0.3365
12	0.5056	37	0.3487	62	0.1023	87	0.1426
13	0.2361	38	0.1280	63	−0.0524	88	−0.2054
14	−0.2348	39	−0.2110	64	−0.1794	89	−0.3622
15	−0.4581	40	−0.3080	65	−0.1506	90	−0.1475
16	−0.2402	41	−0.0980	66	−0.0135	91	0.1975
17	0.1880	42	0.1736	67	0.1098	92	0.2815
18	0.4203	43	0.2451	68	0.1333	93	0.0362
19	0.2518	44	0.0404	69	0.0326	94	−0.2656
20	−0.1416	45	−0.2251	70	−0.1287	95	−0.2595
21	−0.3953	46	−0.2614	71	−0.2059	96	0.0260
22	−0.2578	47	−0.0504	72	−0.1163	97	0.2556
23	0.1495	48	0.2085	73	0.0803	98	0.2187
24	0.4303	49	0.2903	74	0.1935	99	−0.0230
25	0.3123	50	0.0979	75	0.1078	100	−0.2242

$C_0 = 1.2224$

TABLE A.2.3

NORMALIZED SAMPLE SPECTRAL DENSITIES

A. $\beta_1 = -0.25$, $\beta_2 = 0.0625$. (Wold's Sample No. 2)

k	$\frac{2\pi}{C_0} I\left(\frac{2\pi k}{200}\right)$	k	$\frac{2\pi}{C_0} I\left(\frac{2\pi k}{200}\right)$	k	$\frac{2\pi}{C_0} I\left(\frac{2\pi k}{200}\right)$	k	$\frac{2\pi}{C_0} I\left(\frac{2\pi k}{200}\right)$
0	1.5087						
1	1.5029	26	0.5243	51	0.1236	76	0.2538
2	0.5606	27	0.9970	52	1.7227	77	3.8915
3	1.0695	28	0.4683	53	0.1063	78	0.2601
4	3.8996	29	2.2405	54	0.4195	79	0.4844
5	3.3980	30	1.5876	55	0.5291	80	1.9956
6	0.6895	31	0.7753	56	0.3043	81	0.0868
7	0.8402	32	1.1055	57	0.1007	82	0.3921
8	1.7680	33	1.0494	58	1.5100	83	0.1838
9	1.9260	34	6.0570	59	0.3572	84	0.0284
10	0.1815	35	0.5693	60	0.0249	85	0.7412
11	0.7431	36	0.4151	61	2.2647	86	0.1808
12	0.6993	37	0.0128	62	1.2785	87	0.3800
13	3.0737	38	0.1682	63	0.8900	88	0.1344
14	0.9221	39	0.2226	64	1.0273	89	0.0748
15	0.6206	40	0.3264	65	0.5327	90	1.6157
16	5.8050	41	1.6880	66	0.3617	91	0.2269
17	1.2231	42	1.0313	67	0.0274	92	0.3397
18	1.4275	43	0.3604	68	0.8100	93	1.2194
19	1.0515	44	0.3232	69	3.5494	94	0.7040
20	1.2780	45	0.3270	70	1.1472	95	0.5200
21	1.1196	46	0.9212	71	0.3027	96	0.0886
22	1.1128	47	0.9579	72	0.7541	97	0.7174
23	1.7322	48	0.7229	73	0.2701	98	0.5979
24	3.0603	49	0.2063	74	0.2194	99	1.6979
25	0.4891	50	0.1732	75	0.1076	100	0.4220

TABLE A.2.3 (*continued*)

B. $\beta_1 = -0.7$, $\beta_2 = 0.49$. (Wold's Sample No. 2)

k	$\frac{2\pi}{C_0} I\left(\frac{2\pi k}{200}\right)$	k	$\frac{2\pi}{C_0} I\left(\frac{2\pi k}{200}\right)$	k	$\frac{2\pi}{C_0} I\left(\frac{2\pi k}{200}\right)$	k	$\frac{2\pi}{C_0} I\left(\frac{2\pi k}{200}\right)$
0	1.1189						
1	0.9664	26	1.0541	51	0.0877	76	0.0825
2	0.4030	27	1.8778	52	1.1702	77	1.0673
3	0.7823	28	1.0098	53	0.0617	78	0.0662
4	2.6466	29	5.0956	54	0.2595	79	0.1388
5	2.3570	30	3.5765	55	0.3281	80	0.5218
6	0.5149	31	1.7633	56	0.1525	81	0.0210
7	0.5969	32	2.4765	57	0.0426	82	0.1066
8	1.4281	33	2.4593	58	0.7922	83	0.0414
9	1.5064	34	15.0876	59	0.1625	84	0.0074
10	0.1523	35	1.2011	60	0.0080	85	0.1818
11	0.6217	36	0.8276	61	1.0016	86	0.0500
12	0.6459	37	0.0150	62	0.5806	87	0.1026
13	2.8789	38	0.3897	63	0.3768	88	0.0285
14	0.8066	39	0.4910	64	0.3807	89	0.0209
15	0.5380	40	0.5550	65	0.2053	90	0.4074
16	6.0875	41	2.7688	66	0.1410	91	0.0567
17	1.4132	42	1.4481	67	0.0067	92	0.0891
18	1.5177	43	0.5413	68	0.2936	93	0.2865
19	1.2673	44	0.4815	69	1.2444	94	0.1767
20	1.6300	45	0.3843	70	0.3671	95	0.1313
21	1.5202	46	1.0904	71	0.0891	96	0.0178
22	1.7333	47	0.9157	72	0.2535	97	0.1789
23	2.5668	48	0.7222	73	0.0851	98	0.1440
24	4.8883	49	0.2032	74	0.0706	99	0.3831
25	0.7624	50	0.1366	75	0.0374	100	0.1068

TABLE A.2.3 (*continued*)

C. $\beta_1 = -0.9, \beta_2 = 0.81$. (Wold's Sample No. 1)

k	$\frac{2\pi}{C_0} I\left(\frac{2\pi k}{200}\right)$	k	$\frac{2\pi}{C_0} I\left(\frac{2\pi k}{200}\right)$	k	$\frac{2\pi}{C_0} I\left(\frac{2\pi k}{200}\right)$	k	$\frac{2\pi}{C_0} I\left(\frac{2\pi k}{200}\right)$
0	0.1835						
1	0.5806	26	3.6521	51	0.0902	76	0.0538
2	0.5042	27	2.6660	52	0.4664	77	0.0186
3	0.5937	28	0.9321	53	0.0246	78	0.0037
4	0.2787	29	0.6410	54	0.2046	79	0.1264
5	0.0850	30	0.1841	55	0.0311	80	0.0095
6	0.0430	31	4.7018	56	0.0621	81	0.0104
7	0.4200	32	15.7755	57	0.1469	82	0.0102
8	0.2520	33	24.4907	58	0.1169	83	0.0259
9	0.0175	34	6.2684	59	0.0315	84	0.0380
10	0.7423	35	8.8254	60	0.1073	85	0.0065
11	0.1452	36	0.1490	61	0.0048	86	0.0848
12	0.1305	37	7.4089	62	0.0045	87	0.0058
13	0.2365	38	0.0970	63	0.0708	88	0.0471
14	0.0259	39	2.6194	64	0.2145	89	0.0210
15	0.0354	40	1.3477	65	0.0586	90	0.0692
16	1.2747	41	0.2229	66	0.0730	91	0.0013
17	1.0938	42	1.8378	67	0.0349	92	0.0106
18	0.1598	43	0.9971	68	0.1006	93	0.0134
19	0.0107	44	0.2457	69	0.1786	94	0.0117
20	0.0607	45	0.1095	70	0.1164	95	0.0339
21	2.6171	46	0.5497	71	0.0238	96	0.0153
22	0.3760	47	0.5808	72	0.0073	97	0.0466
23	0.2992	48	1.0833	73	0.2072	98	0.0297
24	0.7754	49	0.1481	74	0.0119	99	0.0239
25	0.0146	50	0.3898	75	0.0025	100	0.0039

TABLE A.2.4

NORMALIZED BARTLETT ESTIMATES OF THE SPECTRAL DENSITIES

A. $\beta_1 = -0.25$, $\beta_2 = 0.0625$. (Wold's Sample No. 2)

k	$\frac{2\pi}{C_0}\hat{f}_T\left(\frac{2\pi k}{200}\right)$	k	$\frac{2\pi}{C_0}\hat{f}_T\left(\frac{2\pi k}{200}\right)$	k	$\frac{2\pi}{C_0}\hat{f}_T\left(\frac{2\pi k}{200}\right)$	k	$\frac{2\pi}{C_0}\hat{f}_T\left(\frac{2\pi k}{200}\right)$
0	1.5349						
1	1.5345	26	1.2662	51	0.8224	76	0.7793
2	1.5334	27	1.2455	52	0.8153	77	0.7755
3	1.5316	28	1.2244	53	0.8092	78	0.7713
4	1.5291	29	1.2030	54	0.8041	79	0.7665
5	1.5258	30	1.1813	55	0.7999	80	0.7613
6	1.5217	31	1.1594	56	0.7965	81	0.7558
7	1.5169	32	1.1374	57	0.7940	82	0.7498
8	1.5112	33	1.1155	58	0.7921	83	0.7436
9	1.5048	34	1.0936	59	0.7908	84	0.7372
10	1.4975	35	1.0720	60	0.7900	85	0.7306
11	1.4894	36	1.0507	61	0.7897	86	0.7240
12	1.4805	37	1.0298	62	0.7898	87	0.7173
13	1.4707	38	1.0094	63	0.7900	88	0.7107
14	1.4600	39	0.9896	64	0.7905	89	0.7043
15	1.4484	40	0.9705	65	0.7909	90	0.6982
16	1.4360	41	0.9521	66	0.7914	91	0.6923
17	1.4226	42	0.9346	67	0.7917	92	0.6869
18	1.4084	43	0.9180	68	0.7919	93	0.6819
19	1.3933	44	0.9023	69	0.7918	94	0.6775
20	1.3774	45	0.8877	70	0.7913	95	0.6736
21	1.3607	46	0.8741	71	0.7905	96	0.6704
22	1.3432	47	0.8616	72	0.7892	97	0.6679
23	1.3249	48	0.8501	73	0.7875	98	0.6660
24	1.3060	49	0.8398	74	0.7853	99	0.6649
25	1.2864	50	0.8306	75	0.7826	100	0.6646

TABLE A.2.4 (*continued*)

B. $\beta_1 = -0.7,\ \beta_2 = 0.49$. (Wold's Sample No. 2)

k	$\frac{2\pi}{C_0}\hat{f}_T\left(\frac{2\pi k}{200}\right)$	k	$\frac{2\pi}{C_0}\hat{f}_T\left(\frac{2\pi k}{200}\right)$	k	$\frac{2\pi}{C_0}\hat{f}_T\left(\frac{2\pi k}{200}\right)$	k	$\frac{2\pi}{C_0}\hat{f}_T\left(\frac{2\pi k}{200}\right)$
0	1.2499						
1	1.2506	26	2.2623	51	0.5270	76	0.2781
2	1.2522	27	2.4148	52	0.4735	77	0.2712
3	1.2530	28	2.5912	53	0.4326	78	0.2621
4	1.2515	29	2.7682	54	0.4062	79	0.2497
5	1.2466	30	2.9190	55	0.3936	80	0.2341
6	1.2389	31	3.0173	56	0.3920	81	0.2160
7	1.2314	32	3.0422	57	0.3974	82	0.1969
8	1.2289	33	2.9823	58	0.4058	83	0.1788
9	1.2378	34	2.8375	59	0.4139	84	0.1633
10	1.2648	35	2.6194	60	0.4197	85	0.1519
11	1.3152	36	2.3489	61	0.4223	86	0.1452
12	1.3912	37	2.0529	62	0.4218	87	0.1430
13	1.4909	38	1.7591	63	0.4184	88	0.1447
14	1.6077	39	1.4920	64	0.4128	89	0.1491
15	1.7315	40	1.2691	65	0.4052	90	0.1551
16	1.8497	41	1.0987	66	0.3956	91	0.1618
17	1.9500	42	0.9798	67	0.3839	92	0.1687
18	2.0228	43	0.9037	68	0.3702	93	0.1755
19	2.0639	44	0.8572	69	0.3551	94	0.1821
20	2.0756	45	0.8254	70	0.3393	95	0.1886
21	2.0673	46	0.7952	71	0.3241	96	0.1949
22	2.0538	47	0.7579	72	0.3105	97	0.2004
23	2.0526	48	0.7096	73	0.2994	98	0.2049
24	2.0803	49	0.6515	74	0.2908	99	0.2078
25	2.1489	50	0.5885	75	0.2841	100	0.2087

TABLE A.2.4 (*continued*)

C. $\beta_1 = -0.9,\ \beta_2 = 0.81$. (Wold's Sample No. 1)

k	$\frac{2\pi}{C_0}\hat{f}_T\left(\frac{2\pi k}{200}\right)$	k	$\frac{2\pi}{C_0}\hat{f}_T\left(\frac{2\pi k}{200}\right)$	k	$\frac{2\pi}{C_0}\hat{f}_T\left(\frac{2\pi k}{200}\right)$	k	$\frac{2\pi}{C_0}\hat{f}_T\left(\frac{2\pi k}{200}\right)$
0	0.4574						
1	0.4812	26	1.9682	51	0.2360	76	0.0302
2	0.5128	27	1.7071	52	0.2329	77	0.0398
3	0.4687	28	1.4244	53	0.1873	78	0.0529
4	0.3259	29	1.0239	54	0.1308	79	0.0636
5	0.2040	30	1.7788	55	0.1071	80	0.0535
6	0.2295	31	6.3443	56	0.0990	81	0.0329
7	0.3226	32	13.1094	57	0.1029	82	0.0301
8	0.3489	33	15.6088	58	0.1136	83	0.0375
9	0.3579	34	11.5562	59	0.1064	84	0.0389
10	0.3921	35	6.2179	60	0.0837	85	0.0427
11	0.3584	36	3.9872	61	0.0633	86	0.0480
12	0.2611	37	3.5342	62	0.0655	87	0.0435
13	0.1995	38	2.7475	63	0.1049	88	0.0384
14	0.2327	39	1.9772	64	0.1382	89	0.0439
15	0.4434	40	1.7151	65	0.1105	90	0.0475
16	0.7550	41	1.5282	66	0.0615	91	0.0372
17	0.8461	42	1.2119	67	0.0679	92	0.0228
18	0.7107	43	0.8837	68	0.1192	93	0.0174
19	0.7688	44	0.5833	69	0.1422	94	0.0212
20	1.0923	45	0.4081	70	0.1074	95	0.0289
21	1.2004	46	0.4382	71	0.0662	96	0.0373
22	0.8373	47	0.5520	72	0.0697	97	0.0414
23	0.5144	48	0.5691	73	0.0858	98	0.0371
24	0.9066	49	0.4415	74	0.0659	99	0.0281
25	1.7041	50	0.2890	75	0.0353	100	0.0239

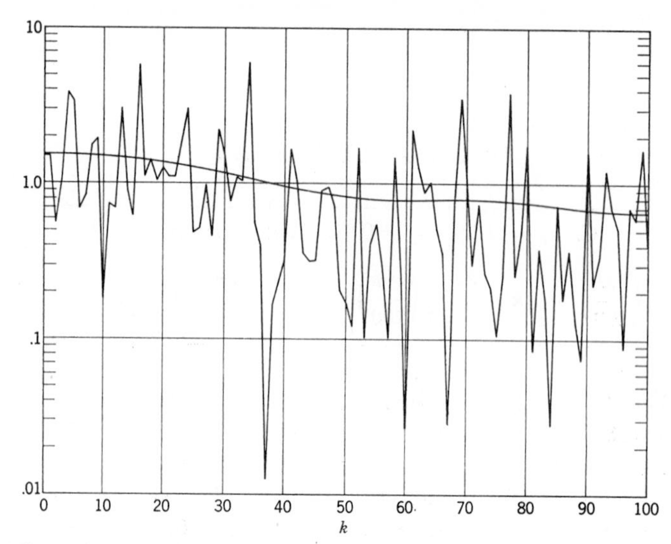

Figure A.2.4. The normalized sample spectral density and normalized Bartlett estimate for the series with $\gamma = 0.25$.

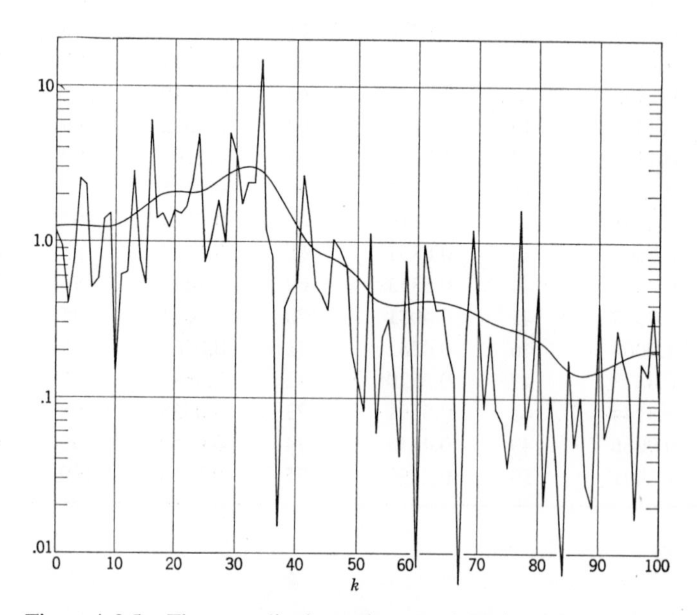

Figure A.2.5. The normalized sample spectral density and normalized Bartlett estimate for the series with $\gamma = 0.7$.

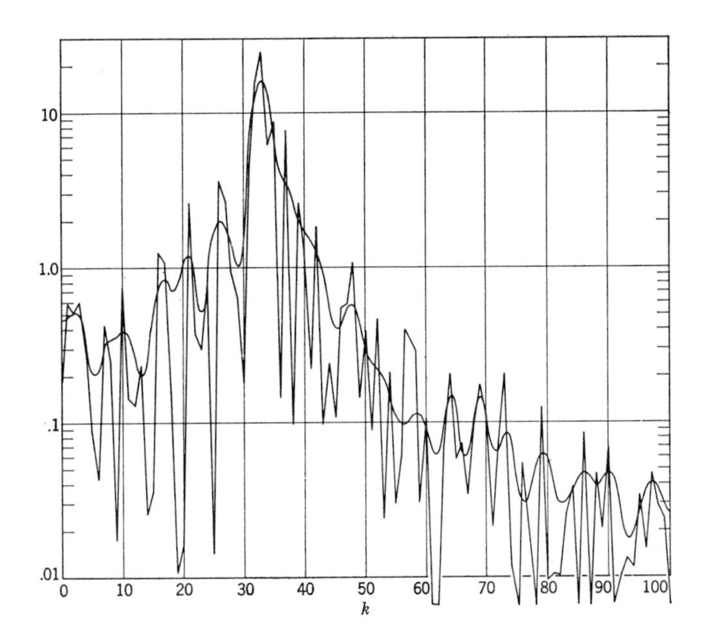

Figure A.2.6. The normalized sample spectral density and normalized Bartlett estimate for the series with $\gamma = 0.9$.

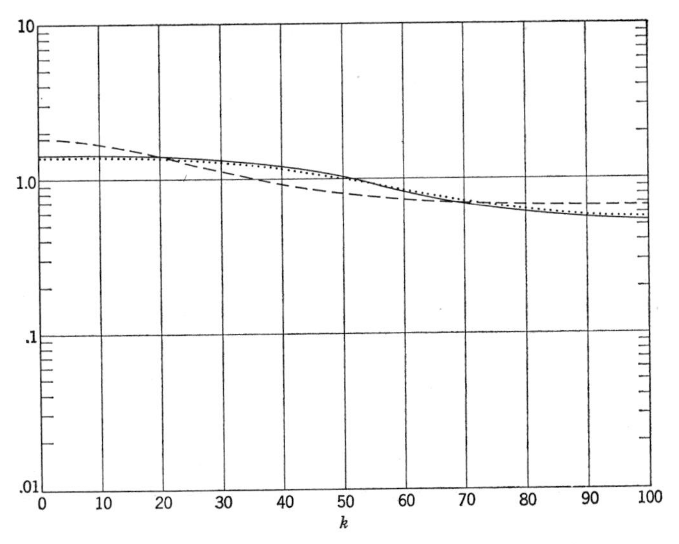

Figure A.2.7. The normalized spectral densities of the process (unbroken line) and two fitted processes when $\gamma = 0.25$.

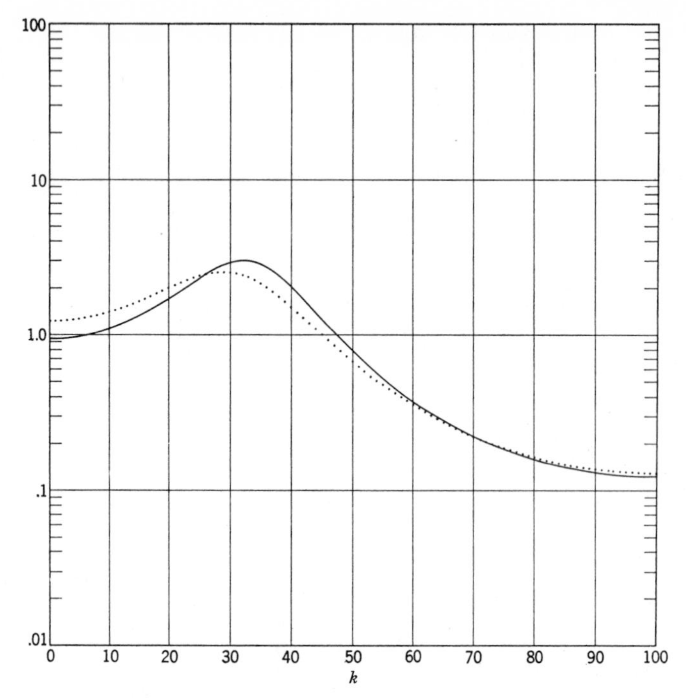

Figure A.2.8. The normalized spectral densities of the process (unbroken line) and the fitted process when $\gamma = 0.7$.

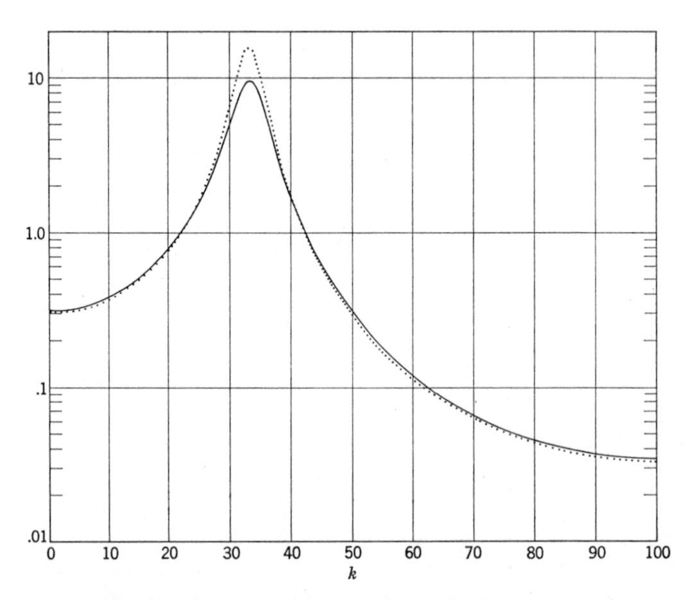

Figure A.2.9. The normalized spectral densities of the process (unbroken line) and the fitted process when $\gamma = 0.9$.

A.3. SUNSPOT NUMBERS

The sunspot numbers were described in Section 5.9. Wolfer's numbers are given in Table A.3.1 for 1749 to 1924. The graph of Waldmeicr's data was given as Figure 5.1. The correlation sequence was graphed in Section 5.9.

The spectrogram was calculated by Schuster (1906) on the basis of semi-annual data for a number of frequencies of the form k/N; the semiannual data were obtained by summing the monthly sunspot numbers over 6-month periods from 1749 to 1901. The results are given in Table A.3.2 and are graphed in Figure A.3.1 on a logarithmic scale. The estimates of the spectral density calculated by Schaerf (1964) for annual data (1749-1924) using Parzen's window (J of Section 9.2.3) and $K = 20$, 40, and 60 are graphed in Figure A.3.2.

TABLE A.3.1

WOLFER'S SUNSPOT NUMBERS, 1749-1924

Year	Wolfer's number	Year	Wolfer's number	Year	Wolfer's number	Year	Wolfer's number
1749	80.9	1793	46.9	1837	138.3	1881	54.3
1750	83.4	1794	41.0	1838	103.2	1882	59.7
1751	47.7	1795	21.3	1839	85.8	1883	63.7
1752	47.8	1796	16.0	1840	63.2	1884	63.5
1753	30.7	1797	6.4	1841	36.8	1885	52.2
1754	12.2	1798	4.1	1842	24.2	1886	25.4
1755	9.6	1799	6.8	1843	10.7	1887	13.1
1756	10.2	1800	14.5	1844	15.0	1888	6.8
1757	32.4	1801	34.0	1845	40.1	1889	6.3
1758	47.6	1802	45.0	1846	61.5	1890	7.1
1759	54.0	1803	43.1	1847	98.5	1891	35.6
1760	62.9	1804	47.5	1848	124.3	1892	73.0
1761	85.9	1805	42.2	1849	95.9	1893	84.9
1762	61.2	1806	28.1	1850	66.5	1894	78.0
1763	45.1	1807	10.1	1851	64.5	1895	64.0
1764	36.4	1808	8.1	1852	54.2	1896	41.8
1765	20.9	1809	2.5	1853	39.0	1897	26.2
1766	11.4	1810	0.0	1854	20.6	1898	26.7
1767	37.8	1811	1.4	1855	6.7	1899	12.1
1768	69.8	1812	5.0	1856	4.3	1900	9.5
1769	106.1	1813	12.2	1857	22.8	1901	2.7
1770	100.8	1814	13.9	1858	54.8	1902	5.0
1771	81.6	1815	35.4	1859	93.8	1903	24.4
1772	66.5	1816	45.8	1860	95.7	1904	42.0
1773	34.8	1817	41.1	1861	77.2	1905	63.5
1774	30.6	1818	30.4	1862	59.1	1906	53.8
1775	7.0	1819	23.9	1863	44.0	1907	62.0
1776	19.8	1820	15.7	1864	47.0	1908	48.5
1777	92.5	1821	6.6	1865	30.5	1909	43.9
1778	154.4	1822	4.0	1866	16.3	1910	18.6
1779	125.9	1823	1.8	1867	7.3	1911	5.7
1780	84.8	1824	8.5	1868	37.3	1912	3.6
1781	68.1	1825	16.6	1869	73.9	1913	1.4
1782	38.5	1826	36.3	1870	139.1	1914	9.6
1783	22.8	1827	49.7	1871	111.2	1915	47.4
1784	10.2	1828	62.5	1872	101.7	1916	57.1
1785	24.1	1829	67.0	1873	66.3	1917	103.9
1786	82.9	1830	71.0	1874	44.7	1918	80.6
1787	132.0	1831	47.8	1875	17.1	1919	63.6
1788	130.9	1832	27.5	1876	11.3	1920	37.6
1789	118.1	1833	8.5	1877	12.3	1921	26.1
1790	89.9	1834	13.2	1878	3.4	1922	14.2
1791	66.6	1835	56.9	1879	6.0	1923	5.8
1792	60.0	1836	121.5	1880	32.3	1924	16.7

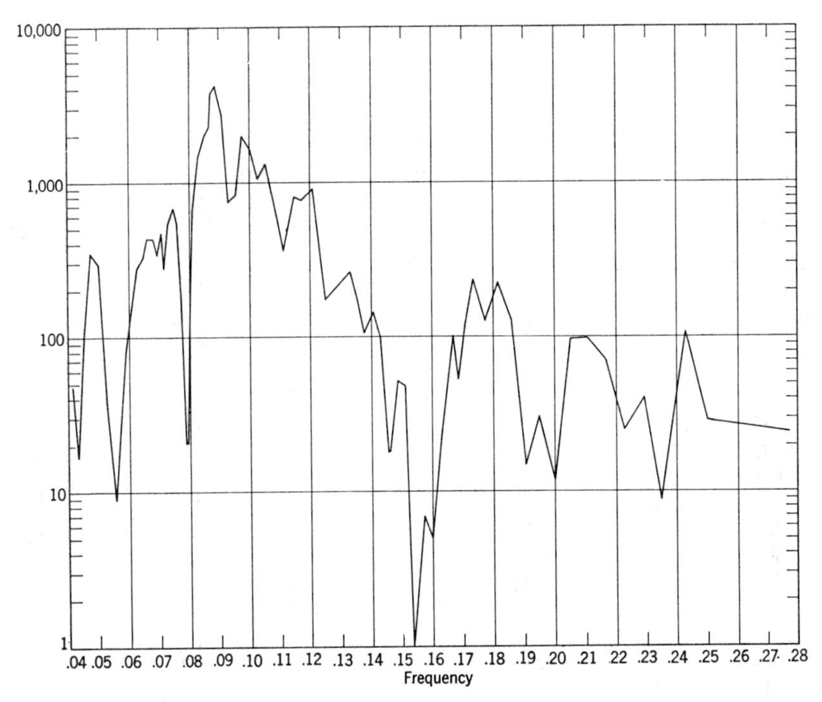

Figure A.3.1. Schuster's spectrogram of sunspot numbers.

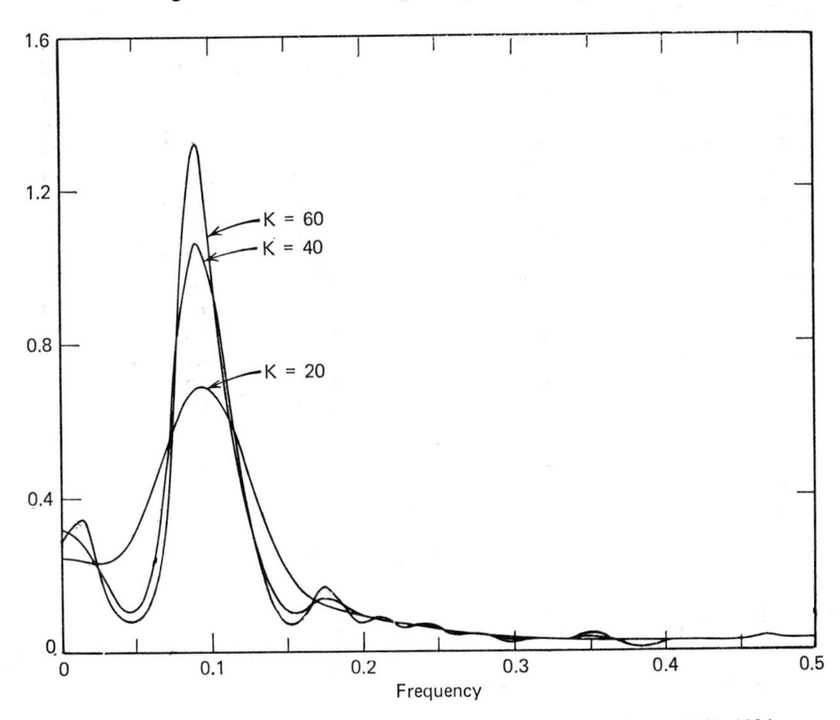

Figure A.3.2. Estimated spectral density for sunspot numbers, 1749–1924.

TABLE A.3.2

SCHUSTER'S SPECTROGRAM FOR SUNSPOTS

Frequency	Period in years	Spectrogram	Frequency	Period in years	Spectrogram
0.0417	24	48	0.1111	9	364
0.0435	23	17	0.1143	8.75	812
0.0455	22	112	0.1176	8.5	770
0.0476	21	349	0.1212	8.25	933
0.0500	20	298	0.1250	8	177
0.0526	19	35	0.1333	7.5	264
0.0556	18	9	0.1356	7.375	173
0.0588	17	83	0.1379	7.25	106
0.0625	16	278	0.1403	7.125	149
0.0645	15.5	340	0.1429	7	99
0.0658	15.25	436	0.1455	6.875	18
0.0667	15	434	0.1481	6.75	52
0.0680	14.75	432	0.1509	6.625	49
0.0690	14.5	342	0.1538	6.5	1
0.0704	14.25	474	0.1568	6.375	7
0.0714	14	278	0.1600	6.25	5
0.0727	13.75	550	0.1633	6.125	24
0.0741	13.5	696	0.1667	6	102
0.0755	13.25	552	0.1686	5.93	54
0.0769	13	198	0.1702	5.875	122
0.0784	12.75	21	0.1739	5.75	236
0.0800	12.5	105	0.1778	5.625	128
0.0816	12.25	675	0.1818	5.5	224
0.0833	12	1464	0.1860	5.375	129
0.0843	11.86	1951	0.1905	5.25	15
0.0851	11.75	2338	0.1951	5.125	31
0.0870	11.5	3700	0.2000	5	12
0.0889	11.25	4230	0.2051	4.875	97
0.0909	11	2724	0.2105	4.75	99
0.0930	10.75	742	0.2162	4.625	70
0.0952	10.5	853	0.2222	4.5	25
0.0976	10.25	2026	0.2286	4.375	41
0.1000	10	1677	0.2353	4.25	9
0.1026	9.75	1050	0.2424	4.125	109
0.1053	9.5	1313	0.2500	4	29
0.1081	9.25	603	0.2778	3.75	24

REFERENCES

A.1. Beveridge (1921), (1922),Granger and Hughes (1969).

A.2. Wold (1965).

A.3. Schaerf (1964), Schuster (1906).

APPENDIX B

Solutions to Selected Problems

Chapter 2, Problem 7. Since $\mathscr{E}(y - Zb) = \mathbf{0}$,

$$\begin{aligned}\mathscr{E}(y - Zb)(b - \beta)' &= \mathscr{E}(y - Zb)b' \\ &= \mathscr{E}(I - ZA^{-1}Z')yy'ZA^{-1} \\ &= (I - ZA^{-1}Z')(\sigma^2 I + Z\beta\beta'Z')ZA^{-1} \\ &= \mathbf{0}.\end{aligned}$$

Chapter 2, Problem 8. The equation $AB = I$ is partitioned in part as

$$\begin{aligned}A_{11}B_{12} + A_{12}B_{22} &= \mathbf{0}, \\ A_{21}B_{12} + A_{22}B_{22} &= I.\end{aligned}$$

From the first equation $B_{12} = -A_{11}^{-1}A_{12}B_{22}$, and (a) follows. Substitution into the second equation yields

$$(A_{22} - A_{21}A_{11}^{-1}A_{12})B_{22} = I.$$

Therefore $A_{22} - A_{21}A_{11}^{-1}A_{12}$ is nonsingular and (b) follows.

Chapter 2, Problem 10. (a) Let B be the matrix of coefficients of the regression of $z_t^{(2)}$ on $z_t^{(1)}$. Then

$$\sum_{t=1}^{T} z_t^{(1)}z_t^{(2)\prime} = \sum_{t=1}^{T} z_t^{(1)}z_t^{(1)\prime}B', \qquad B' = A_{11}^{-1}A_{12}, \quad B = A_{21}A_{11}^{-1},$$

$$\tilde{z}_t^{(2)} = z_t^{(2)} - Bz_t^{(1)} = z_t^{(2)} - A_{21}A_{11}^{-1}z_t^{(1)}.$$

Let $b^{*(1)}$ be the vector of coefficients of the regression of y_t on $z_t^{(1)}$. Then

$$b^{*(1)} = A_{11}^{-1} \sum_{t=1}^{T} z_t^{(1)} y_t,$$

$$\tilde{y}_t = y_t - b^{*(1)\prime} z_t^{(1)} = y_t - \sum_{\tau=1}^{T} y_\tau z_\tau^{(1)\prime} A_{11}^{-1} z_t^{(1)},$$

$$\begin{aligned}
\mathscr{E}\tilde{y}_t &= \beta' z_t - \sum_{\tau=1}^{T} \beta' z_\tau z_\tau^{(1)\prime} A_{11}^{-1} z_t^{(1)} \\
&= \beta' z_t - \sum_{\tau=1}^{T} (\beta^{(1)\prime}\ \beta^{(2)\prime}) \begin{pmatrix} z_\tau^{(1)} \\ z_\tau^{(2)} \end{pmatrix} z_\tau^{(1)\prime} A_{11}^{-1} z_t^{(1)} \\
&= \beta^{(1)\prime} z_t^{(1)} + \beta^{(2)\prime} z_t^{(2)} - \beta^{(1)\prime} \sum_{\tau=1}^{T} z_\tau^{(1)} z_\tau^{(1)\prime} A_{11}^{-1} z_t^{(1)} - \beta^{(2)\prime} \sum_{\tau=1}^{T} z_\tau^{(2)} z_\tau^{(1)\prime} A_{11}^{-1} z_t^{(1)} \\
&= \beta^{(2)\prime} (z_t^{(2)} - A_{21} A_{11}^{-1} z_t^{(1)}) \\
&= \beta^{(2)\prime} \tilde{z}_t^{(2)}.
\end{aligned}$$

(b) The normal equations for $\tilde{b}^{(2)}$ are

$$\text{(i)} \qquad \sum_{t=1}^{T} \tilde{z}_t^{(2)} \tilde{y}_t = \sum_{t=1}^{T} \tilde{z}_t^{(2)} \tilde{z}_t^{(2)\prime} \tilde{b}^{(2)}.$$

From substitution in (i) we find

$$\sum_{t=1}^{T} (z_t^{(2)} - A_{21} A_{11}^{-1} z_t^{(1)}) \left(y_t - \sum_{\tau=1}^{T} y_\tau z_\tau^{(1)\prime} A_{11}^{-1} z_t^{(1)} \right)$$
$$= \sum_{t=1}^{T} (z_t^{(2)} - A_{21} A_{11}^{-1} z_t^{(1)})(z_t^{(2)\prime} - z_t^{(1)\prime} A_{11}^{-1} A_{12}) \tilde{b}^{(2)},$$

which can be written as

$$\text{(ii)} \qquad \sum_{t=1}^{T} z_t^{(2)} y_t - A_{21} A_{11}^{-1} \sum_{t=1}^{T} z_t^{(1)} y_t = (A_{22} - A_{21} A_{11}^{-1} A_{12}) \tilde{b}^{(2)}.$$

However, the normal equations for $b^{(1)}$ and $b^{(2)}$ are

$$A_{11} b^{(1)} + A_{12} b^{(2)} = \sum_{t=1}^{T} z_t^{(1)} y_t,$$

$$A_{21} b^{(1)} + A_{22} b^{(2)} = \sum_{t=1}^{T} z_t^{(2)} y_t.$$

The first equation gives

$$b^{(1)} = -A_{11}^{-1}A_{12}b^{(2)} + A_{11}^{-1}\sum_{t=1}^{T} z_t^{(1)} y_t.$$

Substitution in the second equation gives

$$-A_{21}A_{11}^{-1}A_{12}b^{(2)} + A_{21}A_{11}^{-1}\sum_{t=1}^{T} z_t^{(1)} y_t + A_{22}b^{(2)} = \sum_{t=1}^{T} z_t^{(2)} y_t,$$

which is the same as (ii).

Chapter 3, Problem 5. The expression in the hint is

$$\sum_{t=1}^{T}\left[\sum_{k=0}^{p+1}\binom{p+1}{k} t^k - t^{p+1}\right] = \sum_{k=0}^{p}\binom{p+1}{k}\sum_{t=1}^{T} t^k,$$

and the first sum in the hint is $\sum_{t=2}^{T+1} t^{p+1}$.

Chapter 3, Problem 7. By Problem 5

$$(k+1)\psi_k(T) = (T+1)^{k+1} - 1 - \sum_{r=0}^{k-1}\binom{k+1}{r}\psi_r(T),$$

where $\psi_0(T) = T$. Solve for $\psi_1(T), \ldots, \psi_8(T)$ in order.

Chapter 3, Problem 30. The stated orthogonality requirements for $\phi^*(s)$ yield the $r+1$ homogeneous equations

$$c_0 \sum_{s=-m}^{m} s^{2j} + c_2 \sum_{s=-m}^{m} s^{2j+2} + \cdots + c_{2r} \sum_{s=-m}^{m} s^{2j+2r} = 0, \qquad j = 0, 1, \ldots, r.$$

The coefficient matrix is therefore

$$\sum_{s=-m}^{m}\begin{pmatrix} 1 \\ s^2 \\ \cdot \\ \cdot \\ \cdot \\ s^{2r} \end{pmatrix}(1, s^2, \ldots, s^{2r}),$$

and

$$\sum_{s=-m}^{m}\left(\sum_{i=0}^{r} x_i s^{2i}\right)^2 = 0$$

implies $x_0 = 0$ and $\sum_{i=1}^{r} x_i s^{2i} = 0$, $s = 1, \ldots, m$. If $m \geq r$ the only solution is $x_0 = x_1 = \cdots = x_r = 0$, and the coefficient matrix is positive definite. Thus it is nonsingular and $\phi^*(s) = c_1 s + c_3 s^3 + \cdots + c_{2r-1} s^{2r-1} + s^{2r+1}$. It also follows that $\phi^*_{i,2m+1}(s)$ for i even contains only even powers of s.

Chapter 3, Problem 38. For real numbers d_1 and d_2 and linear operators $\mathcal{O}_1$ and $\mathcal{O}_2$

$$\begin{aligned}(c_1\mathcal{O}_1 + c_2\mathcal{O}_2)(d_1u_t + d_2v_t) &= c_1\mathcal{O}_1(d_1u_t + d_2v_t) + c_2\mathcal{O}_2(d_1u_t + d_2v_t)\\ &= c_1(d_1\mathcal{O}_1u_t + d_2\mathcal{O}_1v_t) + c_2(d_1\mathcal{O}_1u_t + d_2\mathcal{O}_2v_t)\\ &= d_1(c_1\mathcal{O}_1u_t + c_2\mathcal{O}_2u_t) + d_2(c_1\mathcal{O}_1v_t + c_2\mathcal{O}_2v_t)\\ &= d_1(c_1\mathcal{O}_1 + c_2\mathcal{O}_2)u_t + d_2(c_1\mathcal{O}_1 + c_2\mathcal{O}_2)v_t\end{aligned}$$

by (2), (3), and (7).

Chapter 3, Problem 52. By Lemma 3.4.2 $\Delta^{q+1}f(t) = 0$, $t = 1, \ldots, T - q - 1$ since $\Delta^{q+1}f(t)$ for each t involves the values at $q + 2$ points. Since the same operator annihilates the trend over the entire range, the trend is a polynomial of degree (at most) q by Problem 51.

Chapter 3, Problem 57.

$$\begin{aligned}\sum_{i,j,k,l=1}^{n} a_{ij}b_{kl}\mathcal{E}u_iu_ju_ku_l\\ = \kappa_4 \sum_{i=1}^{n} a_{ii}b_{ii} + \sigma^4 \sum_{i=1}^{n} a_{ii} \sum_{i=1}^{n} b_{ii} + 2\sigma^4 \sum_{i,j=1}^{n} a_{ij}b_{ij}.\end{aligned}$$

Apply (39). See Problem 7 of Chapter 8.

Chapter 3, Problem 63. By (39) $\mathcal{E}Q_1/\mathcal{E}Q_2 = \operatorname{tr} \boldsymbol{A}/\operatorname{tr} \boldsymbol{B}$. The variance is

$$\operatorname{Var} \sum_{s,t=1}^{T} \left(a_{st} - \frac{\operatorname{tr} \boldsymbol{A}}{\operatorname{tr} \boldsymbol{B}} b_{st}\right) u_s u_t,$$

and the result follows by direct application of Lemma 3.4.4.

Chapter 3, Problem 64. The random variable

$$\begin{aligned}\sqrt{T}\left[\frac{\boldsymbol{u}'\boldsymbol{A}_T\boldsymbol{u}}{\boldsymbol{u}'\boldsymbol{B}_T\boldsymbol{u}} - \frac{\alpha}{\beta}\right]\\ = \sqrt{T}\left[\frac{\alpha\sigma^2 + (\boldsymbol{u}'\boldsymbol{A}_T\boldsymbol{u} - \alpha\sigma^2)}{\beta\sigma^2 + (\boldsymbol{u}'\boldsymbol{B}_T\boldsymbol{u} - \beta\sigma^2)} - \frac{\alpha}{\beta}\right] = \sqrt{T}\left[\frac{\dfrac{\alpha}{\beta} + \dfrac{\boldsymbol{u}'\boldsymbol{A}_T\boldsymbol{u} - \alpha\sigma^2}{\beta\sigma^2}}{1 + \dfrac{\boldsymbol{u}'\boldsymbol{B}_T\boldsymbol{u} - \beta\sigma^2}{\beta\sigma^2}} - \frac{\alpha}{\beta}\right]\end{aligned}$$

has the limiting distribution of

$$\sqrt{T}\left\{\left[\frac{\alpha}{\beta} + \frac{\boldsymbol{u}'\boldsymbol{A}_T\boldsymbol{u} - \alpha\sigma^2}{\beta\sigma^2}\right]\left[1 - \frac{\boldsymbol{u}'\boldsymbol{B}_T\boldsymbol{u} - \beta\sigma^2}{\beta\sigma^2}\right] - \frac{\alpha}{\beta}\right\}$$

which in turn has the limiting distribution of

$$\sqrt{T}\left\{\frac{u'A_Tu - \alpha\sigma^2}{\beta\sigma^2} - \frac{\alpha}{\beta}\frac{u'B_Tu - \beta\sigma^2}{\beta\sigma^2}\right\} = \frac{\sqrt{T}[u'A_Tu - (\alpha/\beta)u'B_Tu]}{\beta\sigma^2}.$$

Then the statistic under question has a limiting normal distribution with mean 0 and variance

$$\frac{1}{\beta^2\sigma^4}\lim_{T\to\infty} T\,\mathrm{Var}\left(u'A_Tu - \frac{\alpha}{\beta}u'B_Tu\right).$$

The result in the hint follows by direct application of Problem 63 for each fixed T. If $a_{11}^{(T)}/b_{11}^{(T)} = \cdots = a_{TT}^{(T)}/b_{TT}^{(T)}$, then $a_{tt}^{(T)} = kb_{tt}^{(T)}$, $t = 1, \ldots, T$, for some k, and $\alpha = k\beta$. Then

$$a_{tt}^{(T)} - \frac{\alpha}{\beta}b_{tt}^{(T)} = \left(k - \frac{k\beta}{\beta}\right)b_{tt}^{(T)} = 0, \qquad t = 1, \ldots, T,$$

for each T. See Problem 34 of Chapter 8 for a more general result.

Chapter 4, Problem 6. The elements of P may be determined by inspection in $N = MP$, using $e^{i\lambda} = \cos\lambda + i\sin\lambda$. A more direct method uses $\cos(2\pi - \lambda) = \cos\lambda$ and $\sin(2\pi - \lambda) = -\sin\lambda$ to extend the results of (9)-(13). For example (9) becomes

$$\sum_{t=1}^{T}\cos\frac{2\pi j}{T}t\,\cos\frac{2\pi(T-k)}{T}t = \sum_{t=1}^{T}\cos\frac{2\pi j}{T}t\,\cos\frac{2\pi k}{T}t, \qquad 0 \le j, k \le [\tfrac{1}{2}T].$$

By direct multiplication $P = M'N$ and the elements of P are determined with the aid of the indicated extensions of (9)-(13). If T is even, $p_{1T} = 1$, $p_{2t,t} = p_{2t,T-t} = 1/\sqrt{2}$, $p_{2t+1,t} = i/\sqrt{2}$, $p_{2t+1,T-t} = -i/\sqrt{2}$, $t = 1, \ldots, \frac{1}{2}T - 1$, $p_{T,\frac{1}{2}T} = 1$, and all other elements are 0. If T is odd, $p_{1T} = 1$, $p_{2t,t} = p_{2t,T-t} = 1/\sqrt{2}$, $p_{2t+1,t} = i/\sqrt{2}$, $p_{2t+1,T-t} = -i/\sqrt{2}$, $t = 1, \ldots, \frac{1}{2}(T-1)$, and all other elements are 0.

Chapter 4, Problem 18. Use (10)-(13) and (33) to write

$$a_0 = \frac{1}{n}\sum_{t=1}^{n}\bar{y}_t = \bar{y},$$

$$a(kh) = \frac{2}{n}\sum_{t=1}^{n}\bar{y}_t\cos\frac{2\pi k}{n}t,$$

$$b(kh) = \frac{2}{n}\sum_{t=1}^{n}\bar{y}_t\sin\frac{2\pi k}{n}t, \qquad k = 1, \ldots, \tfrac{1}{2}n - 1,$$

$$a_{\frac{1}{2}T} = \frac{1}{n}\sum_{t=1}^{n}\bar{y}_t(-1)^t.$$

Compare (17) and (18) for the case $n = 12$, $h = 3$. Let

$$x^* = \sqrt{\frac{n}{2}}\left(\sqrt{2}a_0, a(1h), b(1h), a(2h), \ldots, b\left(\frac{n-2}{2}h\right), \sqrt{2}a_{\frac{1}{2}T}\right)',$$

$$y^* = (\bar{y}_1, \ldots, \bar{y}_n)'.$$

By (26) and (27) of Section 4.2 $y^* = M^*x^*$, where M^* is orthogonal, and

$$h\sum_{t=1}^{n}(\bar{y}_t - \bar{y})^2 = h[y^{*\prime}y^* - n\bar{y}^2] = h[x^{*\prime}x^* - na_0^2],$$

which is the numerator of (30). To finish we observe, by (15) and the above,

$$\begin{aligned}(T-n)s^2 &= \sum_{t=1}^{T} y_t^2 - T\bar{y}^2 - h\sum_{t=1}^{n}(\bar{y}_t - \bar{y})^2 \\ &= \sum_{t=1}^{n}\sum_{j=0}^{h-1} y_{t+nj}^2 - h\sum_{t=1}^{n}\bar{y}_t^2 \\ &= \sum_{t=1}^{n}\sum_{j=0}^{h-1}(y_{t+nj} - \bar{y}_t)^2.\end{aligned}$$

Chapter 4, Problem 27.

(a)

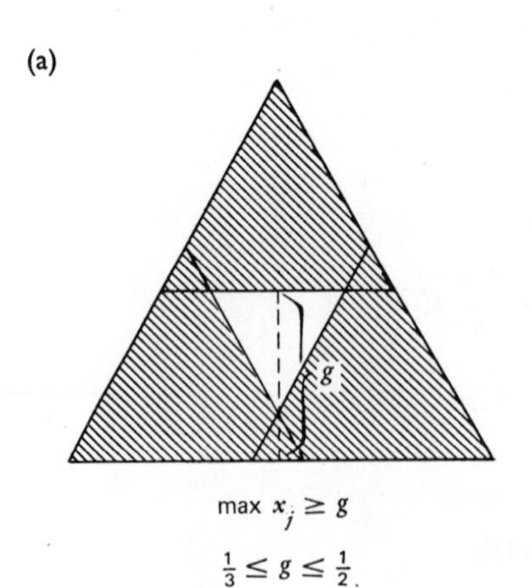

max $x_j \geq g$

$\frac{1}{3} \leq g \leq \frac{1}{2}$.

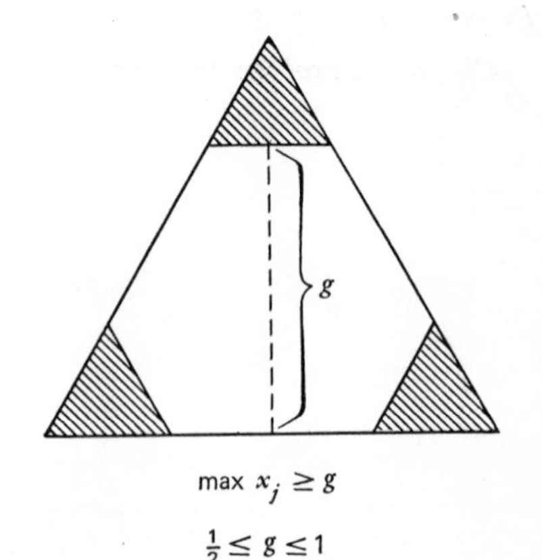

max $x_j \geq g$

$\frac{1}{2} \leq g \leq 1$

(b)

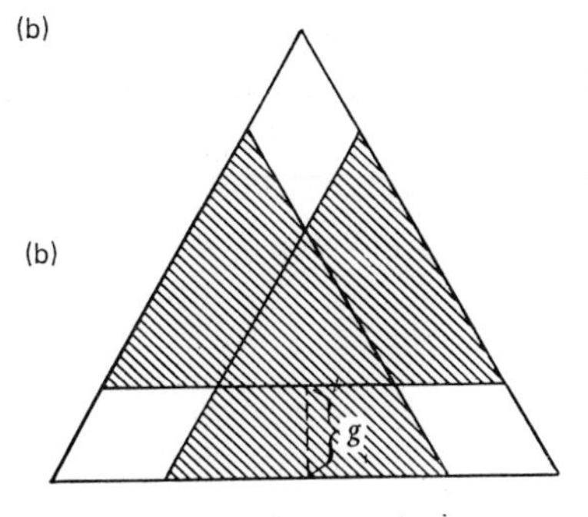

Second largest $x_j \geq g$

$0 \leq g \leq \frac{1}{3}$

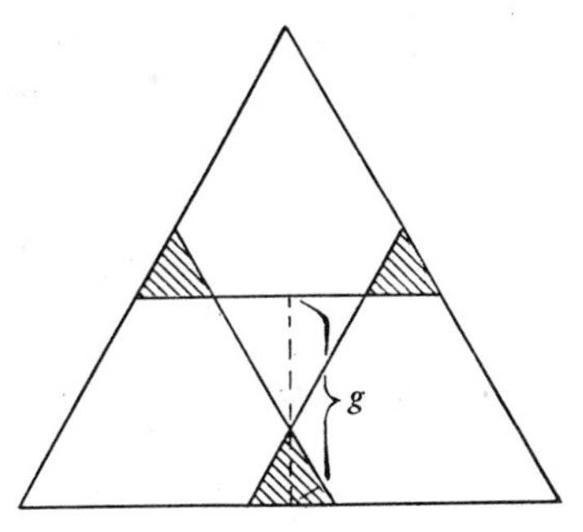

Second largest $x_j \geq g$

$\frac{1}{3} \leq g \leq \frac{1}{2}$

Chapter 4, Problem 28.

(i) $R_0: x_i \leq g,\ i = 1, 2, 3$

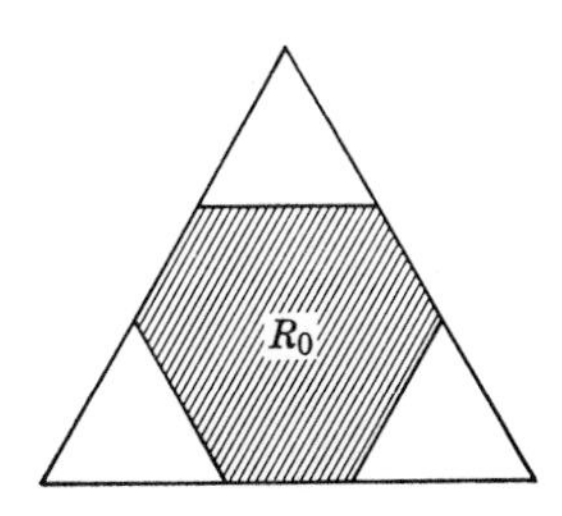

(ii) $R_i: x_i > g,\ x_j \leq g^* x_k,\ x_k \leq g^* x_j,$
i, j, k distinct,

$g^* \sin \theta = \sin(\frac{\pi}{3} - \theta),$

$\theta = \arctan \frac{\sqrt{3}}{2g^* + 1},\ g^* \geq 1$

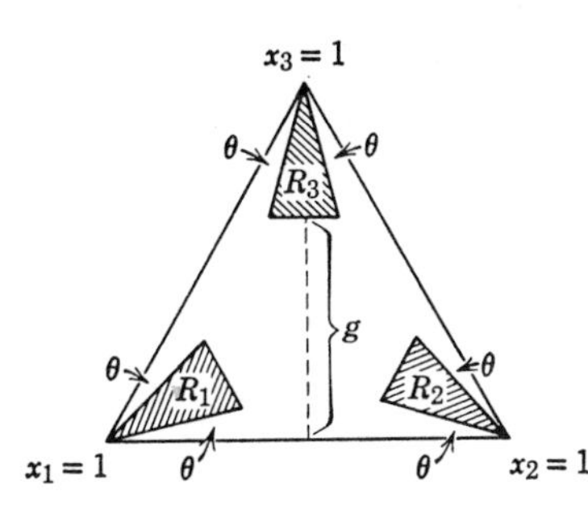

(iii) $R_{ij}: x_i > g,\ x_j > g^* x_k,\ i, j, k$ distinct,

$\theta = \arctan \frac{\sqrt{3}}{2g^* + 1},\ g^* \geq 1$

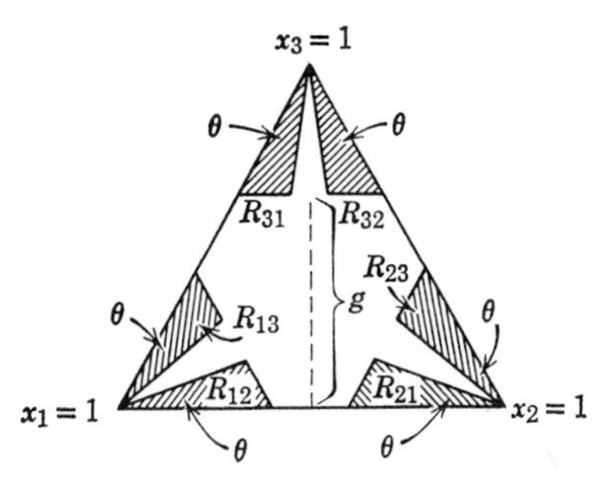

The procedure suggested by Whittle (1952) is to decide $\rho_1 = \rho_2 = \rho_3 = 0$ if $x_i \leq g$, $i = 1, 2, 3$; this region is R_0. Decide $\rho_1 > 0$ if $x_1 > g$; for $g \geq \frac{1}{2}$ this region is $R_1 \cup R_{12} \cup R_{13}$. In this case decide $\rho_2 = \rho_3 = 0$ if $x_2 \leq \bar{g}(x_2 + x_3)$ and $x_3 \leq \bar{g}(x_2 + x_3)$, or equivalently, if $x_2 \leq g^* x_3$ and $x_3 \leq g^* x_2$; this region is R_1. The solution is given for $g \geq \frac{1}{2}$ for ease of illustration. For $\frac{1}{3} < g < \frac{1}{2}$ decide $\rho_1 > 0$ if $x_1 > g$ and $x_1 > x_2$ and $x_1 > x_3$.

Chapter 5, Problem 6. The left-hand side is

$$P_q(t+1)a^{t+1} - P_q(t)a^{t+1} = [P_q(t+1) - P_q(t)]a^{t+1} = P_{q-1}(t)a^{t+1},$$

by (13) of Section 3.4.

Chapter 5, Problem 7. The proof is by induction. For $m = 1$, $(\mathscr{P} - a)a^t = 0$. Assume the result is true for $m = k$. Then

$$(\mathscr{P} - a)^{k+1}P_k(t)a^t = (\mathscr{P} - a)^k(\mathscr{P} - a)P_k(t)a^t = (\mathscr{P} - a)^k P_{k-1}(t)a^{t+1} = 0$$

by Problem 6 and the induction hypothesis.

Chapter 5, Problem 9. The polynomial in $\mathscr{P}$ can be written

$$\sum_{r=0}^{p} \beta_r \mathscr{P}^{p-r} = P_{p-m}(\mathscr{P})(\mathscr{P} - a)^m,$$

where $P_{p-m}(t)$ is a polynomial of degree $p - m$. Then

$$\begin{aligned}\sum_{r=0}^{p} \beta_r w_{t-r} &= \sum_{r=0}^{p} \beta_r \mathscr{P}^{p-r} w_{t-p} \\ &= P_{p-m}(\mathscr{P})(\mathscr{P} - a)^m P_{m-1}(t-p)a^{t-p} \\ &= 0,\end{aligned}$$

by Problem 7.

Chapter 5, Problem 10. We use the method of solution of Problem 9. Then

$$\begin{aligned}\sum_{r=0}^{p} \beta_r w_{t-r} &= P_{p-2m}(\mathscr{P})(\mathscr{P} - \alpha e^{i\theta})^m(\mathscr{P} - \alpha e^{-i\theta})^m[P_{m-1}(t)\alpha^t e^{it\theta} + \bar{P}_{m-1}(t)\alpha^t e^{-it\theta}] \\ &= 0,\end{aligned}$$

by application of Problem 7 twice. Note $P_{m-1}(t)\alpha^t e^{it\theta} + Q_{m-1}(t)\alpha^t e^{-it\theta}$ is also a solution for any polynomial $Q_{m-1}(t)$ of degree $m - 1$, but this solution is real only if $Q_{m-1}(t) = \bar{P}_{m-1}(t)$.

Chapter 5, Problem 17.

$$\Lambda^\tau = \begin{bmatrix} \lambda^\tau & \binom{\tau}{1}\lambda^{\tau-1} & \binom{\tau}{2}\lambda^{\tau-2} & \binom{\tau}{3}\lambda^{\tau-3} & \cdots & \binom{\tau}{n-2}\lambda^{\tau-n+2} & \binom{\tau}{n-1}\lambda^{\tau-n+1} \\ 0 & \lambda^\tau & \binom{\tau}{1}\lambda^{\tau-1} & \binom{\tau}{2}\lambda^{\tau-2} & \cdots & \binom{\tau}{n-3}\lambda^{\tau-n+3} & \binom{\tau}{n-2}\lambda^{\tau-n+2} \\ 0 & 0 & \lambda^\tau & \binom{\tau}{1}\lambda^{\tau-1} & \cdots & \binom{\tau}{n-4}\lambda^{\tau-n+4} & \binom{\tau}{n-3}\lambda^{\tau-n+3} \\ \cdot & \cdot & \cdot & \cdot & & \cdot & \cdot \\ \cdot & \cdot & \cdot & \cdot & & \cdot & \cdot \\ \cdot & \cdot & \cdot & \cdot & & \cdot & \cdot \\ 0 & 0 & 0 & 0 & \cdots & \lambda^\tau & \binom{\tau}{1}\lambda^{\tau-1} \\ 0 & 0 & 0 & 0 & \cdots & 0 & \lambda^\tau \end{bmatrix},$$

where n is the order of the matrix and $\binom{\tau}{j} = 0$ for $j > \tau$. The result may be verified by induction.

Chapter 5, Problem 18. The characteristic polynomial equation $|\Lambda - \nu I| = 0$ is $(\lambda - \nu)^n = 0$, where Λ is of order n. The equation $\Lambda x = \lambda x$, where $x = (x_1, \ldots, x_n)'$, has the solution $x = (x, 0, \ldots, 0)'$, which is unique except for $x \neq 0$.

Chapter 5, Problem 19. The component equations $-\tilde{\mathbf{B}}v = x_i v$, with $v = (v_1, \ldots, v_p)'$, are $-\sum_{r=1}^{p} \beta_r v_r = x_i v_1$, and $v_{r-1} = x_i v_r$, $r = 2, \ldots, p$. If $v_1 = x_i^{p-1}$ the unique solution is $v_r = x_i^{p-r}$, $r = 1, \ldots, p$. That is, to each distinct root there corresponds a unique vector. If $-\tilde{\mathbf{B}} = C\Lambda C^{-1}$ with Λ diagonal and C nonsingular, then $-\tilde{\mathbf{B}}C = C\Lambda$, and $-\tilde{\mathbf{B}}$ has p distinct vectors and hence p distinct roots.

Chapter 5, Problem 20. In (22) $\sum_{j=1}^{p} c_{1j} x_j^\tau c^{j1}$ is the first diagonal element of $C\Lambda^\tau C^{-1} = (-\tilde{\mathbf{B}})^\tau$. It is necessary to show this equals δ_τ in (28) of Section 5.2. Denote the first row of $(-\tilde{\mathbf{B}})^{\tau+1}$ by $(\alpha_{\tau 1}, \alpha_{\tau 2}, \ldots, \alpha_{\tau, p-1}, \alpha_{\tau p})$. Then the first row of $(-\tilde{\mathbf{B}})^{\tau+2} = (-\tilde{\mathbf{B}})^{\tau+1}(-\tilde{\mathbf{B}})$ is $(\alpha_{\tau 2} - \alpha_{\tau 1}\beta_1, \alpha_{\tau 3} - \alpha_{\tau 1}\beta_2, \ldots, \alpha_{\tau p} - \alpha_{\tau 1}\beta_{p-1}, -\alpha_{\tau 1}\beta_p)$. These elements therefore satisfy the same recursion relations as the δ_τ's. [See (14) or (22) of Section 5.2.] They also satisfy the same initial conditions.

Chapter 5, Problem 27. Following the hint, we multiply each side of (31) on the left by C^{-1} and on the right by $(C')^{-1}$. By (10) $C^{-1}\mathbf{B}C = -\Lambda$, so that $A^* - \Lambda A^* \Lambda = \Sigma^*$ follows. If $A^* = (a_{ij}^*)$, $\Sigma^* = (\sigma_{ij}^*)$, and Λ has diagonal elements $\lambda_1, \ldots, \lambda_p$, $a_{ij}^*(1 - \lambda_i\lambda_j) = \sigma_{ij}^*$, $i, j = 1, \ldots, p$. Since the roots of $-\mathbf{B}$ are less than 1 in absolute value A^* is unique, and $A = CA^*C'$ is unique.

Chapter 5, Problem 40. (a) Graph of Deflated Aggregate Disposable Income.

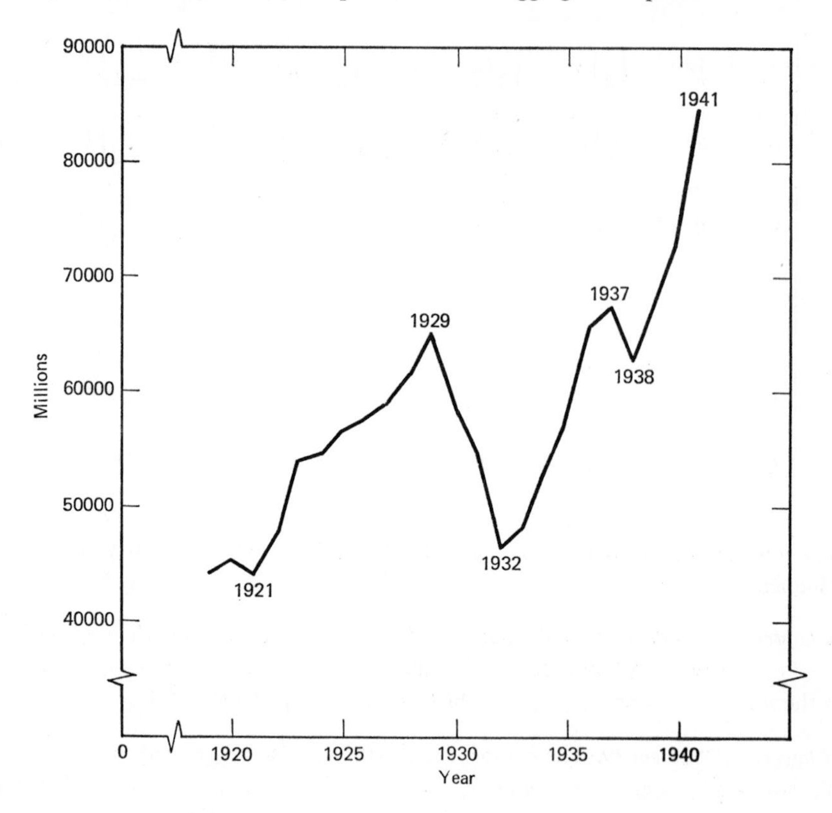

(b) The means of y_t, y_{t-1}, and $y_{t-2}, t = 1, \ldots, 21$, are 59.0980, 57.2485, and 55.8937, respectively. The normal equations for $\hat{\gamma}_2$, $\hat{\beta}_1$, and $\hat{\beta}_2$ are

$$\begin{pmatrix} 770 & 681 & 639 \\ 681 & 1{,}287 & 1{,}050 \\ 639 & 1{,}050 & 1{,}175 \end{pmatrix} \begin{pmatrix} \hat{\gamma}_2 \\ \hat{\beta}_1 \\ \hat{\beta}_2 \end{pmatrix} = - \begin{pmatrix} 821 \\ 1{,}316 \\ 946 \end{pmatrix}.$$

The sum of squares of y_t about its mean is 1,807.38. The estimates are $\hat{\beta}_1 = -1.223$, $\hat{\beta}_2 = 0.5103$, $\hat{\gamma}_1 = -13.13$, $\hat{\gamma}_2 = -0.4095$, and $\hat{\sigma} = 4.045$. (The estimate of σ dividing by "degrees of freedom" $21 - 4 = 17$ is 4.496.) The estimated covariance matrix of $\hat{\gamma}_2$, $\hat{\beta}_1$, and $\hat{\beta}_2$ is $\hat{\sigma}^2$ (or s^2) times

$$\begin{pmatrix} 25.79 & -8.11 & -6.78 \\ -8.11 & 31.25 & -23.52 \\ -6.78 & -23.52 & 33.22 \end{pmatrix}$$

times 10^{-4}. The estimated standard deviations of the estimated coefficients are 0.228, 0.251, and 0.259, respectively, with the use of s.

(c) $\hat{\mu} = 44.7$, $\hat{\delta} = 1.425$.

(d) The associated polynomial equation has roots $0.6115 \pm 0.3693i$. The angle is $31^\circ\, 8'$, which corresponds to a frequency of $31^\circ\, 8'/360^\circ = 0.08648$ or period of 11.57 years; the modulus is 0.7144.

Discussion. The second-order autoregressive model for income can be derived from the economic model

$$Y_t = C_t + I_t + G_t,$$

$$C_t = \alpha_0 + \alpha Y_{t-1} + \alpha^* t,$$

$$I_t = \beta_0 + \beta[C_t + G_t - (C_{t-1} + G_{t-1})] + \beta^* t,$$

$$G_t = \gamma_0 + \gamma^* t,$$

where Y_t, C_t, I_t, and G_t are income, consumption, investment, and government expenditures, respectively, in the year t. Then $\beta_1 = -\alpha(1 + \beta)$ and $\beta_2 = \alpha\beta$. The estimate for α is 0.713, which is the marginal propensity to consume, and the estimate for β is 0.716, which is the component of investment brought out by the change in non-investment expenditures. The fact that the latter coefficient is rather low may be due to the lack of confidence in business conditions during much of the period under observation.

Chapter 6, Problem 13. Substitution of (16) in (20) gives (i), proving the assertion in the hint. Then (i) and (ii) further imply

$$\text{(iii)} \qquad \int_c^\infty (Q_i - \mu) h_i(Q_i \mid Q_0, \ldots, Q_{i-1}; \gamma_i^{(1)})\, dQ_i = 0.$$

By (16) the given conditional distribution has positive mass in each of the intervals $(-\infty, c)$ and (c, ∞); otherwise $\varepsilon_i = 0$ or 1. Then (i) implies $\mu < c$ and (iii) implies the contradictory result $\mu > c$.

Chapter 6, Problem 27. (a) The vector $\mathbf{y} = (y_1, y_2, y_3, y_4)'$ has the normal distribution with mean vector $\mathbf{0}$ and covariance matrix $\mathbf{\Sigma}$ given by

$$\mathbf{\Sigma}^{-1} = \begin{bmatrix} \gamma_0 + \gamma_1 & 0 & 0 & 0 \\ 0 & \gamma_0 + \gamma_1 & 0 & 0 \\ 0 & 0 & \gamma_0 - \gamma_1 & 0 \\ 0 & 0 & 0 & \gamma_0 - \gamma_1 \end{bmatrix}.$$

Then $Y_1 = (\gamma_0 + \gamma_1)(y_1^2 + y_2^2)$ and $Y_2 = (\gamma_0 - \gamma_1)(y_3^2 + y_4^2)$ are independently distributed each as χ^2 with 2 degrees of freedom; the joint density is

$$\tfrac{1}{4} e^{-\frac{1}{2}(Y_1 + Y_2)}, \qquad Y_1, Y_2 \geq 0.$$

The Jacobian of the transformation

$$Y_1 = \tfrac{1}{2}(\gamma_0 + \gamma_1)(Q_0 + Q_1),$$
$$Y_2 = \tfrac{1}{2}(\gamma_0 - \gamma_1)(Q_0 - Q_1),$$

is $\frac{1}{2}(\gamma_0^2 - \gamma_1^2)$, and the joint density of Q_0 and Q_1 is (ii).

(b) The Jacobian of the transformation $Q_0 = Q_0$, $r = Q_1/Q_0$ is Q_0. It follows from (a) that the joint density of Q_0 and r is (iii).

(c) The marginal density of Q_0 is the integral of (iii) over the range of r, namely

$$\frac{\gamma_0^2 - \gamma_1^2}{8} Q_0 e^{-\frac{1}{2}\gamma_0 Q_0} \int_{-1}^{1} e^{-\frac{1}{2}\gamma_1 Q_0 r}\, dr = \frac{\gamma_0^2 - \gamma_1^2}{8} Q_0 e^{-\frac{1}{2}\gamma_0 Q_0} \frac{2}{\gamma_1 Q_0} (e^{\frac{1}{2}\gamma_1 Q_0} - e^{-\frac{1}{2}\gamma_1 Q_0})$$

which is (iv). It may also be written

$$\frac{\gamma_0^2 - \gamma_1^2}{2\gamma_1} e^{-\frac{1}{2}\gamma_0 Q_0} \sinh \tfrac{1}{2}\gamma_1 Q_0, \qquad 0 \le Q_0.$$

(d) The conditional density of r given Q_0 is the quotient of (iii) and (iv).

(e) The marginal density of r is the integral of (iii) over the range of Q_0, namely,

$$\frac{\gamma_0^2 - \gamma_1^2}{8(\gamma_0 + \gamma_1 r)^2} \int_0^\infty [(\gamma_0 + \gamma_1 r)Q_0]^{\frac{1}{2}4-1} e^{-\frac{1}{2}(\gamma_0+\gamma_1 r)Q_0} [(\gamma_0 + \gamma_1 r)dQ_0]$$
$$= \frac{\gamma_0^2 - \gamma_1^2}{8(\gamma_0 + \gamma_1 r)^2} 2^{\frac{1}{2}4}\Gamma(\tfrac{1}{2}4) = \frac{\gamma_0^2 - \gamma_1^2}{2(\gamma_0 + \gamma_1 r)^2}.$$

The integral involves the χ^2-density with 4 degrees of freedom. Note that $\gamma_0 + \gamma_1 r > 0$ since $|\gamma_1| < \gamma_0$ and $|r| \le 1$.

Chapter 6, Problem 37. The characteristic function of u, given by (69), has only simple poles. Suppose $\nu_{m+1} \le R \le \nu_m$ and $H \ge 2$. The density of u is

$$f(u) = \frac{1}{2\pi} \int_{-\infty}^{\infty} e^{-iut} \cdot \frac{dt}{\prod_{j=1}^{H} [1 - 2it(\nu_j - R)]}.$$

We shall determine $f(u)$ for $u \ge 0$ by integrating along a closed contour containing the poles in the lower half of the complex plane.

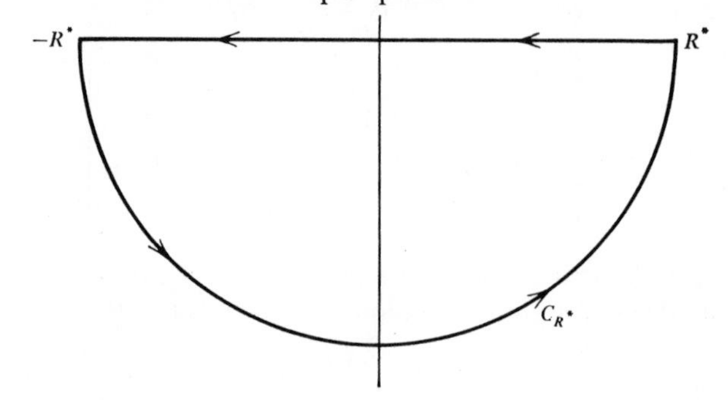

The radius R^* is chosen large enough so that the contour encloses all the poles in the lower half of the complex plane. By the residue theorem

$$\frac{1}{2\pi}\int_{R^*}^{-R^*} e^{-iut} \cdot \frac{dt}{\prod_{j=1}^{H} [1 - 2it(\nu_j - R)]} + \frac{1}{2\pi}\int_{C_{R^*}} e^{-iuz} \cdot \frac{dz}{\prod_{j=1}^{H} [1 - 2iz(\nu_j - R)]}$$

equals $2\pi i$ times the sum of the residues of the integrand at the poles within the contour. Since $\nu_{m+1} \le R \le \nu_m$, poles occur at $z = -\frac{1}{2}i/(\nu_j - R)$, $j = 1, \ldots, m$, which are in the lower half-plane. (If $R = \nu_m$, there are $m - 1$ poles; there is no pole for $j = m$.) We have

$$\frac{1}{1 - 2iz(\nu_j - R)} = \frac{\dfrac{i}{2(\nu_j - R)}}{z + \dfrac{i}{2(\nu_j - R)}}.$$

Then the residue at $z = -\frac{1}{2}i/(\nu_j - R)$ is

$$\frac{i}{4\pi} e^{-\frac{1}{2}u/(\nu_j - R)} \frac{(\nu_j - R)^{H-2}}{\prod_{\substack{k=1\\k\neq j}}^{H} (\nu_j - \nu_k)},$$

and $2\pi i$ times the sum of the residues is

$$\text{(i)} \qquad -\frac{1}{2}\sum_{j=1}^{m} e^{-\frac{1}{2}u/(\nu_j - R)} \frac{(\nu_j - R)^{H-2}}{\prod_{\substack{k=1\\k\neq j}}^{H} (\nu_j - \nu_k)}, \qquad \nu_{m+1} \le R \le \nu_m, \quad m = 1, \ldots, H - 1.$$

Then for $u \geqq 0$, $-f(u)$ equals (i) minus the limit as $R^* \to \infty$ of the integral along C_{R^*}. The latter integral tends to zero, because the numerator of the integrand has absolute value at most 1, the denominator of the integrand is $O(R^{*H})$, and the arc length of C_{R^*} is πR^*. Finally

$$\Pr\{r \ge R\} = \int_0^\infty f(u)\, du$$

$$= \frac{1}{2}\sum_{j=1}^{m} \frac{(\nu_j - R)^{H-2}}{\prod_{\substack{k=1\\k\neq j}}^{H} (\nu_j - \nu_k)} \cdot 2(\nu_j - R), \qquad \nu_{m+1} \le R \le \nu_m, \quad m = 1, \ldots, H - 1,$$

which is (52). To obtain $f(u)$ for $u < 0$ it is necessary to integrate along a closed contour containing the poles in the upper half of the complex plane.

Chapter 6, Problem 69. The derivative of the left-hand side is

$$2\int_0^\pi \frac{dx}{a + b\cos x}$$

and following the hint, we obtain

$$2\int_0^\pi \frac{dx}{(a+b)\cos^2 \frac{1}{2}x + (a-b)\sin^2 \frac{1}{2}x} = 2\int_0^\pi \frac{\sec^2 \frac{1}{2}x\, dx}{(a+b)\left[1 + \dfrac{a-b}{a+b}\tan^2 \frac{1}{2}x\right]}$$

$$= \frac{4}{\sqrt{a^2 - b^2}}\int_0^\infty \frac{dy}{1+y^2} = \frac{2\pi}{\sqrt{a^2 - b^2}}.$$

The derivative of the right-hand side is the same expression. We have shown

$$\text{(i)} \qquad C(a, b) = \int_0^\pi \log(a + b\cos x)\, dx - \pi \log \frac{a + \sqrt{a^2 - b^2}}{2}, \qquad a > b \geq 0,$$

does not depend on the first argument. Let $a = kb > 0$, $k > 1$. Then it follows $C(kb, b) = C^*(b)$, and $C^*(b) = \int_0^\pi \log(k + \cos x)\, dx - \pi \log(k + \sqrt{k^2 - 1})/2$, so that $C(a, b) \equiv C$. As b tends to 0, the left-hand side of (i) remains constant and the right-hand side tends to 0, so that $C(a, b) \equiv 0$.

Chapter 7, Problem 28. The required moving average and autoregressive processes exist by Corollaries 7.5.1 and 7.5.2, respectively. In each case

$$|\sigma(h) - \sigma_m(h)| = \left|\int_{-\pi}^{\pi} e^{i\lambda h}[f(\lambda) - f_m(\lambda)]\, d\lambda\right|$$

$$< \varepsilon \int_{-\pi}^{\pi} d\lambda.$$

Chapter 7, Problem 29. The processes $\{y_t\}$ and $\{w_t\}$ are independent, and by Problem 9 the spectral density of $\{z_t\}$ is

$$\frac{\sigma^2}{2\pi\, |e^{i\lambda} + \beta|^2} + \frac{\tau^2}{2\pi}.$$

The spectral density of $\{z_t^*\}$ is deduced from (51), and the requirements for equality of the two densities are $K\alpha = \beta\tau^2$ and $K(1 + \alpha^2) = \sigma^2 + \tau^2 + \beta^2\tau^2$. Then $K = \beta\tau^2/\alpha$, and α is a root of $\beta\tau^2\alpha^2 - (\sigma^2 + \tau^2 + \beta^2\tau^2)\alpha + \beta\tau^2 = 0$. Both roots are real because

$$(\sigma^2 + \tau^2 + \beta^2\tau^2)^2 - 4(\beta\tau^2)^2 = (\sigma^2 + \tau^2 - \beta^2\tau^2)^2 + 4\sigma^2\beta^2\tau^2 \geq 0.$$

Chapter 7, Problem 33. By Corollary 7.5.3

$$\begin{aligned}
f(\lambda) &= \frac{\sigma^2}{2\pi}\left|1 - \frac{1}{2m+1}\sum_{s=-m}^{m} e^{i\lambda s}\right|^2 \\
&= \frac{\sigma^2}{2\pi}\left|e^{i\lambda m} - \frac{1}{2m+1}\sum_{s=0}^{2m} e^{i\lambda s}\right|^2 \\
&= \frac{\sigma^2}{2\pi}\left|e^{i\lambda m} - \frac{1}{2m+1}\cdot\frac{\sin\lambda(m+\frac{1}{2})}{\sin\frac{1}{2}\lambda}e^{i\lambda m}\right|^2 \\
&= \frac{\sigma^2}{2\pi}\left|1 - \frac{\sin\lambda(m+\frac{1}{2})}{(2m+1)\sin\frac{1}{2}\lambda}\right|^2.
\end{aligned}$$

See (71).

Chapter 8, Problem 2. From the definition (34) and the 4! ways the product $(y_t - \mu)(y_{t+h} - \mu)(y_{t+r} - \mu)(y_{t+s} - \mu)$ may be written it follows that $\nu(h, r, s)$ has 24 symmetries. By (36) $\kappa(h, r, s)$ and $\nu(h, r, s)$ have the same symmetries. The 15 forms not shown are obtained by applying the results of (a) to each of the forms shown in (b).

$$\begin{aligned}
\kappa(h, r, s) &= \kappa(-h, s-h, r-h) = \kappa(r-h, -h, s-h) \\
&= \kappa(r-h, s-h, -h) = \kappa(s-h, -h, r-h) \\
&= \kappa(s-h, r-h, -h); \\
\kappa(h, r, s) &= \kappa(-r, s-r, h-r) = \kappa(h-r, -r, s-r) \\
&= \kappa(h-r, s-r, -r) = \kappa(s-r, -r, h-r) \\
&= \kappa(s-r, h-r, -r); \\
\kappa(h, r, s) &= \kappa(-s, r-s, h-s) = \kappa(h-s, -s, r-s) \\
&= \kappa(h-s, r-s, -s) = \kappa(r-s, -s, h-s) \\
&= \kappa(r-s, h-s, -s).
\end{aligned}$$

Chapter 8, Problem 12.

$$\begin{aligned}
\mathscr{E}(y_t - \mu)^2(y_s - \mu) &= \sum_{r,p,q=-\infty}^{\infty} \gamma_r\gamma_p\gamma_q \mathscr{E} v_{t-r}v_{t-p}v_{s-q} \\
&= \kappa_3 \sum_{r=-\infty}^{\infty} \gamma_r^2\gamma_{r+s-t}.
\end{aligned}$$

Since $\operatorname{Cov}[(y_t - \mu)^2, y_s] = \mathscr{E}(y_t - \mu)^2(y_s - \mu)$,

$$\begin{aligned}
\operatorname{Cov}(\bar{y}, C_0) &= \frac{1}{T^2}\sum_{t,s=1}^{T} \mathscr{E}(y_t - \mu)^2(y_s - \mu) \\
&= \frac{\kappa_3}{T^2}\sum_{t,s=1}^{T}\sum_{r=-\infty}^{\infty} \gamma_r^2\gamma_{r+s-t},
\end{aligned}$$

and (i) follows. Then (i) is in absolute value at most

$$\frac{\kappa_3}{T} \max_r |\gamma_r| \sum_{r=-\infty}^{\infty} |\gamma_r| \sum_{h=-(T-1)}^{T-1} \left(1 - \frac{|h|}{T}\right) |\gamma_{r+h}|.$$

By Lemma 8.3.1 the second summation is bounded by its limit, $\sum_{h=-\infty}^{\infty} |\gamma_h|$, and (ii) follows.

Chapter 8, Problem 17. Under the given assumptions $\text{Cov}(y_t y_s, y_t y_s) = \sigma^4$, $t \neq s$, $\text{Cov}(y_t^2, y_t^2) = 2\sigma^4 + \kappa_4$, and $\text{Cov}(y_t y_s, y_{t'} y_{s'}) = 0$ otherwise. Then [see (76)]

$$\begin{aligned}(2\pi T)^2 \text{Cov}[I(\lambda), I(\lambda')] &= \sum_{t,s,t',s'=1}^{T} e^{i\lambda(t-s)+i\lambda'(t'-s')} \text{Cov}(y_t y_s, y_{t'} y_{s'}) \\ &= T\kappa_4 + \sigma^4 \sum_{t,s=1}^{T} [e^{i(\lambda+\lambda')(t-s)} + e^{i(\lambda-\lambda')(t-s)}] \\ &= T\kappa_4 + \sigma^4 \left[\frac{\sin^2 \frac{1}{2}(\lambda + \lambda')T}{\sin^2 \frac{1}{2}(\lambda + \lambda')} + \frac{\sin^2 \frac{1}{2}(\lambda - \lambda')T}{\sin^2 \frac{1}{2}(\lambda - \lambda')}\right]\end{aligned}$$

for $\lambda \neq \pm\lambda'$. Note $I(\lambda) = I(-\lambda)$. The variances are special cases.

Chapter 8, Problem 26. By (45)

$$\begin{aligned}\sum_{h=-\infty}^{\infty} \sigma^2(h) &= \sigma^4 \sum_{h,s,s'=-\infty}^{\infty} \gamma_s \gamma_{s+h} \gamma_{s'} \gamma_{s'+h} \leq \sigma^4 \max_t |\gamma_t| \sum_{h,s,s'=-\infty}^{\infty} |\gamma_s| \cdot |\gamma_{s+h}| \cdot |\gamma_{s'}| \\ &= \sigma^4 \max_t |\gamma_t| \left(\sum_{s=-\infty}^{\infty} |\gamma_s| \right)^3.\end{aligned}$$

Chapter 9, Problem 21. The expression on the left in the hint is

$$\left| \sum_{r=0}^{2K} \frac{1}{K} g\left(\frac{r-K}{K}\right) e^{idr} \right|,$$

and $g\left(\frac{r-K}{K}\right) = C$ for $A = aK + K < r \leq bK + K = B$ and is 0 otherwise. Let $[x]$ denote the greatest integer less than or equal to x and let $\{x\}$ denote the smallest integer greater than x. Then the above is

$$\frac{C}{K} \left| \sum_{r=\{A\}}^{[B]} e^{idr} \right| = \frac{C}{K} \left| \frac{1 - e^{id([B]-\{A\}+1)}}{1 - e^{id}} \right| \leq \frac{C}{K} \frac{1}{|\sin \frac{1}{2}d|},$$

since $|1 - e^{id}| = 2 |\sin \frac{1}{2}d|$.

Chapter 9, Problem 22. Let

$$\begin{aligned} h_j(x) &= C_j, && a_{j-1} < x \le a_j, \\ &= 0, \qquad \text{elsewhere}, && j = 1, \ldots, n, \end{aligned}$$

so that $h(x) = \sum_{j=1}^{n} h_j(x)$. The result of Problem 21 also holds for $C < 0$ and for $g(x) = C$, $a \le x \le b$, and 0 elsewhere, so that $h(-1) = C_1$ is no restriction. Then the desired limit is

$$\sum_{j=1}^{n} \lim_{K \to \infty} \sum_{r=-K}^{K} \frac{1}{K} h_j\left(\frac{r}{K}\right) e^{idr},$$

and the limit for each j is 0.

Chapter 9, Problem 23. Since $f(x)$ is uniformly continuous in $-1 \le x \le 1$, for any $\varepsilon > 0$ there is a $\delta > 0$ such that $|f(x) - f(y)| < \varepsilon$ if $|x - y| < \delta$. Fix $\varepsilon > 0$ and choose δ and a partition $-1 = a_0 < a_1 < \cdots < a_n = 1$ so that

$$\max_{j=1,\ldots,n} |a_j - a_{j-1}| < \delta.$$

Select a_j' so that $a_{j-1} < a_j' \le a_j$, $j = 1, \ldots, n$. Define $h(x) = f(a_j')$, $a_{j-1} < x \le a_j$, $j = 1, \ldots, n$, and $h(-1) = f(a_1')$. Then $|f(x) - h(x)| < \varepsilon$ in $-1 \le x \le 1$ and

$$\begin{aligned} \left| \sum_{r=-K}^{K} \frac{1}{K} f\left(\frac{r}{K}\right) e^{idr} - \sum_{r=-K}^{K} \frac{1}{K} h\left(\frac{r}{K}\right) e^{idr} \right| &\le \frac{1}{K} \sum_{r=-K}^{K} \left| f\left(\frac{r}{K}\right) - h\left(\frac{r}{K}\right) \right| \\ &< \left(2 + \frac{1}{K}\right) \varepsilon. \end{aligned}$$

By Problem 22

$$\lim_{K \to \infty} \sum_{r=-K}^{K} \frac{1}{K} h\left(\frac{r}{K}\right) e^{idr} = 0$$

for $d \ne k2\pi$, $k = 0, \pm 1, \ldots$, and the result follows because ε is arbitrary.

Chapter 10, Problem 10. For any p-component complex vector $\boldsymbol{c}$ $(\ne \mathbf{0})\boldsymbol{c}'\mathbf{R}(h)\bar{\boldsymbol{c}} = \int_{-\pi}^{\pi} e^{i\lambda h}\, \boldsymbol{c}'d\boldsymbol{M}(\lambda)\bar{\boldsymbol{c}}$, by (64). Since $\mathbf{R}(0)$ is positive definite and $\boldsymbol{M}(\lambda)$ has positive semi-definite increments, $\boldsymbol{c}'\boldsymbol{M}(\lambda)\bar{\boldsymbol{c}}/\boldsymbol{c}'\mathbf{R}(0)\bar{\boldsymbol{c}}$ is a distribution function. Without loss of generality assume $\boldsymbol{c}'\mathbf{R}(0)\bar{\boldsymbol{c}} = 1$. Then $2\pi \int_{-\pi}^{\pi} f(\lambda)\boldsymbol{c}'d\boldsymbol{M}(\lambda)\bar{\boldsymbol{c}}$ is the expectation of the random variable $U = 2\pi f(X)$, where X has distribution function $\boldsymbol{c}'\boldsymbol{M}(\lambda)\bar{\boldsymbol{c}}$. This expectation can also be evaluated as $\int_m^M u\boldsymbol{c}'d\,\mathbf{T}(u)\bar{\boldsymbol{c}}$, where $\boldsymbol{c}'\mathbf{T}(u)\bar{\boldsymbol{c}}$ is the distribution function of U. [See (96) and (97).] The result follows because $\boldsymbol{c}$ is arbitrary.

Bibliography

BOOKS

Anderson, O. (1929), *Die Korrelationsrechnung in der Konjunkturforschung*, Schroeder, Bonn.

Anderson, R. L. (1941), *Serial Correlation in the Analysis of Time Series*, unpublished thesis, Iowa State College, Ames, Iowa.

Anderson, R. L., and E. E. Houseman (1942), *Tables of Orthogonal Polynomial Values Extended to N = 104* (Research Bulletin 297), Agricultural Experiment Station, Iowa State College, Ames, Iowa.

Anderson, T. W. (1958), *An Introduction to Multivariate Statistical Analysis*, John Wiley & Sons, Inc., New York.

Bartlett, M. S. (1966), *An Introduction to Stochastic Processes with Special Reference to Methods and Applications* (Second Edition), Cambridge Univ. Press, Cambridge.

Blackman, R. B., and J. W. Tukey (1959), *The Measurement of Power Spectra from the Point of View of Communications Engineering*, Dover Publications, Inc., New York (printed originally 1958, *Bell System Tech. J.*, **37**, 185–282, 485–569).

Box, George E. P., and Gwilym M. Jenkins (1970), *Time Series Analysis forecasting and control*, Holden-Day, Inc., San Francisco.

Buijs Ballot, C. H. D. (1847), *Les Changements Périodiques de Température*, Utrecht.

Courant, R., and D. Hilbert (1937), *Methoden der Mathematischen Physik*, Springer-Verlag, Berlin.

Cramér, Harald (1946), *Mathematical Methods of Statistics*, Princeton Univ. Press, Princeton, N. J.

Davis, Harold T. (1941), *The Analysis of Economic Time Series* (The Cowles Commission for Research in Economics, Monograph No. 6), Principia Press, Inc., Bloomington, Ind.

Doob, J. L. (1953), *Stochastic Processes*, John Wiley & Sons, Inc., New York.

Draper, N. R., and H. Smith (1966), *Applied Regression Analysis*, John Wiley & Sons, Inc., New York.

Feller, William (1968), *An Introduction to Probability Theory and Its Applications, Vol. I* (Third Edition), John Wiley & Sons, Inc., New York.

Fisher, R. A., and Frank Yates (1963), *Statistical Tables for Biological, Agricultural and Medical Research* (Sixth Edition), Oliver and Boyd Ltd., Edinburgh.

Granger, C. W. J. (1964), in association with M. Hatanaka, *Spectral Analysis of Economic Time Series*, Princeton Univ. Press, Princeton, N. J.

Graybill, Franklin A. (1961), *An Introduction to Linear Statistical Models, Vol. I*, McGraw-Hill Book Co., Inc., New York.

Grenander, Ulf, and Murray Rosenblatt (1957), *Statistical Analysis of Stationary Time Series*, John Wiley & Sons, Inc., New York.

Halmos, Paul R. (1958), *Finite-Dimensional Vector Spaces*, D. Van Nostrand Co., Inc., Princeton, N. J.

Hamming, Richard W. (1962), *Numerical Methods for Scientists and Engineers*, McGraw-Hill Book Co., Inc., New York.

Hannan, E. J. (1960), *Time Series Analysis*, Methuen and Co., Ltd., London, and John Wiley & Sons, Inc., New York.

Hobson, E. W. (1907), *The Theory of Functions of a Real Variable and the Theory of Fourier's Series*, Cambridge Univ. Press, Cambridge.

Jordan, Charles (1939), *Calculus of Finite Differences*, Budapest (reprinted in 1947 by Chelsea Publishing Co., New York).

Kempthorne, Oscar (1952), *The Design and Analysis of Experiments*, John Wiley & Sons, Inc., New York.

Kendall, Maurice G. (1946a), *The Advanced Theory of Statistics, Vol. II*, Charles Griffin and Co., Ltd., London.

Kendall, Maurice G. (1946b), *Contributions to the Study of Oscillatory Time-Series, Nat. Inst. Econ. Soc. Res. Occasional Papers IX*, Cambridge Univ. Press, Cambridge.

Kendall, Maurice G., and Alan Stuart (1961), *The Advanced Theory of Statistics, Vol. 2*, Charles Griffin and Co., Ltd., London, and Hafner Publishing Co., Inc., New York.

Kendall, Maurice G., and Alan Stuart (1966), *The Advanced Theory of Statistics, Vol. 3*, Charles Griffin and Co., Ltd., London, and Hafner Publishing Co., Inc., New York.

Kolmogorov, A. N. (1933), *Grundbegriffe der Wahrscheinlichkeitsrechnung*, Berlin (reprinted in 1946 by Chelsea Publishing Co., New York).

Kuznets, S. (1954), *National Income and Its Composition: 1919–1938*, National Bureau of Economic Research, New York.

Lanczos, Cornelius (1956), *Applied Analysis*, Prentice-Hall, Inc., Englewood Cliffs, N. J.

Laplace, le Marquis P. S. de (1829), *Traité de Mécanique Céleste, Tome Second* (Seconde Édition), Bachelier, Successeur de Mme. Ve. Courcier, Paris.

Lehmann, E. L. (1959), *Testing Statistical Hypotheses*, John Wiley & Sons, Inc., New York.

Loève, Michel (1963), *Probability Theory* (Third Edition), D. Van Nostrand Co., Inc., New York.

Lukacs, Eugene (1960), *Characteristic Functions*, Charles Griffin and Co., Ltd., London, and Hafner Publishing Co., Inc., New York.

Macaulay, F. R. (1938), *Some Theoretical Problems Suggested by the Movements of Interest Rates, Bond Yields and Stock Prices in the United States Since 1856*, National Bureau of Economic Research, New York.

Miller, Kenneth S. (1960), *An Introduction to the Calculus of Finite Differences and Difference Equations*, Henry Holt & Co., New York.

Parzen, Emanuel (1962), *Stochastic Processes*, Holden-Day, Inc., San Francisco.

Plackett, R. L. (1960), *Principles of Regression Analysis*, Clarendon Press, Oxford.

Rao, C. Radhakrishna (1952), *Advanced Statistical Methods in Biometric Research*, John Wiley & Sons, Inc., New York.

Rosenblatt, Murray (1962), *Random Processes*, Oxford Univ. Press, New York.

Scheffé, Henry (1959), *The Analysis of Variance*, John Wiley and Sons, Inc., New York.

Tintner, Gerhard (1940), *The Variate Difference Method*, Principia Press, Inc., Bloomington, Ind.

Tintner, Gerhard (1952), *Econometrics*, John Wiley & Sons, Inc., New York.

Turnbull, H. W., and A. C. Aitken (1952), *An Introduction to the Theory of Canonical Matrices* (Third Edition), Blackie and Son Ltd., London (reprinted in 1961 by Dover Publications, Inc., New York).

United States Bureau of the Census (1955), *Statistical Abstract of the United States*, U.S. Government Printing Office, Washington, D.C.

United States Department of Agriculture (1939), *Agricultural Statistics*, United States Government Printing Office, Washington, D.C.

Wald, A. (1936), *Berechnung und Ausschaltung von Saisonschwankungen*, Springer-Verlag, Vienna.

Waldmeier, M. (1961), *The Sunspot Activity in the Years 1610–1960*, Schulthess & Co., Zürich.

Watson, G. S. (1952), *Serial Correlation in Regression Analysis*, Mimeo Series No. 49, Institute of Statistics, Univ. of North Carolina, Chapel Hill, N. C.

Whittaker, E. T., and G. Robinson (1926), *The Calculus of Observations, A Treatise on Numerical Mathematics* (Second Edition), Blackie & Son Ltd., London.

Whittaker, E. T., and G. N. Watson (1943), *A Course of Modern Analysis*, Cambridge Univ. Press, Cambridge, and The Macmillan Co., New York.

Whittle, Peter (1951), *Hypothesis Testing in Time Series Analysis*, Almqvist and Wiksell Book Co., Uppsala.

Wiener, Norbert (1949), *The Extrapolation, Interpolation and Smoothing of Stationary Time Series with Engineering Applications*, Technology Press of the Massachusetts Institute of Technology, Cambridge, Mass.

Wilks, Samuel S. (1962), *Mathematical Statistics*, John Wiley & Sons, Inc., New York.

Williams, E. J. (1959), *Regression Analysis*, John Wiley & Sons, Inc., New York.

Wold, Herman (1954), *A Study in the Analysis of Stationary Time Series* (Second Edition), Almqvist and Wiksell Book Co., Uppsala.

Wold, Herman O. A. (1965), *Bibliography on Time Series and Stochastic Processes*, Oliver and Boyd Ltd., Edinburgh.

Yaglom, A. M. (1962), *An Introduction to the Theory of Stationary Random Functions* (Revised English Edition, Richard A. Silverman, trans. and ed.), Prentice-Hall, Inc., Englewood Cliffs, N. J.

Zygmund, A. (1959), *Trigonometric Series, Vol. I,* Cambridge Univ. Press, Cambridge.

PAPERS

Akutowicz, Edwin J. (1957), On an explicit formula in least squares prediction, *Math. Scand.*, **5,** 261–266.

Anderson, R. L. (1942), Distribution of the serial correlation coefficient, *Ann. Math. Statist.*, **13,** 1–13.

Anderson, R. L., and T. W. Anderson (1950), Distribution of the circular serial correlation coefficient for residuals from a fitted Fourier series, *Ann. Math. Statist.*, **21,** 59–81.

Anderson, T. W. (1948), On the theory of testing serial correlation, *Skand. Aktuarietidskr.*, **31,** 88–116.

Anderson, T. W. (1959), On asymptotic distributions of estimates of parameters of stochastic difference equations, *Ann. Math. Statist.*, **30,** 676–687.

Anderson, T. W. (1963), Determination of the order of dependence in normally distributed time series, *Proc. Symp. Time Series Anal. Brown Univ.* (Murray Rosenblatt, ed.), John Wiley & Sons, Inc., New York, 425–446.

Anderson, T. W., and Herman Rubin (1950), The asymptotic properties of estimates of the parameters of a single equation in a complete system of stochastic equations, *Ann. Math. Statist.*, **21,** 570–582.

Anderson, T. W., and A. M. Walker (1964), On the asymptotic distribution of the autocorrelations of a sample from a linear stochastic process, *Ann. Math. Statist.*, **35,** 1296–1303.

Bartlett, M. S. (1935), Some aspects of the time-correlation problem in regard to tests of significance, *J. Roy. Statist. Soc.*, **98,** 536–543.

Bartlett, M. S. (1946), On the theoretical specification and sampling properties of autocorrelated time-series, *J. Roy. Statist. Soc. Supp.*, **8,** 27–41, 85–97. [Corrigenda (1948), **10,** 200.]

Bartlett, M. S. (1950), Periodogram analysis and continuous spectra, *Biometrika*, **37,** 1–16.

Bartlett, M. S., and P. H. Diananda (1950), Extensions of Quenouille's test for autoregressive schemes, *J. Roy. Statist. Soc. Ser. B*, **12,** 108–115.

Beveridge, W. H. (1921), Weather and harvest cycles, *Econ. J.*, **31,** 429–452.

Beveridge, W. H. (1922), Wheat prices and rainfall in Western Europe, *J. Roy Statist. Soc.*, **85,** 412–459.

Birnbaum, Allan (1959), On the analysis of factorial experiments without replication, *Technometrics*, **1,** 343–357.

Birnbaum, Allan (1961), A multi-decision procedure related to the analysis of single degrees of freedom, *Ann. Inst. Statist. Math. Tokyo*, **12,** 227–236.

Cave-Browne-Cave, F. E. (1904), On the influence of the time factor on the correlation between the barometric heights at stations more than 1000 miles apart, *Proc. Roy. Soc. London*, **74,** 403–413.

Cheong, H. A., and E. J. Hannan (1968), The asymptotic distribution of spectral estimates, unpublished.

Cooley, James W., Peter A. W. Lewis, and Peter D. Welch (1967), Application of the Fast Fourier Transform to computation of Fourier integrals, Fourier series, and convolution integrals, *IEEE Transactions on Audio and Electroacoustics*, **AU-15,** 79–84.

Cooley, J. W., and J. W. Tukey (1965), An algorithm for the machine calculation of complex Fourier series, *Math. Comput.*, **19,** 297–301.

Cowden, D. J. (1962), Weights for fitting polynomial secular trends, University of North Carolina School of Business Administration Technical Paper No. 4, Chapel Hill, N.C.

Craddock, J. M. (1967), An experiment in the analysis and prediction of time series, *The Statistician*, **17,** 257–268.

Daniels, H. E. (1956), The approximate distribution of serial correlation coefficients, *Biometrika*, **43,** 169–185.

Diananda, P. H. (1953), Some probability limit theorems with statistical applications, *Proc. Cambridge Philos. Soc.*, **49,** 239–246.

Dixon, Wilfrid J. (1944), Further contributions to the problem of serial correlation, *Ann. Math. Statist.*, **15,** 119–144.

Doob, J. L. (1944), The elementary Gaussian processes, *Ann. Math. Statist.*, **15,** 229–282.

Durbin, J. (1959), Efficient estimation of parameters in moving-average models, *Biometrika*, **46,** 306–316.

Durbin, J. (1960a), Estimation of parameters in time-series regression models, *J. Roy. Statist. Soc. Ser. B.*, **22,** 139–153.

Durbin, J. (1960b), The fitting of time-series models, *Rev. Inst. Internat. Statist.*, **28,** 233–244.

Durbin, J. (1963), Trend elimination for the purpose of estimating seasonal and periodic components of time series, *Proc. Symp. Time Series Anal. Brown Univ.* (Murray Rosenblatt, ed.), John Wiley & Sons, Inc., New York, 3–16.

Durbin, J. (1970), An alternative to the bounds test for testing serial correlation in least-squares regression, *Econometrica*, **38,** 422–429.

Durbin, J., and G. S. Watson (1950), Testing for serial correlation in least squares regression. I, *Biometrika*, **37,** 409–428.

Durbin, J., and G. S. Watson (1951), Testing for serial correlation in least squares regression. II, *Biometrika*, **38,** 159–178.

Eicker, F. (1963), Asymptotic normality and consistency of the least squares estimators for families of linear regressions, *Ann. Math. Statist.*, **34**, 447–456.

Fisher, R. A. (1929), Tests of significance in harmonic analysis, *Proc. Roy. Soc. London Ser. A*, **125**, 54–59.

Fisher, R. A. (1940), On the similarity of the distributions found for the test of significance in harmonic analysis, and in Stevens's problem in geometrical probability, *Ann. Eugenics*, **10**, 14–17.

Geisser, Seymour (1956), The modified mean square successive difference and related statistics, *Ann. Math. Statist.*, **27**, 819–824.

Granger, C. W. J., and A. O. Hughes (1969), A new look at some old data: the Beveridge wheat price series, unpublished.

Grenander, Ulf (1954), On the estimation of the regression coefficients in the case of an autocorrelated disturbance, *Ann. Math. Statist.*, **25**, 252–272.

Haavelmo, Trygve (1947), Methods of measuring the marginal propensity to consume, *J. Amer. Statist. Assoc.*, **42**, 105–122.

Hannan, E. J. (1958), The estimation of the spectral density after trend removal, *J. Roy. Statist. Soc. Ser. B*, **20**, 323–333.

Hannan, E. J. (1961), A central limit theorem for systems of regressions, *Proc. Cambridge Philos. Soc.*, **57**, 583–588.

Hannan, E. J. (1963), Regression for time series, *Proc. Symp. Time Series Anal. Brown Univ.* (Murray Rosenblatt, ed.), John Wiley & Sons, Inc., New York, 17–37.

Hannan, E. J. (1964), The estimation of a changing seasonal pattern, *J. Amer. Statist. Assoc.*, **59**, 1063–1077.

Hart, B. J. (1942), Significance levels for the ratio of the mean square successive difference to the variance, *Ann. Math. Statist.*, **13**, 445–447.

Hart, B. J., and John von Neumann (1942), Tabulation of the probabilities for the ratio of the mean square successive difference to the variance, *Ann. Math. Statist.*, **13**, 207–214.

Hoeffding, Wassily, and Herbert Robbins (1948), The central limit theorem for dependent random variables, *Duke Math. J.*, **15**, 773–780.

Hooker, R. H. (1905), On the correlation of successive observations, illustrated by corn prices, *J. Roy. Statist. Soc.*, **68**, 696–703.

Irwin, J. O. (1955), A unified derivation of some well-known frequency distributions of interest in biometry and statistics, *J. Roy. Statist. Soc. Ser. A*, **118**, 389–404.

Jenkins, G. M. (1956), Tests of hypotheses in the linear autoregressive model. II. Null distributions for higher order schemes: non-null distributions, *Biometrika*, **43**, 186–199.

Jenkins, G. M. (1961), General considerations in the analysis of spectra, *Technometrics*, **3**, 133–166.

Kamat, A. R. (1955), Modified mean square successive difference with an exact distribution, *Sankhyā*, **15**, 295–302.

Kantorovich, L. V. (1948), Functional analysis and applied mathematics, *Uspehi Mat. Nauk.*, **3**(6), 89–185.

Karlin, Samuel, and Donald Traux (1960), Slippage problems, *Ann. Math. Statist.*, **31,** 296–324.

Kolmogorov, A. N. (1941a), Stationary sequences in Hilbert space (in Russian), *Bull. Math. Univ. Moscow*, **2**(6), 1–40.

Kolmogorov, A. N. (1941b), Interpolation und Extrapolation von stationären zufälligen Folgen, *Bull. Acad. Sci. U.S.S.R. Ser. Math.*, **5,** 3–14.

Koopmans, Tjalling (1942), Serial correlation and quadratic forms in normal variates, *Ann. Math. Statist.*, **13,** 14–33.

Koopmans, T. C., H. Rubin, and R. B. Leipnik (1950), Measuring the equation systems of dynamic economics, *Statistical Inference in Dynamic Economic Models* (Cowles Commission Monograph No. 10, T. C. Koopmans, ed.), John Wiley & Sons, Inc., New York, 53–237.

Kudo, Akiô (1960), The symmetric multiple decision problems, *Mem. Fac. Sci. Kyushu Univ. Ser. A*, **14,** 179–206.

Lehmann, E. L. (1957), A theory of some multiple decision problems. II, *Ann. Math. Statist.*, **28,** 547–572.

Leipnik, R. B. (1947), Distribution of the serial correlation coefficient in a circularly correlated universe, *Ann. Math. Statist.*, **18,** 80–87.

Madow, William G. (1945), Note on the distribution of the serial correlation coefficient, *Ann. Math. Statist.*, **16,** 308–310.

Magness, T. A., and J. B. McGuire (1962), Comparison of least squares and minimum variance estimates of regression parameters, *Ann. Math. Statist.*, **33,** 462–470.

Mann, H. B., and A. Wald (1943a), On stochastic limit and order relationships, *Ann. Math. Statist.*, **14,** 217–226.

Mann, H. B., and A. Wald (1943b), On the statistical treatment of linear stochastic difference equations, *Econometrica*, **11,** 173–220.

Marsaglia, G. (1954), Iterated limits and the central limit theorem for dependent variables, *Proc. Amer. Math. Soc.*, **5,** 987–991.

Moran, P. A. P. (1947), Some theorems on time series. I, *Biometrika*, **34,** 281–291.

von Neumann, John (1941), Distribution of the ratio of the mean square successive difference to the variance, *Ann. Math. Statist.*, **12,** 367–395.

Olshen, Richard A. (1967), Asymptotic properties of the periodogram of a discrete stationary process, *J. Appl. Prob.*, **4,** 508–528.

Parzen, Emanuel (1957a), A central limit theorem for multilinear stochastic processes, *Ann. Math. Statist.*, **28,** 252–256.

Parzen, Emanuel (1957b), On consistent estimates of the spectrum of a stationary time series, *Ann. Math. Statist.*, **28,** 329–348.

Parzen, Emanuel (1961a), An approach to time series analysis, *Ann. Math. Statist.*, **32,** 951–989.

Parzen, Emanuel (1961b), Mathematical considerations in the estimation of spectra, *Technometrics*, **3,** 167–190.

Quenouille, M. H. (1947), A large-sample test for the goodness of fit of autoregressive schemes, *J. Roy. Statist. Soc.*, **110,** 123–129.

Quenouille, M. H. (1949a), The joint distribution of serial correlation coefficients, *Ann. Math. Statist.*, **20,** 561–571.

Quenouille, M. H. (1949b), Approximate tests of correlation in time-series, *J. Roy. Statist. Soc. Ser. B*, **11,** 68–84.

Quenouille, M. H. (1953), Modifications to the variate-difference method, *Biometrika*, **40,** 383–408.

Rao, M. M. (1960), Estimation by periodogram, *Trabajos Estadíst.*, **11,** 123–137.

Rosenblatt, Murray (1956), Some regression problems in time series analysis, *Proc. Third Berkeley Symposium on Mathematical Statistics and Probability*, (J. Neyman, ed.), *Vol. 1*, Univ. of California Press, Berkeley and Los Angeles, 165–186.

Rosenblatt, M. (1959), Statistical analysis of stochastic processes with stationary residuals, *Probability and Statistics: The Harald Cramér Volume* (Ulf Grenander, ed.) Almqvist and Wiksell Book Co., Stockholm, 246–275.

Rubin, Herman (1945), On the distribution of the serial correlation coefficient, *Ann. Math. Statist.*, **16,** 211–215.

Runge, C. (1903), Über die Zerlegung empirisch gegebener periodischer Funktionen in Sinuswellen, *Z. Math. Phys.*, **48,** 443–456.

Sargan, J. D. (1953), An approximate treatment of the properties of the correlogram and periodogram, *J. Roy. Statist. Soc. Ser. B*, **15,** 140–152.

Schaerf, M. Casini (1964), Estimation of the covariance and autoregressive structure of a stationary time series, Technical Report, Department of Statistics, Stanford University, Stanford, Calif.

Scheffé, Henry (1970), Multiple testing versus multiple estimation. Improper confidence sets. Estimation of directions and ratios, *Ann. Math. Statist.*, **41,** 1–29.

Schuster, A. (1898), On the investigation of hidden periodicities with application to a supposed 26-day period of meteorological phenomena, *Terr. Mag. Atmos. Elect.*, **3,** 13.

Schuster, Arthur (1906), On the periodicities of sunspots, *Philos. Trans. Roy. Soc. London Ser. A*, **206,** 69–100.

Stevens, W. L. (1939), Solution to a geometrical problem in probability, *Ann. Eugenics*, **9,** 315–320.

"Student" (1914), The elimination of spurious correlation due to position in time and space, *Biometrika*, **10,** 179–180.

Tintner, Gerhard (1955), The distribution of the variances of variate differences in the circular case, *Metron*, **17** (3–4), 43–52.

Walker, A. M. (1952), Some properties of asymptotic power functions of goodness-of-fit tests for linear autoregressive schemes, *J. Roy. Statist. Soc. Ser. B*, **14,** 117–134.

Walker, A. M. (1954), The asymptotic distribution of serial correlation coefficients for autoregressive processes with dependent residuals, *Proc. Cambridge Philos. Soc.*, **50,** 60–64.

Walker, A. M. (1961), Large-sample estimation of parameters for moving-average models, *Biometrika*, **48,** 343–357.

Walker, A. M. (1962), Large-sample estimation of parameters for autoregressive processes with moving-average residuals, *Biometrika*, **49,** 117–131.

Walker, A. M. (1965), Some asymptotic results for the periodogram of a stationary time series, *J. Austral. Math. Soc.*, **5,** 107–128.

Walker, A. M. (1968), Large-sample properties of least-squares estimators of harmonic components in a time series with stationary residuals. I. Independent residuals, Technical Report, Department of Statistics, Stanford University, Stanford, Calif.

Walker, G. T. (1914), Correlation in seasonal variation of weather. III. On the criterion for the reality of relationships or periodicities, *Mem. Indian Meteorol. Dept.*, **21**(9), 13–15.

Walker, Gilbert (1931), On periodicity in series of related terms, *Proc. Roy. Soc. London Ser. A*, **131,** 518–532.

Watson, G. S. (1955), Serial correlation in regression analysis. I, *Biometrika*, **42,** 327–341.

Watson, G. S. (1956), On the joint distribution of the circular serial correlation coefficients, *Biometrika*, **43,** 161–168.

Watson, G. S. (1967), Linear least squares regression, *Ann. Math. Statist.*, **38,** 1679–1699.

Watson, G. S., and J. Durbin (1951), Exact tests of serial correlation using non-circular statistics, *Ann. Math. Statist.*, **22,** 446–451.

Watson, G. S., and E. J. Hannan (1956), Serial correlation in regression analysis. II, *Biometrika*, **43**, 436–449.

Whittle, Peter (1952), The simultaneous estimation of a time series harmonic components and covariance structure, *Trabajos Estadíst.*, **3,** 43–57.

Whittle, P. (1954), A statistical investigation of sunspot observations with special reference to H. Alfvén's sunspot model, *The Astrophys. J.*, **120,** 251–260.

Whittle, Peter (1959), Sur la distribution du maximum d'un polynome trigonométrique à coefficients aléatoires, *Colloq. Internat. Centre Nat. Rech. Sci.*, **87,** 173–183.

Young, L. C. (1941), On randomness in ordered sequences, *Ann. Math. Statist.*, **12,** 293–300.

Yule, G. Udny (1927), On a method for investigating periodicities in disturbed series with special reference to Wolfer's sunspot numbers, *Philos. Trans. Roy. Soc. London Ser. A*, **226,** 267–298.

Zyskind, George (1967), On canonical forms, non-negative covariance matrices, and best and simple least squares linear estimators in linear models, *Ann. Math. Statist.*, **38**, 1092–1109.

Index*

*References to books and papers are listed by author for each section at the end of the respective chapter. Therefore, the index does not give author references.

Applied Probability and Statistics (Continued)

LEWIS · Stochastic Point Processes
MANN, SCHAFER and SINGPURWALLA · Methods for Statistical Analysis of Reliability and Life Data
MILTON · Rank Order Probabilities: Two-Sample Normal Shift Alternatives
OTNES and ENOCHSON · Digital Time Series Analysis
RAO and MITRA · Generalized Inverse of Matrices and Its Applications
SARD and WEINTRAUB · A Book of Splines
SEAL · Stochastic Theory of a Risk Business
SEARLE · Linear Models
THOMAS · An Introduction to Applied Probability and Random Processes
WHITTLE · Optimization under Constraints
WILLIAMS · Regression Analysis
WONNACOTT and WONNACOTT · Econometrics
YOUDEN · Statistical Methods for Chemists
ZELLNER · An Introduction to Bayesian Inference in Econometrics

Tracts on Probability and Statistics

BILLINGSLEY · Ergodic Theory and Information
BILLINGSLEY · Convergence of Probability Measures
CRAMÉR and LEADBETTER · Stationary and Related Stochastic Processes
JARDINE and SIBSON · Mathematical Taxonomy
KINGMAN · Regenerative Phenomena
RIORDAN · Combinatorial Identities
TAKACS · Combinatorial Methods in the Theory of Stochastic Processes